Recent Breakthroughs in Basal Ganglia Research

RECENT BREAKTHROUGHS IN BASAL GANGLIA RESEARCH

ERWAN BEZARD
EDITOR

Nova Science Publishers, Inc.
New York

For permission to use material from this book please contact us:
Telephone 631-231-7269; Fax 631-231-8175
Web Site: http://www.novapublishers.com

Library of Congress Cataloging-in-Publication Data

Recent breakthroughs in basal ganglia research / Erwan Bezard (editor).
p. ; cm.
Includes bibliographical references and index.
ISBN 1-59454-880-3
1. Basal ganglia. 2. Basal ganglia--Research.
[DNLM: 1. Basal Ganglia--physiology. 2. Basal Ganglia Diseases--etiology. 3. Basal Ganglia Diseases--therapy. WL 307 R295 2006] I. Bezard, Erwan.
QP383.3.R43 2006
612.8'25--dc22 2006002566

Published by Nova Science Publishers, Inc. ✣ New York

CONTENTS

In: Recent Breakthroughs in Basal Ganglia Research
Editor: Erwan Bezard, p. ix
ISBN: 1-59454-880-3

FOREWORD

When it was first proposed that I edit a book on basal ganglia research, my first reaction was to turn down the invitation since a major problem of our time is the overwhelming number of publications. However, despite the huge number of journals or books, no space really exists where the different fields in basal ganglia research can meet and where weird or purely speculative hypotheses can be elaborated. The present book would hopefully constitute the first of a series offering such a space aimed at promoting cross-fertilization either between fields (e.g. between addiction and drug-induced dyskinesia) or between levels of analysis (e.g. between computation and physiology). Although not all the basal ganglia research is represented in the present volume, it was my primary intention. Each section has been edited by an eminent colleague, A. Benazzouz, T. Boraud, A. Cenci, J. McHaffie, M. Mastumura, W. Meissner, G. Morris and M. Vila, whose efforts should be praised. Others, initially invited as well to cover the missing fields, have not been able to deliver their section because of the increasing missions researchers have to complete. I wish the conceptual novelty of the offered space would convince future contributors about the need of cross-fertilization.

Erwan Bezard
CNRS UMR 5543, University Victor Segalen, Bordeaux, France

In: Recent Breakthroughs in Basal Ganglia Research
Editor: Erwan Bezard, pp. 1-2
ISBN: 1-59454-880-3

PART I: COMPUTATIONAL MODELS OF BG PROCESSING

Thomas Boraud
CNRS UMR 5543, University Victor Segalen,
Bordeaux, France

Why is an electrophysiologist the editor for a computational section? This is because, I definitely think that computational modelling concerns each and every one of us who is working on the topic. New methods for electrophysiological recording have made a significant breakthrough in the basal ganglia (BG) field within the last decade. They have introduced a huge complexity in a field were the consensual theory was based on a "boxes and arrows" scheme of the network. Electrophysiologists are still at the forefront of the collaboration with theoreticians, because to analyse the exponential amount of data multi-electrode recording introduced, they already had to improve their computational background. But anatomists, behavourists, molecular biologists and clinicians are now also concerned (and of course some of them are already involved in the computational approach). In 1994, James Houk, Joel Davis and David Beiser edited a book entitled "models of information processing in the BG" which compiled the most advanced computational approach at that time. It became the bedside book of a whole generation of physiologists in the BG field. It is still unequalled today and this section does not pretend to compete with it. But I think it was just a first step: all the models in this book have been constructed a posteriori from data collected in experiments designed for other purposes. This is a concequence of the common attitude toward computational modelling: experimentalists tend to consider it as an analysis tool one borrows to try to understand data which were not straitforward. Thus results are often ankward or with limited perspectives. It is now time to proceed to the second step and I pleed here for a reversal of the approach. We need to design our experiments in order to test computational model predictions. We have entered a post-"Claude Bernard era" where the link between the idea and the experience should pass through the computational model.

In this section we introduce four computational models of BG processing which use different approaches to address the question of basal ganglia function. The authors have made an effort in order to be understandable by the widest audience and to emphasize the perspectives their models provide.

Kevin Gurney et al. propose the assuption that BG's main role is to solve conflict between independant motivational and sensori-motor units that are physically seperated within the brain but are in competition for behavioural expression. This "action selection" model is the framework of their own work and of the two subsequent chapters of the computational section. The authors descibe their methodology which is structured in two steps: 1) Definition of their model based on an a priori assumption of BG function. 2) Quantification of this assumption by tuning the parameters of the model and verifying that unrealistic parameters decrease its performance.

While the two following chapters adopted two very different methods, in both cases the authors emphasize the "action selection" mechanism that arises from their approach. Leonid Rubchnisky et al. use a conductance based-model of the STN-GPe-GPi microcircuitry and Arthur Leblois et al. propose an analytical approach of the dynamic properties which emerge from the closed or semi-closed architecture of the whole BG-cortex network.

Izar Bar-Gad and Hagaï Bergman conclude this section by moving out of the "action selection" assumption to assume an "information selection" role for the BG network. In their non linear refined variant of the "actor-critic" model, emphasizing the role of dopamine, they suggest that the BG play a role in the extraction of the information transmitted from the cortex through reinforcement driven dimensionality reduction (RDDR).

What, How and Why, is the abc of scientific questioning. Despite the variability of the approaches and the somewhat contradictory conclusions, these models are not mutually exclusive. All of them are "toy models" of the system and thus a reduction (falsification) of the truth to explain the truth. But at least all agree about What the structure performs: the unique architecture of the BG network predestines it to select from the huge variability of cortical generators of diversity. The Why, behaviour (action selection) or information (RDDR)- and the How, local subcircuits or the whole BG-cortex network is now the question to adress by experimentalists in order to test the validity of the various predictions provided by these models.

In: Recent Breakthroughs in Basal Ganglia Research
Editor: Erwan Bezard, pp. 3-15
ISBN: 1-59454-880-3

Chapter 1

A THEORETICAL TREATMENT OF BASAL GANGLIA FUNCTION

K. N. Gurney**, *T. J. Prescott and P. Redgrave
Adaptive Behaviour Research Group, Department of Psychology,
University of Sheffield, S10 2TP, UK.

ABSTRACT

This chapter reviews some of the recent work in our lab on computational modelling of the basal ganglia. The models presented all derive from practicing a single methodology within a common theoretical framework which draws on the ideas of David Marr. In this scheme we start with a computational hypothesis for the basal ganglia (in our case, that it performs 'action selection') and then proceed to mine the neuroscientific data for mechanisms that may support this computation. A successful programme of 'mechanism mining', yielding models that support the given computation, is taken as support of the overall hypothesis. One of the main points of this chapter, therefore, is to advance, using examples, a metatheoretical framework for modelling. Secondly, by presenting a group of models in this way, we are really making the case for a *theory* of basal ganglia function, rather than any particular model *per se*.

1 BUILDING MODELS USING COMPUTATIONAL AND BIOLOGICAL CONSTRAINTS

How should we go about building models of the basal ganglia? Recently, (Gurney et al., 2004a) we have outlined some methodological ideas which attempt to answer this question. In this section, we discuss those parts of that methodology relevant to the models presented here, and describe our hypothesis of basal ganglia function.

One of the desired outcomes of any computational modelling project is to have possible answers to questions such as: what does the brain system of interest do; that is, what computations does it perform and to what end? One approach to answering these questions might be to look carefully at the model input-output relations and try to discern what the system is doing at a sufficiently high level of description. Given the complexity of input-output relations at the level of neural signal processing, this may be at best, difficult, and at worst, intractable. However, an alternative, and more principled approach, is based on the computational analysis proposed originally by David Marr (1982). In this scheme, we posit *a priori* a computational purpose for the basal ganglia. This hypothesised functionality will not, of course, be determined arbitrarily, but derived from empirical considerations about the structure, and physiology of the basal ganglia, together with behaviour implicating its function. Once formulated, such a hypothesis not only supplies an answer to the 'what computation' question above, but also gives the modeller a theoretical stance from which the experimental data may be marshalled. Thus, while a model should be well

* E-mail address: K.gurney@shef.ac.uk

constrained by relevant anatomical, physiological and behavioural data, there is often disagreement in the literature about certain key aspects of the interpretation of this data. For example, the existence of functional medium spiny neuron collaterals in striatum has been the subject of much debate (Jaeger et al., 1994; Tunstall et al., 2002; Koos et al., 2004; Venance et al., 2004) as has the colocalisation of D1 and D2 receptor types in medium spiny neurons (Gerfen et al., 1990; Surmeier et al., 1996; Aizman et al., 2000). There is therefore usually a pressing need to invoke some other constraint in order to tame the 'data mountain' before applying it in a model. This is where a computational hypothesis can be invaluable, for we can now seek to discover whether a model exists that is *consistent* with the data, and that can implement a system that achieves the hypothetical computational function. In particular, we seek to discover suitable neuronal mechanisms within the basal ganglia that support the proposed overall function; we call this process *mechanism mining* (Gurney et al., 2004a).

In our own work we have posited that the main computational purpose of the basal ganglia is one of *action selection* - the resolution of conflicts between independent motivational and sensorimotor units that are physically separated within the brain but are in competition for behavioural expression. Stated informally, it is the problem of how we decide 'what to do next'. This situation is particularly acute if several such units are competing for a common resource as, for example, with the neural systems involved in feeding and drinking, both of which require the use of oral muscles. In other cases, multiple simultaneous actions may be permitted (as, for example, when an animal walks and chews at the same time) provided the actions don't specify simultaneous, incompatible uses of the same muscle group. Competition can also arise in modalities where behavioural expression is indirect, for instance between systems attempting to gain control of cognitive resources. In general, therefore, we suppose that the brain is processing a large number of sensory and cognitive streams or *channels*, each of which may be requesting some action to be taken. The task demanded of the animal, in order for appropriate behaviour to occur, is to suppress the majority of these requests while allowing the expression of only a limited number (in some cases just one). The problem of action selection is clearly crucial to our understanding human behaviour (Duncan, 1995) but also arises within ethological analysis of animal behaviour (McFarland, 1989) (where it is often referred to as the problem of 'behavioural switching' or 'decision-making'). From the biological perspective we can ask the following questions: 1) what is the neural substrate performing action selection? 2) how might this substrate operate functionally? Recently, (Redgrave et al., 1999a) we have reviewed these questions in and argued that, in terms of connectivity and metabolic efficiency, effective action selection favours centralised switching devices and, moreover, that the basal ganglia are well suited to play this role. The link between the basal ganglia and action selection is not original to us, but is a recurring theme in the literature, (see for example Cools, 1980; Mink, 1996; Kropotov and Etlinger, 1999). However, our recent work (Prescott et al., 1999; Redgrave et al., 1999a) has developed this idea as a major unifying hypothesis of basal ganglia function, showing how it relates to known anatomy and physiology, and meets several high level computational requirements of a selection mechanism. Subsequently, we have engaged in a programme of mining the data on basal ganglia anatomy and physiology for mechanisms that support a selective function.

Having posited that the computational purpose of the basal ganglia is one of action selection, we can proceed to test this hypothesis quantitatively in a computational model. That is, under a mechanism mining approach, we seek to discover whether mechanisms instantiated by the essentials of basal ganglia anatomy and physiology will support this putative function. This exercise was successfully carried through in our model of intrinsic basal ganglia function (section 02; Gurney et al., 2001b; Gurney et al., 2004a). Beyond this first step, we now argue that the programme of testing the computational theory of the basal ganglia comprises two main strands. First, the biology of the basal ganglia should be scrutinised further to see if it supports additional mechanisms for selection. That is, additional, biologically realistic features should be added and the model tested for its ability to perform selection. An augmented functionality here is taken as evidence supporting the computational hypothesis, whereas a decrement in performance suggests that the hypothesis may be false. Second, by deliberately introducing non-biological elements, models can be built which act as 'controls'. Thus, under this kind of manipulation, if the computational hypothesis is true, we would expect selection performance to deteriorate. The result of applying these extensions of the methodology are dealt with in the models described in section 03. This section also addresses the problem of refining the selection hypothesis and developing metrics of 'selection performance', in order to test quantitatively whether the impact of modifications to the model.

2 A Model of the Basal Ganglia Constrained by Experimental Data and the Selection Hypothesis

2.1 Biological Basis

Our original model was based on the anatomy and physiology summarised in Figure 1 (specialised for rodent).

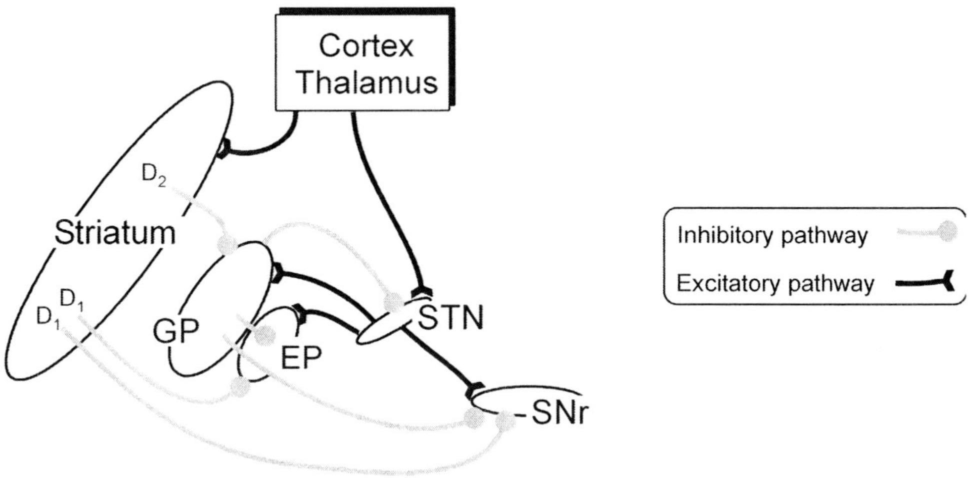

Figure 1. Schematic basal ganglia anatomy showing main pathways intrinsic to the basal ganglia together with cortico-thalamic afferents. Note the two striatal populations of medium spiny neurons have been differentiated by their dopamine receptor type (D1 and D2). These are referred to subsequently as striatum (D1), striatum (D2)

This diagram is not exhaustive in its description of basal ganglia anatomy, but our choice of pathways was guided by the fact that this scheme is also used as the basis for the prevailing functional architecture (Albin et al., 1989) defined in terms of 'direct' and 'indirect' pathways (Figure 2b), and in several other models of basal ganglia (for reviews see Gillies and Arbuthnott, 2000; Prescott et al., 2002).

2.2 Conceptual Foundation of the Original Model

The basic assumption underlying our model was that the brain is processing, in parallel, a large number of sensory, cognitive and motivational streams or channels, each of which may be 'requesting/promoting' different actions to be taken. This channel-based scheme is consonant with the view that basal ganglia comprises a series of afferent and efferent parallel processing streams or loops (Alexander et al., 1986; Hoover and Strick, 1993; Middleton and Strick, 2002). At the systems level, the smallest neuronal population we needed to consider was, therefore, the set of neurons responsible for a single 'channel' within each of the basal ganglia nuclei.

A further assumption was that, implicit in the representation of each action, there is an encoding of its salience or propensity to be selected for execution. In our model, we assumed that channel salience had already been extracted from phasic excitatory input by processes in the basal ganglia input nuclei. The input to the model, therefore, was simply the scalar valued salience of each channel and selection then occurs via selective disinhibition of target structures (Deniau and Chevalier, 1985). In the model, the degree to which this takes place defines another scalar valued signal - the firing rate of population representatives for each channel in the output nuclei. In summary then, large salience signal inputs at striatum and subthalamic nucleus (STN) select for low signal outputs at the entopeduncular nucleus (EP) and substantia nigra pars reticulata (SNr).

2.3 Re-thinking the Basal Ganglia Functional Architecture

We used the computational premise of selection to guide our interpretation of basal ganglia anatomy in functional terms, leading to a new functional architecture. Since we were principally interested at this *systems* level of description in inter-nucleus processing, we sought mechanisms which relied on such interactions rather than those relying on intra-nucleus circuitry. For example, many workers have invoked recurrent inhibition with striatum as a selection mechanism. While our model admits this possibility it does not explicitly invoke this mechanism *per se* (see Gurney et al., 2001b pp412-413).

One inter-nucleus architectural feature that appears to support selection is the diffuse excitation from STN to its targets - GP, SNr/EP (Parent and Hazrati, 1993, 1995)- in combination with more focused inhibition from striatum to the same nuclei. This constitutes an off-centre, on-surround network that can perform a selection function, as noted by Mink and Thach (1993). However, this simple scheme is ambiguous since it is not clear what function the GP serves as an 'output layer' in a feedforward net. We resolved this problem by observing that, while selection could be performed in principle by the complex of striatum(D1), STN, and SNr/EP alone, the relative levels of excitation and inhibition required to achieve this function were only obtained (and indeed guaranteed) by the inhibition supplied by GP (see section 2.5). We therefore hypothesised that the GP acts within a *control pathway* (comprising striatum (D2), STN, and GP) as a source of control signals for the *selection pathway* (striatum (D1), STN, SNr/EP). The new functional architecture described above (Gurney et al., 2001a) is shown in Figure 2a.

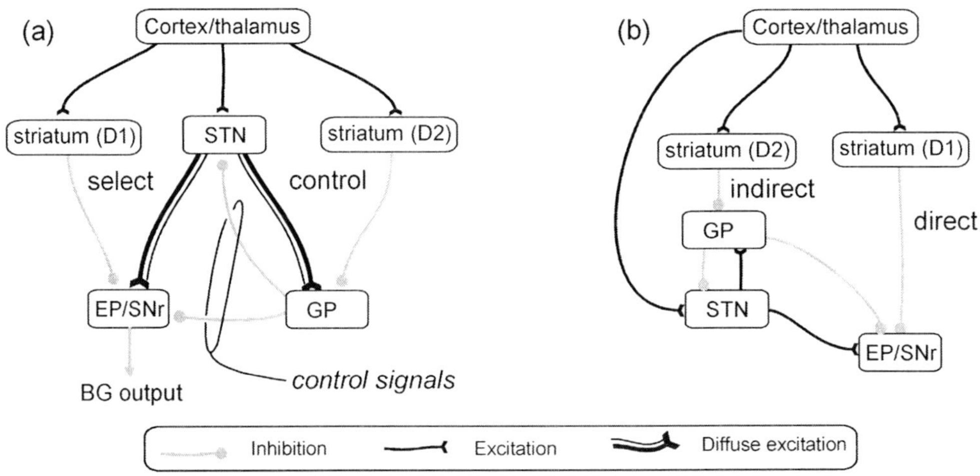

Figure 2. Functional architectures for the basal ganglia (a) The new architecture proposed in (Gurney et al., 2001a). (b) The 'direct/indirect'-pathway architecture of the basal ganglia for comparison.

We stress that this architecture was obtained as a natural consequence of considering system wide mechanisms that address the computational hypothesis of selection. Notice that the new architecture is quite different from the 'direct/indirect' pathway model (Figure 2b) and therefore represents a new hypothesis emerging from our computational theoretic approach.

2.4 Dopamine Modulation

Modulation of neural processing in the basal ganglia by dopaminergic input from the ventral midbrain appears to take place on two time scales (Walters et al., 2000). One is a short-latency phasic response (~100 ms burst) to the onset of biologically significant stimuli (Schultz, 1998; Redgrave et al., 1999b), while the other is a tonic level of activity (1-8 Hz) that can alter for a variety of reasons including pathologies of the substantia nigra pars compacta (SNc), changes in motivational states, and pharmacological interventions (Salamone et al., 1997; Berridge and Robinson, 1998; Horvitz, 2000). Thus far, we have modelled only variations in the tonic level of dopamine.

The action of dopamine within striatum depends on the receptor type at the post-synaptic cell and, even within this constraint, a variety of effects have been reported which make it difficult to discern a clear pattern in the data (Nicola et al., 2000). Nevertheless, there is evidence that dopamine acting via D1-type receptors enhances the effect of salient stimuli (Akkal et al., 1996) but attenuates their effect when activating D2 receptors (Gerfen et al., 1990; Harsing and Zigmond, 1997). This is the scheme we captured in our original model. Finally, we assumed that tonic dopaminergic input to striatum is sufficiently diffuse that it is essentially constant over all channels (Schultz, 1998).

2.5 Result

Figure 3 shows simulation results for a 6 channel model in which only channels 1 and 2 were stimulated (the salience on the remaining channels was set to zero); channel 3 is therefore representative of channels 3-6. Initially the inputs (salience values) are zero but at time t = 1, the salience on channel 1 is increased to 0.4. This results in decreased output y_1 on this channel, and increased output on the other channels. The reduction in y_1 corresponds to a disinhibition of the target structure of channel 1 and, if y_1 is less than some threshold, then channel 1 is deemed has become selected. At t = 2, the salience of channel 2 becomes 0.6. This is sufficient to force y_2 = 0, thereby ensuring selection of channel 2, while pushing y_1 above the tonic value. The output on channel 3 (and all other non-active channels) is increased still further. Thus, if channel 1 were selected prior to t = 2, the selection of channel 2 has caused the interruption of its selection. The next two events are initiated by a transient increase in salience on channel 1 from t = to t = 4.

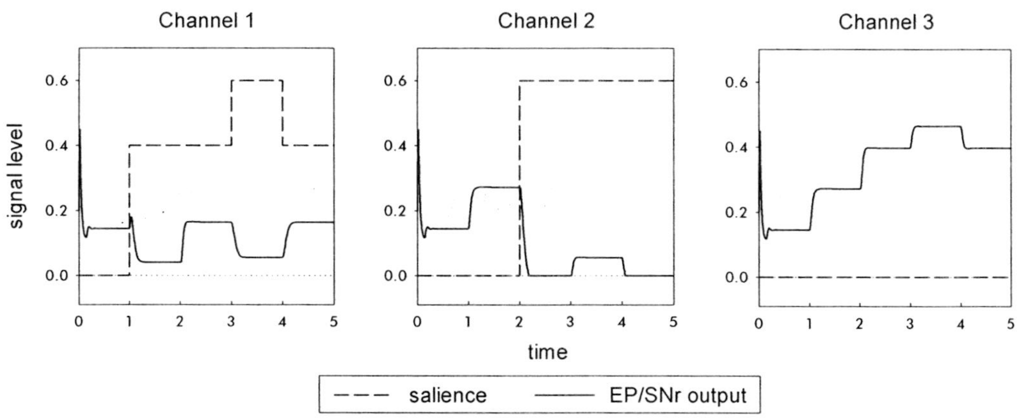

Figure 3. Simulation results from the model described in (Gurney et al., 2001a).

During this time, both channels 1 and 2 have the same, comparatively high salience, but inter-channel competition leads to a higher common output level than would be seen with the same salience on a single channel in isolation.

Figure 4 shows the results of experiments across different levels of simulated dopamine. For a fixed dopamine level, a series of experiments were conducted with a range of salience combinations on two active channels taking values between 0.2 and 1 in steps of 0.1. The protocol was similar to that depicted in Figure 3, but with no transient on channel 1 at t = 4. In each case, a channel was deemed to be 'selected' if it has zero output and the result at equilibrium (when both channels had their maximum salience) was classified as one of no selection, single channel selection or dual channel selection. With no dopamine, neither channel ever gets selected no matter how large the salience values. With an intermediate level of dopamine, saliences below 0.6 fail to initiate selection but, for values greater than or equal to this, channel selection is possible. With closely matched salience values less than 0.9, the inter-channel competition is closely matched and neither channel is selected. However, if both channels are sufficiently salient, they may force simultaneous selection. Turning to the case of high dopamine, a similar general pattern of results occurs but the salience threshold for selection is smaller (0.4) and there is significantly more opportunity for simultaneous channel selection. Thus, our model predicts that one consequence of increased dopamine in the basal ganglia is for selection to be more 'promiscuous'.

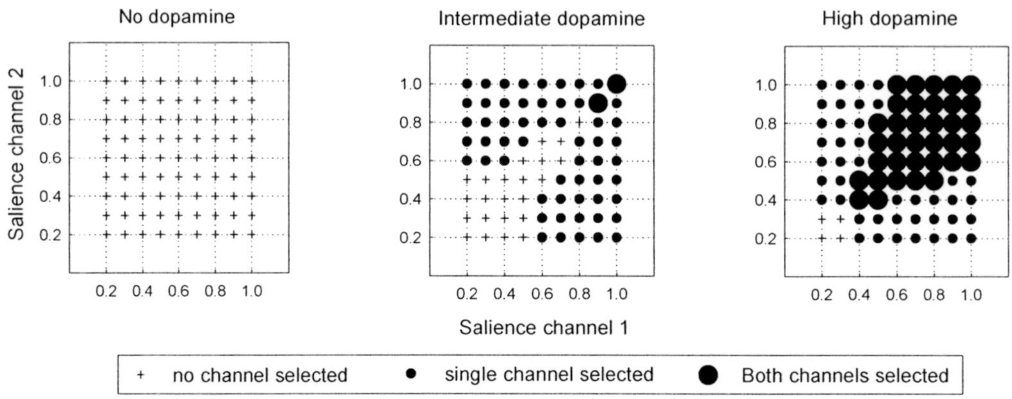

Figure 4. Effect of varying simulated dopamine. Each bubble plot shows the outcomes for a series of pairwise channel competitions

Aside from the simulation studies, aspects of the model were analytically tractable because we chose a piecewise linear form for the leaky-integrator unit output compression function. One important analytic result was that the inhibition provided by GP to STN was just sufficient to ensure that, irrespective of the number of channels in the model, the excitation supplied by STN to SNr/EP was kept within bounds, thereby allowing appropriate selection (as shown in Figure 3). Without this automatic *capacity scaling* the subthalamo-nigral weights would have to be carefully crafted to prevent over-activation of SNr. This automatic gain control function supplied by GP was one of the key features that led us to suppose that GP provides control signals to the rest of the basal ganglia.

3 Introducing Additional Biological Features

Based on the discussion of section 1, we expect that, if the basal ganglia's main computational function is selection, then the incorporation of these additional anatomical constraints would demonstrate an improvement in selection performance. If, in contrast, the basal ganglia is really supporting other functions, then we were misled by the original model by choosing a subset of pathways that unfortunately conspired to demonstrate an erroneous function (that of selection). In this case, we might expect the inclusion of additional biological constraints (not in any way associated with selection *per se*) to have a detrimental effect on the previously observed selection function of the model.

3.1 The New Pathways

To test these ideas, we augmented our original model with the incorporation of three additional anatomical constraints (Gurney et al., 2004b). The first comprises collateral branches of the striatonigral projection - a major part of our selection pathway and the 'direct' pathway in previous models. These collaterals innervate the globus pallidus (Chang et al., 1981; Kawaguchi et al., 1990) - the principal nucleus of our control pathway and the target of the 'indirect' pathway in previous models. Using a large sample of cells, Wu et al. (2000) was able to consolidate previous work demonstrating three types of striatal projection neurons; type I neurons project to GP only; type II project to GP with a collateral to SNr; type III project to GP with collaterals to EP and SNr. This scheme is shown in Figure 5b (Figure 5a shows the projections used in our previous model for comparison).

Further, the anatomical data of Wu et al., (2000) indicate a comparatively weak pallidal projection from types II and III neurons; collaterals of these neurons branch much less profusely in GP than the pallidal projections of type I neurons.

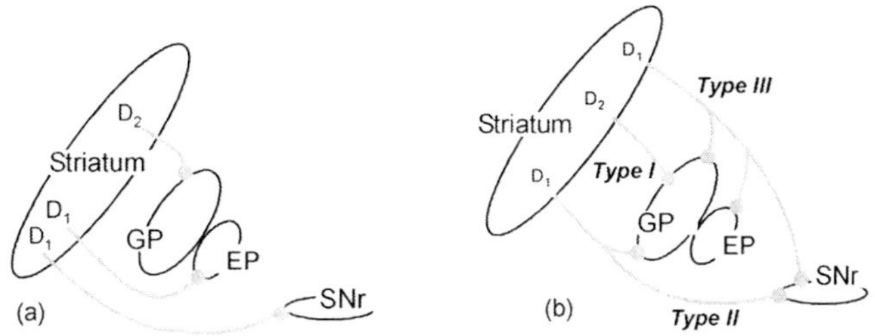

Figure 5. Striatal projections in basal ganglia. (a) Projections incorporated into the original model - this is a subset of the anatomical connectivity shown in Figure 1. (b) Striatal projections incorporating 'direct' pathway striato-pallidal collaterals as classified by Wu et al. (2000).

The other anatomical features we included were the intrinsic axon collaterals within GP (Park et al., 1982; Smith et al., 1998) and SNr (Juraska et al., 1977; Deniau et al., 1982; Mailly et al., 2003) which, because of the inhibitory nature of the parent neurons, are likely to be associated with inhibitory connections. While the evidence for collaterals in EP is sparse (one study Moriizumi et al., 1987 has supplied evidence for them in the cat) we assume EP possesses intrinsic collaterals and continue to consider the output nucleus combination SNr/EP. Local inhibitory collaterals form the basis of a classic neural network mechanism for selective processing which makes the exploration of their functional role in SNr/EP and GP of keen interest. By way of terminology we refer to the model making use of 'Direct' pathway (striato-nigral) collaterals as a *D-model,* one which uses collaterals intrinsic to GP and or SNr/EP as an *I-model,* and one which uses both as a *combined model;* details of all these models are given in (Gurney et al., 2004b).

3.2 Refining the Computational Hypothesis

In order to determine the functional consequences of the new models, it is first necessary to refine the computational concepts underpinning our proposed selection function for basal ganglia. These ideas centre on the ability of the basal ganglia to make transitions (under dopaminergic control) between two modes of operation in which different aspects of selection are emphasised.

Figure 6a shows the outcome of a series of simple dual channel experiments of a similar kind to those yielding the results of Figure 4, obtained with the original model using a moderate value of simulated dopamine.

The results are similar to those in the central panel of Figure 4, except that the outcome classification scheme has been further sub-divided (for details see Gurney et al., 2004b). Thus, the open symbols in Figure 6 correspond to sub-categories of the original case of 'single channel selection' in Figure 4, while dual channel selection is still represented in the same way ('distortion' is a rarely occurring subcategory of the original case of 'non-selection'). Figure 6b shows the results for the same set of experiments but with a high level of simulated dopamine. The main point of contrast is that, with moderate levels of dopamine, the outcome complies well with a *hard selection* regime which is dominated by single channel selection (of whatever variety) whereas at higher dopamine levels, the pattern of results is best described as one of *soft selection* which is dominated by dual channel selection.

The ability of the system to perform hard selection is clearly crucial to it being a candidate for a selection mechanism; we would normally demand that there is a clear 'winner' of the competition for behavioural expression. However, it may be desirable in some circumstances that selection is more 'promiscuous' so that inhibition is removed from multiple channels. Moreover, in a model with noise (possibly mediated by the use of spike train signalling rather than simple firing rates) the conditions for simultaneous selection in the models described here might give rise to rapid switching between channels. This may be a good strategy under some circumstances (e.g early development) for exploring previously unused methods of dealing with novel situations.

These observations led to the following hypothesis which refines the original, generic selection hypothesis of basal ganglia function.

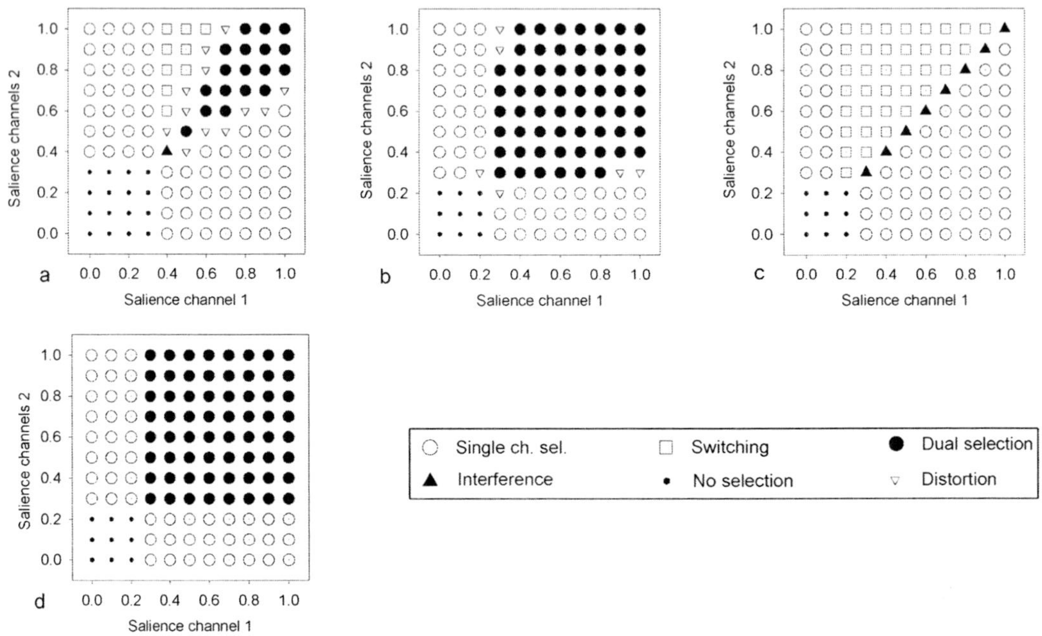

Figure 6. Hard and soft selection. Panels a) and b) are results with the original model at moderate and high dopamine levels respectively. Parts c) and d) correspond to idealised result templates for 'hard' and 'soft' selection regimes respectively. See text for details.

***Refined selection hypothesis*:** The basal ganglia are constructed to perform either hard or soft selection as distinctly as possible under moderate and high levels of dopamine respectively. Each regime must be accessible under a sufficient range of dopamine to allow robust control.

This may be tested using a corollary of the hypothesis itself and the arguments of section 3.

***Corollary of refined selection hypothesis*:** As extra biologically realistic features are added to the model, the ability to support and make transitions between selection regimes will be better supported. On the other hand, the addition of biologically unrealistic pathways should result in a decrement in the performance of these functions.

While this is a perfectly adequate qualitative statement, in order for it to be tested using computational models, it needs to be made quantitative by defining precisely what is meant by 'the ability to support and make transitions between selection regimes' which defines the models 'performance'.

3.2.1 Quantifying the Refined Hypothesis

To quantify model performance it is necessary to be able to measure how well the selection properties of the model correspond to each of the hard and soft selection regimes with a particular level of dopamine. To do this, we determined the match between the pattern of categorical results (see for example, Figure 6a,b) and templates that define idealised performance for each selection regime; these templates are shown in Figure 6c,d respectively. Each level of simulated dopamine is now associated with percentages, P_h, P_s, of individual experimental outcomes (one outcome per grid location on the bubble plots in Figure 6) that match with the hard and soft templates respectively. For example, of the 121 possible outcomes (there are 11 possible salience values on each of channel), 117 of those in the high dopamine condition (Figure 6b) are identical to the corresponding outcomes in the soft selection template of Figure 6d, so that $P_s = 96.7$ %

Template matches were determined for a range of simulated dopamine values. These were defined using the ratio, R_w, of cortico-striatal weights on the connections to the striatum (D1) and striatum (D2) nodes. Figure 7a shows the pattern of results for a combined model. For low to intermediate levels of dopamine (to the left of the vertical dotted line) hard selection is favoured, whereas for higher levels of dopamine (to the right of the line), soft selection is favoured.

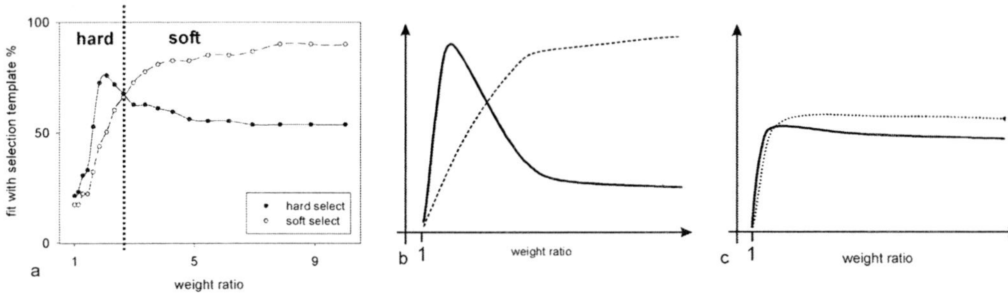

Figure 7. Graphical illustration of the refined computational hypothesis. a) Plots of P_h, P_s, against dopamine-determined weight ratio, R_w, for a combined model. The lines are obtained by cubic spline fitting to the data. b) and c) cartoons of extreme variants of the plots of template fits against R_w. a) a pattern of model results to which better model conformity (than the original model) under realistic modification is supportive of the hypothesis developed in the text. b) a pattern of model results, to which better model conformity under realistic modification would falsify this hypothesis.

The selection hypothesis demands that the data patterning in P_h, P_s, has the following features 1) The existence of large absolute values of P_h, P_s, thereby allowing good access to hard and soft selection regimes. 2) Substantial differences between P_h, and P_s over as much of the range of dopamine modulation as possible, thereby forcing a clear distinction between the model's behaviour in terms of hard/soft selection. 3) A substantial range of dopamine levels over which the hard selection regime may be accessed. This is shown graphically in Figure 7b,c which depict patterns of results displaying these features well, and poorly, respectively. We therefore proceeded to quantify the features 1 – 3 defined above and used them to develop an overall selectivity *merit* which quantified how well the refined hypothesis and its corollary were satisfied. In order to facilitate comparison with the original model, the features were defined in such a way that a merit of zero indicated no change with respect to that model, while positive and negative values showed enhancement and deterioration of the feature respectively. If a feature was not defined for a given model, it was assigned a nominally 'infinite' (actually very large negative) value; this might arise, for example, if there was no separate hard selection regime.

3.3 Results

We focus concentrate here on reporting the behaviour of the overall selectivity merit; full details of the effects of each additional pathway on each of the individual features is given in (Gurney et al., 2004b). Figure 8 shows the merit for a selection of three D-models, three I-models, and a combined model (as defined in section3.1). The D-models are distinguished by the strength of the weighting in the striato-pallidal collateral pathway (one of 0.25, 0.5 or 0.75). The I-models are distinguished according to whether there are collaterals intrinsic to GP or SNr/EP only ('G' and 'S' respectively) or whether they are found in both these nuclei ('S+G'). The combined model used a striato-pallidal weighting of 0.25 and had collaterals in SNr/EP and GP.

3.3.1 Improvements with the Addition of Biologically Realistic Pathways

The D-model with a weak striato-pallidal weighting, shows a performance enhancement. The D-models with increased weighting show progressively worse performance, and that with a weight of 0.75 fails to show a separate hard selection regime (it was therefore assigned a large negative merit as indicated by the downward arrow). In the models, a weight of 0.75 is associated with a comparatively strong pathway (no model weights are larger than 1), and so our modelling result is consistent with the anatomy which indicates a comparatively weak striato-pallidal collateral pathway (Wu et al., 2000).

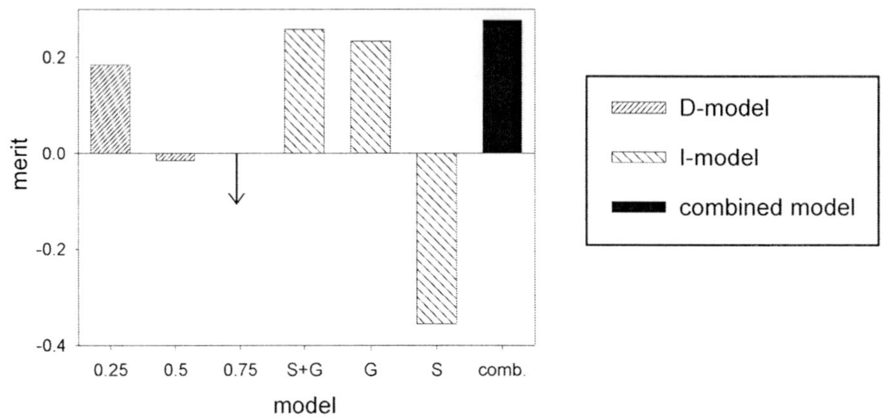

Figure 8. The overall selectivity merits of a series of D- I- and combined models. See text for details.

The I-model (designated 'G') with collaterals in pallidum only confers a selection merit advantage. In contrast, the 'S' model promotes a performance deficit. However, when both collateral sets are combined (in the 'S+G' model) there is not only a performance enhancement, but the increase in merit is slightly more than that in the 'G' model. Thus the collaterals in GP and SNr/EP appear to work in synergy to produce an improvement in selection merit.

The combined model shows the largest increase in selection merit and appears to combine the benefits of both the striato-nigral and intrinsic collaterals. In sum, all the biologically realistic, additional features we have considered result in selection performance enhancement which is supporting evidence for the refined computational hypothesis of selection.

3.3.2 Control Models Show Decrement in Performance

We constructed three models obtained by taking the original model and deliberately adding to it non-biologically plausible connectivity. These models therefore constituted control models in so far as we don't expect there to be a selection advantage (positive selection merit) for any of them; rather we expect negative or zero merit values indicating reduced or unchanged performance. The first such model incorporated an inhibitory pathway from GP to striatum. While there is evidence for a pallido-striatal pathway, it primarily innervates the GABAergic interneurons in striatum (Bevan et al., 1998). Thus the net effect on striatal projection neurons will be excitatory rather than inhibitory and, since it is the projection neurons that are the basis of the model striatum, the pallido-striatal pathway in the model is weighted with the wrong sign for biological realism. The second model incorporated an inhibitory pathway from striatum to subthalamus, while the third used an inhibitory pathway from SNr to striatum. While both these models are realistic with respect to their neurotransmitter effects (they have the correct 'sign') there is no evidence for the existence of such pathways.

Each of the control models was tested with four values 0.25, 0.5, 0.75, 1.0, for the weight magnitude determining the strength of the new pathway; Table 1 shows the resulting merit values. The pallido-striatal and striato-subthalamic models yielded four negative values. However, the nigro-striatal model had two positive values. At first glance this would appear to constitute disconfirmatory evidence for our hypothesis since an apparently unrealistic addition has resulted in a performance improvement. However, there is another way of accounting for positive values in a 'control' model; that is, if the additional pathway on which it is founded is not unrealistic after all.

Table 1. merit-values for the control models. Positive values are highlighted in bold.

	weight magnitude			
Model	**0.25**	0.5	0.75	1.0
GP → Striatum	-0.18	-0.42	-*R*	-*R*
Striatum → STN	-0.03	-0.26	-0.37	-0.7
SNr → Striatum	0.06	0.0	0.11	-R

In respect of the nigro-striatal model, we were therefore prompted to revisit the literature to see if there was any evidence of a return path from SNr to striatum. We were assured that there was certainly no mention of a direct connection, confirming this model as a potential candidate for a control. However, the recent study by Mailly et al (2003) showed that axonal projections of SNr cells innervated SNc and partly occupied the SNc region projecting to the striatal sector from which parent SNR cells receive their afferents, thereby establishing an "indirect nigrostriatonigral circuit" (Mailly et al., 2003). Assuming a facilitatory influence of dopamine in striatum (consistent with its influence via D1 receptors) the net effect of this loop would be to attenuate signals, just as it does in the putative control model. The nigro-striatal model was therefore not a strong candidate for an unrealistic model and so we withdrew it from the control set. We now had two control models supporting the selection hypothesis, and a new model with biologically realistic features (the nigro-striatal model) demonstrating performance enhancement with some weightings, which therefore also supported the hypothesis.

4 Concluding Remarks

We have advocated the use of computational hypotheses in order to organise the potentially overwhelming array of data that may be incorporated during the process of model construction. In particular, we have argued for the hypothesis of 'action selection' as an organising principle in dealing with the basal ganglia. Further, we have used a modelling approach we have dubbed 'mechanism mining' in which, once armed with a computational hypothesis, we go immediately to the data, to seek mechanisms that support this hypothesis. (This is to be placed in contrast with an alternative approach we have termed 'mechanism mapping' Gurney et al., 2004a which first posits a high level algorithm to perform the computation, before attempting to map the mechanisms supporting the algorithm onto the known biological substrate.) We argued that, having established a 'baseline' model that apparently supports the required computation (section 2), the model may then be augmented with additional, known biologically realistic features, to see if these features provide further mechanisms that support the proposed computation. This may necessitate the refinement and quantification of the hypothesis if the effects of the new pathways are to be properly assessed. Improved performance is evidence for the computational hypothesis being true, whereas a decrement in performance is indicative of its inadequacy. By deliberately attempting to construct models with unrealistic pathways, we may further test the hypothesis, since these control models should show a decrement in performance. This methodology was successfully applied in the work reported in (Gurney et al., 2004b) and summarised in section 03.

The refinement of the overall selection computational hypothesis, its quantitative description, and putative mechanisms that support selection, constitute a theoretical framework for basal ganglia function. Individual models with particular pathways are now not the items of primary interest *per se*; rather, the models are tools for testing the theory in a quantitative way. This move from simulation modelling to theory construction constitutes a large programme of multidisciplinary work requiring a combination of mathematical analysis, further simulation modelling, and experimental neuroscience if it is to be carried forward successfully.

References

Aizman O, Brismar H, Uhlen P, Zettergren E, Levey AI, Forssberg H, Greengard P, Aperia A (2000) Anatomical and physiological evidence for D1 and D2 dopamine receptor colocalization in neostriatal neurons. *Nat Neurosci* **3**:226-230.

Akkal D, Burbaud P, Audin J, Bioulac B (1996) Responses of substantia nigra pars reticulata neurons to intrastriatal d1 and d2 dopaminergic agonist injections in the rat. *Neuroscience Letters* **213**:66-70.

Albin RL, Young AB, Penney JB (1989) The functional anatomy of basal ganglia disorders. *Trends in Neurosciences* **12**:366-375.

Alexander GE, DeLong MR, Strick PL (1986) Parallel organization of functionally segregated circuits linking basal ganglia and cortex. *Ann Rev Neurosci* **9**:357-381.

Berridge KC, Robinson TE (1998) What is the role of dopamine in reward: hedonic impact, reward learning, or incentive salience? *Brain Res Brain Res Rev* **28**:309-369.

Bevan MD, Booth PA, Eaton SA, Bolam JP (1998) Selective Innervation of Neostriatal Interneurons by a Subclass of Neuron in the Globus Pallidus of the Rat. *Journal of Neuroscienc* **18**:9438-9452.

Chang HT, Wilson CJ, Kitai ST (1981) Single neostriatal efferent axons in the globus pallidus: a light and electron microscopic study. *Science* **213**:915-918.

Cools AR (1980) Role of the neostriatal dopaminergic activity in sequencing and selecting behavioural strategies: Facilitation of processes involved in selecting the best strategy in a stressful situation. *Behav Brain Res* **1**:361-378.

Deniau JM, Chevalier G (1985) Disinhibition as a basic process in the expression of striatal functions II. The striato-nigral influence on thalamocortical cells of the ventromedial thalamic nucleus. *Brain Research* **334**:227-233.

Deniau JM, Kitai ST, Donoghue JP, Grofova I (1982) Neuronal interactions in the substantia nigra pars reticulata through axon collateral of the projection neurons. *Experimental Brain Research* **47**:105-113.

Duncan J (1995) Attention, intelligence and the frontal lobes. In: The Cognitive Neurosciences (Gazzaniga MS, ed), pp 721-734. Cambridge MA.: MIT Press.

Gerfen CR, Engber TM, Mahan LC, Susel Z, Chase TN, Monsma FJ, Sibley DR (1990) D1 and D2 dopamine receptor regulated gene expression of striatonigral and striatopallidal neurons. *Science* **250**:1429-1431.

Gillies A, Arbuthnott GW (2000) Computational models of the basal ganglia. *Movement Disorders* **15**:762-770.

Gurney K, Prescott TJ, Redgrave P (2001a) A computational model of action selection in the basal ganglia I: A new functional anatomy. *Biological Cybernetics* **84**:401-410.

Gurney K, Prescott TJ, Redgrave P (2001b) A computational model of action selection in the basal ganglia II: Analysis and simulation of behaviour. *Biological Cybernetics* **84**: 411-423.

Gurney K, Prescott TJ, Wickens JR, Redgrave P (2004a) Computational models of the basal ganglia: from robots to membranes. *Trends Neurosci* **27**:453-459.

Gurney K, Wood R, Humphries M, Prescott TJ, Redgrave P (2004b) Testing computational hypotheses of brain systems function using high level models: a case study with the basal ganglia. Network: Computation in Neural Systems.

Harsing LG, Zigmond MJ (1997) Influence of dopamine on GABA release in striatum: evidence for D1-D2 interactions and non-synaptic influences. *Neuroscience* **77**:419-429.

Hoover JE, Strick PL (1993) Multiple output channels in the basal ganglia. *Science* **259**: 819-821.

Horvitz JC (2000) Mesolimbocortical and nigrostriatal dopamine responses to salient non-reward events. *Neuroscience* **96**:651-656.

Jaeger D, Kita H, Wilson CJ (1994) Surround inhibition among projection neurones is weak or nonexistent in the rat neostriatum. *Journal of Neurophysiology* **72**:255-2558.

Juraska JM, Wilson CJ, Groves PM (1977) The substantia nigra of the rat: a Golgi study. *J Comp Neurol* **172**:585-600.

Kawaguchi Y, Wilson JW, Emson P (1990) Projection subtypes of rat neostriatal matrix cells revealed by intracelluar injection of biocytin. *Journal of Neuroscience* **10**:3421-3436.

Koos T, Tepper JM, Wilson CJ (2004) Comparison of IPSCs evoked by spiny and fast-spiking neurons in the neostriatum. *J Neurosci* **24**:7916-7922.

Kropotov JD, Etlinger SC (1999) Selection of actions in the basal ganglia-thalamocortical circuits: review and model. International *Journal of Psychophysiology* **31**:197-217.

Mailly P, Charpier S, Menetrey A, Deniau JM (2003) Three-dimensional organization of the recurrent axon collateral network of the substantia nigra pars reticulata neurons in the rat. *J Neurosci* **23**:5247-5257.

Marr D (1982) Vision: A computational investigation into human representation and processing of visual information. New York: W.H. Freeeman and Co.

McFarland D (1989) Problems of Animal Behaviour. In. New York: Longman Scientific and Technical.

Middleton FA, Strick PL (2002) Basal-ganglia 'projections' to the prefrontal cortex of the primate. *Cereb Cortex* **12**:926-935.

Mink JW (1996) The basal ganglia: Focused selection and inhibition of competing motor programs. *Progress in Neurobiology* **50**:381-425.

Mink JW, Thach WT (1993) Basal ganglia Intrinsic circuits and their role in behavior. *Current Opinion in Neurobiology* **3**:950-957.

Moriizumi T, Nakamura Y, Okoyama S, Kitao Y (1987) Synaptic organization of the cat entopeduncular nucleus with special reference to the relationship between the afferents to entopedunculothalamic projection neurons: an electron microscope study by a combined degeneration and horseradish peroxidase tracing technique. *Neuroscience* **20**:797-816.

Nicola SM, Surmeier DJ, Malenka RC (2000) Dopaminergic modulation of neuronal excitability in the striatum and nucleus acumbens. *Annual Reviews of Neuroscience* **23**:185-215.

Parent A, Hazrati LN (1993) Anatomical aspects of information processing in primate basal ganglia. *Trends in Neuroscience* **16**:111-116.

Parent A, Hazrati LN (1995) Functional anatomy of the basal ganglia 2. The place of subthalamic nucleus and external pallidum in basal ganglia circuitry. *Brain Research Reviews* **20**:128-154.

Park MR, Falls WM, Kitai ST (1982) An intracellular HRP study of the rat globus pallidus. I. Responses and light microscopic analysis. *J Comp Neurol* **211**:284-294.

Prescott TJ, Redgrave P, Gurney K (1999) Layered control architectures in robots and vertebrates. *Adaptive Behaviour* 7:99-127.

Prescott TJ, Gurney K, Redgrave P (2002) The Basal Ganglia: The Handbook of Brain Theory and Neural Networks. In: (Arbib M, ed): MIT Press.

Redgrave P, Prescott TJ, Gurney K (1999a) The basal ganglia: a vertebrate solution to the selection problem? *Neuroscience* **89**:1009-1023.

Redgrave P, Prescott TJ, Gurney K (1999b) Is the short-latency dopamine response too short to signal reward error? *Trends in Neurosciences* **22**:146-151.

Salamone JD, Cousins MS, Synder BJ (1997) Behavioural functions of nucleus accumbens dopamine: empirical and conceptual problems with the anhedonia hypothesi. *Neuroscience and Biobehavioral Reviews* **21**:341-359.

Schultz W (1998) Predictive reward signal of dopamine neurons. *Journal of Neurophysiology* **80**:1-27.

Smith Y, Bevan MD, Shink E, Bolam JP (1998) Microcircuitry of the direct and indirect pathways of the basal ganglia. *Neuroscience* **86**:353-387.

Surmeier DJ, Song WJ, Yan Z (1996) Coordinated expression of dopamine receptors in neostriatal medium spiny neurons. *Journal of Neuroscience* **16**:6579-6591.

Tunstall MJ, Oorschot DE, Kean A, Wickens JR (2002) Inhibitory interactions between spiny projection neurons in the rat striatum. *Journal of Neurophysiology* **88**:1263-1269.

Venance L, Glowinski J, Giaume C (2004) Electrical and chemical transmission between striatal GABAergic output neurones in rat brain slices. *J Physiol* **559**:215-230.

Walters JR, Ruskin DN, Allers KA, Bergstrom DA (2000) Pre- and postsynaptic aspects of dopamine-mediated transmission. *Trends Neurosciences* **10** (supplement):S1471-S1931.

Wu Y, Richard S, parent A (2000) The organization of the striatal output system: a single-cell juxtacellular labeling study in the rat. *Neuroscience Research* **38**:49-62.

In: Recent Breakthroughs in Basal Ganglia Research
Editor: Erwan Bezard, pp. 17-25
ISBN: 1-59454-880-3

Chapter 2

CONDUCTANCE-BASED MODELS OF STN-GPE-GPI CIRCUITS: FROM BIOPHYSICS TO BEHAVIOR†

Leonid L. Rubchinsky*
Department of Mathematical Sciences, IUPUI and Stark Neurosciences Research Institute, Indiana University School of Medicine, Indianapolis, IN 46204
Nancy Kopell#
Center for BioDynamics and Department of Mathematics, Boston University, Boston, MA 02215
Karen A. Sigvardt^
Department of Mathematical Sciences, IUPUI and Stark Neurosciences Research Institute, Indiana University School of Medicine, Indianapolis, IN 46204
Center for Neuroscience and Department of Neurology, University of California, Davis, CA 95616

ABSTRACT

Classical standard models of basal ganglia motor circuits are essentially "box and arrow" models, representing the different parts of the basal ganglia-thalamocortical network and the anatomical connections between them. The power of these models in providing explanations of the role of the basal ganglia in motor control and its dysfunction in basal ganglia disorders such as Parkinson's disease has acknowledged limitations, in particular due to the static nature of the models compared to the dynamic nature of the physiology and pathophysiology of basal ganglia networks. Here we consider conductance-based modeling of basal ganglia circuitry in relation to motor control. This approach opens ways for connecting studies at the basal ganglia at the cellular and behavioral levels. Modeling can explain available experimental and clinical data, and also can help to design new experiments and suggest potential strategies for treatment of basal ganglia disorders. However, given the large number of unknown parameters of these circuits, caution must be applied when drawing conclusions from these modeling efforts.

INTRODUCTION

Physiological mechanisms responsible for the involvement of the basal ganglia (BG) in motor control as well as the pathophysiology responsible for parkinsonian motor symptoms remain unclear. The standard

† This work was supported by 1R01 NS 39121 to KAS.
* Email address: leo@math.iupui.edu
# Email address: nk@bu.edu
^ Email address: kasigvardt@ucdavis.edu

models of BG function (Albin et al., 1989; DeLong, 1990) provided a simple framework for understanding the hypo- and hyperkinetic extrapyramidal movement disorders (e.g., Parkinson's disease (PD) and Huntington's disease). However, many experimental and clinical observations conflict with this model (reviewed by Mink, 1996; Wichmann and DeLong, 1996; Levy et al., 1997). The conceptual framework of this standard "box-and-arrow" model lacks predictive power. Therefore, there is a strong interest in models that may be formalized and thus amenable to analytical or numerical studies; this could better explain available experimental data, predict the outcomes of future experiments and to provide new insights into BG physiology and treatment of BG-associated disorders. The need for biophysically based modeling of the basal ganglia that can adequately address the dynamics associated with different ionic currents present in BG neurons and the spatio-temporal organization of activity within BG nuclei, is widely acknowledged (see, e.g. Wichmann and DeLong, 1996; Bar-Gad and Bergman, 2001; Bergman and Deuschl, 2002). For example, recent beginning of biophysically-based modeling (Terman et al., 2002) examined the dynamics of model networks of GPe and STN neurons and found that the biophysical properties of the neurons can support a variety of spatio-temporal oscillatory activity patterns.

In studies of the relationships between behavior and neural activity in the BG, recordings have revealed many different patterns of activity, differently timed in relation to movement (reviewed by Kimura and Matsumoto, 1997). Based on a series of experimental observations (Georgopoulos et al., 1983; Anderson and Horak, 1985; Mink and Thach, 1991), Mink, (1996) suggested that the function of the BG is to control competing motor programs, inhibiting those that interfere with the voluntary movement being executed, and focusing disinhibition on this desired movement (see also Mink and Thach, 1993; Hikosaka et al., 2000; Boraud et al., 2002). This hypothesis was supported by further experimental observations of pallidal movement-related activity (Turner and Anderson, 1997; Wenger et al., 1999; Boraud et al., 2000) and is a basis for our model. In this context parkinsonian neurodegeneration results in an inability to correctly inhibit unwanted and support wanted motor programs, leading to akinesia, bradykinesia and rigidity.

Modeling of the BG's role in motor control requires a study of how BG networks respond to cortical input to influence thalamocortical circuits responsible for motor control. This paper considers the development and analysis of a minimal biophysically-based model of BG motor control circuitry (see also Rubchinsky et al., 2003). In the construction of a model network for control of motor programs we focused on the pallido-subthalamic subnetwork of BG-thalamo-cortical motor control circuitry. Certainly, a model constructed in this way will be limited in many ways and cannot, of course, describe striatal function. However, it may provide important insight into the mechanisms of simple motor control tasks, such as motor switching, which is critically impaired in PD (e.g. simple opposing movements such as reaching/withdrawing or the pronation/supination of the clinical neurological exam; (Benecke et al., 1987; Weiss et al., 1997). Support of this strategy is grounded in the fact that the pallidum and STN are effective sites for surgical procedures to alleviate the motor symptoms of PD and thus have immediate relation to BG motor control and its impairment in PD. Also providing support of this approach are recent studies which show that during deep brain stimulation (DBS) of STN there is no detectable increase in striatal dopamine (Abosch et al., 2003; Hilker et al., 2003). Similarly, in a study of rodents administered MPTP followed by neuroprotective treatment to recover nigral cells, improvement in motor behavior was not correlated with the improvement of striatal dopaminergic innervation (Crocker et al., 2003).

Besides their relevance for motor control and hypokinetic behavior in PD, GP-STN networks may be involved into parkinsonian tremor generation. Not only has tremor-related activity been observed in STN and GPi of parkinsonian patients, but also experiments with *in vitro* organotypic cultures of STN and pallidal cells (Plenz and Kital, 1999) and numerical experiments with GP-STN networks (Bevan et al., 2002; Terman et al., 2002) have shown the ability of the GP-STN network to exhibit complex oscillatory behavior. Even though the issue of the origin of tremor genesis remains controversial, it is quite probable, that the STN-GP networks must be involved in tremorgenesis, if not responsible for it (Deuschl et al., 2000).

1 Model Network Design

To construct a minimal network with the ability to reproduce basic features of motor program switching, we assume that groups of STN, GPe and GPi neurons are transiently formed by inputs from cortex and striatum to participate in execution of a particular motor program. Other groups of pallidal and subthalamic neurons, at that moment, act to suppress motor programs that may interfere with the desired motion. This pattern of activation during movement is supported by experimental data recorded from non-human primates (e.g. Turner and Anderson, 1997). New movement would require formation of new subsets of neurons within each nucleus. In our model circuit (Fig. 1) we have two subsets of GPe, STN and GPi, each of which represents a transiently formed group of neurons. Input from the cortex and striatum determine which subset in their target structures are to be activated or suppressed. The simple circuit considered here is limited to the control of the minimal number of motor programs - two, the goal being to represent the transient execution of only one motor program at a time ("wanted") and suppression of the other ("unwanted"). The connectivity of the circuit is based on the known anatomy of BG (e.g., reviewed by Wilson, 1998). What we have added to this proposed circuit is the way in which pallidal and STN subsets are connected to one another. The cross-connections were chosen to provide a mechanism for discriminating between cortical signals for the desired movement and inappropriate cortical inputs, and we will show below that other patterns of connectivity lack this mechanism.

In our minimal model of GP-STN circuits, striatal and cortical inputs are represented as external signals to the model circuit with assigned timing targeted to specific subsets of the circuit. These input signals are not derived from biophysical modeling of corticostriatal networks. Instead, we use simple forms of input signals to GP-STN networks. This strategy simplifies the analysis of the model and allows one to study the role of these networks in motor control. The striatum has been studied extensively and as we build more complex models these studies may provide the necessary experimental constraints on the timing and structure of corticostriatal input.

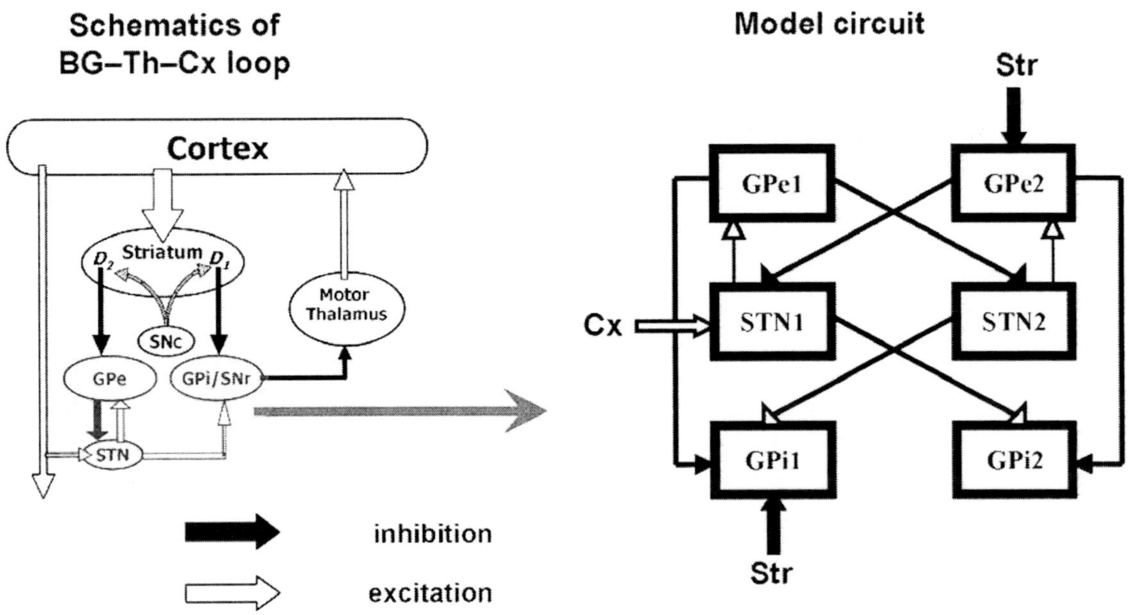

Fig. 1. Model circuit is based on the anatomical organization of basal ganglia - thalamocortical pathways. The depicted pattern of corticostriatal input to the model circuit corresponds to execution of motor program 1 ("wanted") and inhibition of motor program 2 ("unwanted"). The reversed, complementary structure of corticostriatal input will reverse "wanted" and "unwanted" motor programs.

2 Biophysical Properties and Models of Cells

The cellular physiology of STN and GPe has been studied extensively; e.g., STN (Bevan and Wilson, 1999; Beurrier et al., 2000; Bevan et al., 2000) and GPe (Kita and Kitai, 1991; Nambu and Llinas, 1994; Cooper and Stanford, 2000). Those studies gave rise to the formulation of conductance-based models of STN and GPe cells (Terman et al., 2002). Experimental data available on GPi cell physiology is limited. Our GPi model has the same form as the GPe model, but its parameters are adjusted in such a way that the network behavior produces tonic GPi activity with a spiking rate similar to that observed *in vivo*. All these models include standard sodium, potassium and leak currents and incorporate low threshold T-type Ca^{++}-current, high-threshold Ca^{++} current, and Ca^{++}-activated voltage independent after hyperpolarization K^{+}-current. Slow calcium currents enrich the repertoire of cellular dynamics and can produce crucial network dynamics properties. Synaptic connections in the model circuit represent ionotropic glutamatergic and GABAergic synapses. The GPe-STN inhibitory projection is also involves metabotropic $GABA_B$ synapses (Charara et al., 2000; Smith et al., 2001). The addition of these slow-acting synapses can inhibit STN for a sufficiently long time to induce a strong rebound after release from inhibition. Details of the neuronal models and models of synaptic dynamics in the form of coupled ordinary differential equations can be found in (Terman et al., 2002; Rubchinsky et al., 2003).

3 From Biophysics to Behavior. Numerical Experiments on Switching between Motor Programs

3.1 Switching between Motor Programs: Response of the Model Network to Corticostriatal Inputs

A command input to execute a single movement (motor program 1) is modeled as a train of spikes arriving simultaneously in GPe2, STN1 and GPi1 - Input 1 (from t=1000 ms to t=2000 ms, Fig. 2, non-overlapping inputs). This leads to a strong increase in STN1 firing frequency due to the cortical input. In turn, STN firing leads to an increased frequency in GPe1, which inhibits GPi1 and STN2 (so STN2 does not excite GPi1). Striatal input to GPe2 releases STN1 and GPi2 from inhibition and striatal input to GPi1 helps to suppress GPi1 firing. As a result GPi1 ceases to fire spikes and GPi2 firing rate increases. So, during movement 1, GPi1 is silent (releasing the corresponding parts of motor thalamus from inhibition, thus facilitating the desired movement) and GPi2 is overactive, thus inhibiting thalamocortical circuits responsible for "unwanted" motor programs.

Motor program switching is simulated by switching corticostriatal input from one set of network subsets to the complementary one. So Input 2 will drive GPe1, STN2 and GPi2 (from t=2000 ms to t=3000 ms, Fig. 2, non-overlapping inputs). GPi1 fires with a higher than baseline rate when Input 2 is active, while GPi2 has the opposite behavior. As a result, motor program 1 is facilitated and motor program 2 is inhibited during activation of Input 1, and when the input is switched to facilitate motor program 2 and to suppress the previously executed motor program 1, the pattern of GPi activity follows it.

Now consider the dynamics of STN-GP networks under conflicting inputs, i.e., when two inputs, which correspond to conflicting motor programs 1 and 2, coexist for a certain period of time. For numerical simulations of conflicting corticostriatal inputs we consider input signals overlapped in time: Input 1 ends after Input 2 starts (Fig. 2, overlapping inputs; Input 1 ends 500 ms after Input 2 starts). The output of the network is essentially similar to the non-overlapping inputs case. At t=2000 ms the output pattern changes to terminate ongoing movement and promote the new one, even though the input signal for the old movement is still active. This is because high-frequency cortical input to STN2 at t=2000 ms depolarizes STN2, which leads to a strong rebound burst of spikes in STN2 (Fig. 2, activity of STN). This rebound spiking in STN2 excites GPi1 and excites GPe2 (in spite of inhibitory input to GPe2, which is still present). This, in turn, helps to inhibit GPi2 and inhibit STN1 (which, otherwise, would excite GPi2), even though STN1 continues to receive cortical excitation.

3.2 The Influence of Model Network's Parameters and Structure on the Dynamics of Switching

The fast switching relies on certain features of the network parameters. Numerical simulations with a network with weaker $GABA_B$ synapses show the role of the slow metabotropic $GABA_B$ projection from GPe to STN (Fig. 3). The network with weak $GABA_B$ synapses is subject to the same overlapping corticostriatal inputs considered above. The dynamics of GPi are not as extreme as in the example in Fig. 2 because the total amount of inhibition in the network has been changed, but the general pattern of GPi dynamics is easy to see. When there is only Input 1 present, GPi activity facilitates motor program 1 and inhibits motor program 2. However, when the input signals overlap in time both subsets of the BG output nuclei produce high-frequency spiking (Fig. 3). Therefore, both thalamocortical circuits, corresponding to two different motor programs, are inhibited and no movement is performed. This can be interpreted as an akinetic condition. In this case, there is no rebound burst in STN (Fig. 3, activity of STN), which was provided by $GABA_B$ synapses and transient calcium T-current.

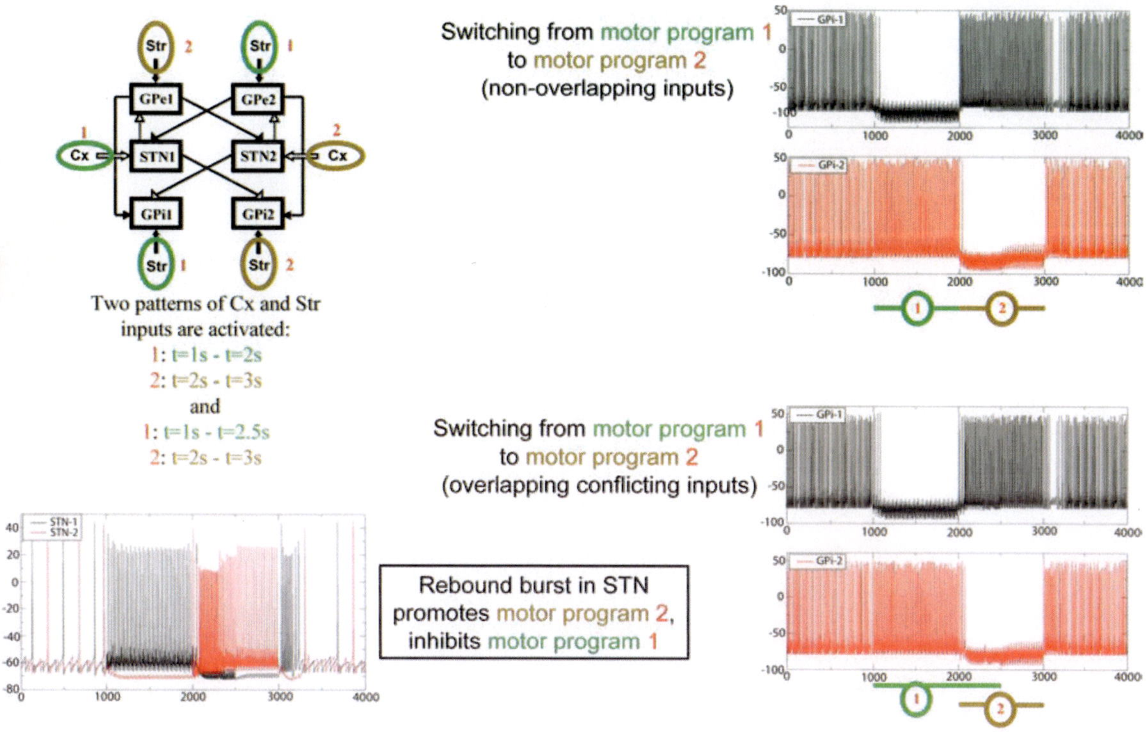

Fig. 2. The response of model network to non-overlapping and overlapping corticostriatal input signals, corresponding to two different motor programs. Green and brown horizontal bars indicate durations of Inputs 1 and 2 respectively. The output of the network (GPi activity) is very similar in both cases.

The structure of the model network is also crucial for performance of motor switching. Addition of reciprocal projections between GPe and STN (in addition to the non-reciprocal links in the network – see Fig. 4) affects this ability. Weak reciprocal links do not affect the performance of the network in response to the overlapping inputs. But the ability to discriminate between conflicting motor programs is gradually lost with the increase of the strengths of reciprocal projections. As a result, the network behavior is similar to the one observed with weak $GABA_B$ synapses (Fig. 3): both GPi segments are activated, which leads to the inhibition of thalamus and prevention of the execution of both old and new motor programs – an akinesia-like condition. Reciprocal links may provide a negative feedback to STN, when STN is being excited by cortex. Therefore the STN firing rate becomes low, there is no rebound burst of activity at the time of initiation of the corticostriatal signal 2 and switching is degraded.

4 Conclusions from Our Model and Their Limitations

The prominent feature of PD motor symptoms is hypokinetic behavior: akinesia and bradykinesia. Akinesia-like behavior is exhibited in the model network when both subsets of GPi become active and significantly inhibit thalamocortical circuits, thus preventing facilitation of any motor program. This situation was observed in a simulation of the disrupted model network under the influence of competing corticostriatal inputs. The model predicts that if the $GABA_B$ GPe to STN projection is weak or absent, then both GPi subsets are active, which is a signature of hypokinetic behavior. A related prediction is that the parkinsonian symptoms should be accompanied by a deficit in the function of metabotropic GABAergic projections from GPe to STN. The introduction into the model network of reciprocal connections between BG nuclei subsets in addition to non-reciprocal connections also lead to akinetic dynamics. In PD there is a loss of specificity of neuronal responses to passive movements in GPi (Filion et al., 1988; Boraud et al., 2000) and increased correlations between activity in distant neural territories (reviewed by Bergman et al., 1998). Action of reciprocal connections (which would be inhibited under normal conditions) may be a mechanism that contributes to this loss of specificity. These predictions from the model can be tested in experiments. Determining if the predicted effects occur in the BG in PD requires studies of changes following dopaminergic degeneration.

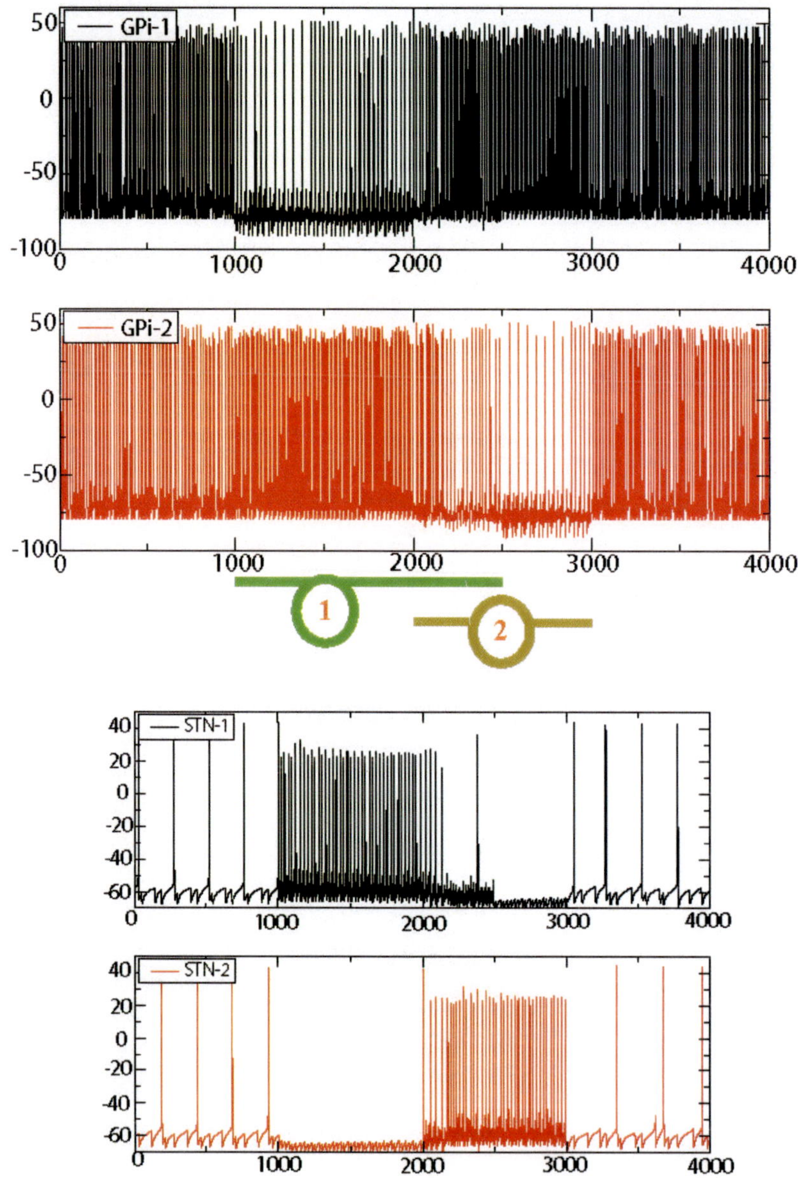

Fig. 3. The dynamics of model network with weak $GABA_B$ GPe-to-STN projection with overlapping corticostriatal inputs, corresponding to two different motor programs. Motor switching is impaired (cf. Fig. 2). Activity of STN2 lacks rebound burst. Green and brown horizontal bars indicate durations of Inputs 1 and 2 respectively. The output of the network with strong reciprocal projections between GPe and STN is very similar to that depicted here.

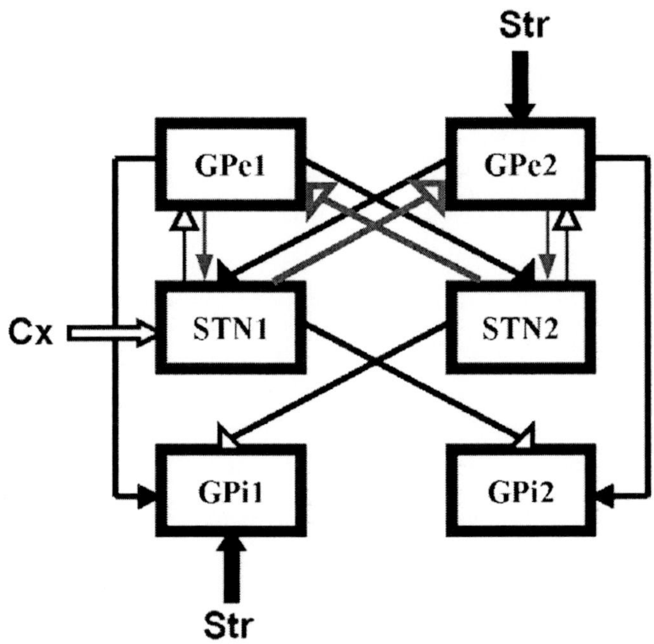

Fig. 4. Addition of reciprocal projections between GPe and STN.

The advantage of such computational dynamical modeling based on known biophysical properties of cells is that such an approach can be matched with the experimental data available and can provide a framework in which to understand the cellular mechanisms of normal and pathological functioning of basal ganglia motor circuits. Such an approach connects motor behavior and the dynamical aspects of basal ganglia function and its pathology with data at the cellular level.

However, the results of the modeling should be considered with caution. As with any model, the correctness of this model depends on the correctness of underlying assumptions (which may be questionable). Since we do not model striatum in a self-consistent way, our model cannot properly describe effects of striatal dynamics, which are surely involved in regulation of motor behavior. Moreover, there are a number of significant phenomena related to BG function that are completely beyond the model considered here (such as reward prediction by nigral neurons, e.g., Schultz, 2002). What this model and like models give us is a possible repertoire of different phenomena, which can arise from the biophysical and anatomical properties used to build this model.

5 What to Expect/Demand from the Future Studies

We would like to make several remarks about the prospects of our model and other models of basal ganglia. Our minimal model describes control of only two different motor programs. We believe that this network is scalable to the large number of pallidal and STN subsets needed to investigate the mechanisms responsible for focused activation and rapid transition among the individual motor programs that accompany the fluid movements during on-going behavior. However, full understanding of BG motor program control will, of course, require more adequate modeling of the spatial and temporal patterns of input into the pallidal/STN networks from the cortex and striatum, which are necessary for the transient formation of the appropriate cell groups within the subthalamo-pallidal networks.

Biophysically-based models should be helpful in design of treatment strategies, as they provide a way to test the outcome of specific treatment procedure *in silica*. The recent paper by Rubin and Terman, (2004) is an example of how conductance-based models may provide new conjectures about mechanisms of deep brain stimulation in treatment of PD. Furthermore, simpler models, which are not based on cellular biophysics and which do not immediately connect behavior and cellular properties, are easier to study both numerically and analytically, and may provide significant insights into BG function or dysfunction, as discussed in other contributions to this book, so we do not refer to these studies here.

Finally, even though motor function is strongly impaired in PD and is the main focus of the present work, PD is also accompanied by an array of cognitive symptoms. The parallels between motor and cognitive deficits (and thus between BG control of motor and cognitive processes) are intriguing (Benecke et al., 1987; Weiss et al., 1997; Franz and Miller, 2002; Shook et al., 2004; see also a discussion by Franz, (2005) in this book). Models similar to those considered here may aid in the exploration of these parallels.

REFERENCES

Abosch A, Kapur S, Lang AE, Hussey D, Sime E, Miyasaki J, Houle S, Lozano AM (2003) Stimulation of the subthalamic nucleus in Parkinson's disease does not produce striatal dopamine release. *Neurosurgery* **53**:1095-1102; discussion 1102-1095.

Albin RL, Young AB, Penney JB (1989) The functional anatomy of basal ganglia disorders. *Trends Neurosci* **12**:366-375.

Anderson ME, Horak FB (1985) Influence of the globus pallidus on arm movements in monkeys. III. Timing of movement-related information. *J Neurophysiol* **54**:433-448.

Bar-Gad I, Bergman H (2001) Stepping out of the box: information processing in the neural networks of the basal ganglia. *Curr Opin Neurobiol* **11**:689-695.

Benecke R, Rothwell JC, Dick JP, Day BL, Marsden CD (1987) Disturbance of sequential movements in patients with Parkinson's disease. *Brain* **110** (Pt 2):361-379.

Bergman H, Deuschl G (2002) Pathophysiology of Parkinson's disease: from clinical neurology to basic neuroscience and back. Mov Disord 17 Suppl 3:S28-40.

Bergman H, Feingold A, Nini A, Raz A, Slovin H, Abeles M, Vaadia E (1998) Physiological aspects of information processing in the basal ganglia of normal and parkinsonian primates. *Trends Neurosci* **21**:32-38.

Beurrier C, Bioulac B, Hammond C (2000) Slowly inactivating sodium current (I(NaP)) underlies single-spike activity in rat subthalamic neurons. *J Neurophysiol* **83**:1951-1957.

Bevan MD, Wilson CJ (1999) Mechanisms underlying spontaneous oscillation and rhythmic firing in rat subthalamic neurons. *J Neurosci* **19**:7617-7628.

Bevan MD, Wilson CJ, Bolam JP, Magill PJ (2000) Equilibrium potential of GABA(A) current and implications for rebound burst firing in rat subthalamic neurons in vitro. *J Neurophysiol* **83**:3169-3172.

Bevan MD, Magill PJ, Terman D, Bolam JP, Wilson CJ (2002) Move to the rhythm: oscillations in the subthalamic nucleus-external globus pallidus network. *Trends Neurosci* **25**:525-531.

Boraud T, Bezard E, Bioulac B, Gross CE (2000) Ratio of inhibited-to-activated pallidal neurons decreases dramatically during passive limb movement in the MPTP-treated monkey. *J Neurophysiol* **83**:1760-1763.

Boraud T, Bezard E, Bioulac B, Gross CE (2002) From single extracellular unit recording in experimental and human Parkinsonism to the development of a functional concept of the role played by the basal ganglia in motor control. *Prog Neurobiol* **66**:265-283.

Charara A, Heilman TC, Levey AI, Smith Y (2000) Pre- and postsynaptic localization of GABA(B) receptors in the basal ganglia in monkeys. *Neuroscience* **95**:127-140.

Cooper AJ, Stanford IM (2000) Electrophysiological and morphological characteristics of three subtypes of rat globus pallidus neurone in vitro. *J Physiol* **527** Pt 2:291-304.

Crocker SJ, Smith PD, Jackson-Lewis V, Lamba WR, Hayley SP, Grimm E, Callaghan SM, Slack RS, Melloni E, Przedborski S, Robertson GS, Anisman H, Merali Z, Park DS (2003) Inhibition of calpains prevents neuronal and behavioral deficits in an MPTP mouse model of Parkinson's disease. *J Neurosci* **23**:4081-4091.

DeLong MR (1990) Primate models of movement disorders of basal ganglia origin. *Trends Neurosci* **13**:281-285.

Deuschl G, Raethjen J, Baron R, Lindemann M, Wilms H, Krack P (2000) The pathophysiology of parkinsonian tremor: a review. *J Neurol* **247** Suppl 5:V33-48.

Filion M, Tremblay L, Bedard PJ (1988) Abnormal influences of passive limb movement on the activity of globus pallidus neurons in parkinsonian monkeys. *Brain Res* **444**:165-176.

Franz E (2005) Converging evidence of the basal ganglia's role in focused action selection and inhibition of competing information. In: Chapter in this book.

Franz EA, Miller J (2002) Effects of response readiness on reaction time and force output in people with Parkinson's disease. *Brain* **125**:1733-1750.

Georgopoulos AP, DeLong MR, Crutcher MD (1983) Relations between parameters of step-tracking movements and single cell discharge in the globus pallidus and subthalamic nucleus of the behaving monkey. *J Neurosci* **3**:1586-1598.

Hikosaka O, Takikawa Y, Kawagoe R (2000) Role of the basal ganglia in the control of purposive saccadic eye movements. *Physiol Rev* **80**:953-978.

Hilker R, Voges J, Ghaemi M, Lehrke R, Rudolf J, Koulousakis A, Herholz K, Wienhard K, Sturm V, Heiss WD (2003) Deep brain stimulation of the subthalamic nucleus does not increase the striatal dopamine concentration in parkinsonian humans. *Mov Disord* **18**: 41-48.

Kimura M, Matsumoto N (1997) Neuronal activity in the basal ganglia. Functional implications. *Adv Neurol* **74**:111-118.

Kita H, Kitai ST (1991) Intracellular study of rat globus pallidus neurons: membrane properties and responses to neostriatal, subthalamic and nigral stimulation. *Brain Res* **564**:296-305.

Levy R, Hazrati LN, Herrero MT, Vila M, Hassani OK, Mouroux M, Ruberg M, Asensi H, Agid Y, Feger J, Obeso JA, Parent A, Hirsch EC (1997) Re-evaluation of the functional anatomy of the basal ganglia in normal and Parkinsonian states. *Neuroscience* **76**: 335-343.

Mink JW (1996) The basal ganglia: focused selection and inhibition of competing motor programs. *Prog Neurobiol* **50**:381-425.

Mink JW, Thach WT (1991) Basal ganglia motor control. III. Pallidal ablation: normal reaction time, muscle cocontraction, and slow movement. *J Neurophysiol* **65**:330-351.

Mink JW, Thach WT (1993) Basal ganglia intrinsic circuits and their role in behavior. *Curr Opin Neurobiol* **3**:950-957.

Nambu A, Llinas R (1994) Electrophysiology of globus pallidus neurons in vitro. *J Neurophysiol* **72**:1127-1139.

Plenz D, Kital ST (1999) A basal ganglia pacemaker formed by the subthalamic nucleus and external globus pallidus. *Nature* **400**:677-682.

Rubchinsky LL, Kopell N, Sigvardt KA (2003) Modeling facilitation and inhibition of competing motor programs in basal ganglia subthalamic nucleus-pallidal circuits. *Proc Natl Acad Sci U S A* **100**:14427-14432.

Rubin JE, Terman D (2004) High frequency stimulation of the subthalamic nucleus eliminates pathological thalamic rhythmicity in a computational model. *J Comput Neurosci* **16**:211-235.

Schultz W (2002) Getting formal with dopamine and reward. *Neuron* **36**:241-263.

Shook S, Franz E, Higginson C, Sigvardt K (2004) Dopamine dependency of cognitive switching operations in Parkinson's patients. Submitted to Neuropsychologia.

Smith Y, Charara A, Paquet M, Kieval JZ, Pare JF, Hanson JE, Hubert GW, Kuwajima M, Levey AI (2001) Ionotropic and metabotropic GABA and glutamate receptors in primate basal ganglia. *J Chem Neuroanat* **22**:13-42.

Terman D, Rubin JE, Yew AC, Wilson CJ (2002) Activity patterns in a model for the subthalamopallidal network of the basal ganglia. *J Neurosci* **22**:2963-2976.

Turner RS, Anderson ME (1997) Pallidal discharge related to the kinematics of reaching movements in two dimensions. *J Neurophysiol* **77**:1051-1074.

Weiss P, Stelmach GE, Hefter H (1997) Programming of a movement sequence in Parkinson's disease. *Brain* **120** (Pt 1):91-102.

Wenger KK, Musch KL, Mink JW (1999) Impaired reaching and grasping after focal inactivation of globus pallidus pars interna in the monkey. *J Neurophysiol* **82**:2049-2060.

Wichmann T, DeLong MR (1996) Functional and pathophysiological models of the basal ganglia. *Curr Opin Neurobiol* **6**:751-758.

Wilson C (1998) Basal Ganglia. In: The Synaptic Organization of the Brain (Shepherd G, ed), pp 329-375. New York: Oxford UP.

In: Recent Breakthroughs in Basal Ganglia Research
Editor: Erwan Bezard, pp. 27-44
ISBN 1-59454-880-3

Chapter 3

BASAL GANGLIA PHYSIOLOGY IS BUILT ON COMPETITION BETWEEN FEEDBACK LOOPS

***A. Leblois*[1,2], *T. Boraud*[2], *H. Bergman*[3,4] *and D. Hansel*[1,4]**
[1] Laboratoire de Neurophysique et de Physiologie
UMR 8119 CNRS, Université René Descartes, Paris 75270, France
[2] Laboratoire de Neurophysiologie, UMR 5543 CNRS,
Université Victor Ségalen, Bordeaux 33076, France
[3] Department of Physiology, The Hebrew University-Hadassah
Medical School, Jerusalem 91120, Israel
[4] Interdisciplinary Center for Neural Computation,
The Hebrew University, Jerusalem, 91904, Israel

Abstract

We have built a simpli ed model of the basal ganglia to investigate their dynamics focussing on how the competion between the positive and negative feedback loops shaped the activity in the cortex and in the GPi in normal an parkinsonian situations. The dynamical variable describing each neuron is its ring rate. The pattern of connections and their polarity are in agreement with anatomical data. Our analysis is performed using analytical calculations and numerical simulations. We nd that the model displays two different regimes, one in which the activity of the populations is stationnary and one in which it oscillates. An important parameter controling the dynamics is the amount of dopamine in the striatum. The oscillating regime could be related to the tremor observed in parkinsonian patients. The change in activity of the various populations during a movement in our model is in agreement with the data of the physiologists.

1 Introduction

A major pathway in the basal ganglia (BG) circuitry leads from most cortical areas to the striatum. Subsequent projections link striatal neurons to the BG output stage, i.e., the GPi (internal part of the pallidum) and SNr (substantia nigra pars reticulata), for simplicity lumped hereafter as GPi. The GABAergic inhibitory projections of GPi neurons control the activity of the excitatory thalamo-cortical networks.

One of the first modern models of the BG (the Albin-DeLong model, Albin et al., 1989; DeLong, 1990) assumes two segregated feed-forward pathways, the so-called direct and indirect pathways, from the striatum to the GPi. The direct pathway is made up of direct GABAergic projections to the GPi. The indirect pathway connects a different population of striatal neurons to the GPi via the GPe (external part of the pallidum) and the STN (subthalamic nucleus). The net effect of the striatum over the GPi is inhibitory for the direct pathway, and excitatory (due to double inhibition) for the indirect pathway. Finally, dopamine modulates the activity of striatal neurons that give rise to the direct and indirect pathways by D1 and D2 excitatory and inhibitory mechanisms, respectively. This classical model of BG received support from the finding of increased firing rates in the STN and GPi in PD (Alexander et al, 1990; DeLong, 1990). Moreover, it was shown that inactivation of these nuclei can improve the motor symptoms in Parkinsonian animals (Bergman et al., 1990; Aziz et al., 1991) and human patients (Benabid et al., 2000).

However, evidence has accumulated that this classical view may be insufficient in several respects. Recent studies indicate that D1 and D2 receptors are co-localized on striatal neurons (Aizman et al., 2000) and that all striatal neurons projecting to GPi also project to GPe (Wu et al., 2000; Bolam et al., 2000). Moreover, dopamine has a major presynaptic effect on the cortico-striatal transmission (Centonze et al., 1999; Calabresi et al., 2000; Kerr and Wickens, 2001) and D1/D2 post synaptic activation cannot merely be described as excitation/inhibition respectively (Nicola et al., 2000). Lesions of the GPi do not only improve the hypokinetic clinical characteristics of PD, but also alleviate hyperkinetic disturbances. In addition, lesions in the thalamus do not lead to Parkinson-like motor symptoms (Obeso et al., 2000). Finally, multi-unit recordings and cross-correlation analysis have shown that, more than the average firing rate, the pattern of activity is modified after dopamine depletion in the GPi (Raz et al, 2000; Hutchison, 1997), STN (Bergman, 1994; Levy, 2000), cortex (Goldberg, 2002) and striatum (Raz, 2001) of PD patient or in animal models the disease. The simple framework of the Albin-Delong model cannot adequately address this dynamic modifications.

The hyper-direct pathway from the cortex to the STN and the GPi, which does not have any specific function in the Albin-Delong model, is now thought to play a primary role in BG dynamics (Nambu et al, 2002). Recent electrophysiological data indicate that this di-synaptic pathway is actually shorter in latency than the direct pathway (Nambu et al, 2000). The motor cortical areas consist of a number of modules, which seem to provide multiple sub-loops inside the BG-cortical network (Nakano et al 2000; Alexander et al, 1990). The primary motor cortex (M1) is embedded in a sub-loop including several pathways from M1 to the motor part of the GPi, through the sensory-motor striatal area (Flaherty and Graybiel 1994; Parent 1995; Kayahara, 1996), and a two synaptic feedback from the same part of the GPi to M1 (Kayahara 1996; Hoover 1999).

Several computational models of the BG-cortical motor network have been developed (see reviews in Beiser et al., 1997; Gillies and Arbuthnott, 2000). However, most of these models were based on the direct/indirect classification of BG pathways, and have overlooked the pre and postsynaptic effects of dopamine. Moreover, the dynamical consequences of the closed loop architecture of the cortico-basal ganglio-cortical network have been poorly explored. Here we suggest a network model of the BG that is based on competition between the two feedback loops, namely the cortico-STN-GPi-thalamo-cortical loop and the cortico-striatal-GPi-thalamo-cortical loop. We show how the response of the cortical population to an external input is modulated by the network and how it depends on the level of striatal dopamine. Moreover, we point out how DA depletion can induce modifications in the pattern of activity in the several populations of the network.

2 Methods

2.1 The Model

Anatomical and electrophysiological data report a clear somatotopic organization of the BG circuitry. We assume in this study that there is a functional segregation along the different pathways. One cortical sub-population is thus embedded in a local network made up of sub-populations of each BG nucleus. Our model circuitry can be thought of as representing one such local network in the motor part of the cortico-basalo-thalamo-cortical circuitry among the set of networks related to a given joint. The network architecture, described in figure 1A, contains five populations of neurons. Three of these populations are excitatory. They correspond to sub-populations of the cortex (Ctx), thalamus (Th) and STN. Two populations are inhibitory. They correspond to sub-populations of the striatum (Str) and GPi. Population α ($\alpha = Ctx, Str, GPi, Th, STN$) consists of N_α neurons. The consequences of the absence of GPe population in the network are addressed in the discussion.

The state of neuron $i = 1, ..., N_\alpha$ in population α is characterized by a smooth rate variable, $m_{i\alpha}$, that measures its instantaneous level of activity. The variable $m_{i\alpha}$ follows the dynamical equation:

$$\tau_\alpha \frac{dm_{i\alpha}}{dt} = -m_{i\alpha} + S_\alpha(I_{i\alpha}) \tag{1}$$

where the function $S_\alpha(x)$ describes the non-linear input-output relationship of the neuron in population α and τ_α is an integration time constant. In all this work, $\tau_\alpha = 5$ ms. The total input on the neuron is denoted by

$I_{i\alpha}$. For simplicity we consider a threshold linear input-output relationship, i.e., $S_\alpha(x) = g_\alpha[x - T_\alpha]_+$ where T_α is the threshold, $g_\alpha > 0$ is a constant and $[x]_+ = x$ for $x > 0$ and 0 otherwise.

We assume an all to all connectivity between two connected populations (see figure 1a). Lateral interactions in the different populations are not incorporated in our model.

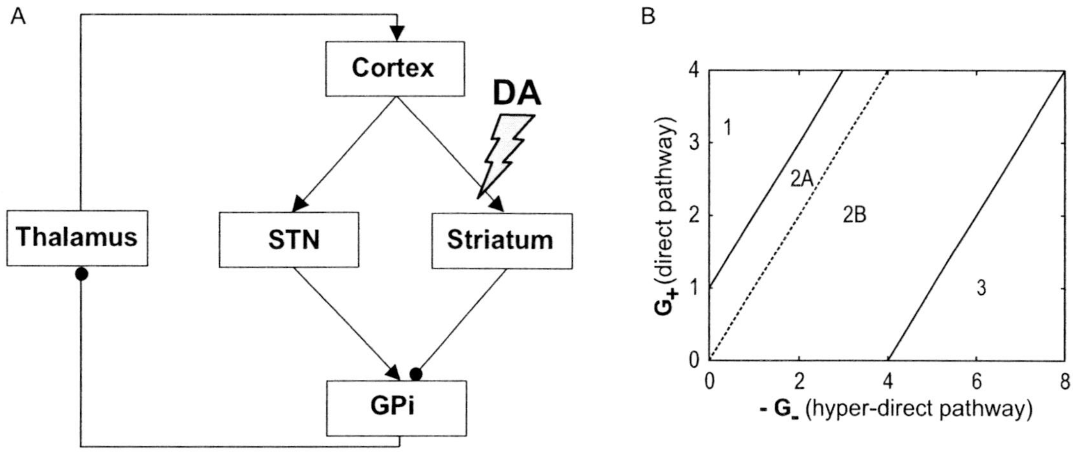

Figure 1: A: The architecture of the model network, made up of five populations: cortex, striatum, STN, GPi, and thalamus. Arrows represent excitatory connections; circles represent inhibitory connections. The connectivity between two populations is all to all. The substantia nigra pars compacta (SNc) is not explicitly considered in the model. However, the influence of dopamine delivered by the SNc dopaminergic neurons on the cortico-striatal synaptic transmission is taken into account in equations 10 and 11 (see methods). B: The different dynamical regimes of the network and the competition between the direct and hyper-direct loops. Region 1: Only open loop states, in which at least one of the populations is silent, are stable. In a broad range of input to cortex, the system display bistability between the states O1 and O2 (see results). Region 2: In this region, the stationary linear state, in which all populations are active, is stable in a wide range of input to cortex. Out of this range, the network is in an open loop state, in which at least one of the populations is silent. We share this domain into 2A, in which the activity in the GPi increases with the input to cortex and 2B, in which the activity in the GPi decreases with the input to cortex. Region 3: The network is either in an open loop state, or, for a broad range of input to cortex, in an oscillatory state.

We will assume that all the connections made by neurons in population α onto neurons in population β have the same strength $G_{\alpha\beta}$ where $\alpha, \beta = Ctx, Str, GPi, Th, STN$. Therefore, the overall pattern of connectivity is characterized by a 5x5 matrix $\mathbf{G}$. The total synaptic input to neuron $i = 1, ..., N_\alpha$ in population α at time t is given by:

$$I_{i\alpha}^{syn}(t) = \sum_\beta \frac{G_{\alpha\beta}}{N_\beta} \sum_{j=1}^{N_\beta} m_{j\beta}(t - \Delta_{\alpha\beta}) \quad (2)$$

where $m_{j\beta}$ is the activity of neuron j in population β and $\Delta_{\alpha\beta}$ a delay in the interactions from population β to population α.

The model circuit can be thought as made of two loops:

- A loop of global positive feedback: Ctx $\longrightarrow$ Str $\longrightarrow$ GPi $\longrightarrow$ Th $\longrightarrow$ Ctx. In this loop the striatum acts on the thalamus through a sequence of two inhibitory sets of connections. Therefore the striatum disinhibits the thalamus. We define $G_+ = G_{StrCtx} G_{GPiStr} G_{ThGPi} G_{CtxTh} > 0$. Because the Ctx $\longrightarrow$ Str $\longrightarrow$ GPi path is usually called direct pathway, this loop will be referred in the following as direct loop.

- A loop of global negative feedback: Ctx $\longrightarrow$ STN $\longrightarrow$ GPi $\longrightarrow$ Th $\longrightarrow$ Ctx. In this loop the STN acts on the thalamus through a sequence of two sets of connections, one excitatory ($STN \longrightarrow$

GPi) and one inhibitory ($GPi \longrightarrow Th$). Therefore it inhibits the thalamus. We define: $G_{-} = -G_{STNCtx}G_{GPiSTN}G_{ThGPi}G_{CtxTh} < 0$. In reference to the hyper-direct pathway from cortex to GPi through the STN (Nambu, 2000), we will call this loop hyper-direct loop in the following.

Finally, the neurons in population α also receive, at time t, an external input $I_{\alpha}^{ext}(t)$. This input is assumed to be the same for all the neurons in the population α.

2.2 Analytical Study

To a large extent the dynamical properties of the model can be studied analytically. For a constant and homogeneous external input on the cortex the dynamical equations of the network, Eqs. (1,2), possess stationary solutions in which all the activities of the neurons are constant in time and are homogeneous across each of the populations: $m_{i\alpha}(t) = m_{\alpha}$. In these states the activities of the neurons are determined by the set of non-linear equations:

$$m_{\alpha} = \left[\sum_{\beta} G_{\alpha\beta} m_{\beta} + I_{\alpha}^{ext} - T_{\alpha}\right]_{+} \tag{3}$$

where $\alpha = Th, STN, Str, GPi$. If one assumes that the total input to each the population is supra-threshold, one finds the solution:

$$\mathbf{m} = [\mathbf{I} - \mathbf{G}]^{-1}\mathbf{I}_{\mathbf{ext}} - \mathbf{T}_{\alpha} \tag{4}$$

This solution is acceptable if all the components of $\mathbf{Gm} + \mathbf{I}^{\mathbf{ext}} - \mathbf{T}$ are positive with $\mathbf{m}$ given by Eq. (4). If this condition is not satisfied another stationary solution of Eq. (3) must exist. It corresponds to a state in which at least one of the populations is subthreshold. Such state will be referred later as an open loop state.

Solutions of Eqs. (1,2) in which the activities of the neurons vary periodically in time can also exist. These solutions cannot be determined analytically in an easy way. We will determine these solutions numerically.

A solution of the dynamical equations is relevant as a possible state of the network if it is stable, i.e., if any small perturbation around it eventually decays at large time. The stability analysis is performed by linearizing the dynamical equations around the solution and solving the equations thus obtained. For a stationary solution it is straightforward to see that the linearized dynamical equations are given by:

$$\tau_{\alpha}\frac{d\delta m_{\alpha}}{dt} = -\delta m_{\alpha} + \sum_{\beta} G_{\alpha\beta}\delta m_{\beta}(t - \Delta_{\alpha\beta}) H\left(\sum_{\beta} G_{\alpha\beta} m_{\beta} + I_{\alpha} - T_{\alpha}\right) \tag{5}$$

where m_{α} are the activities in the stationary state, δm_{α} are the perturbations around this state and $H(x)$, the Heaviside function defined by: $H(x) = 0$ if $x < 0$ and $H(x) = 1$ otherwise. One looks for a solution of these equations of the form:

$$\delta m_{\alpha}(t) \propto e^{\lambda t} \tag{6}$$

That the eigenvalue λ is a solution of the equation:

$$Det\ \ M(\lambda) = 0 \tag{7}$$

where M is a 5x5 matrix given by:

$$M_{\alpha\beta} = -1 - \lambda\tau_{\alpha} + \exp(-\lambda\Delta_{\alpha\beta})\tilde{G}_{\alpha\beta} \tag{8}$$

with $\tilde{G}_{\alpha\beta} = H(\Sigma_{\beta} G_{\alpha\beta} m_{\beta} + I_{\alpha} - T_{\alpha})G_{\alpha\beta}$. The determinant of M can be calculated analytically. However, in general Eq. (7) has to be solved numerically.

Eq. (7) simplifies significantly if one assumes that $\tau_{\alpha} = \tau$ for all α and that the total delays on each of the direct and hyperdirect loops are equal: $\Delta_{STNCtx} + \Delta_{GPiSTN} + \Delta_{ThGPi} + \Delta_{CtxTh} = \Delta_{StrCtx} + \Delta_{GPiStr} + \Delta_{ThGPi} + \Delta_{CtxTh} \equiv \Delta$. In that case one finds that Eq. (7) factorizes:

$$(1 + \lambda\tau)\prod_{k=1}^{4}\left(1 + \lambda\tau + |\tilde{G}|^{\frac{1}{4}} \exp\left(\frac{(2ik + 1 + H(\tilde{G}))\pi}{4} - \lambda\frac{\Delta}{4}\right)\right) = 0 \tag{9}$$

where we have defined: $\tilde{G} = \tilde{G}_{ThGPi}\tilde{G}_{CtxTh}\left(\tilde{G}_{StrCtx}\tilde{G}_{GPiStr} - \tilde{G}_{STNCtx}\tilde{G}_{GPiSTN}\right)$. In particular, for the linear state, $\tilde{G} = G_{+} + G_{-}$. The stationary state is stable if and only if $Re(\lambda) < 0$ for all the solutions of Eq. (9).

2.3 Parameters of the Model

Our network model depends on a large number of parameters. The interactions between neurons are described by six parameters, $G_{\alpha\beta}$. The thresholds of all neurons in a given population α, are assumed to be the same, refered as T_α. Without loss of generality we can assume that the gains of the transfer functions are the same for all the populations. We will take $g = 0.01\ s^{-1}$. We will set all the external inputs but the one to the cortex to zero (Mink, 1996). To simplify the analytical calculations we assume that the characteristic time constants of the neural dynamics are the same for all the populations, $\tau = 5$ ms, and that the time delays $\Delta_{\alpha\beta}$ are identical for all the connections: $\Delta_{\alpha\beta} = \Delta/4$. This latter hypothesis implies that the two loops embedded in the network have the same total delay: Δ. Therefore the properties of the model depend on twelve parameters, all of those except Δ can be constrained to some extent if one requires some specific levels of activity for the five populations.

For most of the discussion of the general properties of the model network we constrain the parameters by requiring that for $I_{Ctx} = 0$ the activity of the different populations are: $m_{Ctx} = 0$, $m_{Str} = 0$, $m_{STN} = 20$ sp/s; $m_{GPi} = 80$ sp/s and $m_{Th} = 20$ sp/s. Other constraints are provided by the requirement that in the linear stationary state the activity of the striatum and the GPi remain at physiological levels (Mink, 1996; Brotchie et al, 1991), i.e. with m_{GPi} in the range $50 - 150$ sp/s and m_{Str} in the range $0 - 30$ sp/s. In particular, this implies that the connection from the striatum on the GPi must be strong. Several set of parameters are compatible with these requirements. A possible set is given in Table I.

Table 1: The fixed parameters of the model. Variable parameters are G_{StrCtx}, $(0 < G_{StrCtx} < 1)$, I_{Ctx}, $(-0.2 < I_{Ctx} < 0.3)$ and T_{Str}.

G_{GPiStr}	-10
G_{ThGPi}	-0.75
G_{ThCtx}	1
G_{STNCtx}	2
G_{GPiSTN}	4
T_{Ctx}	0.1
T_{GPi}	0
T_{STN}	-0.1
T_{Th}	-0.4
I_{Str}	0
I_{GPi}	0
I_{STN}	0
I_{Th}	0

2.4 Numerical Simulations

In a wide range of parameters the network can display stable oscillatory states. It is in principle possible to determine the time evolution of the neural activities combining analytical calculations and numerical solutions of some transcendental equations. However these calculations are not very easy. In this work we study the oscillatory regimes of the network dynamics with numerical simulations. The dynamics was integrated using a standard first order Euler algorithm with a time step $\Delta t = 0.5$ ms. Some simulations with a smaller time step ($\Delta t = 0.05$ ms) were carried out in order to ensure that the dynamics are unchanged. The activity in each population was characterized by its temporal average measured as the instantaneous activity averaged over 1 s and its temporal modulations which was measured by its half peak-to-peak variation averaged over 1 s. A transient of 1 s was discarded prior to averaging.

2.5 The Modulatory Effects of Dopamine

The modulatory effects of dopamine on neuronal excitability is still a matter of debate. It seems as if they could depend crucially on the conditions under which they are studied. It has been shown that some of the dopaminergic terminals on striatal neurons are localized on dendritic spines on which glutamatergic cortico-striatal synapses are also located and it has been proposed that an effect of dopamine would be to enhance synaptic transmission from cortex to striatum. It has been also reported that dopamine activation of D1 receptor may increase the current threshold to firing of spikes by striatal neuron. According to this second effect dopamine depletion makes the firing of spikes by striatal neurons easier.

The first effect is implemented in our model by assuming that the strength of the cortico-striatal synapses decreases with the level of dopamine. Specifically, we take:

$$G_{StrCtx} = \frac{0.85}{1+0.57\exp(-3(D-1))} \tag{10}$$

where D measures the level of dopamine with respect to some reference state ("normal state") for which $D = 1$. We will take the second effect of dopamine depletion into account by changing the threshold of striatal neurons according to the formula:

$$T_{Str} = 0.02\left(1 - \frac{1.1}{1+0.1\exp(-3(D-1))}\right) \tag{11}$$

Therefore, T_{Str} decreases when D increases.

3 Results

3.1 The Direct Feedback Loop

As a starting point of our analysis we consider the dynamics of the direct feedback loop when it is isolated from the rest of the network. This disinhibitory loop goes from the cortex to the GPi through the striatum (direct pathway) and back to the cortex through the thalamus. First we assume that the GPi spontaneous activity is sufficient to inactivate the thalamus. The condition on the (negative) threshold of GPi neurons is the following:

$$T_{GPi} \leq G_{GPiStr}T_{Str} + \frac{T_{Th}}{G_{ThGPi}} \tag{12}$$

If the input to the cortex is sufficiently small, the activity of the striatum is also small. In this case the inhibition of the GPi from the striatum is weak and it can only mildly reduce the activity of the GPi. Since the pallidal neurons display strong spontaneous activity, Eq. (12), the pallidal inhibition on the thalamus can be strong enough to prevent the thalamus to be active. This occurs if the input to cortex is such that:

$$I_{Ctx} < I_{Ctx}^{(1)} \equiv \frac{G_{ThGPi}G_{GPiStr}T_{Str} - G_{ThGPi}T_{GPi} + T_{th}}{G_{ThGPi}G_{GPiStr}G_{StrCtx}} + T_{Ctx} \tag{13}$$

Such a state in which the feedback is suppressed because the thalamus becomes silent is called below the open loop state 1 (state O1). In contrast, if the input to cortex is very strong, the striatum strongly inhibits the GPi that becomes silent. This happens if:

$$I_{Ctx} > I_{Ctx}^{(2)} \equiv \frac{1}{G_{StrCtx}}\left(\frac{I_{GPi}-T_{GPi}}{G_{GPiStr}} - I_{Str} + T_{Str}\right) - G_{CtxTh}(I_{Th} - T_{Th}) \tag{14}$$

In this case, the feedback is suppressed again. We call this state open loop state 2 (state O2). The thalamus, which is completely released from the inhibition of the GPi, is independent of all the inputs to the other populations of the network.

Finally, if $I_{Ctx}^{(1)} < I_{Ctx} < I_{Ctx}^{(2)}$ all the populations in the direct loop are active and the activity of all the populations depend linearly on the inputs to the network (linear state). The formula relating these activities to the model parameters and to the input to the cortex are given in the Appendix. In particular, the activity of the thalamus increases proportionally to the input on the cortex with a gain, $\mathcal{G}$, which depends on the gain of the disinhibitory loop, G_+:

$$\mathcal{G} = \frac{G_{StrCtx} G_{GpStr} G_{ThGp}}{1 - G_+} \tag{15}$$

When G_+ increases from 0 to 1, $\mathcal{G}$ increases. It diverges for $G_+ \rightarrow 1$. Simultaneously, $I_{Ctx}^{(1)}$ increases and $I_{Ctx}^{(2)}$ decreases. They becomes equal for $G_+ \rightarrow 1$. When $G_+ > 1$, the linear state is unstable. Obviously, the two open loop states O1 and O2 are still stable. Moreover, $I_{Ctx}^{(1)} > I_{Ctx}^{(2)}$. This means that the domains of existence of the two open loop states overlap. Therefore, the network displays now bistability in the range, $I_{Ctx}^{(2)} < I_{Ctx} < I_{Ctx}^{(1)}$, of the input to cortex.

If condition, Eq. (12), is not satisfied the spontaneous activity of the GPi is too low to inhibit completely the activity in the thalamus even if the striatum is silent. In this new open loop state, the striatum is silent while the thalamus is active. Since the feedback is still suppressed, it displays the same features as the state O1. For $G_+ < 1$, the two open loop states are stable and exist inside distinct domains of cortical input, while the linear stationary fixed point is stable. For $G_+ > 1$, the linear state exists becomes unstable and the system displays bistability.

A feature of the direct loop is that even if the positive feedback on this loop is very large, the activities in all the populations remain finite. Indeed, the GPi or the thalamus becomes strongly inhibited when the feedback is too strong. Therefore the loop becomes open, preventing the activity of the cortical and thalamic neurons to rise up to saturation of their firing rates. In contrast with a positive feedback loop in which only excitatory populations are connected, a positive feedback loop which uses disinhbition acts as a 'safety valve'. Finally, note that including delays in the interactions between the populations of this loop does not modify its properties.

3.2 The Hyper-direct Feedback Loop

We now consider the hyper-direct loop. It goes from the cortex to the STN, then to the GPi and eventually it goes back to the cortex through the thalamus. Globally it is a negative feedback loop. We consider first the case where the interactions between the populations are not delayed: $\Delta = 0$.

One finds that two dynamical regimes exist for this loop. In the first regime the activity of each population is stationary. This happens if $-4 < G_- < 0$. If the input to the cortex is not sufficiently large so that the cortex is active

$$I_{Ctx} > G_{CtxTh}T_{Th} - G_{ThGPi}G_{CtxTh}T_{GPi} - G_{ThGPi}G_{CtxTh}G_{GPiSTN}T_{STN} \equiv I_{Ctx}^{(3)} \tag{16}$$

but not too large, so that the GPi does not become active enough to inhibit completely the thalamus

$$I_{Ctx} < \frac{G_{ThGPi}T_{GPi} + G_{ThGPi}G_{GPiSTN}T_{STN} - T_{Th}}{G_{ThGPi}G_{STNCtx}G_{GPiStn}} \equiv I_{Ctx}^{(4)} \tag{17}$$

then all populations are active and the loop is closed. In contrast, if the input to the cortex is too small, $I_{Ctx} < I_{Ctx}^{(3)}$, the cortex remains subthreshold. This open loop state with silent cortex is called the rest state. The loop is also open if the input to the cortex is too strong ($I_{Ctx} > I_{Ctx}^{(4)}$). Because the the GPi is so strongly activated that the thalamus is silent. As in the direct loop, this state is called O2.

Stability analysis of the linear state shows that a Hopf bifurcation occurs ar $G_- = -4$. Therefore the loop displays a non-stationary state when $G_- < -4$ and $I_{Ctx}^{(3)} < I_{Ctx} < I_{Ctx}^{(4)}$. In this case, all populations in the loop display stable periodic oscillations. The activities of the neurons are oscillatory and synchronized across the network.

If the delay, Δ, is not zero the qualitative behaviour of the loop is not modified. The bifurcation to oscillations occurs now at a critical value of the gain, $G_c(\Delta)$ which is given by the equation:

$$G_c(\Delta_-, \nu) = \left(\frac{1}{cos(\frac{\pi}{4} + \frac{\Delta\nu}{4})} \right)^4 \tag{18}$$

$$\tau\nu = \tan(\frac{\pi}{4} + \frac{\Delta\nu}{4}) \tag{19}$$

$G_c(\Delta)$ is a decreasing function of Δ. Therefore although a delay is not required for the existence of the bifurcation to oscillations, a delay along the loop enlarges the region of oscillations. Note that the phase shifts between the population activities depend also on Δ.

3.3 The Competition between the Direct and Hyper-direct Loops

We have seen above that the direct and hyper-direct loops display distinct dynamical regimes, according to the strength of their feedback. Here, we will determine and describe the dynamical regimes of the full circuitry. The dynamics of this network depends on the various synaptic strengths through the difference $G = G_+ + G_-$ (see methods). The system displays three different dynamical regimes. We first assume that the delays along the direct and hyper-direct loops are the same. The network is in a linear regime for $-G_c(\Delta) \leq G \leq 1$, where $G_c(\Delta)$ is the solution of Eqs. (18). For the choice of parameters given in Table I, this critical value is obtained for $G_{StrCtx} = .62$. At $G = -G_c$ a bifurcation to oscillations occurs, and the network is in an oscillatory regime in all the domain $G \leq G_c(\Delta)$. In contrast, the network is in a bistable regime for $G > 1$ as it was the case for the disinhibitory loop alone.

In the linear regime the system can either be in the linear state (all the population are active) or in an open loop state, depending on the values of the external inputs. In the oscillatory regime, the system is either in the oscillatory state, in which all populations display oscillatory activity or in an open loop state. In the bistability regime, the system is always in an open loop state, and in some range of external input, it displays bistability between two open loop states, O1 and O2.

The phase diagram of the full network in the $(-G_-, G_+)$ plane is plotted in figure 1B for $\Delta = 0$. It shows how the two loops which have opposing effects on the GPi are competing. For a sufficiently large value of $-G_-$ the network settles in an oscillatory state. A strong enough disinhibition is required to stabilize the network in a steady state. If the disinhibition is strong enough the network can become bistable with two non-linear stable steady states in which one or more populations are inactivated. If $\Delta \neq 0$, the boundary between the linear and the oscillatory regime is shifted, while the other border do not change.

If $\Delta_- \neq \Delta_+$, the boundary between the linear and the oscillatory regime is bent, while the other border do not change.

3.4 The Activities of the Populations

In this section we study in greater detail how the activities of the five populations in the network depend on the external input to the cortex in the three dynamical regimes of the network. The parameters of the network are fixed as in Table I and the balance between the two loops is changed by changing the cortico-striatal synaptic weight, G_{StrCtx}.

3.4.1 The Regime $-G_c(\Delta) \leq G \leq 1$

We first consider the case $T_{Str} > 0$. If the cortical input is sufficiently low, the cortex is silent and the network is in an open loop state called rest state. This happens for $I_{Ctx} < I_{Ctx}^{rest}$ where

$$I_{Ctx}^{rest} = G_{CtxTh}(T_{Th} - G_{ThGPi}(T_{GPi} + G_{GPiSTN}T_{STN} - G_{GPiStr}T_{Str})) + T_{Ctx} \quad (20)$$

The cortex is silent and, obviously, the activity in each population is independent of the input to the cortex. For the chosen parameters these activities are 80 sp/s for the GPi, 20 sp/s for the thalamus and 20 sp/s for the STN. If the input to the cortex is increased, the systems goes into the linear stationary state, in which all populations are active. Depending on the sign of G, two situations occur. Figure 2A shows the activities of the GPi and cortex as a function of the input to cortex in this regime in the two cases $G \geq 0$ and $G \leq 0$. If $G \geq 0$ (or $G_{StrCtx} \geq 0.8$), the disinhibition loop is predominant, and the activity of GPi *decreases* when the input on the cortex increases. This is because the striatum increases its inhibition on the GPi. Subsequently the thalamus increases its activity because it is disinhibited. The network amplifies the activity of the cortex compared to the case in which the cortex would be disconnected from the network. The amplification factor is $1/(1-G) > 1$. If $G < 0$ (ie $G_{StrCtx} < 0.8$), the negative loop is predominant, and the activity of the GPi *increases* when the input to cortex increases. Subsequently the thalamus reduces its activity and the cortex increases its activity less than if it was isolated from the rest of the network. Because of the network, the

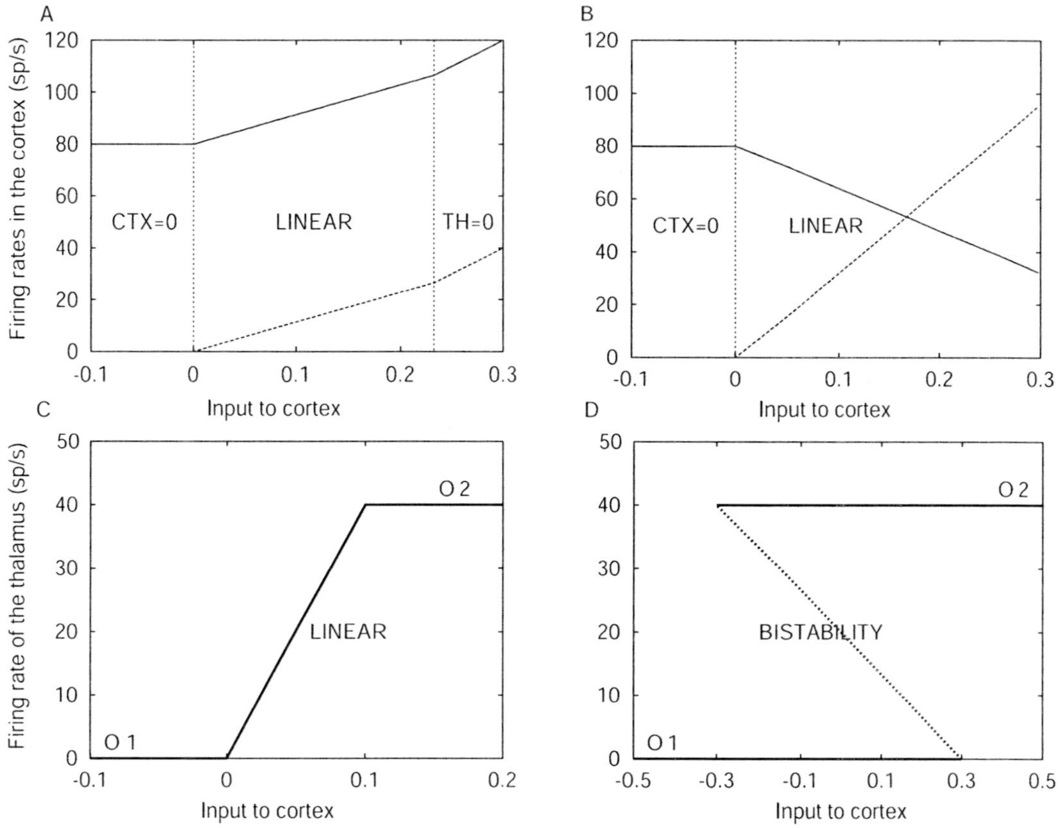

Figure 2: The activities in the network in the linear regime (A, B and C, $-G_c(\Delta) \leq G \leq 1$) and bistable regime (D, $G \geq 1$). All parameters, except the cortico-striatal synaptic strength and the cortical input, are set as in table 1, with $T_{str} = 0$. A, B display the activity in the GPi (solid line) and in the cortex (dashed line) in the linear regime, for $G \leq 0$ (A, with $G_{StrCtx} = 0.7$) and for $G \geq 0$ (B, $G_{StrCtx} = 0.85$). Note that if $G \leq 0$, the GPi activity increases with the input to cortex and the cortical activity is damped by the network. In contrast, if $G \geq 0$, the GPi activity decreases with the input to cortex and the cortical activity is strongly amplified by the network. B, C display the thalamic activity in the linear and bistable regime. C: in the linear regime (for $G_{StrCtx} = 0.85$), the states O1 and O2 (see results) exist in different domains, clearly separated by the stability domain of the linear steady state. D: in the bistable regime, the states O1 and O2 coexist and the linear state is unstable.

activity of the cortex increases linearly with its input with a slope $\beta = 1/(1+|G|) < 1$. When the global feedback G of the network is large and positive and the input to the cortex is sufficiently strong so that the GPi is completely inhibited, the cortex and the thalamus become hyperactive. For the parameters of table I the average firing rates are above 60 sp/s and 40 sp/s for the cortex and the thalamus respectively. Finally, in the domain of parameters where the thalamus is inactivated by the inhibition from the GPi the cortex is effectively isolated from the rest of the network. Therefore its activity increase linearly with its input, with a slope $\beta = 1$.

If $T_{Str} \leq 0$, the states previously described still exist and the activities in the various populations in these states are qualitatively the same. However, a new steady state appears between the rest and the stationary linear steady state, in a narrow range of input to cortex $I^{rest}_{Ctx} < I_{Ctx} < I^{sup}$. For low striatal thresholds, $I^{sup} \approx I^{rest}_{Ctx}$. In this state, the cortex is active and the striatum silent. All populations, except the striatum, display periodic oscillations with a very low amplitude. The cortical activity remains very low.

3.4.2 The Regime $G \leq G_c(\Delta)$

There are two possible open loop states in this regime, the rest state and the state O1. Similarly to what precedes, in these states, the dynamics of the network is trivial. In the rest state ($I_{Ctx} < I_{Ctx}^{rest}$, see Eq. (20)), the cortex is below threshold and the activities in the network are independent of its external input. If $I_{Ctx} > I_c$, where

$$I_c = \frac{G_{ThGPi}T_{GPi} + G_{ThGPi}G_{GPiSTN}T_{STN} - T_{Th} - G_{GPiStr}G_{ThGPi}T_{Str}}{G_{ThGPi}(G_{STNCtx}G_{GPiSTN} - G_{StrCtx}G_{GPiStr})} \tag{21}$$

the thalamus is silent and, in the cortex, the activity increases linearly with the input with a slope $\beta = 1$.

For intermediate levels of input to cortex ($I_{Ctx}^{rest} < I_{Ctx} < I_c$), all populations display oscillatory activity, synchronized over the whole network (see figure 3A). The amplitude of the oscillations in this domain varies with I_{Ctx}. When the cortex receives a small input, the amplitude increases with the input. For sufficiently large I_{Ctx}, the thalamus is transiently silent during a short time window of the oscillation period. The amplitude of the oscillations decreases then with I_{Ctx}. In all the domain of oscillations the average activity in the cortex remains below 10 sp/s. This is substantially smaller than in the linear stationary state for the corresponding values of I_{Ctx} for which the average cortical activity can be as high as $20-40$ sp/s or even larger. A negative striatal threshold increases substantially the amplitude of the modulations of the oscillations. Moreover, it makes bigger the range of input to cortex in which the oscillations occur. Note however that the modulation of the striatal activity remains always small.

When the overall delay on the two sub-loops are equal the frequency of the unstable mode, ν_c, is obtained by solving Eqs. (18). Therefore it depends only on the characteristic time of the dynamics, τ, and on the overall delay, Δ. In particular it is independent of the precise values of the synaptic couplings between the populations, on their thresholds and external inputs. figure 3B shows that this frequency is a decreasing function of Δ. For $\Delta = 0$ and $\tau = 5$ ms, the frequency is $\nu_c \approx 32$ Hz. For $\Delta \to \infty$, ν_c goes to zero like $3\pi/\Delta$. Simulations show that the oscillation frequencies depend almost exclusively on Δ and τ in all the domain where the oscillations occur. In particular, for $\tau = 5$ ms and $\Delta = 20$ ms the frequency of the oscillations is $\nu \approx 12$ Hz in the all range of G_{StrCtx} and I_{Ctx} that we have considered.

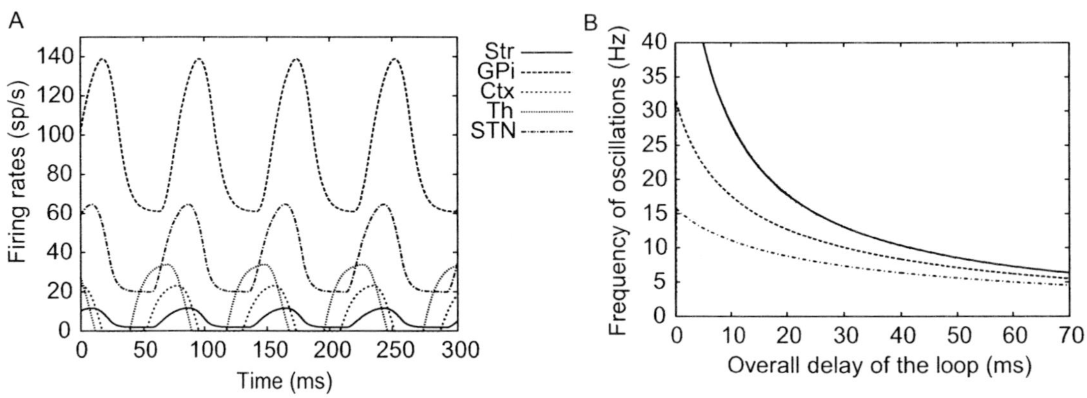

Figure 3: The oscillatory regime. The parameters are the fixed to the values of table 1, with a level of striatal dopamine of $D = 20\%$. A: The activity of the five populations in the oscillatory state, for $I_{ctx} = .135$. B: The frequency of the oscillations as a function of the overall delay Δ for three different values of the integration time constant τ (solid line: $\tau = 2$ ms; dashed line: $\tau = 5$ ms; dash-dotted line: $\tau = 10$ ms).

3.4.3 The Regime $G \geq 1$

For low input to the cortex, the network is in the open loop state O1, in which the thalamus is silent. For high input to cortex, the network is in the open loop state O2, in which the GPi is silent. The dynamics of these two states is trivial since there is no feedback. However, for a broad range of input to the cortex, the network displays bistability between the states O1 and O2. The activity in the thalamus in a wide range of input to cortex is depicted in figure 2d. For low input to cortex, the thalamus is silent and the activity of the cortex increase linearly with its input, with a slope $\beta = 1$. Similarly, if the input to cortex is very strong, the GPi is silent and the cortex increase linearly with its input, with a slope $\beta = 1$. However, in this last case, the cortical activity is high compared to the physiological range. However, in this state, the cortical activity is facilitated by the thalamic spontaneous activity, and thus higher than in the previous case. For intermediate values of the input to cortex, both states coexist and the thalamic, or cortical activity, can jump from a low to a high level of activity, triggered by a transient excitatory current. The range of bistability is given by $I_{Ctx}^{(1)} < I_{Ctx} < I_{Ctx}^{(2)}$, where:

$$I_{Ctx}^{(1)} \equiv \frac{G_{ThGPi}G_{GPiStr}T_{Str} - G_{ThGPi}G_{GPiSTN}T_{STN} - G_{ThGPi}T_{GPi} + T_{th}}{G_{ThGPi}G_{GPiStr}G_{StrCtx}} + T_{Ctx} \tag{22}$$

$$I_{Ctx}^{(2)} \equiv \frac{I_{GPi} - T_{GPi} - G_{GPiStr}(I_{Str} - T_{Str}) + G_{GPiSTN}(I_{STN} - T_{STN})}{G_{GPiStr}G_{StrCtx} - G_{GPiSTN}G_{STNCtx}} - G_{CtxTh}(I_{Th} - T_{Th}) \tag{23}$$

4 Discussion

4.1 The Model

The network we have studied is a toy model of the BG circuits and as such it is based on several simplifications that we now discuss.

Our model includes five populations which correspond to the cortex, the subthalamic nucleus, the internal part of the globus-pallidus, the thalamus and the striatum. The external part of the globus pallidus was not included in the model and therefore only the direct pathway linking the cortex to the GPi was taken into account. This is because we aimed at studying the direct pathway and the effective inhibitory hyper-direct pathway isolated from the rest of the BG network to characterize the specific properties stemming from their competition.

Anatomical studies have shown that lateral connections exist in the striatum (Wickens, 1997; Koos and Tepper, 1999) and in the GPi (Mink, 1996). However, these direct lateral connections might be weak or non-functional as suggested by several electrophysiological studies (Kita, 1993; Jaeger et al, 1994; Tepper et al, 2004), and striatal spiny neurons probably interact through more complex local networks, also involving interneurons (Plenz, 2003). Similarly, the importance and physiological significance of the recurrent excitatory interactions in the STN are still a matter of debate (Wickens, 1997). In this work, we investigate the possible dynamical consequence of the inter-nuclei connectivity pattern in the basal ganglia network, whereas the intra-nuclei connectivity is deliberately neglected. However, we have checked that including lateral interactions in the cortex does not modify substantially the qualitative properties of the network unless they are so strong that the cortex by itself can develop an instability. A more detailed model which will include recurrent excitation together with an additional inhibitory population deserves further investigations.

The dynamics of the network have been described in the framework of simple rate models with threshold-linear transfer functions. This makes the investigation of the model tractable analytically to a large extent. However this prevents us to take into account the intrinsic oscillatory properties and the post-inhibitory rebound properties (Bergman and Deuschl, 2002) of the neurons in the GPi, STN and thalamus. Nevertheless, we believe that this simplification is justified in the present context which is limited to investigate the more general network properties which could contribute to the dynamics of the BG.

The network studied here represents one local network in the motor part of the cortico-basalo-thalamo-cortical circuitry among the set of networks related to a given joint. Several interacting networks, each with a structure similar to the model studied here would correspond to a more complete circuitry related to one joint. The study of such a more comprehensive model is left for further investigations.

4.2 Consequences for the Physiology of the Basal Ganglia

Chevalier and Deniau have proposed that disinhibition of the thalamus by the GPi is a fundamental principle of function in BG (Chevalier et al, 1990). An additional principle which has been proposed by Mink and Thach is that the STN acts as a brake on the activity of the thalamus (Mink et al, 1993). Indeed, an increase

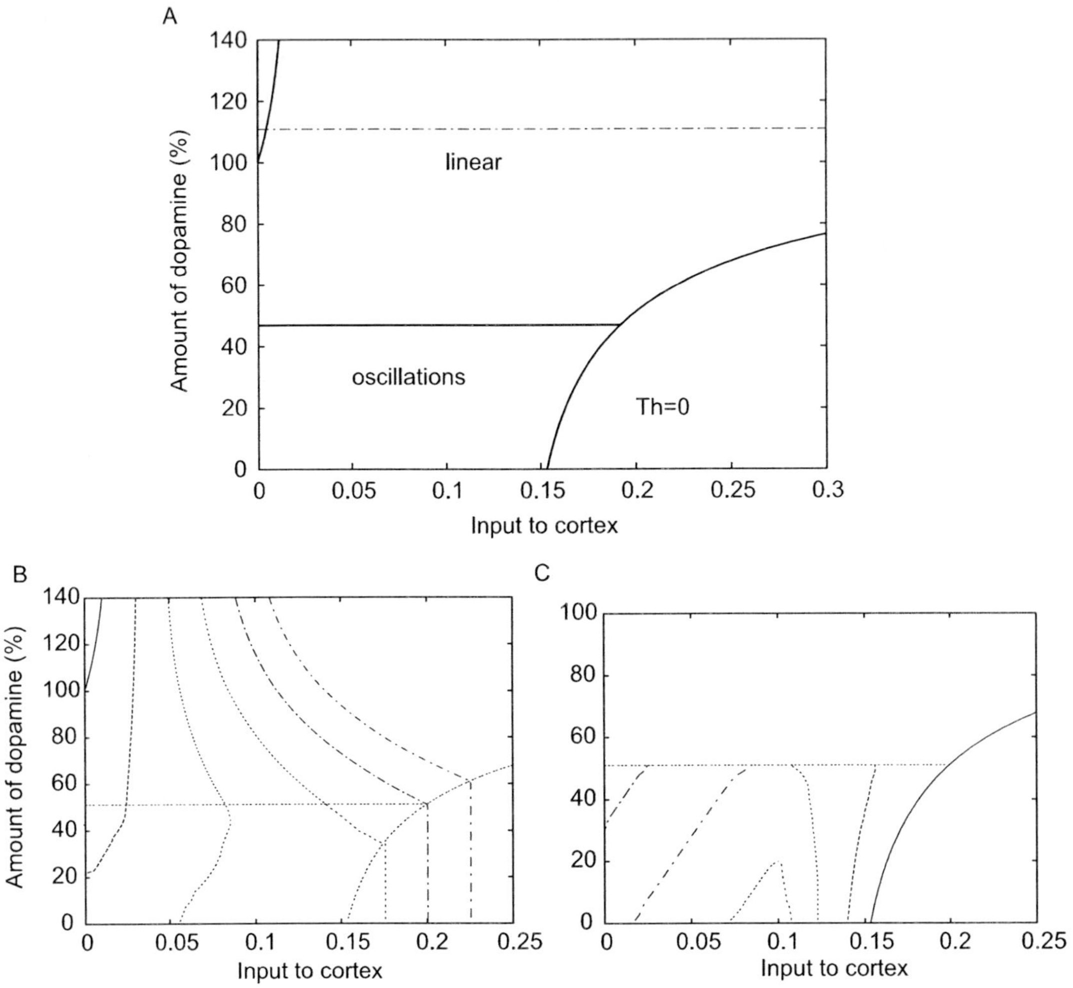

Figure 4: A: The state of the network depending on the external input to the cortex and on the amount of striatal dopamine. Parameters are the ones given in table 1. For levels of dopamine around physiological levels (around 100%), the network is in the linear regime. The cortical activity is either damped (below the dashed line) or enhanced (above the dashed line) by the network in response to a given input on the cortex. For lower dopamine levels, synchronized oscillations can develop in the network, depending on the external input to the cortex. B: Contour plot of the average activity in the cortex. C: Contour plot of the modulation of the activity in the cortex.

in the STN activity makes the pallidal inhibition on the thalamus stronger. These ideas are supported by the anatomy of the BG as well as by several experimental physiological results (Deniau et al, 1985; Mink et al, 1996). Our work is in line with these ideas. These two proposals were formulated originally as principles of feedforward processing of information. The present work generalizes these ideas when feedback is included.

We have analyzed how the closed direct loop which uses disinhibition works as a positive feedback loop. This loop functions as an amplifier of cortical activity provided that the excitatory feedback is not too strong. In a loop which includes only excitatory populations neurons fire at saturation when the positive feedback becomes too strong. We have seen that this is not the case in the disinhibitory loop considered here which functions as a safety valve which prevents the cortical activity to rise too much. The second loop of the network (cortex-STN-GPi-Thalamus-cortex) which has an overall negative feedback dampens the the cortical activity. This contributes to the control of the amplification of the cortical activity.

The phase diagram of the network as a function of the input to cortex, I_{Ctx}, and the dopamine concentration, D, is plotted in figure 4. For physiological levels of dopamine, (normal situation, $D \approx 100\%$), the network is always in the linear regime, unless the cortical input is below rest level (see Eq. 20) and the cortex silent (upper links domain in figure 4). In this regime, the slope of the cortical response to an additional input to cortex strongly depends on the amount of striatal dopamine. For a high level of dopamine (above the dashed line), cortical response is enhanced by the network, whereas for low dopamine levels, the network weakens the cortical response.

Recordings in GPi have demonstrated that about 30% of the neurons increase their activity while 70% of the neurons decrease their activity during movement (Mink and Thach, 1993; Boraud et al., 2000). Our model provides a natural explanation to these facts. If the connectivity in the BG is heterogeneous, one expects that the balance between the two feedbacks will also fluctuate spatially. This will generate a distribution for the sign of the response of the neurons in GPi. If the positive feedback is stronger than the negative one ($G_+ > -G_-$) the neuron decreases its activity when I_{Ctx} increases. In contrast if $G_+ < -G_-$ the activity of the neuron decreases. Finally, for a broad enough distribution of the balance between negative and positive feedback a fraction of the neurons in GPi will display firing rate oscillatory modulations. If this fraction is not too large these oscillations will not be spatially synchronized. This can explain why the activities of about 10% of GPi neurons are found to be modulated in physiological situation (Boraud et al., 2000).

4.3 Consequences for the Pathophysiology of the Basal Ganglia

In Parkinson's disease dopaminergic innervation to the striatum are markedly reduced. Clinical manifestations of the disease are akinesia, rigidity and tremor. One distinguishes between two main forms of the disease, the akineto-rigid form and the tremulus form. Animal models of Parkinson'disease have been developed in the past years (Bergman and Deuschl, 2002). Treatment of monkeys with MPTP (1-methyl-4-phenyl-1,2,3, 6-tetrahydropyridine) induces symptoms which are similar to those of Parkinson disease. It should be noted that depending on the species, MPTP treated monkeys tend to develop either tremulous or non-tremulous form of parkinsonism. However, in contrast to what is commonly observed in man, the tremulous form of parkinsonism is accompanied in general of strong akinesia and rigidity.

Dopamine can have numerous modulatory effects on neural excitability and on synaptic transmissions. Here we have focused on two effects, namely that dopamine enhances cortico-striatal transmission but also increases the threshold to firing of striatal spiny neurons. We have implemented these effects according to Eqs. (10, 11). A natural assumption is that for a movement to be executed the temporal average cortical activity has to increase above some threshold value. For the sake of our discussion we take in the following this threshold to be 15 sp/s. Another assumption that we make below is that tremor is associated with oscillations significantly modulated temporally. Specifically we assume that the threshold to tremor is a modulation level of 10 sp/s (peak-to-peak variation of 20 sp/s).

The hyper-direct feedback loop, Ctx-STN-GPi-Th-Ctx, can develop synchronous oscillations when isolated, if it is sufficiently strong. As we have seen, when this loop is embedded in the all network oscillations can be suppressed provided that cortico-striatal synapses are effective enough. For the parameters of the model studied above this happens for level of dopamine above $D_c \approx 45\%$. If dopamine is depleted below that level the network displays oscillatory activity. Another effect of the dopamine depletion is the reduction of cortical activity for a given input to the cortex (see figure 4B). When D decreases, a higher value of I_{Ctx} is required for the movement. This effect can be related to bradykinesia. As soon as D is under 50% the threshold in the average activity in the cortex required for movement cannot be reached in the all domain of input to the cortex ($I_{Ctx} < I_{Ctx}^{max} = 0.15$). This corresponds to an akinetic state. Simultaneously, the oscillatory activity in the cortex is strongly modulated and it corresponds to tremor for the range of I_{Ctx} where the modulation is above 10 sp/s. The amplitude of oscillations is weak for very low or high input to cortex, whereas they are maximal for intermediate inputs (see figure 4C). The domain of tremor extends toward lower values of I_{Ctx} when the dopamine is more depleted. However, if the input to cortex is strong enough, the thalamus is strongly inhibited by the GPi and the oscillations are suppressed. Note that the frequency of the oscillations for the chosen parameters is about 12 Hz in the whole domain of oscillations.

Variations in the balance between the two effects of the dopamine depletion considered in this work could lead to different forms of Parkinsonian symptomatics. For instance, if the effect on the excitability threshold of the striatal neurons is weak, the temporal modulation of the cortical activity is reduced. This situation could be interpreted as the akinetic-rigid form of Parkinson's disease without tremor. In contrast, if this effect becomes very strong the striatum is more active and it disinhibits more the thalamus. Therefore, the firing in the cortex is facilitated but the oscillations become more modulated. This situation could be interpreted as the tremulous form of Parkinson's disease with limited akinetik-rigid symptoms.

As mentioned above a more complete model of the motor part of the BG includes several sparsely interconnected loops of the form we have studied here but with different balance of negative and positive feedbacks. The effect of dopamine depletion is to increase the fraction of loops for which the positive feedback is small. The fraction of loops in which the GPi neurons increase their activity in relation to movement increases and then the fraction of loops which are oscillating increase.

4.4 Predictions from the Model

A central feature of our model is that it relies on the balance between two loops of opposite feedback. The balance between these two loops depends crucially on the strength of the connections but less on the input to cortex.

The results of our model are compatible with the known effects of the lesions performed in the BG and in the thalamus which reduce or suppress tremor and akinesia. A lesion in the STN tend to reduce or suppress the negative feedback. Therefore, the global feedback of the loop becomes more positive. According to our model, this suppresses the oscillations. However the activity of the cortex can still be amplified by the feedback of the direct loop, allowing movement. After a lesion in the thalamus the cortex does not receive feedback from the BG circuits. This suppresses the oscillations but the activity of the cortex is not amplified anymore (no positive feedback) and a stronger input is required now to induce movement. Therefore this lesion reduces the tremor but does not restore normality in movement. In the same way, the model predicts that a lesion of the GPi stops the oscillatory activity. However, the cortical activity should be enhanced after this lesion.

A prediction of the model is that inactivation in the striatum in a normal monkey should increase the oscillations in the GPi neurons because it cuts the direct loop and therefor make the global feedback strongly negative. However, this should not lead necessarily to tremor if these oscillations are modulated too weakly.

In contrast to the motor part of the BG which includes part of the GPi where a significant fraction of the neurons increase their activity on relation to movement, neurons in the SNr are responding almost always by decreasing their activity (Hikosaka et al, 2000). This can be understood in the framework of our model if one assumes that almost all the neurons in the SNr are part of loops with positive global feedback. This interpretation also implies that in the normal situation the fraction of bursting neurons in the SNr should be much smaller than in the Gpi. It also implies that in MPTP monkeys neurons in the SNr should be less bursty than in the Gpi. This is in agreement with recent experimental results (Wichmann et al, 1999). It is also in line with the fact that in Parkinsonian patients there is no tremor in the eye position (White et al, 1983).

Appendix

Gain of the Loops

$$G_{-} = -G_{STNCtx}G_{GPiSTN}G_{ThGPi}G_{CtxTh}$$

$$G_{+} = G_{StrCtx}G_{GPiStr}G_{ThGPi}G_{CtxTh}$$

$$G = G_{+} + G_{-}$$

Activities of the Populations in the Linear Stationary State

$$m_{Ctx} = \frac{I_{Ctx} - G_{CtxTh}(T_{Th} + G_{ThGPi}(T_{GPi} + G_{GPiSTN}T_{STN} - G_{GPiStr}T_{Str})) - T_{Ctx}}{1-g}$$

$$m_{Str} = \frac{G_{StrCtx}((I_{Ctx} - T_{Ctx}) - G_{CtxTh}(T_{Th} + G_{ThGPi}(T_{GPi} + G_{GPiSTN}T_{STN}))) - (1+G_{+})T_{Str}}{1-g}$$

$$m_{STN} = \frac{G_{STNCtx}((I_{Ctx} - T_{Ctx}) - G_{CtxTh}(T_{Th} + G_{ThGPi}(T_{GPi} + G_{GPiStr}T_{Str})) - (1+G_{-})T_{STN})}{1-g}$$

$$m_{GPi} = \frac{G_{GPiStr}T_{Str} - G_{GPiSTN}T_{STN} + (G_{GPiSTN}G_{STNCtx} - G_{GPiStr}G_{StrCtx})(I_{Ctx} - T_{Ctx})}{1-g} + \frac{(G_{GPiStr}G_{StrCtx} - G_{GPiSTN}G_{STNCtx})G_{CtxTh}T_{Th} - T_{GPi}}{1-g}$$

$$m_{Th} = \frac{G_{ThGPi}(T_{GPi}G_{GPiSTN}T_{STN} - G_{GPiStr}T_{Str}) - T_{Th}}{1-g} - \frac{G_{ThGPi}(G_{GPiSTN}G_{STNCtx} - G_{GPiStr}G_{StrCtx})(I_{Ctx} - T_{Ctx})}{1-g}$$

References

Aizman O, Brismar H, Uhlen P, Zettergren E, Levey AI, Forssberg H, Greengard P, Aperia A (2000) Anatomical and physiological evidence for D1 and D2 dopamine receptor colocalization in neostriatal neurons *Nat Neurosci* **3**: 226-230.

Albin RL, Young AB, Penney JB (1989) The functional anatomy of basal ganglia disorders *Trends Neurosci* **12**: 366-375.

Alexander G E and Crutcher M D (1990) Functional architecture of basal ganglia circuits: neural substrates of parallel processing *Trends in Neurosci.* **13**: 266-71.

Aziz TZ, Peggs D, Sambrook MA, Crossman AR (1991) Lesion of the subthalamic nucleus for the alleviation of 1-methyl-4-phenyl-1,2,3,6-tetrahydropyridine (MPTP)-induced parkinsonism in the primate *Movement Disorders* **6**: 288-293.

Bar-Gad I, Bergman H (2001) Stepping out of the box: information processing in the neural networks of the basal ganglia *Curr Opin Neurobiol* **11**:689-695.

Beiser DG, Hua SE, Houk JC (1997) *Network Models of the Basal Ganglia Cur. Op. in Neurobiol.* **7**: 185-190.

Beiser D G and Houk J C (1998) Model of cortico-basal ganglionic processing: encoding the serial order of sensory events *J. Neurophysiol.* **79**: 3168-3188, 1998

Benabid AL, Krack PP, Benazzouz A, Limousin P, Koudsie A, Pollak P (2000) Deep brain stimulation of the subthalamic nucleus for Parkinson's disease: methodologic aspects and clinical criteria *Neurology* **55**: S40-S44.

Bergman H, Wichmann T, DeLong MR (1990) Reversal of experimental parkinsonism by lesions of the subthalamic nucleus *Science* **249**: 1436-1438.

Bergman H, Wichmann T, Karmon B, DeLong MR (1994) The primate subthalamic nucleus. II. Neuronal activity in the MPTP model of parkinsonism *J Neurophysiol. 1994 Aug*;**72**(2):507-20.

Bergman H and Deuschl G (2002) Pathophysiology of Parkinson's disease: from clinical neurology to basic neuroscience and back *Mov Disord. 2002;17 Suppl* **3**:S28-40. Review.

Berns G S and Sejnowski T J (1998) A computational model of how the basal ganglia produce sequences *J. Cogn. Neurosci.* **10**: 108-121.

Bolam J P, Hanley J J, Booth P A C, and Bevan M D (2000) Synaptic organisation of the basal ganglia *J. Anat.* **196**: 527-542.

Boraud T, Bezard E, Bioulac B, Gross CE (2000) Ratio of inhibited-to-activated pallidal neurons decreases dramatically during passive limb movement in the MPTP-treated monkey *J Neurophysiol* **83**: 1760-1763.

Brotchie J M, Iansek R, Horne M K (1991) Motor function of the monkey globus pallidus. I. Neuronal discharge and parameters of movements *Brain* **114**: 1667-1683.

Calabresi P, Centonze D, Bernardi G (2000) Electrophysiology of dopamine in normal and denervated striatal neurons *Trends in Neurosci.* **23**: S57-S63.

Centonze D, Gubellini P, Picconi B, Calabresi P, Giacomini P, Bernardi G (1999) Unilateral dopamine denervation blocks corticostriatal LTP *J Neurophysiol* **82**: 3575-3579.

Chevalier G and Deniau J M (1990) Disinhibition as basic process in the expression in striatal functions *Trends in Neurosci.* **13**: 277-280.

Contreras Vidal J L and Stelmach G E (1995) A neural model of basal ganglia-thalamocortical relations in normal and parkinsonian movement *Biol. Cybern.* **73**:467-476.

Dayan P, Kakade S, Montague PR (2000) Learning and selective attention *Nat Neurosci. 2000 Nov;3 Suppl*:1218-23. Review.

DeLong MR, Chevalier G (1990) Primate models of movement disorders of basal ganglia origin *Trends Neurosci.* **13**: 281-285.

Deniau JM (1985) Disinhibition as a basic process in the expression of striatal functions. II. The striato-nigral influence on thalamocortical cells of the ventromedial thalamic nucleus. *Brain Res. 1985 May* **20**;334(2):227-33.

Flaherty A W, Graybiel A M (1994) Input-output organisation of the sensorimotor striatum in the squirrel monkey *J. Neurosci.* **14**: 599-610.

Gillies A, Arbuthnott GW (2000) Computational models of the basal ganglia *Movement Disorders* **15**: 762-770.

Gillies A, Willshaw D (2004) Models of the subthalamic nucleus; The importance of intranuclear connectivity *Med Eng Phys. 2004 Nov*;**26**(9):723-32.

Goldberg J A, Boraud T, Maraton S, Haber S N, Vaadia E, Bergman H (2002) Enhanced synchrony among primary motor cortex neurons in the MPTP primate model of Parkinson's disease *J. Neurosci.* **22**(11): 4639-4953

Hikosaka O, Takikawa Y, and Kawagoe R (2000) Role of the basal ganglia in the controle of purposive saccadic eye movements *Physiological reviews* **80**: 953-978.

Hoover J E, Strick P L (1999) The organization of cerebellar and basal ganglia outputs to primary motor cortex as revealed by retrograde transneuronal transport of herpes simplex virus type 1. *J. Neurosci.*

19(4): 1446-1463

Houk J C and Wise S P (1995) Distributed modular architectures linking basal ganglia, cerebelum, and cerebral cortex: their role in planning and controlling action *Cerebral cortex* **5**: 95-110.

Hutchinson W D, Levy R, Dostrovsky J O, Lozano A M, Lang A E (1997) Effects of apomorphine on Globus Pallidus neurons in parkinsonian patients *Annals of Neurology*, **42**: 765-775

Jaeger, D Kita H, Wilson CJ (1994) Surround inhibition among projection neurons is weak or nonexistent in the rat neostriatum *J. Neurophysiol.* **72**: 2555-2558

Kerr JN, Wickens JR (2001) Dopamine D-1/D-5 receptor activation is required for long-term potentiation in the rat neostriatum in vitro *J Neurophysiol* **85**: 117-124.

Kita H (1993) GABAergic circuits of the Striatum In Chemical Signaling in the Basal Ganglia (Arbuthnott, G.W. and Emson, P.C., eds), pp. 51-72, Elsevier

Koòs T and Tepper J M (1999) Inhibitory control of neostriatal projection neurons by GABAergic interneurons *Nature Neurosci.* **2**: 467-472.

Levy R, Hutchison W D, Lozano A M, Dostrovsky J O (2000) High-frequency synchronization of neuronal activity in the subthalamic nucleus of parkinsonian patients with limb tremor *J. Neurosci.* **20**:7766-7775

Mink JW (1996) The basal ganglia: focused selection and inhibition of competing motor programs *Prog Neurobiol* **50**: 381-425.

Mink J W and Thach W T (1993) Basal ganglia intrisic circuits and their role in behavior *Current Opinion in Neurobiol.*, **3**: 950-957.

Nakano K (2000) Neural circuits and topographic organization of the basal ganglia and related regions. *Brain Dev.* **22**:S5-S16

Nambu A, Tokuno H, Hamada I, Kita H, Imanishi M, Akazawa T, Ikeuchi Y, asegawa N (2000) Excitatory cortical inputs to pallidal neurons via the subthalamic nucleus in the monkey. *J. Neurophys.* **84**(1):289-300

Nambu A, Tokuno H, Takada M (2002) Functional significance of the cortico-subthalamo-pallidal 'hyperdirect' pathway. *Neurosci. res.* **43**(2):111-117

Obeso JA, Rodriguez-Oroz MC, Rodriguez M, DeLong MR, Olanow CW (2000) Pathophysiology of levodopa-induced dyskinesias in Parkinson's disease: problems with the current model *Ann Neurol* **47**: S22-S32.

Parent A, Hazrati L N (1995) Functional anatomy of the basal ganglia. I. The cortico-basal ganglia-thalamo-cortical loop *Brain Res. Rev.* **20**: 91.

Plenz, D (2003) When inhibition goes incognito: feedback interactions between spiny projection neurons in striatal function *Trends Neurosci.* **26**, 436-443

Raz A, Vaadia E, and Bergman H (2000) Firing patterns and correlations of spontaneous discharge of pallidal neurons in the normal and the tremulous MPTP-treated vervet model of Parkinsonism *J. of Neurosci.* **20**: 8559-8571.

Raz A, Frechter-Mazar V, Feingold A, Abeles M, Vaadia E, and Bergman H (2001) Activity of pallidal and striatal tonically active neurons is correlated in MPTP-treated monkeys but not in normal monkeys *J. of Neurosci.* **21**: RC128.

Suri R E, Albani C, Glattfelder A H (1997) A dynamic model of motor basal ganglia functions *Biol. Cybern.* **76**: 451-458.

Terman D, Rubin J, Yew A, Wilson C (2002) Activity patterns in a model for the subthalamopallidal network of the basal ganglia *J. Neurosci.* **22**: 296362976.

Tepper JM, Koos T, Wilson CJ (2004) GABAergic microcircuits in the neostriatum *Trends Neurosci. 2004 Nov;***27**(11):662-9.

White OB, Saint-Cyr JA, Tomlinson RD, Sharpe JA (1983) Ocular motor deficits in Parkinson's disease. II. Control of the saccadic and smooth pursuit systems *Brain. Sep;***106** (Pt 3):571-87.

Wichmann T, Bergman H, Starr PA, Subramanian T, Watts RL, DeLong MR. (1999) Comparison of MPTP-induced changes in spontaneous neuronal discharge in the internal pallidal segment and in the substantia nigra pars reticulata in primates *Exp Brain Res. Apr*;**125**(4):397-409

Wickens J (1997) Basal ganglia: structure and computations *Network* **8**: R77-R109.

Wu Y, Richard S, Parent A (2000) The organization of the striatal output system: a single-cell juxtacellular labeling study in the rat *Neurosci Res* **38**: 49-62.

In: Recent Breakthroughs in Basal Ganglia Research
Editor: Erwan Bezard, pp. 45-52
ISBN: 1-59454-880-3

Chapter 4

Reinforcement Driven Nonlinear Dimensionality Reduction in the Multilayer Network of the Basal Ganglia

Izhar Bar-Gad
Gonda Brain Research Center, Bar-Ilan University, Ramat-Gan, Israel
Hagai Bergman*
Center for Neural Computation and Department of Physiology, Hadassah Medical School, The Hebrew University, Jerusalem, Israel

Abstract

Over the last twenty years information processing models of the basal ganglia in health and disease have undergone significant changes which complemented the accumulation of new anatomical, physiological and biochemical findings. A key change was the shift from system level models which used a single rate and connectivity representation for whole nuclei to network level models which emphasize the activity and interaction of elements within the nuclei. We recently suggested the reinforcement driven dimensionality reduction (RDDR) model which emphasizes the role of the basal ganglia in the extraction of information transmitted from the cortex. The basic RDDR model describes a single layer of information processing, the striatum, which receives feedforward input from the cortex and a dopaminergic reinforcement signal. The following manuscript places the RDDR concepts within the more comprehensive framework of consisting of multiple non-linear layers. This expanded model presents further possibilities for complex information processing and encoding in the basal ganglia networks.

Introduction

Over the last twenty years information processing models of the basal ganglia (Beiser et al., 1997; Gillies and Arbuthnott, 2000) in health and disease have undergone significant changes which complemented the accumulation of new anatomical, physiological and biochemical findings. One of the key changes was the shift from system level models, such as the box and arrow models (Albin et al., 1989; DeLong, 1990), which used a single rate and connectivity representation for whole nuclei to network level models which emphasize the different activity of the elements within the nuclei and their interaction. These new models focused on the role of the basal ganglia as an action and sequence selection system (Mink, 1996; Wickens, 1997; Gurney et al., 2004). We recently suggested a novel model, the reinforcement driven dimensionality reduction (RDDR) model (Bar-Gad et al., 2000; Joel et al., 2002; Bar-Gad et al., 2003). This model emphasizes the role of the interaction within and between neurons of the basal ganglia in the

* Correspondence to: Izhar Bar-Gad, Gonda Brain Research Center, Bar-Ilan University, Ramat-Gan, 52900, Israel, Tel: +972-3-5317141 Fax: +972-3-5352184, E-mail address: bargadi@mail.biu.ac.il

extraction of information transmitted from the cortex. The basic RDDR model describes a single layer of information processing, the striatum, which receives feedforward input from the cortex and a reinforcement signal from its dopaminergic inputs. The following manuscript places the RDDR concepts within the more comprehensive framework of the cortico-basal ganglia loop which consists of multiple different layers. This expansion includes incorporation of the multi-layer structure of the network and the different characteristic of the neurons within each layer. The expanded model presents further possibilities for complex information processing and encoding in the basal ganglia networks.

1 Basic Reinforcement Driven Dimensionality Reduction

The basic RDDR model (Bar-Gad et al., 2000; 2003) is an extension of a neural-network based model for performing principal component analysis (PCA; Foldiak, 1989; Kung and Diamantaras, 1990). The network contains two layers (Fig. 1), which represent the cortex (input layer) and the striatum (output layer). The activity function of the neurons in the output layer is linear; i.e. the output is the weighted sum of the feed forward inputs minus the weighted sum of the lateral inputs. The network uses the feed forward network to encode the principal components while the lateral network serves the role of preventing multiple neurons from converging to the same components, thus extending Oja's original PCA model (Oja, 1982) which enables the discovery of the only first principal component. The learning rules are extensions of the normalized Hebbian rule (Oja, 1982) for the feed forward weights and extensions of anti-Hebbian rule for the lateral network. All the learning is local and no global quantities are calculated, thus each weight is affected only by the activation of the pre and post synaptic activities. In the case of lateral connectivity which is asymmetrical, the output of the network will become ordered with the first component (i.e. containing the most information) displayed by the first neuron and the next components displayed in descending order by the next neurons. In contrast, if the connections are symmetrical that information is distributed among the different neurons of the output layer. The RDDR model adds a reinforcement factor to the unsupervised learning performed by the basic PCA networks. This reinforcement signal simulates the dopaminergic input to the striatum from the substantia nigra pars compacta (SNc; Schultz et al., 1997). The modeled reinforcement factor is multiplied by the input and the output of the neuron to create a multi-Hebbian learning algorithm. In a multi-Hebbian learning rule the reinforcement signal regulates the amount of change in the weight of a synaptic connection for a given input/output pair (Reynolds and Wickens, 2002). This type of synaptic modification is in agreement with the electron microscopic studies which indicate that the cortical terminals are often found on the heads of spines and dopaminergic terminals are found nearby, but on the dendritic shafts (Dube et al., 1988; Smith et al., 1994) enabling modification of the synaptic changes occurring in the spine.

2 Nonlinear Multilayer Structure of the Basal Ganglia Cortical Loop

The cortico-basal ganglia loop forms a partially closed loop leading from most parts of the cortex to the input structures of the basal ganglia, the striatum and the sub-thalamic nucleus (STN). The information of the input nuclei is sent via multiple pathways to the output nuclei of the basal ganglia, the globus pallidus internal segment (GPi) and the substantia nigra pars reticulata (SNr). The output nuclei of the basal ganglia send their output to the ventrolateral thalamic nucleus, the ventral anterior thalamic nucleus, the mediodorsal nucleus, the lateral habenular nucleus, the intralaminar (non-specific) thalamic nuclei, and the centromedian and parafasicular nucleus (CM/Pf). These thalamic nuclei send their output mainly to the frontal cortex which itself forms part of the input to the basal ganglia, thus completing and partially closing the loop.

A striking feature of the cortico-basal ganglia loop is the funnel structure observed when looking at the number of neurons within each stage of the loop. The numbers themselves vary greatly depending on the animal and the research methodology. However, the same ratio seems to persist across different studies. The first convergence takes place from the cortex to the input nuclei of the basal ganglia and is of about one order of magnitude (Zheng and Wilson, 2002). A second, and larger reduction in the number of neurons is between the input and output nuclei of the basal ganglia where the factor is of two to three orders of magnitude (Schroder et al., 1975; Thorner et al., 1975; Percheron et al., 1987; Oorschot, 1996). An open question is the number of thalamic neurons receiving projections from the output nuclei of the basal ganglia, and the number of cortical neurons connected to these thalamic neurons. Indirect evidence supports the notion of an increase in the number of neurons in these stages (Arecchi-Bouchhioua et al., 1997; Francois et al., 2002), and therefore a bottle-neck structure of the cortex-basal ganglia-cortex loop.

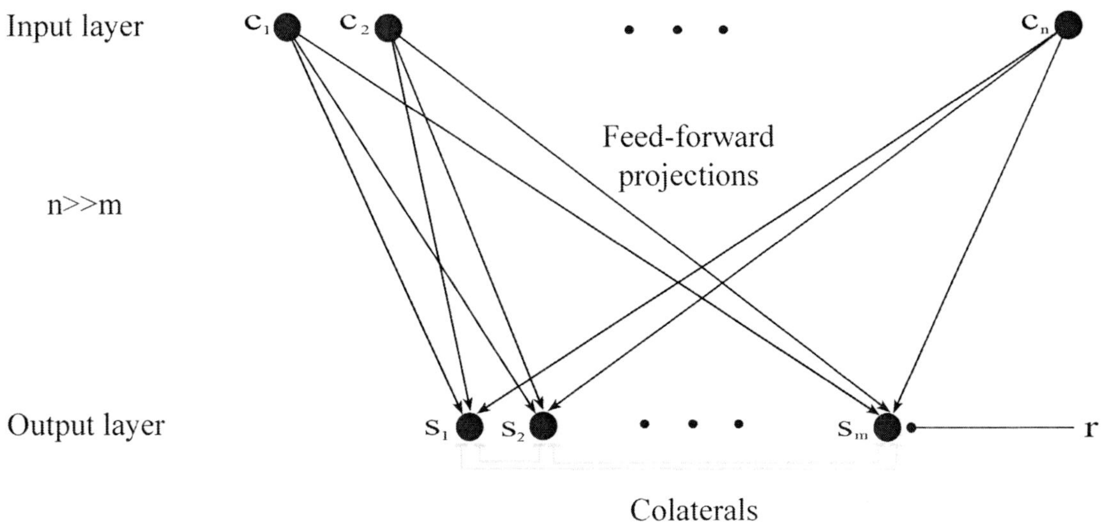

Figure 1: Two layers basic RDDR network scheme. Input elements (C_1, C_2, ..., C_n) are fully connected to the output elements (S_1, S_2, ..., S_m). The output elements are fully laterally connected and in addition receive a scalar input (r).

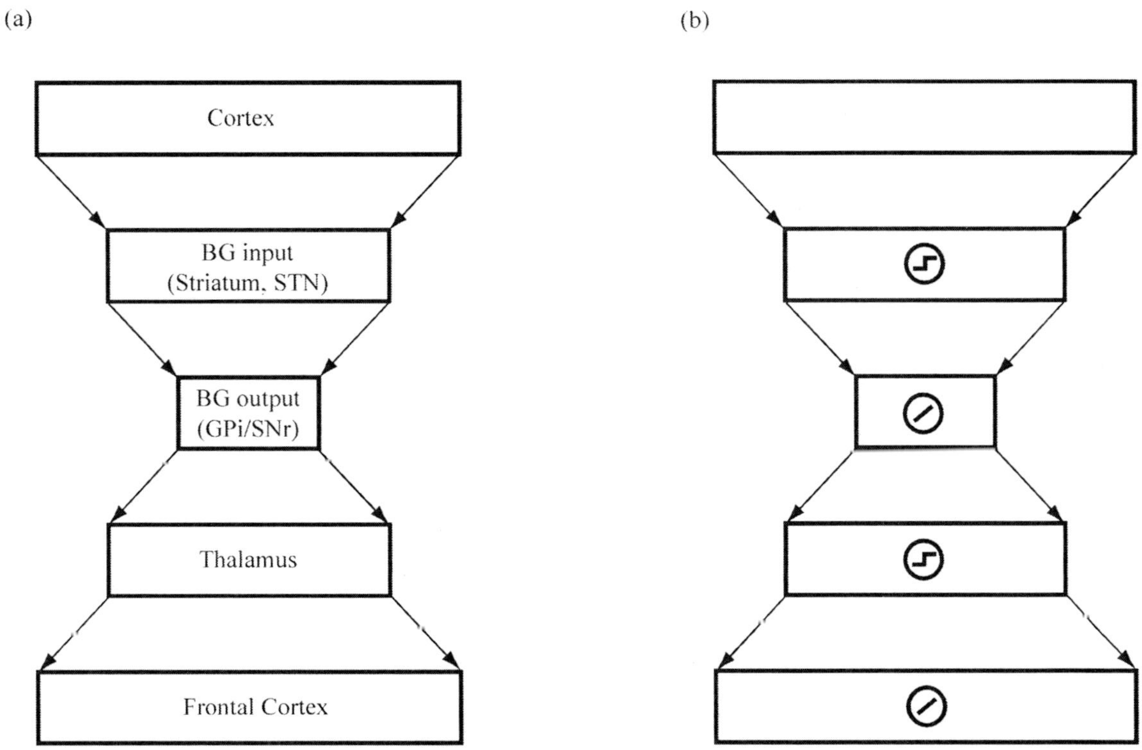

Figure 2: Five layers of the cortico-basal ganglia loop. (a) Scheme of the layers within the cortico-basal ganglia loop. (b) Linearity properties of the different layers in the loop.

3 NONLINEAR MULTILAYER REINFORCEMENT DRIVEN DIMENSIONALITY REDUCTION

The basic RDDR model is primarily a conceptual representation of the basal ganglia in gross lines. The model's simplicity is important for purposes of exploring its main ideas and predictions within the many unknown parameters of a biological system. Expanding the RDDR model to reflect more closely the anatomy and physiology of the basal ganglia enables further understanding of the computational possibilities and the constraints imposed by the biological system.

The basic RDDR model performs PCA by using linear neurons. Neurons with a linear response function transform the sum of their weighted inputs to firing rates, in a linear manner by simple summation. However, neurons are never completely linear as they all have an upper limit which is their maximal firing rate and a lower limit which is zero since neurons cannot switch to negative firing rates. In addition to these obvious upper and lower limits, neurons have an extremely variable response to change in their input even within their response limits. Projection neurons along the different layers of the of the cortico-basal ganglia loop have very different input-output transfer functions ranging from close to linear to highly non-linear. On the one hand, neurons in the globus pallidus have a linear relation between current injected to their soma and their firing rate i.e. linear I-f function (Nakanishi et al., 1990; Kita and Kitai, 1991; Nakanishi et al., 1991; Kita, 1992) and a broad range of firing rates (DeLong, 1972; Miller and DeLong, 1987; Fig. 3). On the other hand, the MSNs of the striatum are non-linear with two basic subthreshold states (Wilson and Kawaguchi, 1996; Stern et al., 1997). Evidence for the occurrence of such two-state membrane potential transitions were further supplied by numerical simulations and electrophysiological recordings in awake monkeys (Kitano et al., 2002). In the 'down' state, MSN neurons have zero firing rate while high firing rates can be reached in the 'up' state (Nisenbaum and Wilson, 1995; Wilson, 1995; Wilson and Kawaguchi, 1996; Stern et al., 1997; Bennet and Wilson, 2000). The dynamic range of these neurons is small, leading mostly to activity in one of the extremes states (Fig. 3).

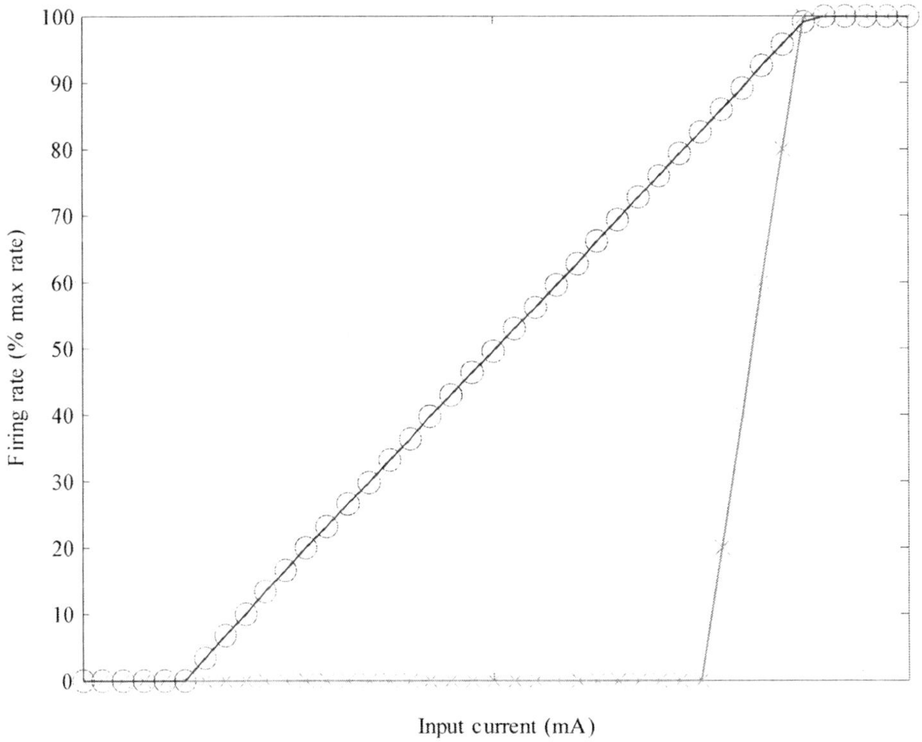

Figure 3: Linearity of neural response. Frequency response of neurons to various input currents. A highly linear neuron (O) resembling projection neurons of the globus pallidus with high firing rate during rest and a broad modulation range. A highly non-linear neuron (X) resembling MSNs of the striatum has a zero firing rate during the down state and a narrow dynamic range during the up state.

The addition of non-linearity to the network enables an increase in the scope of the computation performed by the network. Linear versions of PCA are able to find only linear interactions between the input, and the optimal connections formed are based solely on second order statistics (covariance of the activity). In cases of non-linear interactions between the input elements and higher order statistics of interaction, a linear network will not be able to reduce the dimensionality of the input effectively i.e., the output layer will not maintain the information within the input layer. However, using the same constrained Hebbian learning rules employed in PCA networks with a non-linear transfer function on the neurons (typically a sigmoid) enables learning the structure of complex non-linear inputs embedded within a noisy environment (Oja et al., 1991). The addition of non-linearity elements makes possible the extraction of higher order statistics and the formation of independent components at the output layer, thus enabling the separation of the mixture represented by the inputs into its independent underlying causes (Jutten and Herault, 1991; Karhunen and Joutsensalo, 1994).

An important feature of extremely non-linear computation is generalization. Linear encodings transform different inputs into unique outputs. On the other hand, the extreme non-linear cases of 0/1 transformation (which is equivalent to very sharp sigmoid transfer function) transforms a group of inputs into a single output. The input is thus divided into clusters according to the transformation rules. One such non-linear network was implemented by Carlson, (1990). In his network, the activity of the neurons depends on the neuron's threshold and the transition width. The width defines the neuron's nonlinearity: when the width is 0, activity is highly nonlinear (binary), and when it is large, activity is linear. Both the threshold and the width are adjusted during learning. The output neurons are hierarchically organized (like the APEX linear network and others Kung and Diamantaras, 1990; Rubner and Schulten, 1990). The outcome of this algorithm in the nonlinear regime is a partition of the input into clusters, whereas in the linear case it achieves a linear representation.

The RDDR model may be expanded to a multilayered network which consists of five layers (Fig. 2a). Sequential processing, performed by such a multi-layer dimensionality reduction network, has two main advantages arising from the ability to perform the same computation in several stages (Barlow, 1992):

A single neuron cannot directly access a large portion of the input due to the limit on the number of synapses. A multi-layer structure enables access to a larger portion of the information within the input in a gradual convergence process.

The huge size of the input space impedes efficient learning of its representation due to its sparseness. The problem is therefore broken up into multiple layers, where the first layers learn a limited number of input patterns and subsequent layers learn larger portions of the input space.

In the basal ganglia, each of the two layers gains access to up to 10^3-10^4 input neurons (Yelnik, 2002; Zheng and Wilson, 2002). Theoretically, the combination of the two layers enables the neurons of the output layer (the projection neurons of the GPi) to gain access to a large part of the cortical input (up to 10^7 cortical neurons).

In addition, a multi-layer dimensionality reduction has the ability to perform a different type of processing within each layer. Thus, multi-layer networks enable the combination of linear and non-linear layers within a single network. Oja has studies a network containing a non-linear (encoding) layer followed by a linear (bottleneck) layer performing non-linear and linear PCA respectively (Oja, 1991). The network is able to perform any continuous mapping from the input space to an output layer which is smaller in size. This network is surprisingly similar to the multi-layer structure of the main feed-forward pathway in the basal ganglia (Fig. 2b). Cortical and thalamic inputs reach the MSNs of the striatum, which are non-linear and have a transfer function much like the sigmoid used by Oja. From the striatum the information is passed to the globus pallidus, which is characterized by neurons with a relatively linear transfer function. Oja's work also speculates on the role of two additional layers in decoding information processed by the initial encoding layers. These decoding layers fit nicely within the cortico-basal ganglia-cortico loop as an equivalent to the thalamus and frontal cortex. Further research (DeMers and Cottrell, 1993; Malthouse et al., 1995) on the practical and theoretical implications of such multi-layer non-linear networks has demonstrated their enhanced capabilities in complex feature extraction which cannot be performed by linear networks.

Summary

Experimental data is continuing to prompt new developments in models of the basal ganglia. Data from many fields, including anatomy, physiology, biochemistry, as well as clinical evidence are paving the way for the modeling of additional aspects of the function of these nuclei in health and disease. The RDDR conceptual model was developed to account for the large differences in the neuronal population along the cortico-basal ganglia loop. The model suggests that the basal ganglia perform a compression of the information received from the cortex, based on a reinforcement signal. Thus, the funnel- like structure of the basal ganglia is used to remove the redundancy in the data space spanned by the activity of cortical neurons and to maintain the most important information (from a reward standpoint) to enable efficient planning of new actions. The extension of the model from a two layer linear network to a five layer non-linear network offers a more biologically plausible network which also provides enhanced encoding capabilities.

Acknowledgements

This study was partly supported by a Center of Excellence grant administered by the ISF, the German-Israel Binational Foundation (GIF), and the German-Israeli cooperation in the neurosciences administrated by the Israel Ministry of Science & Technology (MOS) and the Federal Ministry of Education & Research (BMBF).

References

Albin RL, Young AB, Penney JB (1989) The functional anatomy of basal ganglia disorders. *Trends Neurosci* **12**:366-375.

Arecchi-Bouchhioua P, Yelnik J, Francois C, Percheron G, Tande D (1997) Three-dimensional morphology and distribution of pallidal axons projecting to both the lateral region of the thalamus and the central complex in primates. *Brain Res* **754**:311-314.

Bar-Gad I, Morris G, Bergman H (2003) Information processing, dimensionality reduction and reinforcement learning in the basal ganglia. *Prog Neurobiol* **71**:439-473.

Bar-Gad I, Havazelet-Heimer G, Goldberg JA, Ruppin E, Bergman H (2000) Reinforcement-driven dimensionality reduction--a model for information processing in the basal ganglia. *J Basic Clin Physiol Pharmacol* **11**:305-320.

Barlow H (1992) The biological role of the neocortex. In: Information processing in the cortex: Experiments and Theory (Aertsen A, Braitenberg V, eds), pp 54-80: Springer-Verlag.

Beiser DG, Hua SE, Houk JC (1997) Network models of the basal ganglia. *Curr Opin Neurobiol* **7**:185-190.

Bennet B, Wilson C (2000) Synaptology and physiology of neostriatal neurons. In: Brain dynamics and the strital complex (Miller R WJ, ed), pp 111-140. Australia.

Carlson A (1990) Anti-Hebbian learning in a non-linear neural network. *Biol Cybern* **64**:171-176.

DeLong MR (1972) Activity of basal ganglia neurons during movement. *Brain Res* **40**:127-135.

DeLong MR (1990) Primate models of movement disorders of basal ganglia origin. *Trends Neurosci* **13**:281-285.

DeMers D, Cottrell G (1993) Nonlinear dimensionality reduction. *Neural Information Processing Systems* **5**:580-587.

Dube L, Smith AD, Bolam JP (1988) Identification of synaptic terminals of thalamic or cortical origin in contact with distinct medium-size spiny neurons in the rat neostriatum. *J Comp Neurol* **267**:455-471.

Foldiak P (1989) Adaptive Network for Optimal Feature Extraction. In: Proceedings, *International Joint Conference on Neural Networks*, pp 401-405.

Francois C, Tande D, Yelnik J, Hirsch EC (2002) Distribution and morphology of nigral axons projecting to the thalamus in primates. *J Comp Neurol* **447**:249-260.

Gillies A, Arbuthnott G (2000) Computational models of the basal ganglia. *Mov Disord* **15**:762-770.

Gurney K, Prescott TJ, Wickens JR, Redgrave P (2004) Computational models of the basal ganglia: from robots to membranes. *Trends Neurosci* **27**:453-459.

Joel D, Niv Y, Ruppin E (2002) Actor-critic models of the basal ganglia: new anatomical and computational perspectives. *Neural Netw* **15**:535-547.

Jutten C, Herault J (1991) Blind separation of sources, Part I : An adaptive algorithm based on neuromimitic architecture. *Signal Processing* **24**:1-10.

Karhunen J, Joutsensalo J (1994) Representation and Separation of Signals Using Nonlinear PCA Type Learning. *Neural Networks* 7:113-127.

Kita H (1992) Responses of globus pallidus neurons to cortical stimulation: intracellular study in the rat. *Brain Res* **589**:84-90.

Kita H, Kitai ST (1991) Intracellular study of rat globus pallidus neurons: membrane properties and responses to neostriatal, subthalamic and nigral stimulation. *Brain Res* **564**:296-305.

Kitano K, Cateau H, Kaneda K, Nambu A, Takada M, Fukai T (2002) Two-state membrane potential transitions of striatal spiny neurons as evidenced by numerical simulations and electrophysiological recordings in awake monkeys. *J Neurosci* **22**:RC230.

Kung S, Diamantaras K (1990) A neural network learning algorithm for adaptive principal component extraction (APEX). In: *Proceedings, IEEE International Conference on Acoustics, Speech and Signal Processing*, pp 861-864.

Malthouse E, Mah R, Tamhane A (1995) Some theoretical results on nonlinear principal component analysis. In: *Proceedings of the American Control Conference*, pp 744-748.

Miller W, DeLong M (1987) Altered tonic activity of neurons in the globus pallidus and subthalamic nucleus in the primate MPTP model of parkinsonism. In*: The Basal Ganglia II (Carpenter M, Jayaraman A, eds), pp* **415-427***. New York: Plenum Press.*

Mink JW (1996) The basal ganglia: focused selection and inhibition of competing motor programs. *Prog Neurobiol* **50**:381-425.

Nakanishi H, Kita H, Kitai ST (1990) Intracellular study of rat entopeduncular nucleus neurons in an in vitro slice preparation: electrical membrane properties. *Brain Res* **527**:81-88.

Nakanishi H, Kita H, Kitai ST (1991) Intracellular study of rat entopeduncular nucleus neurons in an in vitro slice preparation: response to subthalamic stimulation. *Brain Res* **549**:285-291.

Nisenbaum ES, Wilson CJ (1995) Potassium currents responsible for inward and outward rectification in rat neostriatal spiny projection neurons. *J Neurosci* **15**:4449-4463.

Oja E (1982) A simplified neuron model as a principal component analyzer. *J Math Biol* **15**:267-273.

Oja E (1991) Data Compression, Feature Extraction, and Autoassociation in Feedforward Neural Networks. In: *Artificial Neural Networks* (Kohonen T, Makisara K, Simula O, Kangas J, eds), pp 737-745.

Oja E, Ogawa H, Wangviwattana J (1991) Learning in Nonlinear Constrained Hebbian Networks.

Oorschot DE (1996) Total number of neurons in the neostriatal, pallidal, subthalamic, and substantia nigral nuclei of the rat basal ganglia: a stereological study using the cavalieri and optical disector methods. *J Comp Neurol* **366**:580-599.

Percheron G, Francois C, Yelnik J (1987) Spatial organization and information processing in the core of the basal ganglia. In: *The Basal Ganglia II* (Carpenter M, Jayaraman A, eds), pp 205-226. New York and London: Plenum Press.

Reynolds JN, Wickens JR (2002) Dopamine-dependent plasticity of corticostriatal synapses. *Neural Netw* **15**:507-521.

Rubner J, Schulten K (1990) Development of feature detectors by self-organization. A network model. *Biol Cybern* **62**:193-199.

Schroder K, Hopf A, Lange H, Thorner G (1975) Morphometrisch-statistische Strukturanalysen des Striatum, Pallidum und Nucleus subthalamicus beim Menschen [Morphometrical-statistical structure analysis of human striatum, pallidum and subthalamic nucleus]. *J Hirnforsch* **16**:333-350.

Schultz W, Dayan P, Montague PR (1997) A neural substrate of prediction and reward. *Science* **275**:1593-1599.

Smith Y, Bennett BD, Bolam JP, Parent A, Sadikot AF (1994) Synaptic relationships between dopaminergic afferents and cortical or thalamic input in the sensorimotor territory of the striatum in monkey. *J Comp Neurol* **344**:1-19.

Stern EA, Kincaid AE, Wilson CJ (1997) Spontaneous subthreshold membrane potential fluctuations and action potential variability of rat corticostriatal and striatal neurons in vivo. *J Neurophysiol* 77:1697-1715.

Thorner G, Lange H, Hopf A (1975) [Morphometrical-statistical structure analysis of human striatum, pallidus and subthalamic nucleus. II. Globus pallidus]. *J Hirnforsch* **16**: 401-413.

Wickens J (1997) Basal ganglia: structure and computation. In: Network: *Computation in Neural Systems*, pp R77-R109.

Wilson C (1995) The Contribution of Cortical Neurons to the Firing Pattern of Striatal Spiny Neurons. In: Models of Information *Processing in the Basal Ganglia* (Houk J, Davis J, Beiser D, eds), pp 29-50: MIT Pres.

Wilson CJ, Kawaguchi Y (1996) The origins of two-state spontaneous membrane potential fluctuations of neostriatal spiny neurons. *J Neurosci* **16**:2397-2410.

Yelnik J (2002) Functional anatomy of the basal ganglia. *Mov Disord 17 Suppl* **3**:S15-21.

Zheng T, Wilson CJ (2002) Corticostriatal combinatorics: the implications of corticostriatal axonal arborizations. *J Neurophysiol* **87**:1007-1017.

In: Recent Breakthroughs in Basal Ganglia Research
Editor: Erwan Bezard, pp. 53-54
ISBN: 1-59454-880-3

PART II: THE BASAL GANGLIA: COGNITIVE AND MOTIVATIONAL PROCESSING – FROM ACTION TO VALUE AND BACK

Genela Morris and Hagaï Bergman
Hadassah Medical School – Hebrew University, Jerusalemn, Israel

Inspired by pathologies associated with basal ganglia dysfunction, the first decades of basal ganglia research were dedicated to the role of these nuclei in motor control. Throughout the years, this school of thought has evolved to a well established branch, capable of explaining different pathologies associated with these structures with an increasing resolution of detail. Advances in psychological and computational theory, along with the accumulating evidence of the involvement of elements within the basal ganglia in hedonic and addictive behaviors, has shifted the focus of basal ganglia research. The next stop was the study of basal ganglia involvement in cognitive processing of motivational and rewarding aspects of the surrounding environment. The multidisciplinary study of the brain flourishing in recent years has completed the maneuver. Considerable effort is now dedicated to integrating the extensive knowledge in the two branches. It seems that all lines of work converge to different, often complementary accounts of the basal ganglia as a structure associating environmental cues with action. This association is initiated by the motivational value of the outcomes attributed to the various actions and reinforced by the rewarding value attributed to each such outcome.

The input nucleus of the basal ganglia, the striatum, receives massive neuro-modulatory innervation. Most of the dopamine and acetylcholine found in the brain, as well as substantial amounts of serotonin are co-localized in the striatum. One of the major breakthroughs in linking the basal ganglia stimulus-action association through reward came with the identification of the neuro-modulators targeting the basal ganglia with teaching elements developed in the computational world of reinforcement learning. This view gained support from three lines of study. On the anatomical level, the basal ganglia architecture makes it an ideal candidate to serve as the brain counterpart of an extremely potent computational controller of action known as the actor-critic. System-level physiology has contributed the neural substrate for the critic in this scheme: Recording of dopamine and acetylcholine containing neurons in primate basal ganglia showed that their responses bear remarkable resemblance to the teaching signals predicted by formal learning theory. Recent work has dissociated the roles of dopamine and acetylcholine, and added a third potential complementary teacher, serotonin. Finally, cellular physiology demonstrated the ability of these neurons to mediate learning through their effect on synaptic plasticity.

Control of the actions to be executed in response to environmental input can be accomplished on two levels. Dopamine and acetylcholine have been implicated in both. On the one hand, their reinforcing role provides a tool for changing the association of a particular set of inputs to the desired actions. Such a view of the effect of dopamine on behavior is demonstrated in the first chapter of this section by Wickens et al. who discusses dopamine mediated reinforcement of stimulus-response associations. On the other extreme, we may envision a direct role of immediate control of action. The second chapter presented here by Apicella et al., highlights a potential role of striatal acetylcholine containing interneurons, the physiologically identified tonically active neurons (TANs), in aiding the execution of automatized actions sequences. The following chapter, by Kimura et al., explicitly relates to the dual role of dopamine in action selection and decision making by assigning the different roles to two temporally distinct responses of the dopamine neurons. Finally, in the work presented by Daw et al. the dopamine signal is used both to reinforce particular state-action associations and to choose between them when a dilemma is encountered.

Generally, behaviors can be characterized according to the extent to which they are hard-wired into the neuronal apparatus. On one extreme we find highly standardized, automatic action sequences, which may be viewed as habits. The other extreme is occupied by much more variable modular and complex goal directed behaviors. All authors contributing to this chapter seem to agree that the basal ganglia are involved in choosing, maintaining reinforcing and performing habitual action sequences. In brief, the work of Apicella et al. emphasizes the strong dependence of the signal emitted by the striatal TANs on temporal unpredictability of events. They suggest therefore that this response serves a role in the generation and performance of automatic sequential behaviors. Wickens et al. present evidence for a model in which the striatal projection neurons serve as recognition units for pre-existing templates of stimulus-response associations, perhaps by encoding their value. These associations are reinforced through actor-critic like reinforcement learning. They propose that this value encoding may be impaired in children suffering from attention deficit hyperactivity disorder (ADHD), leading to inappropriate reinforcement of behaviors associated with immediate reward. Kimura et al. discuss two distinct signals emitted by dopamine neurons. The first reports the motivational value of a desired action, and the seconds encodes the error in prediction of reward. The latter signal is implicated in learning in the well established temporal difference (TD) reinforcement learning scheme. The first signal serves a dual purpose. In the immediate sense it is used to initiate approach behavior, and in the long term learning sense it provides a priming signal for the prediction error, controlling its gain. In the last chapter, Daw et al. propose a model for two types of decision making. They start with a thorough discussion of decision making in reinforcement learning. They then go on to elaborate two classes of controllers, differing in the types of behavior that will be executed by them. They suggest that the basal ganglia are an actor-critic network, dedicated to the performance and selection of habitual behaviors. In contrast, goal directed actions require a different form of reinforcement learning control that is based on an internal model of environmental states, actions, outcome and subjective utility and the interplay between them. This controller is proposed to reside in the cortex. Both controllers are proposed be in competition, which is resolved by their relative advantage reliability under various circumstances.

In conclusion, this section presents us with a new look at the role of the basal ganglia in the control of behavior. The proposed view of the basal ganglia as a reinforcement learning network provides new insight into the understanding of this crucial element of the nervous system. We hope that this insight will encourage new approaches to the study of basal ganglia related disorders.

In: Recent Breakthroughs in Basal Ganglia Research
Editor: Erwan Bezard, pp. 55-63
ISBN: 1-59454-880-3

Chapter 5

A Possible Role for Tonically Active Neurons of the Primate Striatum in Learning about Temporal Relationships among Salient Stimuli[#]

Paul Apicella*[*], *Eric Legallet
Laboratoire de Neurobiologie de la Cognition, CNRS, 31 chemin Joseph Aiguier, 13402 Marseille cedex 20, France
Sabrina Ravel
Laboratory of Neuropsychology, National Institute of Mental Health, Bethesda, MD 20892, USA

Abstract

Tonically active neurons (TANs) of the primate striatum are involved in signalling the motivational significance of external stimuli. In a series of experiments in behaving monkeys, we have demonstrated that the response of TANs is dependent on the temporal context in which motivating stimuli are presented, uncertainty about stimulus timing being essential to TAN responsiveness. It therefore appears that TANs might contribute to learning processes involving stimulus predictions mediated by a timekeeping strategy that is built up through extensive practice in a given context. This prediction-dependent effect could be considered as evidence for a role of these neurons in the learning of temporal relationships between motivationally relevant events. We suggest that TANs participate in a mechanism that mediates automatic processing of information about stimulus timing in conditions involving routine and stereotyped behavioral performance.

Introduction

Several lines of evidence suggest that the striatum, the main input structure of the basal ganglia, forms an essential part of the neural circuits that are implicated in the learning, initiation and execution of goal-directed behaviors. However, the exact contribution of this subcortical structure to the processing of information is still controversial. One can get an insight into striatal information processing functions from single neuron recordings in animals performing behavioral tasks. Electrophysiological studies in behaving monkeys have shown that a population of striatal neurons, known as the tonically active neurons or TANs, display phasic variations in discharge rate in response to stimuli having an intrinsic or learned motivational value (Apicella, 2002). These neurons, thought to correspond to the cholinergic interneurons of the striatum, are considered to be an important substrate for reward-based learning (Aosaki et al., 1994; Graybiel, 1995).

[#] This work was supported by Centre National de la Recherche Scientifique, the Biomed II Program of the European Commission, and the French Government (ACI « Neurosciences Integratives et Computationnelles »).
[*] Correspondence: Laboratoire de Neurobiologie de la Cognition, CNRS, 31 chemin Joseph Aiguier, 13402 Marseille cedex 20, France, Phone: 00 33 4 91 16 40 42, Fax: 00 33 4 91 16 42 96, E-mail address: paul.apicella@up.univ-mrs.fr

It was found that the context in which stimuli occur may have a strong modulating effect on TAN responsiveness. In particular, previous research in our laboratory has indicated that the sensitivity of TANs to motivating events can be modified by temporal aspects of stimulus prediction. We have demonstrated that TAN responses are more prevalent when stimuli occur at an unpredicted time than when their timing is predicted. In this review, we will describe findings from our laboratory suggesting that this prediction-dependent effect could be considered as evidence for specific participation by this local neuron system in the learning of the temporal structure of series of events.

1 Temporal Predictability of Reward in a No-task State

The response to a pulse of fruit juice delivered at unpredicted times, unconfounded with other conditioned stimuli, while the monkey remains motionless with no task in progress, is a striking feature for characterizing response properties of TANs (Apicella et al., 1997). These responses mainly consist of short-latency phasic depressions in activity recorded in widespread regions of the caudate nucleus and putamen, with no evidence of preferential localization. They are not tightly correlated with associated motor reactions, i. e., mouth movements to consume the liquid.

We have observed that responses of TANs to the free delivery of liquid were reduced considerably when repetitive rewards were separated by a constant interval of 2 s instead of irregular intervals (Fig. 1). This suggests that a certain degree of temporal regularity in the rate of reward delivery may alter the responsiveness of TANs to this event (Ravel et al., 2001). We interpret this decreased neuronal sensitivity as a consequence of an increase in the monkey's ability to rely on the rate of reward delivery to predict its timing. Additional experiments using time intervals shorter than 2 s have shown that the reduction in TAN responding cannot be attributed simply to a sort of refractoriness when stimuli occur in close temporal proximity (Sardo et al., 2000). Thus, the temporal context in which a motivating stimulus is presented appears to determine its efficacy in eliciting responses from TANs.

2 Influence of an External Signal Predictive of Reward

Our suggestion that the responsiveness of TANs to reward can be lost when the monkey learns to predict its timing may explain why responses to reward are less frequent and weaker in the context of an instrumental task. Following overtraining of a visuo-motor task in which the monkey was required to make an arm-reaching movement to obtain reward, a relative small fraction of TANs responded to reward that was delivered immediately after target contact; instead, the majority of these neurons responded to the visual stimulus triggering the movement (Apicella et al., 1997; Ravel et al., 2001). This lack of responding may be due to the fact that the monkey can use the target contact to predict when rewards occur. On the other hand, the trigger stimulus whose onset time is unpredicted is particularly efficient in eliciting TAN responses. It is possible to decrease or abolish the trigger responses by adding an instruction cue which provides advance information about the timing of the subsequent trigger (Sardo et al., 2000).

To further examine the influence of stimulus prediction, monkeys were subjected to an another testing situation in which reward was automatically delivered after the onset of an external signal, no behavioral reaction on the animal's part being required to obtain it. This is a typical classical conditioning procedure in which the monkey remained motionless and the onset of a visual stimulus (the same light previously used as a movement-triggering stimulus in the instrumental task) was followed by the delivery of reward. Figure 2 shows examples of responsive TANs recorded during the course of learning with this stimulus-reward association. During the initial presentations, TANs responded to both the stimulus and reward (Fig. 2A), but the response to the stimulus rapidly declined with repeated exposure to the stimulus-reward association (Fig. 2B). At later periods of training, responses to the stimulus were restored and reward responses remained present (Fig. 2C). After extensive training, TANs showed little or no response to reward, but they displayed responses to the stimulus that has acquired reward-predicting properties (Fig. 2D). This transfer of TAN responses from the reward to the earlier stimulus acting as a predictor of reward agrees with the findings of other studies (Shimo and Hikosaka, 2001; Ravel et al., 2003). Such an adaptation of TAN responsiveness could contribute to establishing stimulus-reward associations and thus constitutes an electrophysiological correlate of this type of learning.

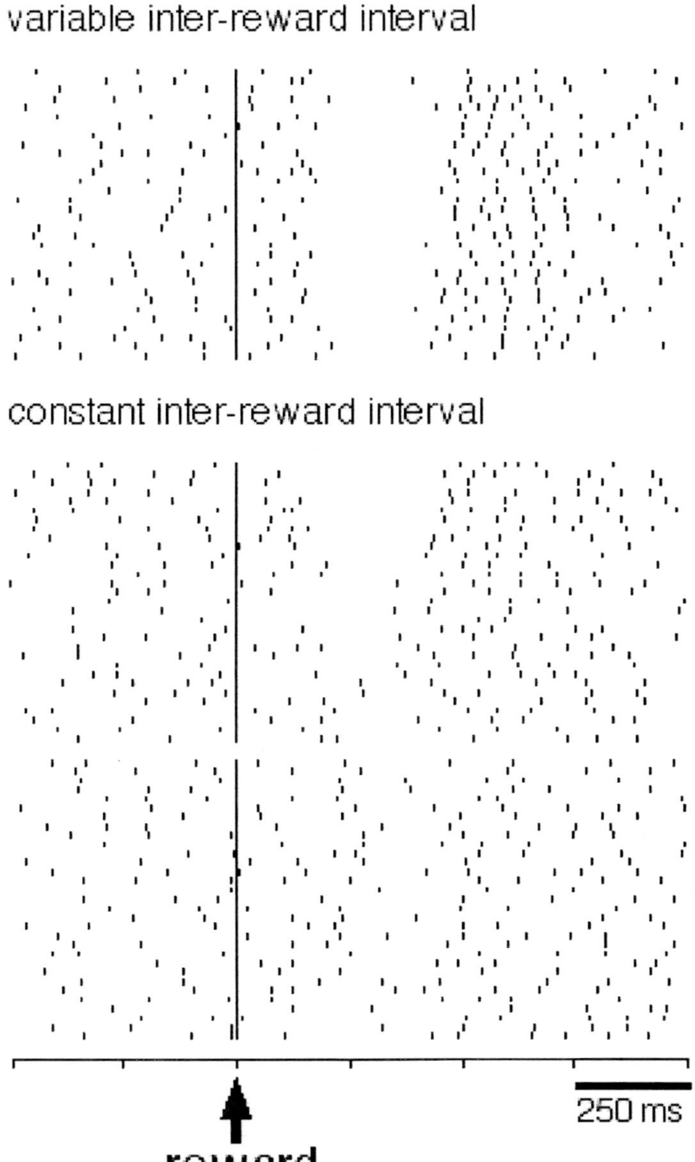

Figure 1. Response of one TAN to repetitive deliveries of liquid outside of a task. The monkey was restrained in a seated position with mechanical stabilization of the head. At unpredictable times, it received a drop of liquid without having to perform any particular behavioral reaction, in the absence of any external signal. Liquid was delivered via a spout positioned directly in front of the monkey's mouth. In the raster displays, each dot represents one spike, and each line of dots a single trial. Raster displays are aligned on the delivery of reward marked by vertical lines. The chronological order of trials is shown from top to bottom in each raster. At the beginning of the test, inter-reward intervals varied from 3.5 to 8.5 s and they were replaced by a fixed 2-s interval. On this example, TAN response subsides after 20-30 repeated deliveries of liquid at the same 2-s interval.

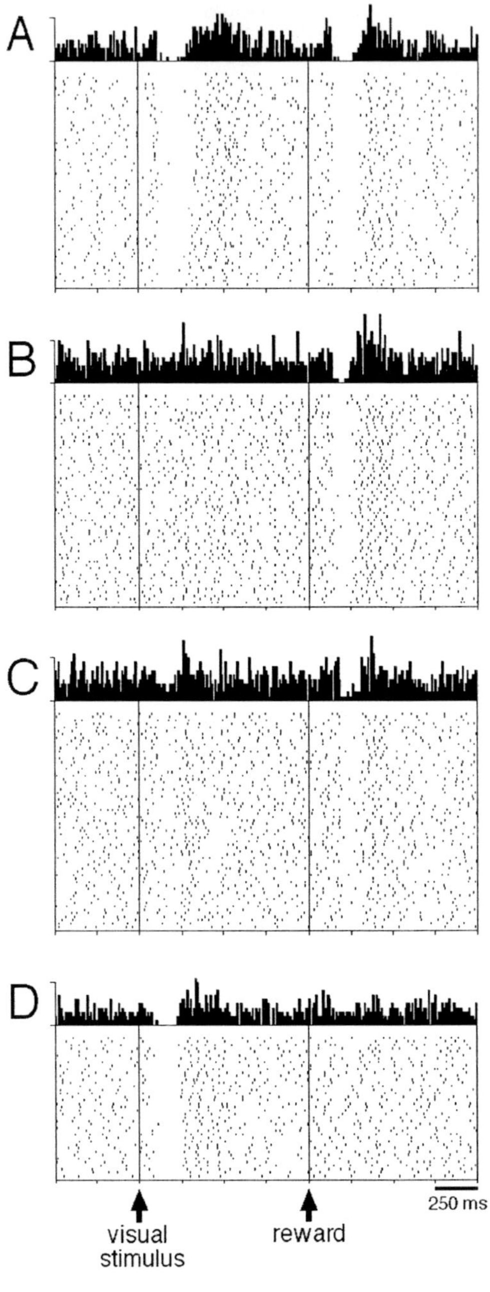

Figure 2. Changes of TAN responses during learning of a stimulus-reward association. Same conventions as in Fig. 1, except that each raster is accompanied with its corresponding peri-event time histogram. Representative responses to the stimulus and reward are shown for 4 TANs (A-D) recorded during successive stages of learning. The monkey previously received training on an instrumental task in which reward was delivered on completion of a movement elicited by a visual stimulus. In a later phase of experimentation, the monkey was subjected to a non-performing condition in which the same visual stimulus preceded the delivery of reward without requiring particular movements. The stimulus was always followed by reward at the end of a fixed interval of 1 s. Vertical calibration: 10 spikes/bin.

3 Influence of Predicted Time Intervals

Our results showed that delivery of reward at the end of a constant time interval does not constitute an optimal condition to elicit TAN responses to reward. It remains to examine if this lack of responsiveness is related to a specific component of stimulus prediction, namely the timing of a salient event. In order to verify that uncertainty about the timing of reward is essential to the response of TANs to this event, we occasionally changed the temporal structure of the non-performing condition in which reward always occurred at the end of a 1-s interval after the visual stimulus. This was done by displacing the delivery of reward toward a later (2 s) or shorter (0.3 s) time. As illustrated in Figure 3, changes in the accustomed duration of the stimulus-reward interval were paralleled by an increase of responses to reward (Apicella et al., 1997; Ravel et al., 2001). It therefore appears that TAN responsiveness is modified in a manner that is consistent with the presumed influence of temporal aspects of stimulus prediction.

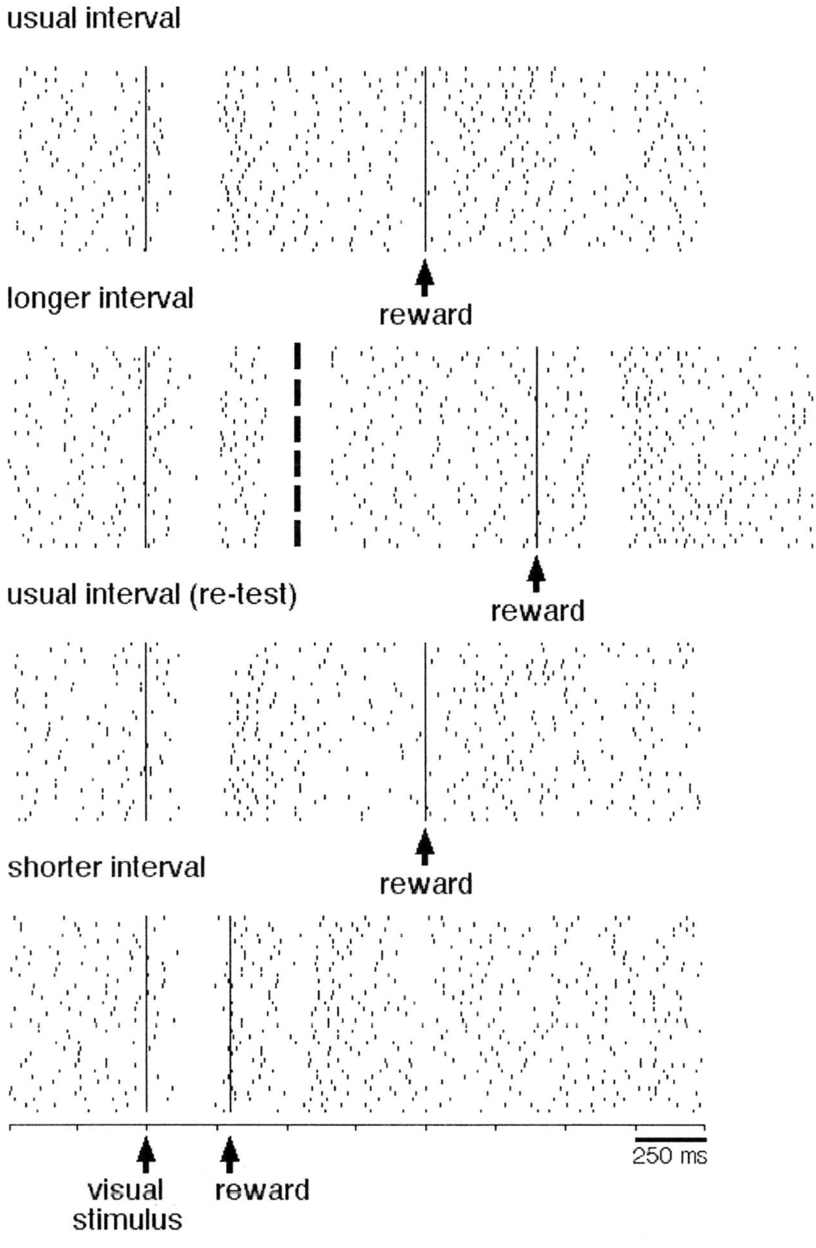

Figure 3. Influence of changing the duration of the stimulus-reward interval in one TAN. Same conventions as in Fig. 1. The same neuron was recorded while the monkey was subjected to a non-performing condition in which a visual stimulus is followed by reward at the end of a 1-s interval. Reward responses became apparent when the interval differed from the accustomed 1-s interval. In contrast, the neuron continued to respond to the visual stimulus regardless of the duration of the stimulus-reward interval. We run these changes in blocks of trials so the monkey can only know which interval is used once he has experienced the first trial in the stimulus-reward sequence of a given block.

4 TIME UNCERTAINTY AND TEMPORAL ORDERING OF TASK EVENTS

As mentioned before, many TANs lack responses to reward when the animal can predict its timing by making a movement in a task performance condition. In order to further examine the influence of temporal predictability of reward in the context of an instrumental task, we devised the so-called surprising reward condition, in which the event that occurs at the beginning of any trial could be either the trigger stimulus or reward (Ravel et al., 2001). On most trials (70 %), the first event was the trigger stimulus, the reward being normally delivered upon execution of the arm-reaching movement; on others (30 %), the first event was the reward automatically delivered in the absence of a predictive stimulus. We also used another version of the surprising reward condition in which an instruction stimulus, serving as a temporal cue, was presented 1.5 s before the occurrence of the first event, regardless of its nature. In this latter condition, the monkey knew the specific time at which an event is expected to take place but did not know exactly which event is about to occur.

Population responses of a sample of TANs recorded in the two versions of the surprising reward condition are shown in Figure 4. Although TANs were not modulated by reward occurring upon correct instrumental responding, the same reward delivered first on some trials led to a strong response of the whole population of TANs. In addition, the magnitude of population responses to the premature reward was weaker in the cued condition than in the uncued condition. This reduction of TAN responses may reflect the certainty at that moment in time that an event will occur, even if there is uncertainty about its identity.

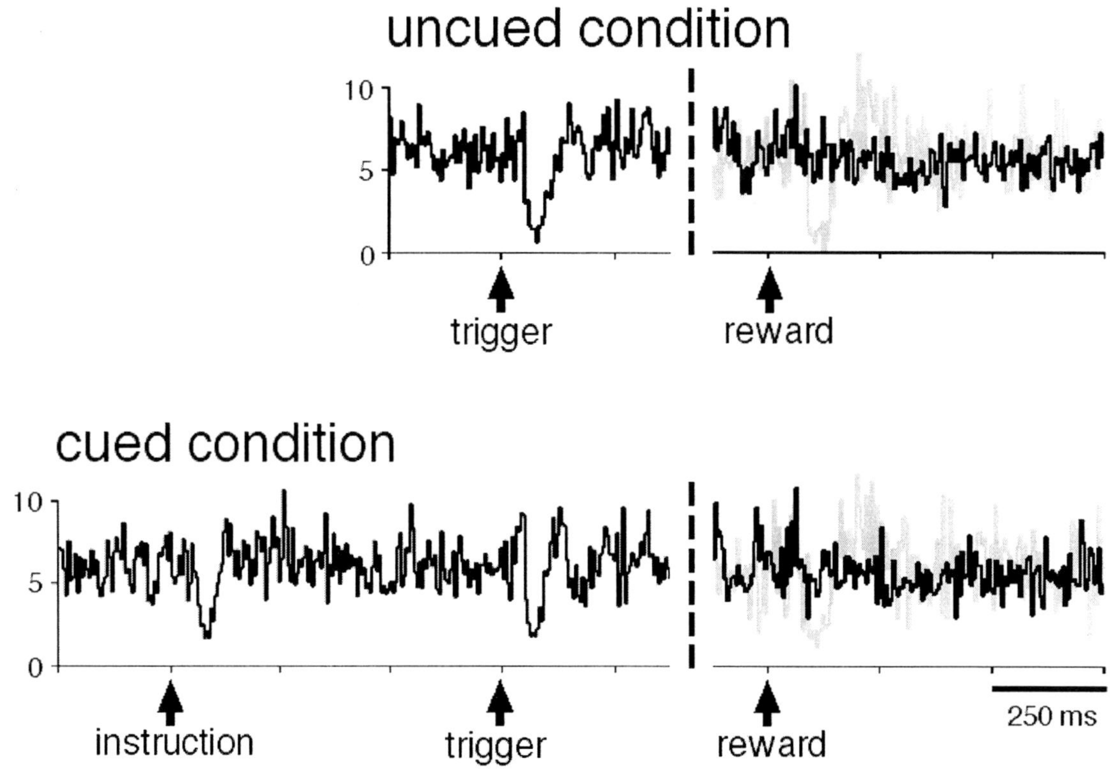

Figure 4. Population responses of TANs to events occurring in two variants of the surprising reward condition. Histograms with black and grey lines indicate the responses to reward delivered normally after target reaching or at the beginning of some trials, respectively. Data were taken from all TANs recorded in both the cued and uncued conditions (N=12), irrespective of their responsiveness to stimuli, and they were separately referenced to the trigger onset and reward. The two types of trials occurred within the same block of trials and were separated off-line for analysis. The change from the uncued condition to the cued condition occurred over two successive blocks of trials. Vertical calibration is in impulses per second. Bin width for histograms is 10 ms.

5 A Role for TANs in Learning and Timing

The data summarized here indicate that the modulation of TAN firing is under the control of the temporal context in which motivating events occur, the sensitivity of TANs to a stimulus being decreased with increasing predictability about the time at which this stimulus will be presented. This prediction-dependent effect is a robust phenomenon observable under different behavioral conditions, such as instrumental task, classically conditioned task and free liquid given outside of a task. This agrees well with the possible contribution of TANs to the learning of temporal regularities in sequential events. There is a widely held view that these neurons are especially concerned with the formation of stimulus-response associations (Aosaki et al., 1994; Ravel et al., 2003; Yamada et al., 2004). Since the hypothesis of (Mishkin et al., 1984), great interest has focused on the role of the striatum in processes underlying procedural learning, such as the acquisition of habits and certain motor skills. Evidence supporting this position is provided by lesion studies in animals (Packard et al., 1989; Hikosaka et al., 1995; Salmon and Butters, 1995; Fernandez-Ruiz et al., 2001), as well as clinical and functional imaging data in humans (Knopman and Nissen, 1991; Knowlton et al., 1996; Jueptner et al., 1997). The contribution of TANs to learning processes may depend on the storage of information about temporal relationships between events. The prediction of the onset time of a stimulus might be executed in an automatic fashion, with a minimum of attention to elapsed time, in a behavioral situation in which each stimulus is precisely timed after extensive practice with the same temporal context. In such a situation, temporal predictions are mediated by some sort of implicit timekeeping strategy that is built up through prolonged training. We propose that TANs could form part of a circuit that selectively mediates automatic processing of temporal relationships between events and thereby participate in the neuronal mechanisms that underlie the performance of routinized behaviors once they are learned.

However, the impact of temporal aspects of stimulus prediction on phasic variations in discharge rate of TANs may depend on the nature of the information available to form expectations and the behavioral requirements of the tasks. It has been recently suggested that the response of these neurons can be related on specific motivational contexts (Shimo and Hikosaka, 2001; Yamada et al., 2004). In addition, the activity of TANs does not appear to reflect the degree of reward predictability (Morris et al., 2004). Also, there are other factors, including arousal and attentional effort (Blazquez et al., 2002), that can interact with stimulus prediction in temporal domain and yet require further research.

It is unclear at present to what extent timing functions are directly carried by striatal circuits under the influence of TANs or involve processing in other brain regions. Discussions of the brain areas important in temporal information processing include the frontal cortex, basal ganglia and cerebellum (Ivry, 1996; Gibbon et al., 1997). Although the relative contributions of these systems to the encoding of temporal information remain to be specified, it is conceivable that both cortical and striatal circuits contain neural mechanisms subserving representations of temporal relationships between events. We speculate that the striatum is especially important for the expression of routinized behaviors that do not require attention to the relation between successive events in a sequence. In addition, a distortion of time perception has been reported in animal with experimentally-induced impaired dopaminergic transmission (Meck, 1996) and in patients with Parkinson's disease (Pastor et al., 1992; Malapani et al., 1998; Riesen and Schnider, 2001), providing evidence that dopaminergic modulation is also necessary for timing functions. In this regard, it has been shown that the sensitivity of midbrain dopaminergic neurons to reward includes not only the occurrence but also the timing of reward (Hollerman and Schultz, 1998).

Conclusion

In conclusion, the results that have been reviewed here provide a framework in which to link response properties of TANs to the expression of goal-directed behaviors containing a learned temporal sequence of events. It seems clear that temporal predictions of stimuli are important for modulating TAN activity. This neuronal system may provide a signal specifying a change in the condition in which timing of sequential events is implicitly represented. This signal would constitute an important mechanism for the adaptation to changing temporal contexts. Although it remains an open question as to how this signal gains access to behavioral outputs, the disruption of a striatal neuron system that is capable of monitoring temporal regularities in environmental conditions may provide a basis for explaining the deficits in programming and controlling behavioral acts of patients with Parkinson's disease, particularly the performance of sequential movements which require appropriate timing (Benecke et al., 1987; Weiss et al., 1997). Our hypothesis may drive a reassessment of the functions of the TANs by adding temporal information processing as an important feature in understanding the role of the striatum in producing goal-directed behavior.

REFERENCES

Aosaki T, Tsubokawa H, Ishida A, Watanabe K, Graybiel AM, Kimura M (1994) Responses of tonically active neurons in the primate's striatum undergo systematic changes during behavioral sensorimotor conditioning. *J Neurosci* **14**:3969-3984.

Apicella P (2002) Tonically active neurons in the primate striatum and their role in the processing of information about motivationally relevant events. *Eur J Neurosci* **16**:2017-2026.

Apicella P, Legallet E, Trouche E (1997) Responses of tonically discharging neurons in the monkey striatum to primary rewards delivered during different behavioral states. *Exp Brain Res* **116**:456-466.

Benecke R, Rothwell JC, Dick JP, Day BL, Marsden CD (1987) Disturbance of sequential movements in patients with Parkinson's disease. *Brain* **110** (Pt 2):361-379.

Blazquez PM, Fujii N, Kojima J, Graybiel AM (2002) A network representation of response probability in the striatum. *Neuron* **33**:973-982.

Fernandez-Ruiz J, Wang J, Aigner TG, Mishkin M (2001) Visual habit formation in monkeys with neurotoxic lesions of the ventrocaudal neostriatum. *Proc Natl Acad Sci U S A* **98**:4196-4201.

Gibbon J, Malapani C, Dale CL, Gallistel C (1997) Toward a neurobiology of temporal cognition: advances and challenges. *Curr Opin Neurobiol* **7**:170-184.

Graybiel AM (1995) Building action repertoires: memory and learning functions of the basal ganglia. *Curr Opin Neurobiol* **5**:733-741.

Hikosaka O, Rand MK, Miyachi S, Miyashita K (1995) Learning of sequential movements in the monkey: process of learning and retention of memory. *J Neurophysiol* **74**:1652-1661.

Hollerman JR, Schultz W (1998) Dopamine neurons report an error in the temporal prediction of reward during learning. *Nat Neurosci* **1**:304-309.

Ivry RB (1996) The representation of temporal information in perception and motor control. *Curr Opin Neurobiol* **6**:851-857.

Jueptner M, Frith CD, Brooks DJ, Frackowiak RS, Passingham RE (1997) Anatomy of motor learning. II. Subcortical structures and learning by trial and error. *J Neurophysiol* **77**:1325-1337.

Knopman D, Nissen MJ (1991) Procedural learning is impaired in Huntington's disease: evidence from the serial reaction time task. *Neuropsychologia* **29**:245-254.

Knowlton BJ, Mangels JA, Squire LR (1996) A neostriatal habit learning system in humans. *Science* **273**:1399-1402.

Malapani C, Rakitin B, Levy R, Meck WH, Deweer B, Dubois B, Gibbon J (1998) Coupled temporal memories in Parkinson's disease: a dopamine-related dysfunction. *J Cogn Neurosci* **10**:316-331.

Meck WH (1996) Neuropharmacology of timing and time perception. *Brain Res Cogn Brain Res* **3**:227-242.

Mishkin M, Malamut B, Bachevalier J (1984) Memories and habits: two neural systems. In: *Neurobiology of Learning and Memory* (G. Lynch JLMaNMW, ed), pp 65-77. New York: Guilford Press.

Morris G, Arkadir D, Nevet A, Vaadia E, Bergman H (2004) Coincident but distinct messages of midbrain dopamine and striatal tonically active neurons. *Neuron* **43**:133-143.

Packard MG, Hirsh R, White NM (1989) Differential effects of fornix and caudate nucleus lesions on two radial maze tasks: evidence for multiple memory systems. *J Neurosci* **9**:1465-1472.

Pastor MA, Artieda J, Jahanshahi M, Obeso JA (1992) Time estimation and reproduction is abnormal in Parkinson's disease. *Brain* **115** Pt 1:211-225.

Ravel S, Legallet E, Apicella P (2003) Responses of tonically active neurons in the monkey striatum discriminate between motivationally opposing stimuli. *J Neurosci* **23**:8489-8497.

Ravel S, Sardo P, Legallet E, Apicella P (2001) Reward unpredictability inside and outside of a task context as a determinant of the responses of tonically active neurons in the monkey striatum. *J Neurosci* **21**:5730-5739.

Riesen JM, Schnider A (2001) Time estimation in Parkinson's disease: normal long duration estimation despite impaired short duration discrimination. *J Neurol* **248**:27-35.

Salmon DP, Butters N (1995) Neurobiology of skill and habit learning. *Curr Opin Neurobiol* **5**:184-190.

Sardo P, Ravel S, Legallet E, Apicella P (2000) Influence of the predicted time of stimuli eliciting movements on responses of tonically active neurons in the monkey striatum. *Eur J Neurosci* **12**:1801-1816.

Shimo Y, Hikosaka O (2001) Role of tonically active neurons in primate caudate in reward-oriented saccadic eye movement. *J Neurosci* **21**:7804-7814.

Weiss P, Stelmach GE, Hefter H (1997) Programming of a movement sequence in Parkinson's disease. *Brain* **120** (Pt 1):91-102.

Yamada H, Matsumoto N, Kimura M (2004) Tonically active neurons in the primate caudate nucleus and putamen differentially encode instructed motivational outcomes of action. *J Neurosci* **24**:3500-3510.

In: Recent Breakthroughs in Basal Ganglia Research
Editor: Erwan Bezard, pp. 65-80
ISBN: 1-59454-880-3

Chapter 6

FRONTOSTRIATAL MECHANISMS IN REINFORCEMENT: IMPLICATIONS FOR ADHD

Jeffery R. Wickens
University of Otago, PO Box 913, Dunedin, New Zealand Department of Anatomy & Structural Biology The Neuroscience Centre
Brian I. Hyland
University of Otago, PO Box 913, Dunedin, New Zealand Department of Physiology, The Neuroscience Centre
Gail Tripp
Department of Psychology, The Neuroscience Centre

ABSTRACT

Alongside an established role in motor activation, the basal ganglia are increasingly recognized as playing a crucial role in integrative behaviour, although the precise nature of this role remains to be elucidated. We consider the hypothesis that the basal ganglia play a particular role in learning on the basis of positive reinforcement. The striatum - the input nucleus of the basal ganglia - is a locus for interaction of neural signals (from the cerebral cortex) reflecting the current situation of the organism, positive reinforcement (from the midbrain dopamine neurons), and behavioural output (via the motor cortex). In this chapter we detail these interactions and consider how they might contribute to the operation of the frontostriatal system during learning. We begin with advances in the basic anatomy and physiology of the striatum, which have clarified many key aspects of its microcircuitry, and the operations it performs. Particular advances have been made in relation to lateral inhibition among the spiny projection neurons in the striatum, the structure and plasticity of corticostriatal afferents, and the role of dopamine. These advances enable us to update a model of corticostriatal cell assemblies involved in learning on the basis of reinforcement. From a clinical perspective, the frontostriatal system is now thought to play an important role in attention-deficit hyperativity disorder (ADHD), and a better understanding of the neurobiology of this system may lead to new insights into the pathophysiology of ADHD and related disorders.

INTRODUCTION

The aim of this chapter is to consider possible basal ganglia mechanisms for learning, and how these might be altered in behavioural disorders. In the past, pathophysiological studies of patients with Parkinson's disease have emphasized the motor functions of the more dorsal parts of the striatum, namely the caudate-putamen. In contrast, the more ventral nucleus accumbens has been associated with the reinforcing effects of drugs of abuse, such as the psychostimulants amphetamine and cocaine. Despite this apparent contrast, the different parts of the striatum (here defined as the caudate, putamen, and accumbens) are remarkably similar in several important respects: (i) the cytology is remarkably constant throughout the striatum; (ii) the physiology of the component neurons from different regions is very similar; and (iii) the dopamine cells projecting to different regions of the striatum display similar activity patterns. While acknowledging that important regional and compartmental differences exist, nevertheless it seems useful to consider whether there is a common input-output operation, based on a canonical circuit, that all parts of the striatum perform. If there is such a common input-output operation, then the major determinant of the function of each striatal subregion is the origin of its corticostriatal inputs and the targets of its outputs. Furthermore, movement and reinforcement functions are inevitably intertwined; reinforcement plays an important role in motor functions of the striatum (Kunig et al., 2000), while motor activation plays an important role in the expression of "higher order" behavioural functions (Wickens, 1990). It is therefore most useful to consider them together.

Rodent, primate and human studies provide evidence that the striatum plays a key role in learning on the basis of positive reinforcement (Packard and Knowlton, 2002; White and McDonald, 2002). In rats, lesions of the dorsal striatum impair acquisition of tasks requiring a win-stay strategy (McDonald and White, 1993), a measure of S-R associative learning (Sage and Knowlton, 2000). Behavioural measures in humans with neurodegenerative diseases of the basal ganglia also provide evidence of a role for the striatum in such learning (Charbonneau et al., 1996; Swainson et al., 2000). There is also a suggestion that parts of the striatum are involved in stimulus-reward associations (eg effect of lesions on place-preference) and delay of reinforcement (Cardinal et al., 2001; 2004; Tanaka et al., 2004).

These insights have important implications for understanding disorders of the frontostriatal system. For example, attention-deficit hyperactivity disorder (ADHD) is a behavioural syndrome of unknown etiology, which is diagnosed on the basis of persistent and developmentally inappropriate levels of overactivity, inattention and impulsivity (American Psychiatric Association, 1994). Several pieces of evidence suggest that the striatum is involved in some aspects of ADHD. For example, differences in the anatomical dimensions of the caudate nucleus (Hynd et al., 1993; Castellanos et al., 1994; 1996; Filipek et al., 1997; Castellanos et al., 2001), decreases in blood flow in regions of the striatum (Lou et al., 1989), and changes in striatal dopamine transporter (DAT) binding (Dougherty et al., 1999) have been described in ADHD. In addition, many researchers agree that ADHD involves altered reinforcement mechanisms, and these may be linked to frontostriatal function. A better understanding of the links between the frontostriatal system and reinforcement learning may lead to important insights into the pathophysiology of ADHD and related disorders.

1 Anatomy and Physiology of Frontostriatal System

The striatum is the major input nucleus of the basal ganglia (Fig 1). Excitatory glutamatergic inputs converge on the striatum from all areas of the cerebral cortex, including sensorimotor, association, prefrontal and limbic regions (McGeorge and Faull, 1989). These are organised in a dorsolateral-ventromedial gradient (Voorn et al., 2004). The dorsolateral striatum receives inputs from sensory and motor cortical areas. The ventromedial striatum (including the nucleus accumbens) receives glutamatergic inputs from the amygdaloid complex and hippocampal formation. The prefrontal cortex projects to an intermediate region (Berendse et al., 1992). The origin of the corticostriatal inputs to each striatum subregion is likely to be a major determinant of the function of that subregion, because the cytology and neurophysiology of the different regions is remarkably similar, suggesting a common operation performed on different sets of inputs.

The principal neuron of the striatum is the medium spiny cell, which constitues the majority of the cells (96 % in the rat, Oorschot, 1996). Cortical inputs make synaptic contact with spiny projection neurons. Hence, the nature of the corticostriatal inputs and the input-output relation of the spiny projection neuron are the crucial determinants of striatal output.

Each spiny neuron receives on the order of 5,000 excitatory synaptic inputs from the cerebral cortex and a similar number from the thalamus. Individual small regions of the cortex terminate in a patchy fashion in the striatum, where they often converge with patches from functionally related areas (Graybiel et al., 1994), and convergence from such areas probably occurs at the level of single spiny cells. Because of its hyperpolarised resting potential and the presence of potassium channels which oppose depolarisation, firing of a spiny cell requires more or less synchronous activation of a number of its afferents.

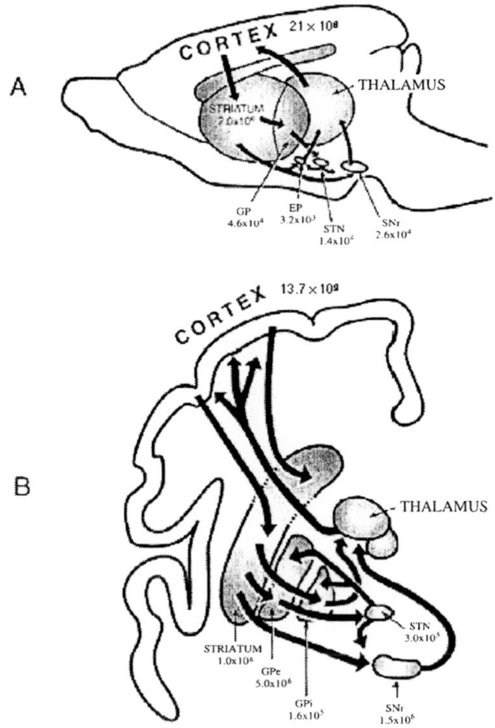

Figure 1. Anatomical organisation of the major structures of the basal ganglia and their interconnections. A. Rat brain, sagittal section. Adapted from Oorschot (1996), with permission of the author. B. Human brain, coronal section. Abbreviations: GP, globus pallidus; GPi, internal segment of globus pallidus; GPe, external segment of globus pallidus; EP, entopeduncular nucleus; STN, subthalamic nucleus; SNr substantia nigra, pars reticulata. Numbers indicate the total number of neurones within each structure (Schroder et al., 1975; Thorner et al., 1975; Braendgaard et al., 1990; Oorschot, 1996; Gredal et al., 2000).

At the cellular level the spiny striatal neurons may function as pattern detectors in a Perceptron-like way. Each cell responds to coactivation of a particular set of corticostriatal inputs, namely that set with strong synaptic connections to the cell in question. The strong connections thus form a template for recognition of all or part of a particular set of corticostriatal neurons in the cortex. Such a set may be thought of as a cortical cell assembly (Braitenberg, 1978) or synfire chain (Abeles, 1991; Abeles et al., 1993; Prut et al., 1998), which represents a particular situation. In the awake animal, firing of spiny striatal neurons probably occurs when a certain cell assembly or synfire chain is activated in the cerebral cortex. Striatal spiny neurons are thus able to act as detectors of particular cortical cell assemblies or repeating patterns of activity in the cerebral cortex.

Physically adjacent spiny neurons, however, probably sample from mostly non-overlapping sets of cortical afferents, and hence detect different parts of an assembly. The degree of overlap is thought to be so small that the input to each cell is probably unique and that sharing of inputs among spiny cells is minimal (Kincaid et al., 1998; Zheng and Wilson, 2002). Each cortical afferent contacts many postsynaptic striatal cells, each of them probably only once. Thus, each single cortical cell provides part of the input pattern for many spiny projection neurons, but is probably combined with a different set of other cortical cells in the total afferent stream to each individual striatal neuron.

The generalisations made above are broadly correct for the whole striatum, but the basic skeleton of spiny projection neurons is also subdivided into neurochemically-distinct compartments. In the primate, "islands" (also called striosomes or patches) of densely-packed cells can be distinguished from a less densely-packed (but considerably more extensive) surrounding "matrix" (Graybiel and Ragsdale, 1978;

Goldman-Rakic, 1982). These two compartments receive different patterns of afferent innervation and make different output connections. For example, spiny neurons in the matrix compartment receive inputs from sensory and motor cortical areas and project to output nuclei such as the globus pallidus (Gerfen, 1984; Gimenez-Amaya and Graybiel, 1990). In the patch compartment spiny neurons receive inputs from specific areas of the frontal cortex (the posterior orbitofrontal/anterior insular cortex and the mediofrontal prelimbic/anterior cingulate cortex, (Eblen and Graybiel, 1995), amygdala and hippocampus and project to the substantia nigra pars compacta, the source of the nigrostriatal dopaminergic system (Gerfen, 1984). In keeping with the principle of a common operation on different inputs and outputs, there are few anatomical and physiological differences between spiny cells in each compartment, although these cells appear to be compartmentally segregated (Kawaguchi et al., 1989). Therefore, we propose that patch and matrix compartments are striatal subsystems that perform the same or similar operations, but on different (and parallel) input-output channels.

2 Spiny Striatal Cells Compute a Value Function on Cortical States

In the awake animal, spiny projection neurons are relatively quiescent cells that fire phasically in a context-dependent way, often in association with movement. The context dependence of their activity is evident in responses related to movements triggered by stimuli, without necessarily responding to the stimuli or movement outside of the task (Kimura, 1986; 1990, 1992). For example, cells in the dorsal striatum that responded to an auditory cue during a movement task usually did not respond to the same cue presented outside the task (Gardiner and Kitai, 1992). On the other hand, dorsolateral cells that responded to movement of the forelimb outside of the instrumental task were not active during lever pressing (Carelli et al., 1997). Similarly, they have been reported to represent learnt associations between particular stimuli and particular responses (Hadj-Bouziane and Boussaoud, 2003).

On the basis of their context dependent response properties it is tempting to suggest that striatal projection neurons with connections to motor centres encode stimulus-response relationships (Graybiel et al., 1994). However, while this is in accordance with stimulus-response habit models, the enormous amount of convergence that takes place in the output nuclei of the basal ganglia places has to be taken into account. For example, in humans although there are on the order of 10^8 neurons in the striatum, there are only 10^5 neurons in the GPi (Schroder et al., 1975; Thorner et al., 1975); in rats, the corresponding numbers are 3 x 10^6 in striatum, with 3 x 10^3 in EP (although there are 2.5 x 10^4 in SNr, Oorschot, 1996). This degree of convergence places important constraints on the degree of response specificity that can be encoded in the output. As shown in Fig 2A, *encoding* stimulus-response connections implies a degree of specificity in output pathways, which seems inconsistent with the degree of convergence in output stations. On the other hand, the *recognition* of specific stimulus-response associations that are stored in the cortex (Fig 2B), requires less specificity in output pathways, which need only report the degree of match or mismatch.

Such a recognition function would be compatible with a role for these cells in the representation of what might be termed the *value* of stimulus-response alternatives to the animal, in other words, the output from the striatum might encode the expectation of a rewarding outcome from the current cortical state Such an expectation could be represented in a relatively low-dimensional output, compatible with the dimensionality reduction possible the basal ganglia (Bar-Gad et al., 2000; Bar-Gad and Bergman, 2001).

There is now ample evidence for such coding of reward expectation in striatal activity. In the rat two related but distinct types of neural activity have recently been described in striatum during navigation of a multiple T-maze (Schmitzer-Torbert and Redish, 2004). Two almost non-overlapping groups of cells were found: maze responsive and reward responsive (defined as firing in the 5 secs after a food reward). Maze responsive cells did not respond consistently to movements such as turning left or right. Responses were highly similar when the same turn sequence was followed, but were not similar when rats took a different route through the same physical location in the environment. These data are consistent with the properties of striatal cells described above, but raise the question of two different populations of cells. It is possible that these two populations correspond to cells in patch and matrix subdivisions of the striatum (White, 1989; White and Hiroi, 1998). We will consider possible roles of patch and matrix compartments later.

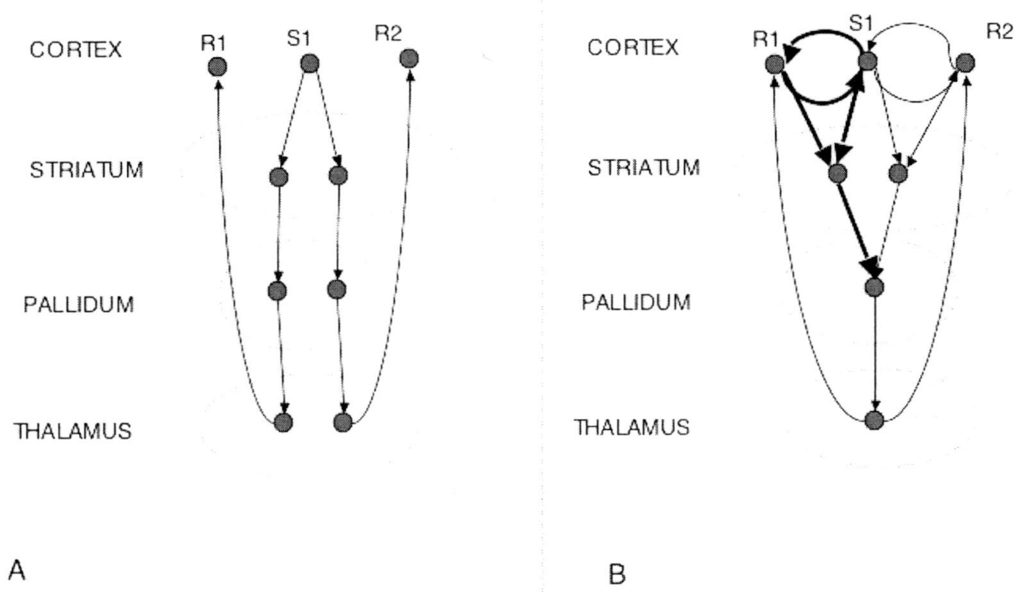

Figure 2. Two different views of the role of the striatum in selection of situation-action assemblies. A. Encoding stimulus-response connections. Connections are made from stimulus (S1) via the striatum to specific cortical representations of alternative responses (R1, R2). This might account for striatal cell involvement in the association of sensory stimuli with a specific response (S1-R1). However, it implies an unrealistically high degree of specificity in output connections, in order to make specific response selections. B. Recognition of specific stimulus-response associations that are stored in the cortex. The degree of match or mismatch controls disinhibition of the cortex. This does not require specificity of connections to outputs. The activity in the S1-R1 assembly is amplified more (indicated by bolder arrows) than the S1-R2 assembly.

In primates, striatal cells respond to primary rewards (Apicella et al., 1991) and in expectation of reward (Schultz et al., 1992). Furthermore, the reward-expectation activity is modified during learning (Tremblay et al., 1998) and is able to encode information about the value of expected rewards, which is incorporated into activity related to the behavioural reaction that leads to the reward (Hassani et al., 2001; Takikawa ct al., 2002).

Studies in which movement requirements are kept constant and the motivational significance of movements is varied also support the hypothesis that striatal output may represent a value function. This seen in tasks in which the cognitive requirements stay the same (memory guided saccade) while motivational significance is varied (Kawagoe et al., 1998). Studies with such tasks have revealed responses that are modulated by expectation of reward, either as enhancement of responses in the rewarded direction (reward-facilitated responses) or enhancement in response to non-rewarded directions (reward-suppressed responses). With each change of the preferred direction, these changes in responses occur rapidly. Such changes are also seen in cells that show a reward-oriented response bias before the presentation of the stimulus (Lauwereyns et al., 2002b; 2002a). Expectation of reward also influences movement-related activity in spiny cells (Kawagoe et al., 1998).

Finally, there is now data from human brain imaging showing that activation of caudate nucleus correlates with reward and reward prediction error (O'Doherty et al., 2003) and parallels the magnitude of subjects behavioural change during learning (Haruno et al., 2004).

To summarise, striatal cells show highly selective responses during behaviour. They seem to encode a complex of stimulus, response, and reward-expectation. In addition, there is a hint that different types of responses may be recorded from cells in patch and matrix compartments. The limited number of output pathways from the striatum argues against a direct role in making specific connections between stimuli and responses. It seems more likely that the striatum encodes the expectation of a rewarding outcome from the current cortical state. We propose that this expectation is encoded as graded activation of spiny projection neurons, such that the greater the striatal output, the greater the expectation of reward.

3 Dopamine, Reinforcement and Dopamine-Dependent Potentiation Underlie Learning of a Value Function

We have previously suggested that striatal cells come to acquire a representation of reward expectation by a three-factor rule for synaptic modification acting at the level of the corticostriatal synapses (Wickens, 1990; 1993b; Wickens and Kotter, 1995). According to this theory, dopamine afferents to the striatum mediate the effects of reinforcement, as proposed by many authors on the basis of diverse evidence. At the cellular level, the action of dopamine is to strengthen particular synapses that were active in the behaviour that led to reward. We now consider more recent results concerning the physical nature of the dopamine signal (that is, its temporal and spatial characteristics) and the circumstances in which this signal is activated by behavioural events.

Dopamine cells fire in different modes, including short bursts, slow regular firing, and non-firing modes (Deniau et al., 1978; Grace and Bunney, 1984a, 1984b; Grace and Onn, 1989; Dai and Tepper, 1998; Paladini and Tepper, 1999; Hyland et al., 2002). The physical character of the dopamine signal resulting from the firing pattern of dopamine cells depends on neurochemical and anatomical properties of the dopamine innervated regions, as well as the temporal structure (tonic or phasic) and degree of synchrony across the population of dopamine cells.

In the striatum, detailed analysis of dopamine release and reuptake mechanisms suggests that each action potential causes dopamine release and immediate overflow of dopamine out of the small synaptic junctions and into the extracellular space (Garris et al., 1994; Venton et al., 2003; Cragg and Rice, 2004; Wickens and Arbuthnott, In Press). Released dopamine diffuses within tens of milliseconds into a sphere of influence extending 5-10 μm radius around each synapse. Dopamine synapses are closely-packed (Groves et al., 1994; Descarries et al., 1996), so during regular asynchronous firing of dopamine cells the spheres of influence of dopamine are likely to coalesce to form a uniform steady-state concentration, as found experimentally (Venton et al., 2003). Phasic firing is likely to produce greater overflow and a momentarily much higher concentration, decaying on a sub-second timescale determined by the dopamine transporter activity (Wickens and Arbuthnott, In Press).

Contrary to what might be expected from Parkinson's disease, recordings of dopamine cell activity in normal animals showed few, if any, responded to movements *per se* (DeLong et al., 1983; Schultz, 1986) Rather, it has emerged from the pioneering studies of Schulz and his co-workers that the phasic responses of dopamine cells preferentially signal stimuli with positive reinforcement or reward value (Ljungberg et al., 1992). As might be expected of cells signalling reward, they do not discriminate between types of primary reward (e.g. food versus drink) or modalities of secondary reward (eg visual versus auditory, Ljungberg et al., 1992). On the other hand they are selective for positive versus negative reinforcement. For example, the phasic response of dopamine cells is not elicited by aversive stimuli (Ungless et al., 2004), and their selectivity for motivationally positive (appetitive) stimuli is also evident in lack of response to stimuli in active avoidance tasks (Mirenowicz and Schultz, 1996).

The responses of dopamine cells show remarkable changes during the course of learning. Their response to primary rewards diminishes as the rewards come to be expected, such as when stimuli have been conditioned to predict such rewards and with repeated exposure the cells gradually come to respond instead to the conditioned stimuli (Schultz et al., 1997; Hollerman and Schultz, 1998). This plasticity may relate to a function for dopamine cells in encoding "errors" in reward prediction. That is, when reward occurs unexpectedly, there is a positive prediction error, and increase in dopamine cell firing, whereas when expected reward does not occur there is a decrease in dopamine cell firing (Schultz, 1998). These response properties match well with features of the prediction error teaching signal of the temporal difference algorithm for reward mediated machine learning (Montague et al., 1996; Schultz et al.,

1997).However, the transfer of dopamine cell responses between primary reinforcers and secondary reinforcers can occur over several trials, during which time responses to several different stimuli may occur in single trials (Takikawa et al., 2004).

Reward-related firing of dopamine cells occurs in phasic bursts consisting of several action potentials within an interval 100-200 ms, on a background of more-or-less regular or tonic firing (Hyland et al., 2002). The phasic bursts probably involve synchronous firing of dopamine cells, as evidenced by the population activity of the dopamine cells (Mirenowicz and Schultz, 1994) and correlated firing of pairs of dopamine cells (Hyland et al., 2002). Thus, in the striatum the reward-related dopamine signal is a subsecond pulse (Phillips et al., 2003), relatively specific in time but spatially widespread, and able to act on the glutamatergic synapses. We now consider the effect of such a dopamine pulse on the efficacy of the corticostriatal glutamatergic synapses, and how this might lead to acquisition of a value function for the cortical state.

As well as being context-dependent, the response properties of striatal projection neurons are highly plastic, and changes in their response properties are associated with several aspects of learning (Miyachi et al., 2002; Schultz et al., 2003). Progressive changes in the responses of these cells have been described during the association of new visual stimuli with established behavioral responses and reward (Tremblay et al., 1998) and during learning of a simple maze (Jog et al., 1999).

Synaptic plasticity in the corticostriatal connections is a probable basis for the changes in response properties that occur in learning. Action potential firing activity in spiny projection neurons is brought about by excitatory inputs from the cerebral cortex and thalamus (Wilson et al., 1983). Thus, changes in the afferent activity or efficacy of these corticostriatal and thalamostriatal inputs probably underlie the specific learning-related changes in striatal cell activity reviewed above. Pulsatile dopamine release evoked during reward-related learning occurs in close proximity to the excitatory corticostriatal and thalamostriatal synapses (Wickens and Arbuthnott, In Press). There has been growing interest in the hypothesis that reward-related release of dopamine controls synaptic plasticity in these inputs.

The mechanism of learning-related actions of dopamine involves, in part, dopamine-dependent synaptic plasticity. Synaptic plasticity in the striatum has been reviewed recently (Reynolds and Wickens, 2002a). Activity dependent plasticity in the corticostriatal pathway appears to obey quite different rules from synapses in the cortex or hippocampus. Conjunction of presynaptic and postsynaptic activity, which in the hippocampus would lead to long-term potentiation (LTP), leads to long-term depression (LTD) in the striatum (Calabresi et al., 1992; 1994; Bonsi et al., 2003). However, such conditions can instead lead to potentiation if the conjunction of presynaptic and postsynaptic activity occurs in close temporal relationship with a pulse of dopamine (Wickens et al., 1996; 2000).

Although much remains to be learned about the rules for synaptic modification in the striatum, there is significant experimental support (Wickens et al., 1996; Reynolds and Wickens, 2000, 2002b) for the following rules:

1. A conjunction of presynaptic cortical and postsynaptic striatal cell activity leads to depression of the active synapses.
2. If a brief pulse of dopamine release occurs coinciding with a pre- and postsynaptic conjunction of activity, this leads to potentiation of the active synapses.

All other combinations of presynaptic cortical activity and postsynaptic striatal activity with dopamine, except for those defined in 1 and 2, produce no potentiation or depression. The potentiation requires dopamine D1 but not D2 receptor activation (Kerr and Wickens, 2001). Similar rules probably apply in the ventral striatum (Floresco et al., 2001) and prefrontal cortex (Gurden et al., 2000). The degree of temporal precision required for the dopamine pulse is on the order of hundreds of milliseconds (Wickens et al., 1996; Wickens, 2000).

It is important to stress the central role of precise timing of the dopamine pulse in these rules. It means that abnormalities in reward processing could occur on a background of what in many respects appeared to be entirely normal gross dopaminergic function (tissue dopamine levels, receptor densities, transporter function, etc.). Reward processing could be disrupted by subtly disturbed functional dynamics of dopamine cell activation pattern or timing, arising from emergent effects of changes at the level of entire neural networks afferent to the dopamine system. Alternatively, alterations in the rate or effectiveness of removal of dopamine from the synaptic cleft might also disrupt such a time-sensitive mechanism.

These rules, in the context of the corticostrial pathway, provide a possible mechanism for acquisition of reward bias in responding of striatal projection neurons. We suggest that the functional significance of this bias is that striatal projection neurons effectively encode the integrated history of reinforcement of actions performed in specific situations.

Selection of actions which lead to maximum reward might then be achieved by competition among striatal projection neurons. Although such competition has been extensively debated, it is now established that lateral inhibition occurs among spiny projection neurons of both dorsal (Tunstall et al., 2002) and ventral (Taverna et al., 2004) striatum. This lateral inhibition is relatively sparse, with each spiny neuron making about 500 inhibitory synaptic contacts with the neurons in its immediate surrounds (Lee et al., 1997), and probably only a few synaptic contacts with any given cell (Koos et al., 2004). This arrangement allows several spiny cells to respond to parts of a large pattern, and as a group to inhibit their neighbours.

We have previously suggested ways in which the output from the striatum might contribute to activation of cortical cell assemblies (Miller and Wickens, 1991; Wickens, 1992, 1993a; Wickens et al., 1994). Spiny neurons in the matrix exert a disinhibitory effect on the cortex, via the thalamus. This positive feedback loop leads to activation of the cortex, but in a manner that is selective for cortical patterns that have previously led to rewards. In the context of corticostriatal loops, inhibitory interactions between spiny neurons within the striatum create a competitive dynamic, which prevents co-activation of incompatible assemblies, and ensures that disinhibition of the most favourable assembly occurs, as illustrated in Fig. 3.

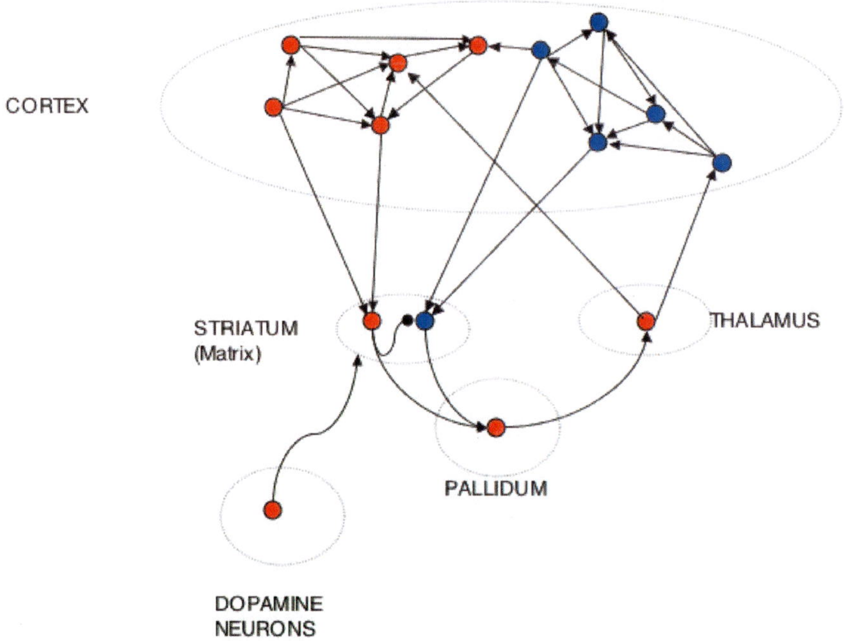

Figure 3. Competition among corticostriatal assemblies. An igniting cortical assembly (red cells) excites a subset of striatal neurones which converge on a particular pallidal cell. The disinhibitory effect of the striatal neurones on the thalamus results in selective amplification of the activity of the assembly, which as it ignites, increases output from the striatum. Inhibitory interactions in the striatum make the activation selective for the assembly with the strongest representation in the striatum. The assembly represented by the other striatal cell shown is less strongly represented in the striatum, and spreading activation from the active cortical cells (red) to the less active cells (blue) would lead to less amplification. The strength of the representation depends on the past history of dopaminergic input, produced in response to a favourable outcome of past activations of the cortical assembly activity, which strengthens the active corticostriatal connections. The overall effect is to make the cell-assembly most often associated with reward in a given situation more likely to ignite in a similar situation in the future.

We also propose that the reinforcement-predicting responses of spiny striatal neurons are used in a second way, as part of the mechanism for learning to predict reinforcement. Although there is great interest in the reward-predicting capabilities of dopamine cell activity, little is understood of the mechanisms that underlie this capability. Although the activity of dopamine cells in particular can fit an appropriately adjusted temporal-difference learning model (Barto, 1994), the neural mechanism that produces such activity has not been identified. Some models for this "adaptive critic" exist, largely based on projections to the dopamine cells from the striatum (Brown et al., 1999; Contreras-Vidal and Schultz, 1999). Although many details remain to be determined, it seems likely that the subset of spiny neurons located in the striosomes, which project to dopamine cells in the midbrain, inhibits dopamine cell firing at the time of expected reinforcement. Other mechanisms are involved in the phasic activation of the dopamine cells, but these are beyond the scope of this chapter.

4 IMPLICATIONS FOR ADHD

Although the striatum is best known for its association with Parkinson's disease and the conspicuous motor symptoms associated with this disease, the foregoing suggests that disorders of the striatum should result in abnormalities of learning and reinforcement. Disorders involving dysfunctional reinforcement mechanisms may therefore be a reflection of striatal pathology. ADHD is an example of such a disorder.

ADHD is a highly prevalent disorder of unknown etiology, which is diagnosed on the basis of persistent and developmentally inappropriate levels of overactivity, inattention and impulsivity (American Psychiatric Association, 1994). Several pieces of evidence suggest that ADHD has a neurological basis. There is a well established familial genetic contribution (Biederman et al., 1992) and a weak association with genetic variations in the dopamine D4 receptor (Swanson et al., 1998) and the dopamine transporter (DAT, Cook et al., 1995). Differences in the anatomical dimensions of the caudate nucleus (Hynd et al., 1993) decreases in blood flow in regions of the striatum (Lou et al., 1989), and changes in striatal DAT binding (Dougherty et al., 1999) have been described in ADHD. Most children with ADHD benefit from medical treatment with methylphenidate (Ritalin), a drug which acts on vesicular release and clearance of brain monoamines including dopamine (Safer and Krager, 1988). Although current knowledge of the pathophysiology of ADHD is very limited (see review by Castellanos and Tannock, 2002), the evidence referred to suggests that the striatum and dopamine are probably involved in some aspects of ADHD. We might on this basis predict abnormal reinforcement mechanisms in children with ADHD, or at least in subgroups with frontostriatal pathology.

Many researchers agree that ADHD altered reinforcement mechanisms may be involved in ADHD, although a wide range of different alterations has been proposed. Wender, (1971, 1972, 1974) argued that children with ADHD have a reduced sensitivity to reward and punishment; a view shared by some others (Haenlein and Caul, 1987). Extending this idea, Barkley, (1989) suggested that children with ADHD do not simply show a reduced sensitivity to reward, but that reward loses its effects more quickly for them; that is, they satiate more rapidly to reinforcers and punishers than other children and they have a reduced sensitivity to partial or delayed reinforcement. In contrast, Douglas argued that children with ADHD are unusually sensitive, not less sensitive, to the effects of reward (Douglas, 1989; Douglas and Parry, 1994). She proposed that children with ADHD show an increased tendency to seek immediate reward, overreact to failures to obtain rewards, and are sometimes more vulnerable to the arousing and distracting effects of rewards. More recently Barkley, (1994, 1997a, 1997b) proposed that children with ADHD are neither more nor less sensitive to reward than other children, nor are they dominated by a tendency to seek immediate reward. Rather, the behaviour of children with ADHD is influenced more by "immediate events and their consequences than by those more distant in time" [(Barkley, 1997a), p.77].

In order to clarify competing accounts of altered reward sensitivity in children with ADHD, Tripp and Alsop, (1999) developed a signal-detection paradigm that provided direct measures of sensitivity to reward. The sensitivity of boys with and without ADHD to differences in reward frequency were compared using this task, in which correct identification of one stimulus was rewarded three times more often than correct identification of the other. Both groups showed a consistent response bias toward the more frequently rewarded alternative, and there were group differences in the stability of this response bias. Importantly, the control children developed a stable response pattern that was governed by their overall history of reward on the task, whereas the ADHD group showed changeable patterns of response bias that were very sensitive to individual instances of reward. This means that children with ADHD were abnormally sensitive to individual instances of reward, rather than the integrated history of reward. This finding is consistent with proposals that children with ADHD are more vulnerable to the arousing and distracting effects of reward (Douglas, 1989; Douglas and Parry, 1994), and that their behaviour is influenced more by immediate events and their consequences (Barkley, 1997a).

Children with ADHD are also particularly sensitive to reward delays. However, there is uncertainty concerning what feature of the delay produces this effect. Sonuga-Barke and colleagues (Sonuga-Barke et al., 1992; Sonuga-Barke, 1994) suggested that hyperactive children were more averse to delays generally. The delay aversion hypothesis has been tested in two recent studies (Kuntsi et al., 2001; Solanto et al., 2001). In both studies children with ADHD or hyperactivity choose a small immediate reward over a larger delayed reward more often than control children. However, neither study assessed preference for immediate reward under all the experimental conditions arranged in the original study (Sonuga-Barke et al., 1992), so these results can also be interpreted as evidence that children with ADHD act to increase reward immediacy. To investigate this hypothesis Tripp and Alsop, (2001) used their signal-detection paradigm to compare sensitivity to pre- and post-reward delays of children with and without combined type ADHD. Correct identification of one stimulus produced an immediate reward and then a 3.5-s delay before the next trial. Correct identification of the other stimulus was associated with a 3.5-s delay before reward was delivered. Children in the ADHD group showed a greater bias toward immediate reward than the controls. This result supports the hypothesis that children with ADHD are unusually sensitive to even brief pre-reward delays.

We suggest that some of the abnormal responses to reinforcement described in children with ADHD may be linked to disorders of the parts of the frontostriatal system we have described above. In particular, failure of the adaptive critic mechanism tentatively identified with the patch compartment of the striatum, may lead to difficulty establishing appropriate reward prediction. This would lead to failure to learn about the secondary reinforcing value of neutral stimuli, and difficulties dealing with discontinuous or delayed reinforcers. Other changes might also account for altered reinforcement processing, for example, including changes in the dopamine transporter and other aspects of dopamine function (Sagvolden et al., In Press). It is not yet possible to define a unique relationship between the abnormal reward sensitivity of children and specific changes in the frontostriatal system.

CONCLUSIONS

Spiny striatal neurons contribute to learning by acquiring responses to cortical states that predict reinforcement. These responses are based on simple pattern-recognition capabilities of spiny neurons operating as feature-detectors. Responses to specific cortical states are acquired by integration of dopamine inputs with cortical inputs and striatal outputs, according to a three-factor rule for synaptic modification which strengthens corticostriatal synapses in an activity and dopamine-dependent way. Thus, the output from the striatum comes to represent a value for the cortical state.

The reinforcement-predicting responses of spiny striatal neurons are used in two ways: Firstly, as part of the mechanism for learning to predict reinforcement ("critic"), a subset of spiny neurons located in the striosomes, which project to dopamine cells in the midbrain, inhibits dopamine cell firing at the time of expected reinforcement. Secondly, as part of the mechanism for selecting and activating the cortical cell assemblies most likely to lead to reinforcement ("actor"), spiny neurons in the matrix exert a disinhibitory effect on the cortex, via the thalamus. Positive feedback between the striatum and frontal cortex facilitates the ignition of cortical assemblies with the greatest value. Inhibitory interactions between spiny neurons within the striatum create a competitive dynamic, which prevents co-activation of incompatible assemblies, and ensures that disinhibition of the most favourable assembly occurs.

In children with ADHD, dysfunction of the frontostriatal system may lead to altered responses to reinforcement. Some of the abnormal responses to reinforcement described in children with ADHD may be linked to failure of the adaptive critic mechanism tentatively identified with the patch compartment of the striatum. Difficulty establishing appropriate reward prediction would lead to failure to learn about the secondary reinforcing value of neutral stimuli, and hence to difficulties dealing with discontinuous or delayed reinforcers.

AKNOWLEDGEMENTS

We wish to acknowledge support from the Health Research Council and the Marsden Fund. JW and GT especially wish to thank the Centre for Advanced Study at the Norwegian Academy of Science and Letters for support, and Fellows of the ADHD research group for helpful discussions.

REFERENCES

Abeles M (1991) Corticonics. Neural circuits in the cerebral cortex. Cambridge: Cambridge University Press.

Abeles M, Bergman H, Margalit E, Vaadia E (1993) Spatiotemporal firing patterns in the frontal cortex of behaving monkeys. *J Neurophysiol* **70**:1629-1638.

American Psychiatric Association (1994) DSM-IV: Diagnostic and Statistical Manual of Mental Disorders, 4th ed. In: (APA, ed). Washington DC.

Apicella P, Ljungberg T, Scarnati E, Schultz W (1991) Responses to reward in monkey dorsal and ventral striatum. *Exp Brain Res* **85**:491-500.

Bar-Gad I, Bergman H (2001) Stepping out of the box: information processing in the neural networks of the basal ganglia. *Curr Opin Neurobiol* **11**:689-695.

Bar-Gad I, Havazelet-Heimer G, Goldberg JA, Ruppin E, Bergman H (2000) Reinforcement-driven dimensionality reduction--a model for information processing in the basal ganglia. *J Basic Clin Physiol Pharmacol* **11**:305-320.

Barkley RA (1989) The problem of stimulus control and rule-governed behavior in attention deficit disorder with hyperactivity. In: Attention deficit disorder; current concepts and emerging trends in attentional and behavioural disorders of childhood. (Bloomingdale LM, Swanson, J., ed), pp 203-228. Oxford: Pergamon Press.

Barkley RA (1994) Impaired delayed responding: A unified theory of attention-deficit hyperactivity disorder. In: Disruptive behavior disorders in children (Routh DH, ed), pp 11-57. New York: Plenun.

Barkley RA (1997a) Behavioral inhibition, sustained attention, and executive functions: Constructing a unifying theory of ADHD. *Psychological Bulletin* **121**:65-94.

Barkley RA (1997b) ADHD and the nature of self-control. New York: Guilford.

Barto AG (1994) Adaptive critics and the basal ganglia. In: Models of information processing in the basal ganglia (Houk JC, Davis JL, Beiser DG, eds), pp 215-232. Cambridge: MIT Press.

Berendse HW, Galis-de Graaf Y, Groenewegen HJ (1992) Topographical organization and relationship with ventral striatal compartments of prefrontal corticostriatal projections in the rat. *J Comp Neurol* **316**:314-347.

Biederman J, Faraone SV, Keenan K, Benjamin J, Krifcher B, Moore C, Sprich-Buckminster S, Ugaglia K, Jellinek MS, Steingard R, et al. (1992) Further evidence for family-genetic risk factors in attention deficit hyperactivity disorder. Patterns of comorbidity in probands and relatives psychiatrically and pediatrically referred samples. *Archives of General Psychiatry* **49**:728-738.

Bonsi P, Pisani A, Bernardi G, Calabresi P (2003) Stimulus frequency, calcium levels and striatal synaptic plasticity. *Neuroreport* **14**:419-422.

Braendgaard H, Evans SM, Howard CV, Gundersen HJ (1990) The total number of neurons in the human neocortex unbiasedly estimated using optical disectors. *J Microsc* **157** (Pt 3):285-304.

Braitenberg V (1978) Cell assemblies in the cerebral cortex. In: Theoretical approaches to complex systems (Heim R, Palm G, eds), pp 171-188. Berlin: Springer.

Brown J, Bullock D, Grossberg S (1999) How the basal ganglia use parallel excitatory and inhibitory learning pathways to selectively respond to unexpected rewarding cues. *J Neurosci* **19**:10502-10511.

Calabresi P, Pisani A, Mercuri NB, Bernardi G (1994) Post-receptor mechanisms underlying striatal long-term depression. *Journal of Neuroscience* **14**:4871-4881.

Calabresi P, Maj R, Pisani A, Mercuri NB, Bernardi G (1992) Long-term synaptic depression in the striatum: physiological and pharmacological characterization. *Journal of Neuroscience* **12**:4224-4233.

Cardinal RN, Winstanley CA, Robbins TW, Everitt BJ (2004) Limbic corticostriatal systems and delayed reinforcement. *Ann N Y Acad Sci* **1021**:33-50.

Cardinal RN, Pennicott DR, Sugathapala CL, Robbins TW, Everitt BJ (2001) Impulsive choice induced in rats by lesions of the nucleus accumbens core. *Science* **292**:2499-2501.

Carelli RM, Wolske M, West MO (1997) Loss of lever press-related firing of rat striatal forelimb neurons after repeated sessions in a lever pressing task. *J Neurosci* **17**: 1804-1814.

Castellanos FX, Tannock R (2002) Neuroscience of attention-deficit/hyperactivity disorder: the search for endophenotypes. *Nat Rev Neurosci* **3**:617-628.

Castellanos FX, Giedd JN, Eckburg P, Marsh WL, Vaituzis AC, Kaysen D, Hamburger SD, Rapoport JL (1994) Quantitative morphology of the caudate nucleus in attention deficit hyperactivity disorder. *Am J Psychiatry* **151**:1791-1796.

Castellanos FX, Giedd JN, Berquin PC, Walter JM, Sharp W, Tran T, Vaituzis AC, Blumenthal JD, Nelson J, Bastain TM, Zijdenbos A, Evans AC, Rapoport JL (2001) Quantitative brain magnetic resonance imaging in girls with attention- deficit/hyperactivity disorder. *Arch Gen Psychiatry* **58**:289-295.

Castellanos FX, Giedd JN, Marsh WL, Hamburger SD, Vaituzis AC, Dickstein DP, Sarfatti SE, Vauss YC, Snell JW, Lange N, Kaysen D, Krain AL, Ritchie GF, Rajapakse JC, Rapoport JL (1996) Quantitative brain magnetic resonance imaging in attention-deficit hyperactivity disorder. *Arch Gen Psychiatry* **53**:607-616.

Charbonneau D, Riopelle RJ, Beninger RJ (1996) Impaired incentive learning in treated Parkinson's disease. *Can J Neurol Sci* **23**:271-278.

Contreras-Vidal JL, Schultz W (1999) A predictive reinforcement model of dopamine neurons for learning approach behavior. *J Comput Neurosci* **6**:191-214.

Cook EH, Jr., Stein MA, Krasowski MD, Cox NJ, Olkon DM, Kieffer JE, Leventhal BL (1995) Association of attention-deficit disorder and the dopamine transporter gene. *American Journal of Human Genetics* **56**:993-998.

Cragg SJ, Rice ME (2004) DAncing past the DAT at a DA synapse. *Trends Neurosci* **27**:270-277.

Dai M, Tepper JM (1998) Do silent dopaminergic neurons exist in rat substantia nigra in vivo? *Neuroscience* **85**:1089-1099.

DeLong MR, Crutcher MD, Georgopoulos AP (1983) Relations between movement and single cell discharge in the substantia nigra of the behaving monkey. *J Neurosci* **3**:1599-1606.

Deniau JM, Hammond C, Riszk A, Feger J (1978) Electrophysical properties of identified output neurons of the rat substantia nigra (Pars Compacta and Pars Reticulata): Evidences for the existence of branched neurons. *Brain Research* **32**:409-422.

Descarries L, Watkins KC, Garcia S, Bosler O, Doucet G (1996) Dual character, asynaptic and synaptic, of the dopamine innervation in adult rat neostriatum: a quantitative autoradiographic and immunocytochemical analysis. *J Comp Neurol* **375**:167-186.

Dougherty DD, Bonab AA, Spencer TJ, Rauch SL, Madras BK, Fischman AJ (1999) Dopamine transporter density in patients with attention deficit hyperactivity disorder. *Lancet* **354**:2132-2133.

Douglas VI (1989) Can Skinnerian theory explain attention deficit disorder?-A reply to Barkley. In: Attention deficit disorder; current concepts and emerging trends in attentional and behavioural disorders of childhood. (Bloomingdale LM, Swanson, J., ed), pp 235-254. Oxford: Pergamon Press.

Douglas VI, Parry PA (1994) Effects of reward and non-reward on frustration and attention in attention deficit disorder. *Journal of Abnormal Child Psychology* **22**:281-302.

Eblen F, Graybiel AM (1995) Highly restricted origin of prefrontal cortical inputs to striosomes in the macaque monkey. *Journal of Neuroscience* **15**:5999-6013.

Filipek PA, Semrud-Clikeman M, Steingard RJ, Renshaw PF, Kennedy DN, Biederman J (1997) Volumetric MRI analysis comparing subjects having attention-deficit hyperactivity disorder with normal controls. *Neurology* **48**:589-601.

Floresco SB, Blaha CD, Yang CR, Phillips AG (2001) Modulation of hippocampal and amygdalar-evoked activity of nucleus accumbens neurons by dopamine: cellular mechanisms of input selection. *Journal of Neuroscience* **21**:2851-2860.

Gardiner TW, Kitai ST (1992) Single-unit activity in the globus pallidus and neostriatum of the rat during performance of a trained head movement. *Exp Brain Res* **88**:517-530.

Garris PA, Ciolkowski EL, Pastore P, Wightman RM (1994) Efflux of dopamine from the synaptic cleft in the nucleus accumbens of the rat brain. *J Neurosci* **14**:6084-6093.

Gerfen CR (1984) The neostriatal mosaic: compartmentalization of corticostriatal input and striatonigral output systems. *Nature* **311**:461-464.

Gimenez-Amaya JM, Graybiel AM (1990) Compartmental origins of the striatopallidal projection in the primate. *Neuroscience* **34**:111-126.
Goldman-Rakic PS (1982) Cytoarchitectonic heterogeneity of the primate neostriatum: Subdivisions into island and matrix cellular compartments. *Journal of Comparative Neurology* **205**:398-413.
Grace AA, Bunney BS (1984a) The control of firing pattern in nigral dopamine neurons: single spike firing. *J Neurosci* **4**:2866-2876.
Grace AA, Bunney BS (1984b) The control of firing pattern in nigral dopamine neurons: burst firing. *J Neurosci* **4**:2877-2890.
Grace AA, Onn SP (1989) Morphology and electrophysiological properties of immunocytochemically identified rat dopamine neurons recorded in vitro. *J Neurosci* **9**:3463-3481.
Graybiel AM, Ragsdale CW (1978) Histochemically distinct compartments in the striatum of human, monkey and cat demonstratedby acetylcholinesterate staining. *Proceedings of the National Academy of Sciences (USA)* **75**:5723-5726.
Graybiel AM, Aosaki T, Flaherty AW, Kimura M (1994) The basal ganglia and adaptive motor control. *Science* **265**:1826-1831.
Gredal O, Pakkenberg H, Karlsborg M, Pakkenberg B (2000) Unchanged total number of neurons in motor cortex and neocortex in amyotrophic lateral sclerosis: a stereological study. *J Neurosci Methods* **95**:171-176.
Groves PM, Linder JC, Young SJ (1994) 5-hydroxydopamine-labeled dopaminergic axons: three-dimensional reconstructions of axons, synapses and postsynaptic targets in rat neostriatum. *Neuroscience* **58**:593-604.
Gurden H, Takita M, Jay TM (2000) Essential role of D1 but not D2 receptors in the NMDA receptor-dependent long-term potentiation at hippocampal-prefrontal cortex synapses in vivo. *J Neurosci* **20**:RC106.
Hadj-Bouziane F, Boussaoud D (2003) Neuronal activity in the monkey striatum during conditional visuomotor learning. *Exp Brain Res* **153**:190-196.
Haenlein M, Caul WF (1987) Attention deficit disorder with hyperactivity: A specific hypothesis of reward dysfunction. *Journal of the American Academy of Child and Adolescent Psychiatry* **26**:356-362.
Haruno M, Kuroda T, Doya K, Toyama K, Kimura M, Samejima K, Imamizu H, Kawato M (2004) A neural correlate of reward-based behavioral learning in caudate nucleus: a functional magnetic resonance imaging study of a stochastic decision task. *J Neurosci* **24**:1660-1665.
Hassani OK, Cromwell HC, Schultz W (2001) Influence of expectation of different rewards on behavior-related neuronal activity in the striatum. *J Neurophysiol* **85**:2477-2489.
Hollerman JR, Schultz W (1998) Dopamine neurons report an error in the temporal prediction of reward during learning. *Nat Neurosci* **1**:304-309.
Hyland BI, Reynolds JN, Hay J, Perk CG, Miller R (2002) Firing modes of midbrain dopamine cells in the freely moving rat. *Neuroscience* **114**:475-492.
Hynd GW, Hern KL, Novey ES, Eliopulos D, Marshall R, Gonzalez JJ, Voeller KK (1993) Attention deficit-hyperactivity disorder and asymmetry of the caudate nucleus. *Journal of Child Neurology* **8**:339-347.
Jog MS, Kubota Y, Connolly CI, Hillegaart V, Graybiel AM (1999) Building neural representations of habits. *Science* **286**:1745-1749.
Kawagoe R, Takikawa Y, Hikosaka O (1998) Expectation of reward modulates cognitive signals in the basal ganglia. *Nat Neurosci* **1**:411-416.
Kawaguchi Y, Wilson CJ, Emson PC (1989) Intracellular recording of identified neostriatal patch and matrix spiny cells in a slice preparation preserving cortical inputs. *Journal of Neurophysiology* **62**:1052-1068.
Kerr JND, Wickens JR (2001) Dopamine D-1/D-5 receptor activation is required for long-term potentiation in the rat neostriatum in vitro. *Journal of Neurophysiology* **85**:117-124.
Kimura M (1986) The role of primate putamen neurons in the association of sensory stimuli with movement. *Neuroscience Research* **3**:436-443.
Kimura M (1990) Behaviorally contingent property of movement-related activity of the primate putamen. *Journal of Neurophysiology* **63**:1277-1296.
Kimura M (1992) Behavioral modulation of sensory responses of primate putamen neurons. *Brain Research* **578**:204-214.

Kincaid AE, Zheng T, Wilson CJ (1998) Connectivity and convergence of single corticostriatal axons. *Journal of Neuroscience* **18**:4722-4731.

Koos T, Tepper JM, Wilson CJ (2004) Comparison of IPSCs evoked by spiny and fast-spiking neurons in the neostriatum. *J Neurosci* **24**:7916-7922.

Kunig G, Leenders KL, Martin-Solch C, Missimer J, Magyar S, Schultz W (2000) Reduced reward processing in the brains of Parkinsonian patients. *Neuroreport* **11**:3681-3687.

Kuntsi J, Oosterlaan J, Stevenson J (2001) Psychological mechanisms in hyperactivity: I Response inhibition deficit, working memory impairment, delay aversion, or something else? *Journal of Child Psychology and Psychiatry* **42**:199-210.

Lauwereyns J, Watanabe K, Coe B, Hikosaka O (2002a) A neural correlate of response bias in monkey caudate nucleus. *Nature* **418**:413-417.

Lauwereyns J, Takikawa Y, Kawagoe R, Kobayashi S, Koizumi M, Coe B, Sakagami M, Hikosaka O (2002b) Feature-based anticipation of cues that predict reward in monkey caudate nucleus. *Neuron* **33**:463-473.

Lee T, Kaneko T, Shigemoto R, Nomura S, Mizuno N (1997) Collateral projections from striatonigral neurons to substance P receptor-expressing intrinsic neurons in the striatum of the rat. *J Comp Neurol* **388**:250-264.

Ljungberg T, Apicella P, Schultz W (1992) Responses of monkey dopamine neurons during learning of behavioral reactions. *Journal of Neurophysiology* **67**:145-163.

Lou HC, Henriksen L, Bruhn P, Borner H, Nielsen JB (1989) Striatal dysfunction in attention deficit and hyperkinetic disorder. *Archives of Neurology* **46**:48-52.

McDonald RJ, White NM (1993) A triple dissociation of memory system: hippocampus, amygdala, and dorsal striatum. *Behavioral Neuroscience* **107**:3-22.

McGeorge AJ, Faull RL (1989) The organisation of the projections from the cerebral cortex to the striatum in the rat. *Neuroscience* **29**:503-537.

Miller R, Wickens JR (1991) Corticostriatal cell assemblies in selective attention and in representation of predictable and controllable events. *Concepts in Neuroscience* **2**:65-95.

Mirenowicz J, Schultz W (1994) Importance of unpredictability for reward responses in primate dopamine neurons. *Journal of Neurophysiology* **72**:1024-1027.

Mirenowicz J, Schultz W (1996) Preferential activation of midbrain dopamine neurons by appetitive rather than aversive stimuli. *Nature* **379**:449-451.

Miyachi S, Hikosaka O, Lu X (2002) Differential activation of monkey striatal neurons in the early and late stages of procedural learning. *Exp Brain Res* **146**:122-126.

Montague PR, Dayan P, Sejnowski TJ (1996) A framework for mesencephalic dopamine systems based on predictive Hebbian learning. *Journal of Neuroscience* **16**:1936-1947.

O'Doherty JP, Dayan P, Friston K, Critchley H, Dolan RJ (2003) Temporal difference models and reward-related learning in the human brain. *Neuron* **38**:329-337.

Oorschot DE (1996) Total number of neurons in the neostriatal, pallidal, subthalamic, and substantia nigral nuclei of the rat basal ganglia: A stereological study using the Cavalieri and optical disector methods. The *Journal of Comparative Neurology* **366**:580-599.

Packard MG, Knowlton BJ (2002) Learning and memory functions of the basal ganglia. *Annu Rev Neurosci* **25**:563-593.

Paladini CA, Tepper JM (1999) GABA(A) and GABA(B) antagonists differentially affect the firing pattern of substantia nigra dopaminergic neurons in vivo. *Synapse* **32**:165-176.

Phillips PE, Stuber GD, Heien ML, Wightman RM, Carelli RM (2003) Subsecond dopamine release promotes cocaine seeking. *Nature* **422**:614-618.

Prut Y, Vaadia E, Bergman H, Haalman I, Slovin H, Abeles M (1998) Spatiotemporal structure of cortical activity: properties and behavioral relevance. *J Neurophysiol* **79**:2857-2874.

Reynolds JN, Wickens JR (2002a) Dopamine-dependent plasticity of corticostriatal synapses. *Neural Netw* **15**:507-521.

Reynolds JNJ, Wickens JR (2000) Substantia nigra dopamine regulates synaptic plasticity and membrane potential fluctuations in the rat neostriatum, in vivo. *Neuroscience* **99**:199-203.

Reynolds JNJ, Wickens JR (2002b) Self-stimulation and synaptic plasticity. In: The Basal Ganglia VII (Nicholson L, ed).

Safer DJ, Krager JM (1988) A survey of medication treatment for hyperactive/inattentive students. *Journal of the American Medical Association* **260**:2256-2258.

Sage JR, Knowlton BJ (2000) Effects of US devaluation on win-stay and win-shift radial maze performance in rats. *Behav Neurosci* **114**:295-306.

Sagvolden T, Johansen EB, Aase H, Russell VA (In Press) A dynamic developmental theory of Attention-Deficit /Hyperactivity Disorder (ADHD) predominantly hyperactive/impulsive and combined subtypes. Behavioral and Brain Sciences.

Schmitzer-Torbert N, Redish AD (2004) Neuronal activity in the rodent dorsal striatum in sequential navigation: separation of spatial and reward responses on the multiple T task. *J Neurophysiol* **91**:2259-2272.

Schroder KF, Hopf A, Lange H, Thorner G (1975) Morphometrisch-statistische Strukturanalysen des Striatum, Pallidum und Nucleus subthalamacus beim Menschen. I Striatum. *Journal fur Hirnforschung* **16**:333-350

Schultz W (1986) Responses of midbrain dopamine neurons to behavioral trigger stimuli in the monkey. *J Neurophysiol* **56**:1439-1461.

Schultz W (1998) Predictive reward signal of dopamine neurons. J Neurophysiol 80:1-27.

Schultz W, Dayan P, Montague PR (1997) A neural substrate of prediction and reward. *Science* **275**:1593-1599.

Schultz W, Tremblay L, Hollerman JR (2003) Changes in behavior-related neuronal activity in the striatum during learning. *Trends Neurosci* **26**:321-328.

Schultz W, Apicella P, Scarnati E, Ljungberg T (1992) Neuronal activity in monkey ventral striatum related to the expectation of reward. *J Neurosci* **12**:4595-4610.

Solanto MV, Abikoff H, Sonuga-Barke E, Schachar R, Logan GD, Wigal T, Hechtman L, Hinshaw S, Turkel E (2001) The ecological validity of delay aversion and response inhibition as measures of impulsivitiy in AD/HD: A supplement to the NIMH multimodal treatment study of AD/HD. *Journal of Abnormal Child Psychology* **29**:215-228.

Sonuga-Barke EJ, Taylor E, Sembi S, Smith J (1992) Hyperactivity and delay aversion--I. The effect of delay on choice. *Journal of Child Psychology & Psychiatry & Allied Disciplines* **33**:387-398.

Sonuga-Barke EJS (1994) Annotation: On dysfunction and function in psychological theories of childhood disorder. *Journal of Child Psychology and Psychiatry* **35**:801-815.

Swainson R, Rogers RD, Sahakian BJ, Summers BA, Polkey CE, Robbins TW (2000) Probabilistic learning and reversal deficits in patients with Parkinson's disease or frontal or temporal lobe lesions: possible adverse effects of dopaminergic medication. *Neuropsychologia* **38**:596-612.

Swanson JM, Sunohara GA, Kennedy JL, Regino R, Fineberg E, Wigal T, Lerner M, Willians L, LaHoste GL, Wigal S (1998) Association of the dopamine receptor D4 (DRD4) gene with a refined phenotype of attention deficit hyperactivity disorder (ADHD): a family-based approach. *Molecular Psychiatry* **3**:38-41.

Takikawa Y, Kawagoe R, Hikosaka O (2002) Reward-dependent spatial selectivity of anticipatory activity in monkey caudate neurons. *J Neurophysiol* **87**:508-515.

Takikawa Y, Kawagoe R, Hikosaka O (2004) A possible role of midbrain dopamine neurons in short- and long-term adaptation of saccades to position-reward mapping. *J Neurophysiol* **92**:2520-2529.

Tanaka SC, Doya K, Okada G, Ueda K, Okamoto Y, Yamawaki S (2004) Prediction of immediate and future rewards differentially recruits cortico-basal ganglia loops. *Nat Neurosci* **7**:887-893.

Taverna S, van Dongen YC, Groenewegen HJ, Pennartz CM (2004) Direct physiological evidence for synaptic connectivity between medium-sized spiny neurons in rat nucleus accumbens in situ. *J Neurophysiol* **91**:1111-1121.

Thorner G, Lange H, Hopf A (1975) Morphometrisch-statistische Strukturanalysen des Striatum, Pallidum und Nucleus Subthalamics beim Menschen. II Pallidum. *Journal fur Hirnforschung* **16**:401-413.

Tremblay L, Hollerman JR, Schultz W (1998) Modifications of reward expectation-related neuronal activity during learning in primate striatum. *J Neurophysiol* **80**:964-977.

Tripp EG, Alsop B (1999) Sensitivity to reward frequency in boys with attention-deficit hyperactivity disorder. *Journal of Child Clinical Psychology* **28**:366-375.

Tripp G, Alsop B (2001) Sensitivity to reward delay in children with attention deficit hyperactivity disorder (ADHD). *J Child Psychol Psychiat* **42**:691-698.

Tunstall MJ, Oorschot DE, Kean A, Wickens JR (2002) Inhibitory interactions between spiny projection neurons in the rat striatum. *Journal of Neurophysiology* **88**:1263-1269.

Ungless MA, Magill PJ, Bolam JP (2004) Uniform inhibition of dopamine neurons in the ventral tegmental area by aversive stimuli. *Science* **303**:2040-2042.

Venton BJ, Zhang H, Garris PA, Phillips PE, Sulzer D, Wightman RM (2003) Real-time decoding of dopamine concentration changes in the caudate-putamen during tonic and phasic firing. *J Neurochem* **87**:1284-1295.

Voorn P, Vanderschuren LJ, Groenewegen HJ, Robbins TW, Pennartz CM (2004) Putting a spin on the dorsal-ventral divide of the striatum. *Trends Neurosci* **27**:468-474.

Wender PH (1971) Minimal brain dysfunction syndrome in children. New York: Wiley.

Wender PH (1972) The minimal brain dysfunction syndrome in children. *J Nervous and Mental Disorders* **155**:55-71.

Wender PH (1974) Some speculations concerning a possible bichemical basis of minimal brain dysfunction. *Life Science* **14**:1605-1621.

White NM (1989) A functional hypothesis concerning the striatal matrix and patches: Mediation of S-R memory and reward. *Life Sciences* **45**:1943-1957.

White NM, Hiroi N (1998) Preferential localization of self-stimulation sites in striosomes/patches in the rat striatum. *Proc Natl Acad Sci U S A* **95**:6486-6491.

White NM, McDonald RJ (2002) Multiple parallel memory systems in the brain of the rat. *Neurobiol Learn Mem* **77**:125-184.

Wickens JR (1990) Striatal dopamine in motor activation and reward-mediated learning. Steps towards a unifying model. *Journal of Neural Transmission* **80**:9-31.

Wickens JR (1992) The contribution of the striatum to cortical function. In: Information Processing in the Cortex (Aertsen A, Braitenberg V, eds), pp 271-284. Berlin: Springer.

Wickens JR (1993a) Corticostriatal interactions in neuromotor programming. *Human Movement Science* **12**:17-35.

Wickens JR (1993b) A theory of the striatum. Oxford: Pergamon Press.

Wickens JR (2000) Dopamine regulation of synaptic plasticity in the neostriatum: a cellular model of reinforcement. In: Brain dynamics and the striatal complex (Miller R, Wickens JR, eds), pp 65-76: Harwood Academic.

Wickens JR, Kotter R (1995) Cellular models of reinforcement. In: Models of Information Processing in the Basal Ganglia. (Houk JC, Davis JL, Beiser DG, eds), pp 187-214: M.I.T. Press.

Wickens JR, Arbuthnott G (In Press) Structural and functional interactions in the striatum at the receptor level. In: Classical Transmitters, part I. Dopamine (Dunnett, Bentivoglio, Björklund, Hökfelt, eds).

Wickens JR, Hyland B, Anson G (1994) Cortical cell assemblies: A possible mechanism for motor programs. *Journal of Motor Behaviour* **26**:66-82.

Wickens JR, Begg AJ, Arbuthnott GW (1996) Dopamine reverses the depression of rat cortico-striatal synapses which normally follows high frequency stimulation of cortex in vitro. *Neuroscience* **70**:1-5.

Wilson CJ, Chang HT, Kitai ST (1983) Disfacilitation and long-lasting inhibition of neostriatal neurons in the rat. *Experimental Brain Research* **51**:227-235.

Zheng T, Wilson CJ (2002) Corticostriatal combinatorics: the implications of corticostriatal axonal arborizations. *Journal of Neurophysiology* **18**:4722-4731.

In: Recent Breakthroughs in Basal Ganglia Research
Editor: Erwan Bezard, pp. 81-89
ISBN: 1-59454-880-3

Chapter 7

MOTIVATIONAL SIGNALS OF DOPAMINE NEURONS

Minoru Kimura**, *Kazuki Enomoto, Naoyuki Matsumoto
Neurophysiology, Kyoto Prefectural University of Medicine, Kyoto
Takemasa Satoh
Tottori University, Medical School, Yonago
Tatsuo Sato
Physiology, Kyoto University Medical School, Kyoto Japan

ABSTRACT

Rewards are critically involved in the processes of decision-making and behavioral learning. The midbrain dopamine (DA) containing neurons are major neural substrate for the reward mechanisms, because discharge rates of DA neurons after the occurrence of reward after subjects performed behavioral actions are quantitatively correlated with errors of reward expectation (REEs). On the other hand, magnitudes of responses of DA neurons to reward-predictive stimuli are negatively correlated with subjects' reaction time to work for reward. Thus, the DA responses have dual and correlated roles: the one is motivating subjects toward a goal through reward, and the other is working as efficient gain controller of coding REEs. Therefore, an essential role of the DA signals in the striatum seems to give reward value, subjective desirability, to information of states and actions transferred through cortico-striate topographically organized pathways.

Key words: dopamine neuron; learning; teaching signal; reward prediction error; trial and error; decision-making

INTRODUCTION

Rewards such as food, sex and money are critically involved in the processes of decision-making (Herrnstein and Vaughn, 1980; Arnauld and Nichole, 1982) and behavioral learning (Thorndike, 1911; Hull, 1943; Rescorla and Wagner, 1972). The midbrain dopamine containing (DA) neurons are major neural substrate for the reward mechanisms.

Recording of DA neuron activity of behaving monkey showed that DA neurons increase discharges when a reward occurs unexpectedly, decrease discharges when expected reward is withheld and maintain a baseline discharge rate when reward is retrieved as expected (Schultz et al., 1997; Schultz, 1998). The following studies demonstrated that DA neuron activity reflects discrepancy between a level of expectation of reward contingent with particular external event or an animal's action and actually obtained reward, that

* Address for correspondence: Neurophysiology, Kyoto Prefectural University of Medicine, Graduate School, Kawaramachi-Hirokoji, Kamigyo-ku, Kyoto 602-8566 Japan. TEL: 81 75 251-5313, FAX: 81 75 241-1499. Email address: mkimura@koto.kpu-m.ac.jp

is reward expectation error (REE) (Fiorillo et al., 2003; Satoh et al., 2003; Morris et al., 2004; Nakahara et al., 2004).

On the other hand, a substantial body of evidence suggests involvement of DA systems in the processes of motivation (Robbins and Everitt, 1996; Koepp et al., 1998; Salamone and Correa, 2002; Wise, 2002), in switching attentional and behavioral selections to salient stimuli that underlie associative learning (Redgrave et al., 1999; Spanagel and Weiss, 1999). It has been well documented that DA neurons show phasic activations by a wide variety of salient stimuli including novel and high intensity stimuli (Jacobs, 1986; Schultz and Romo, 1987; Ljungberg et al., 1992; Horvitz et al., 1997). In this article, we propose a new role of motivational signal of DA neurons in decision-making, action and learning, based upon our recent findings.

1 Dopamine Neurons Encode Errors of Reward Expectation in Decision-Making Task

Macaque monkeys were trained to perform decision-making task under discrete levels of reward expectation (Satoh et al., 2003). When a start-cue LED appeared on a wood panel in front of monkeys, monkeys depressed a hold button (Figure 1A). After the hold button was depressed for about 1 s, three green LEDs were simultaneously turned on. These were target buttons and only one of them was a correct button. If monkeys depressed the correct button, high tone (1 kHz) beep occurred, as a positive reinforcer, and reward water was delivered. But in the case of incorrect choices, low tone (300 Hz) beep occurred, as a negative reinforcer, and no reward was delivered. In the next trial, monkeys remembered previously chosen incorrect buttons and chose one of the remaining buttons. Once a correct button was chosen, monkeys could obtain reward two more times by choosing a previously found correct button. Thus, the trials in a single block were divided into two epochs. The first epoch was the trial-and-error epoch where the monkey searched for the correct button on a trial and error basis. Three types of trials occurred: trials in which the monkeys selected the correct button at the first, second or third choice in a single block (N1, N2 and N3, respectively). The second epoch was the repetition epoch in which the monkeys selected the known correct button in two successive trials after they had found the correct button during the trial-and-error epoch. Two types of trials occurred: the first and the second trials at the repetition epoch (R1 and R2, respectively).

Activity of single DA neurons was recorded in the midbrain of two monkeys (Figure 1B). Two-thirds of DA neurons responded to positive reinforcer by increasing their discharge rate from baseline rate of 4 to 5 spikes/s, while they responded to negative reinforcer after incorrect choices by decrease of discharge rate. Figure 2 A plots average magnitudes of increase and decrease of discharge rate in 52 DA neurons recorded in single monkey after positive and negative reinforcers respectively. The magnitudes of DA neuron responses to reinforcers were the maximum in N1 trials in which level of reward expectation was lowest, whereas they were minimum in R1 and R2 trials in which reward expectation was highest at almost 100 %. Superimposition of REEs on the magnitude of DA neuron responses to reinforcers demonstrated that the responses to reinforcers precisely encode REEs. Negative REEs were also encoded by the decrease of discharge rate, although the gain of encoding by decreasing discharge rate is much smaller than that for positive reinforcer. These results were confirmed in the trial-by-trial analysis of the DA neuron activity (Figure 2B).

The findings that the responses of DA neurons to reinforcers after an animal's choices precisely encode REEs provide a solid experimental support to the models of reinforcement learning. These models suggest that DA neurons transmit REEs as reinforcement signals derived from a sum of the reward predictions at successive times that act like temporal derivatives (TD) and reward received to the target in the dorsal and ventral striatum and frontal cortices (Sutton, 1988; Barto, 1995; Houk et al., 1995; Montague et al., 1996; Schultz et al., 1997; Suri and Schultz, 1998; Sutton and Barto, 1998). Coding of positive REEs by increase of DA neuron spikes, thus increase of DA release, would facilitate adaptive changes of synaptic transmission related to reward-based learning in the target structures. The coding negative REEs by DA neurons might contribute to extinction or unlearning of actions, like the case of teaching signals (climbing fiber activity) in the cerebellar learning (Medina et al., 2002).

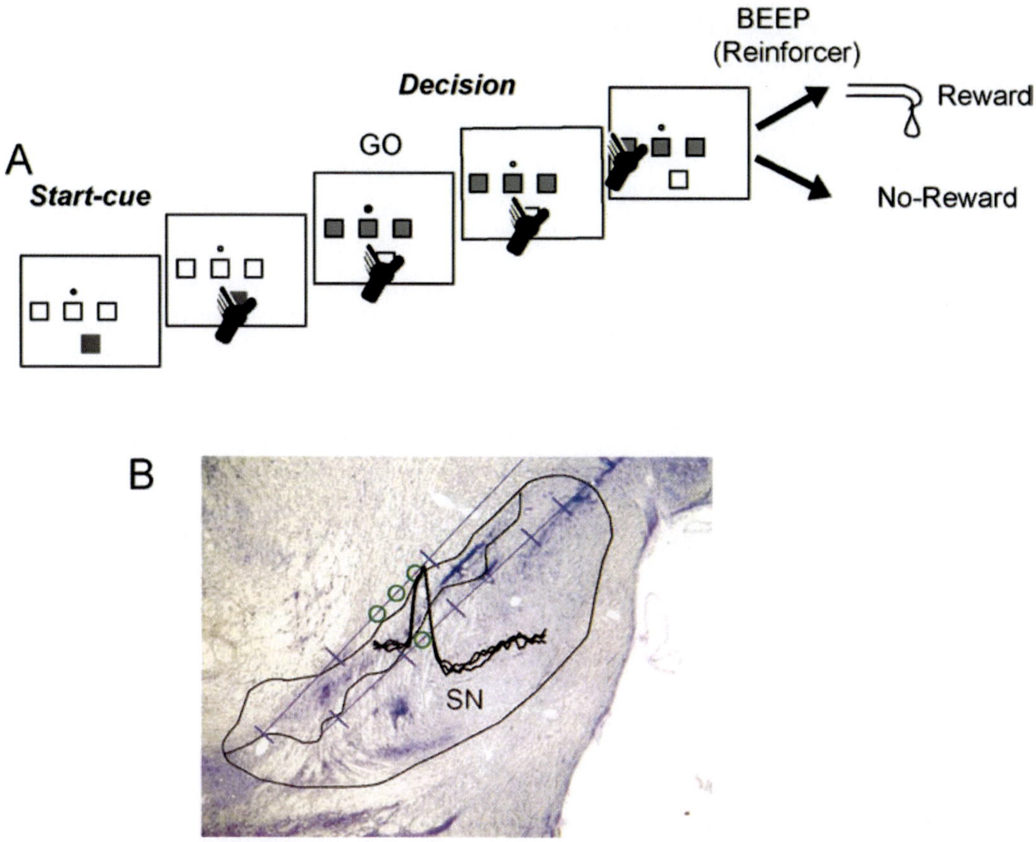

Figure 1. Reward-based decision task and DA neuron activity. A, Time chart of events during the task. B, Superimposition of DA neuron spikes on a Nissl-stained histology section at the recording sites in the midbrain. SN: substantia nigra. Circles and bars indicate recording sites of DA neurons and non-DA neurons. (Modified from Satoh et al. 2003)

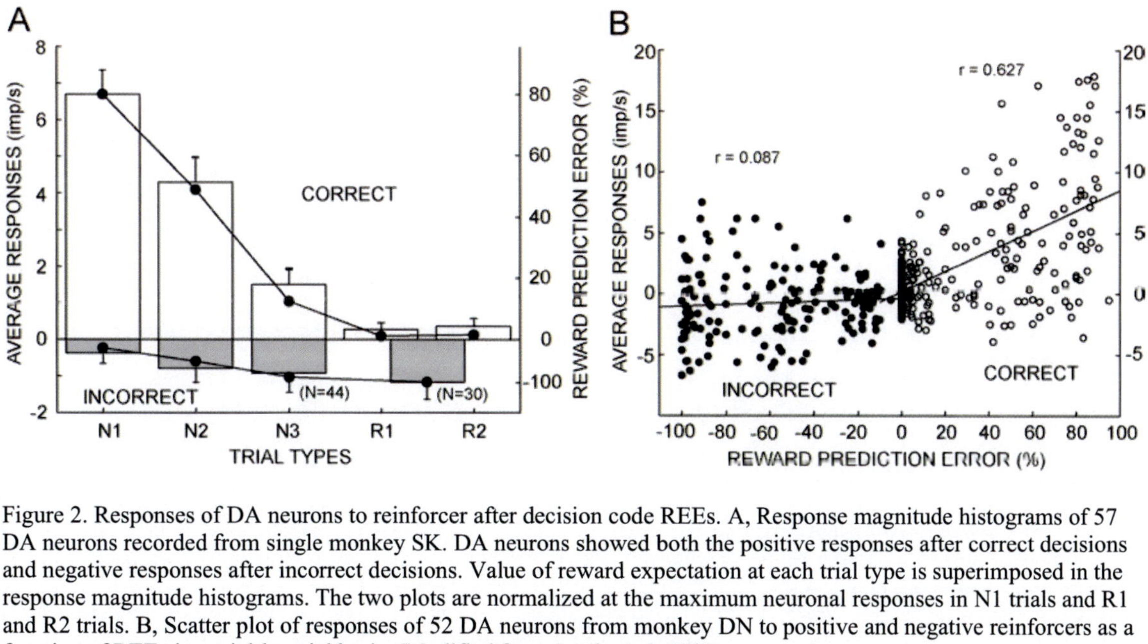

Figure 2. Responses of DA neurons to reinforcer after decision code REEs. A, Response magnitude histograms of 57 DA neurons recorded from single monkey SK. DA neurons showed both the positive responses after correct decisions and negative responses after incorrect decisions. Value of reward expectation at each trial type is superimposed in the response magnitude histograms. The two plots are normalized at the maximum neuronal responses in N1 trials and R1 and R2 trials. B, Scatter plot of responses of 52 DA neurons from monkey DN to positive and negative reinforcers as a function of REEs in a trial-by-trial basis. (Modified from Satoh et al. 2003)

2 Motivational Signals of Dopamine Neurons as Incentive to Work for Reward

About a half of DA neurons responded not only to reinforcers after decisions but also to task start cue, conditioned stimulus (CS), in each trial. Magnitudes of the responses to CS were related to the level of reward expectation in each trial, smaller at low reward expectation and larger at higher expectation, but there were considerable discrepancy between the two especially in the repetition trials with highest reward expectation (Satoh et al., 2003). We tested an alternative hypothesis that the responses to a CS may reflect animal's motivation to work for a reward. We used a time for monkeys to depress the start button (reaction times) after it (CS) was presented as an index of how much the monkeys were motivated to work at the trial, because reaction time is well-known behavioral measures reflecting levels of motivation (Konorski, 1967; Shidara et al., 1998; Watanabe et al., 2001; Kobayashi et al., 2002; Takikawa et al., 2002). To dissociate an involvement of reward expectation in the CS responses from that of motivation, we studied the correlation of reaction time and magnitude of DA neuron response within single trial types in which monkeys performed the trial with consistent level of reward expectations. We found that, within single trial type with identical level of reward expectation, the magnitudes of the CS responses were larger at the trials with shorter reaction time for monkeys to depress hold button after appearance of CS, while they were smaller at the trials with longer reaction times (Figure 3A). There was a significant negative correlation between the CS response magnitudes and reaction times (R1 trials, $r = -0.465$, $p < 0.001$).

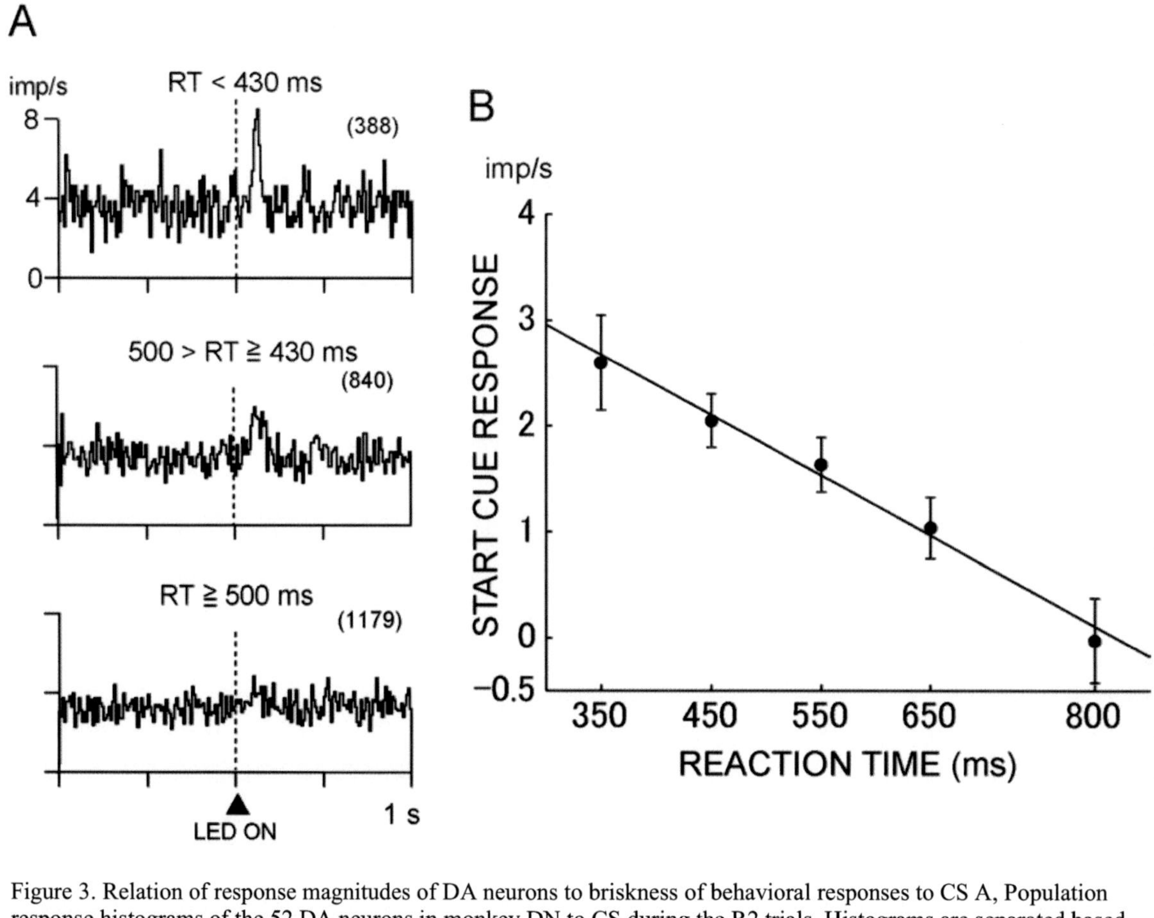

Figure 3. Relation of response magnitudes of DA neurons to briskness of behavioral responses to CS A, Population response histograms of the 52 DA neurons in monkey DN to CS during the R2 trials. Histograms are separated based upon the trials with short, middle and long reaction times to CS. The figures in parenthesis are the number of trials involved in each histogram. RT: reaction time. B, Correlation of magnitude of neural responses to CS in 52 DA neurons in monkey DN to reaction times to depress the start button after CS. The trials were classified into 5 groups based on the reaction times, and the mean and SEM of DA neuron responses in these groups of trials are plotted. (Modified from Satoh et al. 2003)

Pavlovian incentive theories suggest that if a stimulus becomes associated with primary reward, not only the Pavlovian association between the stimulus and a conditioned response occurs, but also a motivational transformation occurs. That is, the stimulus takes on specific motivational, incentive, properties that were originally possessed only by the primary reward itself (Bolles, 1972; Bindra, 1978; Toates, 1986; Dickinson and Balleine, 1994). Because the responses to CS were negatively correlated with reaction times at trials with identical level of reward expectation, it appears that the responses to CS might reflect participation in the processes of motivation. A similar negative correlation was recently reported between the magnitude of positive responses of the pedunculopontine tegmental nucleus (PPTN) neurons to the fixation point onset (CS) for an eye movement task and the reaction time of eye fixation after the CS (Kobayashi et al., 2002). In addition, the magnitude of PPTN neuron responses was positively correlated with the correct performance rate. These results suggested that the PPTN system might be involved in the processes of motivational and attentional control of movement and in the neuronal mechanisms for reinforcement learning (Dormont et al., 1998; Brown et al., 1999; Kobayashi et al., 2002; Kobayashi et al., 2004). Monosynaptic axonal projections from the PPTN to DA neurons in the substantia nigra pars compacta have been demonstrated (Futami et al., 1995). Thus, the PPTN is a strong candidate for a brain structure that supplies midbrain DA neurons with Pavlovian incentive-related signals.

The short latency DA neuron responses to behaviorally salient stimuli, such as reward, aversive and novel stimuli, have been proposed to mediate terminating current behavioral selection and opening new ones, 'switching' (Cools, 1980; Robbins and Sahakian, 1983; Redgrave et al., 1999). Tail pinch, aversive stimulus, causes a decrease of discharges in the identified DA neurons in the ventral tegmental area (VTA) of anesthetized rat, while non-dopaminergic VTA neurons showed increase of discharges by aversive stimulus (Ungless et al., 2004).

Consistent observations about short latency activation of DA neurons are first behavioral events predictive of reward. Thus, basic role of the DA neuron responses to the reward predictive events appears to motivate subjects toward a goal through reward. This role must be especially crucial not only for early process of learning but also for orienting reaction to the salient events, while REEs should play a major role in refinement of learning in the later processes of learning (Satoh et al., 2003). This view is not inconsistent with incentive theories (Bolles, 1972; Bindra, 1978; Toates, 1986; Dickinson and Balleine, 1994; Berridge and Robinson, 2003).

3 Motivational Signals of Dopamine Neurons as Gain Controller of Coding Reward Expectation Error

What kind of roles does the simultaneous coding of motivational properties and REEs by single DA neuron activity play? To address this issue, we studied the relation of responses to CS and those to positive reinforcers (high tone beep) within single trial type. The responses in all N2 trials in single monkeys were quantitatively examined. Responses to CS were positively correlated with those to positive reinforcers (monkey DN, $r = 0.234$, 52 neurons and monkey SK, $r = 0.524$, 56 neurons, $p<0.001$). The results, thus, support an interesting view that the number of DA neuron spikes encoding REEs, gain of coding REEs, appears to be modulated by the responses to CS that reflect levels of motivation. For instance, rate of learning could be faster when animals are highly motivated because of stronger activation of DA neurons, and thus larger amount of DA release, while slower when less motivated, even at identical REEs as a consequence of an action. This suggests a new and richer model for DA neurons as teaching signals in reinforcement learning than currently proposed. On the other hand, it is consistent with the theory of classical conditioning in which rate of learning is assumed to be under the influence of factors, such as attention or motivation (Rescorla and Wagner, 1972; Dickinson, 1980). The progress of learning (ΔP) is expresses as a product of learning constant (α) and reward expectation error (δ). Thus, we propose that the responses of DA

$$\Delta P = \alpha\delta$$

neurons to first reward-predictive stimuli have dual and correlated roles: the one is motivating subjects toward a goal through reward, and the other is working as efficient gain controller of coding REEs.

Animals would approach or escape from those stimuli that gain reinforcing efficacy by means of their association with appetitive or aversive stimuli, conditioned reinforcers. This process must play fundamental but distinct roles for behavioral decisions and learning from reinforcement process. This view was supported by the observation that the responses to CS and to reinforcer behaved independently during learning. The gain of coding the motivation estimated by reaction times did not change during learning, suggesting the invariance of the mechanism linking the incentive attribution to CS with DA release in the dorsal and ventral striatum and frontal cortices during learning. In contrast, there was a remarkable elevation of the gain for encoding REEs by DA neuron spike density, thus DA release during learning.

4 Can the Delayed Reinforcement Signals of Dopamine Work as Teaching Signal in the Striatum?

Learning theories suppose that teaching signals of DA neurons are used for modifying cortico-striate synaptic transmission in the striatum (Barto, 1995; Montague et al., 1996; Schultz et al., 1997; Brown et al., 1999). The unexpectedly obtained reward after an action activates DA neurons and increase transient DA release in the striatum that leads to reinforce selectively the synaptic transmission related to planning the action. Actually, long-term potentiation of cortico-striate EPSP occurs in a dopamine D1 receptor-dependent mechanisms (Calabresi et al., 2000; Reynolds et al., 2001; Wickens et al., 2003). On the other hand, the reinforcement signals of DA neurons arrive at the striatum always with delays, after action-related synaptic activity occurred. In some cases, the delay could be as long as the order of seconds. It was assumed that activation of striate neurons might leave the traces of patterns of synaptic activation for a short period of time, eligibility traces, in terms of, for instance, intracellular or intra-dendritic calcium increase through which delayed DA reinforcement might enable long-term potentiation to occur (Houk et al., 1995; Wickens and Kotter, 1995).

Recently, the delayed reinforcement hypothesis was tested experimentally by Wickens and his colleagues (Bergman et al., 2005). It was shown that precise temporal relationship is required for the dopamine-dependent induction of long-term potentiation. If DA pulses that induce long-term potentiation when applied simultaneously with pre- and post-synaptic conjunction of striatal neuron activity are delayed by as little as 500 ms, long-term depression occurs instead. This strict temporal requirement argues against a cellular eligibility trace. Does the short time window for the synaptic plasticity mean restricted contribution of reinforcement functions of dopamine? Further studies involving quantitative, cellular and behavioral experiments are necessary for this critically important issue about the functional roles of dopamine.

On the other hand, short latency responses of DA neurons to reward-predictive stimuli could enable striatal neurons to 'prepare' for dopamine-dependent modification of action-related cortico-striate synaptic barrage. It is well-know that D1 receptor activation induces activation of intracellular molecular cascade (Kebabian and Greengard, 1971; Fienberg et al., 1998; Greengard et al., 1998) which is supposed to be a cellular basis of the mechanisms. Thus, from an initial trial-and-error, exploratory phase to later implicit, exploitation phase of learning, the DA neuron responses to salient external stimuli, such as reward-predictive events, seem to play a pivotal role in both behavioral guidance to reward and in learning.

Concluding Remarks

In this article, we put our major emphasis on the coding of motivation and REEs by DA neurons. Motivation and REEs are discrete neuronal information, although they are used as correlated neural codes in most efficient way. Precise coding of REEs is a basis as teaching signal supported by learning theories. We propose that the signals of motivation evoked by the first reward-predictive cue may have dual and correlated roles. First, it is to motivate subjects acting toward a goal through obtaining rewards. This role must be especially crucial for initial phase of learning in new environment, and for orienting reaction to the salient, reward-predictive events. Second, the number of DA neuron spikes encoding REEs, gain of coding REEs, appears to be modulated by the responses to CS that reflect levels of motivation. In other words, even at identical REEs, rate of learning could be faster when animals are highly motivated because of stronger activation of DA neurons, and thus larger amount of DA release, while slower when less motivated. This suggests a new and richer model for DA neurons as teaching signals in reinforcement learning than currently proposed.

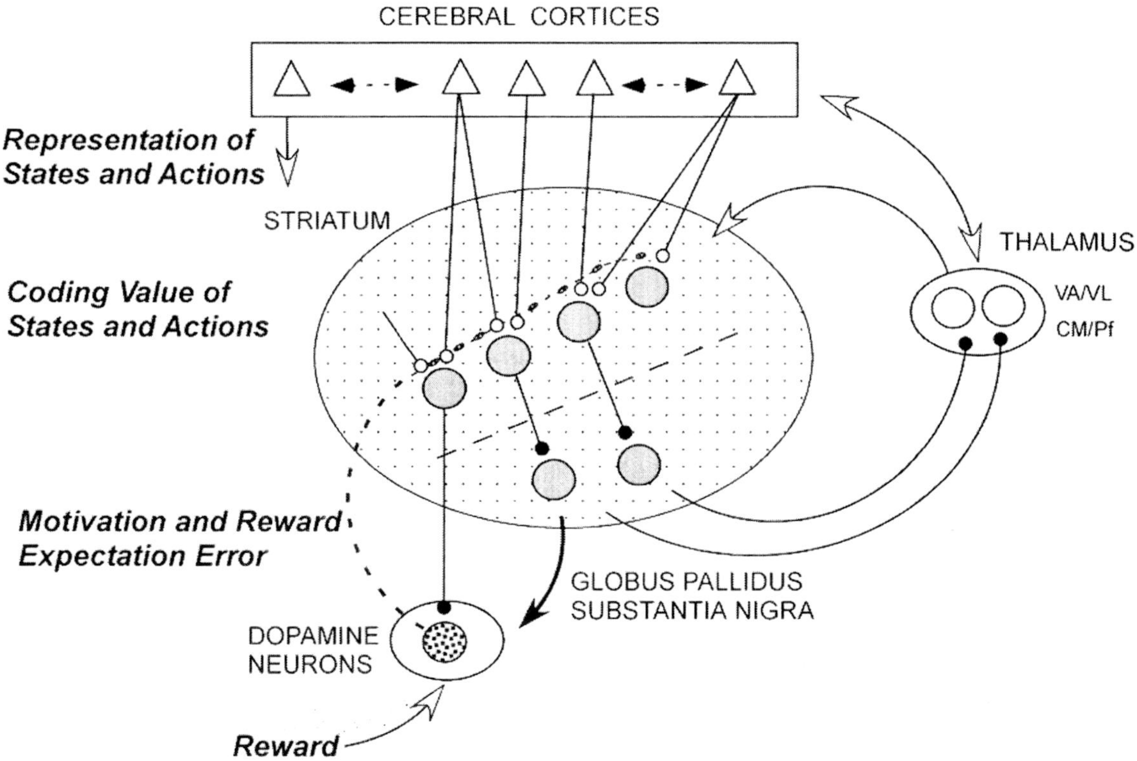

Figure 4. Schematic illustration of roles of DA neurons in reward-based decision-making and action selection through the neuronal circuit of the basal ganglia and cerebral cortex.

Through integrating external sensory information, action repertoire and biological significance, past, current and future states and actions are represented in the neuronal circuits in the cerebral cortex. The DA signals of motivation and REEs, as evaluation signal, are transmitted to their major target of the striatum, limbic system and cerebral frontal cortex. Therefore, essential role of the DA signals in the striatum seems to give reward value, subjective desirability, to information of states and actions transferred through cortico-striate topographically organized pathways. The valued information of states and actions in the striatum would, then, be processed for selecting highest valued action as the processes of decision-making and action planning through projecting globus pallidus, substantia nigra-thalamo-cortical loops (Figure 4).

Acknowledgements

This study was supported by a Grant-in-aid for Scientific Research on Priority Areas (C)-Advanced Brain Science Project and (B) from the Ministry of Education, Culture, Sports, Science and Technology of Japan. We thank Harue Matsuda and Ryoko Sakane for their technical assistance.

References

Arnauld A, Nichole P (1982) The Art of Thinking: Port-Royal Logic. Indianapolis: Bobbs-Merrill.

Barto AG (1995) Adaptive critics and the basal ganglia. In: *Models of Information Processing in the Basal Ganglia* (Houk JC, Davis JL, Beiser DG, eds), pp 215-232. Cambridge: The MIT Press.

Bergman H, Bolam P, Graybiel AM, Kimura K, Plenz D, Seung S, Surmeier DJ, Wickens JR (2005) Microcircuits, molecules and motivated behaviour: microcircuits in the striatum. In: Microcircuits: The interface between neurons and global brain function (Grillner S, ed).

Berridge KC, Robinson TE (2003) Parsing reward. *Trends Neurosci* **26**:507-513.

Bindra D (1978) How adaptive behavior is produced: a perceptual-motivation alternative to response reinforcement. *Behav Brain Sci* **1**:41-91.

Bolles RC (1972) Reinforcement, expectancy, and learning. *Psychol Rev* **79**:394-409.

Brown J, Bullock D, Grossberg S (1999) How the basal ganglia use parallel excitatory and inhibitory learning pathways to selectively respond to unexpected rewarding cues *J Neurosci* **19**:10502-10511.

Calabresi P, Gubellini P, Centonze D, Picconi B, Bernardi G, Chergui K, Svenningsson P, Fienberg AA, Greengard P (2000) Dopamine and cAMP-regulated phosphoprotein 32 kDa controls both striatal long-term depression and long-term potentiation, opposing forms of synaptic plasticity. *J Neurosci* **20**:8443-8451.

Cools AR (1980) Role of the neostriatal dopaminergic activity in sequencing and selecting behavioural strategies: facilitation of processes involved in selecting the best strategy in a stressful situation. *Behav Brain Res* **1**:361-378.

Dickinson A (1980) Contemporaly Animal Learning Theory. Cambridge: Cambridge Univ. Press.

Dickinson A, Balleine B (1994) Motivational control of goal-directed action. *Anim Learn Behav* **22**:1-18.

Dormont JF, Conde H, Farin D (1998) The role of the pedunculopontine tegmental nucleus in relation to conditioned motor performance in the cat. I. Context-dependent and reinforcement-related single unit activity. *Exp Brain Res* **121**:401-410.

Fienberg AA, Hiroi N, Mermelstein PG, Song W, Snyder GL, Nishi A, Cheramy A, O'Callaghan JP, Miller DB, Cole DG, Corbett R, Haile CN, Cooper DC, Onn SP, Grace AA, Ouimet CC, White FJ, Hyman SE, Surmeier DJ, Girault J, Nestler EJ, Greengard P (1998) DARPP-32: regulator of the efficacy of dopaminergic neurotransmission. *Science* **281**:838-842.

Fiorillo CD, Tobler PN, Schultz W (2003) Discrete coding of reward probability and uncertainty by dopamine neurons. *Science* **299**:1898-1902.

Futami T, Takakusaki K, Kitai ST (1995) Glutamatergic and cholinergic inputs from the pedunculopontine tegmental nucleus to dopamine neurons in the substantia nigra pars compacta. *Neurosci Res* **21**:331-342.

Greengard P, Nairn AC, Girault JA, Ouimet CC, Snyder GL, Fisone G, Allen PB, Fienberg A, Nishi A (1998) The DARPP-32/protein phosphatase-1 cascade: a model for signal integration. *Brain Res Brain Res Rev* **26**:274-284.

Herrnstein RJ, Vaughn WJ (1980) The allocation of individual behavior. In: *Limits to Action* (Staddon JER, ed), pp 143-176. New York: Academic.

Horvitz JC, Stewart T, Jacobs BL (1997) Burst activity of ventral tegmental dopamine neurons is elicited by sensory stimuli in the awake cat. *Brain Res* **759**:251-258.

Houk JC, Adams JL, Barto AG (1995) A model of how the basal ganglia generate and use neural signals that predict reinforcement. In: *Models of Information Processing in the Basal Ganglia* (Houk JC, Davis JL, Beiser DG, eds), pp 249-270. Cambridge: The MIT Press.

Hull CL (1943) Principles of behavior, an introduction to behavior theory. New York: D. Appleton-Century Co.

Jacobs BL (1986) Single unit activity of brain monoamine-containing neurons in freely moving animals. *Ann N Y Acad Sci* **473**:70-77.

Kebabian JW, Greengard P (1971) Dopamine-sensitive adenyl cyclase: possible role in synaptic transmission. *Science* **174**:1346-1349.

Kobayashi Y, Inoue Y, Isa T (2004) Pedunculo-pontine control of visually guided saccades. *Prog Brain Res* **143**:439-445.

Kobayashi Y, Inoue Y, Yamamoto M, Isa T, Aizawa H (2002) Contribution of pedunculopontine tegmental nucleus neurons to performance of visually guided saccade tasks in monkeys. *J Neurophysiol* **88**:715-731.

Koepp MJ, Gunn RN, Lawrence AD, Cunningham VJ, Dagher A, Jones T, Brooks DJ, Bench CJ, Grasby PM (1998) Evidence for striatal dopamine release during a video game. *Nature* **393**:266-268.

Konorski J (1967) Integrative activity of the brain. Chicago: University of Chicago Press.

Ljungberg T, Apicella P, Schultz W (1992) Responses of monkey dopamine neurons during learning of behavioral reactions. *J Neurophysiol* **67**:145-163.

Medina JF, Nores WL, Mauk MD (2002) Inhibition of climbing fibres is a signal for the extinction of conditioned eyelid responses. *Nature* **416**:330-333.

Montague PR, Dayan P, Sejnowski TJ (1996) A framework for mesencephalic dopamine systems based on predictive Hebbian learning. *J Neurosci* **16**:1936-1947.

Morris G, Arkadir D, Nevet A, Vaadia E, Bergman H (2004) Coincident but distinct messages of midbrain dopamine and striatal tonically active neurons. *Neuron* **43**:133-143.

Nakahara H, Itoh H, Kawagoe R, Takikawa Y, Hikosaka O (2004) Dopamine neurons can represent context-dependent prediction error. *Neuron* **41**:269-280.

Redgrave P, Prescott TJ, Gurney K (1999) Is the short-latency dopamine response too short to signal reward error? *Trends Neurosci* **22**:146-151.

Rescorla RA, Wagner AR (1972) Current research and theory. In: Classical Conditioning II (Black AH, Prokasy WF, eds), pp 64-99. New York: Appleton Century Crofts.

Reynolds JN, Hyland BI, Wickens JR (2001) A cellular mechanism of reward-related learning. *Nature* **413**:67-70.

Robbins TW, Sahakian BJ (1983) In: *Stimulants: Neurochemical, Behavioural and Clinical Perspectives* (Creese I, ed), pp 301-338: Plenum Press.

Robbins TW, Everitt BJ (1996) Neurobehavioural mechanisms of reward and motivation. *Curr Opin Neurobiol* **6**:228-236.

Salamone JD, Correa M (2002) Motivational views of reinforcement: implications for understanding the behavioral functions of nucleus accumbens dopamine. *Behav Brain Res* **137**:3-25.

Satoh T, Nakai S, Sato T, Kimura M (2003) Correlated coding of motivation and outcome of decision by dopamine neurons. *J Neurosci* **23**:9913-9923.

Schultz W (1998) Predictive reward signal of dopamine neurons. *J Neurophysiol* **80**:1-27.

Schultz W, Romo R (1987) Responses of nigrostriatal dopamine neurons to high-intensity somatosensory stimulation in the anesthetized monkey. *J Neurophysiol* **57**:201-217.

Schultz W, Dayan P, Montague PR (1997) A neural substrate of prediction and reward. *Science* **275**:1593-1599.

Shidara M, Aigner TG, Richmond BJ (1998) Neuronal signals in the monkey ventral striatum related to progress through a predictable series of trials. *J Neurosci* **18**:2613-2625.

Spanagel R, Weiss F (1999) The dopamine hypothesis of reward: past and current status. *Trends Neurosci* **22**:521-527.

Suri RE, Schultz W (1998) Learning of sequential movements by neural network model with dopamine-like reinforcement signal. *Exp Brain Res* **121**:350-354.

Sutton RS (1988) Learning to predict by the method of temporal differences. *Mach Learn* **3**:9-44.

Sutton RS, Barto AG (1998) Reinforcement Learning. Cambridge: The MIT press.

Takikawa Y, Kawagoe R, Itoh H, Nakahara H, Hikosaka O (2002) Modulation of saccadic eye movements by predicted reward outcome. *Exp Brain Res* **142**:284-291.

Thorndike EL (1911) Animal intelligence. New York: The Macmillan Company.

Toates F (1986) Motivational systems. Cambridge, UK: Cambridge University Press.

Ungless MA, Magill PJ, Bolam JP (2004) Uniform inhibition of dopamine neurons in the ventral tegmental area by aversive stimuli. *Science* **303**:2040-2042.

Watanabe M, Cromwell HC, Tremblay L, Hollerman JR, Hikosaka K, Schultz W (2001) Behavioral reactions reflecting differential reward expectations in monkeys. *Exp Brain Res* **140**:511-518.

Wickens J, Kotter R (1995) Cellular models of reinforcement. In: *Models of Information Processing in the Basal Ganglia* (Houk JC, Davis JL, Beiser DG, eds), pp 187-214. Cambridge: The MIT Press.

Wickens JR, Reynolds JN, Hyland BI (2003) Neural mechanisms of reward-related motor learning. *Curr Opin Neurobiol* **13**:685-690.

Wise RA (2002) Brain reward circuitry: insights from unsensed incentives. *Neuron* **36**:229-240.

In: Recent Breakthroughs in Basal Ganglia Research
Editor: Erwan Bezard, pp. 91-106
ISBN: 1-59454-880-3

Chapter 8

ACTIONS, POLICIES, VALUES, AND THE BASAL GANGLIA

Nathaniel D. Daw, Yael Niv and Peter Dayan
Gatsby Computational Neuroscience Unit,
University College London

ABSTRACT

The basal ganglia are widely believed to be involved in the learned selection of actions. Building on this idea, reinforcement learning (RL) theories of optimal control have had some success in explaining the responses of their key dopaminergic afferents. While these *model-free* RL theories offer a compelling account of a range of neurophysiological and behavioural data, they offer only an incomplete picture of action control in the brain. Psychologists and behavioural neuroscientists have long appealed to the existence of at least two separate control systems underlying the learned control of behaviour. The *habit* system is closely identified with the basal ganglia, and we associate it with the model-free RL theories. The other system, more loosely localised in prefrontal regions and without such a detailed theoretical account, is associated with cognitively more sophisticated *goal-directed actions*. On the critical issue of which system determines the ultimate output when they disagree, there is a wide range of experimental results and sparse theoretical underpinning. Here, we extend the RL account of neural action control by first interpreting goal-directed actions in terms of an alternative *model-based* strategy for RL. Then, by considering the relative *uncertainties* of model-free and model-based controllers, we offer a new and more comprehensive account of the confusing experimental results about how the systems trade off control. Our theory offers a more sharply delineated view of the contributions of the basal ganglia to learned behavioural control.

INTRODUCTION

The basal ganglia, and specifically the dorsal striatum, are widely believed to be involved in aspects of the learned selection of actions (Packard and Knowlton, 2002). This belief has motivated, and in turn been reinforced by, a concerted effort to understand basal ganglia function in terms of theoretical ideas from adaptive optimal control, and particularly a branch of this called reinforcement learning (RL) (Sutton and Barto, 1998). Optimal control is particularly challenging, and therefore particularly interesting, in sequential tasks such as mazes, in which the rewarding or punishing consequences of a decision may take time to play out.

One method from RL, the temporal difference (TD) learning algorithm (Sutton, 1988), in association with the so-called *actor-critic* model, has been specially important for neural modeling. This is because a key signal in the algorithm, the TD prediction error, accurately captures many aspects of the phasic activity of midbrain dopamine neurons in animals performing appetitive tasks (Houk et al., 1995; Montague et al., 1996; Schultz et al., 1997). Amongst other places, these neurons deliver dopamine to the ventral and dorsal striatum and their afferents, where they likely control plasticity (Wickens et al., 1996). In this framework (Montague et al., 1996; O'Doherty et al., 2004), dopaminergically controlled plasticity in the ventral striatum and its afferents has been associated with the critic, which learns predictions of long term future reward. Dopaminergically controlled plasticity in the dorsal striatum has been associated with the actor, which specifies the subjects' action choices.

There are many algorithmic routes toward *optimal* behaviour. Insight into the actual structure of control is better provided by studying their characteristic forms of *sub*optimality. Indeed, a striking conjunction between theory and experiment arises when subjects are exposed to shifts in the circumstances or contingencies of a task, for instance, to motivational manipulations that devalue food reward by shifting the subject from hunger to satiation[1]. In this case, when the dorsal striatum appears to be in control of the choices of actions, the behaviour of the subjects is indeed suboptimal in that it fails to respect the change. Instead, subjects persist in performing actions that bring about rewards in which they are motivationally uninterested. A simple, so-called *model-free* form of the actor-critic has exactly the same problem, because it bases its decisions on *stored* predictions that require relearning for them to be revised in the light of the new values of the outcomes. This characteristic is not mere happenchance – rather it critically reflects the way the actor-critic uses such predictions to ameliorate the difficulty of choosing between actions whose consequences for reward or punishment might otherwise be obscure due to delay.

The results of numerous behavioural experiments (reviewed in Dickinson, 1985, 1994; Dickinson and Balleine, 2002) employing such manipulations show that only some classes of behaviour are devaluation-insensitive in this manner, and thus that the actor-critic is at best an incomplete theory of action selection. Other behaviours react instantly to revaluation treatments, without the need for relearning. The two classes of behaviour are operationally defined as being devaluation-insensitive or *habitual* and devaluation-sensitive or *goal-directed* (since devaluations change a subject's goals). In some cases in which a sequence of multiple actions has to be performed to attain reward, experiments have even demonstrated behavioural inconsistencies, with subjects performing the first action in the chain, but then omitting the second action for a reward that is not desired.

The complete set of rather rococo behavioural distinctions has led psychologists to theories involving multiple parallel routes to action (Dickinson, 1985, 1994; Dickinson and Balleine, 2002; Kilcross and Blundell, 2002), and has also inspired some modelling (Dayan, 2002; Dayan and Balleine, 2002). However, there has not yet been a full computational investigation of the goal-directed route to action, or into the competition between these parallel routes.

In this chapter, we first consider goal-directed control in terms of *model-based* methods of RL. Rather than storing long-term values, these methods anticipate the long-term consequences of actions by searching in a *forward model* of the task contingencies. Along with other authors, we locate the major substrate of goal-directed control in the prefrontal cortex. Model-free and model-based controllers thus have distinct anatomical substrates and employ different strategies for choosing between actions whose consequences might take time to unfold. We then propose a theory of action choice that arbitrates between the two controllers when their suggested actions disagree, proposing that competition for the control of behaviour is based on the relative reliability of the information upon which each system depends. This offers a new account for a body of seemingly puzzling behavioural and neuroscientific data, and a better delimited view of the contributions of the basal ganglia to learned behaviour.

1 Decision Strategies in Reinforcement Learning

In a typical RL setting, an agent is placed in an environment (such as a maze or an operant chamber), and allowed repeatedly to take actions (such as pressing a lever) and observe their consequences (such as gaining food pellets). The goal is defined in terms of learning to choose actions that optimise some long-term measure of future utility. As in a maze, where early decisions can have a huge impact on the speed with which the goal is ultimately attained, the chief difficulty is tracking the consequences of each action over an extended timescale.

The most straightforward strategy for optimal control is to build an internal *world model* detailing the immediate consequences of actions, and then to search this model recursively through many stages to anticipate and evaluate the long-term consequences of any particular choice. In psychological terms, such a model is a form of the cognitive map stressed by (Tolman, 1932), and studied by many others. It consists of three components: transitions, outcomes, and utilities. The first defines how actions induce transitions

[1] There are potentially important differences between devaluation achieved by such motivational shifts and devaluation achieved by, for instance, food aversion treatments. In this chapter, we suppress this distinction for simplicity.

between situations (called *states* of the world), *ie,* the probability of going from one state to another when a particular action is performed (going from room A to room B when turning right). The next is the mappings of states or actions to affectively important outcomes (such as the particular food available at some location in a maze). The last is the mapping from each outcome to its subjective affective utility, a scalar measure of desirability[2]. Crucially, this can change depending on the animal's state of deprivation (for instance, a food pellet will be more desirable when an animal is hungry than when it is full). A long-term measure of utility can simply be the sum, over a sequence of states, of the immediate utilities received in each.

Although building or learning a model is straightforward, using (*ie* searching) it to infer the best action is not. For instance, in a maze, in order to decide which action to take in one room, one has to use the world model to search through many different paths in an expanding tree of future possibilities, adding up expected utilities to determine the value expected for each path. Clearly, comparing the consequences of many long sequences of behaviour is practically impossible, and therefore an alternative strategy is essential. Starting as far back as the seminal contribution of (Samuels, 1959), two main ideas have been suggested (distinct though often used together), namely *pruning* and *caching.* Pruning involves prioritising paths and exploring only a subset. Except in special cases, pruning introduces inaccuracy, since prizes or pitfalls may be left undiscovered.

The alternative to pruning is caching, *ie* storing anticipated results of the search. In the actor-critic, a temporal-difference algorithm (Sutton, 1988) is used to learn cached *values* (long-term utilities) for each state. These can be used, in turn, to learn a cached *policy* indicating which action is best in each state (Barto et al., 1983; Sutton, 1984; Barto, 1995)[3]. The policy itself can be represented in a number of ways; in this paper, we consider a version of the actor-critic called advantage learning (Baird, 1994), which specifies a policy by retrieving from storage exactly the quantities that tree search builds on the fly. Specifically, advantage learning caches quantities called 'advantages' for each action in each state. The advantage of a particular action in a particular state represents the additional long-term value expected for taking that action over the baseline expectation reflected in the state's cached value; the sum of the state value and the advantage is the predicted long-term value for taking the action in the state.

Advantage learning, like other actor-critic methods, is *model-free* in that caching obviates the need for building and searching in a model. However, the key benefit from caching, that expansion of the search tree is unnecessary, is also its key limitation. If the task is changed, for instance by altering either the transitions (*eg,* blocking or opening routes in a maze, as studied by Tolman, (1932) or the immediate utilities (*eg,* by shifting the motivational state of rats pressing a lever for food, from hungry to sated), then the cached values and policy will not change with it, unless there is an explicit opportunity for relearning of each relevant value (which typically requires multiple experiences of the entire trajectory from decision to outcome). By contrast, decisions derived from search in the full tree are immediately sensitive to changes in any of the transitions or utilities given only local experience with the changed aspects themselves, since the tree search will encounter and take into account the new contingencies.

Model-free and model-based controllers actually live at two ends of a spectrum. Elaborations of the sort explored in Dayan, (2002) embed some, but not all, features of a learned model into a model-free controller. It is also common to combine both caching and pruning, i.e. to substitute cached values for unexplored paths in partial tree search (Samuels, 1959). In RL it is conventional to explore such variants. However, we interpret the data on animal behavioural choice as suggesting that a different strategy is at work, involving the cooperation and competition of (at least) two separate and simultaneously active controllers, one model-free (and caching) and the other model-based (and perhaps using pruning). We discuss the substantial evidence implicating the basal ganglia in instantiating the model-free component, and the rather more flimsy evidence implicating the prefrontal cortex in the model-based component in section 3. RL theory does not offer extensive guidance as to how to combine multiple simultaneous controllers when they disagree. However, there is a body of work in areas such as multisensory integration arguing that combination should be based on relative accuracy or certainty (Lindley, 1985; Yuille and Bültoff, 1996; Dayan et al., 2000; Ernst and Banks, 2002). That is, circumstances that promote one system over the other should do so in virtue of their differential impact on the reliability of each system.

---

[2] RL treatments normally collapse the last two components, outcome identities and utilities, into a single function mapping states to *rewards*; we separate them to allow for a clearer treatment of motivational manipulation, and in the light of some lesion data (Blundell et al., 2003).

[3] Only the policy ultimately determines the decisions. However, in methods like actor-critic, it is essential to learn the values of states in order to learn the optimal policy.

In the rest of this chapter, we first link particular forms of model-free (advantage learning) and model-based (tree-search) controllers to habitual and goal-directed action choice respectively. We then construct a theory of the uncertainties of the two sorts of controllers, and use it to account for, and re-interpret, animal behavioural data. Finally, we review the evidence about their neural realizations and the complexities of their interaction.

2 Controller Competition

The difference between model-based and model-free controllers in the light of task changes mirrors venerable ideas in psychology about multiple routes to action (reviewed in Dickinson and Balleine, 2002). As we mentioned, psychologists distinguish between habitual responses, characterised by insensitivity to manipulations of the outcome utility, and goal-directed actions, which are sensitive to such manipulations. That literature takes a somewhat subtly different view than we have described about the information relied on by each system. There, goal-directed actions are assumed to be controlled by associations between responses and their outcomes ('R-O' associations); habits are envisioned to be outcome-independent because they instead rely on associations between stimuli and responses ('S-R'). The RL quantities discussed above (policies, values, etc.) broadly parallel this associative learning terminology (with states in place of stimuli; actions for responses, etc.), but they are more general and precise. For instance, it is not clear how the concept of the R-O association applies when multiple actions and outcomes may occur sequentially or interleaved. This could be problematic, for instance, in extending to such situations a previous account of habit formation (Dickinson, 1985), which relies on the strength of R-O contingencies.

Both goal-directed and habitual action are forms of instrumental behaviour. We will also briefly discuss another class of behaviour, known as Pavlovian responses. Rather than being a chosen arbitrary action (such as a lever press), Pavlovian behaviours are thought to be emitted automatically when an animal anticipates reinforcement, as in the famous example of Pavlov's dog salivating in expectation of food. Pavlovian responses can compete with, and thereby complicate the analysis of, instrumental choices. As with instrumental choices, some Pavlovian responses seem to be sensitive to motivational manipulations, whilst others are insensitive (*eg* Holland and Rescorla, 1975; Holland and Straub, 1979). Though we lack the space to develop the notion completely, we suggest that the same underlying computational framework applies to Pavlovian as well as to instrumental behaviour.

The difference between habitual and goal-directed controllers has led to a confusing wealth of experimental results on the way that actions are chosen when they disagree. In a typical such experiment (illustrated in Figure 1), an animal is placed in a deprived motivational state (*eg* hunger), and trained to perform an action such as pressing a lever, in order to obtain a motivationally relevant outcome such as food pellets. Normally, obtaining reward actually requires a sequence of actions: pressing the lever to release food, and pushing a flap on a food magazine to retrieve it. After training, the utility of the outcome is devalued. This can be accomplished, for instance, by feeding the animal until it is sated (a motivational shift), or by conditioning aversion to the food by pairing it with drug-induced illness. The animal is then tested to see whether it will still perform the actions associated with the devalued food. Importantly, this test is performed in extinction – without reward delivery – to ensure that any change in behaviour is attributable to existing knowledge about the outcome and its utility rather than to new experience with either[4].

In some circumstances – especially for actions that have been moderately trained – devaluation indeed reduces performance of the action relative to non-devalued controls (Figure 1a,b; Adams and Dickinson, 1981; Balleine and Dickinson, 1992). Consistent with the folk-psychological maxim that repetition breeds habits, with increased training, goal-directed actions often become devaluation-insensitive, *ie*, habitual (Figure 1c; Adams, 1982; Dickinson et al., 1995). However, there are a number of additional interacting factors, and hitherto no satisfactory explanation uniting them. For instance, magazine entries (the action more proximal to reward) can resist habitisation even while leverpresses become habitual (Killcross and Coutureau, 2003; Figure 1a). Devaluation sensitivity also persists despite overtraining for actions trained in more complex tasks involving multiple possible responses and outcomes (Colwill and Rescorla, 1985;

[4] Of course, the reward omission itself causes a reduction in responding; to control for this, responding is compared to another group that is also tested in extinction but without having experienced devaluation of the outcome.

Holland, 2004). Table 1 summarises some of the important factors influencing devaluation sensitivity in experiments.

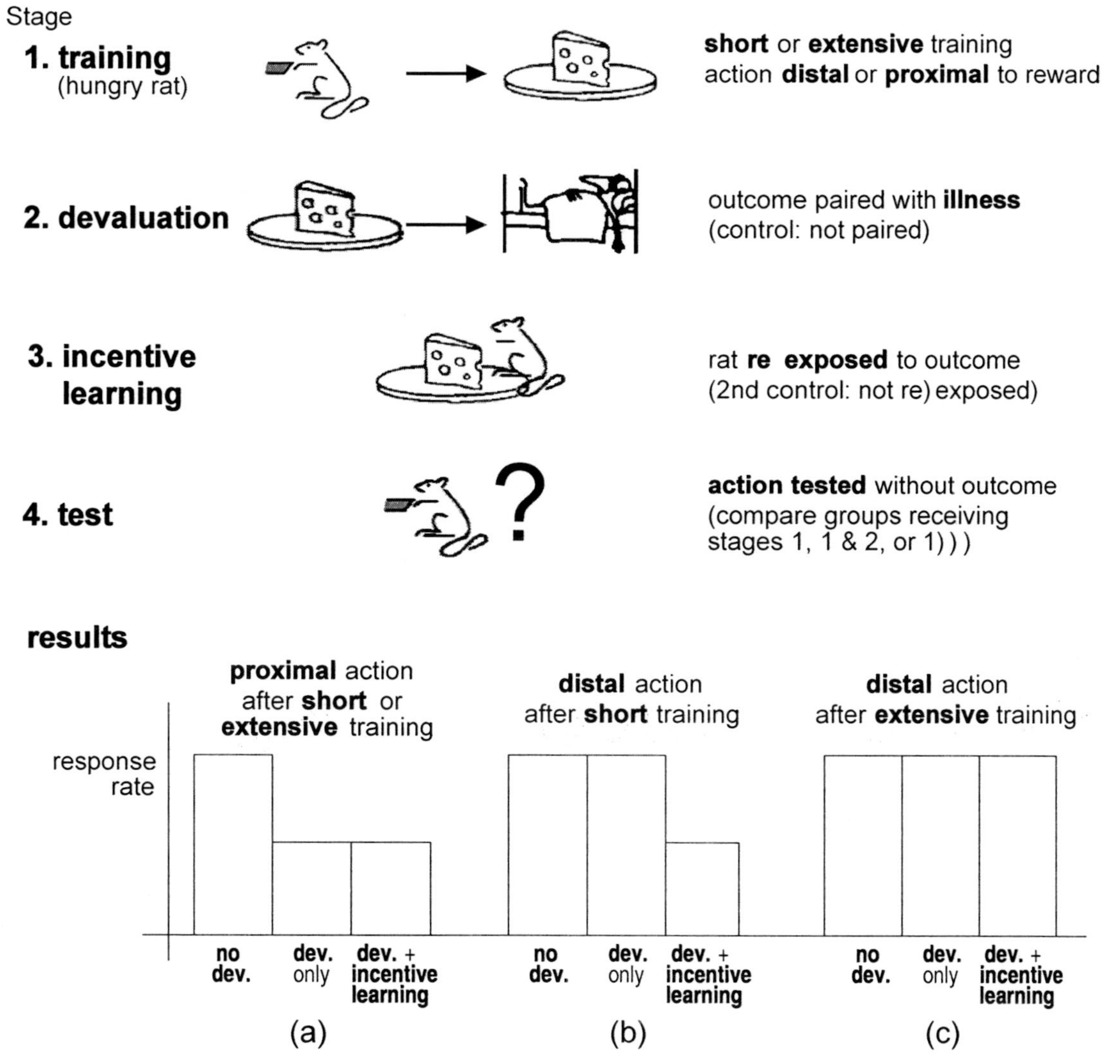

Figure 1: The stages in a typical devaluation experiment, together with a qualitative illustration of the results (based on data from, e.g., Balleine et al., 1995; Dickinson et al., 1995). Drawings from Dickinson and Balleine, 1994.

There is a further phenomenon that complicates interpreting these data with a two-controller model. If the reward is relatively novel (in particular, if it has never been experienced in the devalued state, *eg* while sated or after pairing with illness), then the outcome-sensitive actions subdivide further. Some actions are affected *immediately* by devaluation (Figure 1a), while other actions are affected by devaluation *only if* the animal is exposed to the outcome in the devalued state prior to the instrumental test (Balleine and Dickinson, 1991; Balleine, 1992; Figure 1b). This extra experience is called *incentive learning*, under the assumption (which we challenge below) that the re-exposure allows learning about the reduced utility ('incentive') of the outcome. Note that both of these categories differ from habits, which are usually devaluation-insensitive regardless of incentive learning (Dickinson et al., 1995; Figure 1c). Remarkably, both profiles of devaluation sensitivity can be observed simultaneously in different behaviours occurring in sequence. For instance, in classic experiments by Balleine, (1992), magazine flap responses were immediately affected by devaluation, but contemporaneously recorded leverpresses were sensitive only after re-exposure. Similarly, in an experiment in which an animal had to carry out a sequence of two instrumental actions (a chain-pull and a leverpress) to obtain reinforcement, the proximal action was instantly devaluation-sensitive while the distal one required incentive learning (Balleine et al., 1995).

Table 1: Some factors promoting devaluation sensitivity vs. insensitivity in behavioural experiments.

devaluation sensitivity	devaluation insensitivity
moderate training	extensive training
multiple responses/outcomes	single response/outcome
incentive learning	no incentive learning
action proximal to reward	action distal from reward

We now offer an account of this apparent proliferation of behavioural categories, suggesting that they arise directly from the nature of the competition between the goal-directed and habitual controllers.

2.1 Uncertainty-Based Arbitration

As we have described it, decision-making within both the caching and tree-search systems depends on predicted action values. That is, the two systems employ *different* methods for estimating the *same* underlying quantities. The two estimation methods may therefore be more or less accurate under different circumstances. Here we propose that the experimental circumstances promoting devaluation sensitivity or insensitivity (Table 1) are those for which model search or caching, respectively, are relatively more accurate.

More concretely, we assume that arbitration between the two systems is based on the relative accuracy of their estimation methods. This raises the question of how these can be judged – it is known that such accuracies are extremely difficult to quantify (Singh and Dayan, 1998; Kearns and Singh, 1999, 2000). One measure that arises naturally in many cases of competition in machine learning and neuroscience is *uncertainty* (Lindley, 1985; Yuille and Bültoff, 1996; Dayan et al., 2000; Ernst and Banks, 2002). Both systems' estimates of the action values have uncertainty (defined as the variances, *ie* expected squared errors, in the estimates) because the controllers have only limited experience on top of impoverished prior information[5]. We assume that the controllers themselves estimate the uncertainty in their action value predictions. This is computationally very difficult, but various approximations have been proposed (Dearden et al., 1998; 1999; Mannor et al., 2004 in press). The actor-critic can cache uncertainty information together with the cached values and advantages (Dearden et al., 1998). Instead of caching values, the model-based system stores an estimated world model, which is itself uncertain, and uses it to generate projected trajectories through the tree and accumulate action values over them. Uncertainty in the action values can be accumulated during this search based on the uncertainty in the estimates of the transitions, outcomes and utilities (Dearden et al., 1999; Mannor et al., 2004 in press). This process is illustrated for a simple leverpressing task in Figure 2a, which shows how search proceeds forward through a sequence of anticipated states, and their values and uncertainties then propagate back up the tree.

The critical question for uncertainty-based competition is what makes one system more or less uncertain than the other. The tree based system stores all relevant information about past experience in its world model, and an exhaustive tree search can optimally incorporate all this information in constructing a value estimate. The caching system avoids searching in a world model, a shortcut that necessitates that new information be incorporated into its state estimates by a more haphazard and inefficient process of *bootstrapping*, or training each value estimate in terms of other estimates. Thus early in training (when data are scarce) cached values are only distantly related to their ultimate settings (Singh and Dayan, 1998), and the more data-efficient tree system has an advantage in accuracy. This advantage is enhanced in more complex, data-intensive tasks, such as when there are multiple responses and outcomes, but wanes as experience accumulates (Figure 2b, top).

[5] We distinguish *internal* from *external* uncertainty. External uncertainty results from actual stochasticity in the world (as when a lever-press pays off with 50 % probability). Internal uncertainty is the Bayesian counterpart to the confidence interval in classical statistics, quantifying ignorance rather than stochasticity (as when an animal cannot reliably distinguish between payoff rates between 40 % and 60 % due to having only observed a few lever-presses). In differentially assessing the two controllers, it is internal uncertainty that is important.

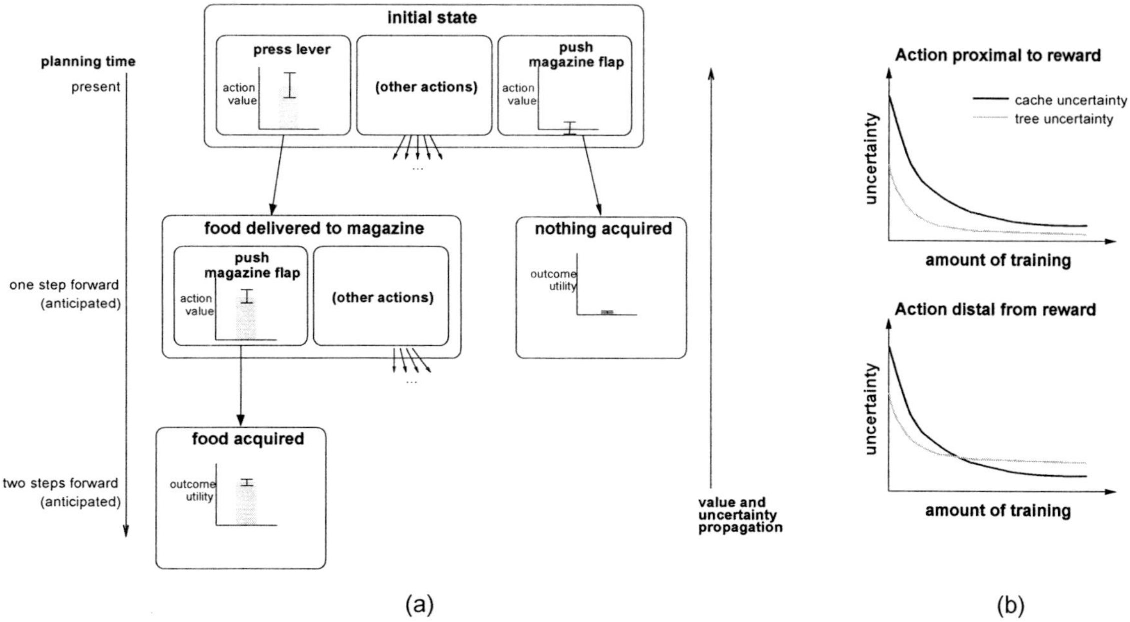

Figure 2: (a) Value and uncertainty propagation in tree search. Illustrated is a simple instrumental conditioning task in which a rat must press a lever and then push a flap, opening a food magazine to receive reward. Two paths through the tree (of many) are explored, and derived action values are shown with error bars representing uncertainty. Not directly depicted is uncertainty about the transitions (which state will result from an action), which is one reason that uncertainty about values grows moving up the tree, away from the rewards. (b) Uncertainty as a function of training for the cache and tree-search systems. Above, for actions proximal to reward, the tree is consistently more reliable than the cache, because it is more data-efficient, though the magnitude of this advantage declines with training. Below, for actions more distal from reward, the tree system is subject to additional uncertainty as a result of accumulated computational inaccuracy, allowing the cache system to surpass its reliability after some training.

Why should the caching system ever dominate, if an *unlimited* tree-search system is strictly more accurate (Kearns and Singh, 1999) ? We assume that the tree search system has limitations of its own, for instance, that it is not capable of exhaustively searching all paths through the tree to derive accurate action values. In this case, it must rely on some approximation such as pruning (exploring only a subset of paths) or using a simplified approximation to the probability distribution over future states resulting from each contemplated action. This will introduce inaccuracy, which will compound over each step of search. Since the caching system requires no such approximation, its estimates would then be relatively advantaged for actions farther from the goal (*ie*, those that the goal-directed system must use deeper search to evaluate; Figure 2b, bottom). Which system is more reliable at evaluating any particular action then depends on the balance of the effects of inexperience and task complexity (favouring the tree system early in training and in complex tasks) against search depth (favouring the habit system for actions more distal from reward). Such a scheme unites and explains most of the experimental results concerning factors influencing outcome sensitivity, summarised in Table 1.

It remains for us to account for the effect of incentive learning. The key observation is that *uncertainty* in the habit system's value estimates is also cached, and thus (analogous to the value estimates themselves) insensitive to changes in confidence about the outcome utility. We assume that a devaluation without re-exposure immediately reduces the goal-directed system's utility estimate, and more importantly, for outcomes that have not been experienced in the new devalued state, that the change is also accompanied by increased uncertainty about the estimate of the new utility. Through the search process, the goal-directed system would immediately compute reduced values for actions leading to the devalued state, but also *increased uncertainties* about those values. This introduces an additional factor favouring the habit system, since because of caching, its uncertainties are (incorrectly) unaffected by devaluation. For actions that are only weakly under goal directed control (*eg*, more distal ones), devaluation without re-exposure can thus tip the balance in favour of the habit system. Performance of these actions would then be insensitive to devaluation, until a re-exposure treatment restored certainty in the goal-directed system's utility estimate, allowing it to reassert control. In contrast, the goal-directed system would retain control of those actions it most staunchly dominates (such as those more proximal to reward, or in tasks with multiple actions and outcomes) allowing them to adjust to devaluation without need for re-exposure. This pattern is consistent with the experimental data, explaining the difference between distal and proximal responsivity patterns depicted in Figure 1a,b (Balleine et al., 1995) and similar differences in the effect of incentive learning when there are multiple versus single actions and outcomes (Balleine and Dickinson, 1992; Rescorla, 1994).

3 The Neural Substrate

We have so far concentrated mainly on a psychological picture of the controllers and their interaction. We now turn to the neural substrate, first of the two separate systems, and then their arbitration.

The core data anchoring the actor-critic model in the basal ganglia are recordings of midbrain dopamine neurons in primates engaged in appetitive learning tasks (Schultz, 1992, 1998). The phasic responses of these neurons – to unpredicted but not predicted primary rewards, for example, and to cues signalling reward – closely resemble the TD error signal for reward prediction, which is used in the actor-critic to learn both values and policies. The correspondence between the TD signal and the neuronal responses has been worked out in detail (Houk et al., 1995; Montague et al., 1996; Schultz et al., 1997; Kakade and Dayan, 2002; Daw et al., 2003). Expanding from this identification of dopamine as a teaching signal, the actor-critic model proposes that dopamine targets – particularly in the striatum – are the sites of value and policy learning (Houk et al., 1995; Montague et al., 1996; Suri and Schultz, 2001; Dayan, 2002). Although much debated (Horvitz, 2000; Ungless et al., 2004), there is evidence suggesting that phasic dopamine is involved in appetitive but not aversive learning (Mirenowicz and Schultz, 1996). It has been, however, suggested that dorsal raphe serotonin complements this function in aversive learning, playing a similar role for predictions of punishment (Daw et al., 2002).

The separation between the two learning functions in the actor-critic parallels a fundamental division in the functional anatomy of the basal ganglia. Although there are many finer distinctions, a division of the striatum into dorsal and ventral subregions is well established anatomically, pharmacologically, physiologically, and even in functional neuroimaging. (For extensive references, see Voorn et al., 2004, who propose that the distinction is more accurately described as dorsolateral versus ventromedial.) The ventral striatum receives (and indirectly reciprocates) projections from limbic structures such as orbitofrontal cortex, and is implicated in reward and motivation (Cardinal et al., 2002a). Much of the dorsal striatum, by contrast, is connected with motor cortices and is itself implicated in learned motor control (Alexander et al., 1986; Packard and Knowlton, 2002). Dopamine neurons are analogously grouped, into the ventral tegmental area (VTA) and substantia nigra pars compacta (SNc) (Joel and Weiner, 2000).

The actor-critic model thus suggests that the dopamine projection from VTA, which targets ventral striatum and other limbic areas, supports the learning of state values, while that from SNc to dorsal striatum supports policy learning (Figure 3). An early observation influential in this identification is the little apparent difference between the activities of VTA and SNc dopamine neurons (Schultz et al., 1993), despite their association with functionally quite disparate striatal territories. This negative finding is expected under the actor-critic, since values and policies share a common TD error signal. This overall functional mapping recently received more direct support in rodent neurophysiological (Daw, 2003) and human fMRI experiments (O'Doherty et al., 2004) explicitly designed to exercise the two learning functions differentially. This proposal is also more consistent with anatomical data reviewed recently by Joel et al. (Joel et al., 2002) than is an alternative proposed mapping of actor and critic functions to different striatal subterritories (Houk et al., 1995).

Lesions of the dorsolateral striatum seem to abolish habitual control (Yin et al., 2004), as would be expected under the model. That is, even extensively trained actions show devaluation sensitivity in animals with such lesions. According to the model, the values cached in the *ventral* striatum are associated with states (and not actions), and so they would also be appropriate to drive Pavlovian responses (which typically anticipate outcomes whose delivery is dependent on the state but not the animal's actions) as well as to support instrumental learning more dorsally. Indeed, lesions implicate the ventral striatum in some Pavlovian responses and in Pavlovian-instrumental interactions (Hall et al., 2001; Cardinal et al., 2002b). Though this has been spottily tested, under our model, these are expected to be the devaluation-*insensitive* aspects of Pavlovian conditioning, since they originate from cached values. (As with instrumental choices, we expect devaluation-*sensitive* Pavlovian responses to originate in a model-based system.)

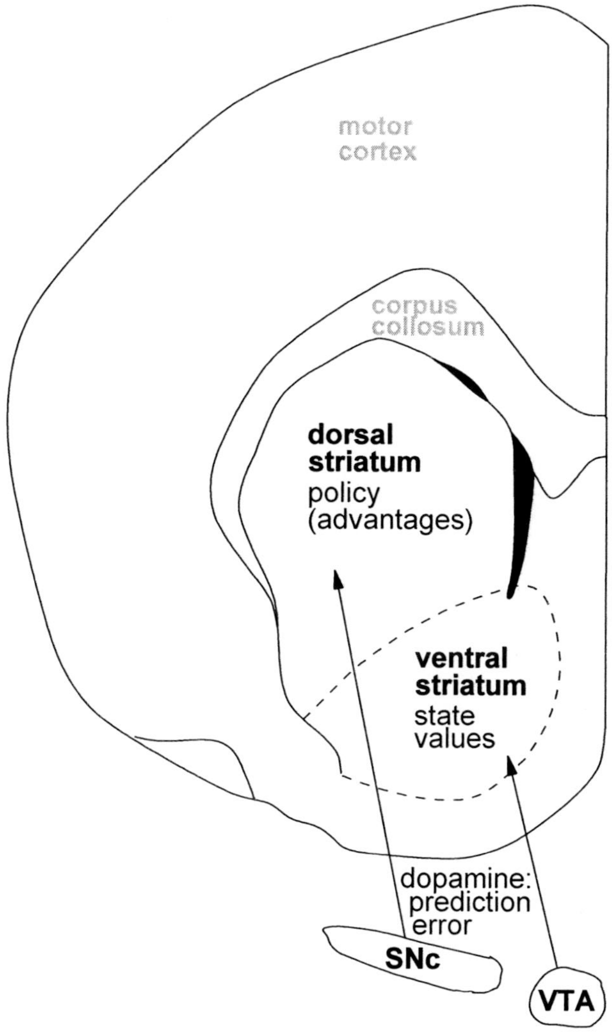

Figure 3: Suggested anatomical substrates of actor/critic model in midbrain and striatum6.

In contrast to the rather well specified habitual controller, there is much less data on the nature and substrates of outcome-sensitive, goal-directed actions (Dickinson, 1985, 1994). Broadly, one would expect this sort of planning function to be associated with prefrontal cortex, for a host of neuropsychological and neurophysiological reasons (Owen, 1997; Miller and Cohen, 2001; Clark et al., 2004). This general impression is supported by more specific studies of instrumental behaviour: Lesions of rat prelimbic cortex (a subarea of medial prefrontal cortex) abolish devaluation sensitivity for undertrained behaviours that would normally be goal-directed (Balleine and Dickinson, 1998; Corbit and Balleine, 2003; Killcross and Coutureau, 2003). The full pattern of results suggests involvement of this area in representation or evaluation of the world model.

As described in Section 2, model-based planning also relies directly on information about outcomes and their utilities. Electrophysiological, lesion and imaging data implicate areas such as the orbitofrontal region of prefrontal cortex (Schoenbaum et al., 1998, Tremblay and Schultz, 1999; O'Doherty et al., 2000), gustatory insular cortex (Balleine and Dickinson, 1998, 2000), and basolateral amygdala (Schoenbaum et al., 1998; Blundell et al., 2001) in learning and representing these. Notably, lesions in all of these areas also abolish devaluation sensitivity (Balleine and Dickinson, 1998; Gallagher et al., 1999; Balleine and Dickinson, 2000; Balleine et al., 2003; Blundell et al., 2003), and further behavioural assays (*eg,* Blundell

[6] Figure adapted from *The Rat Brain in Stereotaxic Coordinates, Compact Third Edition*, Paxinos and Watson, copyright 1996, reprinted with permission from Elsevier.

et al., 2001, 2003) support the interpretation that, in terms of our theory, these areas are involved in the mapping from states to outcomes, or outcomes to utilities, rather than in the tree itself.

The lesion results we have discussed have broader implications for the organisation of control, in that they demonstrate the apparent possibility of *separately* disabling either the goal-directed or habitual controller, leaving the other intact. The results suggest a measure of independence between the systems, and indicate that both systems are capable of specifying actions even when they would not normally be in control (for instance that the habit system can control even moderately trained behaviour if the goal-directed system is disabled).

There is no direct evidence as to the neural basis of arbitration between the systems, though there have been suggestions that this function may be integrated into either system rather than being anatomically separate. Thus Redgrave and colleagues (Redgrave et al., 1999) propose that the basal ganglia may select between control systems (as well as between habitual actions). The converse view is that the prefrontal planning system performs the arbitration, by overriding or gating habitual control. Killcross and Coutureau, (2003) deploy this suggestion to explain their finding (paradoxical under the conventional anatomical mapping that we have described) that lesions of rat infralimbic cortex (another subarea of medial prefrontal cortex) disrupt control by the habit system, rather than by the goal-directed system (Coutureau and Killcross, 2003; Killcross and Coutureau, 2003).

4 Discussion

We have offered a new framework for interpreting behavioural and physiological evidence regarding multiple routes to action, which contacts and enriches theories of RL and its neural substrates. We presented a bipartite model of action control, in which the predominant actor-critic model of dopaminergic and basal ganglia function (which is seen as the model of habitual control) is accompanied by a second, cortical system, which is capable of learning a model of the world and searching in it to plan actions. When the controllers differ in the actions they prefer, we have suggested that their relative certainties control which action gets executed, and have used this to explain the wealth of data as to circumstances promoting or inhibiting devaluation sensitivity, including the puzzling findings of incentive learning. Also, unlike some previous models in which goal-directed and habitual control are rather entangled (Dickinson, 1994; Dayan, 2002), the new model is consonant with lesion studies that suggest that the two behavioural systems are relatively dissociable and independent (Coutureau and Killcross, 2003; Killcross and Coutureau, 2003; Yin et al., 2004).

The standard account of incentive learning is different from the one we have presented. We view the goal-directed system as always being capable of immediate devaluation. The need for further learning arises because re-exposure restores control to this system over the habitual system. By contrast, the standard view (Balleine, 1992; Balleine, 2000; Dickinson and Balleine, 2002) holds that the relevant actions are under the control of the goal-directed system throughout the incentive learning treatment, with re-exposure changing the value ascribed within this system. Immediate devaluation sensitivity for magazine entries is then explained by assuming that these are actually not controlled by the goal-directed or habitual systems, but are instead Pavlovian responses, controlled by a *separate,* Pavlovian evaluator whose utilities are immediately devaluable (Dickinson and Dawson, 1987). A main feature of our theory is the elimination of this appeal to a third system. Our account of these data is thus more parsimonious, and also more directly applicable to Balleine et al.'s, (1995) demonstration of immediate devaluation even for unambiguously instrumental responses.

Various experimental studies could arbitrate between the two views of these issues. Notably, because we view non-devaluability prior to incentive learning as resulting from habitual rather than goal-directed control, we predict that lesions disabling the habit system (Killcross and Coutureau, 2003) should render instrumental actions immediately susceptible to devaluation, eliminating the need for re-exposure. (The standard view predicts no such effect.) Further, in our view, the crucial aspect of re-exposure in an incentive learning experiment is its effect on *uncertainty* about the utility of an outcome rather than, as normally assumed, on the estimate of the utility itself. This may explain why incentive learning is dopamine-independent (Dickinson et al., 2000) – other, Pavlovian experiments testing manipulations of uncertainty instead implicate another neuromodulator, acetylcholine (Holland, 1997; Dayan and Yu, 2001).

We suggest that some of the same brain areas implicated in those studies may also play a role in incentive learning.

The model also has testable ramifications for the effects of devaluation on the responses of dopamine neurons. Dopamine responses to cues originate in the cached values, and should thus be insensitive to post-training changes in outcome value (*eg,* due to shifts in deprivation state). In contrast, dopamine responses to primary reward may be directly sensitive to devaluation, since they are based on the reward's observed utility. Neither of these predictions has been directly tested. However, the magnitude of the dopamine response to a cue has been shown to be correlated with the latency of a behavioural response triggered by that cue (Satoh et al., 2003), and this has been taken as evidence that the dopamine responses carry motivational information. Our theory would indeed expect correlations between trial-to-trial fluctuations in dopamine responding and behavioural vigour (for habitual actions), since in advantage learning, both are controlled by the cached values. What has not been established is what systematic factors (if any) drove the fluctuations measured in the experiment (Satoh et al., 2003); one possibility is that the dopamine response is modulated by the animal's overall *drive,* which is a more generalised form of motivation that is independent of outcome expectancy and thus could be reported even by a system ignorant of the identity or utility of the outcome.

In common with other theories, our account of the goal-directed system is sketchy. The existing data are radically insufficient to verify whether devaluation-sensitive behaviours are actually products of a full, model-based search system, due to the extreme simplicity of the tasks used. There are many different algorithms for model learning and planning, and many shortcuts or hybrids that have advantages over either the actor/critic or full model search. Some of the earliest work on RL suggested that pruning and caching be combined, with cached values standing in for unexplored subtrees (Samuels, 1959); more recently, uncertainty has been used to decide where to make this substitution (Baum and Smith, 1997). Applied to the model, these ideas would suggest a rich interaction between the goal-directed and habitual controllers – perhaps richer than the relative independence suggested by the lesion studies. Another hybrid is the successor representation (Dayan, 1993), which was used in a previous RL model of goal-directed action (Dayan, 2002). This algorithm uses caching, but of expected *outcome identities* rather than policies or values. Because of this, it can adapt immediately to outcome devaluations such as those in the experiments discussed here, but not to manipulations changing the transition structure of the task (as by rearranging walls in a maze) that would invalidate the cached outcomes. The latter sort of manipulation was pioneered by Tolman, (1932), but has seen little testing in modern times. Understanding the capabilities and limitations of the goal-directed system will require much more experimentation studying how organisms can react to different sorts of contingency changes in more complicated sequential decision tasks.

Also, we are at the earliest stages of understanding the factors governing uncertainty in the two systems and the way this influences arbitration. For instance, an alternative to our suggestion that inaccuracy accumulates due to approximate tree search is that whether exact or approximate, tree search incurs costs of energetic or computational resources (again accumulating over each step). This could, for deeper searches, outweigh the cost of potentially reduced utility that would result from abandoning the search altogether and falling back on the less accurate caching system. A distinct possibility for differential uncertainty that would favour the cache is that the two systems might be tuned to differing expectations about how quickly the estimated aspects of the world (transitions, utilities) are *changing.* The possibility of change is an ongoing source of uncertainty, and a system that assumed faster change would be more uncertain asymptotically but quicker to adapt in situations when the world actually changed (as for the Kalman filter, Anderson and Moore, 1979). The goal-directed system, whose comparative advantage is most apparent when there is only little data, could have a built-in bias for transience, and therefore high asymptotic uncertainties, whereas the habit system, which anyway requires more substantial training, might have a built-in bias for stability and correspondingly lower asymptotic uncertainties. Finally, although we have spoken as though arbitration is all-or-nothing, it is a straightforward and reasonable extension to assume behaviour is based on both controllers' predicted values, with contributions weighted by their uncertainty.

Another crucial issue for future work is connecting the well-understood theory of discrete, sequential decisions (which we have drawn on here as an idealisation) with the realistic experimental situation, in which animals are free to emit responses in continuous time and analysis typically focuses on the *rates* with which they do so. A more careful treatment of these issues will be required before we can address a further factor known to influence devaluation, which has to do with how payoffs are scheduled as a function of responding (interval vs. ratio schedules, Dickinson et al., 1983). This issue is, at present, better explained under a previous account of habit formation (Dickinson, 1985).

Though we have not specifically discussed the circumstances under which Pavlovian responses should be sensitive to devaluation (Holland and Rescorla, 1975; Holland and Straub, 1979), we expect the same considerations to apply as for instrumental action. The predictions triggering Pavlovian responses, like those driving choices, might originate either from model search or from cached values, with corresponding sensitivity or insensitivity to outcome devaluation. By eliminating the appeal to a separate and qualitatively different Pavlovian motivational system, our model opens the possibility of a unified account of Pavlovian and instrumental behaviours, in which both share a common forward model and common cached values. A major task for the future is to appraise this model in the light of the (equally complex) data on devaluation sensitivity in Pavlovian conditioning. Do the same factors impact Pavlovian as instrumental devaluation? Are the systems really shared? And if not, how do they interact? Of special relevance will be two paradigms that couple Pavlovian and instrumental conditioning (and also involve dopamine and ventral striatum), Pavlovian-instrumental transfer (Lovibond, 1983) and conditioned reinforcement (Williams and Dunn, 1994).

Finally, what of the basal ganglia? We suggest that the new theory delimits more precisely the role of the dorsal striatum in habitual behaviour, and also the roles of dopamine and of the ventral striatum's value predictions in learning those habits. We have not discussed dopamine's additional on-line, energising effect on behaviour (as in Pavlovian-instrumental transfer Dickinson et al., 2000). This likely also involves values learned in the ventral striatum, if only through their influence over the dopamine projection to the dorsal habit areas (Haber et al., 2000). Rounding out the cornucopia of open concerns is the observation that the evaluation of search trees in the goal-directed system may tax working memory (Corbit and Balleine, 2003). This suggests the unexplored possibility that even the goal-directed system may be affected by dopamine (and by striatal influences on it), which is thought to have a specific role in prefrontal mnemonic function (Williams and Goldman-Rakic, 1995; Cohen et al., 2002).

ACKNOWLEDGEMENTS

We are grateful to Tony Dickinson, Peter Holland, Daphna Joel, and Maneesh Sahani for helpful discussions, and to Rudolf Cardinal, Aaron Courville, Daphna Joel, and Genela Morris for thoughtful comments on an earlier draft. The authors are supported by the Gatsby Foundation, the EU BIBA project (PD), a Royal Society USA Research Fellowship (ND), and the Interdisciplinary Center for Neural Computation (YN).

REFERENCES

Adams CD (1982) Variations in the sensitivity of instrumental responding to reinforcer devaluation. *Quarterly Journal of Experimental Psychology* **34***B*:77-98.

Adams CD, Dickinson A (1981) Instrumental responding following reinforcer devaluation. *Quarterly Journal of Experimental Psychology* **33***B*:109-112.

Alexander GE, DeLong MR, Strick PL (1986) Parallel organization of functionally segregated circuits linking basal ganglia and cortex. *Annu Rev Neurosci* **9**:357-381.

Anderson BDO, Moore JB (1979) Optimal Filtering. NJ: Prentice-Hall.

Baird LC (1994) Reinforcement learning in continuous time: Advantage updating. In: Proceedings of the International Conference on Neural Networks. Orlando, FL.

Balleine B (1992) Instrumental performance following a shift in primary motivation depends on incentive learning. *J Exp Psychol Anim Behav Process* **18**:236-250.

Balleine B, Dickinson A (1991) Instrumental performance following reinforcer devaluation depends upon incentive learning. *Quarterly Journal of Experimental Psychology* **43***B*:279-296.

Balleine B, Dickinson A (1992) Signalling and incentive processes in instrumental reinforcer devaluation. *Quarterly Journal of Experimental Psychology* **45***B*:285-301.

Balleine BW (2000) Incentive processes in instrumental conditioning. In: *Handbook of Contemporary Learning Theories* (Klein RRMaSB, ed), pp 307-366. Mahwah, NJ: Lawrence Erlbaum.

Balleine BW, Dickinson A (1998) Goal-directed instrumental action: contingency and incentive learning and their cortical substrates. *Neuropharmacology* **37**:407-419.
Balleine BW, Dickinson A (2000) The effect of lesions of the insular cortex on instrumental conditioning: evidence for a role in incentive memory. *J Neurosci* **20**:8954-8964.
Balleine BW, Killcross AS, Dickinson A (2003) The effect of lesions of the basolateral amygdala on instrumental conditioning. *J Neurosci* **23**:666-675.
Balleine BW, Garner C, Gonzalez F, Dickinson A (1995) Motivational control of heterogeneous instrumental chains. *Journal of Experimental Psychology: Animal Behavior Processes* **21**:203-217.
Barto AG (1995) Adaptive critics and the basal ganglia. In: *Models of Information Processing in the Basal Ganglia* (J. C. Houk JLD, and D. G. Beiser, ed), pp 215-232. Cambridge, MA: MIT Press.
Barto AG, Sutton RS, Anderson CW (1983) Neuron-like adaptive elements that can solve difficult learning control problems. *IEEE Transactions on Systems, Man, and Cybernetics SMC*-**13**:834-846.
Baum EB, Smith WD (1997) A Bayesian approach to relevance in game playing. *Artificial Intelligence* **97**:195-242.
Blundell P, Hall G, Killcross S (2001) Lesions of the basolateral amygdala disrupt selective aspects of reinforcer representation in rats. *J Neurosci* **21**:9018-9026.
Blundell P, Hall G, Killcross S (2003) Preserved sensitivity to outcome value after lesions of the basolateral amygdala. *J Neurosci* **23**:7702-7709.
Cardinal RN, Parkinson JA, Hall J, Everitt BJ (2002a) Emotion and motivation: the role of the amygdala, ventral striatum, and prefrontal cortex. *Neurosci Biobehav Rev* **26**: 321-352.
Cardinal RN, Parkinson JA, Lachenal G, Halkerston KM, Rudarakanchana N, Hall J, Morrison CH, Howes SR, Robbins TW, Everitt BJ (2002b) Effects of selective excitotoxic lesions of the nucleus accumbens core, anterior cingulate cortex, and central nucleus of the amygdala on autoshaping performance in rats. *Behav Neurosci* **116**: 553-567.
Clark L, Cools R, Robbins TW (2004) The neuropsychology of ventral prefrontal cortex: decision-making and reversal learning. *Brain Cogn* **55**:41-53.
Cohen JD, Braver TS, Brown JW (2002) Computational perspectives on dopamine function in prefrontal cortex. *Curr Opin Neurobiol* **12**:223-229.
Colwill RM, Rescorla RA (1985) Instrumental responding remains sensitive to reinforcer devaluation after extensive training. *Journal of Experimental Psychology: Animal Behavior Processes* **11**:520-526.
Corbit LH, Balleine BW (2003) The role of prelimbic cortex in instrumental conditioning. *Behav Brain Res* **146**:145-157.
Coutureau E, Killcross S (2003) Inactivation of the infralimbic prefrontal cortex reinstates goal-directed responding in overtrained rats. *Behav Brain Res* **146**:167-174.
Daw ND (2003) Reinforcement Learning Models of the Dopamine System and Their Behavioral Implications. In: PhD thesis, School of Computer Science: Carnegie Mellon University.
Daw ND, Kakade S, Dayan P (2002) Opponent interactions between serotonin and dopamine. *Neural Netw* **15**:603-616.
Daw ND, Courville AC, Touretzky DS (2003) Timing and partial observability in the dopamine system. In: *Advances in Neural Information Processing Systems*, pp 99-106. Cambridge, MA: MIT Press.
Dayan P (1993) Improving generalisation for temporal difference learning: The successor representation. *Neural Computation* **5**:613-624.
Dayan P (2002) Motivated reinforcement learning. In: Advances in Neural Information Processing Systems (T. Dietterich SB, and Z. Ghahramani, ed), pp 11-18. Cambridge, MA: MIT Press.
Dayan P, Yu A (2001) ACh, uncertainty, and cortical inference. *Advances in Neural Information Processing Systems* **14**:189-196.
Dayan P, Balleine BW (2002) Reward, motivation, and reinforcement learning. *Neuron* **36**:285-298.
Dayan P, Kakade S, Montague PR (2000) Learning and selective attention. *Nat Neurosci* **3** Suppl:1218-1223.
Dearden R, Friedman N, Russell SJ (1998) Bayesian Q-learning. In: Proceedings of the 15th National Conference on Artificial Intelligence (AAAI), pp 761-768.
Dearden R, Friedman N, Andre D (1999) Model based Bayesian exploration. In: *Proceedings of the 15th Conference on Uncertainty in Artificial Intelligence (UAI)*, pp 150-159.
Dickinson A (1985) Actions and habits - the development of behavioural autonomy. Philosophical Transactions of the Royal Society of London Series B - *Biological Sciences* **308**:67-78.
Dickinson A (1994) Instrumental conditioning. San Diego: Academic Press.

Dickinson A, Dawson GR (1987) Pavlovian processes in the motivational control of instrumental performance. *Quarterly Journal of Experimental Psychology* **39***B*:201-213.

Dickinson A, Balleine B (1994) Motivational control of goal-directed action. *Animal Learning and Behavior* **22**:1-18.

Dickinson A, Balleine B (2002) The role of learning in motivation. In: *Learning, Motivation and Emotion.* (Gallistel CR, ed). Wiley, New York: Stevens' Handbook of Experimental Psychology (3rd ed.).

Dickinson A, Nicholson DJ, Adams CD (1983) The effect of the instrumental contingency on susceptibility to reinforcer devaluation. *Quarterly Journal of Experimental Psychology* **35***B*:249-263.

Dickinson A, Smith J, Mirenowicz J (2000) Dissociation of Pavlovian and instrumental incentive learning under dopamine antagonists. *Behav Neurosci* **114**:468-483.

Dickinson A, Balleine B, Watt A, Gonzalez F, Boakes RA (1995) Motivational control after extended instrumental training. *Animal Learning and Behavior* **23**:197-206.

Ernst MO, Banks MS (2002) Humans integrate visual and haptic information in a statistically optimal fashion. *Nature* **415**:429-433.

Gallagher M, McMahan RW, Schoenbaum G (1999) Orbitofrontal cortex and representation of incentive value in associative learning. *J Neurosci* **19**:6610-6614.

Haber SN, Fudge JL, McFarland NR (2000) Striatonigrostriatal pathways in primates form an ascending spiral from the shell to the dorsolateral striatum. *J Neurosci* **20**:2369-2382.

Hall J, Parkinson JA, Connor TM, Dickinson A, Everitt BJ (2001) Involvement of the central nucleus of the amygdala and nucleus accumbens core in mediating Pavlovian influences on instrumental behaviour. *Eur J Neurosci* **13**:1984-1992.

Holland PC (1997) Brain mechanisms for changes in processing of conditioned stimuli in Pavlovian conditioning: Implications for behavior theory. *Animal Learning and Behavior* **25**:373-399.

Holland PC (2004) Relations between Pavlovian-instrumental transfer and reinforcer devaluation. *Journal of Experimental Psychology: Animal Behavior Processes* **30**: 104-117.

Holland PC, Rescorla RA (1975) The effect of two ways of devaluing the unconditioned stimulus after first- and second-order appetitive conditioning. *Journal of Experimental Psychology: Animal Behavior Processes* **1**:355-363.

Holland PC, Straub JJ (1979) Differential effect of two ways of devaluing the unconditioned stimulus after Pavlovian appetitive conditioning. *Journal of Experimental Psychology: Animal Behavior Processes* **5**:65-78.

Horvitz JC (2000) Mesolimbocortical and nigrostriatal dopamine responses to salient non-reward events. *Neuroscience* **96**:651-656.

Houk JC, Adams JL, Barto AG (1995) A model of how the basal ganglia generate and use neural signals that predict reinforcement. In: *Models of Information Processing in the Basal Ganglia* (J. C. Houk JLD, and D. G. Beiser, ed), pp 249-270. Cambridge, MA: MIT Press.

Joel D, Weiner I (2000) The connections of the dopaminergic system with the striatum in rats and primates: an analysis with respect to the functional and compartmental organization of the striatum. *Neuroscience* **96**:451-474.

Joel D, Niv Y, Ruppin E (2002) Actor-critic models of the basal ganglia: new anatomical and computational perspectives. *Neural Netw* **15**:535-547.

Kakade S, Dayan P (2002) Dopamine: Generalization and bonuses. *Neural Networks* **15**: 549-559.

Kearns M, Singh S (1999) Finite-sample rates of convergence for Q-learning and indirect methods. *Advances in Neural Information Processing Systems* **11**:996-1002.

Kearns M, Singh S (2000) Bias-variance error bounds for temporal difference updates. In: *Proceedings of the 13th Annual Conference on Computational Learning Theory*, pp 142-147.

Kilcross S, Blundell P (2002) Associative representations of emotionally significant outcomes. In: *Emotional Cognition: From Brain to Behaviour* (Oaksford SMaM, ed), pp 35-73. Amsterdam: John Benjamins.

Killcross S, Coutureau E (2003) Coordination of actions and habits in the medial prefrontal cortex of rats. *Cereb Cortex* **13**:400-408.

Lindley DV (1985) Reconciliation of discrete probability distributions. In: *Bayesian Statistics* (J. M. Bernado MHD, D. V. Lindley, and A. F. M. Smith, ed), pp 375-390. Amsterdam: Elsevier Science Publishers B.V.

Lovibond PF (1983) Facilitation of instrumental behavior by a Pavlovian appetitive conditioned stimulus. *Journal of Experimental Psychology: Animal Behavior Processes* **9**:225-247.

Mannor S, Simester D, Sun P, Tsitsiklis JN (2004 (in press)) Bias and variance in value function estimation. In: *Proceedings of the 21st International Conference on Machine Learning (ICML).*
Miller EK, Cohen JD (2001) An integrative theory of prefrontal cortex function. *Annu Rev Neurosci* **24**:167-202.
Mirenowicz J, Schultz W (1996) Preferential activation of midbrain dopamine neurons by appetitive rather than aversive stimuli. *Nature* **379**:449-451.
Montague PR, Dayan P, Sejnowski TJ (1996) A framework for mesencephalic dopamine systems based on predictive Hebbian learning. *J Neurosci* **16**:1936-1947.
O'Doherty J, Dayan P, Schultz J, Deichmann R, Friston K, Dolan RJ (2004) Dissociable roles of ventral and dorsal striatum in instrumental conditioning. *Science* **304**:452-454.
O'Doherty J, Rolls ET, Francis S, Bowtell R, McGlone F, Kobal G, Renner B, Ahne G (2000) Sensory-specific satiety-related olfactory activation of the human orbitofrontal cortex. *Neuroreport* **11**:893-897.
Owen AM (1997) Cognitive planning in humans: neuropsychological, neuroanatomical and neuropharmacological perspectives. *Prog Neurobiol* **53**:431-450.
Packard MG, Knowlton BJ (2002) Learning and memory functions of the Basal Ganglia. *Annu Rev Neurosci* **25**:563-593.
Redgrave P, Prescott TJ, Gurney K (1999) The basal ganglia: a vertebrate solution to the selection problem? *Neuroscience* **89**:1009-1023.
Rescorla RA (1994) A note on depression of instrumental responding after one trial of outcome devaluation. *Quarterly Journal of Experimental Psychology* **47***B*:27-37.
Samuels AL (1959) Some studies in machine learning using the game of checkers. IBM *Journal of Research and Development* **3**:210-229.
Satoh T, Nakai S, Sato T, Kimura M (2003) Correlated coding of motivation and outcome of decision by dopamine neurons. *J Neurosci* **23**:9913-9923.
Schoenbaum G, Chiba AA, Gallagher M (1998) Orbitofrontal cortex and basolateral amygdala encode expected outcomes during learning. *Nat Neurosci* **1**:155-159.
Schultz W (1992) Activity of dopamine neurons in the behaving primate. In: *Seminars in the Neurosciences*, pp 129-138.
Schultz W (1998) Predictive reward signal of dopamine neurons. *J Neurophysiol* **80**:1-27.
Schultz W, Apicella P, Ljungberg T (1993) Responses of monkey dopamine neurons to reward and conditioned stimuli during successive steps of learning a delayed response task. *J Neurosci* **13**:900-913.
Schultz W, Dayan P, Montague PR (1997) A neural substrate of prediction and reward. *Science* **275**:1593-1599.
Singh SP, Dayan P (1998) Analytical mean squared error curves in Temporal Difference learning. *Machine Learning* **32**:5-40.
Suri RE, Schultz W (2001) Temporal difference model reproduces anticipatory neural activity. *Neural Comput* **13**:841-862.
Sutton RS (1984) Temporal Credit Assignment in Reinforcement Learning. In: PhD thesis. Amherst, MA: University of Massachusetts.
Sutton RS (1988) Learning to predict by the method of temporal differences. *Machine Learning* **3**:9-44.
Sutton RS, Barto AG (1998) Reinforcement Learning: An Introduction. Cambridge, MA: MIT Press.
Tolman EC (1932) Purposive Behavior in Animals and Men. New York: Appleton-Century-Crofts.
Tremblay L, Schultz W (1999) Relative reward preference in primate orbitofrontal cortex. *Nature* **398**:704-708.
Ungless MA, Magill PJ, Bolam JP (2004) Uniform inhibition of dopamine neurons in the ventral tegmental area by aversive stimuli. *Science* **303**:2040-2042.
Voorn P, Vanderschuren LJ, Groenewegen HJ, Robbins TW, Pennartz CM (2004) Putting a spin on the dorsal ventral divide of the striatum. *Trends Neurosci* **27**:468-474.
Wickens JR, Begg AJ, Arbuthnott GW (1996) Dopamine reverses the depression of rat corticostriatal synapses which normally follows high-frequency stimulation of cortex in vitro. *Neuroscience* **70**:1-5.
Williams BA, Dunn R (1994) Conditioned reinforcement: Neglected or outmoded explanatory construct? *Psychonomic Bulletin and Review* **1**:457-475.
Williams GV, Goldman-Rakic PS (1995) Modulation of memory fields by dopamine D1 receptors in prefrontal cortex. *Nature* **376**:572-575.
Yin HH, Knowlton BJ, Balleine BW (2004) Lesions of dorsolateral striatum preserve outcome expectancy but disrupt habit formation in instrumental learning. *Eur J Neurosci* **19**:181-189.

Yuille AL, Bültoff HH (1996) Bayesian decision theory and psychophysics. New York: Cambridge University Press.

In: Recent Breakthroughs in Basal Ganglia Research
Editor: Erwan Bezard, pp. 107
ISBN: 1-59454-880-3

PART III: THE PEDUNCULOPONTINE TEGMENTAL NUCLEUS AND THE BASAL GANGLIA

Masaru Matsumura
Neuroscience and Neurosurgery, Chuo Gunma Neurosurgery Hospital 64-1, Nakao-machi, Takasaki, Gunma, 370-0001, Japan

It was at the International Basal Ganglia Society Meeting, Nemunosato, Mie, Japan in 1995, that Prof. S. T. Kitai proposed to invite the pedunculopontine tegmental nucleus (Ppn) as a member of the basal ganglia. The dense connections between the basal ganglia and this nucleus had been well documented, however, the majority of the attendees accepted to welcome this nucleus as "a cousin" or "a brother or a sister" of the basal ganglia. During the last decade, several groups performed enthusiastic experiments of the Ppn and the position of this nucleus has been determined.

The Ppn had been well examined anatomically and physiologically in rodents, however, the studies in non-human primates had been rare. Matsumura and his group was the first to study the physiological and behavioral aspects of the Ppn in monkeys. They recorded the single-cell activity in awake monkeys and presented how this nucleus contributes to the voluntary limb movements. Their results are reviewed and a circuitry of the basal ganglia with the Ppn is discussed in his chapter.

In the course of their study, they found a unique, in other word, a paradoxical feature of this nucleus in relation to the Parkinson's disease. A kainic acid-induced lesioning of the Ppn resulted in parkinsonism. But the neurotoxin, MPTP, failed to induce further parkinsonism in monkeys which had received Ppn lesions. This paradox is discussed in the second chapter.

In the third chapter, one of the classical feature of the Ppn, the control of locomotion and postural muscle tone is discussed in relation to the basal ganglia. The authors argue for a role of basal ganglia projections to the Ppn in the automatic control of movement that occurs in conjunction with voluntary control processes. Sustained output from the basal ganglia to the thalamocortical loop and the brainstem may regulate the degree of freedom of movements by controlling the excitability of the target systems.

The Ppn has been believed to be a simple relay between the cerebral cortex and the spinal cord, acting as a brainstem control center for automated motor behaviors, such as locomotion. Today, there are another concerns that the Ppn functions as a modulatory center that receives the excitatory drive from the cortex and controls the activity of the basal ganglia and thalamus, possibly influencing cognitive behaviors including motor learning and reward system, as well as voluntary movement control.

Together with the thalamus and the somatomotor cortex, the Ppn stays as a member of the basal ganglia-related family

In: Recent Breakthroughs in Basal Ganglia Research
Editor: Erwan Bezard, pp. 109-121
ISBN: 1-59454-880-3

Chapter 9

THE BASAL GANGLIA CONTROL OF LOCOMOTION AND POSTURAL MUSCLE TONE

Kaoru Takakusaki**, *Kazuya Saitoh,
Hirohumi Harada and Makoto Kashiwayanagi
Department of Physiology, Asahikawa Medical College, Midorigaoka-Higashi, 2-1, Asahikawa, Japan

ABSTRACT

Here we hypothesize a conceptual framework that may provide a better understanding of the role of a GABAergic basal ganglia-brainstem system throughout the mesopontine tegmentum in the automatic integration of postural muscle tone and locomotion that occurs in conjunction with voluntary control processes. Excitability of the muscle tone inhibitory system and the locomotion executing system in the brainstem-spinal cord can be steadily balanced by a net excitatory cortical input and a net inhibitory basal ganglia input to these systems. Based on this framework, we propose that dysfunction of the basal ganglia-brainstem system together with that of cortico-basal ganglia loop underlies the pathogenesis of motor disturbances expressed in basal ganglia diseases.

INTRODUCTION

Marsden, (1982) has hypothesized that "the basal ganglia are responsible for the autonomic execution of learned motor plans". This suggests that the basal ganglia are involved in both the process of a learned motor act and expression of the automatic execution of movements. Voluntary movements are always associated with unconscious automatic control processes. For example, the initiation and termination of locomotion, and avoiding obstacles during locomotion, are volitional processes that require accurate control, and part of this control is automatic. Similarly, we are largely unaware of the automatic control of rhythmic limb movements, postural muscle tone, and the postural reflexes that accompany locomotion. The fact that each aspect of locomotion is seriously impaired in Parkinsonian patients (Murray et al., 1978; Morris et al., 1994) indicates that the basal ganglia must play a crucial role in integrating the volitional and automatic aspects of the control of posture and locomotion.

The basal ganglia and cerebellar loops with the motor areas of the cerebral cortex are involved in the volitional control of movement (Middleton and Strick, 2000). In addition, the basal ganglia contribute to the planning and execution of voluntary movements via a series of parallel basal ganglia thalamocortical loops (Alexander and Crutcher, 1990; DeLong, 1990; Turner and Anderson, 1997). However the role of the basal ganglia in the automatic aspects of regulating muscle tone and rhythmic limb movements during

* Address all corresponding to Kaoru Takakusaki, M.D., Ph.D. Department of Physiology, Asahikawa Medical College, Midorigaoka-Higashi 2-1, Asahikawa, 078-8510, Japan. Tel: +81-166-68-2331 / Fax: +81-166-68-2339, E-mail address: kusaki@asahikawa-med.ac.jp

locomotion is less understood and infrequently discussed. Recently emphasis has been given to the importance of the direct basal ganglia outflow to the brainstem regions (Garcia-Rill, 1991; Inglis and Winn, 1995; Reese et al., 1995; Pahapill and Lozano, 2000; Takakusaki et al., 2003b; Mena-Segovia et al., 2004; Takakusaki et al., 2004a) where fundamental neuronal networks for controlling muscle tone and locomotion (Grillner, 1981; Mori, 1987; Garcia-Rill, 1991; Rossignol, 1996) are located. Because the pedunculopontine tegmental nucleus (PPN) and its adjacent areas include the midbrain locomotor region (MLR; Grillner, 1981) and a region inhibiting muscle tone (Lai and Siegel, 1990; Takakusaki et al., 2003b; 2004c; 2004d), it is critical to establish how the basal ganglia-brainstem system controls muscle tone and locomotion.

This review considers a conceptual framework that may provide a better understanding of the role of the GABAergic basal ganglia–brainstem system in the integration of postural muscle tone and locomotion. We first consider the neural bases of the control of postural muscle tone and locomotion, and their regulation by GABAergic output from the basal ganglia. In the second part, the general principles that underline expression of motor behaviors played by the basal ganglia output are considered. Finally the functional significance of this system in relation to the pathogenesis of motor disturbances in basal ganglia disorders is discussed.

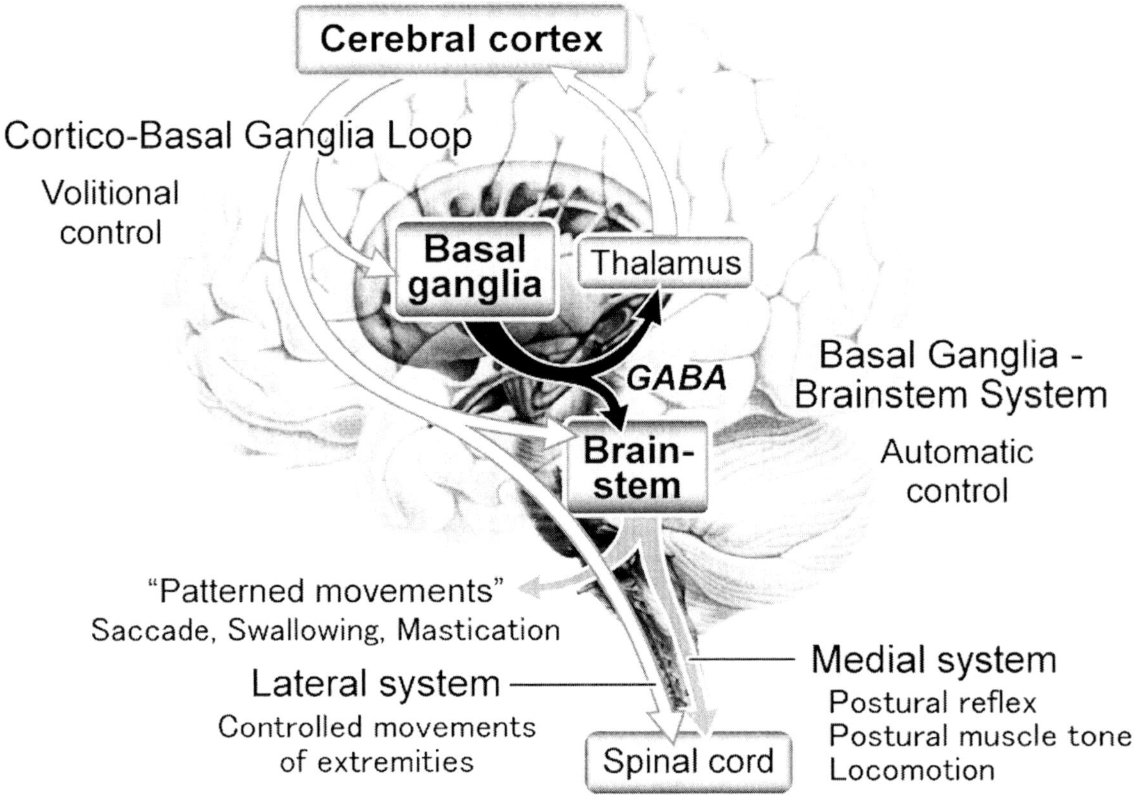

Figure 1. Volitional and automatic control of locomotor movements. The GABAergic output of the basal ganglia to the thalamocortical neurons and the brainstem neurons integrate the volitional and automatic control processes of movements. See text for further explanation. Abbreviations: GABA, gamma-amino-butyric acid.

1 A General Framework for the Control of Movements by the Basal Ganglia

Hikosaka et al., (2000) have proposed that by using two kinds of output the basal ganglia have two ways to control movements. One method of controlling movement would be via the thalamocortical networks, and the other via brainstem motor networks. The outputs from the basal ganglia are schematically illustrated in Fig.1. The basal ganglia output to the cerebral cortex through the thalamocortical loop may be responsible for the volitional control of movements (Middleton and Strick, 2000). Precisely controlled movements of the extremities can be achieved via the corticospinal tract. On the other hand, particular patterns of movements such as a saccade, mastication, vocalization, swallowing, and locomotion, are thought to be generated by specific neuronal networks in the brainstem and spinal cord (see Grillner, 2003). The basal ganglia output to the networks in the brainstem and the spinal cord for these movements could be involved in the automatic control that accompanies the voluntary movements. In this article, we emphasize the importance of GABAergic basal ganglia projections, via the substantia nigra pars reticulata (SNr), to the mesopontine tegmentum for integration of the control of postural muscle tone and locomotion.

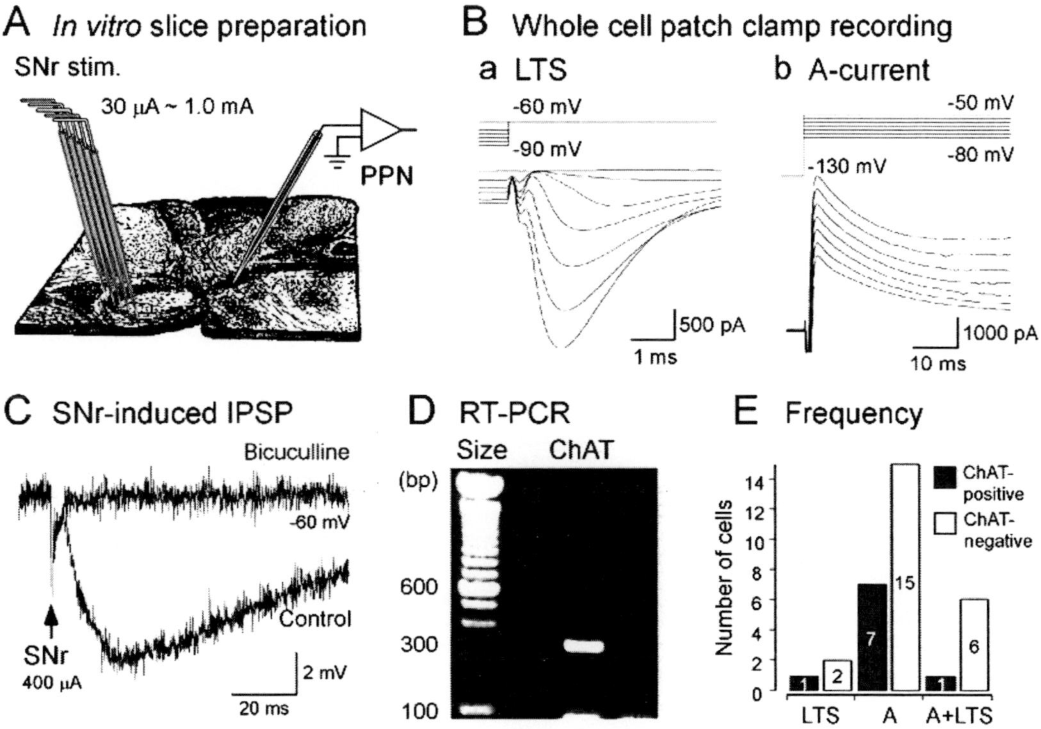

Figure 2. Evidence of GABAergic inhibitory effects upon PPN neurons. A. The experimental setup for the *in vitro* rat brainstem slice preparation. Multiple stimulating electrodes were placed on the SNr area. A recording patch pipette was inserted into the PPN area. B. Electrophysiological identification of a PPN neuron. (a) In the voltage clamp recordings, transient inward currents, low-threshold Ca^{2+} spikes (LTS), were displayed by applying a test pulse-potential of –60 mV after a prepulse-potential that was varied between –90 to –60 mV. (b) Transient outward currents (A-currents) were also observed following a hyperpolarizing prepulse of up to –130 mV which was then stepped to a series of potentials from –80 to –50 mV. C. Stimulation of the SNr (indicated by an arrow) induced an IPSP in the presence of antagonists for ionotropic glutamate receptors (50 μM AP-5 and 10 μM CNQX). The IPSP was diminished after an application of bicuculline (30 μM). D. Choline acetyltransferase (ChAT) mRNA was detected with a single-cell RT-PCR procedure. E. The frequency of the ChAT mRNA positive PPN neurons (hatched bar) and negative PPN neurons (open bar) in relation to the electrophysiological phenotypes. Abbreviations: IPSP, inhibitory postsynaptic potential; SNr, substantia nigra pars reticulata.

1.1 Basal Ganglia Outflow to the Brainstem

The basal ganglia are more highly interconnected with the mesopontine tegmentum, including the PPN, than with any other brain region (Mena-Segovia et al., 2004). The PPN receives efferents of the basal ganglia, and particularly from the SNr in rats (Beckstead et al., 1979; Spann and Grofova, 1991) and cats (Moriizumi et al., 1988). In the PPN cholinergic (Armstrong et al., 1983; Rye et al., 1987; Spann and Grofova, 1992) and non-cholinergic neurons are intermingled. The latter include glutamatergic (Clements et al., 1991), GABAergic (Ottersen and Storm-Mathisen, 1984; Krosaka et al., 1988), and peptidergic (e.g.,

Substance P; Vincent et al., 1983) neurons. It has been suggested that the nigrotegmental efferents use GABA as a neurotransmitter and have terminals preferentially on non-cholinergic neurons rather than on cholinergic neurons (Grofova and Zhou, 1998). Saitoh et al., (2003) recently demonstrated *in vitro* rat brain slice preparation (Fig.2A), that GABAergic efferents, possibly from the SNr, induced inhibitory effects upon PPN neurons (Fig.2C). The electrophysiological membrane properties of the PPN neurons were characterized by the presence of either low threshold calcium spikes (LTS), or transient outward current (A-current), or both (Fig.2B; Kang and Kitai, 1990; Takakusaki and Kitai, 1997; Saitoh et al., 2003). Stimulation of the SNr induced monosynaptic IPSPs which were diminished by an application of bicuculline, a $GABA_A$ receptor antagonist (Fig.2C). A single-cell, RT-PCR amplification technique, revealed that approximately 30 % of the neurons were cholinergic in nature (Fig.2D–E). These findings suggest that both cholinergic and non-cholinergic neurons in the PPN receive GABAergic efferents from the SNr.

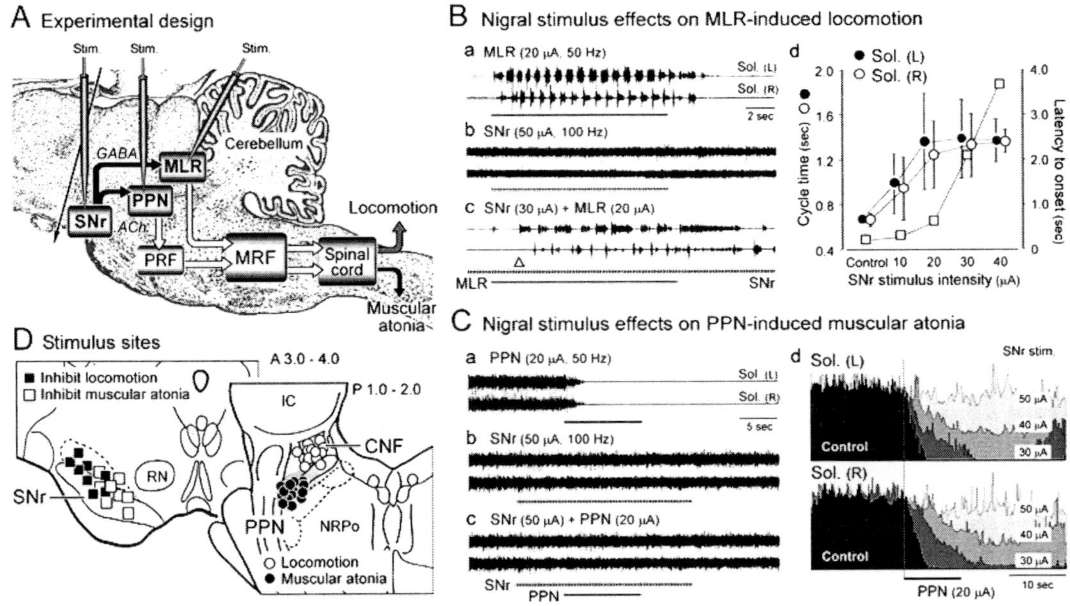

Figure 3. Nigral stimulus effects upon locomotion and muscle tone suppression. A. Experimental diagram. Electrical stimuli were delivered to the MLR, the PPN and the SNr. B. SNr stimulus effects upon MLR-induced locomotion. (a) Locomotion on a moving treadmill induced by stimulating the MLR. (b) SNr stimulation did not change the level of muscle tone. (c) Conditioning SNr stimuli, which were delivered throughout the recording period, reduced step cycles, delayed the onset (indicated by open arrowhead) and disturbed rhythmic alteration of limb movements during locomotion. (d) Changes in locomotor parameters. By increasing the stimulus intensity, the cycle time (filled and open circles) of locomotion was increased and the locomotor onset (hatched squares) was delayed. C. SNr stimulus effects upon PPN-effects. (a) PPN stimulation (20 μA) suppressed muscle tone. (b) SNr stimulation did not alter the level of muscle tone. (c) The PPN-effect was abolished by conditioning SNr stimuli. (d) the EMG activity of the left (upper panel) and right (lower panel) soleus muscles were rectified and integrated. PPN stimulation abolished muscle tone within 5 seconds (control). However, SNr stimulation with 30 μA delayed the onset of the PPN-induced atonia. PPN stimulation reduced the level of muscle tone, but did not abolished it when the SNr was stimulated with 40 μA. PPN stimulation did not alter the level of muscle tone when SNr stimuli with 50 μA were delivered. D. Right panel. Optimal stimulus sites for evoking locomotion (open circles) and muscular atonia (filled circles) on a coronal plane of the brainstem. Most of the locomotion sites were located in the cuneiform nucleus (CNF), and most of the muscular atonia stimulus sites were located in the ventrolateral part of the PPN. The hatched area in the PPN indicates where cholinergic neurons were abundantly distributed. Left panel. Optimal nigral stimulus sites in the SNr for suppression of the MLR-induced locomotion (open squares) and the PPN-induced muscular atonia (filled squares). Results in B-D are modified from Takakusaki et al. (2003a). Abbreviations: A, anterior; ACh, acetylcholine; IC, inferior colliculus; MLR, midbrain locomotor region; MRF, medullary reticular formation; NRPo, nucleus reticularis pontis oralis; P, posterior; PPN, pedunculopontine tegmental nucleus; PRF, pontine reticular formation; Sol, soleus muscle.

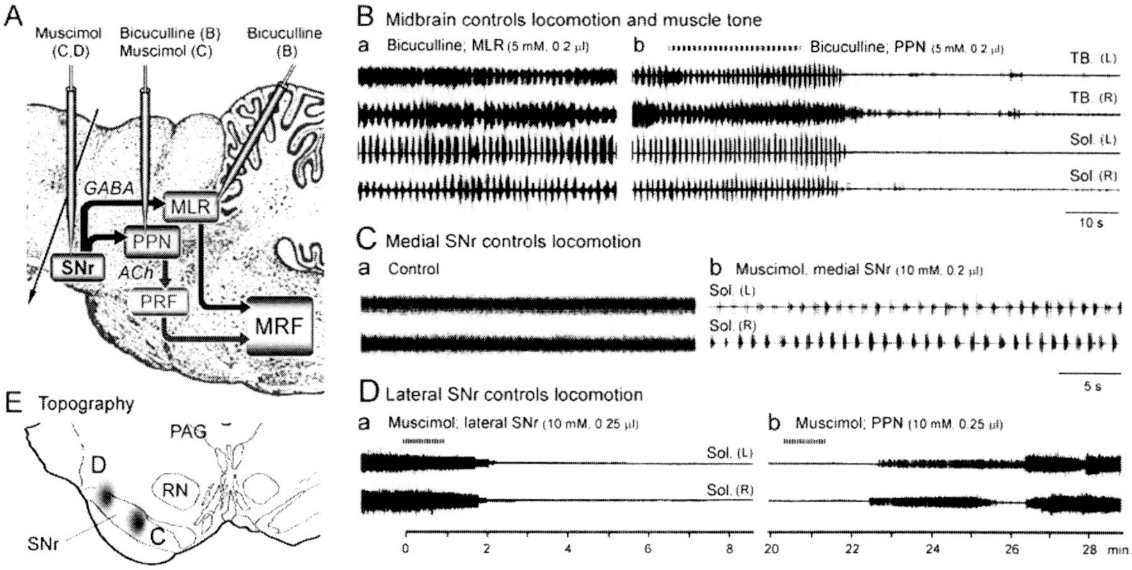

Figure 4. GABAergic nigrotegmental projections control locomotion and muscle tone. A. Experimental diagram. Chemical stimulation was delivered to the SNr, the MLR and the PPN. B. (a) Quadrupedal locomotion observed at 15 min after injecting bicuculline into the MLR. (b) Another injection of bicuculline into the PPN in the same cat suppressed the locomotion. A dashed line above the recording indicates the period of the injection. Muscle activity was recorded bilaterally from the triceps brachial (TB) muscles and soleus muscles. C. (a) Tonic contraction appeared bilaterally in the soleus muscles. (b) Inactivation of the medial SNr by a microinjection of muscimol generated locomotor activity. D. (a) Inactivation of the lateral SNr by injecting muscimol bilaterally abolished soleus muscle activity. (b) Twenty minutes after the SNr muscimol injection, muscimol was injected into the PPN. The level of muscle tone was re-established approximately 6 min after the PPN muscimol injection. E. Muscimol injection sites in the SNr.

1.2 Brainstem Areas Responsible for Generating Locomotion and Muscle Tone Suppression

There are two functionally identified regions in the lateral part of the mesopontine tegmentum; one is the MLR, and the other is a muscle tone inhibitory region (Fig.3A and D). In decerebrate cat preparation, where the striatum, thalamus, and cerebral cortex were removed but the SNr was preserved, repetitive electrical stimuli applied to the MLR elicited locomotion on a moving treadmill (Fig.3Ba). Repetitive electrical stimuli which were delivered to the inhibitory region in the PPN suppressed postural muscle tone which was due to decerebrate rigidity (Fig.3Ca). In Fig. 3D the effective sites for locomotion (open circles) and muscle tone suppression (filled circles) are shown on a coronal plane of the brainstem (right panel in Fig.3D). The locomotor evoking sites were dorsomedially located to the inhibitory sites. The former were mostly located in the cuneiform nucleus (CNF) and the latter were located in the ventrolateral part of the PPN where cholinergic neurons are abundant.

At least two major pathways descend from the MLR (Grillner, 1981; Mori, 1987; Rossignol, 1996). One is via reticulospinal neurons in the medial medullary reticular formation (MRF), and the other is via the pontomedullary locomotor strip which descends through the dorsolateral funiculus. These two pathways activate the locomotor central pattern generator (CPG) to produce the rhythm of locomotion. Signals from the MLR may also activate muscle tone facilitatory systems such as the raphespinal and coerulospinal tracts (see Fig.5A; Mori, 1987; Lai and Siegel, 1990). Cortical projections to the MLR have not been reported. It is possible that cortical signals are polysynaptically mediated through the subthalamic locomotor region (Rossignol, 1996). Because stimulation of the SLR elicited locomotion, even after bilateral lesions in the MLR, effects by which SLR stimulation could be mediated by direct projections from the SLR to the MRF (Rossignol, 1996). It has been reported that a patient with a lesion in the dorsal part of mesopontine tegmentum could not stand and walk, even though the patient did not have any pyramidal signs (Masdeu et al., 1994). This indicates that an MLR exists in the mesopontine tegmentum of humans.

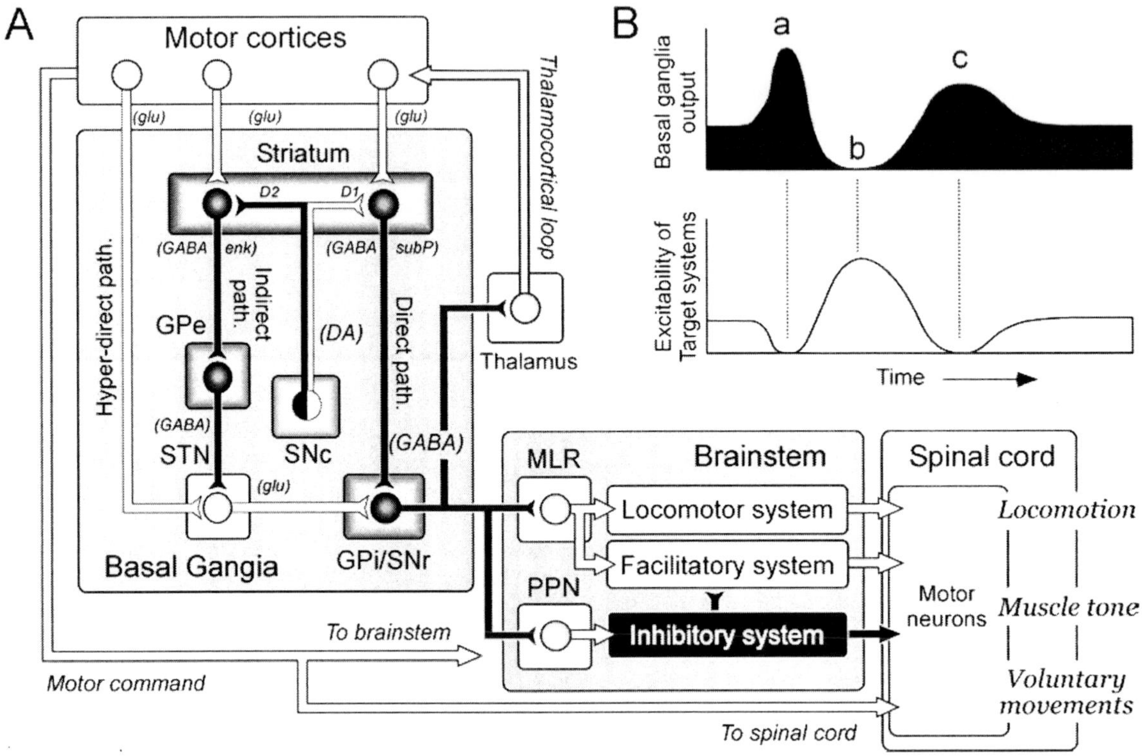

Figure 5. Possible neural connections between basal ganglia-thalamocortical loop and brainstem motor systems. A. A composite model of basal ganglia motor circuits and brainstem-spinal pathways for controlling saccades, locomotion, muscle tone and voluntary movements. B. Basal ganglia modulation of target motor systems. Upper and lower schematic graphs indicate the level of the basal ganglia output and the excitability of target systems, respectively. See text for further explanation. Abbreviations: D1, dopamine 1 receptors; D2, dopamine 2 receptors; DA, dopamine; enk, enkephaline; glu, glutamate; GPe, external segment of the globus pallidus; GPi, internal segment of the globus pallidus; SC, superior colliculus; SNc, substantia nigra pars compacta; STN, subthalamic nucleus; Sub P, substance P.

Neuroanatomical studies have shown that the PPN has descending projections to the pontine reticular formation (PRF, Mitani et al., 1988; Lai et al., 1993) and the medial part of the MRF (Shiromani et al., 1990; Lai et al., 1993). It has also been shown that either electrical or chemical stimulation applied to each reticular formation area suppressed muscle tone in a decerebrate cat preparation (see Takakusaki et al., 2004b). Because the PPN-induced atonia was blocked by injecting atropine sulfate into the PRF (Takakusaki et al., 2003b), the PPN effects may be mediated by the activation of the cholinoceptive PRF neurons. These neurons, in turn, may excite reticulospinal neurons in the MRF (Takakusaki et al., 1993; 1994; Fig.3A). The γ-motoneurons and interneurons intercalated in reflex pathways may possibly be suppressed in parallel (Takakusaki et al., 2001; 2003a). The inhibitory system also reduces the activity of coerulospinal neurons which comprise one of the descending excitatory systems (Mileykovskiy et al., 2000). Muscle tone suppression could therefore be induced by an activation of the inhibitory system together with an inactivation of the excitatory systems (see Fig.5A). The inhibitory system possibly regulates muscle tone during wakefulness and induces muscular atonia during rapid eye movement (REM) sleep (see discussion in Takakusaki et al., 2004a). Patients with a lesion in the dorsolateral mesopontine tegmentum have been reported to retain their muscle tone during REM sleep ("REM without atonia"; Culebras and Moore, 1989; Kimura et al., 2000). This observation indicates the presence of a muscle tone inhibitory region in the human midbrain.

1.3 Basal Ganglia Control of Locomotion and Muscle Tone

How do the SNr projections to the MLR/PPN control locomotion and muscle tone? To answer this question, we must examine the effects of stimulation of the SNr on MLR/PPN-induced movements (Fig.3C-D). Stimulation of the SNr alone did not alter muscular activity (Fig.3Bb). However stimuli applied to the medial part of the SNr reduced the number of MLR-activated step cycles, increased the duration of the stance phase, and disrupted the rhythmic alternation of limb movements (Fig.3Bc). Stimulation of the SNr at a higher strength eventually stopped MLR-activated locomotion (not illustrated). Furthermore, the onset of the locomotion was delayed by SNr stimuli of progressively increasing strength (Fig.3Bd). Consequently, the SNr-MLR projection affects both the steady state (e.g., postural control and rhythmic limb movements) and dynamic state (e.g., initiation and termination) of locomotion. On the other hand, conditioning stimuli applied to the lateral part of the SNr attenuated and blocked the PPN-induced muscular atonia (Fig.3Ca-c). A superimposition of the rectified soleus EMGs (Fig.3Cd) clearly demonstrated that, the PPN stimulation completely suppressed muscle tone within a few seconds. However, the latency to the muscular atonia was delayed and the degree of muscle tone suppression was reduced as the SNr stimulus strength was progressively increased from 30 to 50 μA. It should be noted that there was a partial functional topography in the SNr (left panel in Fig.3D). The neurons in the lateral SNr (filled squares) may control muscle tone, while those in the medial SNr (open squares) may control locomotion, via GABAergic nigrotegmental projections to the PPN and the MLR.

To further reveal whether the nigrotegmental projections were GABAergic in nature, a small amount (0.10–0.25 μl) of $GABA_A$ receptor agonist (muscimol) and antagonist (bicuculline) was injected into the mesopontine tegmentum and the SNr, so that changes in muscular activity could be examined (Fig.4A). An injection of bicuculline into the MLR elicited locomotion on a moving treadmill (Fig.4Ba). Successive injections of bicuculline into the ventrolateral PPN inhibited the locomotion and suppressed muscle tone (Fig.4Bb). In addition, a muscimol injection into the medial part of the SNr ("C" in Fig.4E) induced locomotion (Fig.4Cb; Takakusaki et al., 2004d). It was also observed that an injection of muscimol into the lateral part of the SNr ("D" in Fig.4E) reduced the level of muscle tone (Fig.4Da), and this was re-established by succeeding muscimol injections into the PPN (Fig.4Db). This evidence indicates that the locomotor system and the muscle tone inhibitory system are under tonic inhibitory control, from the GABAergic projections, from the SNr to the MLR/PPN (Fig.4A). Accordingly, our opinion is that the basal ganglia achieve control of postural muscle tone and locomotion by a combined inhibition/disinhibition of both the muscle tone inhibitory system, and the locomotion executing system, via the GABAergic nigrotegmental projections.

2 How Do the Basal Ganglia Control Movements?

2.1 The Basal Ganglia Motor Circuits and the Brainstem-Spinal Cord Motor Systems

The excitability of basal ganglia output from the SNr and internal segment of the globus pallidus (GPi) is thought to be regulated by hyper-direct, direct, and indirect, pathways in the basal ganglia circuits (Fig.5; Alexander and Crutcher, 1990; DeLong, 1990; Hikosaka et al., 2000; Nambu, 2004). Usually neurons in the GPi/SNr could inhibit target neurons in the thalamus and brainstem with their tonic, high background activity, thus preventing unnecessary movements (Fig.1; Hikosaka et al., 2000). To initiate movements, motor commands from the motor cortices first increase the basal ganglia output by an activation of the hyper-direct pathway from the cerebral cortices to the subthalamic nucleus (STN) so that the excitability of the target system would be further reduced (Nambu, 2004; Fig.5Ba). Signals via the direct pathway from the striatum to the GPi/SNr remove this sustained inhibition, resulting in a disinhibition of the target motor systems (Hikosaka et al., 2000; Fig.5Bb). The phasic activity of GABAergic output neurons in the striatum, which are mostly silent, interrupts the tonic GPi/SNr inhibition, and movements are allowed to occur (Hikosaka, 1989). Finally, signals through the indirect pathway, involving the external segment of the globus pallidus (GPe) and the STN, can further re-enhance the inhibition of the target systems (Fig.5Bc). This sequential information processing would enhance the temporal contrast of the excitability of the target motor system. Accordingly an enhancement of tonic inhibition and a release from the inhibition (disinhibition) are key mechanisms of the basal ganglia control of movements. Moreover, the sequential processing ensures that only the selected motor program is initiated, executed, and terminated, at the appropriate times, whereas other competing programs are cancelled.

The above mechanisms may act on brainstem networks, including the muscle tone control system and locomotor system, in addition to thalamocortical networks. Therefore these brainstem systems could be combined with basal ganglia motor circuits (Fig.5A). In the combined model, the basal ganglia output controls the MLR for locomotion and the inhibitory region in the PPN for muscle tone. Because the MLR and the PPN receive GABAergic input from the SNr, the locomotor system and muscle tone control system may be regulated by the hyper-direct, direct, and indirect pathways. When locomotor movement is being prepared, tonic activity of SNr neurons would continuously inhibit both systems. When a trigger signal occurred the hyper-direct pathway would enhance the inhibition. Then the direct pathway would release the activity of these systems, resulting in an initiation of locomotion that would be followed by a smooth reduction of the level of muscle tone. To terminate the locomotion, the direct pathway would inhibit each system, resulting in a cessation of rhythmic locomotor movements and an accompanying increase in the level of muscle tone. A parallel organization from the SNr to the MLR/PPN would be therefore assist regulation of the level of muscle tone which was appropriate for the initiation and termination of locomotion.

2.2 The Importance of Sustained Basal Ganglia Inhibition upon the Target Motor Systems; the Basal Ganglia May Control the Degree of Freedom of Movements

Cortico-basal ganglia loops are involved with the motor behavior that requires volition, cognition, and attention (Brooks, 1995; Middleton and Strick, 2000). The majority of motor cortical neurons significantly altered their discharge properties when a walking subject has to accurately overcome obstacles (Drew et al., 1996). This accuracy requires a precise visuomotor coordination (Georgopoulos and Grillner, 1989). Therefore cortical processing is required for the volitional aspects of locomotion. The cortico-basal ganglia loop could help to serve this purpose. In contrast, a BG-BS system seems required for the automatic regulation of postural muscle tone and rhythmic limb movements during locomotion. The motor cortices have projections to the PPN (Matsumura et al., 2000) and to the pontomedullary reticular formation (PMRF, Matsuyama and Drew, 1997). Therefore the muscle tone control system and the locomotor system can be controlled, in parallel, by a combined input to the brainstem of net inhibition from the basal ganglia and net excitation from the motor cortex (Fig. 6A).

Given the above consideration, the motor cortical neurons that receive basal ganglia output may control the velocity and the amount of voluntary movement (Turner and Anderson, 1997; ordinate on the left of the graph in Fig.6Ba). The GABAergic basal ganglia output to the MLR reduced the drive from the MLR, resulting in disruption of the activity of CPGs in the spinal cord. Basal ganglia efferents to the MLR may therefore control the locomotor pattern (ordinate on the right). In addition, basal ganglia efferents to the PPN may determine the level of muscle tone via the muscle tone control systems (abscissa). Because the basal ganglia output is variable in a normal condition, the degree of freedom for the amount and the velocity of movement, the locomotor velocity, and the muscle tone, can be large. Each parameter can take any of the coordinates within the frame in Fig.6Ba. Consequently, a sustained basal ganglia output to each target system may control the degree of freedom of the excitability of these systems during ongoing movements. For example, when a subject needs to adapt to a heavy load during walking, the subject may unconsciously select an appropriate gait pattern which is associated with a higher level of muscle tone and slower walking speed. Such a gait pattern could be realized by an increase in sustained SNr output to the PPN and the MLR. This would result in a decrease in the excitabilities of the muscle tone inhibitory system and locomotor rhythm generating system (Fig.6Bb). To the present, emphasis has not been given to the importance of sustained inhibitory output from the basal ganglia during ongoing saccadic and voluntary arm-hand movements. However, we emphasize a crucial role for the sustained output from the basal ganglia to the target motor systems (for example, the PPN and MLR) for controlling the steady-state of ongoing movements. These ongoing movements would include the maintenance of postural muscle tone and rhythmic limb movements during locomotion. We suggest that the basal ganglia may control the degree of freedom of movements by modulating the excitability of target motor systems.

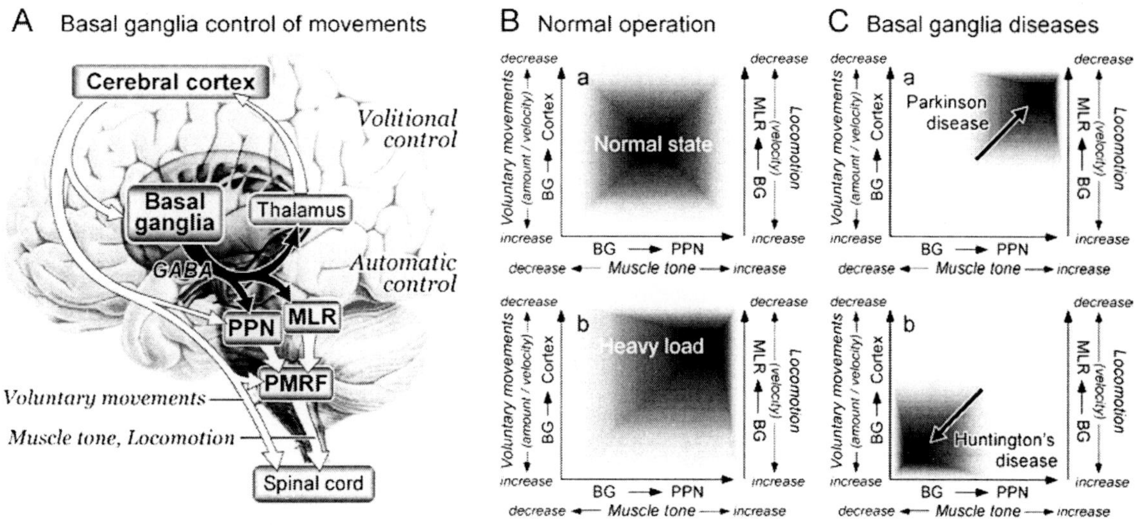

Figure 6. A hypothetical model for the control of movements by the basal ganglia. A. GABAergic basal ganglia projections to the thalamocortical neurons may be involved in the volitional control of movements, while those to the MLR and the PPN may be responsible for the automatic control processes of locomotor movements and postural muscle tone. B. Normal operation of the basal ganglia for control of voluntary movements, locomotion and muscle tone. C. Motor disturbances in basal ganglia diseases such as Parkinson disease (a) and Huntington's chorea (b). See text for explanation. Abbreviations: BG, Basal ganglia; PMRF, pontomedullary reticular formation.

3 Pathophysiology of Motor Disturbances of Basal Ganglia Diseases

The GABAergic basal ganglia output is thought to be overactive in Parkinson disease (Wichmann and DeLong, 1996, 2003). If the schemas in Figs.5 and 6 are appropriate, an excessive GABAergic inhibition upon thalamocortical neurons may decrease the velocity and amount of movement (bradykinesia and hypokinesia, respectively). An increase in basal ganglia inhibition, together with a decrease in cortical excitation of the PPN, may increase the level of muscle tone (hypertonus). Similarly, an excessive inhibition of the MLR and a decrease in cortical excitation of the brainstem reticular formation may elicit gait failure. Additionally, less activity of the premotor cortex may disturb the motor programming required for precise gait control (Hanakawa et al., 1999; Pahapill and Lozano, 2000). As a result, the degree of freedom for each movement would be restricted, and the frame, which indicates the degree of freedom for movement, will be smaller and move to the upper right (Fig.6Ca). Gait disturbances, including delays in gait onset (frozen gait), an increase in the stance phase in locomotor cycles, and a decrease in locomotor velocity, are observed in Parkinsonian patients (Murray et al., 1978; Morris et al., 1994; Pahapill and Lozano, 2000). Because these gait failures resemble the locomotor pattern induced by SNr stimulation (Fig. 3; Takakusaki et al., 2003b), a dysfunction of the BG-BS system could be the primary basis for Parkinson disease-induced gait impairments. Moreover, muscular rigidity (hypertonus), which is one of the most prominent symptoms of Parkinson disease, could be the result of inhibition of the muscle tone inhibitory system. Namely, muscular rigidity could be interpreted in terms of a loss of inhibition to α- and γ-motoneurons.

In contrast, the reduction of output from the basal ganglia in Huntington's chorea may increase movement (hyperkinesias) and decrease muscle tone (hypotonus). The frame would be restricted and move to the lower left for this disease (Fig.6Cb). From these considerations, we suggest that a BG-BS contributes to an automatic control of movements that occur in conjunction with voluntary control processes. Moreover, the output of the basal ganglia would determine the degree of freedom of each movement, and a restriction of the degree of freedom could exist in the background of Parkinson and Huntington's diseases. Consequently, dysfunction of the BG-BS system together with that of the cortico-basal ganglia loop may underlie the pathogenesis of the motor disturbances in these basal ganglia diseases.

Dystonia is a syndrome characterized by abnormal postures, muscle spasms, and tremor, due to involuntary co-contraction of muscle agonists and antagonists. Dystonia is task specific, and patients only develop muscular co-contraction when performing skilled movements such as writing (van der Kamp et al., 1989). By using positron emission tomography an inappropriate over-activity of the basal ganglia projections to the premotor and dorsal prefrontal cortex has been observed (Brooks, 1995). However the activity of the primary sensorimotor and caudal premotor cortices is rather attenuated (Hutchins et al., 1988). Most studies, by contrast, have not found abnormalities in the brainstem. This evidence suggests that the activity of the BG-BS system and that of the cortico-basal ganglia loop are controlled separately in dystonia.

The non-motor functions, such as cognitive and psychotic processes, are impaired in patients with Parkinson disease (Taylor et al., 1986; Graybiel, 1995; Mellers et al., 1995; Hikosaka et al., 2000) and Huntington's disease (McHugh and Folsten, 1975). There are also abnormalities in the wake-sleep states in Parkinsonian patients (Rye et al., 1999; Bliwise et al., 2000; Eisensehr et al., 2001). The PPN is considered as a part of the ascending reticular activation system (ARAS; Jones, 1991; Steriade, 1996; Takakusaki et al., 1996; Garcia-Rill, 1997) through its ascending projections to the non-specific thalamic nuclei, indicating that this nucleus is involved in the maintenance of the non-motor functions (Steriade, 1996; Rye, 1997; Maloney et al., 1999; Koyama and Sakai, 2000; Datta, 2002). Because a loss of cholinergic PPN neurons has been observed in Parkinson disease (Hirsch et al., 1987; Zweig et al., 1987; Jellinger, 1988; Zweig et al., 1989), a dysfunction of cholinergic projection systems from the PPN could contribute to the deficiencies of both the motor and non-motor functions in this disease.

4 Concluding Comments

The present article argues for a role of basal ganglia projections to the mesopontine tegmentum, including the PPN, in the automatic control of movement that occurs in conjunction with voluntary control processes. Sustained output from the basal ganglia to the thalamocortical loop and the brainstem may regulate the degree of freedom of movements by controlling the excitability of the target systems. We propose that dysfunction of the basal ganglia-brainstem system, together with that of the cortico-basal ganglia loop, underlies the pathogenesis of motor disturbances which are expressed in basal ganglia diseases.

Acknowledgments

This study was supported by the Japanese Grants-in-Aid for Scientific Research (C) and RISTEX of JST (Japan Science and Technology Agency) and a grant from the Uehara Memorial Foundation to KT.

References

Alexander GE, Crutcher MD (1990) Functional architecture of basal ganglia circuits: neural substrates of parallel processing. *Trends Neurosci* **13**:266-271.

Armstrong DA, Saper CB, Levey AI, Winer BH, Terry RD (1983) Distribution of cholinergic neurons in the rat brain demonstrated by the immunohistochemical localization of choline acethyltransferase. *J Comp Neurol* **216**:53-68.

Beckstead RM, Domesick VB, Nauta WJ (1979) Efferent connections of the substantia nigra and ventral tegmental area in the rat. *Brain Res* **175**:191-217.

Bliwise DL, Willians ML, Irbe D, Ansari FP, Rye DB (2000) Inter-rater reliability for identification of REM sleep in Parkinson's disease. *Sleep* **23**:671-676.

Brooks DJ (1995) The role of the basal ganglia in motor control: contributions from PET. *J Neurol Sci* **128**:1-13.

Clements JR, Toth DD, Highfield DA, Grant SJ (1991) Glutamate-like immunoreactivity is present within cholinergic neurons of the laterodorsal tegmental and pedunculopontine nuclei. *Adv Exp Med Biol* **295**:127-142.

Culebras A, Moore JT (1989) Magnetic resonance findings in REM sleep behavior disorder. *Neurology* **39**:1519-1523.

Datta S (2002) Evidence that REM sleep is controlled by the activation of brain stem pedunculopontine tegmental kainate receptor. *J Neurophysiol* **87**:1790-1798.

DeLong MR (1990) Primate models of movement disorders of basal ganglia origin. *Trends Neurosci* **13**:281-285.

Drew T, Jiang W, Kably B, Lavoie S (1996) Role of the motor cortex in the control of visually triggered gait modifications. *Can J Physiol Pharmacol* **74**:426-442.

Eisensehr I, v Lindeiner H, Jager M, Noachtar S (2001) REM sleep behavior disorder in sleep-disordered patients with versus without Parkinson's disease: is there a need for polysomnography? *J Neurol Sci* **186**:7-11.

Garcia-Rill E (1991) The pedunculopontine nucleus. *Prog Neurobiol* **36**:363-389.

Garcia-Rill E (1997) Disorders of the reticular activating system. *Med Hypotheses* **49**: 379-387.

Georgopoulos AP, Grillner S (1989) Visuomotor coordination in reaching and locomotion. *Science* **245**:1209-1210.

Graybiel AM (1995) Building action repertoires: memory and learning functions of the basal ganglia. *Curr Opin Neurobiol* **5**:733-741.

Grillner S (1981) Control of locomotion in bipeds, tetrapods, and fish. In: *The Nervous System II* (Brooks VB, ed), pp 1179-1236. Bethesda: Am. Physiol. Soc. Press.

Grillner S (2003) The motor infrastructure: from ion channels to neuronal networks. *Nat Rev Neurosci* **4**:573-586.

Grofova I, Zhou M (1998) Nigral innervation of cholinergic and glutamatergic cells in the rat mesopontine tegmentum: light and electron microscopic anterograde tracing and immunohistochemical studies. *J Comp Neurol* **395**:359-379.

Hanakawa T, Katsumi Y, Fukuyama H, Honda M, Hayashi T, Kimura J, Shibasaki H (1999) Mechanisms underlying gait disturbance in Parkinson's disease: a single photon emission computed tomography study. *Brain* **122** (Pt 7):1271-1282.

Hikosaka O (1989) Role of basal ganglia in initiation of voluntary movements. In: *Dynamic Interactions in Neural Networks: Models and Data.* (M.A. Arbib and Amari S, ed), pp 153-167. New York: Springer-Verlag.

Hikosaka O, Takikawa Y, Kawagoe R (2000) Role of the basal ganglia in the control of purposive saccadic eye movements. *Physiol Rev* **80**:953-978.

Hirsch EC, Graybiel AM, Duyckaerts C, Javoy-Agid F (1987) Neuronal loss in the pedunculopontine tegmental nucleus in Parkinson disease and in progressive supranuclear palsy. *Proc Natl Acad Sci U S A* **84**:5976-5980.

Hutchins KD, Martino AM, Strick PL (1988) Corticospinal projections from the medial wall of the hemisphere. *Exp Brain Res* **71**:667-672.

Inglis WL, Winn P (1995) The pedunculopontine tegmental nucleus: where the striatum meets the reticular formation. *Prog Neurobiol* **47**:1-29.

Jellinger K (1988) The pedunculopontine nucleus in Parkinson's disease, progressive supranuclear palsy and Alzheimer's disease. *J Neurol Neurosurg Psychiatry* **51**:540-543.

Jones BE (1991) Paradoxical sleep and its chemical/structural substrates in the brain. *Neuroscience* **40**:637-656.

Kang Y, Kitai ST (1990) Electrophysiological properties of pedunculopontine neurons and their postsynaptic responses following stimulation of substantia nigra reticulata. *Brain Res* **535**:79-95.

Kimura K, Tachibana N, Kohyama J, Otsuka Y, Fukazawa S, Waki R (2000) A discrete pontine ischemic lesion could cause REM sleep behavior disorder. *Neurology* **55**: 894-895.

Koyama Y, Sakai K (2000) Modulation of presumed cholinergic mesopontine tegmental neurons by acetylcholine and monoamines applied iontophoretically in unanesthetized cats. *Neuroscience* **96**:723-733.

Krosaka T, Tauchi M, Dahl J (1988) Cholinergic neurons containing GABA-like and/or glutamic acid decarboxylase-like immunoreactivities in various brain regions of the rat. *Expl Brain Res* **70**:605-617.

Lai YY, Siegel JM (1990) Muscle tone suppression and stepping produced by stimulation of midbrain and rostral pontine reticular formation. *J Neurosci* **10**:2727-2734.

Lai YY, Clements JR, Siegel JM (1993) Glutamatergic and cholinergic projections to the pontine inhibitory area identified with horseradish peroxidase retrograde transport and immunohistochemistry. *J Comp Neurol* **336**:321-330.

Maloney KJ, Mainville L, Jones BE (1999) Differential c-Fos expression in cholinergic, monoaminergic, and GABAergic cell groups of the pontomesencephalic tegmentum after paradoxical sleep deprivation and recovery. *J Neurosci* **19**:3057-3072.

Marsden CD (1982) The mysterious motor function of the basal ganglia: the Robert Wartenberg Lecture. *Neurology* **32**:514-539.

Masdeu JC, Alampur U, Cavaliere R, Tavoulareas G (1994) Astasia and gait failure with damage of the pontomesencephalic locomotor region. *Ann Neurol* **35**:619-621.

Matsumura M, Nambu A, Yamaji Y, Watanabe K, Imai H, Inase M, Tokuno H, Takada M (2000) Organization of somatic motor inputs from the frontal lobe to the pedunculopontine tegmental nucleus in the macaque monkey. *Neuroscience* **98**:97-110.

Matsuyama K, Drew T (1997) Organization of the projections from the pericruciate cortex to the pontomedullary brainstem of the cat: a study using the anterograde tracer Phaseolus vulgaris-leucoagglutinin. *J Comp Neurol* **389**:617-641.

McHugh PR, Folsten MF (1975) Psychiatric syndromes of Huntington's chorea: a clinical and phenomenologic study. In: Psychiatric aspects of neurological disease (Beston F. BD, ed), pp 267-286. New York: Grune & Stratton.

Mellers JD, Quinn NP, Ron MA (1995) Psychotic and depressive symptoms in Parkinson's disease. A study of the growth hormone response to apomorphine. *Br J Psychiatry* **167**:522-526.

Mena-Segovia J, Bolam JP, Magill PJ (2004) Pedunculopontine nucleus and basal ganglia: distant relatives or part of the same family? *Trends Neurosci* **27**:585-588.

Middleton FA, Strick PL (2000) Basal ganglia and cerebellar loops: motor and cognitive circuits. *Brain Res Brain Res Rev* **31**:236-250.

Mileykovskiy BY, Kiyashchenko LI, Kodama T, Lai YY, Siegel JM (2000) Activation of pontine and medullary motor inhibitory regions reduces discharge in neurons located in the locus coeruleus and the anatomical equivalent of the midbrain locomotor region. *J Neurosci* **20**:8551-8558.

Mitani A, Ito K, Hallanger AE, Wainer BH, Kataoka K, McCarley RW (1988) Cholinergic projections from the laterodorsal and pedunculopontine tegmental nuclei to the pontine gigantocellular tegmental field in the cat. *Brain Res* **451**:397-402.

Mori S (1987) Integration of posture and locomotion in acute decerebrate cats and in awake, freely moving cats. *Prog Neurobiol* **28**:161-195.

Moriizumi T, Nakamura Y, Tokuno H, Kitao Y, Kudo M (1988) Topographic projections from the basal ganglia to the nucleus tegmenti pedunculopontinus pars compacta of the cat with special reference to pallidal projection. *Exp Brain Res* **71**:298-306.

Morris ME, Iansek R, Matyas TA, Summers JJ (1994) The pathogenesis of gait hypokinesia in Parkinson's disease. *Brain* **117** (Pt 5):1169-1181.

Murray MP, Sepic SB, Gardner GM, Downs WJ (1978) Walking patterns of men with parkinsonism. *Am J Phys Med* **57**:278-294.

Nambu A (2004) A new dynamic model of the cortico-basal ganglia loop. *Prog Brain Res* **143**:461-466.

Ottersen OP, Storm-Mathisen J (1984) Glutamate- and GABA-containing neurons in the mouse and rat brain, as demonstrated with a new immunocytochemical technique. *J Comp Neurol* **229**:374-392.

Pahapill PA, Lozano AM (2000) The pedunculopontine nucleus and Parkinson's disease. *Brain* **123** (Pt 9):1767-1783.

Reese NB, Garcia-Rill E, Skinner RD (1995) The pedunculopontine nucleus--auditory input, arousal and pathophysiology. *Prog Neurobiol* **47**:105-133.

Rossignol S (1996) Neural control of stereotypic limb movements. In: Handbook of Physiology (Rowell LB, Shepherd, J.T., ed), pp 173-216. New York: Oxford University Press.

Rye DB (1997) Contributions of the pedunculopontine region to normal and altered REM sleep. *Sleep* **20**:757-788.

Rye DB, Saper CB, Lee HJ, Wainer BH (1987) Pedunculopontine tegmental nucleus of the rat: cytoarchitecture, cytochemistry, and some extrapyramidal connections of the mesopontine tegmentum. *J Comp Neurol* **259**:483-528.

Rye DB, Johnston LH, Watts RL, Bliwise DL (1999) Juvenile Parkinson's disease with REM sleep behavior disorder, sleepiness, and daytime REM onset. *Neurology* **53**:1868-1870.

Saitoh K, Hattori S, Song WJ, Isa T, Takakusaki K (2003) Nigral GABAergic inhibition upon cholinergic neurons in the rat pedunculopontine tegmental nucleus. *Eur J Neurosci* **18**:879-886.

Shiromani PJ, Lai YY, Siegel JM (1990) Descending projections from the dorsolateral pontine tegmentum to the paramedian reticular nucleus of the caudal medulla in the cat. *Brain Res* **517**:224-228.

Spann BM, Grofova I (1991) Nigropedunculopontine projection in the rat: an anterograde tracing study with phaseolus vulgaris-leucoagglutinin (PHA-L). *J Comp Neurol* **311**: 375-388.

Spann BM, Grofova I (1992) Cholinergic and non-cholinergic neurons in the rat pedunculopontine tegmental nucleus. *Anat Embryol (Berl)* **186**:215-227.

Steriade M (1996) Arousal: revisiting the reticular activating system. *Science* **272**:225-226.

Takakusaki K, Kitai ST (1997) Ionic mechanisms involved in the spontaneous firing of tegmental pedunculopontine nucleus neurons of the rat. *Neuroscience* **78**:771-794.

Takakusaki K, Kohyama J, Matsuyama K (2003a) Medullary reticulospinal tract mediating a generalized motor inhibition in cats: III. Functional organization of spinal interneurons in the lower lumbar segments. *Neuroscience* **121**:731-746.

Takakusaki K, Kohyama J, Matsuyama K, Mori S (1993) Synaptic mechanisms acting on lumbar motoneurons during postural augmentation induced by serotonin injection into the rostral pontine reticular formation in decerebrate cats. *Exp Brain Res* **93**:471-482.

Takakusaki K, Shimoda N, Matsuyama K, Mori S (1994) Discharge properties of medullary reticulospinal neurons during postural changes induced by intrapontine injections of carbachol, atropine and serotonin, and their functional linkages to hindlimb motoneurons in cats. *Exp Brain Res* **99**:361-374.

Takakusaki K, Shiroyama T, Yamamoto T, Kitai ST (1996) Cholinergic and noncholinergic tegmental pedunculopontine projection neurons in rats revealed by intracellular labeling. *J Comp Neurol* **371**:345-361.

Takakusaki K, Kohyama J, Matsuyama K, Mori S (2001) Medullary reticulospinal tract mediating the generalized motor inhibition in cats: parallel inhibitory mechanisms acting on motoneurons and on interneuronal transmission in reflex pathways. *Neuroscience* **103**:511-527.

Takakusaki K, Habaguchi T, Saitoh K, Kohyama J (2004a) Changes in the excitability of hindlimb motoneurons during muscular atonia induced by stimulating the pedunculopontine tegmental nucleus in cats. *Neuroscience* **124**:467-480.

Takakusaki K, Saitoh K, Harada H, Kashiwayanagi M (2004b) Role of basal ganglia-brainstem pathways in the control of motor behaviors. *Neurosci Res* **50**:137-151.

Takakusaki K, Oohinata-Sugimoto J, Saitoh K, Habaguchi T (2004c) Role of basal ganglia-brainstem systems in the control of postural muscle tone and locomotion. *Prog Brain Res* **143**:231-237.

Takakusaki K, Habaguchi T, Ohtinata-Sugimoto J, Saitoh K, Sakamoto T (2003b) Basal ganglia efferents to the brainstem centers controlling postural muscle tone and locomotion: a new concept for understanding motor disorders in basal ganglia dysfunction. *Neuroscience* **119**:293-308.

Takakusaki K, Saitoh K, Harada H, Okumura T, Sakamoto T (2004d) Evidence for a role of basal ganglia in the regulation of rapid eye movement sleep by electrical and chemical stimulation for the pedunculopontine tegmental nucleus and the substantia nigra pars reticulata in decerebrate cats. *Neuroscience* **124**:207-220.

Taylor AE, Saint-Cyr JA, Lang AE (1986) Frontal lobe dysfunction in Parkinson's disease. The cortical focus of neostriatal outflow. *Brain* **109** (Pt 5):845-883.

Turner RS, Anderson ME (1997) Pallidal discharge related to the kinematics of reaching movements in two dimensions. *J Neurophysiol* **77**:1051-1074.

van der Kamp W, Berardelli A, Rothwell JC, Thompson PD, Day BL, Marsden CD (1989) Rapid elbow movements in patients with torsion dystonia. *J Neurol Neurosurg Psychiatry* **52**:1043-1049.

Vincent SR, Satoh K, Armstrong DM, Fibiger HC (1983) Substance P in the ascending cholinergic reticular system. *Nature* **306**:688-691.

Wichmann T, DeLong MR (1996) Functional and pathophysiological models of the basal ganglia. *Curr Opin Neurobiol* **6**:751-758.

Wichmann T, DeLong MR (2003) Pathophysiology of Parkinson's disease: the MPTP primate model of the human disorder. *Ann N Y Acad Sci* **991**:199-213.

Zweig RM, Jankel WR, Hedreen JC, Mayeux R, Price DL (1989) The pedunculopontine nucleus in Parkinson's disease. *Ann Neurol* **26**:41-46.

Zweig RM, Whitehouse PJ, Casanova MF, Walker LC, Jankel WR, Price DL (1987) Loss of pedunculopontine neurons in progressive supranuclear palsy. *Ann Neurol* **22**:18-25.

In: Recent Breakthroughs in Basal Ganglia Research
Editor: Erwan Bezard, pp. 123-135
ISBN: 1-59454-880-3

Chapter 10

THE PEDUNCULOPONTINE TEGMENTAL NUCLEUS AND VOLUNTARY MOVEMENT

Masaru Matsumura*

Neuroscience and Neurosurgery, Chuo Gunma Neurosurgery Hospital 64-1, Nakao-machi, Takasaki, Gunma, 370-0001, Japan

ABSTRACT

Electrophysiological studies of the pedunculopontine tegmental nucleus (PPN) in monkeys revealed that the PPN neurons facilitate limb movements in conjunction with the action of the basal ganglia, thalamus, and cerebellum. Anatomical studies revealed that the PPN is activated by direct inputs from multiple motor-related cerebral cortices. The PPN receives inhibitory inputs from the basal ganglia. In turn, it sends excitatory outputs extensively through its divergent projections to the basal ganglia and thalamus.

I hypothesize that the PPN is hypoactive in parkinsonism, receiving an excessive inhibition from the basal ganglia. When the PPN loses its excitatory output to the substantia nigra, the nigro-striatal dopaminergic system fails and results in akinesia and rigidity. The PPN functions as a modulatory center that receives an excitatory drive from the cortex and controls the activity of nigral dopamine neurons, possibly influencing cognitive behaviors including motor learning and reward system, as well as voluntary movement control

INTRODUCTION

The pedunculopontine tegmental nucleus (PPN) is a cluster of cells that is located in the caudal mesencephalic tegmentum extending from the caudal border of the red nucleus to the parabrachial nucleus. Olsewski and Baxter defined this nucleus in the human brainstem cytoarchitecturally (Olszewski and Baxter, 1982). In the human brain, the PPN is bounded on its lateral side by fibers of the medial lemniscus and on its medial side by fibers of the superior cerebellar peduncle and its decussation. Rostrally, the anterior aspect of the PPN contacts the dorsomedial aspect of the posterolateral portion of the substantia nigra, while the retrorubral fields borders it dorsally. The most dorsal aspect of the PPN is bounded caudally by the pontine cuneiform and subcuneiform nuclei and ventrally by the pontine reticular formation. The most caudal pole of the PPN is adjacent to neurons of the locus coeruleus (Olszewski and Baxter, 1982). The PPN consits of a neurochemicaly and morphologically heterogenous population of neurons. Two subdivisions of this nucleus have been recognized on the basis of cell density. The pars compacta of the PPN is located within the caudal half of the nucleus in its dorsal aspect. Cells of the pars dissipatus of the PPN, on the other hand, are distributed sparsely within its rostral half of the nucleus. The former is reported to contain >90 % of cholinergic neurons and the latter to contain less cholinergic neurons (Mesulam et al., 1989). The latter is known to contain a considerable number of glutamatergic neurons (Lavoie and Parent, 1994a). Additional neuronal types known in the PPN include a type of dopaminergic

* Tel. +81-27-363-6161 / fax. +81-27-363-6620 E-mail address: masaru_matsu@mail.wind.ne.jp

neurons, a type of noradrenergic neurons, and a small group of GABAergic interneurons (see review by Pahapill and Lozano, 2000). Anatomical studies have revealed dense connections between the PPN and the basal ganglia (Shink et al., 1997). In primates, the PPN receives massive inhibitory afferents from the substantia nigra pars reticulata and the internal pallidum, both of which constitute the output stations of the basal ganglia (Harnois and Filion, 1980, 1982; Hazrati and Parent, 1991; Saitoh et al., 2003). In turn, the PPN projects profusely to the substantia nigra pars compacta and the subthalamic nucleus (Tokuno et al., 1988; Lavoie and Parent, 1994a; Charara et al., 1996). These ascending projections are known to be cholinergic and glutamatergic and project to both ipsilaterally and contralaterally (Lavoie and Parent, 1994b; Parent and Hazrati, 1995). Additionally, thalamic projections (Hallanger et al., 1987; Pare et al., 1988) and striatal projections (Parent et al., 1983; Nakano et al., 1990) have also been reported to exist bilaterally. These bilateral projections are proposed to play an important role in the inter-limb coordination and in the locomotion (Matsumura et al., 1995). As for the lower motor system, the PPN sends descending fibers to the ventromedial portions of the pontomedullary reticular formation, which in turn gives a rise to the reticulospinal tract (Jackson and Crossman, 1983; Garcia-Rill, 1986; Rye et al., 1987; Garcia-Rill and Skinner, 1988; Rye et al., 1988; Spann and Grofova, 1989; Skinner et al., 1990; Garcia-Rill, 1991).

Given the diverse connections with many areas of the brain and the spinal cord, it is not surprising that so many functions associated with the PPN have been reported. These are the control of the sleep-wake cycle, locomotor activity, muscle tone, incentive motivation, biting and a gnawing, antinociception, gating of the startle reflex, and cognitive and auditory processing (Garcia-Rill, 1986; Steckler et al., 1994; Inglis and Winn, 1995; Reese et al., 1995; Takakusaki et al., 2004).

Concerning the motor functions, emphasis has been laid on the role of the PPN in the locomotion and the control of a muscle tone. These results are based on the studies with rodents or felines, however, recent studies in primates suggest that the PPN is more concerned with the voluntary movements (Matsumura et al., 1996, 1997; Kobayashi et al., 2001; Kobayashi et al., 2002; Matsumura et al., 2002) and parkinsonism (Kojima et al., 1997; Matsumura et al., 2000; Takada et al., 2000). A neuronal loss in the PPN in patients with Parkinson's disease has been reported (Hirsch et al., 1987; Jellinger, 1988; Zweig et al., 1989). These reports strongly suggest that the PPN may act as an important relay in the basal ganglia circuitry in the control of motor behavior. However, the neuronal mechanism of this nucleus in motor behaviors is not still clear. And the number of reports of the experimental results on PPN in non-human primates are small.

In this article, I review a series of my experiments on the roles of PPN in voluntary movements. I sought to elucidate the functional role of the PPN in the control of voluntary movements using macaque monkeys. I propose a model of basal ganglia-circuitry which includes connections between the basal ganglia and PPN.

1 There Exist Different Types of Neurons of the PPN

In the PPN area of macaque monkeys, neurons were recorded in a awake condition and classified into two types: one with a wide depolarizing spike (a duration of the negative phase >360 μsec) and one with a narrow depolarizing spike (a duration of the negative phase <360 μsec; Figure 1; Matsumura et al., 1996, 1997). Generally, "wide" neurons had a lower firing rate and were tonically active. "Narrow" neurons had a higher firing rate and were phasically active. Thirty-eight percent of the neurons were of "wide" type neurons and 62 % were of "narrow" type neurons. There was no apparent segregation between the spike width and either distribution of neurons or spontaneous firing rate and between the firing rate and activity pattern during the motor behaviors (Matsumura et al., 1997; Kobayashi et al., 2002). The significance of the different types of the depolarization is not still clear.

Anatomical studies suggested that the PPN contains cholinergic and non-cholinergic, most of which are known to be glutamatergic, neurons (Lavoie and Parent, 1994a, 1994b; Futami et al., 1995; Charara et al., 1996; Smith et al., 1996). Electrophysiological studies of the PPN neurons of rodents *in vitro* revealed the existence of three different types of neurons according to their membrane properties (Kang and Kitai, 1990). This is similar to the reported neurons *in vivo* experiments in rats (Scarnati et al., 1987). In the decerebrate cats, extracellular recording has also revealed the presence of three separate populations of neurons (Garcia-Rill and Skinner, 1988). These include two groups of tonically-active neurons, they are called 'on' cells and 'off' cells. The third group of neurons presents a bursting firing pattern, and is termed 'bursters'. Recently, Takakusaki et al. classified cholinergic neurons of the PPN into two types, one with a short-duration spike and the other with a long-duration spike (Takakusaki and Kitai, 1997; Takakusaki et al., 1997). Interestingly, their results in a slice preparation are similar to our findings in monkeys in an awake condition. In terms of task-related neuronal activity during the arm movements, we could not find any dissociation between the two groups of neurons. Kobayashi et al. also analyzed the width of action potential of PPN neurons in monkeys and also failed to detect any segregation between the spike width and activity pattern during the saccade task (Kobayashi et al., 2002). Cholinergic and non-cholinergic projections would provide various tonic and phasic events of behavioral states (Steriade et al., 1988; Takakusaki et al., 1996). The existence of heterogeneous groups of neurons suggests that PPN contains modules with different functions, however, the neurochemical identity and the functional property of these neurons remain to be proved.

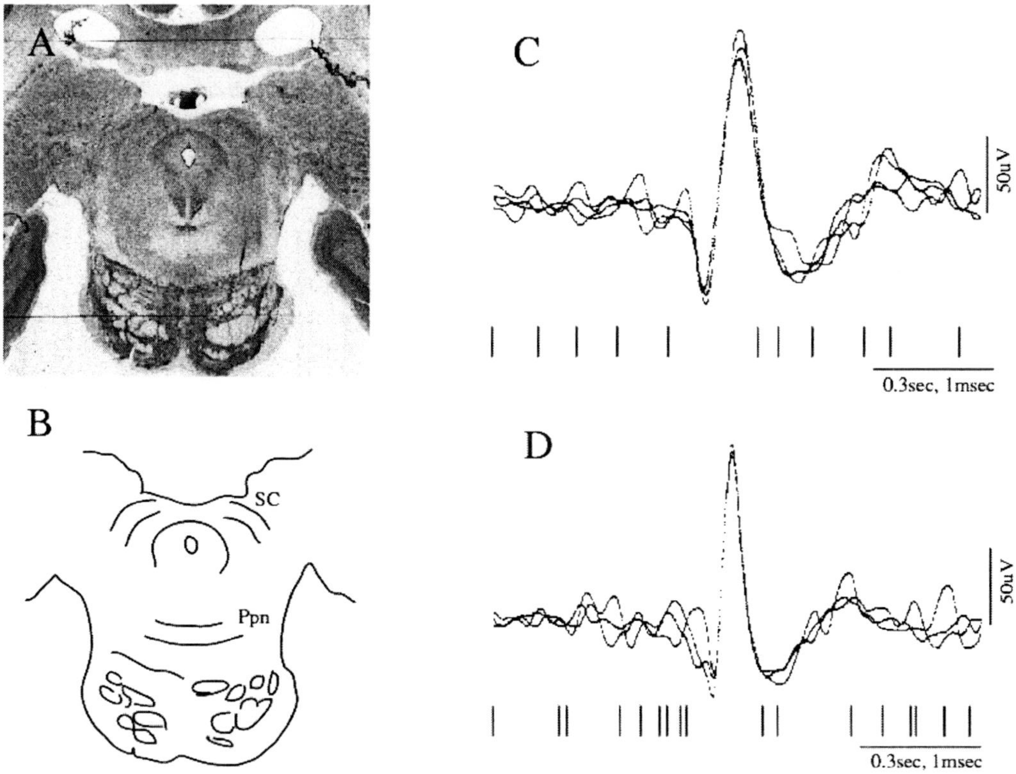

Figure 1. A Nissl-stained frontal section through the PPN in a monkey (A). Histological drawing of a section shown as A (B). SC; superior colliculus, PPN; pedunculopontine tegmental nucleus. Traces of action-potential waveforms (above) and raster displays (below) of the spontaneous firing of a neuron with a wide action potential (C) and those of a neuron with a narrow action potential (D). Each vertical bar corresponds a spike.

2 Activity Related to Voluntary Movements in the Neurons of the PPN

The neuronal activity of the PPN was recorded while a monkey was performing a trained lever-pull task with a hand (Matsumura et al., 1996, 1997). At the time of lever-off movement, neurons showed changes of the firing rate that started usually before the event (less than 100 ms) and lasted for two to several hundreds milliseconds. Nearly half of the PPN neurons changed their activity in response to voluntary arm movements (Figure 2). The onsets of changes were well related to the movement onset than to the presentation of a cue that was given before the movement onset. An increase in the firing rate occurred more frequently than a decrease of the firing rate of the PPN neurons. Changes were observed in both types of neurons (types of the neurons are shown in Figure 1). Neurons showed changes of the firing rate that started usually before the event (less than 100 ms) and lasted across the movement. This suggests that the PPN activity is closely related to the initiation and execution of movement.

Activity related to the arm movements often occurred while the animal was manipulating the lever both with the ipsilateral hand and with the contralateral hand to the side of the recording neurons. The patterns of changes during the contralateral task and the ipsilateral task were mostly identical, however, it was sometimes observed that an increase of the firing rate occurred on one of the task and a decrease occurred on the other side of the task. Responsive neurons were often found in the medio-caudal area of the PPN. However, they were found in the whole area of the PPN.

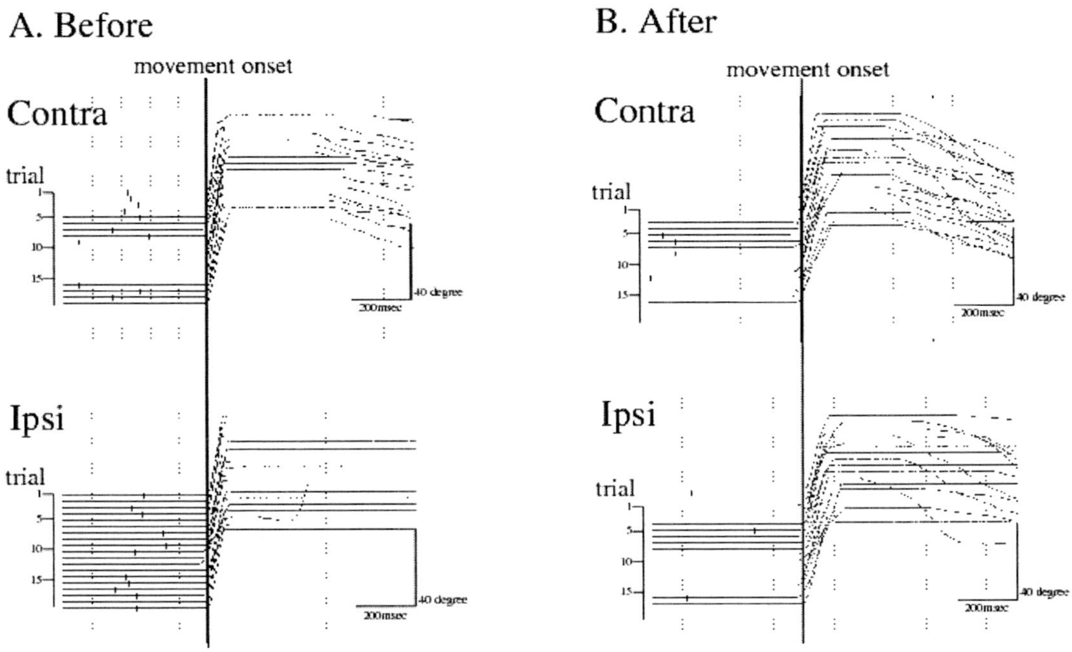

Figure 2. Neuronal activity of a PPN neuron during arm movement of the side contralateral to the recorded neuron (Contra) and that of during arm movement of the side ipsilateral to the recorded neuron (Ipsi). An increase of the neuronal firing is observed around the movement onsets in both conditions. Traces of lever position (top), histograms (middle) and raster displays (bottom) of activity of a single neuron are aligned with the movement onset (arrow). More than ten consecutive trials are presented. A short bar on each trace of lever position represents the time of trigger presentation.

In the PPN of primates, saccade-related neuronal activity has been reported (Kobayashi et al., 2002). PPN neurons changed their firing rate in relation to voluntary saccades. The changes were either an increase or a decrease. In cats, Dormont et al. has reported that PPN neurons showed a change of firing rate during operantly conditioned performance of a lever-release movement of a forelimb (Dormont et al., 1998). They also reported the effects of reversible inactivation of PPN. In their study, muscimol (a γ-aminobutyric acid (GABA) agonist) injection induced prolonged inter-trial intervals and arrest of performance, however, the reaction times were not affected (Conde et al., 1998). Their results were similar to ours (see the section 6 of this paper), however, the difference in species as well as the difference of the behavioral task should be considered for the evaluation of the results. Their work as well as our work showed that the role of PPN is to facilitate the motor system. The connections of PPN with the basal ganglia seems to be developed well in primates and the descending connections from PPN to the lower brainstem seems to be more prominent in felines than in primates, however further anatomical and physiological studies are required.

3 Bilateral Response during Voluntary Movements in the Neurons of the PPN

My study indicated that the PPN contained neurons that changed their activity during both the contralateral and the ipsilateral arm movements (Matsumura et al., 1996; Watanabe et al., 1996; Matsumura et al., 1997). The present task required the animal a single joint movement rather than an involvement of a broad set of muscles. Also to be noticed, the neuronal activity during the contralateral arm movement and that of ipsilateral arm movement were rarely identical, however, the pattern of the neuronal activity were similar. The pallido-tegmental projections were found to be bilateral (Hazrati and Parent, 1991). The substantia nigra pars reticulata is also reported to send fibers to the contralateral and ipsilateral PPN (Spann and Grofova, 1991). The efferent projections of the PPN to the sunstantia nigra pars compacta and to the thalamus were also found both contralaterally and ipsilaterally. However, the contralateral projections were reported to be less abundant (Jackson and Crossman, 1983; Steriade et al., 1988; Lavoie and Parent, 1994b). The evidence of the bilateral PPN projections to the subthalamic nucleus has been presented for monkeys (Lavoie and Parent, 1994b). The deep cerebellar nuclei send axon collaterals to the ipsilateral cholinergic and non-cholinergic neurons of the PPN on their course up to the contralateral thalamus (Hazrati and Parent, 1991). It is reported that the PPN neurons also project to the contralateral PPN neurons (Lavoie and Parent, 1994b).

The existence of these bilateral connections may underlie the facts that neurons of the PPN changed their activity during both the contralateral and the ipsilateral arm movements. The PPN may contribute to the coordination of the upper limb movement on the two sides in conjunction with the action of the basal ganglia, the thalamus, the cerebellum, and the PPN itself of the contralateral side.

4 The Origin the Motor-Related Signals of the PPN

The existence of the activity related to voluntary movements in the PPN neurons suggests that the PPN exerts control influences on voluntary movements of various body parts by being closely associated with other brain structures. My group recently reported that the excitatory input from the PPN is crucial to activate dopaminergic neurons in the substantia nigra pars compacta, thereby being involved not only in somatic motor control, but also in cognitive behaviors such as motor learning and the reward system (Kojima et al., 1997). To address the issue on the functional role of the PPN in primates, two fundamental anatomical questions are raised as to (1) where excitatory input comes from to activate the PPN neurons, and (2) how somatotopical representations are arranged in the PPN. With respect to the first question, the most likely origin of the excitatory input to the PPN is the cerebral cortex. It was also critical to investigate the distribution patterns in the PPN of cortical inputs representing individual body parts. No data, however, had so far been available on the cortical projections that terminate within regions corresponding to the PPN, except that the existence of such projections arising from the primary motor cortex was briefly reported (Hartmann von Monakow et al., 1979; Moon Edley and Graybiel, 1983). Injections of anterograde tracers into the sensorimotor regions of the medial pallidum of squirrel monkeys led to a field of labeled terminals that was strictly confined to a sector of the PPN located more rostrally and medially than the PPN pars compacta (Shink and Smith, 1995).

Thus, we made an attempt to analyze the organization of somatic motor inputs from multiple motor-related areas of the frontal lobe to the PPN (Matsumura et al., 2000). The anterograde axonal tracing was performed in the macaque monkeys by injecting biotinylated dextran amine (BDA) and wheat germ agglutinin-conjugated horseradish peroxidase (WGA-HRP) in to the following cortical areas: the primary motor cortex, the supplementary and presupplementary motor areas, the dorsal and ventral divisions of the premotor cortex and the frontal eye field. Somatotopical representations in each motor-related area were defined with intracortical microstimulation mapping.

The bundles of anterogradely labeled fibers which emerged from the injection sites of BDA and WGA-HRP were seen to descend ipsilaterally through the internal capsule and, then, the cerebral peduncle at the levels of the forebrain and midbrain. At the pontine level, part of the fibers left the cerebral peduncle and coursed dorsally or dorso-laterally towards the PPN. Through all of the injection cases examined, the terminal labeling in the PPN was restricted to the side ipsilateral to the injection site. No clear rostrocaudal topography in terminal label distribution was detected in the PPN. Accumulations of terminal label were the most prominent at the level of the trochlear nucleus where the major portion of the PPN was situated. In this area of the PPN, we often recorded neurons that changed their activity during the arm movements (Matsumura et al., 1997). The terminal labeling was found more frequently in the dorsal aspect than in the ventral aspect to the superior cerebellar peduncle (Figure 3).

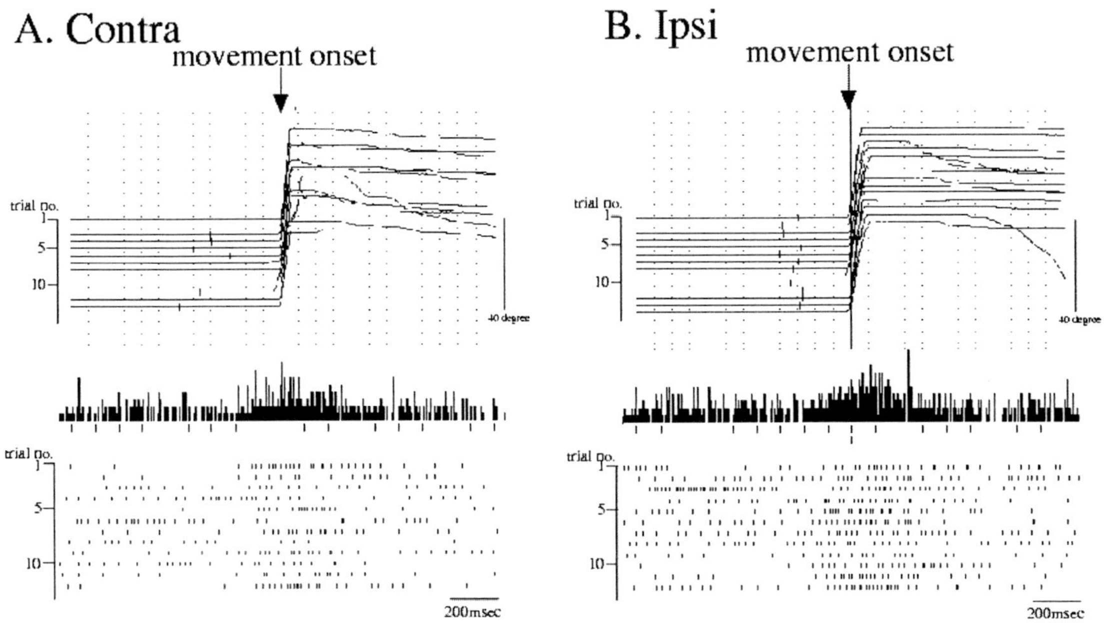

Figure 3. Distributions of terminal labeling in the PPN after injections of anterograde tracers into the distal hindlimb area (A), distal forelimb area (B), and orofacial area (C) of the primary motor cortex in monkeys. The PPN area is presented with the contour of the superior cerebellar peduncle. Note that projection fibers from the distal hindlimb area of the primary motor cortex terminate predominantly in the dorso-lateral part of the PPN (A). The projections from the forelimb area of the primary motor cortex are found in the central part of the PPN (B). Slight medially located are the terminals from the orofacial areas of the primary motor cortex (C). The distribution pattern of the corticotegmental inputs from the orofacial, forelimb, and hindlimb representations of the primary motor cortex considerably overlap, although these inputs tend to be arranged orderly from medial to lateral within the PPN (D).

The projection fibers from the distal hindlimb area of the primary motor cortex terminate predominantly in the dorso-lateral part of the PPN (Figure 3A). The projections from the forelimb area of the primary motor cortex are found in the central part of the PPN (Figure 3B). Slight medially located are the terminals from the orofacial areas of the primary motor cortex (Figure 3C). Corticotegmental inputs from orofacial, forelimb, and hindlimb representations of the primary motor cortex tended to be arranged orderly from medial to lateral in the PPN (Figure 3D). However, the distribution areas of these inputs considerably overlapped. The main terminal zones from the forelimb regions of the primary motor cortex, the supplementary and presupplementary motor areas, and the dorsal and ventral divisions of the premotor cortex appeared to overlap largely in the middle aspect of the PPN. Finally, the corticotegmental input from the frontal eye field was scattered over the PPN. Thus, the PPN is likely to receive partly separate but essentially convergent cortical inputs not only from multiple motor-related areas representing the same body part, but also from multiple regions representing diverse body parts. The somatotopical representations are intermingled rather than segregated in the PPN (Matsumura et al., 2000).

This study provided anatomical evidence that the PPN receives direct cortical inputs from multiple motor-related areas of the frontal lobe in non-human primates. In view of the fact that the general neurotransmission from the cerebral cortical areas to the subcortical structures is mediated by glutamate, such excitatory inputs from the motor-related areas are crucial to activate PPN neurons. Thus, the activity of PPN neurons is likely to be regulated by a cooperative but opposite action of excitatory and inhibitory inputs from the cortical motor-related areas and the basal ganglia, respectively.

5 Does Inactivation of PPN Produce any Disability in Motor Behavior?

It is now established that PPN receives projections from the output stations of the basal ganglia, i.e., the internal pallidum and the substantia nigra pars reticulata, which are known to be abnormally hyperactive in the parkinsonian state (Mitchell et al., 1989; Filion and Tremblay, 1991; Matsumura et al., 1995). Electrical stimulation of the substantia nigra pars reticulata in a slice preparation evoked inhibitory postsynaptic potentials in the neurons of the PPN. These inhibitory postsynaptic potentials were abolished with a GABA-A receptor antagonist, bicuculline (Saitoh et al., 2003). The behavioral consequence of this condition, however, had never been studied in an experimental animal.

We examined the effects of a reversible blockade of PPN by injecting muscimol in a macaque monkey which was trained to perform a lever-pull task with an arm (Matsumura et al., 2002). A small amount of muscimol was injected locally into PPN area, where neurons that changed their activity related to arm movements were recorded. After the injection, lever-pull movements became slow for the arm of the contralateral side. Ten to fifteen minutes after the start of the injection, the velocity of the lever movement decreased and the latency increased (Figure 4). Soon after, the symptom appeared for both sides of the arms. After 30 minutes, the monkey often presented a difficulty in holding the lever at the maximum position and returned the lever to the neutral position soon after the reward delivery. The monkey was alert and motivated well on the task within a couple of hours after the injection. When a large dose of muscimol (two to three μl) was injected, a hypokinesia on the limbs of the contralateral side was more apparent and the monkey could not continue the task. A circling (toward the contralateral side, 1-3 turns/min.) was observed occasionally outside the task.

PPN receives inputs from the internal pallidum, which is known to be abnormally hyperactive in the parkinsonian state (Mitchell et al., 1989; Filion and Tremblay, 1991). PPN sends projections to the substantia nigra pars compacta (Tokuno et al., 1988; Futami et al., 1995), whose cell loss is the major source in the parkinson's disease. These projection neurons of the PPN are reported to be excitatory and their excitation is mediated by both glutamatergic and cholinergic transmission (Futami et al., 1995). Additionally, these connections are mostly bilateral, however, the ipsilateral connections are predominant.

My group previously reported that unilateral kainic acid-induced lesion in the PPN resulted in contralateral hemiparkinsonism (Kojima et al., 1997). The monkeys presented general hypokinesia, a flexed posture and a bradykinesia on the contralateral limbs. The hemiparkinsonian symptoms observed after PPN destruction might be ascribed to a decrease in the nigrostriatal neuron activity due to excitatory input ablation. Injecting muscimol into the unilateral PPN area during the trained arm movements, we observed slowness and prolonged reaction time of voluntary fast arm movements. A decrease of general motor behavior after unilateral microinjection of muscimol into the PPN was also reported by other groups (Aziz et al., 1998; Munro-Davies et al., 1999; Nandi et al., 2002).

The inactivation of PPN might have resulted in a decreased activity of dopaminergic neurons of the substantia nigra pars compacta. Using a *in vivo*-voltammetry technique, a decrease of a current intensity of dopamine in the striatum was confirmed after a unilateral injection of kainic acid into the PPN in monkeys, confirming a decreased activity of dopaminergic neurons of the substantia nigra pars compacta (The results are preliminary).

As stated above, in parkinsonian monkeys, as well as in parkinsonian patients, abnormally high activity of neurons in the internal pallidum have been reported. This might inactivate the neurons of PPN through its GABAergic projection to the PPN. Thus, the decreased activity of the neurons in the PPN may underlie the pathophysiology of Parkinson's disease. Actually, in the postmortem study of the brains of the patients with Parkinson's disease revealed neuronal loss in the PPN (Hirsch et al., 1987; Jellinger, 1988; Zweig et al., 1989). The current surgical treatment of Parkinson's disease by lesioning or high frequency stimulation of the subthalamic nucleus or the internal segment of the globus pallidus has been shown to be effective. It has been suggested that part of this beneficial effect of subthalamic and internal pallidal lesions is achieved by reducing the descending GABAergic inhibitory influence from the internal pallidum to the PPN region (Aziz et al., 1998). However, axial symptoms like gait, akinesia and postural disturbance do not respond as well as tremor, rigidity and bradykinesia (Lozano and Lang, 1998). Recently, it was demonstrated that microinjection of GABA receptor A antagonist substance, bicuculline, into the PPN of non-human primates rendered parkinsonian with 1-methyl-4-phenyl-1,2,3,6-tetrahydropyridine (MPTP) resulted in significant improvement of akinesia (Nandi et al., 2002). This finding opens up new possibilities in the management of akinesia, not just in Parkinson's disease but also in related disorders like multi-system atrophy and progressive nuclear palsy (Nandi et al., 2002).

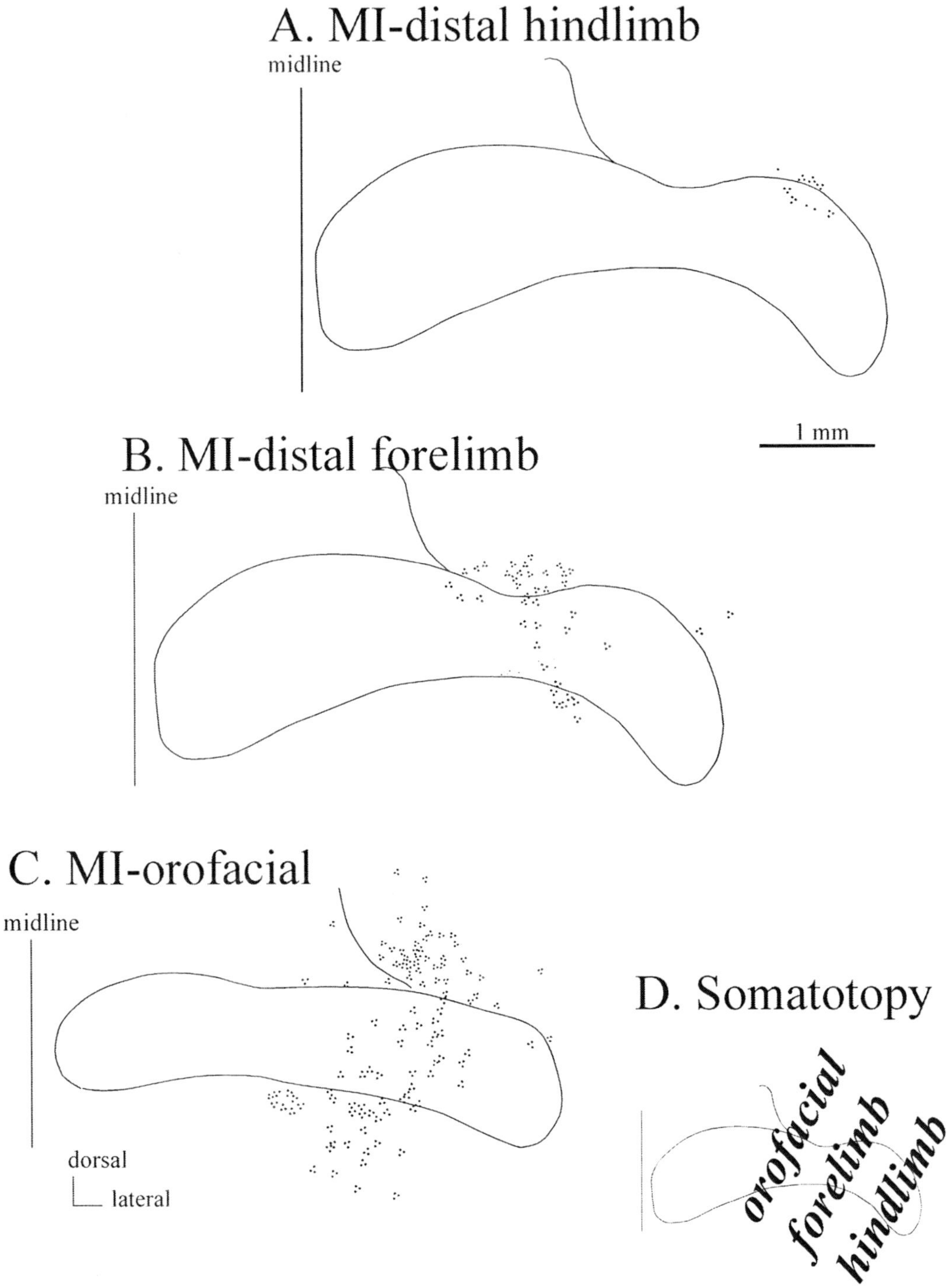

Figure 4. Traces of lever position recorded before (A) and 15 minutes after (B) an injection of muscimol into PPN area. After an injection, the reaction times increased and the velocity of the lever movement decreased. Each trace is aligned with the movement onset (vertical bar). Contra; The monkey was performing the arm movement of the contralateral side to the muscimol injection in to the PPN area. Ipsi; The monkey was performing the arm movement of the ipsilateral side to the muscimol injection into the PPN area. A short bar on each trace of lever position represents the time of trigger presentation.

6 Basal-Ganglia Circuitry and the PPN Circuitry

The current network models of basal ganglia circuitry explain pathological conditions, including Parkinson's disease, as resulting from an imbalance in the activity level between the direct and indirect pathways from the striatum to the output structures of the basal ganglia, i.e., the internal pallidum and the substantita nigra pars reticulata (Albin et al., 1989; DeLong, 1990; Alexander, 1994; Parent and Hazrati, 1995; Wichmann and DeLong, 1996). The usefulness of this model has been accepted widely, however, most of them only show the increase or decrease of the neural activities of basal ganglia, thalamus and the cortex. The descending projections from the basal ganglia to the brainstem, i.e., the PPN and the superior colliculus has less been discussed. To achieve more realistic understanding of the neuronal circuitry, especially on the pathophysiology of the Parkinson's disease, the neuronal connections between the PPN and the substantia nigra pars compacta should be paid attention.

The PPN is activated by the cortical motor inputs and likely exert modulatory actions extensively through its divergent projections to the basal ganglia and thalamus. The major projections are directed to the substantia nigra pars reticulata, subthalamic nucleus and the thalamus. The PPN can be in a strategic position to modulate, at multiple sites through its divergent outputs, the activity of the functional loop involving the somatic motor cortex, the basal ganglia, and the thalamus (Figure 5).

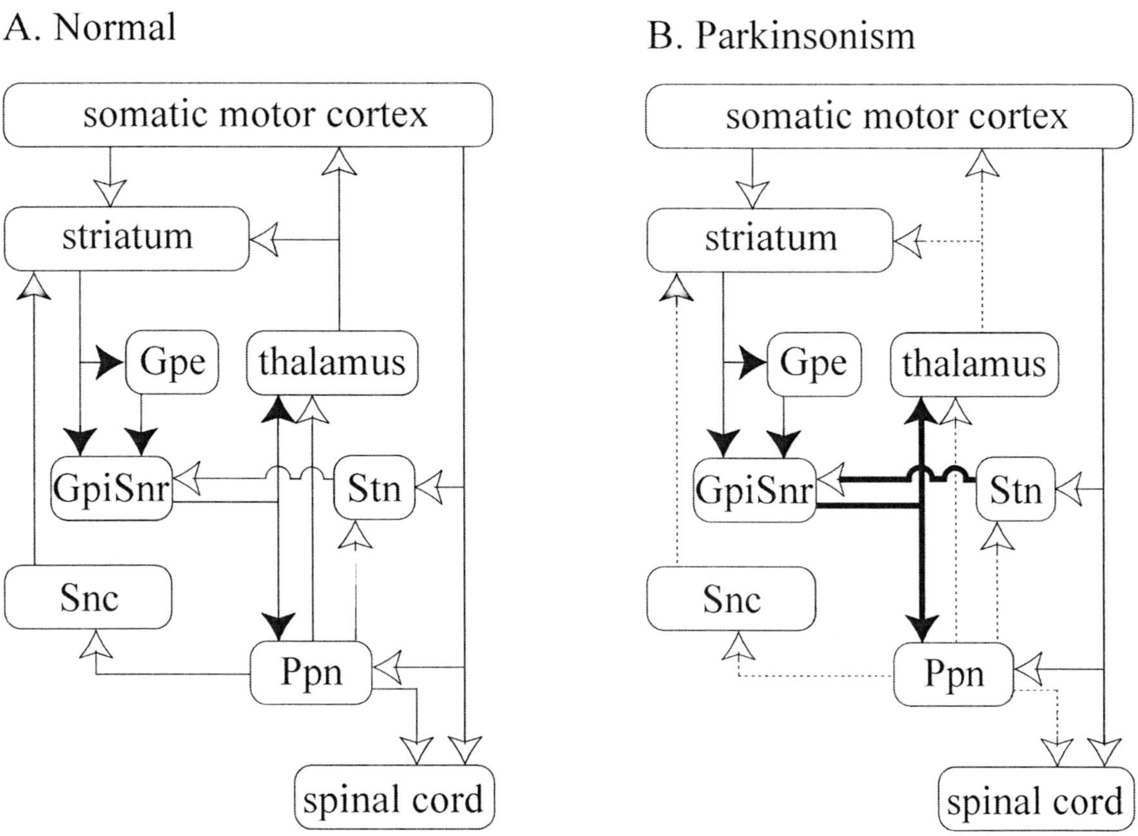

Figure 5. Functional connectivity within the basal ganglia-thalamocortical circuit with the PPN. A, Normal. The PPN receives excitatory inputs from the multiple motor-related cerebral cortices and inhibitory inputs from the basal ganglia, i.e., the internal segment of the globus pallidus and the substantia nigra pars reticulata (GpiSnr). In turn, it sends an excitatory output to the substantia nigra pars compacta (Snc) and a descending output to the brain stem. B. Parkinsonism. Receiving the excessive inhibition from the basal ganglia (GpiSnr) the PPN is hypoactive. Thus, the outputs from the PPN to the Snc decrease. This results in a hypo-activity of the nigrostriatal system. The corticospinal system and the descending system of the PPN are also hypoactive. This results in the poverty of movement, i.e., akinesia. (Open arrows, excitatory projections, Filled arrows, inhibitory projections. Solid line, normal, Interrupted line, hypoactive, Bold line, hyperactive. Gpe; external pallidum, Gpi: internal pallidum, PPN;pedunculopontine tegmental nucleus, Snc; substantia nigra pars compacta, Stn; subthalamic nucleus).

Two major concepts on the functional role of the PPN are conceivable. One is that the PPN is a relay between the cerebral cortex and the spinal cord, acting as a brainstem control center for interlimb coordination in locomotion and bimanual motor behavior. The other concerns that the PPN functions as a modulatory center that receives the excitatory drive from the cortex and controls the activity of nigral dopamine neurons, possibly influencing cognitive behaviors including motor learning and reward system, as well as voluntary movement control. Together with the findings presented in this paper, the latter concept is more favorable.

ACKNOWLEDGEMENTS

The author wish to thank his colleagues for their cooperation in the studies presented in this paper. They are Jun Kojima, Yoshio Yamaji, Hisamasa Imai, Atsushi Nambu, Masahiko Inase, Masahiko Takada, Hironobu Tokuno, Katsushige Watanabe and Chihiro Ohye.

REFERENCES

Albin RL, Young AB, Penney JB (1989) The functional anatomy of basal ganglia disorders. *Trends Neurosci* **12**:366-375.

Alexander GE (1994) Basal ganglia-thalamocortical circuits: their role in control of movements. *J Clin Neurophysiol* **11**:420-431.

Aziz TZ, Davies L, Stein J, France S (1998) The role of descending basal ganglia connections to the brain stem in parkinsonian akinesia. *Br J Neurosurg* **12**:245-249.

Charara A, Smith Y, Parent A (1996) Glutamatergic inputs from the pedunculopontine nucleus to midbrain dopaminergic neurons in primates: Phaseolus vulgaris-leucoagglutinin anterograde labeling combined with postembedding glutamate and GABA immunohistochemistry. *J Comp Neurol* **364**:254-266.

Conde H, Dormont JF, Farin D (1998) The role of the pedunculopontine tegmental nucleus in relation to conditioned motor performance in the cat. II. Effects of reversible inactivation by intracerebral microinjections. *Exp Brain Res* **121**:411-418.

DeLong MR (1990) Primate models of movement disorders of basal ganglia origin. *Trends Neurosci* **13**:281-285.

Dormont JF, Conde H, Farin D (1998) The role of the pedunculopontine tegmental nucleus in relation to conditioned motor performance in the cat. I. Context-dependent and reinforcement-related single unit activity. *Exp Brain Res* **121**:401-410.

Filion M, Tremblay L (1991) Abnormal spontaneous activity of globus pallidus neurons in monkeys with MPTP-induced parkinsonism. *Brain Res* **547**:142-151.

Futami T, Takakusaki K, Kitai ST (1995) Glutamatergic and cholinergic inputs from the pedunculopontine tegmental nucleus to dopamine neurons in the substantia nigra pars compacta. *Neurosci Res* **21**:331-342.

Garcia-Rill E (1986) The basal ganglia and the locomotor regions. *Brain Res* **396**:47-63.

Garcia-Rill E (1991) The pedunculopontine nucleus. *Prog Neurobiol* **36**:363-389.

Garcia-Rill E, Skinner RD (1988) Modulation of rhythmic function in the posterior midbrain. *Neuroscience* **27**:639-654.

Hallanger AE, Levey AI, Lee HJ, Rye DB, Wainer BH (1987) The origins of cholinergic and other subcortical afferents to the thalamus in the rat. *J Comp Neurol* **262**:105-124.

Harnois C, Filion M (1980) Pallidal neurons branching to the thalamus and to the midbrain in the monkey. *Brain Res* **186**:222-225.

Harnois C, Filion M (1982) Pallidofugal projections to thalamus and midbrain: a quantitative antidromic activation study in monkeys and cats. *Exp Brain Res* **47**:277-285.

Hartmann von Monakow K, Akert K, Künzle H (1979) Projections of precentral and premotor cortex to the red nucleus and other midbrain areas in Macaca fascicularis. *Exp Brain Res* **34**:91-105.

Hazrati LN, Parent A (1991) Contralateral pallidothalamic and pallidotegmental projections in primates: an anterograde and retrograde labeling study. *Brain Res* **567**:212-223.

Hirsch EC, Graybiel AM, Duyckaerts C, Javoy-Agid F (1987) Neuronal loss in the pedunculopontine tegmental nucleus in Parkinson disease and in progressive supranuclear palsy. *Proc Natl Acad Sci U S A* **84**:5976-5980.

Inglis WL, Winn P (1995) The pedunculopontine tegmental nucleus: where the striatum meets the reticular formation. *Prog Neurobiol* **47**:1-29.

Jackson A, Crossman AR (1983) Nucleus tegmenti pedunculopontinus: efferent connections with special reference to the basal ganglia, studied in the rat by anterograde and retrograde transport of horseradish peroxidase. *Neuroscience* **10**:725-765.

Jellinger K (1988) The pedunculopontine nucleus in Parkinson's disease, progressive supranuclear palsy and Alzheimer's disease. *J Neurol Neurosurg Psychiatry* **51**:540-543.

Kang Y, Kitai ST (1990) Electrophysiological properties of pedunculopontine neurons and their postsynaptic responses following stimulation of substantia nigra reticulata. *Brain Res* **535**:79-95.

Kobayashi Y, Saito Y, Isa T (2001) Facilitation of saccade initiation by brainstem cholinergic system. *Brain Dev* **23** Suppl 1:S24-27.

Kobayashi Y, Inoue Y, Yamamoto M, Isa T, Aizawa H (2002) Contribution of pedunculopontine tegmental nucleus neurons to performance of visually guided saccade tasks in monkeys. *J Neurophysiol* **88**:715-731.

Kojima J, Yamaji Y, Matsumura M, Nambu A, Inase M, Tokuno H, Takada M, Imai H (1997) Excitotoxic lesions of the pedunculopontine tegmental nucleus produce contralateral hemiparkinsonism in the monkey. *Neurosci Lett* **226**:111-114.

Lavoie B, Parent A (1994a) Pedunculopontine nucleus in the squirrel monkey: projections to the basal ganglia as revealed by anterograde tract-tracing methods. *J Comp Neurol* **344**:210-231.

Lavoie B, Parent A (1994b) Pedunculopontine nucleus in the squirrel monkey: cholinergic and glutamatergic projections to the substantia nigra. *J Comp Neurol* **344**:232-241.

Lozano AM, Lang AE (1998) Pallidotomy for Parkinson's disease. *Neurosurg Clin N Am* **9**:325-336.

Matsumura M, Watanabe K, Ohye C (1996) Neuronal activity of monkey pedunculopontine tegmental nucleus area. I. Activity related to voluntary arm movements. In: Basal Ganglia V (Ohye C, Kimura, M., and McKenzie, J., ed), pp 209-215. New York: Plenum Press.

Matsumura M, Watanabe K, Ohye C (1997) Single-unit activity in the primate nucleus tegmenti pedunculopontinus related to voluntary arm movement. *Neurosci Res* **28**: 155-165.

Matsumura M, Watanabe K, Ohye C (2002) Effects of reversible blockade of pedunculopontine tegmental nucleus on a voluntary arm movement in monkeys. In: Basal Ganglia VI (Graybiel AM DM, ed). New York: kluwer Academic/Plenum Pub.

Matsumura M, Tremblay L, Richard H, Filion M (1995) Activity of pallidal neurons in the monkey during dyskinesia induced by injection of bicuculline in the external pallidum. *Neuroscience* **65**:59-70.

Matsumura M, Nambu A, Yamaji Y, Watanabe K, Imai H, Inase M, Tokuno H, Takada M (2000) Organization of somatic motor inputs from the frontal lobe to the pedunculopontine tegmental nucleus in the macaque monkey. *Neuroscience* **98**:97-110.

Mesulam MM, Geula C, Bothwell MA, Hersh LB (1989) Human reticular formation: cholinergic neurons of the pedunculopontine and laterodorsal tegmental nuclei and some cytochemical comparisons to forebrain cholinergic neurons. *J Comp Neurol* **283**:611-633.

Mitchell IJ, Clarke CE, Boyce S, Robertson RG, Peggs D, Sambrook MA, Crossman AR (1989) Neural mechanisms underlying parkinsonian symptoms based upon regional uptake of 2-deoxyglucose in monkeys exposed to 1-methyl-4-phenyl-1,2,3,6-tetrahydropyridine. *Neuroscience* **32**:213-226.

Moon Edley S, Graybiel AM (1983) The afferent and efferent connections of the feline nucleus tegmenti pedunculopontinus pars compacta. *J Comp Neurol* **217**:187-215.

Munro-Davies LE, Winter J, Aziz TZ, Stein JF (1999) The role of the pedunculopontine region in basal-ganglia mechanisms of akinesia. *Exp Brain Res* **129**:511-517.

Nakano K, Hasegawa Y, Tokushige A, Nakagawa S, Kayahara T, Mizuno N (1990) Topographical projections from the thalamus, subthalamic nucleus and pedunculopontine tegmental nucleus to the striatum in the Japanese monkey, Macaca fuscata. *Brain Res* **537**:54-68.

Nandi D, Aziz TZ, Giladi N, Winter J, Stein JF (2002) Reversal of akinesia in experimental parkinsonism by GABA antagonist microinjections in the pedunculopontine nucleus. *Brain* **125**:2418-2430.

Olszewski J, Baxter D (1982) Cytoarchitecture of the Human Brain Stem., 2nd ed. Edition.

Pahapill PA, Lozano AM (2000) The pedunculopontine nucleus and Parkinson's disease. *Brain* **123** (Pt 9):1767-1783.

Pare D, Smith Y, Parent A, Steriade M (1988) Projections of brainstem core cholinergic and non-cholinergic neurons of cat to intralaminar and reticular thalamic nuclei. *Neuroscience* **25**:69-86.

Parent A, Hazrati LN (1995) Functional anatomy of the basal ganglia. I. The cortico-basal ganglia-thalamo-cortical loop. *Brain Res Brain Res Rev* **20**:91-127.

Parent A, Mackey A, De Bellefeuille L (1983) The subcortical afferents to caudate nucleus and putamen in primate: a fluorescence retrograde double labeling study. *Neuroscience* **10**:1137-1150.

Reese NB, Garcia-Rill E, Skinner RD (1995) The pedunculopontine nucleus--auditory input, arousal and pathophysiology. *Prog Neurobiol* **47**:105-133.

Rye DB, Saper CB, Lee HJ, Wainer BH (1987) Pedunculopontine tegmental nucleus of the rat: cytoarchitecture, cytochemistry, and some extrapyramidal connections of the mesopontine tegmentum. *J Comp Neurol* **259**:483-528.

Rye DB, Lee HJ, Saper CB, Wainer BH (1988) Medullary and spinal efferents of the pedunculopontine tegmental nucleus and adjacent mesopontine tegmentum in the rat. *J Comp Neurol* **269**:315-341.

Saitoh K, Hattori S, Song WJ, Isa T, Takakusaki K (2003) Nigral GABAergic inhibition upon cholinergic neurons in the rat pedunculopontine tegmental nucleus. *Eur J Neurosci* **18**:879-886.

Scarnati E, Proia A, Di Loreto S, Pacitti C (1987) The reciprocal electrophysiological influence between the nucleus tegmenti pedunculopontinus and the substantia nigra in normal and decorticated rats. *Brain Res* **423**:116-124.

Shink E, Smith Y (1995) Differential synaptic innervation of neurons in the internal and external segments of the globus pallidus by the GABA- and glutamate-containing terminals in the squirrel monkey. *J Comp Neurol* **358**:119-141.

Shink E, Sidibe M, Smith Y (1997) Efferent connections of the internal globus pallidus in the squirrel monkey: II. Topography and synaptic organization of pallidal efferents to the pedunculopontine nucleus. *J Comp Neurol* **382**:348-363.

Skinner RD, Kinjo N, Henderson V, Garcia-Rill E (1990) Locomotor projections from the pedunculopontine nucleus to the spinal cord. *Neuroreport* **1**:183-186.

Smith Y, Charara A, Parent A (1996) Synaptic innervation of midbrain dopaminergic neurons by glutamate-enriched terminals in the squirrel monkey. *J Comp Neurol* **364**:231-253.

Spann BM, Grofova I (1989) Origin of ascending and spinal pathways from the nucleus tegmenti pedunculopontinus in the rat. *J Comp Neurol* **283**:13-27.

Spann BM, Grofova I (1991) Nigropedunculopontine projection in the rat: an anterograde tracing study with phaseolus vulgaris-leucoagglutinin (PHA-L). *J Comp Neurol* **311**: 375-388.

Steckler T, Inglis W, Winn P, Sahgal A (1994) The pedunculopontine tegmental nucleus: a role in cognitive processes? *Brain Res Brain Res Rev* **19**:298-318.

Steriade M, Pare D, Parent A, Smith Y (1988) Projections of cholinergic and non-cholinergic neurons of the brainstem core to relay and associational thalamic nuclei in the cat and macaque monkey. *Neuroscience* **25**:47-67.

Takada M, Matsumura M, Kojima J, Yamaji Y, Inase M, Tokuno H, Nambu A, Imai H (2000) Protection against dopaminergic nigrostriatal cell death by excitatory input ablation. *Eur J Neurosci* **12**:1771-1780.

Takakusaki K, Kitai ST (1997) Ionic mechanisms involved in the spontaneous firing of tegmental pedunculopontine nucleus neurons of the rat. *Neuroscience* **78**:771-794.

Takakusaki K, Shiroyama T, Kitai ST (1997) Two types of cholinergic neurons in the rat tegmental pedunculopontine nucleus: electrophysiological and morphological characterization. *Neuroscience* **79**:1089-1109.

Takakusaki K, Shiroyama T, Yamamoto T, Kitai ST (1996) Cholinergic and noncholinergic tegmental pedunculopontine projection neurons in rats revealed by intracellular labeling. *J Comp Neurol* **371**:345-361.

Takakusaki K, Saitoh K, Harada H, Kashiwayanagi M (2004) Role of basal ganglia-brainstem pathways in the control of motor behaviors. *Neurosci Res* **50**:137-151.

Tokuno H, Moriizumi T, Kudo M, Nakamura Y (1988) A morphological evidence for monosynaptic projections from the nucleus tegmenti pedunculopontinus pars compacta (TPC) to nigrostriatal projection neurons. *Neurosci Lett* **85**:1-4.

Watanabe K, Matsumura M, Ohye C (1996) Neuronal activity of monkey pedunculopontine tegmental nucleus area. II. Activity related to load application on working arms. In: Basal Ganglia V (Ohye C, Kimura, M., and McKenzie, J., ed), pp 249-257. New York: Plenum Press.

Wichmann T, DeLong MR (1996) Functional and pathophysiological models of the basal ganglia. *Curr Opin Neurobiol* **6**:751-758.

Zweig RM, Jankel WR, Hedreen JC, Mayeux R, Price DL (1989) The pedunculopontine nucleus in Parkinson's disease. *Ann Neurol* **26**:41-46.

In: Recent Breakthroughs in Basal Ganglia Research
Editor: Erwan Bezard, pp. 137-145
ISBN: 1-59454-880-3

Chapter 11

INVOLVEMENT OF PPN IN PARKINSON'S DISEASE: PRODUCTION VS PROTECTION

Masahiko Takada*
Tokyo Metropolitan Institute for Neuroscience, Fuchu, Tokyo 183-8526, Japan.
Masaru Matsumura
Chuo Gunma Neurosurgery Hospital, Takasaki, Gunma 370-0001, Japan.

ABSTRACT

The pedunculopontine tegmental nucleus (PPN) and its linkage with the basal ganglia have been implicated in motor control. The PPN is the major origin to provide excitatory input to the substantia nigra pars compacta, thus exerting modulatory effects on the activity of dopaminergic nigrostriatal neurons. Loss or dysfunction of the dopaminergic neurons is well known to cause Parkinson's disease. In this review, we first tested the hypothesis that the decrease in nigrostriatal neuron activity due to PPN lesions might produce parkinsonian motor signs. Second, we paradoxically addressed an issue that excitatory input ablation with PPN lesions might protect against the onset of parkinsonian motor signs induced by 1-methyl-4-phenyl-1,2,3,6-tetrahydropyridine (MPTP). This is based on the proposal that an excitatory mechanism mediated by glutamatergic neurotransmission could participate in the pathogenesis of neurodegenerative disorders

INTRODUCTION

Emphasis has long been placed on the involvement of the pedunculopontine tegmental nucleus (PPN) in motor control. One of the most essential factors to accept such a notion is strong linkage between the PPN and the basal ganglia. Many pieces of anatomical data have shown that the PPN is interconnected with structures of the basal ganglia, including the globus pallidus, the subthalamic nucleus (STN), and the substantia nigra (see Garcia-Rill, 1986, 1991; Kojima et al., 1997; Takada et al., 2000). Of particular importance is that the PPN provides major excitatory input to the substantia nigra pars compacta (SNc) (see Kojima et al., 1997; Takada et al., 2000). Parent and his colleagues (Lavoie and Parent, 1994b; Charara et al., 1996; Smith et al., 1996) have reported in the monkey that a large number of PPN neurons containing acetylcholine and glutamate project to the SNc, and that glutamate-enriched axon terminals of PPN neurons make asymmetrical synaptic contact with dopaminergic neurons in the SNc. Moreover, an *in vitro* intracellular recording study using rat slice preparations has demonstrated that excitatory postsynaptic responses to PPN stimulation that are elicited in SNc dopamine neurons are suppressed by anti-cholinergic and anti-glutamatergic agents (Futami et al., 1995). Given that the PPN is the origin that sends the excitatory input almost exclusively to the SNc (see Kojima et al., 1997; Takada et al., 2000), such a

* All correspondences to : Masahiko Takada, Tokyo Metropolitan Institute for Neuroscience, 2-6, Musashidai, Fuchu, Tokyo 183-8526, Japan. Tel. +81-42-325-3881 / Fax. +81-42-321-8678. E-mail address: takada@tmin.ac.jp

projection seems critical to govern the activity of SNc dopamine neurons. Degeneration of these dopaminergic neurons is well known to cause Parkinson's disease that is characterized by abnormal motor signs, such as akinesia, rigidity, and tremor (Albin et al., 1989; DeLong, 1990). It is the consensus that the pathogenesis of Parkinson's disease is dopamine deficiency in the striatum due to the selective loss of SNc neurons (Albin et al., 1989; DeLong, 1990). If nigrostriatal neuron activity is dampened by PPN lesions, then parkinsonian motor signs may be produced. In the first part of the present review, we have tested this hypothesis by monitoring behavioral changes in monkeys in which the PPN was removed (see Kojima et al., 1997). Paradoxically, a proposal has been made that an excitatory mechanism mediated by glutamatergic neurotransmission can participate in the pathogenesis of neurodegenerative disorders (Choi, 1988; Olney, 1988; Klockgether and Turski, 1989; Olney, 1989). In the second part of the present review, we have challenged to this issue by investigating whether excitatory input ablation with PPN lesions may prevent nigrostriatal neuron death induced by 1-methyl-4-phenyl-1,2,3,6-tetrahydropyridine (MPTP), to protect monkeys against the onset of parkinsonian motor signs (see Takada et al., 2000).

1 Production of Parkinsonism by PPN Lesions

In Japanese monkeys (*Macaca fuscata*), the PPN that had been identified electrophysiologically was lesioned unilaterally through microinjection of kainic acid thereinto, and the changes in motor behaviors were observed. For evaluating parkinsonian motor signs, the rating scale of Smith et al., (1993) was adopted here. The PPN-lesioned monkeys underwent a series of tests that include tremor (score: 0–3), posture (0–2), gait (0–4), bradykinesia (0–4), balance (0–2), gross motor skills (0–3), and defence reaction (0–2). The severity of parkinsonism was graded by a summation of each score (maximum: 20). Within 24–48 hr after PPN lesioning, mild to moderate levels of flexed posture and hypokinesia occurred in the upper limb contralateral to the lesion. The monkeys usually became reluctant to utilize the flexed upper limb, for example, when reaching for food. Similar but milder hemiparkinsonian symptoms appeared in the lower limb on the side contralateral to the lesion. Muscular rigidity also seemed evident in the upper and lower limbs, although it was difficult to evaluate exactly by manual examination alone. No resting tremor was observed in any of the PPN-lesioned monkeys. The total score of parkinsonian motor signs was 4–8 at the peak stage for each monkey (Fig. 1). Such hemiparkinsonian symptoms were gradually ameliorated and, the monkeys returned to normal within 10–20 days after PPN lesioning (Fig. 1).

The extent of PPN lesions was histologically defined as a substantial number of neurons were eliminated from the PPN on the lesioned side. The process of recovery from hemiparkinsonism generated by unilateral PPN lesions depended on the size of the lesions; larger lesions resulted in severer levels of motor disturbances with longer durations (compare Monkeys 4–6 with Monkeys 1–3, Fig. 1). No nigrostriatal neurons were affected, because there were no marked changes in tyrosine hydroxylase (TH) immunoreactivity of SNc dopamine neurons.

Together with the previous anatomical work showing the existence of the direct excitatory (i.e., cholinergic and glutamatergic) projection from the PPN to the SNc (Scarnati et al., 1984; 1986; Beninato and Spencer, 1987; Clarke et al., 1987; Scarnati et al., 1987; Beninato and Spencer, 1988; Lee et al., 1988; Tokuno et al., 1988; Gould et al., 1989; Bolam et al., 1991; Di Loreto et al., 1992; Lavoie and Parent, 1994b; Futami et al., 1995; Oakman et al., 1995; Charara et al., 1996; Takakusaki et al., 1996), pharmacophysiological studies indicate that PPN neurons modulate the activity of dopaminergic nigrostriatal neurons through an excitatory mechanism (Niijima and Yoshida, 1988; Hernandez-Lopez et al., 1992; Blaha and Winn, 1993; Hernandez-Lopez et al., 1994; Gongora-Alfaro et al., 1996; Chapman et al., 1997). In addition, no likely origin other than the PPN has so far been reported for providing powerful excitatory input to the nigrostriatal neurons, except that the SNc receives some glutamatergic inputs from the cerebral cortex and the STN (Carter, 1982; Kita and Kitai, 1987; Sakai, 1988). Thus, parkinsonian motor signs produced by PPN lesions might be ascribed to a decrease in nigrostriatal neuron activity due to excitatory input ablation. It would be interesting that the PPN-lesioned monkeys restore almost completely from the motor impairments at 10–20-day postoperative periods. In view of the fact that the PPN innervates the SNc bilaterally with an ipsilateral predominance (Lavoie and Parent, 1994a), such a compensatory event may be explained by an increased input from the contralateral PPN. The actual mechanism underlying the improvement of hemiparkinsonian symptoms still remains a mystery.

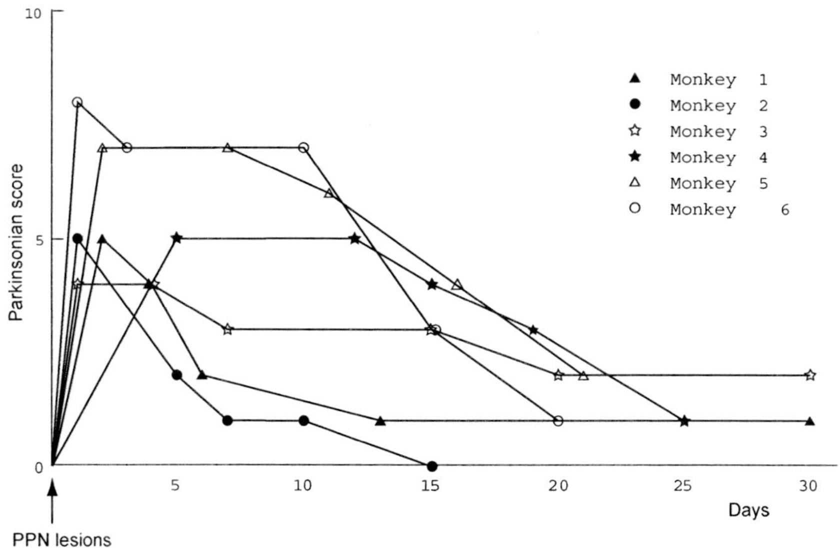

Figure 1. Time-dependent changes in the parkinsonian score after PPN lesioning (Monkeys 1–6).

2 PROTECTION AGAINST MPTP-INDUCED PARKINSONISM BY PPN LESIONS

Unilateral lesions of the PPN were made in Japanese monkeys, as described above. These PPN-lesioned monkeys exhibited parkinsonian motor signs in the upper and lower limbs contralateral to the lesion. In each monkey, the total parkinsonian score was maximally 4–8, and the motor impairments relevant to hemiparkinsonism were amended day by day until the parkinsonian score was reduced to 0–2 within 10–20 days after PPN lesioning (Fig. 2A). Then, the monkeys were given systemic administration of MPTP repeatedly for 1–3 months (every 3–4 days; 0.3–0.6 mg/kg per shot). The alterations in motor behaviors in the MPTP-treated monkeys of which PPN had been lesioned (P7–P12) were evaluated in comparison with those in the normal monkey group (P1–P3) and the MPTP-treated monkey group with the PPN intact (P4–P6).

In monkeys rendered parkinsonian by MPTP treatment (P4–P6), motor disturbances such as akinesia, muscular rigidity, and postural instability were readily developed at the cumulative dose of 1.2 mg/kg. Finally, every motor function was seriously affected at cumulative doses of 1.5–2.4 mg/kg as the total parkinsonian score reached 15 or 16 (Table 1; see also P5 in Fig. 2A). In remarkable contrast, the PPN-lesioned monkeys displayed no drastic behavioral changes when MPTP was administered at cumulative doses of 1.5–2.4 mg/kg. For some monkeys to exhibit parkinsonian motor signs as severely as P4–P6, much larger doses of MPTP (3.0–3.6 mg/kg) were necessary. As shown in Figure 2A, only slight or no parkinsonian signs at all appeared in P11 and P12 which were exposed over 3 months to extremely large doses of MPTP (7.2–10.8 mg/kg).

After monitoring of the behavioral changes, histological analysis was performed to examine the distribution of dopaminergic nigrostriatal neurons by means of TH immunohistochemistry. In the MPTP-treated monkey group with the PPN intact (P4–P6), the number of TH-immunoreactive neuronal cell bodies was greatly decreased in the SNc, as compared to that in the normal monkey group (P1–P3; Table 1). By contrast, a significantly large number of TH-immunoreactive SNc cells were left in the PPN-lesioned monkey group (P7–P12; Table 1). In these monkeys, TH-immunoreactive neurons were located more frequently on the side ipsilateral to PPN lesions than contralaterally (Table 1). Similar results were obtained for the density of TH-immunoreactive axon terminals in the striatum (mainly in the putamen; Table 1). While TH-immunoreactive striatal terminals were drastically reduced in the MPTP-treated monkey group with the PPN intact (P4–P6), those in the PPN-lesioned monkey group remained to considerable extent (P7–P12) (Table 1), except for a few symptomatic monkeys. In all of the PPN-lesioned monkeys, TH immunoreactivity was higher in the striatum ipsilateral than contralateral to PPN lesions. The significant correlations were detected between the final parkinsonian score, the number of TH-positive nigral cells, and the density of TH- positive striatal terminals (Fig. 2B–D).

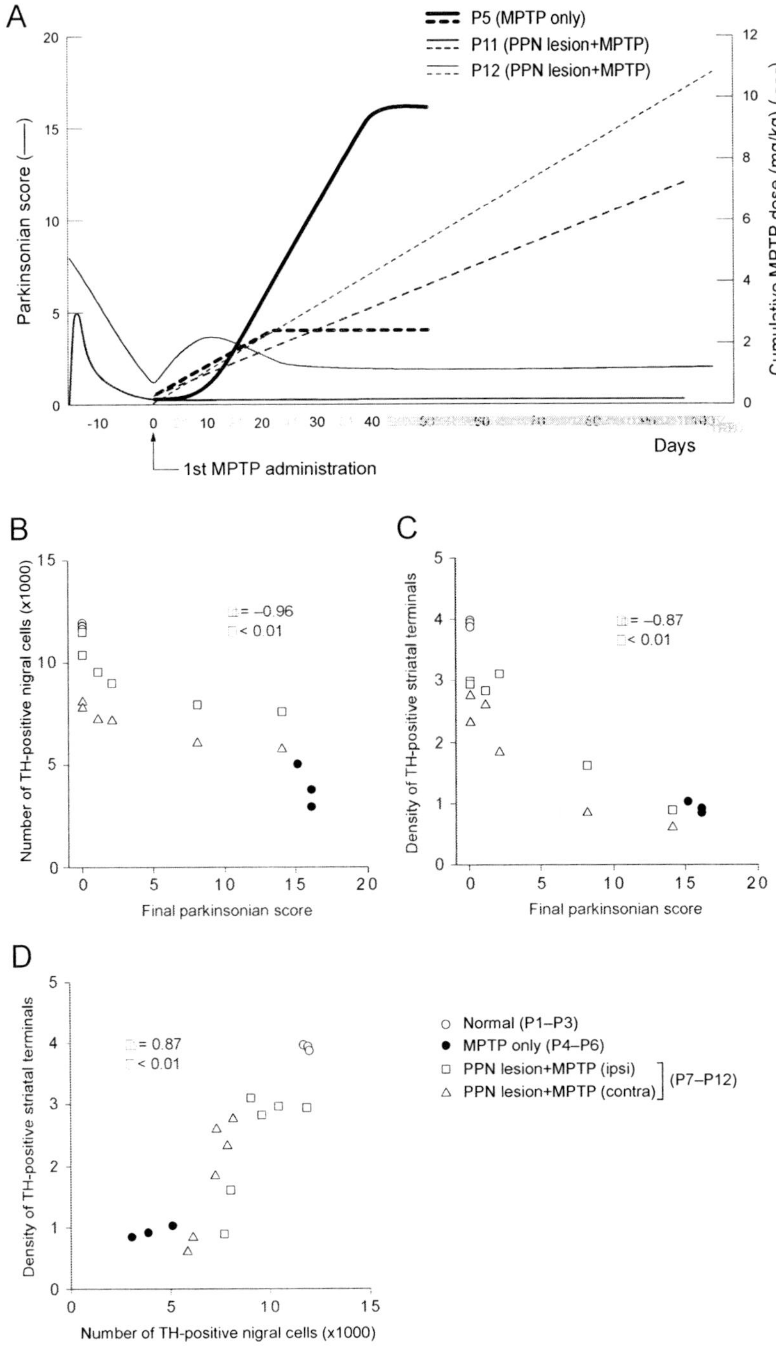

Figure 2. (A) Long-term suppression of parkinsonian motor signs after MPTP treatment in two PPN-lesioned monkeys (P11, P12), in comparison with an MPTP-treated monkey with the PPN intact (P5). Early increases in the parkinsonian scores in P11 and P12 are due to transient hemiparkinsonism produced by PPN lesions. Note that despite large cumulative doses of MPTP, the parkinsonian scores in these PPN-lesioned monkeys display only a slight or no increase at all. (B–D) Correlations between the final parkinsonian score, the number of TH-positive nigral cells, and the density of TH-positive striatal terminals. Normal, normal control monkeys (P1–P3); MPTP only, MPTP-treated monkeys with the PPN intact (P4–P6); PPN lesion+MPTP, MPTP-treated monkeys with the PPN lesioned (P7–P12); ipsi, ipsilateral to the lesion; contra, contralateral to the lesion. The rank correlation coefficient (r_s) and significance level (P) are indicated in each panel. For densitometric analysis of TH-positive striatal terminals, see Takada et al. (2000).

Table 1

	Symptom-onset day	Symptom-onset MPTP dose (mg/kg)	Final parkinsonian score	Number of TH-positive nigral cells	Relative density of TH-positive striatal terminals
Normal (P1–P3)			0.0 0.0	11810 152	3.93 0.47
MPTP only (P4–P6)	15.0 4.0	1.2 0.0	15.7 0.6	3960 1041	0.94 0.10
PPN lesion+MPTP (P7–P12)	ND	ND	4.2 5.7	9396 1550 (ipsi) 7076 935 (contra)	2.41 0.92 (ipsi) 1.84 0.92 (contra)

In view of the fact that the PPN is the major origin of glutamatergic input to dopaminergic nigrostriatal neurons, such an excitatory input seems crucial to activate them (Scarnati et al., 1984; 1986; 1987; Tokuno et al., 1988; Di Loreto et al., 1992; Futami et al., 1995; Kojima et al., 1997; Scarnati and Florio, 1997; Bezard and Gross, 1998). Based on the concept that excitation mediated by excitatory amino acids is involved in neurodegenerative events (Choi, 1988; Olney, 1988; Klockgether and Turski, 1989; Olney, 1989), removal of the glutamatergic input by PPN lesions might protect the nigrostriatal neurons against MPTP-induced neurotoxic insults. Here we have revealed that excitatory input ablation with PPN lesions does suppress or delay the onset of MPTP neurotoxicity to the nigrostriatal neurons. In comparison with the MPTP-treated monkey group with the PPN intact, nigrostriatal cell loss and parkinsonian motor deficits were largely prevented or attenuated in the PPN-lesioned monkey group.

Of particular interest is that the PPN lesions *per se* without MPTP treatment produce mild to moderate levels of hemiparkinsonism-like motor impairments that occur in the contralateral limbs (Kojima et al., 1997). This may probably be ascribable to the decreased activity of nigrostriatal neurons due to excitatory input ablation. In each of the PPN-lesioned monkeys, the motor impairments appeared transiently until 10–20 days after PPN lesioning. Our findings indicate that nigral excitation mediated by the glutamatergic input that originates from the PPN is responsible for the dopaminergic neurotoxicity of MPTP. In this work, it is important to elucidate the laterality of the nigrostriatal pathway that might predominantly be affected by a combination of unilateral PPN lesion and systemic MPTP administration. If, in the PPN-lesioned monkey with MPTP treatment, the nigrostriatal function was normally maintained on the side ipsilateral to the lesion, then parkinsonian motor signs and nigrostriatal cell loss should be restricted to the side ipsilateral or contralateral to the lesion, respectively. In fact, significant differences between the sides ipsilateral and contralateral to the lesion were histologically detected in both the number of surviving dopaminergic SNc cells and the density of remaining dopaminergic striatal terminals. Such laterality, however, was not so evident from a behavioral viewpoint, although in many of the PPN-lesioned monkeys, the limbs utilized preferentially after MPTP treatment were switched from the ipsilateral to the contralateral side to the lesion.

Another source for glutamatergic input to the SNc is the STN (Kita and Kitai, 1987; Smith et al., 1990; Parent and Hazrati, 1995; Kang and Futami, 1999). It has been demonstrated that chronic inactivation of the STN also exerts a neuroprotective effect on the SNc in a rat model of Parkinson's disease (Piallat et al., 1996, 1999). Contrary to the protective action of glutamatergic input ablation against nigrostriatal cell death, pharmacological blockade of glutamatergic input to the SNc with intranigral injection of the glutamate receptor antagonist, kynurenic acid, has been shown to accelerate the onset of parkinsonism in monkeys who are still asymptomatic after MPTP administration (Bezard et al., 1997; Bezard and Gross, 1998). This may be due to a reduced driving force on nigrostriatal neurons, as PPN lesions alone induce parkinsonian motor impairments (Kojima et al., 1997).

Given a complex interplay between glutamate and dopamine in the striatum, their functional interactions have been implicated in the pathophysiology of Parkinson's disease (Carlsson and Carlsson, 1990; Nieoullon and Kerkerian-Le Goff, 1992; Greenamyre, 1993; Di Chiara et al., 1994; Schmidt, 1995; Starr, 1995a, 1995b; Lange et al., 1997). In addition, glutamate receptor antagonists have been reported to synergize with the conventional antiparkinsonian agent, L-DOPA (Greenamyre, 1993; Schmidt, 1995; Starr, 1995a, 1995b; Lange et al., 1997). Thus, we suggest here that the possible glutamate–dopamine interactions may occur at the level of the SNc as well as of the striatum. Based on the fact that not only glutamate, but also acetylcholine is contained in PPN neurons projecting to the SNc (Scarnati et al., 1986; Woolf and Butcher, 1986; Beninato and Spencer, 1987; Clarke et al., 1987; Rye et al., 1987; Beninato and Spencer, 1988; Lee et al., 1988; Gould et al., 1989; Bolam et al., 1991; Di Loreto et al., 1992; Lavoie and Parent, 1994b; Futami et al., 1995; Oakman et al., 1995; Charara et al., 1996; Smith et al., 1996; Takakusaki et al., 1996), a cooperative action of both glutamatergic and cholinergic inputs on the excitatory mechanism of dopaminergic nigrostriatal neurons cannot as yet be ruled out. Anatomical studies have revealed that most of large neurons in the PPN are cholinergic (Woolf and Butcher, 1986; Rye et al., 1987; Spann and Grofova, 1992; Lavoie and Parent, 1994c; Takakusaki et al., 1996; Kojima et al., 1997), and that, among these cholinergic neurons, large and medium-sized glutamatergic neurons are intermingled (Lavoie and Parent, 1994c). It has also been shown that glutamate and acetylcholine are coexpressed in a subclass of single PPN neurons, and that glutamatergic and cholinergic inputs from the PPN converge onto single neurons in the SNc (Lavoie and Parent, 1994b, 1994c; Futami et al., 1995). Moreover, a constellation of pharmacophysiological studies indicate that acetylcholine increases the activity of nigrostriatal neurons via the nicotinic and/or muscarinic acetylcholine receptors and, thus, elevates the release of dopamine from their axon terminals (Lichtensteiger et al., 1976; 1982; Clarke et al., 1985; 1987; Niijima and Yoshida, 1988; Hernandez-Lopez et al., 1992; Blaha and Winn, 1993; Chapman et al., 1997).

Glutamate hyperactivity in the STN and its output targets (i.e., the globus pallidus and the substantia nigra) is now regarded as an important mechanism underlying the cardinal symptoms that characterize Parkinson's disease (Mitchell et al., 1989; Palombo et al., 1990; Guridi et al., 1993; Klockgether et al., 1994; Obeso et al., 1998). It has, in fact, been elucidated that inactivation of overactive STN neurons exerts a major antiparkinsonian effect, as evidenced by the finding that experimental ablation of the STN reverses all of the principal parkinsonian signs induced by MPTP (Bergman et al., 1990; Aziz et al., 1991; 1992). Such an improved understanding of the pathophysiology of Parkinson's disease has led to a renaissance of stereotaxic surgery, including pallidotomy and deep brain stimulation in the STN (Guridi et al., 1993; Klockgether et al., 1994; Obeso et al., 1998). The present data suggest that nigral excitation driven by the PPN may also be involved in the pathophysiology of Parkinson's disease. Surgical or pharmacological manipulations of the PPN-derived excitatory input to dopaminergic nigrostriatal neurons could provide a new basis for therapeutic strategies directed towards protection against Parkinson's disease.

References

Albin RL, Young AB, Penney JB (1989) The functional anatomy of basal ganglia disorders. *Trends Neurosci* **12**:366-375.

Aziz TZ, Peggs D, Sambrook MA, Crossman AR (1991) Lesion of the subthalamic nucleus for the alleviation of 1-methyl-4-phenyl-1,2,3,6-tetrahydropyridine (MPTP)-induced parkinsonism in the primate. *Mov Disord* **6**:288-292.

Aziz TZ, Peggs D, Agarwal E, Sambrook MA, Crossman AR (1992) Subthalamic nucleotomy alleviates parkinsonism in the 1-methyl-4-phenyl-1,2,3,6-tetrahydropyridine (MPTP)-exposed primate. *Br J Neurosurg* **6**:575-582.

Beninato M, Spencer RF (1987) A cholinergic projection to the rat substantia nigra from the pedunculopontine tegmental nucleus. *Brain Res* **412**:169-174.

Beninato M, Spencer RF (1988) The cholinergic innervation of the rat substantia nigra: a light and electron microscopic immunohistochemical study. *Exp Brain Res* **72**:178-184.

Bergman H, Wichmann T, DeLong MR (1990) Reversal of experimental parkinsonism by lesions of the subthalamic nucleus. *Science* **249**:1436-1438.

Bezard E, Gross CE (1998) Compensatory mechanisms in experimental and human parkinsonism: towards a dynamic approach. *Prog Neurobiol* **55**:93-116.

Bezard E, Boraud T, Bioulac B, Gross CE (1997) Compensatory effects of glutamatergic inputs to the substantia nigra pars compacta in experimental parkinsonism. *Neuroscience* **81**:399-404.

Blaha CD, Winn P (1993) Modulation of dopamine efflux in the striatum following cholinergic stimulation of the substantia nigra in intact and pedunculopontine tegmental nucleus-lesioned rats. *J Neurosci* **13**:1035-1044.

Bolam JP, Francis CM, Henderson Z (1991) Cholinergic input to dopaminergic neurons in the substantia nigra: a double immunocytochemical study. *Neuroscience* **41**:483-494.

Carlsson M, Carlsson A (1990) Interactions between glutamatergic and monoaminergic systems within the basal ganglia--implications for schizophrenia and Parkinson's disease. *Trends Neurosci* **13**:272-276.

Carter CJ (1982) Topographical distribution of possible glutamatergic pathways from the frontal cortex to the striatum and substantia nigra in rats. *Neuropharmacology* **21**:379-383.

Chapman CA, Yeomans JS, Blaha CD, Blackburn JR (1997) Increased striatal dopamine efflux follows scopolamine administered systemically or to the tegmental pedunculopontine nucleus. *Neuroscience* **76**:177-186.

Charara A, Smith Y, Parent A (1996) Glutamatergic inputs from the pedunculopontine nucleus to midbrain dopaminergic neurons in primates: Phaseolus vulgaris-leucoagglutinin anterograde labeling combined with postembedding glutamate and GABA immunohistochemistry. *J Comp Neurol* **364**:254-266.

Choi DW (1988) Glutamate neurotoxicity and diseases of the nervous system. *Neuron* **1**:623-634.

Clarke PB, Hommer DW, Pert A, Skirboll LR (1987) Innervation of substantia nigra neurons by cholinergic afferents from pedunculopontine nucleus in the rat: neuroanatomical and electrophysiological evidence. *Neuroscience* **23**:1011-1019.

Clarke PBS, Hommer DW, Pert A, Skirboll LR (1985) Electrophysiological actions of nicotine on substantia nigra single units. *Br J Pharmacol* **85**:827-835.

DeLong MR (1990) Primate models of movement disorders of basal ganglia origin. Trends *Neurosci* **13**:281-285.

Di Chiara G, Morelli M, Consolo S (1994) Modulatory functions of neurotransmitters in the striatum: ACh/dopamine/NMDA interactions. *Trends Neurosci* **17**:228-233.

Di Loreto S, Florio T, Scarnati E (1992) Evidence that non-NMDA receptors are involved in the excitatory pathway from the pedunculopontine region to nigrostriatal dopaminergic neurons. *Exp Brain Res* **89**:79-86.

Futami T, Takakusaki K, Kitai ST (1995) Glutamatergic and cholinergic inputs from the pedunculopontine tegmental nucleus to dopamine neurons in the substantia nigra pars compacta. *Neurosci Res* **21**:331-342.

Garcia-Rill E (1986) The basal ganglia and the locomotor regions. *Brain Res* **396**:47-63.

Garcia-Rill E (1991) The pedunculopontine nucleus. *Prog Neurobiol* **36**:363-389.

Gongora-Alfaro JL, Hernandez-Lopez S, Martinez-Fong D, Flores G, Aceves J (1996) Circling behavior elicited by cholinergic transmission in the substantia nigra pars compacta: involvement of nicotinic and muscarinic receptors. *Neuroscience* **71**:729-734.

Gould E, Woolf NJ, Butcher LL (1989) Cholinergic projections to the substantia nigra from the pedunculopontine and laterodorsal tegmental nuclei. *Neuroscience* **28**:611-623.

Greenamyre JT (1993) Glutamate-dopamine interactions in the basal ganglia: relationship to Parkinson's disease. *J Neural Transm Gen Sect* **91**:255-269.

Guridi J, Luquin MR, Herrero MT, Obeso JA (1993) The subthalamic nucleus: a possible target for stereotaxic surgery in Parkinson's disease. *Mov Disord* **8**:421-429.

Hernandez-Lopez S, Gongora-Alfaro JL, Martinez-Fong D, Aceves J (1992) A cholinergic input to the substantia nigra pars compacta increases striatal dopamine metabolism measured by in vivo voltammetry. *Brain Res* **598**:114-120.

Hernandez-Lopez S, Gongora-Alfaro JL, Martinez-Fong D, Rosales MG, Aceves J (1994) Cholinergic stimulation of rostral and caudal substantia nigra pars compacta produces opposite effects on circling behavior and striatal dopamine release measured by brain microdialysis. *Neuroscience* **62**:441-447.

Kang Y, Futami T (1999) Arrhythmic firing in dopamine neurons of rat substantia nigra evoked by activation of subthalamic neurons. *J Neurophysiol* **82**:1632-1637.

Kita H, Kitai ST (1987) Efferent projections of the subthalamic nucleus in the rat: light and electron microscopic analysis with the PHA-L method. *J Comp Neurol* **260**:435-452.

Klockgether T, Turski L (1989) Excitatory amino acids and the basal ganglia: implications for the therapy of Parkinson's disease. *Trends Neurosci* **12**:285-286.

Klockgether T, Loschmann PA, Wullner U (1994) New medical and surgical treatments for Parkinson's disease. *Curr Opin Neurol* **7**:346-352.

Kojima J, Yamaji Y, Matsumura M, Nambu A, Inase M, Tokuno H, Takada M, Imai H (1997) Excitotoxic lesions of the pedunculopontine tegmental nucleus produce contralateral hemiparkinsonism in the monkey. *Neurosci Lett* **226**:111-114.

Lange KW, Kornhuber J, Riederer P (1997) Dopamine/glutamate interactions in Parkinson's disease. *Neurosci Biobehav Rev* **21**:393-400.

Lavoie B, Parent A (1994a) Pedunculopontine nucleus in the squirrel monkey: projections to the basal ganglia as revealed by anterograde tract-tracing methods. *J Comp Neurol* **344**:210-231.

Lavoie B, Parent A (1994b) Pedunculopontine nucleus in the squirrel monkey: cholinergic and glutamatergic projections to the substantia nigra. *J Comp Neurol* **344**:232-241.

Lavoie B, Parent A (1994c) Pedunculopontine nucleus in the squirrel monkey: distribution of cholinergic and monoaminergic neurons in the mesopontine tegmentum with evidence for the presence of glutamate in cholinergic neurons. *J Comp Neurol* **344**:190-209.

Lee HJ, Rye DB, Hallanger AE, Levey AI, Wainer BH (1988) Cholinergic vs. noncholinergic efferents from the mesopontine tegmentum to the extrapyramidal motor system nuclei. *J Comp Neurol* **275**:469-492.

Lichtensteiger W, Felix D, Lienhart R, Hefti F (1976) A quantitative correlation between single unit activity and fluorescence intensity of dopamine neurones in zona compacta of substantia nigra, as demonstrated under the influence of nicotine and physostigmine. *Brain Res* **117**:85-103.

Lichtensteiger W, Hefti F, Felix D, Huwyler T, Melamed E, Schlumpf M (1982) Stimulation of nigrostriatal dopamine neurones by nicotine. *Neuropharmacology* **21**:963-968.

Mitchell IJ, Clarke CE, Boyce S, Robertson RG, Peggs D, Sambrook MA, Crossman AR (1989) Neural mechanisms underlying parkinsonian symptoms based upon regional uptake of 2-deoxyglucose in monkeys exposed to 1-methyl-4-phenyl-1,2,3,6-tetrahydropyridine. *Neuroscience* **32**:213-226.

Nieoullon A, Kerkerian-Le Goff L (1992) Cellular interactions in the striatum involving neuronal systems using "classical" neurotransmitters: possible functional implications. *Mov Disord* **7**:311-325.

Niijima K, Yoshida M (1988) Activation of mesencephalic dopamine neurons by chemical stimulation of the nucleus tegmenti pedunculopontinus pars compacta. *Brain Res* **451**:163-171.

Oakman SA, Faris PL, Kerr PE, Cozzari C, Hartman BK (1995) Distribution of pontomesencephalic cholinergic neurons projecting to substantia nigra differs significantly from those projecting to ventral tegmental area. *J Neurosci* **15**:5859-5869.

Obeso JA, Guridi J, Alvarez L, Macias R, Linazasoro G (1998) Ablative surgery for Parkinson's disease. In: Parkinson's Disease and Movement Disorders, 3rd edn Edition (Jankovic JaT, E., ed), pp 1049-1064. Baltimore, MD: Williams & Wilkins.

Olney JW (1988) Revelations in excitotoxicology. In: *Frontiers in Excitatory Amino Acid Research.* (Cavalheiro EA, Lehmann, J. and Turski, L., ed), pp 589-596. New York: Alan R. Liss.

Olney JW (1989) Excitotoxicity and N-methyl-D-aspartate receptors. *Drug Dev Res* **17**:299-319.

Palombo E, Porrino LJ, Bankiewicz KS, Crane AM, Sokoloff L, Kopin IJ (1990) Local cerebral glucose utilization in monkeys with hemiparkinsonism induced by intracarotid infusion of the neurotoxin MPTP. *J Neurosci* **10**:860-869.

Parent A, Hazrati LN (1995) Functional anatomy of the basal ganglia. II. The place of subthalamic nucleus and external pallidum in basal ganglia circuitry. *Brain Res Brain Res Rev* **20**:128-154.

Piallat B, Benazzouz A, Benabid AL (1996) Subthalamic nucleus lesion in rats prevents dopaminergic nigral neuron degeneration after striatal 6-OHDA injection: behavioural and immunohistochemical studies. *Eur J Neurosci* **8**:1408-1414.

Piallat B, Benazzouz A, Benabid AL (1999) Neuroprotective effect of chronic inactivation of the subthalamic nucleus in a rat model of Parkinson's disease. *J Neural Transm Suppl* **55**:71-77.

Rye DB, Saper CB, Lee HJ, Wainer BH (1987) Pedunculopontine tegmental nucleus of the rat: cytoarchitecture, cytochemistry, and some extrapyramidal connections of the mesopontine tegmentum. *J Comp Neurol* **259**:483-528.

Sakai ST (1988) Corticonigral projections from area 6 in the raccoon. *Exp Brain Res* **73**:498-504.

Scarnati E, Florio T (1997) The pedunculopontine nucleus and related structures: functional organization. In: *The Basal Ganglia and New Surgical Approaches for Parkinson's Disease*, Advances in Neurology (Obeso JA, DeLong, M.R., Ohye, C. and Marsden, C.D., ed), pp 97-110. Philadelphia, PA: Lippincott-Raven.

Scarnati E, Campana E, Pacitti C (1984) Pedunculopontine-evoked excitation of substantia nigra neurons in the rat. *Brain Res* **304**:351-361.

Scarnati E, Proia A, Campana E, Pacitti C (1986) A microiontophoretic study on the nature of the putative synaptic neurotransmitter involved in the pedunculopontine-substantia nigra pars compacta excitatory pathway of the rat. *Exp Brain Res* **62**:470-478.

Scarnati E, Proia A, Di Loreto S, Pacitti C (1987) The reciprocal electrophysiological influence between the nucleus tegmenti pedunculopontinus and the substantia nigra in normal and decorticated rats. *Brain Res* **423**:116-124.

Schmidt WJ (1995) Balance of transmitter activities in the basal ganglia loops. *J Neural Transm Suppl* **46**:67-76.

Smith RD, Zhang Z, Kurlan R, McDermott M, Gash DM (1993) Developing a stable bilateral model of parkinsonism in rhesus monkeys. *Neuroscience* **52**:7-16.

Smith Y, Hazrati LN, Parent A (1990) Efferent projections of the subthalamic nucleus in the squirrel monkey as studied by the PHA-L anterograde tracing method. *J Comp Neurol* **294**:306-323.

Smith Y, Charara A, Parent A (1996) Synaptic innervation of midbrain dopaminergic neurons by glutamate-enriched terminals in the squirrel monkey. *J Comp Neurol* **364**:231-253.

Spann BM, Grofova I (1992) Cholinergic and non-cholinergic neurons in the rat pedunculopontine tegmental nucleus. *Anat Embryol (Berl)* **186**:215-227.

Starr MS (1995a) Antiparkinsonian actions of glutamate antagonists--alone and with L-DOPA: a review of evidence and suggestions for possible mechanisms. *J Neural Transm Park Dis Dement Sect* **10**:141-185.

Starr MS (1995b) Glutamate/dopamine D1/D2 balance in the basal ganglia and its relevance to Parkinson's disease. *Synapse* **19**:264-293.

Takada M, Matsumura M, Kojima J, Yamaji Y, Inase M, Tokuno H, Nambu A, Imai H (2000) Protection against dopaminergic nigrostriatal cell death by excitatory input ablation. *Eur J Neurosci* **12**:1771-1780.

Takakusaki K, Shiroyama T, Yamamoto T, Kitai ST (1996) Cholinergic and noncholinergic tegmental pedunculopontine projection neurons in rats revealed by intracellular labeling. *J Comp Neurol* **371**:345-361.

Tokuno H, Moriizumi T, Kudo M, Nakamura Y (1988) A morphological evidence for monosynaptic projections from the nucleus tegmenti pedunculopontinus pars compacta (TPC) to nigrostriatal projection neurons. *Neurosci Lett* **85**:1-4.

Woolf NJ, Butcher LL (1986) Cholinergic systems in the rat brain: III. Projections from the pontomesencephalic tegmentum to the thalamus, tectum, basal ganglia, and basal forebrain. *Brain Res Bull* **16**:603-637.

In: Recent Breakthroughs in Basal Ganglia Research
Editor: Erwan Bezard, pp. 147-150
ISBN: 1-59454-880-3

PART IV: DO BASAL GANGLIA AND CEREBELLAR LOOPS COMMUNICATE IN MOTOR CONTROL AND MOVEMENT DISORDERS ?

Wassilios Meissner
Department of Neurology, CHU Pellegrin,
University of Bordeaux, France

The basal ganglia and the cerebellum are implicated in motor control. Both neuronal networks receive afferents from different cortical areas and project back to these areas via the thalamus thereby forming distinct functional loops. The classical view of functional anatomy assumes these loops to be strictly segregated at the level of the motor thalamus (Schell and Strick, 1984; Alexander et al., 1986), while some overlap of the terminal fields of basal ganglia and cerebellar loops has been described for different cortical areas that are involved in motor control (Hoover and Strick, 1993; Middleton and Strick, 1994; Hoover and Strick, 1999). Recent studies using multiple labeling techniques have confirmed the segregation within the motor thalamus, but have also suggested the existence of overlapping and interdigitating projections in the transition zone between bordering nuclei (Rouiller et al., 1994; Sakai et al., 1996, 1999, 2002). Thus, pallidothalamic and cerebellothalamic fibers might converge on thalamocortical or thalamostriatal projection neurons within these transition zones. In turn, such a convergence could have functional implications in physiological and pathophysiological motor control by modulating not only the activity of thalamocortical, but also of thalamostriatal projection neurons. Beyond a possible convergence of pallidothalamic and cerebellothalamic fibers in the motor thalamus, a strong projection from the VPLo to the dorsolateral striatum has been described in macaques (McFarland and Haber, 2000). This motor thalamic nucleus mainly receives its afferents from the deep cerebellar nuclei, suggesting that cerebellar loops could modulate the activity of the same striatal output circuits through ventral thalamic and cortical motor areas. Cerebellar loops may thereby provide important information to the basal ganglia for proper motor control. Furthermore, the intralaminar thalamic nuclei, in particular the centromedian and parafascicular nucleus that belong to the ascending reticulo-thalamo-cortical activating system, project to the striatum (Sadikot et al., 1990; Fenelon et al., 1991; Sadikot et al., 1992; Sidibe and Smith, 1999), and also to other basal ganglia nuclei (for review see Herrero et al., 2002; Lanciego et al., 2004). The intralaminar thalamic nuclei receive direct and indirect afferents from the basal ganglia and the deep cerebellar output nuclei (Sakai and Patton, 1993; Sidibe et al., 2002). They supply the striatum with information about behaviorally significant sensory events (Matsumoto et al., 2001) and in this way the cerebellum may provide important information to the basal ganglia.

With respect to motor control, the basal ganglia seem to be involved in selecting movements based on internal cues, while the cerebellum is expected to contribute to the triggering and guiding of externally cued movements (Mushiake and Strick, 1993, 1995). When it comes to motor thalamus nuclei that are expected to receive either pallidal or cerebellar inputs, the distinction between internally generated and visually triggered movement is not as obvious. In this respect, sixty percent of neurons in the oral portion of the ventral posterior lateral nucleus (VPLo), which mainly receives cerebellar input, increase their firing rate when a monkey performs an internally generated movement (van Donkelaar et al., 1999, 2000).

Moreover, fifty percent of the neurons in the oral portion of the ventral lateral nucleus, which mainly receive pallidal input, increase their firing rate when a monkey performs a visually triggered movement. Thus, it is tempting to speculate that the functional segregation in motor control between basal ganglia and cerebellar loops may be restricted to some specific nuclei of the motor thalamus, while others may integrate simultaneously basal ganglia and cerebellar inputs. In addition to motor control, it has been shown that basal ganglia and cerebellum are also involved in non-motor tasks such as mental imagery, processing of sensory information, attention, planning and language (for review see Doya, 2000). However, it is still a matter of debate whether their physiological role is involved in switching attentional set, coding prediction errors, operating as an internal timing system or optimizing movements by using sensory feedback (Jueptner and Weiller, 1998; Rao et al., 2001; Dreher and Grafman, 2002).

Major motor symptoms of Parkinson's disease (PD) are bradykinesia, rigidity, tremor and postural instability. While bradykinesia is expected to result from abnormal basal ganglia activity (for review see DeLong, 1990), tremor and postural instability are presumably not related to basal ganglia dysfunction. This implies the involvement of other neuronal networks. Among these, the cerebellum and associated loops seem to be promising candidates. In this regard, studies in humans using [^{18}F] fluorodeoxyglucose and positron emission tomography have shown that levodopa infusion reduces increased resting-state metabolic activity in the cerebellum of PD patients bilaterally (Feigin et al., 2001), while high frequency stimulation of the internal pallidum as well as subthalamotomy increases the resting-state metabolic activity in the cerebellum bilaterally (Fukuda et al., 2001; Su et al., 2001). A comparison of the regional cerebral blood flow by using single photon emission computed tomography showed a significant overactivation of the ipsilateral cerebellum during repetitive hand movements in akinetic PD patients. This overactivation normalized after levodopa treatment (Rascol et al., 1997). Finally, a study that investigated the neural basis for impaired time reproduction in PD by using functional magnetic resonance imaging (fMRI) technique found a decreased cerebellar signal in a paced-finger tapping task irrespective if there were "on" or "off" levodopa (Elsinger et al., 2003). These results suggest the implication of cerebellar loops in the pathophysiology of PD. However, the exact meaning of these findings remains a mystery. One could imagine that changes in the activity of cerebellar loops could be involved in mechanisms that compensate for basal ganglia dysfunction in PD patients, thereby delaying the onset of PD symptoms. This idea is supported by the results of animal experiments suggesting that structures outside the basal ganglia might compensate for the nigral dopamine loss (Bezard et al., 2001; 2003). However, it remains unknown how these mechanisms intervene in the course of PD. Do cerebello-thalamo-striatal loops modulate the abnormal activity of the basal ganglia that follows the degeneration of nigral dopamine neurons? Are interactions between basal ganglia and cerebellar loops in the thalamus or cortex relevant compensatory mechanisms?

The advances in electrophysiological techniques allowing simultaneous recordings of many neurons in related networks and the establishment of fMRI are crucial for a more thorough characterization and integration of possible interactions between cortex, basal ganglia, cerebellum and their related circuits in the future. Of course, a better understanding of these interactions would have direct implications for the development of new therapeutic strategies in PD and other movement disorders. The following chapter will focus on aspects of basal ganglia and cerebellum involvement in motor and non-motor control. At first, Catherine Elsinger will present recent data based on functional neuroimaging that will allow to gain more insight into how basal ganglia and cerebellum might differentially contribute to behaviors that require ongoing control of timing and working memory functions under physiological and pathophysiological conditions. In the second part of her contribution she will discuss levodopa-induced changes on basal ganglia and cerebellar activity and finally provide an integrative theory of the role of cortical and subcortical networks in temporal processing. Then, Paul van Donkelaar and Chris Miall will present recent findings based on extracellular electrophysiological recordings in the motor thalamus of non-human primates that were trained to perform a visually triggered or internally generated movement. Their results are at odds with the classical view of functional anatomy and support the idea that there is much less functional segregation between projections from the basal ganglia and the cerebellum at the level of the motor thalamus as was once thought.

REFERENCES

Alexander GE, DeLong MR, Strick PL (1986) Parallel organization of functionally segregated circuits linking basal ganglia and cortex. *Annu Rev Neurosci* **9**:357-381.

Bezard E, Gross CE, Brotchie JM (2003) Presymptomatic compensation in Parkinson's disease is not dopamine-mediated. *Trends Neurosci* **26**:215-221.

Bezard E, Crossman AR, Gross CE, Brotchie JM (2001) Structures outside the basal ganglia may compensate for dopamine loss in the presymptomatic stages of Parkinson's disease. *Faseb J* **15**:1092-1094.

DeLong MR (1990) Primate models of movement disorders of basal ganglia origin. *Trends Neurosci* **13**:281-285.

Doya K (2000) Complementary roles of basal ganglia and cerebellum in learning and motor control. *Curr Opin Neurobiol* **10**:732-739.

Dreher JC, Grafman J (2002) The roles of the cerebellum and basal ganglia in timing and error prediction. *Eur J Neurosci* **16**:1609-1619.

Elsinger CL, Rao SM, Zimbelman JL, Reynolds NC, Blindauer KA, Hoffmann RG (2003) Neural basis for impaired time reproduction in Parkinson's disease: an fMRI study. *J Int Neuropsychol Soc* **9**:1088-1098.

Feigin A, Fukuda M, Dhawan V, Przedborski S, Jackson-Lewis V, Mentis MJ, Moeller JR, Eidelberg D (2001) Metabolic correlates of levodopa response in Parkinson's disease. *Neurology* **57**:2083-2088.

Fenelon G, Francois C, Percheron G, Yelnik J (1991) Topographic distribution of the neurons of the central complex (centre median-parafascicular complex) and of other thalamic neurons projecting to the striatum in macaques. *Neuroscience* **45**:495-510.

Fukuda M, Mentis MJ, Ma Y, Dhawan V, Antonini A, Lang AE, Lozano AM, Hammerstad J, Lyons K, Koller WC, Moeller JR, Eidelberg D (2001) Networks mediating the clinical effects of pallidal brain stimulation for Parkinson's disease: a PET study of resting-state glucose metabolism. *Brain* **124**:1601-1609.

Herrero MT, Barcia C, Navarro JM (2002) Functional anatomy of thalamus and basal ganglia. *Childs Nerv Syst* **18**:386-404.

Hoover JE, Strick PL (1993) Multiple output channels in the basal ganglia. *Science* **259**:819-821.

Hoover JE, Strick PL (1999) The organization of cerebellar and basal ganglia outputs to primary motor cortex as revealed by retrograde transneuronal transport of herpes simplex virus type 1. *J Neurosci* **19**:1446-1463.

Jueptner M, Weiller C (1998) A review of differences between basal ganglia and cerebellar control of movements as revealed by functional imaging studies. *Brain* **121** (Pt 8): 1437-1449.

Lanciego JL, Gonzalo N, Castle M, Sanchez-Escobar C, Aymerich MS, Obeso JA (2004) Thalamic innervation of striatal and subthalamic neurons projecting to the rat entopeduncular nucleus. *Eur J Neurosci* **19**:1267-1277.

Matsumoto N, Minamimoto T, Graybiel AM, Kimura M (2001) Neurons in the thalamic CM-Pf complex supply striatal neurons with information about behaviorally significant sensory events. *J Neurophysiol* **85**:960-976.

McFarland NR, Haber SN (2000) Convergent inputs from thalamic motor nuclei and frontal cortical areas to the dorsal striatum in the primate. *J Neurosci* **20**:3798-3813.

Middleton FA, Strick PL (1994) Anatomical evidence for cerebellar and basal ganglia involvement in higher cognitive function. *Science* **266**:458-461.

Mushiake H, Strick PL (1993) Preferential activity of dentate neurons during limb movements guided by vision. *J Neurophysiol* **70**:2660-2664.

Mushiake H, Strick PL (1995) Pallidal neuron activity during sequential arm movements. *J Neurophysiol* **74**:2754-2758.

Rao SM, Mayer AR, Harrington DL (2001) The evolution of brain activation during temporal processing. *Nat Neurosci* **4**:317-323.

Rascol O, Sabatini U, Fabre N, Brefel C, Loubinoux I, Celsis P, Senard JM, Montastruc JL, Chollet F (1997) The ipsilateral cerebellar hemisphere is overactive during hand movements in akinetic parkinsonian patients. *Brain* **120** (Pt 1):103-110.

Rouiller EM, Liang F, Babalian A, Moret V, Wiesendanger M (1994) Cerebellothalamocortical and pallidothalamocortical projections to the primary and supplementary motor cortical areas: a multiple tracing study in macaque monkeys. *J Comp Neurol* **345**:185-213.

Sadikot AF, Parent A, Francois C (1990) The centre median and parafascicular thalamic nuclei project respectively to the sensorimotor and associative-limbic striatal territories in the squirrel monkey. *Brain Res* **510**:161-165.

Sadikot AF, Parent A, Smith Y, Bolam JP (1992) Efferent connections of the centromedian and parafascicular thalamic nuclei in the squirrel monkey: a light and electron microscopic study of the thalamostriatal projection in relation to striatal heterogeneity. *J Comp Neurol* **320**:228-242.

Sakai ST, Patton K (1993) Distribution of cerebellothalamic and nigrothalamic projections in the dog: a double anterograde tracing study. *J Comp Neurol* **330**:183-194.

Sakai ST, Inase M, Tanji J (1996) Comparison of cerebellothalamic and pallidothalamic projections in the monkey (Macaca fuscata): a double anterograde labeling study. *J Comp Neurol* **368**:215-228.

Sakai ST, Inase M, Tanji J (1999) Pallidal and cerebellar inputs to thalamocortical neurons projecting to the supplementary motor area in Macaca fuscata: a triple-labeling light microscopic study. *Anat Embryol (Berl)* **199**:9-19.

Sakai ST, Inase M, Tanji J (2002) The relationship between MI and SMA afferents and cerebellar and pallidal efferents in the macaque monkey. *Somatosens Mot Res* **19**:139-148.

Schell GR, Strick PL (1984) The origin of thalamic inputs to the arcuate premotor and supplementary motor areas. *J Neurosci* **4**:539-560.

Sidibe M, Smith Y (1999) Thalamic inputs to striatal interneurons in monkeys: synaptic organization and co-localization of calcium binding proteins. *Neuroscience* **89**:1189-1208.

Sidibe M, Pare JF, Smith Y (2002) Nigral and pallidal inputs to functionally segregated thalamostriatal neurons in the centromedian/parafascicular intralaminar nuclear complex in monkey. *J Comp Neurol* **447**:286-299.

Su PC, Ma Y, Fukuda M, Mentis MJ, Tseng HM, Yen RF, Liu HM, Moeller JR, Eidelberg D (2001) Metabolic changes following subthalamotomy for advanced Parkinson's disease. *Ann Neurol* **50**:514-520.

van Donkelaar P, Stein JF, Passingham RE, Miall RC (1999) Neuronal activity in the primate motor thalamus during visually triggered and internally generated limb movements. *J Neurophysiol* **82**:934-945.

van Donkelaar P, Stein JF, Passingham RE, Miall RC (2000) Temporary inactivation in the primate motor thalamus during visually triggered and internally generated limb movements. *J Neurophysiol* **83**:2780-2790.

In: Recent Breakthroughs in Basal Ganglia Research
Editor: Erwan Bezard, pp. 151-161
ISBN: 1-59454-880-3

Chapter 12

NEURAL MECHANISMS SUBSERVING THE PROCESSING OF TIME

*Catherine L. Elsinger**
Research, Neurognostics, Inc., 10437 Innovation Drive, Suite 309, Milwaukee, WI, 53226;
Department of Neurology, Medical College of Wisconsin

ABSTRACT

Humans are remarkably proficient at learning large repertoires of skills, nearly all aspects of which require precise timing and sequencing of actions. Central to controlling action are planning operations, which encompass a broad range of cognitive processes that allow us to anticipate events, select movements, specify their ordering, and control actions online. The relative contribution of the basal ganglia and cerebellum to timing and sensory integration processes has been long debated. In this chapter evidence is presented in support of an integrative theory describing the selective roles that the fronto-striatal and cerebello-cortical circuits play in precise timekeeping.

INTRODUCTION

The classical notion that the cerebellum and the basal ganglia are dedicated exclusively to motor control is under dispute given increasing evidence of their involvement in non-motor function. There is abundant evidence from physiological, behavioral, and functional imaging studies to support the hypothesis that both the cerebellum and the basal ganglia are involved in actions that require precise timing processes. However, the degree to which cortical and sub-cortical structures participate in time perception or time reproduction processes has been under-evaluated due to a lack of alternative theories that would enable us to understand how these brain structures function concurrently in the context of perceptual-motor tasks. This chapter describes converging evidence from recent functional neuroimaging investigations that provide a framework for describing how the basal ganglia and cerebellum may differentially contribute to behaviors that require ongoing control of timing and working memory functions. The chapter begins with a brief description of a cognitive model of timekeeping and the two classic paradigms used to test temporal processing and the neural systems underlying these abilities in humans. This is followed by a brief review of recent evidence in support of an integrative theory of the contribution of cortical and subcortical structures to temporal processing.

* Tel: 414-727-7950 (x120); Fax: 414-727-7952; E-mail address: celsinger@neurognostics.com

1 Cognitive Model of Temporal Processing

Contemporary research posits the existence of an internal timekeeping system that is independent of sensory or motor feedback (Wing and Kristofferson, 1973; Ivry and Keele, 1989; Sergent et al., 1993) . The influential scalar theory of timing has been the prevailing model of a temporal processing system (Gibbon et al., 1984). Three stages are distinguished in the scalar theory; the clock mechanism measures duration, a memory mechanism compares the current duration with a duration stored in reference memory, and a decision mechanism selects a response. In this model of temporal processing, control of timing (the internal clock) is implemented by a pacemaker mechanism that represents time through the accumulation of pulses. The accumulation of these internal pulses is controlled by a switch hypothesized to be influenced by attentional processes. Once the representation of interval duration is formulated, it is passed through working memory where it can be maintained in accord with existing goals. Over time, interval representations are stored in long-term memory. Lastly, during the decision process, pulse counts collected by the accumulator representing the current interval are compared to the representation stored in memory. In this model of prospective timing, perception of time can be influenced by the level of attention associated with starting and stopping pulses to the accumulator (the switch), as is the case if pulses are lost and a subjective interval is thus shortened. Time perception can also be influenced by memory processes, based on the rate at which information from the accumulator is encoded into working memory. At the decision-making level, attentional or strategic factors can influence response thresholds that determine when the current time is sufficiently close to the remembered time. The three component processes (clock, memory, decision) of temporal processing have been studied using a variety of tasks. The two most widely used tasks are the time perception task and time reproduction task described below.

2 Testing Models of Temporal Processing

In the time perception procedure, a standard interval to be timed is presented, followed by a comparison interval which is slightly shorter or longer than the standard (on the order of milliseconds). Subjects are required to make a decision about whether the comparison interval is shorter or longer in duration than the standard interval. Mean accuracy is a measure of the degree to which the representation of time stored in memory reflects the actual physical time. The closer the comparison interval is to the standard, the more difficult it is for subjects to discriminate between the new interval and the standard.

The most commonly used time reproduction procedure, the paced-finger tapping (PFT) task, requires that subjects tap their index finger in response to the presentation of a series of tones, separated by a constant interval (synchronization condition), typically in the sub-second range. The auditory pacing stimulus is then discontinued, and the subject continues to tap at the same pace (continuation condition). During this latter condition when the tone is absent, it is possible to assess central timing because timing competency is dependent on an internal representation of the target interval duration. It is important to note that central timing mechanisms are likely to be engaged during synchronization as well as continuation. However, the extent to which the clock mechanism functions in the same way, or is affected by attentional processes related to additional sensory information, is unclear. The advantage of examining the continuation phase is simply that subjects have to access a representation of an interval stored in memory in order to perform the task accurately. Relying on continual refreshing of the target interval or anticipation of coincident events cannot be used as a strategy in the continuation phase. In the PFT task, accuracy and variability in reproducing the inter-response interval (IRI) are measures of interest.

Versions of both the time perception and the PFT task have been used extensively in healthy adults in order to examine the neural systems involved in temporal processing (Ivry and Hazeltine, 1995; Ivry, 1996; Rao et al., 1997; Harrington et al., 1998a; Harrington and Haaland, 1999; Rao et al., 2001). Scalar timing theory predicts that clock, memory, and decision processes should be represented by specific patterns of accuracy and variability, and can thus be influenced differentially by the introduction of pharmacological interventions and/or existing pathology.

The neural systems that support different component processes of time perception and reproduction have been under dispute. Furthermore, our understanding of how clock, memory, and other processes affect time perception and time reproduction performance is far from complete. Temporal processing is comprised of multiple processes, making it difficult to separate the effects of the clock on performance from the effects of other processes, such as working memory, attention, or decision making. This is especially true of the PFT task, in which mechanisms underlying increased or decreased timing variability are difficult to ascertain. Time discrimination paradigms have similar problems, although they can be more analytic depending on how the behavioral data are decomposed. For instance, a recent study using a time perception task sought to dissociate neural mechanisms underlying encoding of time intervals from those involved during decision making processes (Harrington et al., 2004c). By manipulating the difficulty of the task, and independently examining measures of duration encoding accuracy (bisection point) and temporal processing sensitivity (coefficient of variation), they were better able to isolate neural systems that mediate interval encoding (timekeeping) from systems engaged during overt decision-making (interval comparison) processes.

Functional imaging techniques employing both PFT and time discrimination tasks have been used to investigate timing behaviors in healthy populations (Rao et al., 1997; 2001; Harrington et al., 2004c) and in patients with neuropathology affecting structures hypothesized to be involved in timing (Ivry et al., 1988; Ivry and Keele, 1989; Feigin et al., 2001; Elsinger et al., 2003; Harrington et al., 2004b; Paulsen et al., 2004). The goal of these studies has been to identify the neural systems that are not only involved, but *essential* for timekeeping processes. In the following two sections, evidence in support of two possible neural mechanisms underlying timekeeping processes is reviewed.

3 The Cerebellum as a Mechanism Underlying Timekeeping

In a series of experiments primarily involving patients who have damage to the cerebellum, Ivry and colleagues (Ivry et al., 1988; Ivry and Keele, 1989; Ivry, 1996) have suggested that the lateral cerebellum and its primary output, the dorsal dentate nucleus, are involved in central timekeeping operations. In these experiments patients show impairment when they must estimate or reproduce the duration of an interval lasting several hundred milliseconds. Harrington and colleagues (Harrington et al., 2004a; 2004b) cite two reasons why this evidence alone does not conclusively implicate the cerebellum as a central timekeeping mechanism. First, subjects examined in the studies cited above have included patients with cerebellar atrophy, which is rarely restricted to the cerebellum. Damage to other areas of the cerebral cortex can produce deficits in timing across many timing ranges (several hundred milliseconds to several seconds; Harrington et al., 1998b; Casini and Ivry, 1999), making it difficult to differentiate deficits in timekeeping mechanisms from cognitive processes that support timing (e.g., attention, working memory). Secondly, these studies have typically used only one timing test (e.g., time perception or time reproduction), or have tested subjects on only one time interval. This methodological limitation precludes the identification of a timekeeping mechanism that supports timing across different tasks and across many interval ranges. In a recent study, Harrington and colleagues (Harrington et al., 2004b) addressed the limitations of past research using a well-described population of 21 cerebellar stroke patients and 30 healthy age- and education-matched controls performing both time perception and time reproduction tasks. They hypothesized that if cerebellar damage disrupts a central timekeeping operation, performance deficits should be observed in time reproduction *and* time perception tasks. Significant impairments in time perception were not found in patients with cerebellar damage. Moreover, an analysis of the individual subject data showed that few patients exhibited both time perception and time reproduction deficits at both standard intervals (Harrington et al., 2004a). Impairments were observed only during time reproduction (increased clock variability) in patients with damage to the middle and superior lobules of the cerebellum. In order to differentiate deficits specific to a timekeeper from those associated with cognitive processes that influence timekeeping, time reproduction performance was correlated with independent measures of working memory, attention, processing speed, and sensory discrimination. Harrington et al., (2004b) found that increased variability in time reproduction was correlated with poor performance on the working memory task. Because cerebellar damage did not affect both time reproduction *and* time perception performance, their results do not support a role for the cerebellum as the central timekeeper. Instead, they conclude that deficits in timing movements may be a function of disruption in the acquisition of sensory and cognitive information crucial to task performance. If patients with damage to the middle and superior lobules of the cerebellum are showing deficits in working memory performance, it is possible that the deficits they exhibit in time reproduction reflect a deficit in maintaining an effective representation of temporal duration in working memory, and do not reflect a deficit in timing (clock mechanism) per se.

4 The Basal Ganglia's Role in Timekeeping Processes

Converging evidence from studies in animals and humans suggests that timing depends not on the cerebellum, but on activation of striato-thalamo-cortical circuits (Alexander et al., 1986; Malapani et al., 1998; Meck and Benson, 2002; Matell et al., 2003; Hinton and Meck, 2004). The basal ganglia were initially thought to be solely involved in motor functions because patients with disorders of the striatum (e.g., Parkinson's disease and Huntington's disease) showed predominately motor symptoms. However, research has shown that patients with Parkinson's disease (PD), demonstrate abnormal performance on the PFT, characterized by decreased accuracy and variability changes, supporting the idea that the basal ganglia may play a critical role in motor timing (Ivry and Keele, 1989; Pastor et al., 1992; Pastor and Artieda, 1996). Pathological changes in Parkinson's disease include a loss of nigral dopaminergic neurons projecting to the dorsal putamen (Brooks et al., 1990). The major output of the dorsal putamen is the ventrolateral thalamus, which in turn projects to the supplementary motor area (SMA; Alexander et al., 1986) . Unmedicated PD patients exhibit faster mean IRIs than control subjects on the PFT (Ivry and Keele, 1989; Elsinger et al., 2003) and demonstrate greater IRI variability than controls and medicated PD patients (O'Boyle et al., 1996). Other studies indicate that medicated PD patients tap faster and show greater variability than controls at intervals between 167 and 600 ms (Konczak et al., 1997; Harrington and Haaland, 1999).

There is evidence from data collected in healthy adults to support the idea that timed movements are dependent on multiple interrelated neural systems that include the basal ganglia. In an fMRI study using 300 or 600 ms pacing intervals, Rao et al. (1997) found that the PFT task activated the left sensorimotor cortex and right cerebellum (dorsal dentate nucleus) during both synchronization and continuation conditions. During the continuation condition, which stresses internal timekeeping processes, the medial premotor system (SMA, left putamen, and left ventrolateral thalamus) was activated. The continuation condition also activated a region within the right inferior frontal gyrus and superior temporal gyrus. These findings suggest that timed movements are dependent on three interrelated neural systems; one that is involved in explicit timing (putamen, ventrolateral thalamus, SMA), another that mediates auditory working memory (inferior frontal gyrus, superior temporal gyrus), and a third that involves sensorimotor processing (dorsal dentate nucleus, sensorimotor cortex). Likewise, in an event-related fMRI study examining time perception performance in normal adults, Rao et al. (2001) found that activation of the basal ganglia (putamen and caudate nucleus) occurred early, during active encoding of the interval to be timed, while activation of the cerebellum was observed much later, during motor execution. These results are supported by findings from another functional imaging study by Ferrandez et al. (2003) using a time perception task (paired with an intensity discrimination task and control task). Greater activation was observed in the putamen and SMA during the duration discrimination task only. Activation in the cerebellum arose during performance of *both* the duration and intensity discrimination tasks, suggesting that the cerebellum is engaged during attention to, or integration of, sensory information (duration or intensity) in preparation for a motor response. Taken together, findings from the functional imaging studies reviewed above support the basal ganglia's role as central timekeeper, and the cerebellum's potential role in sensorimotor integration and/or attentional processes.

There is also evidence from functional imaging studies in patients with pathology to support the role of medial frontostriatal circuits in timing. In a recent fMRI study (Elsinger et al., 2003), PD patients and healthy age-matched controls were imaged while performing the PFT task. PD patients underwent two imaging sessions, one 'on' and the other 'off' dopamine supplementation. Relative to controls, PD patients were less accurate in performing the continuation task and showed greater performance variability in both synchronization and continuation conditions relative to controls (see Figure 1). No PFT performance differences were observed between on and off medication states in PD patients, despite significantly greater motor symptoms on the Unified Parkinson's Disease Rating Scale (UPDRS) in the off medication state. Other studies have reported increased performance variability in PD patients on medication (Konczak et al., 1997; Harrington and Haaland, 1999). However, it is entirely possible that the absence of performance differences between the on and off medication state was due to the fact that more than half of the patients continued with dopamine agonist therapy. It will be critical in follow-up studies for patients to be completely off their dopaminergic therapies in order to rule out the possibility that performance was enhanced by agonist therapies alone.

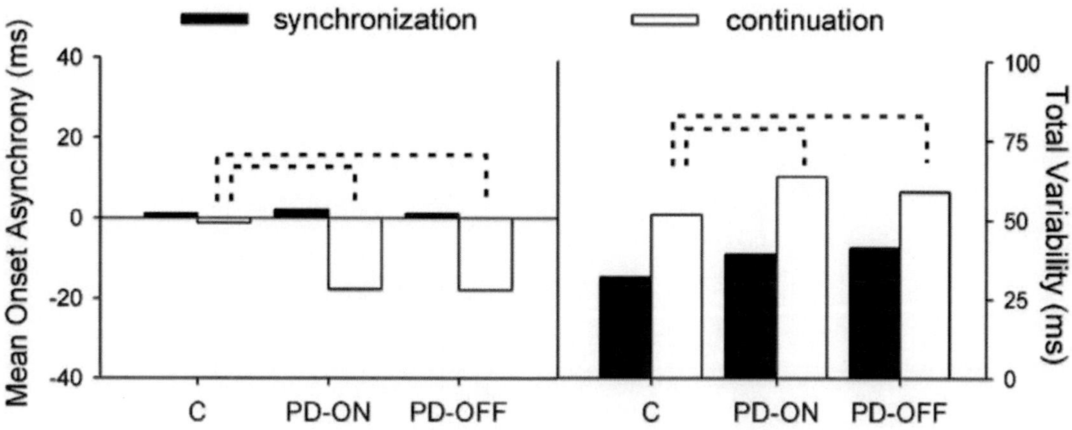

Figure 1. Mean deviation from the 600ms target interval (left panel) and total variability (right panel) as a function of task condition (*synchronization, continuation*) for the controls (C) PD ON, and PD OFF groups. Reprinted from the Journal of International Neuropsychological Society, 9 (7), Elsinger, C.L., Rao, S.M., Zimbelman, J., Reynolds, N., Blindauer, K., Hoffman, R., Neural basis for impaired time reproduction in Parkinson's disease, 1088-1098, 2003.

During performance of the synchronization task, fMRI results for the Controls showed robust activation of the left sensorimotor cortex (SMC), bilateral superior temporal gyrus (STG), and right cerebellum, consistent with findings in young adults (Rao et al., 1997). The PD patients, both on and off medication, showed a reduced spatial extent of activation relative to the Controls, with the only region of overlap occurring in the STG. In the SMC, the area of activation for the PD group (on and off) was located more caudal to that observed in the Control subjects. Levodopa had relatively little effect on the activation observed during performance of the synchronization task.

In the previous fMRI study in young adults (Rao et al., 1997), the continuation condition, which taxes the internal timekeeping system, resulted in additional activation of the medial premotor circuit (SMA, putamen, and thalamus). The older subjects in this study activated the SMA, but not the putamen or thalamus. Activation of the putamen and thalamus, however, was observed in the PD patients when they were on, but not off, medication (see Figure 2). The SMA was not activated in the PD patients while off medication; however, activation was observed during the on state, albeit smaller in spatial extent than the Controls. In addition, the PD patients on medication had more rostral activation of the SMC, similar to the maps of the Control subjects and unlike that of the patients when off medication. Activation was not observed in the cerebellum of the PD patients either on or off medication.

To summarize, there are four key findings to report from this study. First, PD patients demonstrated impairments on the PFT, characterized by a reduction in the IRI during the continuation condition and an increase in total variability during both the synchronization and continuation conditions. Second, PD patients demonstrated an overall reduction in brain activity relative to controls, most notably in the sensorimotor cortex and cerebellum during both the synchronization and continuation conditions, and in the SMA during the continuation condition. Third, despite improvements in the UPDRS motor scale with levodopa treatment, no differences in PFT accuracy and variability were observed in PD patients between the on or off levodopa states. Fourth, despite the absence of a behavioral improvement on the PFT, PD patients taking levodopa exhibited a partial normalization in their brain activation patterns relative to control subjects, characterized by increased activity in medial frontostriatal circuits (SMA, thalamus, and putamen) during the continuation condition only. These findings are discussed below in the context of the broader literature.

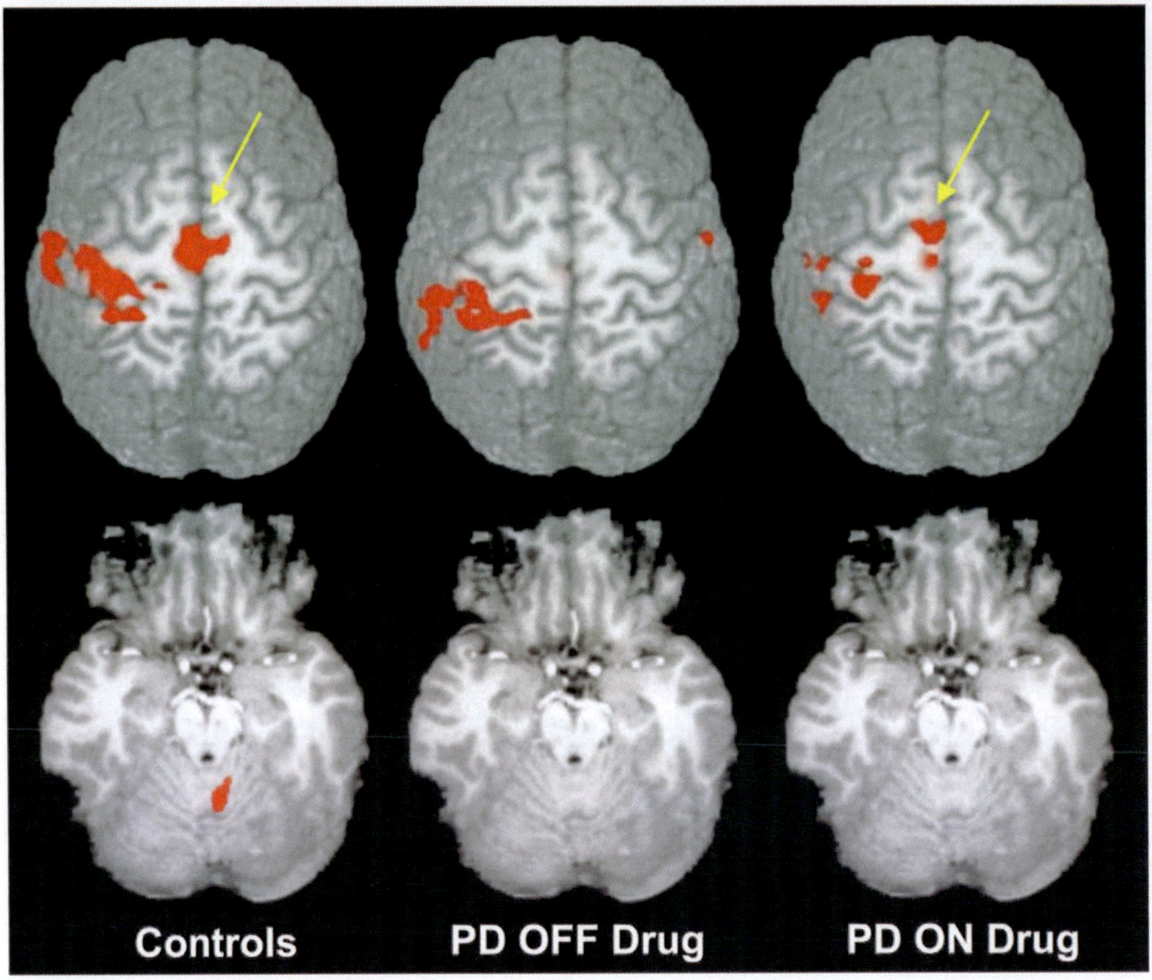

Figure 2. Red areas indicate activity during the continuation task relative to rest. Yellow arrow points to SMA activation in upper panel. Lower panel shows activation in cerebellum for controls only. Adapted from Journal of International Neuropsychological Society, 9 (7), Elsinger, C.L., Rao, S.M., Zimbelman, J., Reynolds, N., Blindauer, K., Hoffman, R., Neural basis for impaired time reproduction in Parkinson's disease, 1088-1098, 2003.

5 Levodopa-Induced Changes in Motor Timing Performance

The behavioral findings from the Elsinger et al., (2003) study are consistent with past reports of increased variability and IRI shortening in PD patients both on and off medications (Ivry and Keele, 1989; O'Boyle et al., 1996; Konczak et al., 1997; Harrington et al., 1998a), and consistent with the idea that PD patients underestimate time intervals due to an altered internal clock (Pastor et al., 1992). Warren Meck, (1996) has suggested that the basal ganglia functions as a clock-counter system, whereby dopaminergic neurons of the substantia nigra terminating on D2 receptors operate as pacemaker units, with pulses accumulating in the dorsal striatum. As PD patients experience a loss of these critical neurons in the substantia nigra, this would alter the number of pulses sent to dorsal striatum, thereby shortening interval estimation at the encoding level.

6 Levodopa-Induced Changes in Neural Activity

The second important finding in the Elsinger et al., (2003) was the overall reduction in brain activity in the sensorimotor cortex and cerebellum during both the synchronization and continuation tasks. The cerebellar region (dorsal dentate nucleus) activated in both young (Rao et al., 1997) and older healthy adults (Elsinger et al., 2003) has been theorized to play a role in sensorimotor integration (Penhune et al., 1998) through supervision and regulation of relevant sensory and cognitive input from the cortex (Desmond et al., 1997), and interpretation and coordination of temporal patterns of incoming sensory stimuli with motor output (Thach, 1998). The lack of cerebellar activity observed in PD patients during both the synchronization and continuation condition (unmodified by the reinstatement of drug), combined with increased response variability (relative to controls) might suggest that the cerebellum is involved in regulating clock processes. An alternative interpretation is that the cerebellum is necessary for integrating sensorimotor information in preparation for a motor response. Activation in the SMC during the PD off state was located more posteriorly than that of controls. This suggests that PD patients may be relying on sensory information from key-presses as feedback to guide upcoming finger movements, perhaps in a compensatory manner. However, given the absence of activity in the cerebellum, patients were not able to make use of this information to enhance performance. The finding that decreased cerebellar activation resulted in significantly increased variability during both the synchronization and continuation task, and not the continuation task exclusively, is consistent with the findings of Harrington et al., (2004b) that deficits in movement timing and response preparation may be due to disruption in the acquisition and maintenance of sensory and cognitive information crucial to task performance and not deficits in timekeeping processes per se.

Additional support for the role of frontal-striatal circuitry in timing comes from an fMRI study in patients with Huntington's disease (HD), another neurodegenerative disorder affecting the basal ganglia. Similar to the pathology of PD, the striatum deteriorates, but rather than resulting in a poverty of movement (as is the case in PD) HD results in an excess of uncontrolled motor activity. Prior to the onset of motor symptoms however, as the striatum shows signs of atrophy time perception abilities show significant impairment (Paulsen et al., 2004). In the Paulsen et al. study, HD patients and healthy controls performed a time perception task (paired with a pitch perception task and control task). Patients in the HD group were subdivided into two groups based on estimated years to diagnosis of manifest disease (Close $<$ 12 years; Far $\geq$ 12 years). Disease onset age was estimated using a regression equation based on number of trinucleotide CAG repeats in the gene that encodes the protein huntingtin (gene IT-15) on chromosome 4 (The Huntington's Disease Collaborative Research Group, 1993). Impairment was observed during the time perception task for Close participants relative to Far participants and Controls, but no differences were observed between Far participants and Controls. Close participants demonstrated reductions in caudate volume bilaterally, while Far participants did not. Functional imaging results demonstrated that the behavioral deficit observed in Close HD patients was associated with significantly decreased activation in the basal ganglia (caudate/thalamus) and preSMA/cingulate region relative to controls. Far HD patients showed an intermediate level of activation in the caudate and thalamus, but an increase in activation in medial structures (anterior cingulate and SMA) relative to the Close HD and Control participants. Enhanced cerebellar activation was not observed in the time minus control or the pitch minus control comparison in healthy controls. No group differences at all were observed in the pitch minus control comparison. This result supports the hypothesis that the basal ganglia are critical for timing processes, while the cerebellum may play a role in processes other than explicit timing.

7 Levodopa-Induced Clinical Improvements and Cognitive Impairments

An important finding in the study described above was that despite improvements in the UPDRS motor scale with levodopa treatment, no differences in PFT accuracy and variability were observed in PD patients between the on or off levodopa states (Elsinger et al., 2003). This finding is consistent with other studies which report improvements in clinical outcome measures (i.e., UPDRS-motor score), while observing no improvement or inconsistent improvement in other measures of perceptual motor behavior (Feigin et al., 2001; 2002). Eidelberg, Feigin and colleagues (Feigin et al., 2001; 2002) examined the relationship between levodopa administration and improvement in functioning and/or metabolic changes in Parkinson's disease. In these studies, the authors used PET imaging to measure resting-state regional brain metabolism and motor activation responses as a function of levodopa administration in patients with PD. They found that metabolic changes (at rest) with levodopa correlated with clinical improvement (UPDRS motor scales, and reduced tremor) and resulted in decreased metabolic activity in the putamen, thalamus, cerebellum and primary motor cortex. In other words, clinical improvement during the resting condition was related to the degree of suppression or modulation observed in cortico-striato-pallido-thalamocortical networks as a function of levodopa administration (Feigin et al., 2001). In a second experiment (Feigin et al., 2002), PD patients were scanned using H^2O/PET while they performed a predictable paced sequence of reaching movements both on and off levodopa medication. The 'timed' response task required subjects to move a cursor on a digitizing table while hand position and target locations were displayed on a computer screen. Target locations were eight positions arranged in a circle (45° apart), with a central circle representing the start area. Targets always appeared in a predictable counterclockwise order, and were presented at constant time intervals in synchrony with a 160 ms tone. Subjects were instructed to anticipate the tone such that they initiated the movement prior to the appearance of the target circle and the movement reversal (back to center) occurred in synchrony with the tone. Dependent measures were movement time (movement onset to reversal) and spatial error (shortest distance of reversal point to center target). The authors found that dopaminergic supplementation improved some aspects of behavioral performance, and worsened others. Moreover, changes in behavioral performance were associated with very specific patterns of increases in functional activation in the basal ganglia and cerebellum. Administration of levodopa improved movement initiation (movement timing) which was correlated with increased activation in the putamen bilaterally, and the left thalamus. Conversely, increased activation in the cerebellar vermis was correlated with accuracy (increased spatial error). The authors note that increases in activation were not observed in the SMA and sensorimotor cortex as a function of levodopa administration. This is not surprising given that this task involved movements that were guided by external cues (both visual and auditory). The sensorimotor integration requirements of this task are greater than those employed in a typical PFT task, thus spatial error (i.e. increased variability) in this task could reflect difficulty with precisely integrating visual *and* auditory information about distance to be traveled with the timing and motor requirements of the task. A principal concern (as in the Elsinger et al. study (2003), is that although patients refrained from taking levodopa medication, they continued with their dopamine agonist therapy. The authors speculate that levodopa administration may interfere with the activity of cerebellocortical pathways, which leads to impairments in motor execution (i.e. increased spatial error). This is consistent with increased variability in performance parameters observed in the study described earlier (Elsinger et al., 2003). Additional support for this idea comes from an imaging study evaluating the effect of pallidal stimulation on spatial and temporal movement accuracy (Fukuda et al., 2001). Using PET at rest and during performance of the motor task used in the Feigin et al. studies (2001; 2002), Fukuda and colleagues examined the mechanism of motor improvement induced by deep brain stimulation (DBS) of the internal segment of the globus pallidus (GPi) in PD patients in a medication-free state. They found that stimulation of the GPi improved movement onset time (reduced timing errors) as well as spatial errors. Stimulation was also associated with increases in activation in the SMA, and left sensorimotor cortex. Unlike results observed with the introduction of levodopa, GPi stimulation resulted in a positive correlation between motor performance and activation in the sensorimotor cortex, thalamus and contralateral cerebellum. Improvements in spatial accuracy were correlated with increases in both cerebellar hemispheres and in the left sensorimotor cortex. Notably, because task parameters (movement frequency, amplitude) were kinematically controlled across DBS conditions, increases in activation responses cannot be attributed to changes in task parameters as a function of DBS intervention.

Other studies have reported that PD patients off medications show increased activation in the cerebellum as a compensatory mechanism for poorly functioning basal ganglia-thalamo-cortical circuitry (Rascol et al., 1997). Although it is possible that cerebello-cortical pathways may be engaged to compensate for deficits in frontostriatal circuitry, methodological concerns limit the conclusions that can be drawn from this work. In this particular study SPECT imaging was used to assess regional cerebral blood flow induced by the execution of a finger-to-thumb opposition task in PD patients. Movement rate was "matched" for the on/off states. Unfortunately, the on/off state of levodopa medication was controlled as a between-subjects factor, precluding a direct comparison of brain activation patterns induced by the motor task within a single subject on and off medication. It is possible that the increased cerebellum activation in the off condition was due to increased dependence on sensory information from finger movements. In the absence of dopamine intervention cerebello-cortical pathways may not have been disrupted, allowing the afferent somatosensory information to guide movements.

8 An Integrative Theory of the Role of Cortical and Subcortical Systems in Timing

Taken together, evidence from functional imaging studies in healthy adults, and adults with neuropathology, support the role of cortico-striatal systems in timekeeping behavior. How do we develop a framework for describing how the basal ganglia and cerebellum may differentially contribute to behaviors that require ongoing control of timing and working memory functions? Further, in the context of pathology of the basal ganglia, do cerebello-cortical circuits compensate for reduced striatal function?

Factors such as early learning and motor adaptation phases of learning, where adjustment and monitoring of sensory-motor information is critical, have been hypothesized to be the role of cortico-cerebellar systems (Doyon et al., 2003; Wu et al., 2004). Evidence suggests that the cerebellum may modulate attention and sensory responsiveness, which are necessary preparatory actions that optimize the successful acquisition and analysis of relevant sensory information during anticipation of an expected stimulus or during exploration of a novel environment (Allen et al., 1997; Courchesne and Allen, 1997). Anticipation of imminent sensory events improves the likelihood that cognitive resources will be directed effectively, with the goal of facilitating the coordination of incoming sensory information with motor responses. If we assume that the principal role of the basal ganglia is as the central timing mechanism, or clock, the cerebellum can be viewed as a mechanism which predicts the internal conditions for a particular motor activity. That is, given the strong physiological interconnectivity the cerebellum has with basal ganglia, cerebrum, limbic system, brain stem and spinal cord (Braitenberg, 2002), it is ideally placed to integrate sensory information from multiple sources thereby facilitating preparatory processes for action. This theory predicts that a novel complex task will lead to strong cerebellar activation, but once the task has been learned, activation in the cerebellum may decrease. This idea is supported by results showing that patterns of cerebellar activation change as a function of extended practice (Parsons et al., 2004). In particular, Parsons et al. speculate that declining activity in the cerebellum during early stages of learning may be due to a relatively quick optimization of sensorimotor input (determining internal state) in addition to more gradual adjustments that take place for monitoring other processes which relate to planning spatiotemporal properties of movements (preparatory processes for action).

In the context of this theory, it is possible to reconcile seemingly discrepant findings in the timing literature described above. If dopaminergic interventions negatively affect cerebello-thalamo-cortical circuits, function of the cerebellum will be altered. In the context of a task that requires precise timekeeping processes, this would disrupt the acquisition of sensory and cognitive information crucial to task performance and preclude the maintenance of an effective representation of temporal information in working memory. Thus, precise timekeeping is dependent on frontal-striatal circuitry for proper clock functions, and cerebello-cortical circuitry is required for integration and optimization of sensory information which facilitates preparatory processes for perceptual motor behavior.

Acknowledgments

The author would like to thank Stephen M. Rao, Deborah L. Harrington, and Sean Hinton for their substantial contributions to the experiments described in this chapter. The study by Elsinger et al. (2003) was funded by grants from the Charles Dana Foundation, Milwaukee Foundation's Parsons Fund, and the National Institutes of Health (to S.M.R.), the Medical College of Wisconsin General Clinical Research Center, and the W.M. Keck Foundation.

References

Alexander G, DeLong M, PL S (1986) Parallel organization of functionally segregated circuits linking basal ganglia and cortex. In: *Annual Review of Neuroscience* (Cowan W, ed), pp 357-381. Washington, DC: Society for Neuroscience.

Allen G, Buxton RB, Wong EC, Courchesne E (1997) Attentional activation of the cerebellum independent of motor involvement. *Science* **275**:1940-1943.

Braitenberg V (2002) In defense of the cerebellum. *Ann N Y Acad Sci* **978**:175-183.

Brooks DJ, Ibanez V, Sawle GV, Quinn N, Lees AJ, Mathias CJ, Bannister R, Marsden CD, Frackowiak RS (1990) Differing patterns of striatal 18F-dopa uptake in Parkinson's disease, multiple system atrophy, and progressive supranuclear palsy. *Ann Neurol* **28**:547-555.

Casini L, Ivry RB (1999) Effects of divided attention on temporal processing in patients with lesions of the cerebellum or frontal lobe. *Neuropsychology* **13**:10-21.

Courchesne E, Allen G (1997) Prediction and preparation, fundamental functions of the cerebellum. *Learn Mem* **4**:1-35.

Desmond JE, Gabrieli JD, Wagner AD, Ginier BL, Glover GH (1997) Lobular patterns of cerebellar activation in verbal working-memory and finger-tapping tasks as revealed by functional MRI. *J Neurosci* **17**:9675-9685.

Doyon J, Penhune V, Ungerleider LG (2003) Distinct contribution of the cortico-striatal and cortico-cerebellar systems to motor skill learning. *Neuropsychologia* **41**:252-262.

Elsinger CL, Rao SM, Zimbelman JL, Reynolds NC, Blindauer KA, Hoffmann RG (2003) Neural basis for impaired time reproduction in Parkinson's disease: an fMRI study. *J Int Neuropsychol Soc* **9**:1088-1098.

Feigin A, Fukuda M, Dhawan V, Przedborski S, Jackson-Lewis V, Mentis MJ, Moeller JR, Eidelberg D (2001) Metabolic correlates of levodopa response in Parkinson's disease. *Neurology* **57**:2083-2088.

Feigin A, Ghilardi MF, Fukuda M, Mentis MJ, Dhawan V, Barnes A, Ghez CP, Eidelberg D (2002) Effects of levodopa infusion on motor activation responses in Parkinson's disease. *Neurology* **59**:220-226.

Ferrandez AM, Hugueville L, Lehericy S, Poline JB, Marsault C, Pouthas V (2003) Basal ganglia and supplementary motor area subtend duration perception: an fMRI study. *Neuroimage* **19**:1532-1544.

Fukuda M, Mentis M, Ghilardi MF, Dhawan V, Antonini A, Hammerstad J, Lozano AM, Lang A, Lyons K, Koller W, Ghez C, Eidelberg D (2001) Functional correlates of pallidal stimulation for Parkinson's disease. *Ann Neurol* **49**:155-164.

Gibbon J, Church RM, Meck WH (1984) Scalar timing in memory. *Ann N Y Acad Sci* **423**:52-77.

Harrington D, Lee R, Boyd L, Rapcsak S, Knight R (2004a) Reply to: Evaluating the role of the cerebellum in temporal processing: beware of the null hypothesis. *Brain* **127**:E14.

Harrington DL, Haaland KY (1999) Neural underpinnings of temporal processing: a review of focal lesion, pharmacological, and functional imaging research. *Rev Neurosci* **10**:91-116.

Harrington DL, Haaland KY, Hermanowicz N (1998a) Temporal processing in the basal ganglia. *Neuropsychology* **12**:3-12.

Harrington DL, Haaland KY, Knight RT (1998b) Cortical networks underlying mechanisms of time perception. *J Neurosci* **18**:1085-1095.

Harrington DL, Lee RR, Boyd LA, Rapcsak SZ, Knight RT (2004b) Does the representation of time depend on the cerebellum? Effect of cerebellar stroke. *Brain* **127**:561-574.

Harrington DL, Boyd LA, Mayer AR, Sheltraw DM, Lee RR, Huang M, Rao SM (2004c) Neural representation of interval encoding and decision making. *Brain Res Cogn Brain Res* **21**:193-205.

Hinton SC, Meck WH (2004) Frontal-striatal circuitry activated by human peak-interval timing in the supra-seconds range. *Brain Res Cogn Brain Res* **21**:171-182.

Ivry R, Keele S (1989) Timing functions of the cerebellum. *J Cog Neurosci* **1**:136-152.

Ivry RB (1996) The representation of temporal information in perception and motor control. *Curr Opin Neurobiol* **6**:851-857.

Ivry RB, Hazeltine RE (1995) Perception and production of temporal intervals across a range of durations: evidence for a common timing mechanism. *J Exp Psychol Hum Percept Perform* **21**:3-18.

Ivry RB, Keele SW, Diener HC (1988) Dissociation of the lateral and medial cerebellum in movement timing and movement execution. *Exp Brain Res* **73**:167-180.

Konczak J, Ackermann H, Hertrich I, Spieker S, Dichgans J (1997) Control of repetitive lip and finger movements in Parkinson's disease: influence of external timing signals and simultaneous execution on motor performance. *Mov Disord* **12**:665-676.

Malapani C, Rakitin B, Levy R, Meck WH, Deweer B, Dubois B, Gibbon J (1998) Coupled temporal memories in Parkinson's disease: a dopamine-related dysfunction. *J Cogn Neurosci* **10**:316-331.

Matell MS, Meck WH, Nicolelis MA (2003) Interval timing and the encoding of signal duration by ensembles of cortical and striatal neurons. *Behav Neurosci* **117**:760-773.

Meck WH (1996) Neuropharmacology of timing and time perception. Brain Res Cogn *Brain Res* **3**:227-242.

Meck WH, Benson AM (2002) Dissecting the brain's internal clock: how frontal-striatal circuitry keeps time and shifts attention. *Brain Cogn* **48**:195-211.

O'Boyle DJ, Freeman JS, Cody FW (1996) The accuracy and precision of timing of self-paced, repetitive movements in subjects with Parkinson's disease. *Brain* **119** *(Pt 1)*:51-70.

Parsons M, Harrington D, Rao S (2004) Distinct neural systems underlie learning visuomotor and spatial representations of motor skills. *Hum Brain Mapp* **24**:229-247.

Pastor M, Artieda J (1996) Involvement of the basal ganglia in timing perceptual and motor tasks. In: *Time, internal clocks and movement* (Pastor M, Artieda J, eds), pp 235-255: Elsevier Science B.V.

Pastor MA, Artieda J, Jahanshahi M, Obeso JA (1992) Time estimation and reproduction is abnormal in Parkinson's disease. *Brain* **115** *Pt 1*:211-225.

Paulsen JS, Zimbelman JL, Hinton SC, Langbehn DR, Leveroni CL, Benjamin ML, Reynolds NC, Rao SM (2004) fMRI biomarker of early neuronal dysfunction in presymptomatic Huntington's Disease. *AJNR Am J Neuroradiol* **25**:1715-1721.

Penhune VB, Zattore RJ, Evans AC (1998) Cerebellar contributions to motor timing: a PET study of auditory and visual rhythm reproduction. *J Cogn Neurosci* **10**:752-765.

Rao SM, Mayer AR, Harrington DL (2001) The evolution of brain activation during temporal processing. *Nat Neurosci* **4**:317-323.

Rao SM, Harrington DL, Haaland KY, Bobholz JA, Cox RW, Binder JR (1997) Distributed neural systems underlying the timing of movements. *J Neurosci* **17**:5528-5535.

Rascol O, Sabatini U, Fabre N, Brefel C, Loubinoux I, Celsis P, Senard JM, Montastruc JL, Chollet F (1997) The ipsilateral cerebellar hemisphere is overactive during hand movements in akinetic parkinsonian patients. *Brain* **120** *(Pt 1)*:103-110.

Sergent V, Hellige JB, Cherry B (1993) Effects of responding hand and concurrent verbal processing on time-keeping and motor-implementation processes. *Brain Cogn* **23**:243-262.

Thach WT (1998) A role for the cerebellum in learning movement coordination. *Neurobiol Learn Mem* **70**:177-188.

The Huntington's Disease Collaborative Research Group (1993) A novel gene containing a trinucleotide repeat that is expanded and unstable on Huntington's disease chromosomes. The Huntington's Disease Collaborative Research Group. *Cell* **72**:971-983.

Wing A, Kristofferson A (1973) Response delays and the timing of discrete motor responses. *Percept Psychophys* **14**:5-12.

Wu T, Kansaku K, Hallett M (2004) How self-initiated memorized movements become automatic: a functional MRI study. *J Neurophysiol* **91**:1690-1698.

In: Recent Breakthroughs in Basal Ganglia Research
Editor: Erwan Bezard, pp. 163-169

ISBN: 1-59454-880-3

Chapter 13

Functional Segregation and Integration in the Motor Thalamus

Paul van Donkelaar
Department of Human Physiology, University of Oregon, USA
R. Chris Miall
Behavioral Brain Sciences Centre, School of Psychology, University of Birmingham, U.K.

Abstract

The motor thalamus receives input from both the basal ganglia and cerebellum and projects to a number of different motor areas in the cerebral cortex. This chapter reviews recent research which attempts to provide insight into the extent to which these different projections are functionally segregated. This research demonstrates that there is some overlap in function for the basic context under which movements occur, but a unique contribution of basal ganglio-thalamo-cortical pathways to the integration of movement and reward-related activity.

Summary

The motor thalamus receives input from both the basal ganglia and cerebellum and subsequently projects to a number of different motor areas in the cortex. The relative segregation of these two ascending pathways has led to the notion that they make different contributions to the control of movement. In this chapter, we review evidence that supports this contention as well as some which suggests that the functional anatomy at the level of the thalamus is not as distinct as was once thought.

The control of movement requires processing in a variety of structures throughout the central nervous system including the basal ganglia and cerebellum. The fact that these two subcortical structures make different contributions to motor control is clear from the characteristic deficits that arise when each is damaged. Individuals with Parkinson's disease resulting from degeneration of the dopamine-producing neurons in the substantia nigra pars reticulata have a great deal of difficulty producing movements based on internal cues (Jackson et al., 1995; Morris et al., 1996; Ketcham et al., 2003). By contrast, individuals with damage to the cerebellum display deficits in generating movements triggered and guided by external sensory cues (van Donkelaar and Lee, 1994; Crowdy et al., 2000; Morton and Bastian, 2004). Interestingly, when these two groups of patients attempt movements engaging the undamaged subcortical pathway their deficits are substantially reduced. Thus, Parkinson's patients produce much quicker and more coordinated movements when provided with external sensory cues engaging cerebellar processing (Azulay et al., 1999) and patients with cerebellar damage generate less variable motor output when asked to do so based on internal cues (van Donkelaar and Lee, 1994).

The basal ganglia and cerebellum each send ascending projections to a number of different motor cortical areas via the ventrolateral or motor thalamus. The motor thalamus itself is composed of several different subnuclei each of which receives varying levels of input from both the basal ganglia and cerebellum. Originally, it was thought that these ascending projections were completely segregated and fairly straightforward: the basal ganglia projects via the oral portion of the ventral lateral nucleus (VLo) and the parvocellular portion of the ventral anterior nucleus (VApc) mainly to the supplementary motor area and the motor cortex; by contrast, the cerebellum projects via the ventral posterolateral nucleus (VPLo) and area X of the thalamus mainly to the premotor and motor cortices. More recently, this pattern of connectivity has been re-evaluated and it has become clear that it is much more complicated than originally thought. These complications are made worse by the fact that different groups working on the neuroanatomy of these ascending projections use different nomenclature and have different criteria for segregating the nuclei within the motor thalamus. Based on current knowledge, there appears to be graded input from both the basal ganglia and cerebellum to each of the subnuclei in the motor thalamus. Within each of these subnuclei, however, the projections from the basal ganglia and cerebellum appear to be interdigitated but still segregated. Indeed, it has been demonstrated that in transitional zones between adjacent nuclei, both basal ganglia and cerebellar efferents are located in close proximity to the same thalamocortical neuron (Sakai et al., 1996). In other words, a cluster of cells that receive input from the basal ganglia may be located adjacent to a cluster receiving input from the cerebellum. Thus, instead of the idea that basal ganglia and cerebellar inputs are completely segregated across different subnuclei in the motor thalamus the more parsimonious explanation is that the relative number or volume of clusters receiving from the basal ganglia and cerebellum is modulated across the different thalamic subnuclei.

This pattern of interdigitation and segregation is carried forward at the level of the cortex. Thus, each cortical motor area appears to receive projections from both the basal ganglia and cerebellum in varying proportions, yet the receiving cells remain more or less segregated. In particular, none of the neuroanatomical studies which have simultaneously injected label into both the cerebellum and basal ganglia have found evidence of doubly-stained cells in the receiving subnuclei of the thalamus. A similar claim can be made for the subsequent thalamocortical projections.

This fine-grained anatomical segregation suggests that function may be segregated as well within each of the thalamic subnuclei and motor cortical areas. In a series of recent experiments (van Donkelaar et al., 1999, 2000) we attempted to address this issue by recording from and temporarily inactivating cells in each of the thalamic subnuclei that receive basal ganglia and cerebellar projections in macaque monkeys trained to make limb movements. We examined the extent to which each subnucleus contributed to the planning, initiation, control, and reward related aspects of task performance. To further differentiate the functional specificity of each subnucleus we trained the monkeys to perform the limb movements in response to the appearance of a visual target (a visually-triggered condition) or to initiate the movement based on their own internal cues (an internally-generated condition). Thus, in the former condition an external stimulus provided information about when and where to move, whereas in the latter condition the animal decided when to move on their own.

Our results demonstrated that the four thalamic subnuclei we tested showed a continuum of functional preferences for either the visually-triggered or internally-generated movements that mirrored the pattern of connectivity from the basal ganglia and cerebellum. In particular, Area X, which receives a substantial projection from the portion of the cerebellar dentate nucleus that is involved in the visual control of movement (Mushiake and Strick, 1993) contained a large proportion of cells (~50 %) that displayed activity only in the visually-triggered condition (i.e., the activity was present during visually-triggered trials but not in the internally-generated trials). Moreover, when this area was temporarily inactivated with lignocaine the animal displayed markedly increased reaction times and trajectory abnormalities during the visually-triggered trials that were not present when the internally-generated movements were made. In contrast to this pattern of results, the exact opposite finding was obtained in VApc, which receives a large input from that portion of the globus pallidus that has been shown to be involved in the planning and control of internally-generated movements (Mushiake and Strick, 1995). Here a large proportion of cells (50 %) were active only during internally-generated movements; moreover, the animal displayed slower and smaller movements in this condition but not the visually-triggered condition when this subnucleus was temporarily inactivated with lignocaine. Thus, this pattern of results suggests that Area X is mainly concerned with the planning and control of visually-triggered movements, and VApc with internally-generated movements.

The results from VLo and VPLo fell in between these two extremes, as the proportional input from the basal ganglia and cerebellum would predict. In particular, the most common cells (~50-60 %) recorded from in these two subnuclei were equally active in both visually-triggered and internally-generated tasks.

The only characteristic that differentiated these two subnuclei was the fact that the second largest group of cells in each area had activity that corresponded with the putative functional input from the basal ganglia or cerebellum respectively. Specifically, about 30 % of the cells in VPLo responded only in the visually-triggered task, whereas approximately the same proportion in VLo responded only in the internally-generated task. When VPLo was inactivated with lignocaine, the animal became hemiplegic in the contralateral limb for ~5-30 minutes. When movements were produced they were slower and more variable in the visually-triggered task but relatively normal in the internally-generated task. A similar, though less severe, deficit limited to the initiation of the visually-triggered movements was observed following inactivation of VLo. Taken together, the results from VPLo and VLo are contrary to what one would predict if there was a strict functional segregation between the basal ganglia and cerebellar inputs to the motor thalamus. Rather, cells in both of these subnuclei appear to contribute to movements in general regardless of the specific context under which they are produced. This implies that some portions of the ascending projections from both the cerebellum and basal ganglia, relaying through the thalamus and onto the motor areas of the cortex, are functionally concerned with the most general aspects of movement programming common to all contexts. In this regard it may be informative to re-examine some of the deficits associated with cerebellar damage and Parkinson's disease. In particular, there are certain similarities in the motor deficits in these two disease states – slowness of movement initiation and completion – that are often overlooked so that the differences can be highlighted. A more intriguing idea may be that both cerebellar and Parkinson's patients have a common deficit in planning and controlling movement that is exacerbated when either external or internal cues, respectively, are used to drive the movement. This is certainly consistent with the findings from our studies which show that there is a progressive modulation in the extent to which each thalamic subnucleus contributes to movements based on external or internal cues.

Another general aspect of movement performance is related to the motivation to complete the task. Clearly, if someone is not motivated to perform a task then it is less likely to be completed than if they are motivated. The difference in effort shown by a child cleaning their room versus playing a video game comes to mind. At some level in the brain the activity related to task performance must interact with activity related to the motivation for that task, which itself is linked to the expectation of reward.

The basal ganglia are ideally situated to perform this integrative function (Schultz, 1998). The ventral striatum and caudate in particular have been shown to contain separate groups of cells that become activated in relation to motor output or reward delivery (Schultz et al., 1995). Indeed, behaviorally task related ventral striatal and caudate cells can be modulated by reward expectancy (Hollerman et al., 1998; Kawagoe et al., 1998). But to date few cells have been reported that explicitly combine motor and reward related information. One possibility is that these signals are integrated further along the pathway connecting the ventral striatum and caudate to the globus pallidus and thalamus (Selemon and Goldman-Rakic, 1990; Hedreen and DeLong, 1991; Strick et al., 1995).

In the course of measuring the movement-related thalamic activity that we have described above (van Donkelaar et al., 1999, 2000), we noticed that a large proportion of cells in the basal ganglia-, but not the cerebellar-receiving portion of the motor thalamus were active after the movement was completed, while the reward was being delivered. These cells were initially defined as movement related if their activity was modulated when the animal reached for a food reward. Further probing of their responses to passive arm movements, postural adjustments or licking/chewing movements ensured that only cells with clear arm movement-related activity were subsequently tested.

Because the main task was designed to contrast external and internally triggered actions, reward delivery and the monkey's return movement were nearly simultaneous and, thus, the activity described above could be due to either event. To gain insight into this potential confound, in one animal we tested 20 cells from VLo and VApc and 11 from Area X and VPLo while the animal was pulled approximately 30 cm away from the experimental apparatus. From this position, he could still see the target and could see but not reach the manipulandum used to generate the reaching movements. The experimenter then held the manipulandum and performed the reaching movement for which the animal received the normal food reward. Figure 1 shows the activity of a cell from VPLo during active movement (Fig. 1a) and during passive observation of the movement (Fig. 1b). In both cases a reward was delivered at the end of the movement, yet the activity of the cell was modulated only in the active condition. Figures 1d and 1e display a cell from VLo that continued to fire during the reward delivery period even when no overt movement was performed by the animal. Clearly, this cell was activated during both movement generation and reward delivery. This trend was confirmed for the group of cells recorded from each subnucleus by comparing the activity in the active movement and passive observation conditions prior to, during, and after the movement. This analysis revealed that cells in the basal ganglia-receiving portion of the motor thalamus responded differently across the 3 time periods as a function of the movement condition (Fig. 1f). Activity during the pre-movement period in the passive observation condition was lower than during all three periods in the active generation condition. By contrast, the activity during the post-movement period in the passive observation condition was not significantly different from that observed in the active generation condition. This interaction was not present, however, for the cerebellar-receiving cells (Fig. 1c). Thus, the cells we recorded in the basal ganglia-receiving subnuclei appear to code for both movement generation and reward delivery, whereas cerebellar-receiving cells do not.

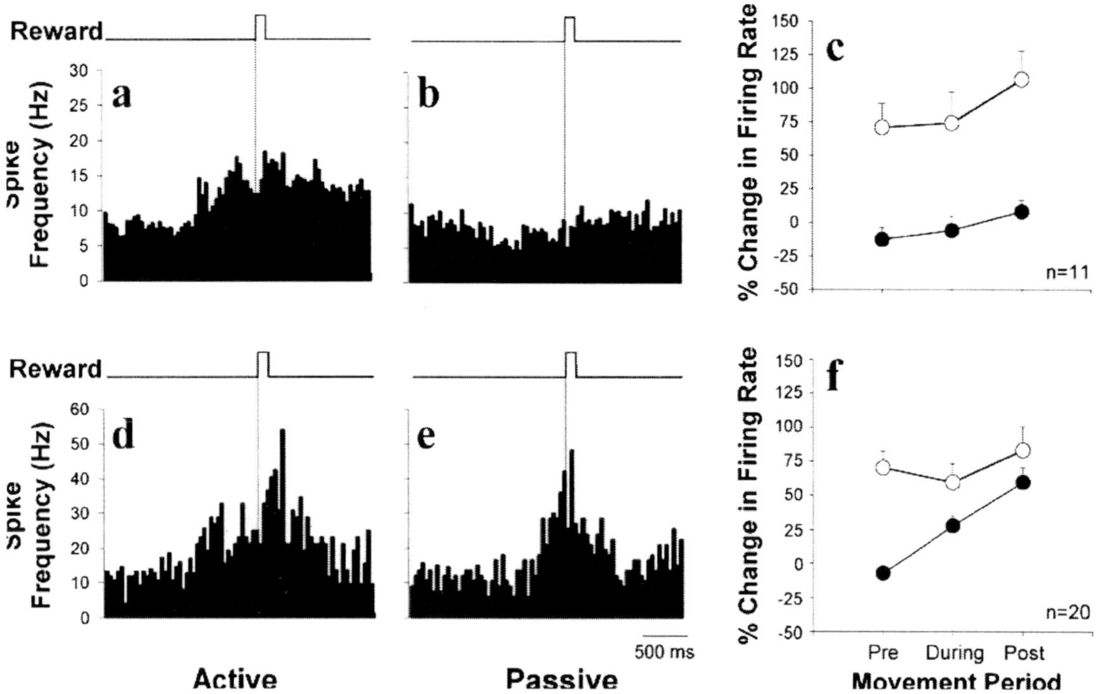

Figure 1. A cell in cerebellar-receiving motor thalamus active during and after active movement (a) but unmodulated during observation of the experimenter performing the movement (b). A cell from basal ganglia-receiving motor thalamus displaying increased firing rates during both active movement (d) and passive observation (e). Group mean change in firing rate (*error bars*, between cell SE) in the active movement (open symbols) and passive observation (solid symbols) conditions for cells in cerebellar-receiving (c) and basal ganglia-receiving territories (f).

To further characterize this putative reward-related activity, we trained a third animal to alternately push and pull the lever between near and far target zones. The task was similar to the visually-triggered reaching task described above, except that a variable hold period of at least 500 ms was introduced upon capture of the target, before reward was signaled with a tone and food delivery. In alternate blocks the animal was rewarded at the near or far target position, to dissociate hand position and reward. Critically, on a proportion of randomly interleaved trials, success was not rewarded with either a tone or food delivery.

Figure 2a displays the activity of a typical basal ganglia-receiving cell that possessed both movement and reward related activity. When the reward was withheld the late burst of activity was no longer present (Fig. 2b). Of 49 cells tested in this paradigm, 28 (57 %) showed significant modulation in activity during both the movement and reward periods. The activity in a much smaller number of cells was related purely to the movement (4/49 – 8 %), purely to the reward (3/49 – 6 %) or purely to licking responses (2/49 – 4 %). Of the 28 movement-reward related cells, 8 showed responses that could not be fully dissociated from the licking activity. The remaining 20/49 (41 %) movement-reward cells were not lick-related. Thus, the majority of movement related cells that we tested in the basal ganglia-receiving portion of the thalamus in this paradigm appeared to combine movement and reward signals. These data demonstrate that in addition to carrying movement related activity these cells also encode some aspect of reward occurrence. By contrast, cells in the cerebellar-receiving portion of the motor thalamus did not respond to rewards.

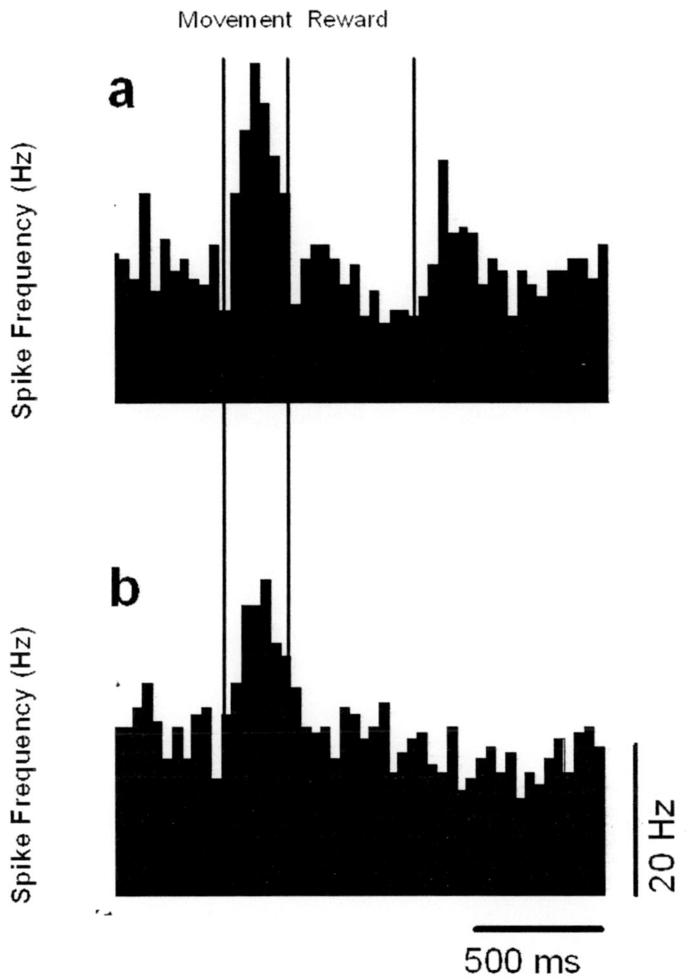

Figure 2. A basal ganglia-receiving thalamic cell with both movement and reward related activity. (a) Movement for reward; (b) movement without reward. Vertical lines: 1st, average movement onset; 2nd, movement offset; 3rd, reward delivery.

Given that we observed reward-related activity in a site that is typically implicated in movements encoding, the question arises concerning its source. Sidibé and colleagues (Sidibe et al., 1997) have recently demonstrated that the rostromedial pole and the dorsal one-third of the internal segment of the globus pallidus (GPi) receive input, respectively, from the ventral striatum and caudate. These latter two parts of the basal ganglia contain cells that carry reward-related signals (Schultz et al., 1995; Hollerman et al., 1998; Kawagoe et al., 1998; Schultz, 1998). Moreover, Sidibé and coworkers (Sidibe et al., 1997) have shown that the ventral striatal- and caudate-receiving portions of the GPi project to several different thalamic nuclei including those that have previously been shown to possess movement-related activity (Anderson and Turner, 1991; Strick et al., 1995). They could therefore provide the reward-related signals we observed. Alternatively, the reward signal could arrive via corticothalamic projections from prefrontal cortical areas (Preuss and Goldman-Rakic, 1987) that have also been shown to possess reward-related activity (e.g., Inoue et al., 1985; Leon and Shadlen, 1999; Hikosaka and Watanabe, 2000). Thus, there appears to be a neuroanatomical substrate for the functional convergence of movement and reward-related activity in the basal ganglia-receiving portion of the motor thalamus.

This convergence represents an explicit mechanism by which reward information can be integrated into signals encoding movement generation. To our knowledge, this combination of disparate information has not been demonstrated in previous research addressing the relationship between movement and reward processing. A number of studies have shown clear modulation of behavior-related activity by reward prediction or expectation both within the basal ganglia (Hollerman et al., 1998; Kawagoe et al., 1998) and some of its cortical targets (Shima and Tanji, 1998; Tremblay and Schultz, 1999). Cells that respond in this manner appear to provide information as to whether a specific behavior is likely to be rewarded given past experience. Thus, when combined with the results from the present study, a network of cortical and subcortical sites emerges by which movement and reward signals interact.

Because single cells in the basal ganglia-receiving portion of the thalamus code for both intentional arm movements and for rewards, this structure may be the key area that integrates movement and motivational information and, thus, provide the key link that drives goal-directed task performance. This may help to explain why the pathway linking the basal ganglia to the thalamus is intimately involved in internally-generated movements – such movements require an internally-generated signal that motivates the subject to initiate the response. Such a signal is less relevant during externally-triggered movements where the appearance of the target provides an initial motivating signal that helps to drive the response. It may also provide some insight into the "poverty" of movement that is typically observed in Parkinson's disease and "abulia" or lack of will that often occurs following caudate lesions (Bhatia and Marsden, 1994) – if the thalamic cells that integrate movement and motivation signals receive deficient input from the basal ganglia, then the patient may feel unmotivated to make the effort to move. Indeed, in our inactivation studies we noted that infusion of lignocaine into VLo led to a reduction in the number of attempts to move during each block of trials. However, unlike with VPLo inactivation this did not appear to be due to hemiplegia because the animals were able to move their limbs spontaneously outside of the trained experimental tasks. This is consistent with the notion that VLo cells integrate movement information with signals associated with the motivation to move derived in this case from the rewards provided to the animal. Taken together, this evidence provides new insight into the role of the basal ganglia-receiving portion of the thalamus in the linking of movement and motivation – something that is vital to normal task performance. It also clearly demonstrates a functional dissociation between the basal ganglia- and cerebellar-receiving parts of the motor thalamus.

Conclusion

In conclusion, our experiments have demonstrated that certain parts of the motor thalamus are functionally segregated whereas other parts contribute equivalently to movements under different contexts. The segregation is most obvious in the movement-related activity of cells in Area X and VApc and in the effects of inactivation in these areas. It is much less apparent, although still present to a certain extent, in VPLo and VLo. Indeed, the fact that the majority of cells in these latter two thalamic subnuclei respond equally in movements that are visually-triggered and internally-generated indicates that there is much less functional segregation between these components of the projections from the basal ganglia and cerebellum as was once thought.

Acknowledgments

The authors would like to thank Richard E. Passingham, John F. Stein, and Erika Lorencz for their substantial contributions to the experiments described in this chapter. This research was funded by grants from the Wellcome Trust and the Medical Research Council to R.C. Miall and R.E. Passingham.

References

Anderson ME, Turner RS (1991) Activity of neurons in cerebellar-receiving and pallidal-receiving areas of the thalamus of the behaving monkey. *J Neurophysiol* **66**:879-893.

Azulay JP, Mesure S, Amblard B, Blin O, Sangla I, Pouget J (1999) Visual control of locomotion in Parkinson's disease. *Brain* **122** (Pt 1):111-120.

Bhatia KP, Marsden CD (1994) The behavioural and motor consequences of focal lesions of the basal ganglia in man. *Brain* **117** (Pt 4):859-876.

Crowdy KA, Hollands MA, Ferguson IT, Marple-Horvat DE (2000) Evidence for interactive locomotor and oculomotor deficits in cerebellar patients during visually guided stepping. *Exp Brain Res* **135**:437-454.

Hedreen JC, DeLong MR (1991) Organization of striatopallidal, striatonigral, and nigrostriatal projections in the macaque. *J Comp Neurol* **304**:569-595.

Hikosaka K, Watanabe M (2000) Delay activity of orbital and lateral prefrontal neurons of the monkey varying with different rewards. *Cereb Cortex* **10**:263-271.

Hollerman JR, Tremblay L, Schultz W (1998) Influence of reward expectation on behavior-related neuronal activity in primate striatum. *J Neurophysiol* **80**:947-963.

Inoue M, Oomura Y, Aou S, Nishino H, Sikdar SK (1985) Reward related neuronal activity in monkey dorsolateral prefrontal cortex during feeding behavior. *Brain Res* **326**: 307-312.

Jackson SR, Jackson GM, Harrison J, Henderson L, Kennard C (1995) The internal control of action and Parkinson's disease: a kinematic analysis of visually-guided and memory-guided prehension movements. *Exp Brain Res* **105**:147-162.

Kawagoe R, Takikawa Y, Hikosaka O (1998) Expectation of reward modulates cognitive signals in the basal ganglia. *Nat Neurosci* **1**:411-416.

Ketcham CJ, Hodgson TL, Kennard C, Stelmach GE (2003) Memory-motor transformations are impaired in Parkinson's disease. *Exp Brain Res* **149**:30-39.

Leon MI, Shadlen MN (1999) Effect of expected reward magnitude on the response of neurons in the dorsolateral prefrontal cortex of the macaque. *Neuron* **24**:415-425.

Morris ME, Iansek R, Matyas TA, Summers JJ (1996) Stride length regulation in Parkinson's disease. Normalization strategies and underlying mechanisms. *Brain* **119** (Pt 2):551-568.

Morton SM, Bastian AJ (2004) Prism adaptation during walking generalizes to reaching and requires the cerebellum. *J Neurophysiol* **92**:2497-2509.

Mushiake H, Strick PL (1993) Preferential activity of dentate neurons during limb movements guided by vision. *J Neurophysiol* **70**:2660-2664.

Mushiake H, Strick PL (1995) Pallidal neuron activity during sequential arm movements. *J Neurophysiol* **74**:2754-2758.

Preuss TM, Goldman-Rakic PS (1987) Crossed corticothalamic and thalamocortical connections of macaque prefrontal cortex. *J Comp Neurol* **257**:269-281.

Sakai ST, Inase M, Tanji J (1996) Comparison of cerebellothalamic and pallidothalamic projections in the monkey (Macaca fuscata): a double anterograde labeling study. *J Comp Neurol* **368**:215-228.

Schultz W (1998) Predictive reward signal of dopamine neurons. *J Neurophysiol* **80**:1-27.

Schultz W, Apicella P, Romo R, Scarnati E (1995) Context-dependent activity in primate striatum reflecting past and future behavioral events. In: *Models of Information Processing in the Basal Ganglia* (Houk J, Davis J, Beiser D, eds), pp 11-28. Cambridge, MA: MIT Press.

Selemon LD, Goldman-Rakic PS (1990) Topographic intermingling of striatonigral and striatopallidal neurons in the rhesus monkey. *J Comp Neurol* **297**:359-376.

Shima K, Tanji J (1998) Role for cingulate motor area cells in voluntary movement selection based on reward. *Science* **282**:1335-1338.

Sidibe M, Bevan MD, Bolam JP, Smith Y (1997) Efferent connections of the internal globus pallidus in the squirrel monkey: I. Topography and synaptic organization of the pallidothalamic projection. *J Comp Neurol* **382**:323-347.

Strick P, Dum R, Picard N (1995) Macro-organization of the circuits connecting the basal ganglia with the cortical motor areas. In: *Models of Information Processing in the Basal Ganglia* (Houk J, Davis J, Beiser D, eds), pp 117-130. Cambridge, MA: MIT Press.

Tremblay L, Schultz W (1999) Relative reward preference in primate orbitofrontal cortex. *Nature* **398**:704-708.

van Donkelaar P, Lee RG (1994) Interactions between the eye and hand motor systems: disruptions due to cerebellar dysfunction. *J Neurophysiol* **72**:1674-1685.

van Donkelaar P, Stein JF, Passingham RE, Miall RC (1999) Neuronal activity in the primate motor thalamus during visually triggered and internally generated limb movements. *J Neurophysiol* **82**:934-945.

van Donkelaar P, Stein JF, Passingham RE, Miall RC (2000) Temporary inactivation in the primate motor thalamus during visually triggered and internally generated limb movements. *J Neurophysiol* **83**:2780-2790.

In: Recent Breakthroughs in Basal Ganglia Research
Editor: Erwan Bezard, pp. 171
ISBN: 1-59454-880-3

PART V: OCCULOMOTRICITY AND BASAL GANGLIA

John G. McHaffie

Department of Neurobiology and Anatomy, Wake Forest University School of Medicine, Winston Salem, N.C. 27157-1010, USA

Evolving notions of the behavioral roles of the basal ganglia allude to its involvement in a variety of cognitive and sensorimotor processes well beyond what was envisioned for the extrapyramidal motor system in the not too distant past. Essential to understanding these 'new' functional roles is the perspective that many basal ganglia-related processes, including action selection, attention, decision making, reward, and sensory guidance of movement, to name a few, are best viewed as operations of networks involving interactions of the basal ganglia with other structures rather than information processing within the basal ganglia *per se*. It is with this perspective in mind that the chapters appearing in this section on visuomotor aspects of basal ganglia were selected. The fundamental belief is that garnering new insights into how visual cues ultimately give rise to purposeful, goal-directed behaviors can be achieved only by considering the function of visuomotor structures within the context of the basal ganglia architecture, and, as a corollary, that input from these structures is critical for shaping basal ganglia activity itself.

Each of the chapters addresses issues central to contemporary basal ganglia research. One influential hypothesis suggests that the fundamental contribution of the basal ganglia to behaviors is in its role as an action selection mechanism, whereby specific intrinsic pathways serve to select one motor program while complementary pathways inhibit competing demands. While this view has been successfully incorporated into basal ganglia models, physiological support for global inhibition in the visuomotor domain has been lacking. Recent work by McHaffie and colleagues, however, reveals that the basal ganglia shape subcortical visuomotor activity in the midbrain superior colliculus bilaterally via such mechanisms. Interestingly, these data provide novel insights into the mechanisms of cortically-induced visual hemineglect, a phenomenon not traditionally associated with the basal ganglia. A second compelling question addresses the potential role of dopaminergic neurotransmission in reinforcement learning. One strategy for decoding the information content of the important short-latency phasic dopaminergic signals would be to determine the nature of sensory inputs to the ventral midbrain dopaminergic neurons. Recent data from Redgrave and colleagues suggest that subcortical visual systems involving the superior colliculus are likely to be the predominant source of short latency visual input onto these neurons. Thus, not only is visuomotor activity in the superior colliculus modulated by basal ganglia outputs, but the visual activity in the superior colliculus provides critical sensory input to dopaminergic neurons which are so essential to normal basal ganglia function. Finally, the concept of closed cortico-basal ganglia-thalamo-cortical loops has been one of the most influential issues in contemporary basal ganglia research. Yet, despite its conceptual relevance, the physiological ramifications of such a loop architecture have received little experimental attention. Contemporary work by Stanford and colleagues in oculomotor regions of central thalamus suggests that thalamic visuomotor neurons have the capacity to dynamically represent the contextual cues necessary for choosing between competing stimuli in a visual orienting task; it is suggested that the evolution of activity relating to target/saccade choice could be one function of a closed "oculomotor loop" that links cortex and basal ganglia.

In: Recent Breakthroughs in Basal Ganglia Research
Editor: Erwan Bezard, pp. 173-184

ISBN: 1-59454-880-3

Chapter 14

SUPERIOR COLLICULUS MEDIATED VISUOMOTOR FUNCTION DEPENDS ON COUPLING GLOBAL INHIBITION AND FOCAL DISINHIBITION

John G. McHaffie*[*], *Huai Jiang and Barry E. Stein
Department of Neurobiology and Anatomy, Wake Forest University School of Medicine, Winston Salem, N.C. 27157-1010, USA

ABSTRACT

The transformation of sensory information into adaptive motor behaviors is premised on the appropriate allocation of motor resources in spite of ongoing, and often conflicting, behavioral demands. Understanding how this selection process occurs is paramount in current models of basal ganglia function. We summarize here how the basal ganglia utilize the complementary processes of focal disinhibition and global inhibition conveyed by segregated populations of substantia nigra pars reticulata neurons to bilaterally coordinate the sensory activity of superior colliculus neurons on each side of the brain. Together, these two complementary processes ensure accurate movements of the head and eyes to salient targets despite the presence of potentially distracting stimuli. Moreover, understanding how the basal ganglia coordinate the inter-hemispheric control of visuomotor activity may have important consequences for understanding visual hemineglect

INTRODUCTION

The pivotal role of superior colliculus (SC) output neurons in visuomotor behaviors is facilitated by inputs from a wide variety of cortical and subcortical sensory and motor structures (Edwards et al., 1979; Huerta and Harting, 1984). These inputs converge and integrate with the intrinsic circuitry of the SC to form map-like representations of contralateral visual, auditory and somatosensory space as well as a motor representation of contralateral eye/head movements (Stein and Meredith, 1993). The various maps are in spatial register with one another and they, in turn, are in register with the motor map. In fact, many SC neurons are involved in multiple maps: they not only receive inputs from one or more sensory modalities and, thus, have two or more overlapping sensory receptive fields (Stein and Meredith, 1993), but they also exhibit motor responses that are involved in shifting gaze toward their receptive field locations. The specific direction and amplitude of the gaze shift depends on the locus and extent of activity within the SC topography (Lee et al., 1988). For example, activation in rostral SC, where central sensory space is represented, produces very small shifts of gaze. In contrast, activation of progressively more caudal regions,

[*] Address correspondence : John McHaffie, Ph.D. Department of Neurobiology and Anatomy, Wake Forest University School of Medicine Medical Center Boulevard, Winston-Salem, N.C. 27157-1010 Phone: (336) 716-9078 / Fax: (336) 716-4534. E-mail address: mchaffie@wfubmc.edu

where progressively more temporal regions of contralateral sensory space are represented, evokes progressively larger contralateral gaze shifts. Similarly, activation of the medial aspect of the structure, where upper sensory space is represented, evokes gaze shifts with upward components whereas activation of the lateral portion of the structure, where lower sensory space is represented, evokes gaze shifts with downward components. This SC 'motor error' signal then is conveyed to regions of the brainstem that activate the motor nuclei innervating the extraocular and neck muscles (Munoz and Guitton, 1985). Consequently, the factors that shape the overall pattern of activity within the SC are critical for determining the metrics of the resultant gaze shift and its accuracy.

One of the best known modulators of SC output neurons' activity is the substantia nigra, pars reticulata (SNr); specifically, its ipsilateral nigrocollicular projection. Their axons terminate in a discontinuous patch-like pattern within the deep SC (Graybiel, 1978; Beckstead, 1981; Behan et al., 1987; Harting et al., 1988), where they establish direct synaptic contact with SC output neurons (May and Hall, 1984; Karabelas and Moschovakis, 1985; Tokuno and Nakamura, 1987; Williams and Faull, 1988). Because these ipsilateral, or 'uncrossed,' nigrocollicular neurons are GABA-ergic (Chevalier et al., 1981), their high rates of spontaneous activity (Hikosaka and Wurtz, 1983; Chevalier et al., 1984; Chevalier et al., 1985; Joseph and Boussaoud, 1985) tonically suppresses their SC target neurons (Chevalier et al., 1984). Prior to a gaze shift, however, there is a temporary cessation of activity in a subset of nigrocollicular neurons (Hikosaka and Wurtz, 1983; Joseph and Boussaoud, 1985), which phasically releases a corresponding subset of SC output neurons from inhibition. This facilitates activity in a region of the SC corresponding to the location of the target in space and results in a gaze shift appropriate to position the retinae on the target. This process of 'focal disinhibition' is not only the fundamental neural process by which the basal ganglia influence SC-mediated visuomotor behaviors but also has proven to be the basic template by which they interact with other motor-related structures (Chevalier and Deniau, 1990).

Contemporary studies, however, have suggested that basal ganglia-mediated *inhibition* of target structures may also be involved in the coordination of motor behaviors (Alexander et al., 1986; Alexander and Crutcher, 1990). Thus, when a given motor behavior is to be elaborated (via disinhibition), competing activity that might otherwise interfere with the intended movements must be suppressed. Physiological observations of excitatory responses in the globus pallidus, pars interna, the output nuclei of the basal ganglia involved with the axial musculature (Georgopoulos et al., 1983; Anderson et al., 1985; Mitchell et al., 1987; Brotchie et al., 1991; Mink and Thach, 1991), prompted Mink, (1996) to articulate a model of limb movement control whereby these two complementary processes of focal disinhibition and global inhibition are used to coordinate the agonist and antagonist muscles. Similar notions also have been incorporated into computer models of the basal ganglia (Gurney et al., 2004) wherein the neuroarchitecture underlying these two processes has been embodied as an action selection mechanism (Redgrave et al., 1999; Gurney et al., 2001; 2004). Together, such physiological and modeling data have fostered the emergent view of the basal ganglia in which its fundamental contribution to overt behavior is its ability to select one motor program while simultaneously suppressing competing ones. Such a neural plan is particularly appealing because of its elegant simplicity and broad applicability. Surprisingly, however, the relevance of this view to SC-mediated visuomotor behavior has been questioned due to the absence of strong empirical evidence for the presence of global inhibition in its circuitry (Mink, 1996; Anderson et al., 2001).

With these hypotheses in mind, we sought to determine if, in fact, there was a single general plan by which the basal ganglia affects both skeletomotor and visuomotor functions and, thus, whether basal ganglia-mediated global inhibition plays a significant role in regulating SC-mediated visuomotor activity. One likely source of this inhibition, should it exist, is the crossed nigrocollicular projection (Beckstead, 1981). Widespread and powerful inhibition derived from this input has been used as an effective means of explaining the visual hemineglect that results from large unilateral lesions of visual cortex despite the fact that the ipsilateral SC remains intact albeit non functional (Wallace et al., 1989, 1990). Curiously, however, the actual physiological properties of crossed nigrocollicular neurons have never been assayed: it had simply been assumed that their physiological properties would mimic those of their better-known uncrossed counterparts (Wallace et al., 1990). However, we have determined recently that this assumption is incorrect: crossed nigrocollicular neurons have a unique physiological profile that contrasts markedly with their ipsilateral counterparts in order to shape SC activity bilaterally (Jiang et al., 2003).

1 METHODS

We used both electrophysiological and neuroanatomical techniques to detail the features of crossed SNr neurons. In chronically instrumented ketamine-anesthetized cats, single-neuron extracellular recordings were made from visually-responsive neurons in the left SNr that could be antidromically activated from arrays of electrodes in the ipsilateral and contralateral SC. By also recording through the antidromic stimulating electrodes, we were able to roughly determine the correspondence between the visual receptive fields of nigrocollicular neurons and neurons in SC regions that they were targeting. A similar array of stimulating electrodes was implanted in the left extraprimary visual cortex to determine whether nigrocollicular neurons also receive information from the ipsilateral cortex (see Jiang et al., 2003 for details). After mapping the SNr neurons' visual receptive fields, controlled visual stimuli were used to evaluate how these neurons respond to natural visual stimuli.

In order to determine whether the crossed nigrocollicular projection was GABA-ergic, wheatgerm agglutinin- horseradish peroxidase (WGA-HRP) was injected into the SC, and the SNr was coprocessed for retrograde WGA-HRP and glutamic acid decarboxylase immunohistochemistry. We also sought to determine whether axons from the crossed nigrocollicular neurons contact SC output neurons by injecting biotinylated dextran (BDA) into the left SNr and co-processing the right SC for anterogradely transport BDA and parvalbumin immunohistochemistry (see Jiang et al., 2003 for details).

2 RESULTS

Data were obtained from 59 SNr neurons that met established electrophysiological criteria for antidromic activation (Fuller and Schlag, 1976) and were, therefore, classified as nigrocollicular. They fell into two groups: one activated antidromically from the left SC and thus defined as 'uncrossed' (n = 43, 73 %) and another (n = 16, 27 %) activated antidromically from the right SC and defined as 'crossed.' No SNr neurons were found that could be activated antidromically from both colliculi, suggesting that the crossed and uncrossed nigrocollicular projections arise from segregated neuronal populations. Consistent with previous anatomical studies (Beckstead, 1981; Wallace et al., 1990), crossed neurons tended to be clustered more anterior in the SNr whereas uncrossed neurons were more broadly distributed. All SNr neurons in our sample were orthodromically activated from extrastriate visual cortex, indicating that activity in visual cortex can influence SC neurons via trans-synaptic pathways that course through the basal ganglia.

Uncrossed neurons in our sample displayed characteristics typical of nigrocollicular neurons. They had high rates of spontaneous activity (mean = 36.81 +/- 18.08 Hz; range = 11.06 - 86.26 Hz) and relatively small visual receptive fields (mean = 876 +/- 443.11^{o2}; range = 340-5225) whose centers were restricted to the contralateral hemifield. The receptive fields of these neurons were in close spatial correspondence with those of ipsilateral SC neurons at the effective antidromic stimulation site(s). Thus, uncrossed neurons and their SC targets had receptive fields that were centered in contralateral space and in topographic register with each other (n = 43/43; 100 %). All uncrossed neurons encountered (n = 43/43; 100 %) were phasically inhibited by visual stimuli. Figure 1 illustrates the salient features of an individual uncrossed neuron that characterize the physiological profile of the population.

By contrast, crossed neurons displayed a very different physiological profile. All crossed neurons (n = 16/16; 100 %) were phasically excited by visual stimuli, indicating that they respond to visual stimuli in a manner opposite that of their uncrossed counterparts. Similarly, the spontaneous activity of crossed neurons (mean = 12.52 +/- 10.29 Hz; range = 0.12 - 40 Hz) was far lower than that of uncrossed neurons ($p<<$ 0.001; 2-tailed T-test). Indeed, some of them exhibited no spontaneous activity throughout the entire period in which they were studied. This general paucity of spontaneous activity may explain why these neurons have not been studied previously. They are very difficult to localize and most would not have been detected in the present studies had not antidromic electrical stimuli been used to search for them. The paucity of spontaneous activity in these neurons is also a key to their presumptive functional role: assuming that they, like their uncrossed counterparts, are GABA-ergic (see below), the absence of ongoing activity enables them to *phasically* inhibit their target neurons in the opposite SC.

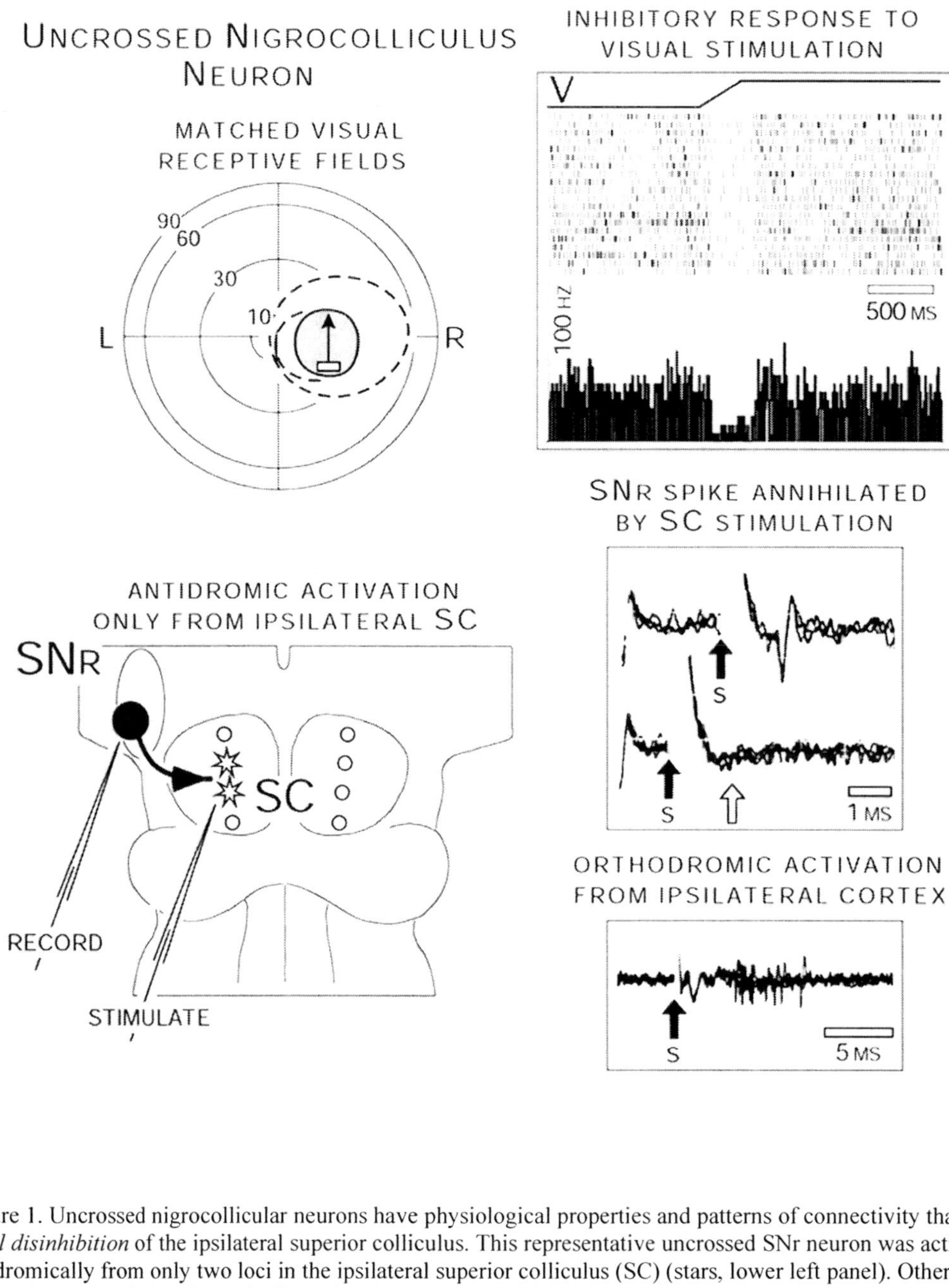

Figure 1. Uncrossed nigrocollicular neurons have physiological properties and patterns of connectivity that facilitate *focal disinhibition* of the ipsilateral superior colliculus. This representative uncrossed SNr neuron was activated antidromically from only two loci in the ipsilateral superior colliculus (SC) (stars, lower left panel). Other stimulating electrodes (open circles) in the ipsilateral and contralateral SC were ineffective. The neuron's *small* contralateral visual receptive field (gray shaded oval, upper left panel) is depicted in a diagram of visual space. Note that the SNr neuron's receptive field is in the *same* hemifield and in spatial register with the multiunit receptive fields of SC neurons (unfilled dotted ovals) recorded at the two effective collicular stimulation sites. As shown in the upper right panel, this uncrossed neuron had a *high* rate of spontaneous activity that was *inhibited* by a visual stimulus moved upward through its receptive field. The annihilation of a spontaneously occuring spike (open arrow, middle right panel) during the collision test confirmed antidromic activation from the ipsilateral SC. In the lower right panel, orthodromic activation from cortical stimulating electrodes confirmed that this neuron also received input from the ipsilateral extrastriate visual cortex, presumably via multisynaptic routes coursing through the basal ganglia. Adapted with permission from Jiang et al., 2003.

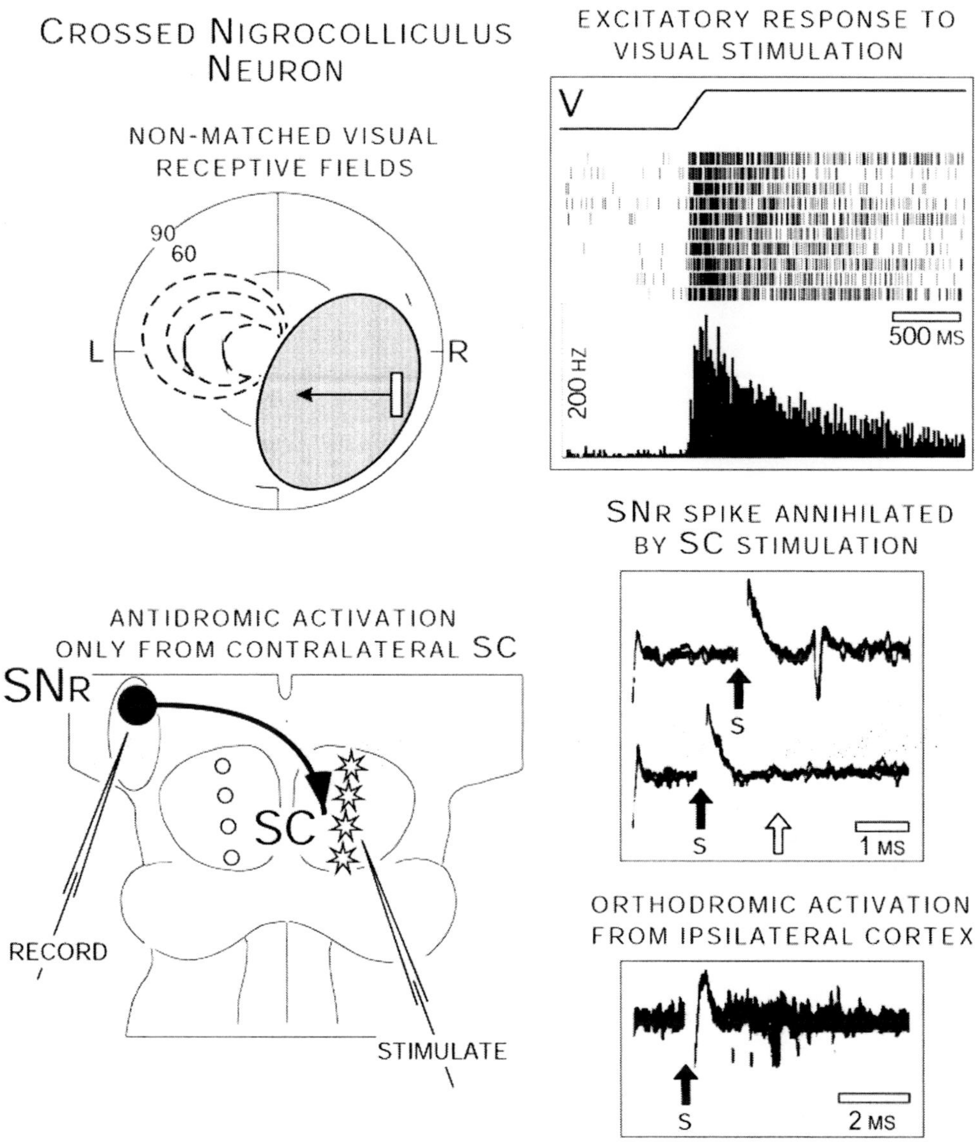

Figure 2. Crossed nigrocollicular neurons have physiological properties and patterns of connectivity that facilitate *global inhibition* of the contralateral superior colliculus. In contrast to the uncrossed SNr neuron shown in figure 1, this representative crossed SNr neuron was activated antidromically from all four stimulating electrodes in the contralateral SC (stars, lower left panel) but not from any electrode in the ipsilateral SC (open circles). The neuron's *large* contralateral visual receptive field (gray shaded oval, upper left panel) was in the hemifield *opposite* to those of SC neurons (unfilled dotted ovals) recorded at the four effective collicular stimulation sites. As shown in the upper right panel, this crossed neuron had a *low* rate of spontaneous activity that was *increased* by a visual stimulus moved through its receptive field, with the period of excitation outlasting stimulus presentation. The annihilation of a spontaneously occuring spike (open arrow, middle right panel) during the collision test confirmed antidromic activation from the contralateral SC. In the lower right panel, orthodromic activation from cortical stimulating electrodes confirmed that this crossed neuron also received input from the ipsilateral extrastriate visual cortex. This neuron, and others like it, were preferentially located in the more rostral aspects of SNr. Adapted with permission from Jiang et al., 2003.

It is also of substantial interest to note that the visual receptive fields of crossed neurons (mean = 7499 +/- 3068.46^{o2}; range = 3080 - 12100^{o2}) were not only significantly larger than those of uncrossed neurons (p<< 0.001; 2-tailed T-test) but they were not in spatial register with SC neurons at the effective antidromic stimulation sites in the opposite SC. In every instance (n = 16/16; 100 %), the centers of crossed neurons' receptive fields were in the hemifield opposite those recorded at the effective SC antidromic activation site(s). Furthermore, many (n= 14/16; 87 %) crossed neurons were activated antidromically from all (n=4) SC stimulation sites, suggesting that, when activated, crossed neurons can phasically exert global inhibition over their target neurons in the opposite SC. Figure 2 illustrates the salient features of an individual crossed neuron that characterize the physiological profile of the population.

In order to determine if, as suspected, crossed neurons utilize the same neurotransmitter as their uncrossed counterparts (Chevalier et al., 1981), WGA-HRP was injected into one SC, and the tissue co-processed for glutamic acid decarboxylase (GAD) immunohistochemistry, an enzymatic marker for GABA-ergic neurons. The somas of all crossed nigrocollicular neurons retrogradely-labeled with WGA-HRP also were immunoreactive for the GAD antibody (Figure 3A), thus providing strong evidence that they, like their uncrossed counterparts, are GABA-ergic (Chevalier et al., 1981). Consistent with our difficulty in isolating crossed neurons in electrophysiological experiments, counts of retrogradely labeled SNr neurons revealed an uncrossed-to-crossed neuron ratio of approximately 12:1, similar to numbers reported previously (Beckstead, 1981).

For crossed neurons to significantly influence SC visuomotor behaviors, their axons should terminate on the deep laminae neurons that project to the brainstem and spinal cord. In the cat, SC neurons projecting to the brainstem are known to express the calcium-binding protein parvalbumin (Mize et al., 1991). Thus, we sought to determine if crossed nigrocollicular projections terminate on parvalbumin-containing SC neurons. To evaluate this possibility, we combined injections of the anterograde tracer biotinylated dextran amine (BDA) in the rostral SNr with parvalbumin immunohistochemistry. BDA-labeled crossed nigrocollicular fibers were seen coursing deep to the ipsilateral SC and followed the dorsolateral contour of the periaqueductal gray before crossing the midline and entering the opposite SC. Their deep and medial trajectory, combined with the low stimulus current employed for antidromic activation (100-300 μA) via SC stimulating electrodes, are likely reasons for why crossed neurons were not activated antidromically from the ipsilateral SC. Crossed fibers were distributed in the intermediate and deep laminae of the contralateral SC, and labeled boutons were observed in close apposition to the soma and proximal dendrites of parvalbumin immunopositive neurons (Figure 3B). These anatomical data strongly suggest that crossed neurons establish direct contact with contralateral SC output neurons controlling shifts of gaze (Munoz and Guitton, 1985).

The consequences of activating crossed nigrocollicular neurons on the visual responsiveness of their contralateral SC target neurons was then evaluated. This was accomplished by recording sensory evoked activity through the SC 'stimulating' electrodes while orthodromically stimulating the SNr through the previous 'recording' electrode. The recordings were made from the SC electrode having the lowest antidromic threshold for each crossed nigrocollicular neuron while stimulating at the recording site of that crossed nigrocollicular neuron (Figure 4). Following control tests with no SNr stimulation, presentation of the visual stimulus was concatenated with microstimulation delivered through the contralateral SNr electrode at the peak of the control visual response. In all such paired recordings (n = 4), a single, brief (100 μs), low intensity (100- 300 μA) pulse delivered through the SNr electrode abolished SC evoked activity for periods of approximately 30 ms, suggesting that crossed nigrocollicular neurons can exert potent inhibitory influences on the visual responses of contralateral SC neurons.

3 Discussion

Our present electrophysiological observations reveal a striking dichotomy between the properties of crossed nigrocollicular neurons and those described for their uncrossed counterparts (Hikosaka and Wurtz, 1983; Chevalier et al., 1984; 1985; Joseph and Boussaoud, 1985). Taken together, the data suggest that the basal ganglia interacts with the SC by coupling focal disinhibition and global inhibition to coordinate neural activity bilaterally in the subcortical circuitry critical for visuomotor behaviors. The differing responsiveness of these two segregated populations of GABA-ergic nigrocollicular neurons, one excited and another inhibited by visual stimuli, likely reflects the differential convergence of the parallel and antagonistic direct and indirect pathways that course through the basal ganglia (Alexander et al., 1986; Alexander and Crutcher, 1990; Mink and Thach, 1991). Working in synchrony, these two complementary nigrocollicular pathways facilitate visuomotor activity in one colliculus (via the uncrossed pathway) while simultaneously inhibiting activity in the other colliculus (via the crossed pathway).

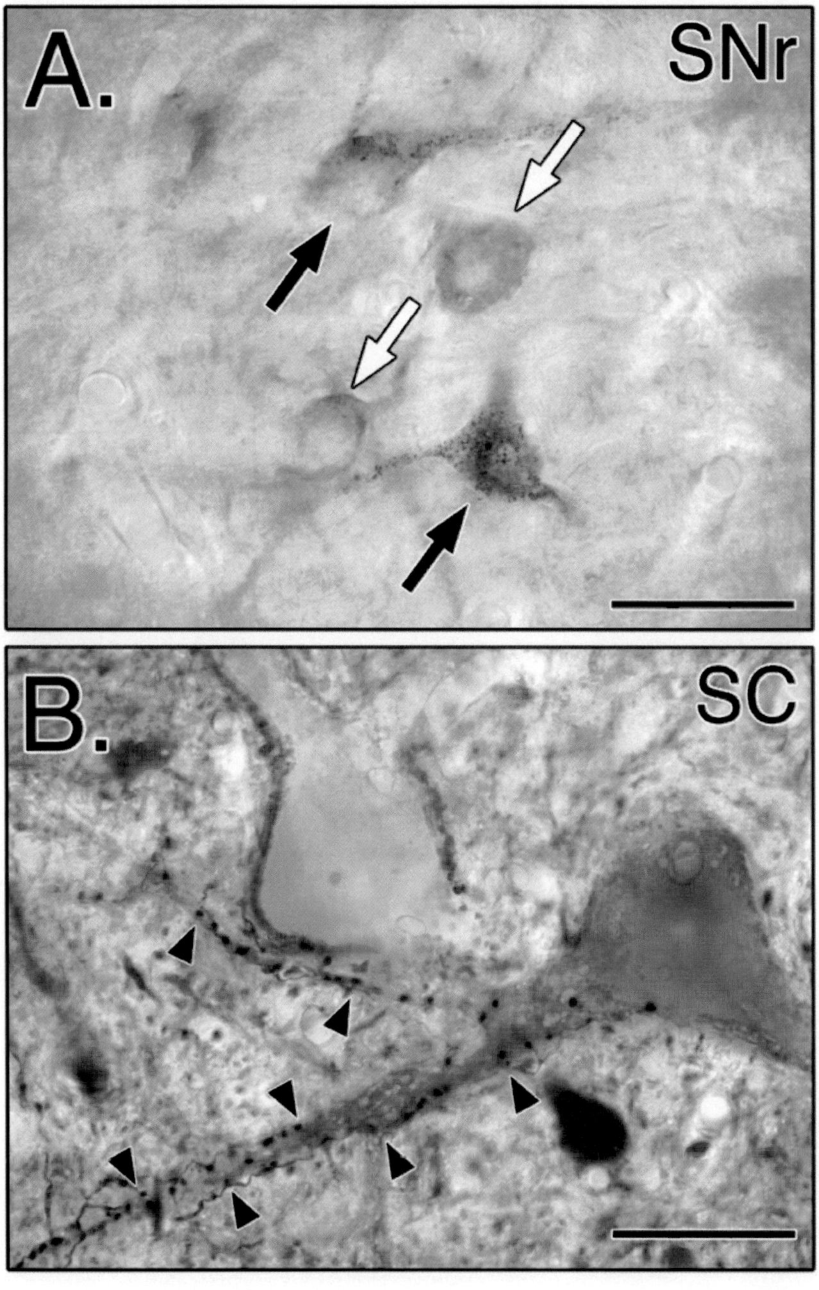

Figure 3. Crossed nigrocollicular neurons projecting to the contralateral SC are, like their uncrossed counterparts, GABA-ergic and terminate on presumptive SC output neurons in the contralateral colliculus. A section through the left SNr (A) shows single-labeled GAD-positive neurons (white arrows) and double-labeled GAD- and HRP-positive neurons (black arrows) following WGA-HRP injections into the right SC. All retrogradely labeled neurons contralateral to the injection site also were positive for GAD. Following a BDA injection into the left SNr, a section through the right SC (B) reveals BDA-labeled axonal swellings, indicative of synaptic sites, in close apposition to SC dendrites (arrowheads) labelled with parvalbumin, a calcium binding protein preferentially distributed in SC output neurons. Scale bars are 50 μm. Adapted with permission from Jiang et al., 2003.

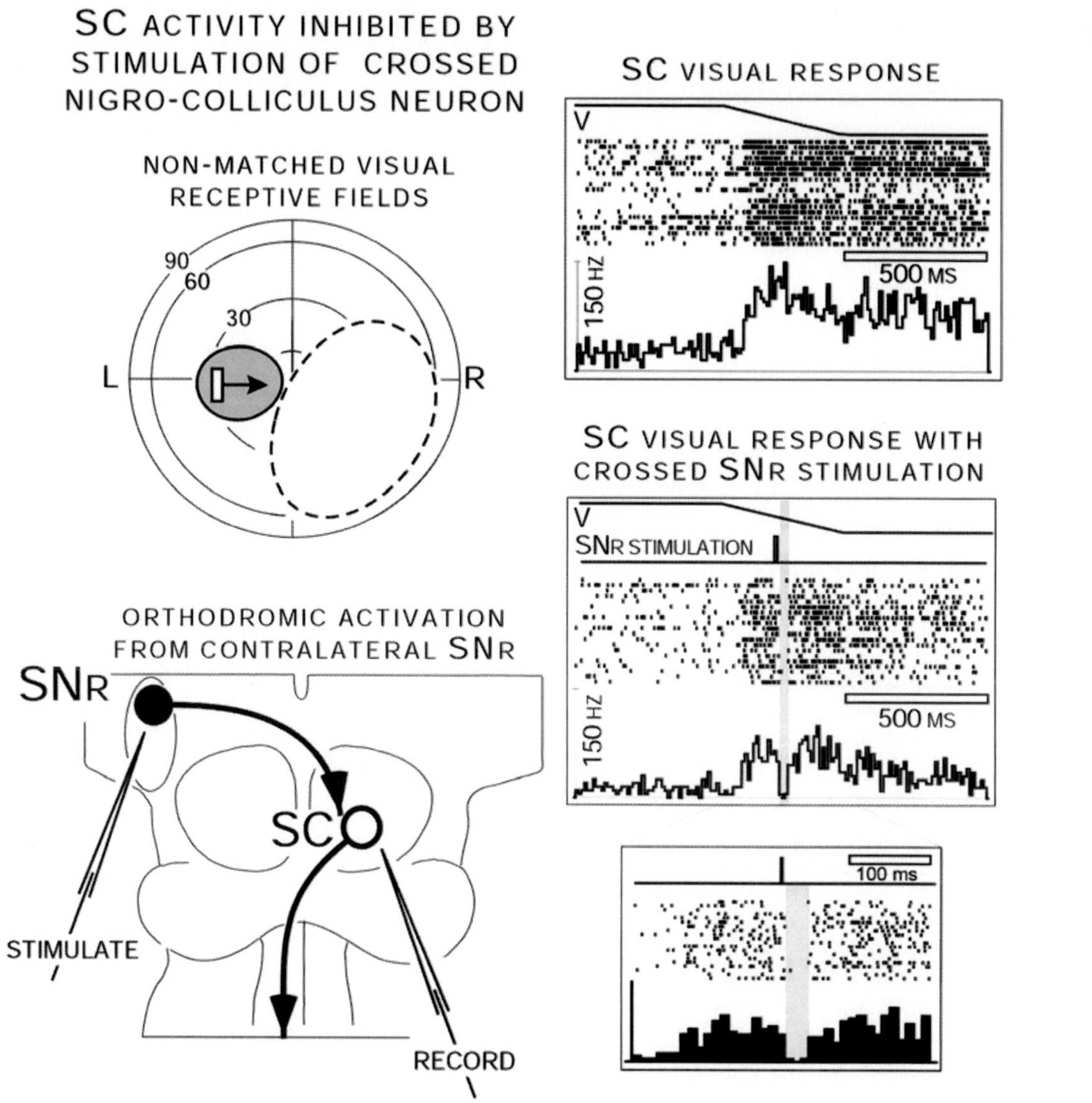

Figure 4. Activation of crossed nigrocollicular neurons inhibits their SC target neurons in the contralateral colliculus. Visual activity was recorded in the right SC through the stimulating electrode that had previously produced the lowest threshold for antidromically activating the crossed nigrocollicular neuron shown in figure 2. The receptive field at this SC stimulation site was in left visual space (gray shaded oval), whereas the receptive field of the crossed neuron at the SNr stimulation site was in the opposite hemifield (unfilled dotted oval). The upper right panel shows the response of the SC neuron to a rightward moving visual stimulus (the most effective direction of movement). Following this control series, the visual stimulus was then concatenated with brief (0.1 ms) low intensity (100 μA) microstimulation through the previous SNr recording electrode (middle right panel). Note that SNr stimulation inhibited the visual response (shaded gray area, shown at higher temporal resolution in the inset, lower right panel) and that the duration of the inhibition (approximately 30 ms) long outlasted the duration of the electrical stimulus. Adapted with permission from Jiang et al., 2003.

In addition to their opposite physiological responsiveness to visual stimuli, each of these nigrocollicular populations possesses distinct physiological properties that facilitate bilateral control of midbrain visuomotor activity. For example, uncrossed neurons always have comparatively small receptive fields centered in the same hemifield and are spatially congruent with the receptive fields of their target SC neurons (Hikosaka and Wurtz, 1983; Joseph and Boussaoud, 1985). This spatial register renders uncrossed neurons particularly well-suited to focally disinhibit topographically appropriate regions of the ipsilateral SC. In contrast, the receptive fields of crossed neurons are considerably larger, far larger than those of most SC visual neurons (Stein and Meredith, 1993). Moreover, the receptive fields of crossed neurons always are centered in the hemifield opposite those of their target SC neurons; this, and their broad innervation territories in contralateral SC, allows them to suppress activity within widespread territories of the target SC. Despite such physiological differences, however, crossed neurons are, like uncrossed neurons, GABA-ergic (Chevalier et al., 1981) and establish direct contact with presumptive collicular output neurons in the contralateral SC (May and Hall, 1984; Karabelas and Moschovakis, 1985); thus, they are well positioned to inhibit activity in the nonselected SC, thereby keeping it from influencing brainstem motor areas. Despite the comparatively small numbers of crossed neurons (Beckstead, 1981), the positioning of crossed nigrocollicular boutons on or near the neuronal soma ensures that their influence is potent. Our physiological demonstration that even brief electrical activation of crossed neurons results in profound, long-duration suppression of visually-evoked activity in the opposite SC is consistent with these observations.

The significance of such an organizational scheme in the control of visuomotor behavior can best be appreciated by considering notions of how gaze shifts are coded by brainstem motor centers. Brainstem integrators of collicular information are believed to compute appropriate movement vectors by averaging the activity from a population of SC neurons (Lee et al., 1988). It has been assumed, however, that this neuronal population is restricted to a single SC, specifically, the one phasically disinhibited by uncrossed nigrocollicular activity. Nevertheless, recent experiments indicate that aberrant activity between the two colliculi leads to inappropriate shifts of orientation, suggesting that activity from *both* colliculi is integrated into this computation (Karabelas and Moschovakis, 1985; Wang and Redgrave, 1997). Thus, the active suppression of extraneous neural activity associated with potential distractors in the non-selected SC is just as necessary for achieving an accurate gaze shift as the facilitation of activity associated with the target in the selected SC. This notion is consistent with prior observations that neurons in the two colliculi fire in a "push-pull" fashion during spontaneous eye movements: when neurons in one SC become active, those in the other are rendered silent (Infante and Leiva, 1986). The crossed nigrocollicular pathways, therefore, provide an extrinsic substrate for synchronizing intercollicular activity, independent of any contribution from intrinsic SC commissural pathways (Behan, 1985). Such interhemispheric coordination may be particularly important in visual search strategies in the natural environment, where the eyes are continually redirected between the hemifields.

The notion that the basal ganglia couples global inhibition with focal disinhibition is not without precedence. Global inhibition is recognized as an essential component in action selection, where competing behavioral demands are suppressed while desired programs are elaborated (Redgrave et al., 1999). More specifically, the prevalence of excitatory responses (and, thus, target inhibition) in the globus pallidus, pars interna, during limb movements has been suggested to underlie the coordination of agonist and antagonist musculature during movements (Mink, 1996). Although the lack of excitatory responses in the SNr has prompted suggestions that limb and visuomotor movements are organized differently (Mink, 1996; Anderson et al., 2001), the present finding (and recent reports of excitation of premotor activity in behaving monkeys (Handel and Glimcher, 2000; Sato and Hikosaka, 2002) suggests that it is crucial to visuomotor behaviors. These results are consistent with a single operational scheme by which the basal ganglia controls the expression of overt behaviors although the segregation of excitatory responses in crossed nigrocollicular neurons may reflect the biological constraints imposed by the visual system.

The differing responsiveness of crossed and uncrossed neurons to visual stimuli likely reflects the differential convergence of the parallel and antagonistic pathways that course through the basal ganglia (Alexander et al., 1986; Alexander and Crutcher, 1990). For example, the phasic inhibition observed in uncrossed neurons is due to the direct striatonigral pathway (Chevalier et al., 1985). Thus, phasic activation of normally quiescent striatonigral GABA-ergic neurons by a visual stimulus will produce the phasic inhibition characteristic of the uncrossed neurons. Although the specific pathway(s) that produce the phasic excitation of crossed neurons is unknown, it is likely due to those transbasal ganglia circuits that ultimately modulate the subthalamus, the source of excitatory input onto SNr neurons (Kita and Kitai, 1987).

One of the intriguing implications of this organizational scheme is that it provides a potential physiological basis for the Sprague Effect, a paradoxical case of the amelioration of a lesion-induced deficit by a second lesion (Sprague, 1966). Contralateral visual neglect produced by large unilateral visual cortex lesions can be "reversed" by lesions that affect crossed nigrocollicular neurons (Sprague, 1966; Wallace et al., 1989, 1990). According to the present data, eliminating the crossed nigrocollicular pathway removes the widespread inhibition conveyed by these neurons onto the cortically-deafferented SC, which presumably reinstates responsiveness in that SC. This, then, would permit previously ineffectual visual inputs to once again initiate SC activity and SC-mediated orientation behaviors. Thus, the amelioration of hemineglect can be interpreted as the disruption of an essential component of the circuitry that is normally used by the basal ganglia to coordinate visuomotor activity in the two halves of the neuraxis. This is consistent with the developing appreciation of the basal ganglia as a critical structure in the manifestation of human hemineglect syndromes (Karnath, 2001).

ACKNOWLEDGEMENTS

Supported by NIH grants NS35008, NS22543, NS36916

REFERENCES

Alexander GE, Crutcher MD (1990) Functional architecture of basal ganglia circuits: neural substrates of parallel processing. *Trends Neurosci* **13**:266-271.

Alexander GE, DeLong MR, Strick PL (1986) Parallel organization of functionally segregated circuits linking basal ganglia and cortex. *Ann Rev Neurosci* **9**:357-381.

Anderson ME, Ruffo M, Buford JA, Inase M (2001) Pallidal and cortical determinates of thalamic activity. In: *Basal Ganglia and Thalamus in Health and Movement Disorders* (Kultas-Ilinsky K, Ilinsky IA, eds), pp 93-104. New York: Kluwer Academic/Plenum Publishers.

Anderson ME, Horak FB, Aug. JoN, Messages L, Library CrbtC, Abstract (1985) Influence of the globus pallidus on arm movements in monkeys. III. Timing of movement-related information. *J Neurophysiol* **54**:433-448.

Beckstead RM (1981) A comparison of the intranigral distribution of nigrotectal neurons labelled with horseradish peroxidase in the monkey, cat and rat. *J Neurosci* **1**:121-125.

Behan M (1985) An EM-autoradiographic and EM-HRP study of the commissural projection of the superior colliculus in the cat. *J Comp Neurol* **234**:105-116.

Behan M, Lin CS, Hall WC (1987) The nigrotectal projection in the cat: an electron microscope autoradiographic study. *Neurosci* **21**:529-539.

Brotchie P, Iansek R, Horne MK (1991) Motor function of the monkey globus pallidus. 1. Neuronal discharge and parameters of movement. *Brain* **114**:1667-1683.

Chevalier G, Deniau JM (1990) Disinhibition as a basic process in the expression of striatal functions. *Trends Neurosci* **13**:277-280.

Chevalier G, Vacher S, Deniau JM (1984) Inhibitory nigral influence on tectospinal neurons, a possible implication of basal ganglia in orienting behaviour. Exp Brain Res 53:320-326.

Chevalier G, Thierry AM, Shibazaki T, Feger J (1981) Evidence for a GABAergic inhibitory nigrotectal pathway in the rat. *Neurosci Lett* **21**:67-70.

Chevalier G, Vacher S, Deniau JM, Desban M (1985) Disinhibition as a basic process in the expression of striatal function. I. The striato-nigral influence on tecto-spinal/tecto-diencephalic neurons. *Brain Res* **334**:215-226.

Edwards SB, Ginsburg CL, Henkel CK, Stein BE (1979) Sources of subcortical projections to the superior colliculus in the cat. *J Comp Neurol* **184**:309-330.

Fuller JH, Schlag JD (1976) Determination of antidromic excitation by the collision test: Problems of interpretation. *Brain Res* **112**:283-298.

Georgopoulos AP, DeLong MR, Crutcher MD (1983) Relations between parameters of step-tracking movements and single cell discharge in the globus pallidus and subthalamic nucleus of the behaving monkey. *J Neurosci* **3**:1586-1598.

Graybiel AM (1978) Organisation of the nigrotectal connection: An experimental tracer study in the cat. *Brain Res* **143**:339-348.
Gurney K, Prescott TJ, Redgrave P (2001) A computational model of action selection in the basal ganglia. I. A new functional anatomy. *Biol Cybern* **84**:401-410.
Gurney K, Prescott TJ, Wickens JR, Redgrave P (2004) Computational models of the basal ganglia: from robots to membranes. *Trends Neurosci* **27**:453-459.
Handel A, Glimcher PW (2000) Contextual modulation of substantia nigra pars reticulata neurons. *J Neurophysiol* **83**:3042-3048.
Harting JK, Huerta MF, Hashikawa R, Weber JT, D.P. VL (1988) Neuroanatomical studies of the nigrotectal projection in the cat. *J Comp Neurol* **278**:615-631.
Hikosaka O, Wurtz RH (1983) Visual and oculomotor function of monkey substantia nigra pars reticulata. IV. Relation of substantia nigra to superior colliculus. *J Neurophysiol* **49**:1285-1301.
Huerta MF, Harting JK (1984) The mammalian superior colliculus: Studies of its morphology and connections. In: *Comparative Neurology of the Optic Tectum* (Vanegas H, ed), pp 687-773. New York: Plenum.
Infante C, Leiva J (1986) Simultaneous unitary neuronal activity in both superior colliculi and its relation to eye movements in the cat. *Brain Res* **381**:390-392.
Jiang H, Stein BE, McHaffie JG (2003) Opposing basal ganglia processes shape midbrain visuomotor activity bilaterally. *Nature* **423**:982-986.
Joseph JP, Boussaoud D (1985) Role of the cat substantia nigra pars reticulata in eye and head movements. 1. Neural activity. *Exp Brain Res* **57**:286-296.
Karabelas AB, Moschovakis AK (1985) Nigral inhibitory termination of efferent neurons of the superior colliculus: an intracellar horseradish peroxidase study in the cat. *J Comp Neurol* **239**:309-329.
Karnath HO (2001) New insights into the functions of the superior temporal cortex. *Nat Rev Neurosci* **2**:568-576.
Kita H, Kitai ST (1987) Efferent projections of the subthalamic nucleus in the rat: light and electron microscopic analysis with the PHA-L method. *J Comp Neurol* **260**:435-452.
Lee C, Rohrer WH, Sparks DL (1988) Population coding of saccadic eye movements by neurons in the superior colliculus. *Nature* **332**:357-360.
May PJ, Hall WC (1984) Relationships between the nigrotectal pathway and the cells of origin of the predorsal bundle. *J Comp Neurol* **226**:357-376.
Mink JW (1996) The basal ganglia: focused selection and inhibition of competing motor programs. *Prog Neurobiol* **50**:381-425.
Mink JW, Thach WT (1991) Basal ganglia motor control. I. Nonexclusive relation of pallidal discharge to five movement modes. *J Neurophysiol* **65**:273-300.
Mitchell SJ, Richardson RT, Baker FH, DeLong MR (1987) The primate globus pallidus: neuronal activity related to direction of movement. 68:491-505.
Mize RR, Jeon CJ, Butler GD, Luo Q, Emson PC (1991) The calcium binding protein calbindin-D 28K reveals subpopulations of projection and interneurons in the cat superior colliculus. *J Comp Neurol* **307**:417-436.
Munoz DP, Guitton D (1985) Tectospinal neurons in the cat have discharges coding gaze position error. *Brain Res* **341**:184-188.
Redgrave P, Prescott TJ, Gurney K (1999) The basal ganglia: a vertebrate solution to the selection problem. *Neurosci* **89**:1009-1023.
Sato M, Hikosaka O (2002) Role of primate substantia nigra pars reticulata in reward-oriented saccadic eye movement. *J Neurosci* **22**:2363-2373.
Sprague JM (1966) Interactions of cortex and superior colliculus in mediation of visually guided behavior in the cat. *Science* **153**:1544-1547.
Stein BE, Meredith MA (1993) The Merging of the Senses. Cambridge, MA: The MIT Press.
Tokuno H, Nakamura Y (1987) Organization of the nigrotectospinal pathway in the cat: a light and electron microscopic study. *Brain Res* **436**:76-84.
Wallace SF, Rosenquist AC, Sprague JM (1989) Recovery from cortical blindness mediated by destruction of nontectotectal fibres in the commisure of the superior colliculus in the cat. *J Comp Neurol* **284**:429-450.
Wallace SF, Rosenquist AC, Sprague JM (1990) Ibotenic acid lesions of the lateral substantia nigra restore visual orientation behaviour in the hemianopic cat. *J Comp Neurol* **296**: 222-252.

Wang S, Redgrave P (1997) Microinjections of muscimol into lateral superior colliculus disrupt orienting and oral movements in the formalin model of pain. *Neurosci* **81**:967-988.

Williams MN, Faull RLM (1988) The nigrotectal projection and tectospinal neurons in the rat. A light and electron microscopic study demonstrating a monosynaptic nigral input to identified tectospinal neurons. *Neurosci* **25**:533-562.

In: Recent Breakthroughs in Basal Ganglia Research
Editor: Erwan Bezard, pp. 185-199
ISBN: 1-59454-880-3

Chapter 15

SENSORY CONTROL OF DOPAMINERGIC NEURONES

P. Redgrave*, V. Coizet, E. Dommett, E. Comoli, and P. G. Overton
Department of Psychology, University of Sheffield, Sheffield, S10 2TP, UK

ABSTRACT

A current influential view is that short-latency sensory responses of ventral midbrain dopaminergic neurones signal the occurrence of unpredicted rewarding stimuli. This hypothesis requires that sensory systems afferent to dopaminergic neurones should have the capacity to discriminate sensory events that are both unexpected and in some way associated with reward. Surprisingly, little is known about the source of rapid sensory input to dopaminergic neurones. Consequently, recent anatomical and electrophysiological experiments are reviewed which show, at least in the case of visual stimuli, that the midbrain superior colliculus is the principal, if not sole source of short-latency sensory input to the ventral midbrain. However, the perceptual properties of collicular neurones are largely restricted to detecting the appearance, disappearance or movement of transient visual stimuli. This response profile is therefore likely to place important constraints on the range of sensory events that can be signalled by dopaminergic neurones at short latency.

INTRODUCTION

There is widespread agreement that the ascending dopamine projections that originate in the ventral midbrain play a critical role in motivation and reinforcement learning. However, the exact nature of their contribution remains unclear (Salamone, 1994; Robbins and Everitt, 1996; Berridge and Robinson, 1998; Schultz, 1998; Horvitz, 2000; Schultz and Dickinson, 2000). A current influential view is that phasic releases of dopamine within forebrain target structures provide reward prediction error signals that are used to adjust future response probabilities (Montague et al., 1996; Schultz, 1998, 2004). An important part of the evidence on which this view is based involves the detection of short-latency, short-duration responses in midbrain dopaminergic (DA) neurones following unexpected presentations of behaviourally significant stimuli (Figure 1). For example, immediately following unexpected primary rewards (Figure 1B), and arbitrary stimuli classically conditioned by association with primary rewards (Figure 1C), DA neurones typically exhibit a burst of impulses. The response comprises a short latency (50-110 ms), short duration (<200 ms) burst of 3-6 impulses, which is superimposed on spontaneous, low level spike activity (1-9 spikes/s; Overton and Clark, 1997). The latency and duration of this sensory-elicited burst of firing are relatively invariant and largely independent of the modality of the eliciting stimulus.

* Corresponding Author Prof. Peter Redgrave. Dept Psychology, University of Sheffield, Sheffield S10 2TP, UK. E-mail address: P.Redgrave@sheffield.ac.uk

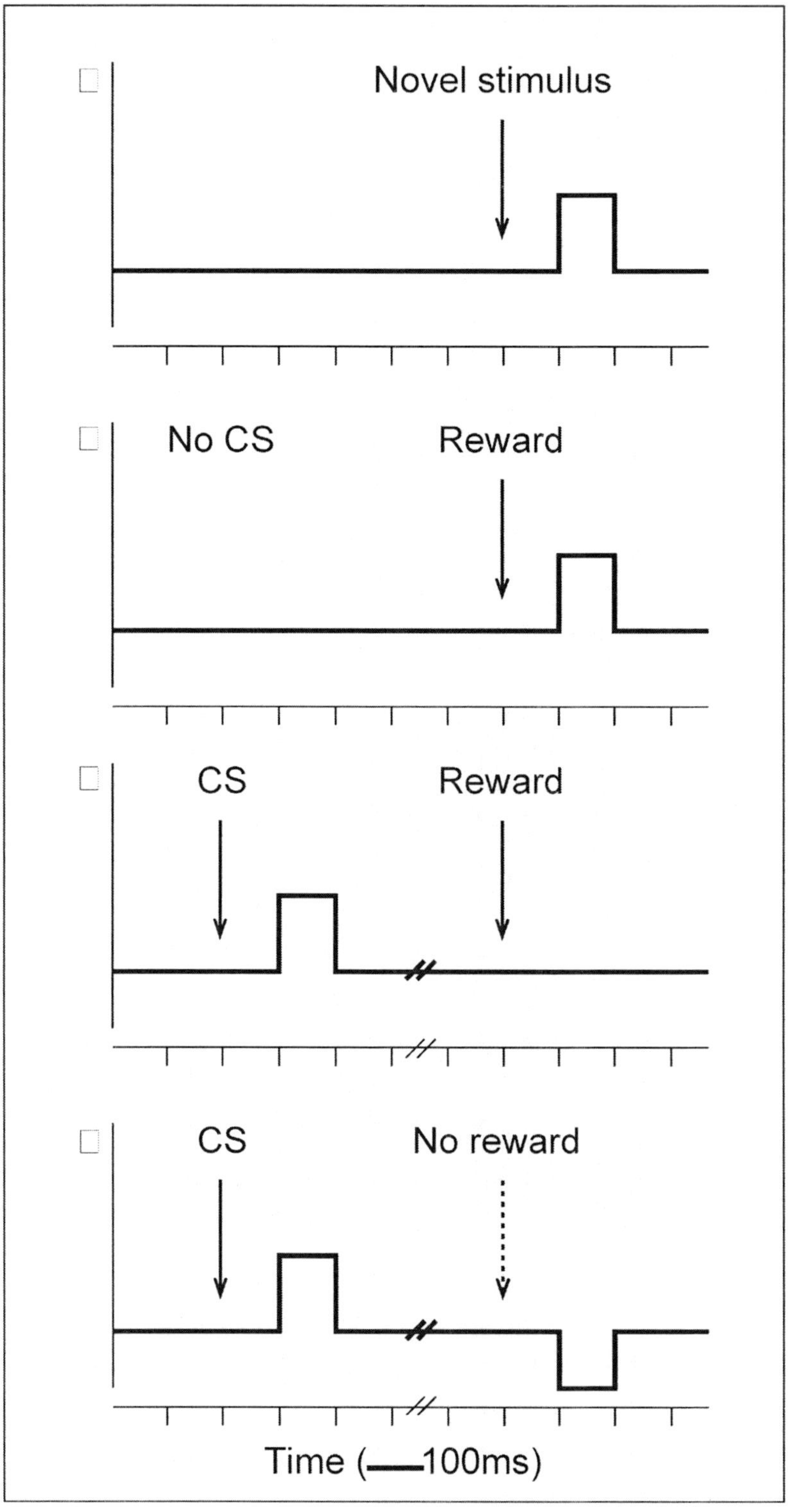

Figure 1: A schematic illustration of dopaminergic (DA) cell responses to salient stimuli in different experimental conditions (Schultz, 1998). a. The unexpected presentation of a novel stimulus evokes a burst of firing in a significant proportion of DA cells with a latency and duration of approximately 100 ms. b. A similar response is elicited by unexpected primary rewards. c. When a conditioned stimulus (CS) reliably predicts a primary reward, the burst of activity in DA cells transfers to the conditioned stimulus. d. After conditioning, if an anticipated (predicted) reward is not delivered, there is a short pause (~100 ms) in the baseline activity approximately 100 ms after the anticipated time of delivery. Reproduced with permission from (Redgrave et al., 1999).

An intriguing feature of the short latency DA response is its capacity to change over time with repeated stimulus presentation and with changes in experimental context (Schultz, 1998). For example, the initially robust response to novelty (Figure 1A) habituates rapidly when the stimulus is repeated in the absence of reinforcement. Interestingly, habituation of the DA neuronal response seems to correlate with the diminishing capacity of the stimulus to elicit behavioural interrupt and subsequent orienting. If a primary reward is presented in an increasingly predictable manner, it also fails to evoke a phasic DA response. However, the short latency DA response will gradually shift to an unexpected neutral stimulus, such as a light or tone, which reliably predicts the delivery of a primary reward (Figure 1C). After conditioning, if the reward is withheld after presentation of the predicting stimulus, a depression in the spontaneous activity of the DA neurones is reliably observed (Figure 1D). This pause in ongoing DA activity seems to have a latency of 50-100 ms after the time of expected reward delivery and lasts a further ~100 ms. Finally, extensive over-training produces a gradual attenuation of neuronal responses as performance becomes highly stable and automatised (Schultz, 1998). Mainly on the basis of these observations, it has been proposed by Schultz and his colleagues (Schultz et al., 1997; Contreras-Vidal and Schultz, 1999; Suri and Schultz, 1999), and others (Montague et al., 1996; Brown et al., 1999; Kakade and Dayan, 2000; Suri, 2001; BarGad et al., 2003), that the short latency DA response represents a reward prediction error signal which is relayed to neural mechanisms in the forebrain capable of associative learning. This view receives support from computational investigations in which simulated DA neuronal responses have been shown to be an effective 'teaching signal' in several models of reinforcement learning (Schultz and Dickinson, 2000; Suri, 2002; BarGad et al., 2003).

While the reward prediction error hypothesis continues to receive wide support, we (Redgrave et al., 1999; Coizet et al., 2003; Comoli et al., 2003) and others (Horvitz, 2000) have argued that there are outstanding difficulties with this view which have yet to be addressed. Firstly, the stereotyped short-latency, short-duration response of DA neurones is apparently not confined to reward and reward-related stimuli. Other biologically salient events including novel stimuli (Schultz, 1998; Horvitz, 2000), intense sensory phenomena (Schultz, 1998) and in some cases aversive stimuli (Guarraci and Kapp, 1999) are equally effective at eliciting the response. The apparent difficulty of DA neurones to distinguish stimuli associated specifically with reward may be related to the fact that the typical onset latency of a saccadic eye movement, which brings an unexpected event onto the fovea, is in the range 180-200 ms (80-110 ms for express saccades; Sparks, 1986; Grantyn, 1988) (Figure 2). With DA phasic response latencies almost always in the range 70-100 ms, the relevant perceptual classification of sensory stimuli must therefore occur prior to, or during, foveation. Moreover, if the function of foveation is to bring a stimulus onto the region of retina maximally connected to neural systems involved in feature extraction and object recognition (Thorpe and Fabre-Thorpe, 2001), the short latency phasic DA responses must occur on the basis of pre-attentive stimulus processing. If this analysis is correct, it must be assumed that if the DA neurones signal reward prediction error, the computations required to generate this signal would have to be conducted *before* the animal switches its gaze to see fully what the stimulus is.

If it is the case that DA neurones exhibit responses to a wider class of stimuli than those unambiguously related to reward, and do so prior to the onset of orienting saccadic eye-movements, then it is possible that the phasic DA signal may have a broader remit than informing target structures of unexpected reward errors alone. The research issue we have therefore been addressing for the past few years is to try and discover exactly what the remit of the short-latency phasic DA signal is. Our starting point was the assumption that information content of the DA signal will be determined partly by the integrative properties of the neurones themselves, and partly by the perceptual properties of their afferent sensory systems. An important step in decoding the DA signal would therefore be to identify, and then elucidate, the perceptual properties of sensory projections afferent to DA neurones.

Surprisingly, comparatively little is known about the source of sensory input to DA neurones (Schultz, 1998). Thus, in the case of visual stimuli (widely used as conditioned stimuli in the electrophysiological experiments reviewed above), it must be assumed that one or both of the two major targets of the retinal ganglion cells, the geniculostriate projection system or the midbrain superior colliculus (SC), are involved. An important clue as to which of the 'two visual systems' (Schneider, 1969) is responsible for relaying information to DA neurones at short latency may be gained by considering their relative transmission and processing latencies. In the thalamocortical visual circuits, which are responsible for feature detection and object recognition, the latencies are generally in the range ~60 ms in V1 to ~130 ms in pre-frontal cortical visual areas (Thorpe and Fabre-Thorpe, 2001). However, direct projections from these circuits to regions of the ventral midbrain that contain DA neurones originate mainly from prefrontal regions at the end of the processing chain, which have the longest (>100 ms) response latencies (Carter, 1982; Sesack and Pickel, 1992; Naito and Kita, 1994; Tong et al., 1996). It is therefore unlikely that the geniculo-cortical systems are in a position to provide the short latency (<100 ms) route whereby visual information related to unexpected events gains access to mesencephalic DA cells.

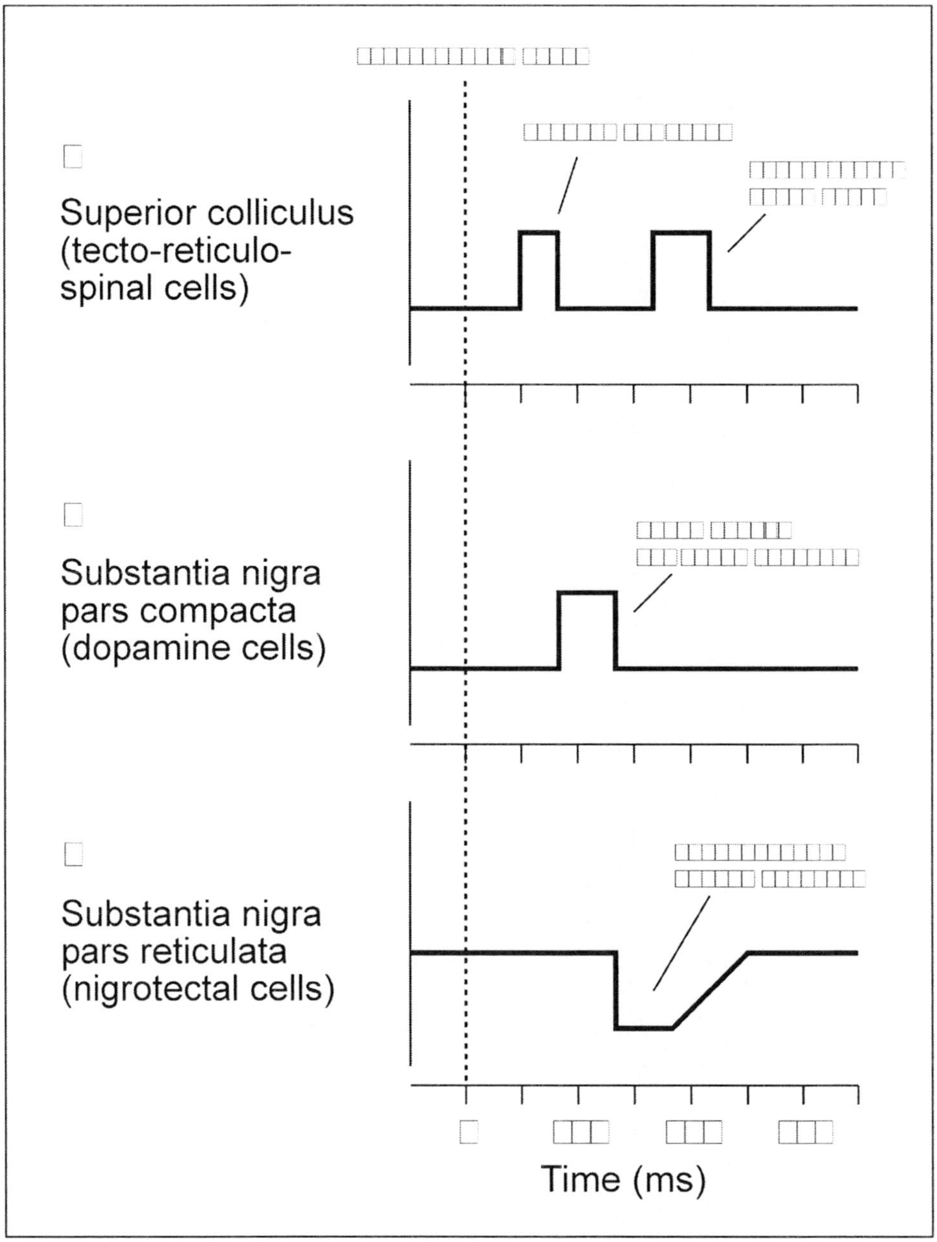

Figure 2: Short latency dopamine responses occur prior to saccadic eye movements that bring unexpected events onto the fovea. a. An unexpected visual event typically elicits a short latency sensory reaction (~50 ms) and a longer latency (>150 ms) pre-saccadic motor burst in primate superior colliculus (Jay and Sparks, 1987). b. The typical latency of the response of dopaminergic cells to unexpected stimuli is 70-100 ms (Schultz, 1998). c. A significant visual stimulus elicits a disinhibitory output response from the basal ganglia which generally coincides with the pre-saccadic motor burst in collicular target neurones (Hikosaka and Wurtz, 1983) i.e. >150 ms. Reproduced with permission from (Redgrave et al., 1999).

In contrast, the reported response latencies of cells in the midbrain SC to visual stimuli are reliably less than those of DA cells in the ventral midbrain (i.e. 40-50 ms; cf. 70-100 ms for DA neurones; Goldberg and Wurtz, 1972; Wurtz and Goldberg, 1972; Jay and Sparks, 1987). In view of the comparative visual response latencies of the geniculostriate and subcortical visual systems we have proposed that the retinotectal projection is the most plausible route by which visual information is relayed to DA cells. To test this hypothesis, we have conducted a series of anatomical and electrophysiological experiments.

1 ANATOMICAL EXPERIMENTS

1.1 Anterograde Tract Tracing

According to available data at the time, our expectation was that the projection from the SC to the ventral midbrain was likely to be indirect. The SC was known to have efferent projections to the subthalamic nucleus (Tokuno et al., 1994) and pedunculopontine nucleus (Redgrave et al., 1987), both of which have direct contacts with DA cells in the ventral midbrain (Deniau et al., 1978; Bolam et al., 1991; Smith and Grace, 1992; Lokwan et al., 1999). Even more indirect routes were possible via the midline and intralaminar thalamus (Krout et al., 2001) and amygdala, which also projects to DA neurones (Gonzales and Chesselet, 1990; Doron and LeDoux, 1999). We were therefore surprised when our injections of the anterograde tracers *Phaseolus vulgaris* leucoagglutinin (PHA-L) or biotinylated dextran amine (BDA) into the rat SC revealed a significant, previously unreported, direct projection to the ipsilateral substantia nigra and ventral tegmental area (Comoli et al., 2003; Figure 3). We found that the 'tectonigral pathway' descends vertically from the superior colliculus through the underlying midbrain reticular formation then curves rostrally to enter the caudal pole of substantia nigra pars compacta (Figure 3a). One of the reasons it had not previously been noted might be that most histology is done on coronal sections and this projection is best visualised in sections cut sagittally. Tectonigral fibres run the full rostro-caudal length of pars compacta (Figure 3b), with some fibres turning medially to enter the ventral tegmental area at rostral levels.

Throughout pars compacta (Figure 3c) and the ventral tegmental area (Figure 3d) numerous anterogradely labelled axonal boutons were seen in close association with TH-positive perikarya and dendrites, as well as in adjacent regions devoid of TH immunostaining (Figures 3b, c and d). Tectonigral fibres and boutons were also seen in regions of the caudal substantia nigra pars reticulata that contain the ventral group of dopaminergic neurones (Gonzalez-Hernandez and Rodriguez, 2000). Anterograde labelling was sparse or absent from regions of the pars reticulata devoid of DA neurones. Structures dorsally adjacent to the substantia nigra pars compacta, including zona incerta, also contained anterogradely labelled boutons and axons. At the light microscopic level both PHA-L and BDA produced qualitatively and quantitatively similar anterograde labelling. However, in some cases, cell bodies of neurones in the substantia nigra pars reticulata were also labelled with BDA. This raised the possibility that some of the boutons that were evident in pars compacta may derive from local collaterals of retrogradely labelled nigrotectal neurones. To investigate this possibility and verify that anterogradely labelled processes made synaptic contacts with DA neurones, we proceeded to examine the tectonigral pathway using electron microscopy.

At least three classes of anterogradely labelled synaptic terminals were observed. (i) Boutons of medium size (0.5-1.5 μm in diameter) characteristically formed asymmetric (presumed excitatory) synapses with both DA (tyrosine hydroxylase (TH) -positive) and non-DA (TH-negative) dendrites. (ii) Medium-sized boutons (0.5-1.0 μm in diameter) that formed symmetric (presumed inhibitory) synapses with TH-positive and TH-negative dendrites. (iii) Larger labelled terminals (up to 2.5 μm in diameter) that also formed symmetric synapses with both TH-positive and TH-negative structures. The larger boutons had the morphological appearance of terminals formed by the local axon collaterals of the substantia nigra pars reticulata output neurones (Damlama et al., 1993) and were present only in those animals in which the tracer injections also gave rise to retrograde labelling in pars reticulata. These results suggested that the direct tectonigral projection has both excitatory and inhibitory components that are likely to originate from different cell populations in the superior colliculus.

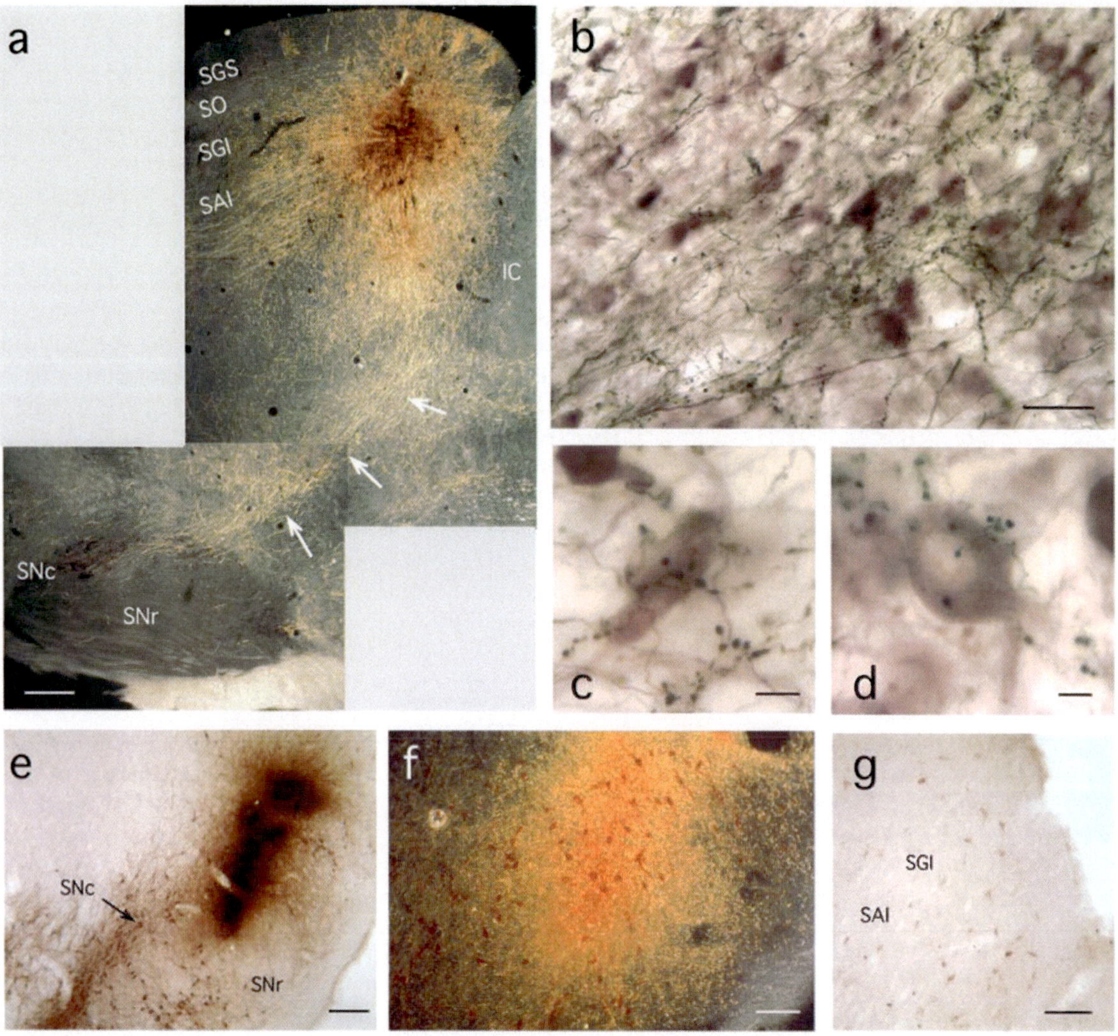

Figure 3: The tectonigral projection revealed by anterograde and retrograde tract tracing. a. A darkfield photomontage of a parasagittal section of the rat midbrain illustrating an injection site of the anterograde tracer BDA in the caudal intermediate layers of the superior colliculus, with ventrally projecting tectonigral fibres (white arrows) directed to tyrosine hydroxylase (TH)-positive regions (purple label) of the substantia nigra pars compacta. b. Tectonigral fibres, with numerous boutons, coursing rostrally (from top right to bottom left) in a parasagittal section, running between TH-positive cells (purple label) in pars compacta. c. Tectonigral boutons in close association with a TH-immunopositive neurone in pars compacta. Note also the anterogradely labelled terminals in adjacent TH-negative regions. d. Tectonigral boutons apposed to a TH-positve perikaryon in the ventral tegmental area. e. An injection site of the retrograde tracer CTb in the lateral substantia nigra pars compacta in coronal section. f. A darkfield photomicrograph of an adjacent section showing the injection site (orange label) was centred on a region containing numerous TH-positive cells (purple). g. Cells retrogradely labelled with CTb in the lateral intermediate layers of the superior colliculus in coronal section. *Abbreviations:* Collicular layers - SGS, stratum griseum superficiale; SO stratum opticum; SGI, stratum griseum intermediale; SAI, stratum album intermediale. Substantia nigra - SNc, pars compacta; SNr, pars reticulata. Calibration bars: a = 0.5 mm; b = 30 μm; c and d = 5 μm; e = 200 μm; f and g = 100 μm. Reproduced with permission from (Comoli et al., 2003).

1.2 Retrograde Tract Tracing

In order to identify cells in the superior colliculus that give rise to direct tectonigral fibres, small quantities of the retrograde tracer, cholera toxin subunit B, was injected into the substantia nigra pars compacta. The tissue was also processed to reveal TH-immunoreactive structures (Figures 3e and f). Retrogradely labelled neurones were concentrated in the intermediate layers of the superior colliculus, although scattered cells were observed in the adjacent deep and optical layers (Figure 3g). Labelled neurones appeared as a relatively uniform population of small cells (mean diameter = 10.78 ± 0.48 μm). The localisation of retrogradely labelled cells in different layers of the colliculus, in combination with the anterograde tracing data, provided bi-directional tracing evidence for a tectonigral projection. Both the anterograde and retrograde tracing data also raise the possibility that the tectonigral projection might comprise several functionally diverse components.

In summary, the results of these anatomical experiments demonstrated a significant, previously unreported anatomical projection from a primary visual structure directly to the pars compacta region of the substantia nigra in which DA neurones are located. This relay is therefore in a position to transmit short latency visual information throughout dopamine-containing regions of the ventral midbrain. To test this hypothesis, we have conducted a series of electrophysiological experiments to determine whether the superior colliculus acts as a critical relay for short-latency components of visual-evoked responses recorded locally in the substantia nigra pars compacta.

2 Electrophysiological Experiments

2.1 Experimental Stimulation of the Superior Colliculus

Our first step was use direct chemical stimulation of the SC to explore the possibility of functional connections to the ventral midbrain (Coizet et al., 2003). The strategy used was to obtain simultaneous electrophysiological recordings from the SC and substantia nigra pars compacta, in urethane anaesthetised rats, during chemical activation of the SC with the GABA antagonist bicuculline. The disinhibitory effect of injections of this drug into the SC of anaesthetised rats is known to activate local neurones. This activation is often characterised by periods of intense burst firing (Figure 4). Neurones recorded from substantia nigra and the ventral tegmental area were classified as putative DA (~40 %) or putative non-DA (~ 60 %) on the basis of anatomical and electrophysiological criteria (Grace and Bunney, 1983). In almost half the cases, chemical stimulation of the SC evoked a corresponding increase in neural activity in the ventral midbrain. This excitatory effect was apparent in both DA and non-DA neurones. In approximately 10 % of cases, SC activation elicited a reliable suppression of activity.

The close relationship between activity in the SC and that of DA neurones in the ventral midbrain is illustrated in Figure 5. During the pre-injection baseline period, the SC multi-unit record exhibited short periods of increased activity (bursts) approximately every 2 s (~0.5 Hz) (top trace Figure 5a), while at the same time a DA neurone in the ventral tegmental area fired only infrequently (Figure 5a bottom trace). Following an injection of bicuculline into the SC, the local burst pattern became markedly more pronounced and intense (top trace Figure 5b). This local disinhibitory action of bicuculline in the SC was sufficient to elicit single or double spikes in the DA neurones that coincided perfectly with the bursts in the SC (bottom trace Figure 5b). Later in the recording session, the burst pattern in the SC became more variable with some bursts lasting considerably longer (top trace Figure 5c).

The differences in SC burst durations were mirrored in the ventral tegmental area by corresponding changes in the number of spikes/burst exhibited by the DA neurone (bottom trace Figure 5c). The results of this study (Coizet et al., 2003) demonstrated clearly close functional connections between the SC and both DA and non-DA neuronal elements in the substantia nigra pars compacta and the ventral tegmental area. These findings were entirely consistent with the anatomical evidence described above which revealed that both excitatory and inhibitory components of the tectonigral projection make connections with both DA and non-DA neurones. The next phase of our investigation proceeded to test whether the SC represents a critical relay in the transmission of short-latency visual information to the DA neurones.

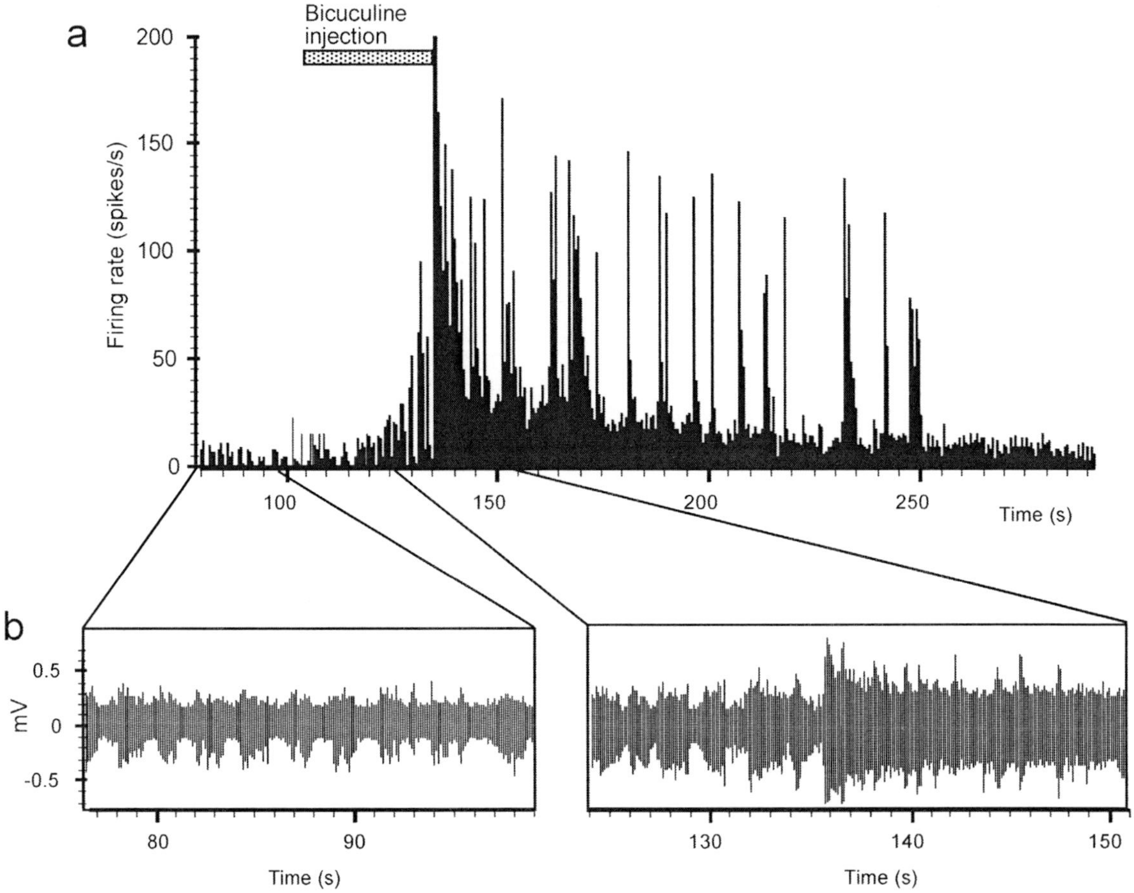

Figure 4: Effect of bicuculline (BIC) on neurones recorded from the SC. a. The rate histogram (1 s bins) shows a typical SC multi-unit response to a nearby injection of BIC. Note that throughout the response there was intense bursting separated by periods of relative inactivity. b. Expansion of two periods of the waveform trace from which the rate histogram in (a) was derived. The clear low frequency oscillatory pattern of burst firing (<1.0 Hz) in the pre-injection period was disrupted by the intense disinhibitory activity evoked by the injection of BIC. Time scales are referenced to the beginning of the recording session and shaded bars represent the injection period. Reproduced with permission from (Coizet et al., 2003).

2.2 Effect of Collicular Manipulations on Visual Evoked Activity in Substantia Nigra

The first stage of these experiments was to determine the relative latencies of visually evoked potentials (VEPs) recorded simultaneously from the superficial layers of the SC and the substantia nigra pars compacta of urethane-anaesthetised rats (Comoli et al., 2003). Field potential signal averaging revealed short-latency, whole field light flash-evoked responses in both structures (Figure 6). Both initial onset and peak latencies were reliably shorter for VEPs recorded from the SC, as would be predicted if the SC acts as relay for visual information destined for the ventral midbrain.

Secondly, we reasoned that if the SC transmits short latency visual signals to the substantia nigra pars compacta, then local manipulations of SC neurotransmission should produce appropriate modulations of nigral VEPs. Disinhibitory microinjections of bicuculline next to the SC recording electrode increased both the amplitude of local VEPs, and the VEP responses recorded from the substantia nigra pars compacta. Conversely, both SC and nigral VEPs were reliably decreased by microinjections of the local anaesthetic lidocaine into the SC. These results established the SC as a powerful regulator of short-latency visual input to the substantia nigra pars compacta.

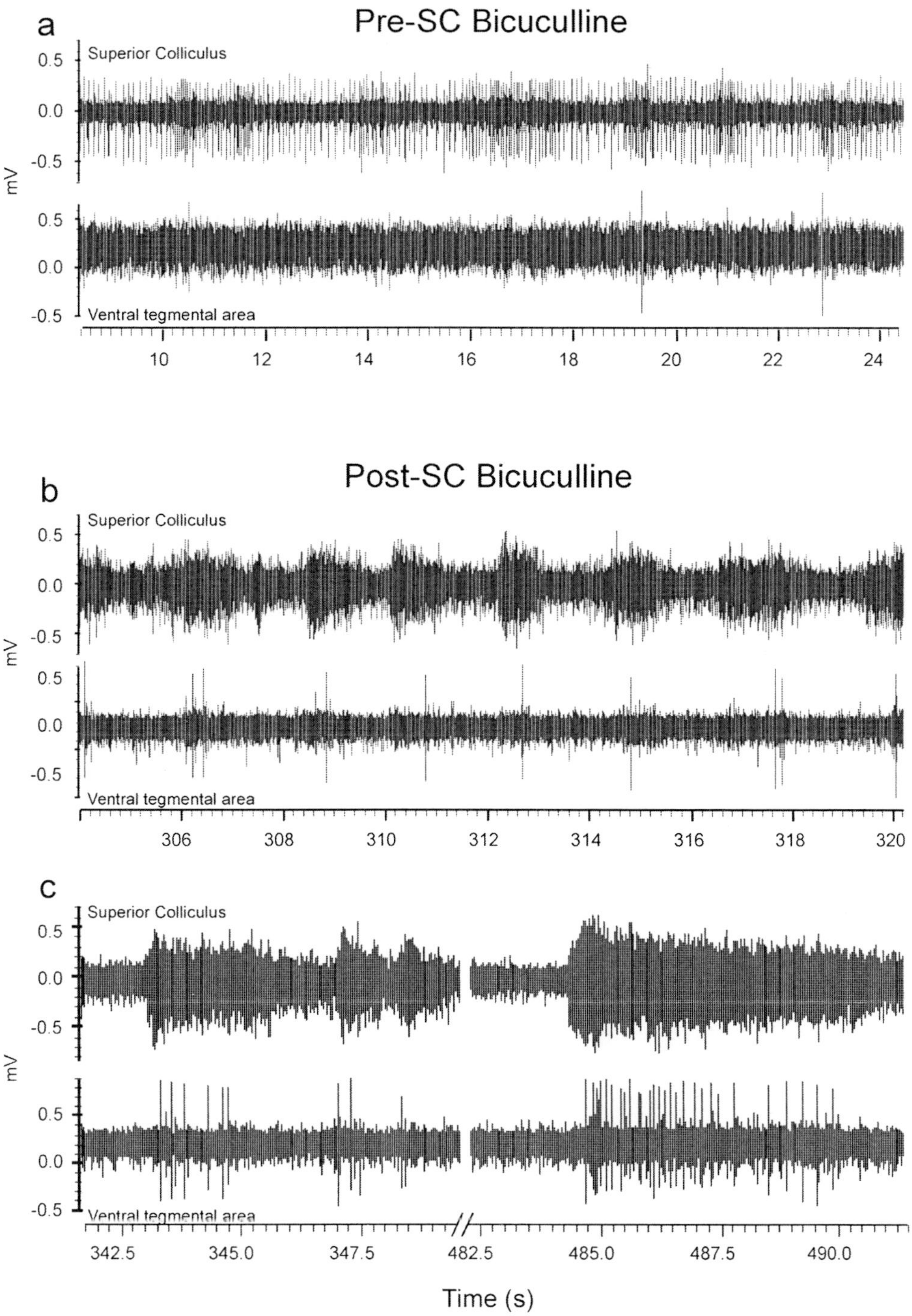

Figure 5: Chemical stimulation of the superior colliculus (SC) induced correlated activity in the ventral tegmental area (VTA). a. Before the injection of bicuculline (BIC), the SC waveform (top trace) exhibited spontaneous multi-unit firing, punctuated by short periods of increased activity (bursts) approximately every 2 s (~0.5 Hz). During this time, a VTA broad-spike neurone fired only infrequently (e.g. 2 spikes in the latter half of the bottom trace). b. Following a 200 nl injection of BIC, the pattern of local burst firing in the SC increased in intensity and separation (top trace). In this state, each burst in the SC was sufficient to elicit closely associated single or double spikes from the VTA neurone (bottom trace). c. Later in the session, variation in the duration of bursts recorded from the SC (top trace) were reflected in corresponding changes in the numbers of spikes/burst recorded from the dopaminergic neurone in the VTA (bottom trace). In each case time scales are referenced to the beginning of the recording session. Reproduced with permission from (Coizet et al., 2003)

Thirdly, we then tested whether the integrity of the SC is a necessary condition for the transmission of visual information to the ventral midbrain. Baseline VEPs were recorded from both SC and nigral electrodes in urethane-anaesthetised rats. Subsequent aspiration of the ipsilateral visual cortex confirmed the powerfully suppressive effect that unilateral cortical ablation has on SC visual processing, as reported widely by others (Sprague, 1966; Jiang et al., 2003). We also noted that cortical ablation induced a correlated suppression of visual processing in pars compacta. Since it has been reported that application of bicuculline to the superior colliculus can reverse the inhibitory effects of cortical lesions (Ciaramitaro et al., 1997) and potentiate local VEPs, a swab soaked in bicuculline was applied to the surface of the SC exposed by the cortical ablation. Following this treatment, and in the complete absence of ipsilateral visual cortex, the bicuculline rapidly re-instated VEPs recorded both locally in the SC *and* in the substantia nigra. These results effectively dissociated the potential involvement of cortical and midbrain visual systems in the provision of short latency visual input to the substantia nigra pars compacta.

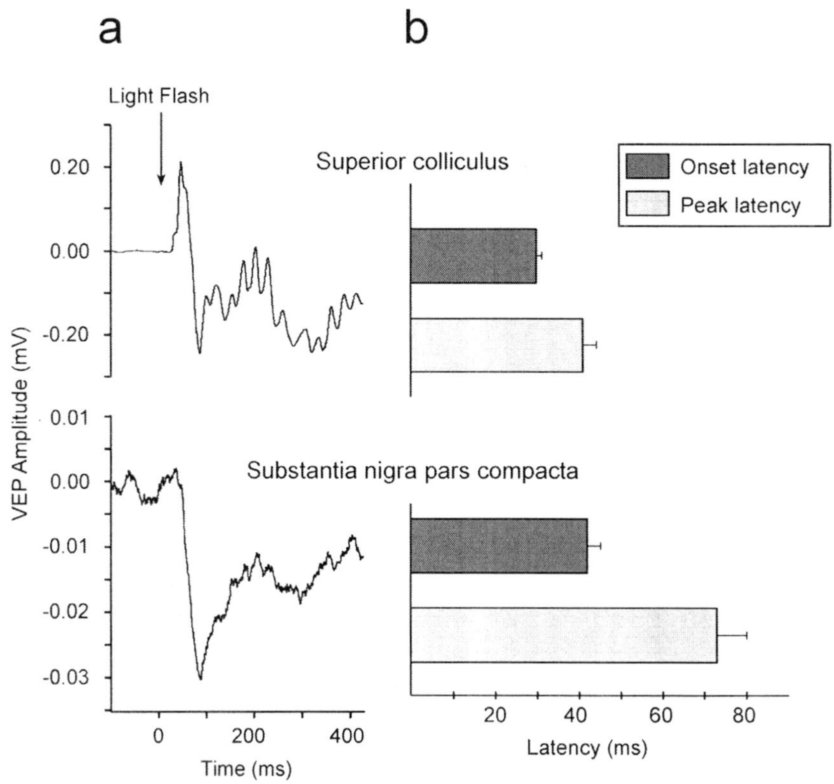

Figure 6: Response latencies of visually evoked field potentials (VEPs) in the superior colliculus and the substantia nigra pars compacta. a. A typical example of collicular and nigral VEPs from an individual subject. The two dotted lines indicate initial onset and peak latencies respectively for the collicular VEP. b. Group data (n=12) showing that both the mean onset latencies and peak latencies were shorter for collicular VEPs. Reproduced with permission from (Comoli et al., 2003).

Having demonstrated that a functioning SC is essential for visually evoked field potentials to be recorded in substantia nigra pars compacta, it remained for us to show that the SC is critical for short latency visual activation of DA neurones themselves. The forth stage of our electrophysiological investigations (Overton et al., 2003) was therefore based on a recent report that anaesthetic-related suppression of visual sensitivity in the intermediate and deep layers of the SC can be restored temporarily by local injections of disinhibitory agents such as bicuculline (Katsuta and Isa, 2003). This experimental paradigm provided us with the opportunity to render a non-working system (anaesthetic-related disruption of the retino-tecto-nigral relay) temporarily operational again by a specific and localised manipulation (a local injection of bicuculline into the SC). This study (Overton et al., 2003) can therefore be viewed as a mirror image of a normal lesion/inactivation investigation in which a specific manipulation is used either permanently or temporarily to disable a working system. The prediction for our investigation was that the bicuculline-associated restoration of visual sensitivity to neurones in the intermediate and deep layers of the SC would confer visual sensitivity to DA neurones in substantia nigra.

The first stage in this study used simultaneous multiunit recording from the intermediate and deep SC, and single unit recording from DA neurones in substantia nigra or the ventral tegmental area in anaesthetised animals to reveal that, in all cases (N=35), neither the superior colliculus nor midbrain DA neurones were responsive to a whole-field light flash. Following a disinhibitory microinjection bicuculline into the SC (directly adjacent to the recording electrode), local SC neurones became sensitive to the light flash. When the SC became responsive to the light stimulus, a large proportion (~85 %) of DA neurones also exhibited clear short-latency responses to the light. More than half of these (~60 %) were excited by the visual stimulus, while the ongoing activity of the others (~40 %) was inhibited. Again the response latencies of neurones in the superior colliculus were reliably shorter than those of the simultaneously recorded DA neurones. Control experiments involving a comparable enhancement of visual processing in the striate cortex, produced by direct application of bicuculline to the surface of visual cortex, left DA neurones unresponsive to the light stimulus. These results again reveal the retino-tecto-nigral route as the predominant means for transmitting visual information to DA neurones in the ventral midbrain. A further important aspect of the methodology used in this study was that the whole-field light flash was entirely predictable (every 2 s, for 100's of trials). Measures of response magnitude throughout the experimental sessions revealed no signs of the rapid habituation reported by others (Schultz, 1998), of either excitatory or inhibitory DA responses. Rather, the magnitude of the responses of DA neurones appeared more accurately to reflect the waxing and waning of bicuculline's disinhibitory effect in the SC. This provides further evidence that the state of inhibition/disinhibition in the SC is perhaps one of the critical determinants of the DA response to visual stimuli.

3 FUNCTIONAL IMPLICATIONS

The present series of experiments has established a direct projection from a primary retino-recipient structure, the midbrain SC, to DA containing regions of the ventral midbrain and that the functional state of the SC is critical for the transmission of short latency visual information to DA neurones. This work therefore establishes the retino-tecto-nigral projection as the most likely source of short latency visual input to both DA and non-DA neurones in the ventral midbrain.

3.1 Excitatory and Inhibitory Components of the Tectonigral Projection

An important aspect of both the anatomical and electrophysiological experiments discussed above is the suggestion that the tectonigral pathway may have several functionally distinguishable components. First, our electron microscopy data revealed that tectonigral boutons form both asymmetric (presumed excitatory) and symmetric (presumed inhibitory) synapses in substantia nigra pars compacta. Moreover, it is likely that the different types of contact originate from different cell populations in the SC. This supposition is consistent with the discovery of retrogradely labelled tectonigral neurones in different layers of the SC. Anatomical, physiological and behavioural analyses of the SC suggest a modular architecture (Mana and Chevalier, 2001) in which multisensory (Stein and Meredith, 1993) and contextual inputs from cortical (Mana and Chevalier, 2001; Tsumori et al., 2001) and limbic sources (Heilman, 1994; Cameron et al., 1995) are directed to anatomically and functionally dissociable output channels (Dean et al., 1989; Redgrave et al., 1993; Mana and Chevalier, 2001). Many of these functional units are differentially distributed both within and across the laminar structure of the SC (Mana and Chevalier, 2001). It is therefore likely that tectonigral cells of origin in different SC regions will be subject to qualitatively different influences. This conclusion is supported by our electrophysiological investigations that consistently showed local disinhibitory chemical activation of the superior colliculus was associated with both excitatory and inhibitory effects on the activity of ventral midbrain neurones.

The discovery of possible excitatory and inhibitory components in a direct tectonigral projection suggests that this pathway could mediate both the positive *and* negative short-latency modulation of DA neurones by visual stimuli (Schultz, 1998). Activation of DA neurones could be determined by relevant pre-attentive visual signals being directed to an excitatory component of the tectonigral projection.

Alternatively, the profuse afferent connections to the SC from regions of the brain associated with expectation (e.g. cortex; Basso, 1998 and basal ganglia; Hikosaka et al., 1989) and timing (basal ganglia and cerebellum Hikosaka et al., 1989; Dreher and Grafman, 2002) suggest that sufficient information may be present in the SC to enable the absence of an expected event to be detected at short latency. If the results of such a computation were directed to an inhibitory component of the tectonigral projection, the temporary suppression of DA neuronal activity observed shortly after the omission of an expected reward (Schultz, 1998) could be explained.

3.2 Visual Processing in the Superior Colliculus

In unanaesthetised animals, intermediate and deep layer neurones of the superior colliculus (Stein and Meredith, 1991) and midbrain DA neurones (Schultz, 1998) are both exquisitely sensitive to unexpected novel visual events, but habituate rapidly if stimuli become predictable or are not maintained by association with primary reward (Wurtz and Albano, 1980; Sparks, 1986). In our final electrophysiological study (Overton et al., 2003) it was therefore significant that when the collicular relay was experimentally 'unblocked' and consequently resistant to habituation, DA activation also showed consistent responses to predictable stimuli over many consecutive trials. From this it may be inferred that, in normally awake preparations, mechanisms that modulate sensory processing in the SC (Ikeda and Hikosaka, 2003) would be in a position to regulate the often-reported habituation and reward-related dishabituation of DA neurones to neutral visual stimuli (Schultz, 1998). If this reasoning is correct, the capacity of DA neurones to distinguish between various classes of visual event, and modify their responses on the basis of reinforcement contingencies will depend, in large part, on the visual processing capacities of the SC (Wurtz and Albano, 1980). In most vertebrate species, the SC is sensitive to stimuli that exhibit a restricted range of properties, which are represented according to a well-defined retino-topic map (Wurtz and Albano, 1980; Sparks, 1986). Thus on the one hand, most visually responsive cells are transiently activated 40-60 ms after the appearance or disappearance of a stimulus in a specific region of the visual field. Within this region, they are also exquisitely sensitive to moving stimuli and respond maximally to movement in the range of 0.5-30 °/s. On the other hand, SC cells respond poorly, if at all, to the contrast, velocity, wavelength or geometric configuration of visual stimuli. The SC therefore appears to be a structure which is specialised for analysing 'where' an unpredicted stimulus is in the external world, with a limited capacity for detecting 'what' it is (Schneider, 1969).

Analyses of the stimulus-processing properties of SC cells provides much of the functional evidence supporting the widely accepted conclusion that the SC is a mechanism in the vertebrate brain responsible for redirecting the sense organs and attention to unexpected, biologically salient novel events (Wurtz and Albano, 1980; Sparks, 1986; Grantyn, 1988; Dean et al., 1989). The SC's specific processing capacities are not, however, what one would necessarily expect of a mechanism who's function was to distinguish the many different classes of rewarding events from the rest of the visual milieu. If the reward prediction error hypothesis of DA function (Montague et al., 1996; Schultz and Dickinson, 2000) is to apply, it must be assumed that afferent and necessarily pre-saccadic and pre-attentive stimulus processing, performed in large part by the SC, can do this. Within the context of highly structured experimental protocols and with extensively trained subjects, there is now good evidence that DA neurones can indeed discriminate familiar, and in most cases spatially predictable but temporally unexpected stimuli that signify rewards with differential probability and different magnitude (Fiorillo et al., 2003; Morris et al., 2004). Consistent with these experiments are those reporting that the initial sensory properties of visual cells in the SC can be modulated by behavioural variables: in particular, a response enhancement is observed if the stimulus is to be the target of a rewarded saccade (Goldberg and Wurtz, 1972; Ikeda and Hikosaka, 2003). However, the extent to which the pre-attentive perceptual capacities of this evolutionarily ancient subcortical structure can be configured to distinguish specific classes of visual event in more ecologically valid experimental contexts, where events can be spatially as well as temporally unpredictable, remains to be explored. Whatever the outcome, the answer will provide an important constraint on the information that can be signalled by the phasic activity of DA neurones.

ACKNOWLEDGEMENTS

This work was partly funded by Wellcome Trust grants to PR (059735) and PO (062742), and an MRC Programme Grant (PR).

REFERENCES

BarGad I, Morris G, Bergman H (2003) Information processing, dimensionality reduction and reinforcement learning in the basal ganglia. *Prog Neurobiol* **71**:439-473.

Basso MA (1998) Cognitive set and oculomotor control. *Neuron* **21**:665-668.

Berridge KC, Robinson TE (1998) What is the role of dopamine in reward: hedonic impact, reward learning, or incentive salience? *Brain Res Brain Res Rev* **28**:309-369.

Bolam JP, Francis CM, Henderson Z (1991) Cholinergic input to dopaminergic neurons in the substantia nigra: a double immunocytochemical study. *Neurosci* **41**:483-494.

Brown J, Bullock D, Grossberg S (1999) How the basal ganglia use parallel excitatory and inhibitory learning pathways to selectively respond to unexpected rewarding cues. *J Neurosci* **19**:10502-10511.

Cameron AA, Khan IA, Westlund KN, Cliffer KD, Willis WD (1995) The efferent projections of the periaqueductal gray in the rat: a phaseolus vulgaris leucoagglutinin study .1. ascending projections. *J Comp Neurol* **351**:568-584.

Carter CJ (1982) Topographical distribution of possible glutamatergic pathways from the frontal cortex to the striatum and substantia nigra in rats. *Neuropharmacology* **21**: 370-383.

Ciaramitaro VM, Todd WE, Rosenquist AC (1997) Disinhibition of the superior colliculus restores orienting to visual stimuli in the hemianopic field of the cat. *J Comp Neurol* **387**:568-587.

Coizet V, Comoli E, Westby GWM, Redgrave P (2003) Phasic activation of substantia nigra and the ventral tegmental area by chemical stimulation of the superior colliculus: an electrophysiological investigation in the rat. *Eur J Neurosci* **17**:28-40.

Comoli E, Coizet V, Boyes J, Bolam JP, Canteras NS, Quirk RH, Overton PG, Redgrave P (2003) A direct projection from superior colliculus to substantia nigra for detecting salient visual events. *Nat Neurosci* **6**:974-980.

Contreras-Vidal JL, Schultz W (1999) A predictive reinforcement model of dopamine neurons for learning approach behavior. *J Comput Neurosci* **6**:191-214.

Damlama M, Bolam JP, Tepper JM (1993) Axon collaterals of pars reticulata projection neurons contact nigral dopaminergic neurons. *Soc Neurosci Abs* **19**:1432.

Dean P, Redgrave P, Westby GWM (1989) Event or emergency ? Two response systems in the mammalian superior colliculus. *Trends Neurosci* **12**:137-147.

Deniau JM, Hammond C, Chevalier G, Feger J (1978) Evidence for branched subthalamic nucleus projections to substantia nigra, entopeducular nucleus and globus pallidus. *Neurosci Lett* **9**:117-121.

Doron NN, LeDoux JE (1999) Organization of projections to the lateral amygdala from auditory and visual areas of the thalamus in the rat. *J Comp Neurol* **412**:383-409.

Dreher JC, Grafman J (2002) The roles of the cerebellum and basal ganglia in timing and error prediction. *Eur J Neurosci* **16**:1609-1619.

Fiorillo CD, Tobler PN, Schultz W (2003) Discrete coding of reward probability and uncertainty by dopamine neurons. *Science* **299**:1898-1902.

Goldberg ME, Wurtz RH (1972) Activity of superior colliculus in behaving monkey. II. Effect of attention on neuronal responses. *J Neurophysiol* **35**:560-574.

Gonzales C, Chesselet M-F (1990) Amygdalonigral pathway: An anterograde study in the rat with Phaseolus vulgaris leucoagglutinin (PHA-L). *J Comp Neurol* **297**:182-200.

Gonzalez-Hernandez T, Rodriguez M (2000) Compartmental organization and chemical profile of dopaminergic and GABAergic neurons in the substantia nigra of the rat. *J Comp Neurol* **421**:107-135.

Grace AA, Bunney BS (1983) Intracellular and extracellular electrophysiology of nigral dopaminergic neurons-1. Identification and characterization. *Neuroscience* **10**:301-315.

Grantyn R (1988) Gaze control through superior colliculus: structure and function. In: *Neuroanatomy of the oculomotor system* (Buttner-Ennever JA, ed), pp 273-333. Amsterdam: Elsevier.

Guarraci FA, Kapp BS (1999) An electrophysiological characterization of ventral tegmental area dopaminergic neurons during differential pavlovian fear conditioning in the awake rabbit. *Behav Brain Res* **99**:169-179.

Heilman KM (1994) Emotion and the brain - a distributed modular network mediating emotional experience. In: (Zaidel DW, ed), pp 139-158. 525 B Street, Suite 1900, San Diego, CA 92101-4495: Academic Press Inc.

Hikosaka O, Wurtz RH (1983) Visual and oculomotor function of monkey substantia nigra pars reticulata. I. Relation of visual and auditory responses to saccades. *J Neurophysiol* **49**:1230-1253.

Hikosaka O, Sakamoto M, Usui S (1989) Functional properties of monkey caudate neurons III. Activities related to expectation of target and reward. *J Neurophysiol* **61**:814-831.

Horvitz JC (2000) Mesolimbocortical and nigrostriatal dopamine responses to salient non-reward events. *Neuroscience* **96**:651-656.

Ikeda T, Hikosaka O (2003) Reward-dependent gain and bias of visual responses in primate superior colliculus. *Neuron* **39**:693-700.

Jay MF, Sparks DL (1987) Sensorimotor integration in the primate superior colliculus. I Motor convergence. *J Neurophysiol* **57**:22-34.

Jiang H, Stein BE, McHaffie JG (2003) Opposing basal ganglia processes shape midbrain visuomotor activity bilaterally. *Nature* **423**:982-986.

Kakade S, Dayan P (2000) Dopamine bonuses. *NIPS* **12**:131-137.

Katsuta H, Isa T (2003) Release from GABA(A) receptor-mediated inhibition unmasks interlaminar connection within superior colliculus in anesthetized adult rats. *Neurosci Res* **46**:73-83.

Krout KE, Loewy AD, Westby GWM, Redgrave P (2001) Superior colliculus projections to midline and intralaminar thalamic nuclei of the rat. *J Comp Neurol* **431**:198-216.

Lokwan SJA, Overton PG, Berry MS, Clark D (1999) Stimulation of the pedunculopontine tegmental nucleus in the rat produces burst firing in A9 dopaminergic neurons. *Neuroscience* **92**:245-254.

Mana S, Chevalier G (2001) Honeycomb-like structure of the intermediate layers of the rat superior colliculus: Afferent and efferent connections. *Neuroscience* **103**:673-693.

Montague PR, Dayan P, Sejnowski TJ (1996) A framework for mesencephalic dopamine systems based on predictive Hebbian learning. *J Neurosci* **16**:1936-1947.

Morris G, Arkadir D, Nevet A, Vaadia E, Bergman H (2004) Coincident but distinct messages of midbrain dopamine and striatal tonically active neurons. *Neuron* **43**:133-143.

Naito A, Kita H (1994) The cortico-nigral projection in the rat - an anterograde tracing study with biotinylated dextran amine. *Brain Res* **637**:317-322.

Overton PG, Clark D (1997) Burst firing in midbrain dopaminergic neurons. *Brain Res Rev* **25**:312-334.

Overton PG, Dommett EJ, Coizet V, Redgrave P (2003) Visually-evoked responses in dopaminergic neurones in the anaesthetised rat following chemical activation of the superior colliculus. In: *Abstract viewer/Itinerary planner Online* (704.13 PN, ed). Wahington, DC: Society for Neuroscience.

Redgrave P, Mitchell IJ, Dean P (1987) Descending projections from the superior colliculus in rat: a study using orthograde transport of wheatgerm-agglutinin conjugated horseradish peroxidase. *Exp Brain Res* **68**:147-167.

Redgrave P, Westby GWM, Dean P (1993) Functional architecture of rodent superior colliculus: relevance of multiple output channels. In: *Progress in Brain Research* (Hicks TP, Molotchnikoff S, Ono T, eds), pp 69-77. Amsterdam: Elsevier.

Redgrave P, Prescott TJ, Gurney K (1999) Is the short latency dopamine response too short to signal reward error ? *Trends Neurosci* **22**:146-151.

Robbins TW, Everitt BJ (1996) Neurobehavioural mechanisms of reward and motivation. *Curr Opin Neurobiol* **6**:228-236.

Salamone JD (1994) The involvement of nucleus accumbens dopamine in appetitive and aversive motivation. *Behav Brain Res* **61**:117-133.

Schneider GE (1969) Two visual systems. *Science* **163**:895-901.

Schultz W (1998) Predictive reward signal of dopamine neurons. *J Neurophysiol* **80**:1-27.

Schultz W (2004) Neural coding of basic reward terms of animal learning theory, game theory, microeconomics and behavioural ecology. *Curr Opin Neurobiol* **14**:139-147.

Schultz W, Dickinson A (2000) Neuronal coding of prediction errors. *Annu Rev Neurosci* **23**:473-500.

Schultz W, Dayan P, Montague PR (1997) A neural substrate of prediction and reward. *Science* **275**:1593-1599.

Sesack SR, Pickel VM (1992) Prefrontal cortical efferents in the rat synapse on unlabeled neuronal targets of catecholamine terminals in the nucleus accumbens septi and on dopamine neurons in the ventral tegmental area. *J Comp Neurol* **320**:145-160.

Smith ID, Grace AA (1992) Role of the subthalamic nucleus in the regulation of nigral dopamine neuron activity. *Synapse* **12**:287-303.

Sparks DL (1986) Translation of sensory signals into commands for control of saccadic eye movements: role of the primate superior colliculus. *Physiol Rev* **66**:118-171.

Sprague JM (1966) Interaction of cortex and superior colliculus in the mediation of visually guided behavior in the cat. *Science* **153**:1544-1547.

Stein BE, Meredith MA (1991) Functional organization of the superior colliculus. In: *Electrophysiology of Vision*. (Leventhal AG, ed), pp 85-110. Hampshire, U.K.: Macmillan Press.

Stein BE, Meredith MA (1993) The merging of the senses. Cambridge, MA.: The MIT Press.

Suri RE (2001) Anticipatory responses of dopamine neurons and cortical neurons reproduced by internal model. *Exp Brain Res* **140**:234-240.

Suri RE (2002) TD models of reward predictive responses in dopamine neurons. *Neural Netw* **15**:523-533.

Suri RE, Schultz W (1999) A neural network model with dopamine-like reinforcement signal that learns a spatial delayed response task. *Neuroscience* **91**:871-890.

Thorpe SJ, Fabre-Thorpe M (2001) Seeking categories in the brain. *Science* **291**:260-263.

Tokuno H, Takada M, Ikai Y, Mizuno N (1994) Direct projections from the deep layers of the superior colliculus to the subthalamic nucleus in the rat. *Brain Res* **639**:156-160.

Tong ZY, Overton PG, Clark D (1996) Stimulation of the prefrontal cortex in the rat induces patterns of activity in midbrain dopaminergic neurons which resemble natural burst events. *Synapse* **22**:195-208.

Tsumori T, Yokota S, Ono K, Yasui Y (2001) Organization of projections from the medial agranular cortex to the superior colliculus in the rat: a study using anterograde and retrograde tracing methods. *Brain Res* **903**:168-176.

Wurtz RH, Goldberg ME (1972) The primate superior colliculus and the shift of visual attention. *Invest Ophthalmol* **11**:441-450.

Wurtz RH, Albano JE (1980) Visual-motor function of the primate superior colliculus. *Ann Rev Neurosci* **3**:189-226.

In: Recent Breakthroughs in Basal Ganglia Research
Editor: Erwan Bezard, pp. 201-211
ISBN: 1-59454-880-3

Chapter 16

PERSPECTIVES ON SUBCORTICAL CONTRIBUTIONS TO HIGHER-ORDER ASPECTS OF VISUO-OCULOMOTOR CONTROL

Terrence R. Stanford
Wake Forest University School of Medicine

ABSTRACT

Nearly two decades ago, Alexander and colleagues proposed that the neural architecture linking the basal ganglia and cortex comprised a series of parallel circuits (Alexander et al., 1986). As outlined originally, this group of circuits consisted of a motor, an oculomotor, a prefrontal, and a limbic loop, with each originating in a different region of frontal cortex and remaining topographically segregated throughout their course within the basal ganglia and thalamus enroute to returning to their cortical site of origin. Since the time of this influential review, anatomical evidence has continued to grow in support of the existence of reentrant circuits involving frontal cortex, basal ganglia, and thalamus (Lynch et al., 1994; Middleton & Strick, 2002; Kelly & Strick, 2004).

Though the concept of parallel, segregated loops linking the cortex and subcortex was originally proposed and substantiated for the basal ganglia, it now seems clear that a similar closed-loop architecture may characterize cortico-cerebellar interactions as well. Again largely on the strength of anatomical evidence, the existence of topographically segregated loops relating to a range of functions from purely motor to purely cognitive have been suggested (Middleton & Strick, 1997, 2001; Mason et al., 2000; Dum & Strick, 2002, 2003; Kelly & Strick, 2003).

The postulate of a canonical loop circuit suggests that this common neural architecture may provide a general solution to a computational problem shared by sensorimotor and more cognitive domains. Yet, while the anatomical evidence for closed loop architectures linking frontal cortex with the basal ganglia and with the cerebellum is compelling, there is yet little known about precisely what computations these loops are necessary for performing. This chapter considers reentrant circuits in the specific context of oculomotor control and reviews some recent evidence suggesting a role for these loops in specifying the goals of saccadic eye movements in a context-dependent manner. In particular, the chapter will consider recent single-neuron neurophysiological results from primate central thalamic regions that serve as the penultimate synapses in the aforementioned cortico-subcortical loops. The data, which demonstrate that neurons within primate central thalamus participate in the process selecting saccadic goals from competing alternatives, provide empirical support for the idea that cortico subcortical interactions are an essential part of visuo-oculomotor selection circuitry.

1 TARGET AND RESPONSE SELECTION FOR VISUAL ORIENTING

In a typically complex and dynamic real-world environment, choices or decisions (see Schall, 2005 for discussion of the distinction) about where to look (or equivalently, what to look at), are governed both by the physical properties (bottom up) of the objects in the scene (e.g., color, brightness, size) and by a host of internal factors (top down) relating to the current goals of the animal. While bottom up factors would tend to promote gaze shifts to objects or events that stand out in some way (e.g., abrupt onset, bright color), this tendency can be mitigated by top down factors that specify a particular goal for the current circumstances. For example, if one is searching for his/her car keys, gaze shifts may be biased toward the most probable locations (e.g., the countertop) or to objects that possess certain physical qualities (e.g., highly reflective). In the laboratory, many neurophysiological studies have reported neural correlates of visual target and/or saccadic response selection in the activity of single neurons recorded from rhesus monkeys engaged in some variant of an oculomotor choice task. There are now a large number of excellent review articles on the matter of target selection and the closely related phenomenon of perceptual decision-making in nonhuman primates. I refer the interested reader to several of these for a more thorough treatment of this rapidly advancing area (Gold and Shadlen, 2001; Glimcher, 2001, 2002, 2003, Glimcher & Rustichini, 2004; Reddi, 2001; Schall, 2001, 2002, 2004; Sugrue et al., 2005). For the purposes of this chapter, however, I'll provide a brief introduction to the topic by considering some representative neurophysiological findings, mainly from the primate frontal eye fields (FEF), an extensively studied visuo-oculomotor region and the origin of the 'oculomotor loop' as originally defined by Alexander and colleagues, (Alexander et al., 1986; 1990).

For some of the earliest demonstrations of 'selection-related' activity in frontal cortex, Schall and colleagues recorded from FEF neurons while monkeys performed a simple visual 'pop-out' task in which the objective was to locate and generate a saccade to a feature (color) singleton, generally a green (or red) stimulus embedded in an array of red (or green) distracters (Schall et al., 1995; Thompson et al., 1996, 1997). Performance on this task can be logically viewed as a serial process, requiring the monkey first to differentiate the target from the multiple distracter stimuli and then to plan and execute an accurate saccade to foveate it. The first stage involves a perceptual discrimination (identifying what is present), while the latter requires selecting the appropriate response (determining what to do about it). From the work of Schall and colleagues, it appears that different FEF neurons participate to varying degrees in the perceptual discrimination and motor planning (and execution) phases of task performance (Thompson et al., 1996; 1997, Murthy et al., 2001; see Schall, 2002 for review). A neuron thought to have a role in the perceptual discrimination typically responds with a transient burst linked to the appearance of a stimulus in its response field, followed by a period of sustained activity. When recorded during performance of the feature singleton or 'pop-out' task, the initial onset phase of the stimulus-linked response is indistinguishable for a target or distracter in the response field. However, over the course of approximately 150 ms, the neuron's activity evolves to discriminate between a target and a distracter and, once achieved, this difference in firing level is usually maintained until a saccade to the target is initiated (Thompson et al., 1996; Murthy et al., 2001; see Schall, 2002 for review). Neurons involved in visual selection have activity levels that correlate with the time of perceptual discrimination and are presumed to reflect explicit knowledge of the target's identity (time of target detection – see Thompson et al., 1996 for details). Upon detecting the target, other FEF neurons appear to participate in preparing the response to look at the target. Such neurons have activity levels that correlate with the time of saccade onset (Sato et al., 2003; see Schall, 2002 for review). Thus, sequential stages of visual orientation (goal identification and response planning) appear to be represented in at least partially segregated populations of FEF neurons.

As the term implies, 'pop-out' is a highly efficient form of visual search requiring relatively little perceptual discrimination time. Naturally, search efficiency can be reduced by making targets and distracters more similar and thus diminishing the tendency for the target to 'pop-out' perceptually (e.g., green versus yellow-green rather than green versus red; see Bichot and Schall, 1999). Even more dramatic increases in visual search times may be observed when the target is defined not by one feature (e.g., color), but by a conjunction of features (e.g., red AND square) such that targets and distracters can have one or more features in common (e.g., **red** circle or green **square**). The key finding from such manipulations, and the one that implicates an FEF neuron as participating in the perceptual discrimination, is that the time required to resolve targets and distracters increases in proportion to the difficulty of the perceptual discrimination (Bichot et al., 2001; Sato et al., 2001).

2 Distributed Cortical Architecture for Visuo-Oculomotor 'Selection'

The basic principles of neural target/response selection, described above for neurons in the FEF, appear to generalize to neurons in other visuocortical domains and for other visuo-oculomotor tasks as well. For example, employing a task that required monkeys to judge the direction of a motion stimulus, Kim & Shadlen (1999) reported analogous results for neurons in both the FEF and in the dorsolateral prefrontal cortex (DLPFC); when the perceptual strength (i.e., relative motion coherence) of the motion signal was varied, the time required for neurons to 'judge' direction was inversely related to the difficulty of the discrimination (see Gold & Shadlen, 2001 for review).

Analogous observations from different cortical regions suggest that the process of linking specific stimuli to specific actions is not domain specific nor for that matter the workings of a strict functional hierarchy. Instead, it is generally thought that these functions are supported by parallel processing over a highly recurrent cortical network (See Houk & Wise, 1995; Pierrott-Deseilligny et al., 1995; Gaymard et al., 1998; Lynch & Tian, in press, for reviews). As outlined in Fig. 1, the FEF is reciprocally connected with a number of other frontal cortical regions. These areas, which in addition to the DLPFC, include the supplementary eye fields (SEF), contain neurons that display analogous visual and saccade-related activations and which are active over a comparable time frame as are those in FEF (Hanes et al., 1995; Russo & Bruce, 1996). In addition to frontal regions, the cortical visuo-oculomotor network comprises reciprocal links with parietal regions (not shown), most notably, a parietal eye field (LIP) located in the lateral bank of the intraparietal sulcus (see Lynch & Tian, in press for recent review). Finally, a still broader network would include reciprocal links to several extrastriate visual regions (Lynch & Tian, in press) Though primarily if not exclusively sensory, these latter regions process information about visual stimulus attributes, such as motion (MT, MST) and color (V4) – stimulus features upon which choices/decisions are based.

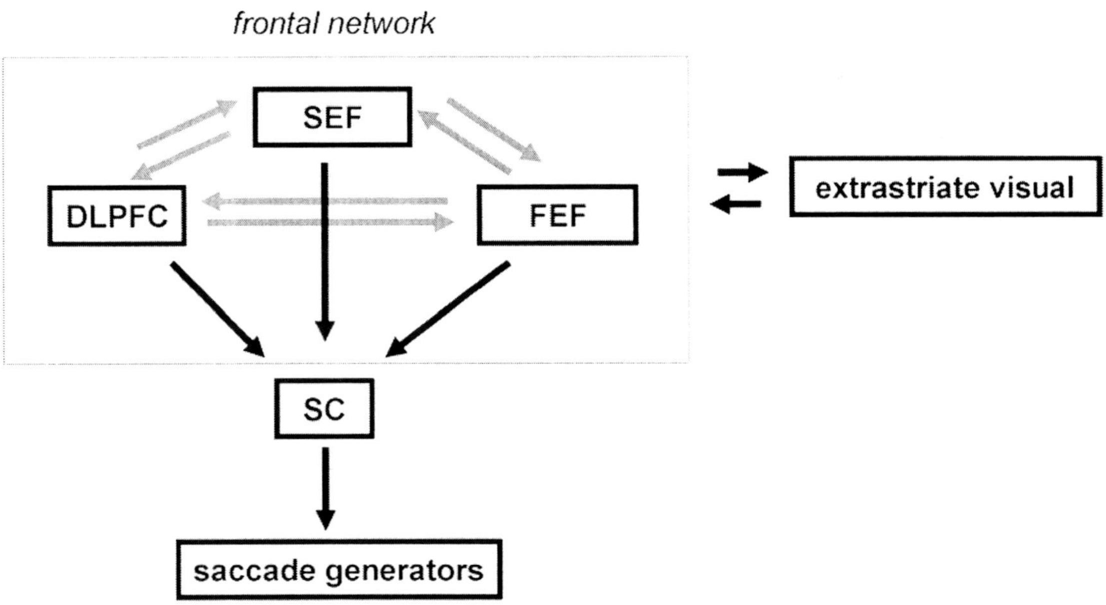

Figure 1. Schematic illustrating the reciprocal cortico-cortico connections of frontal cortical areas involved in visuo-oculomotor control. Gray arrows illustrate reciprocal connections between the frontal eye fields (FEF), supplementary eye fields (SEF), and dorsolateral prefrontal cortex (DLPFC). Reciprocity between frontal and extrastriate visual regions is noted by bi-directional arrows (black) while descending projections from frontal cortex to the superior colliculus (SC) in the midbrain are indicated (black unidirectional arrows). See text for details.

Given access to information about visual features (e.g., motion, color) from extrastriate regions, the fronto-parietal network, as outlined above, would appear to contain all of the requisite elements for effecting visuomotor transformations like those required to select targets in a visual search task. Then, having identified the visual goal, the results of the selection process could then be translated into a motor plan and/or voluntary motor command to be conveyed to downstream structures. As shown in Fig 1, the frontal-cortical network in primates provides several independent output projections to the superior colliculus (SC), where the motor commands for saccade vector (amplitude and direction) are represented topographically in the form of a two-dimensional map (see Sparks, 1986; Stanford, 2004 for reviews). From the SC, saccadic commands are relayed to the pontine and medullary saccade generator circuitry responsible for controlling saccade dynamics (see Sparks & Mays, 1990 for review).

3 Expanding the Visuo-Oculomotor 'Selection' Network to Subcortical Structures

In principle, most aspects of the selection of visual stimuli and the choosing of appropriate motor responses could be carried out by processing local to cortex and the result subsequently passed on to subcortical effector systems. However, anatomical evidence, along with theoretical and logical considerations, suggest that subcortical systems could play a more active role in shaping these signals. Again, considering the protracted period of time over which cortical selection/decision-related signals have been shown to evolve, it is intuitively apparent that the information conveyed in closed loops connecting cortical and subcortical structures, such as the 'oculomotor loop' defined by Alexander and colleagues, could have a profound impact on their development. Indeed, recent conceptualizations of the cortico-basal ganglia loop system argue that it is an architecture ideally suited for solving the general problem of action selection (Redgave et al., 1999) (see below), of which visual target/saccade selection is a special case. Moreover, many recent studies have now documented the influence of reward expectation on visuomotor signaling pertaining to response selection in both frontal and parietal cortices (Platt & Glimcher, 1999; Dorris & Glimcher, 2004; Sugrue et al., 2004; See Gold & Shadlen, 2002; Glimcher & Rustichini, 2004; Sugrue et al., 2005 for recent reviews). The observation that, during visuomotor task performance, the influence of relatively low-level motivational systems is well integrated with those mediating relatively high-level visuo-oculomotor control implicates subcortical dopamine systems along with the familiar cortico-basal ganglia-thalamo-cortical loops (Alexander et al., 1986, 1990; See Schultz 2001, 2002 for reviews). While this is perhaps the clearest case for the influence of subcortical systems on cortical selection mechanisms, the potential contributions of closed loops involving the cerebellum and potentially even the SC might also be considered in this context. Figure 2, from left to right, outlines the essential connections of each of these putative closed loops (See Sommer, 2003 for review).

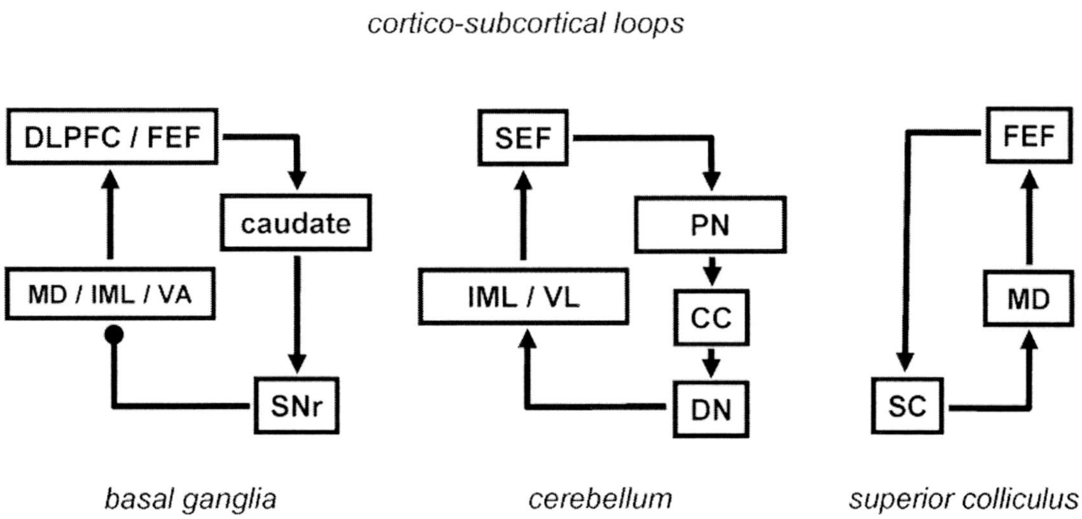

Figure 2. Schematic representation of putative closed cortico-subcortical loops. **Left:** Cortico-basal ganglia-thalamo-cortical loops including the "oculomotor loop" as originally defined by Alexander and colleagues. **Middle:** Cortico-cerebello-thalamo-cortical loop. **Right:** Cortico-colliculo-thalamo-cortical loop. See text for details. **Abbreviations: CC:** cerebellar cortex, **DLPFC:** dorsolateral prefrontal cortex, **DN:** dentate nucleus, **FEF:** frontal eye fields, **IML:** internal medullary lamina, **MD:** mediodorsal nucleus, **PN:** pontine nuclei, **SC:** superior colliculus, **SNr:** substantia nigra pars reticulata, **VA:** ventral anterior nucleus, **VL:** ventral lateral nucleus.

Though highly suggestive arguments could be made on anatomical and logical grounds, the fact remains that physiological data relevant to the role of these regions in target/response selection is in short supply. Strick and colleagues have reported that cerebellar dentate neurons respond in association with cognitively demanding reaching tasks (Mushiake & Strick, 1993). They also note that, along with motor deficits, gross cerebellar lesions lead to cognitive impairments in humans (see Middleton & Strick, 2000 for review). Neural and clinical data relating to the basal ganglia are more abundant. Early single-unit studies established that subsets of neurons within the caudate nucleus and the SNr maintain spatial information in the absence of a visible target when monkeys are required to make saccades to remembered locations (Hikosaka & Wurtz, 1983; Hikosaka et al., 1989). This, along with deficits in performing memory-guided saccades following experimental lesions of the caudate nucleus, has led to the suggestion that activity within a basal ganglia thalamocortical loops contributes to sustaining working memory representations in prefrontal cortex (see Hikosaka et al., 2000 for review). Perhaps more relevant to the issue of target selection, Hikosaka and colleagues have shown dramatic shifts in the spatial preferences of visually-responsive caudate neurons in a task that varied the association between reward magnitude and stimulus/response location (Kawagoe et al., 1998).

In contrast to those in cerebellum and basal ganglia, neurons in SC have been extensively studied and have been shown to exhibit many of the properties of cortical neurons when examined in the context of target selection and or perceptual decision-making tasks (Glimcher & Sparks, 1992; Horwitz & Newsome, 1999; McPeek & Keller, 2002; Horwitz et al., 2004; see Gold 2004, Keller et al., 2005 for reviews). Given that the SC receives feedforward input from all of the aforementioned cortical visuo-oculomotor areas (see Fig. 2), this is neither surprising nor compelling evidence that the SC is anything more than a passive recipient of information, its activity a reflection but not a participant in a cortically-mediated process. At issue is whether or not the SC neurons that display selection-related activity are part of a closed loop with frontal cortex (see Fig. 2; right). Recent studies have established that SC neurons do provide ascending information (via MD of thalamus) to the FEF, but on the basis of reversible lesions, it is argued that these signals function primarily as an efference copy of the SC motor command (see Sommer, 2003 for review; Sommer & Wurtz, 2004a, b) – a signal critical for updating visual representations consequent to eye movements, not for selecting the targets of impending saccades. Nevertheless, the possibility that information relevant to selecting the target or the impending response to the target is conveyed within this transthalamic pathway cannot be ruled out. Along these lines, McPeek & Keller (2004) recently reported deficits in target selection, but not saccade execution, following reversible inactivation of the SC. Even more recently, Newsome and colleagues (Muller et al., 2005) concluded that signals (artificially produced via microstimulation) originating within the SC can focus spatial attention in the absence of eye movements. Insofar as the deployment of attention requires the activation of cortical networks, this finding is consistent with ascending influences from the SC to cortex having an impact on visuomotor cognition.

4 Visuo-Oculomotor 'Selection' Signals in the Central Thalamus

With their principle outputs directed to cortex, and as an obligatory link in each of the putative cortico-subcortical loops outlined in Fig. 2, several nuclei within central thalamus could provide information critical to understanding the nature of the information subcortical structures provide to cortex. Unlike the other components of the putative loops, the basal ganglia, cerebellum, and SC, which have both ascending and descending outputs, the projections of the thalamic nuclei in question are principally, if not exclusively, ascending and are those that convey the influence of ascending pathways onto cortical neurons. Yet, despite their crucial anatomical position, until recently virtually nothing was known about the task-related activations of neurons within the central thalamic nuclei that serve as the penultimate synapse in basal ganglia, cerebellar, and collicular loops with cortex (see Sommer 2003 for review). In primates, early data consisted entirely of the seminal study by John Schlag and Madeleine Schlag-Rey, published in the form of companion papers more than two decades ago (Schlag-Rey & Schlag, 1984; Schag & Schlag-Rey, 1984). Though limited in scope, having been conducted in monkeys performing spontaneous eye movements, these studies nevertheless revealed the existence of signals related to both visual stimuli and saccadic eye movements. Only recently, have a few groups begun to reexamine these areas using more specific tasks suitable for examining the neural signals that may link stimulus and response in these regions (see Sommer, 2003 for a recent review).

With regard to stimulus/response selection, our laboratory has examined the activity of neurons within these regions, which include nuclei of the internal medullary lamina (IML) and paralaminar regions of the ventral anterior (VA), ventral lateral (VL) and mediodorsal nuclei (MD). Having first observed that, like neurons in other visuo-oculomotor areas, thalamic neurons are active during all phases of a simple delayed saccade task (Wyder et al., 2003), we sought to determine if such neurons could carry information relevant to the selection of visual targets. To do so, neurons were recorded while monkeys performed a 'choice' task in which a saccade to the correct stimulus was required to obtain a reward. On each trial, a monkey was required to choose between two stimuli on the basis color (red or green), with the 'correct' choice on any given trial indicated by the color of an independent cue. The result, reported by Wyder et al., (2004), indicated that thalamic neurons can combine information about the sensory stimulus with information about the behavioral relevance of that stimulus. That is, like their cortical counterparts, such neurons were capable of discriminating between a rewarded target and an irrelevant distracter in their response field.

Adapted from Wyder et al., (2004), Figure 3 illustrates the 'selection-related' activations of two different central thalamic neurons (A & B – See Legend for details). For each of these neurons, it is shown that activity for the same physical stimulus differs considerably depending on the identity of the stimulus as either the target or the distracter. Thus, prior to the receiving the informative cue, the stimulus in the response field was equally likely to become a target or distracter and the activity of these neurons was accordingly ambivalent. However, after the cue, activity evolved to reflect the change in behavioral status, increasing if the stimulus was identified as the target and decreasing if identified as a distracter. It is important to emphasize that presentation of the cue changed only the behavioral significance of the eccentric stimuli and not their physical appearance.

The ability of thalamic neurons to differentiate between a target and distracter is analogous to the selectivity shown by neurons within the FEF and other visuo-oculomotor areas. As discussed by Wyder et al., (2004), if one takes into consideration differences in task design across studies, the time course over which thalamic activity evolves is likely to be comparable to that observed within cortical visuo-oculomotor areas and, at a minimum, overlaps considerably with cortical task-related activations. The demonstration of 'selection-related' activity in the thalamic regions that constitute the final relay for closed cortical loops (see Fig. 2) satisfies the minimal requirement for the argument that such loop circuits play a role in shaping cortical 'selection-related' signals.

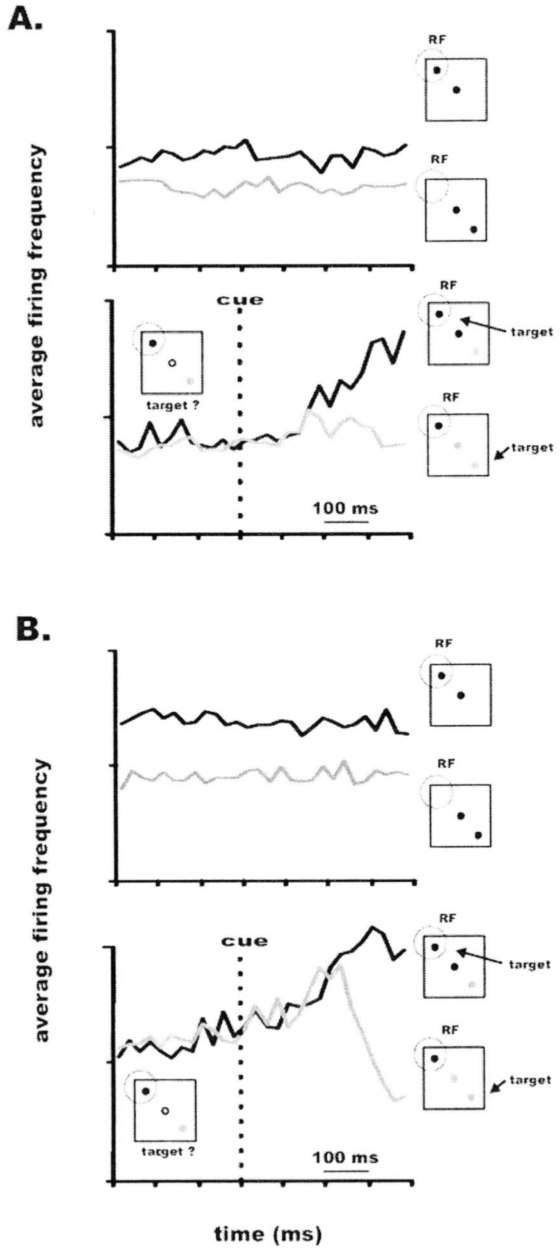

Figure 3. Activity of two different central thalamic neurons that differentiate between a target and distracter. For each neuron, (A & B), average firing frequency is plotted as function of time during performance of a single target control task (top panel of A & B) and a 2-stimulus choice task (bottom panel of A & B). Square panes depict key spatial and temporal aspects of the tasks indicating the position of visual stimuli (black or gray dots) in relation to the receptive field (circle) of the neuron. In the choice task, each trial began with the presentation of a yellow central fixation stimulus (indicted by an unfilled circle at center of pre-cue task panes on left) which the monkey had to look to within 500 ms. After a delay, two eccentric stimuli were illuminated, one red (depicted in black) and one green (depicted in gray). The two stimuli were always of equivalent eccentricity and differed in direction by 180 degrees. During the before-cue period (left of vertical dotted line), each stimulus was a *potential* saccade target. After a delay the central fixation light changed color, randomly, to either red (black) or green (gray) cueing the monkey to the identity of the eventual saccade target (color match) and distracter (non-match). The monkey was required to maintain fixation throughout the after-cue period until offset of the central fixation stimulus after which a saccade to the target could be made to obtain reward (not shown). Note that prior to presentation of the cue (vertical dotted line), the neurons respond similarly. However, upon presentation of the cue, in each case activity evolves, increasing or decreasing to reflect the identity of the receptive field stimulus as either a target (black traces) or a distracter (gray traces), respectively.

5 Future Considerations, which Loops, why Loops?

The single-unit data from central thalamus establishes the existence of a neural correlate for a relatively high-order visuo-oculomotor function, but neither establishes a causal relationship nor pinpoints the particular circuit or circuits within which such activity develops. As noted by Wyder et al., (2004) neurons with similar activity profiles were observed in several thalamic nuclei, including those thought to receive input from the basal ganglia (VA) and cerebellar (VL) circuits and there is yet no strong evidence for a functional segregation. However, in recognition that this is but one study and thus a small sample, it is premature to draw any conclusions in this regard. Future studies using antidromic and orthodromic stimulation methods to examine functional connectivity or the introduction of artificial signals using microstimulation and or reversible inactivation to probe the causal relationship between activity and behavior (see Stanford, 2004 for discussion) will prove valuable to address these issues.

In closing, it's worth considering briefly why closed loop architectures linking cortex to subcortical structures would even be of value in the context of target/response selection. As noted above, there remains relatively little evidence to suggest that the fronto-parietal network could not solve this problem alone. However, there may be certain advantages to 'farming out' parts of the problem to a parallel acting subcortical system whose architecture may be well suited for solving some aspect of the problem. Indeed, Redgrave and colleagues (Redgrave et al., 1999) advance a compelling argument that reentrant loops involving cortex and basal ganglia are an ideal substrate for 'selecting' actions from a slate of competing alternatives (see Mink, 1996 for a similar conceptual framework). The examples cited by Redgrave and colleagues consider complex behaviors involving competition between high-level behavioral systems (i.e., appetitive; fight or flight), however the mechanisms of conflict resolution, which employ the basal ganglia output to suppress unwanted actions and permit (via disinhibition) context-appropriate actions, is readily applied to the problem of choosing a visual target/saccade response selection. It will be interesting to observe if the proposed interplay between permissive and suppressive signals turns out to be an integral component of the 'switch' that initiates the target/distracter selection process apparent in the activities of neurons throughout the proposed cortical-subcortical network.

Acknowledgments

Supported by NIH/NEI EY12389.

References

Alexander GE, DeLong MR, Strick PL (1986) Parallel organization of functionally segregated circuits linking basal ganglia and cortex. *Annu Rev Neurosci* **9**:357-381.

Alexander GE, Crutcher MD, DeLong MR (1990) Basal ganglia-thalamocortical circuits: parallel substrates for motor, oculomotor, "prefrontal" and "limbic" functions. *Prog Brain Res* **85**:119-146.

Bichot NP, Schall JD (1999) Saccade target selection in macaque during feature and conjunction visual search. *Vis Neurosci* **16**:81-89.

Bichot NP, Thompson KG, Chenchal Rao S, Schall JD (2001) Reliability of macaque frontal eye field neurons signaling saccade targets during visual search. *J Neurosci* **21**:713-725.

Dorris MC, Glimcher PW (2004) Activity in posterior parietal cortex is correlated with the relative subjective desirability of action. *Neuron* **44**:365-378.

Dum RP, Strick PL (2003) An unfolded map of the cerebellar dentate nucleus and its projections to the cerebral cortex. *J Neurophysiol* **89**:634-639.

Dum RP, Li C, Strick PL (2002) Motor and nonmotor domains in the monkey dentate. *Ann N Y Acad Sci* **978**:289-301.

Gaymard B, Ploner CJ, Rivaud S, Vermersch AI, Pierrot-Deseilligny C (1998) Cortical control of saccades. *Exp Brain Res* **123**:159-163.

Glimcher PW (2001) Making choices: the neurophysiology of visual-saccadic decision making. *Trends Neurosci* **24**:654-659.

Glimcher P (2002) Decisions, decisions, decisions: choosing a biological science of choice. *Neuron* **36**:323-332.
Glimcher PW (2003) The neurobiology of visual-saccadic decision making. *Annu Rev Neurosci* **26**:133-179.
Glimcher PW, Sparks DL (1992) Movement selection in advance of action in the superior colliculus. *Nature* **355**:542-545.
Glimcher PW, Rustichini A (2004) Neuroeconomics: the consilience of brain and decision. *Science* **306**:447-452.
Gold JI (2004) Through the looking glass. Focus on "representation of an abstract perceptual decision in macaque superior colliculus". *J Neurophysiol* **91**:1936-1937.
Gold JI, Shadlen MN (2001) Neural computations that underlie decisions about sensory stimuli. *Trends Cogn Sci* **5**:10-16.
Gold JI, Shadlen MN (2002) Banburismus and the brain: decoding the relationship between sensory stimuli, decisions, and reward. *Neuron* **36**:299-308.
Hanes DP, Thompson KG, Schall JD (1995) Relationship of presaccadic activity in frontal eye field and supplementary eye field to saccade initiation in macaque: Poisson spike train analysis. *Exp Brain Res* **103**:85-96.
Hikosaka O, Wurtz RH (1983) Visual and oculomotor functions of monkey substantia nigra pars reticulata. III. Memory-contingent visual and saccade responses. *J Neurophysiol* **49**:1268-1284.
Hikosaka O, Sakamoto M, Usui S (1989) Functional properties of monkey caudate neurons. I. Activities related to saccadic eye movements. *J Neurophysiol* **61**:780-798.
Hikosaka O, Takikawa Y, Kawagoe R (2000) Role of the basal ganglia in the control of purposive saccadic eye movements. *Physiol Rev* **80**:953-978.
Horwitz GD, Newsome WT (1999) Separate signals for target selection and movement specification in the superior colliculus. *Science* **284**:1158-1161.
Horwitz GD, Batista AP, Newsome WT (2004) Representation of an abstract perceptual decision in macaque superior colliculus. *J Neurophysiol* **91**:2281-2296.
Houk JC, Wise SP (1995) Distributed modular architectures linking basal ganglia, cerebellum, and cerebral cortex: their role in planning and controlling action. *Cereb Cortex* **5**:95-110.
Kawagoe R, Takikawa Y, Hikosaka O (1998) Expectation of reward modulates cognitive signals in the basal ganglia. *Nat Neurosci* **1**:411-416.
Keller EL, Lee KM, McPeek RM (2005) Readout of higher-level processing in the discharge of superior colliculus neurons. *Ann N Y Acad Sci* **1039**:198-208.
Kelly RM, Strick PL (2003) Cerebellar loops with motor cortex and prefrontal cortex of a nonhuman primate. *J Neurosci* **23**:8432-8444.
Kelly RM, Strick PL (2004) Macro-architecture of basal ganglia loops with the cerebral cortex: use of rabies virus to reveal multisynaptic circuits. *Prog Brain Res* **143**:449-459.
Kim JN, Shadlen MN (1999) Neural correlates of a decision in the dorsolateral prefrontal cortex of the macaque. *Nat Neurosci* **2**:176-185.
Lynch JC, Tian JR (in press) Cortico-cortical networks and cortico-subcortical loops for the higher control of eye movements. In: *Progress in brain research: Neuroanatomy of the oculomotor system* (Buttner-Ennever J, ed).
Lynch JC, Hoover JE, Strick PL (1994) Input to the primate frontal eye field from the substantia nigra, superior colliculus, and dentate nucleus demonstrated by transneuronal transport. *Exp Brain Res* **100**:181-186.
Mason A, Ilinsky IA, Maldonado S, Kultas-Ilinsky K (2000) Thalamic terminal fields of individual axons from the ventral part of the dentate nucleus of the cerebellum in Macaca mulatta. *J Comp Neurol* **421**:412-428.
McPeek RM, Keller EL (2002) Saccade target selection in the superior colliculus during a visual search task. *J Neurophysiol* **88**:2019-2034.
McPeek RM, Keller EL (2004) Deficits in saccade target selection after inactivation of superior colliculus. *Nat Neurosci* **7**:757-763.
Middleton FA, Strick PL (1996) Basal ganglia and cerebellar output influences non-motor function. *Mol Psychiatry* **1**:429-433.
Middleton FA, Strick PL (1997) New concepts about the organization of basal ganglia output. *Adv Neurol* **74**:57-68.

Middleton FA, Strick PL (1997) Dentate output channels: motor and cognitive components. *Prog Brain Res* **114**:553-566.
Middleton FA, Strick PL (2000) Basal ganglia output and cognition: evidence from anatomical, behavioral, and clinical studies. *Brain Cogn* **42**:183-200.
Middleton FA, Strick PL (2001) Cerebellar projections to the prefrontal cortex of the primate. *J Neurosci* **21**:700-712.
Middleton FA, Strick PL (2002) Basal-ganglia 'projections' to the prefrontal cortex of the primate. *Cereb Cortex* **12**:926-935.
Mink JW (1996) The basal ganglia: focused selection and inhibition of competing motor programs. *Prog Neurobiol* **50**:381-425.
Muller JR, Philiastides MG, Newsome WT (2005) Microstimulation of the superior colliculus focuses attention without moving the eyes. *Proc Natl Acad Sci U S A* **102**:524-529.
Murthy A, Thompson KG, Schall JD (2001) Dynamic dissociation of visual selection from saccade programming in frontal eye field. *J Neurophysiol* **86**:2634-2637.
Mushiake H, Strick PL (1993) Preferential activity of dentate neurons during limb movements guided by vision. *J Neurophysiol* **70**:2660-2664.
Pierrot-Deseilligny C, Rivaud S, Gaymard B, Muri R, Vermersch AI (1995) Cortical control of saccades. *Ann Neurol* **37**:557-567.
Platt ML, Glimcher PW (1999) Neural correlates of decision variables in parietal cortex. *Nature* **400**:233-238.
Reddi BA (2001) Decision making: the two stages of neuronal judgement. *Curr Biol* **11**:R603-606.
Redgrave P, Prescott TJ, Gurney K (1999) The basal ganglia: a vertebrate solution to the selection problem? *Neuroscience* **89**:1009-1023.
Russo GS, Bruce CJ (1996) Neurons in the supplementary eye field of rhesus monkeys code visual targets and saccadic eye movements in an oculocentric coordinate system. *J Neurophysiol* **76**:825-848.
Sato T, Murthy A, Thompson KG, Schall JD (2001) Search efficiency but not response interference affects visual selection in frontal eye field. *Neuron* **30**:583-591.
Sato TR, Watanabe K, Thompson KG, Schall JD (2003) Effect of target-distractor similarity on FEF visual selection in the absence of the target. *Exp Brain Res* **151**:356-363.
Schall JD (2001) Neural basis of deciding, choosing and acting. *Nat Rev Neurosci* **2**:33-42.
Schall JD (2002) The neural selection and control of saccades by the frontal eye field. *Philos Trans R Soc Lond B Biol Sci* **357**:1073-1082.
Schall JD (2004) On building a bridge between brain and behavior. *Annu Rev Psychol* **55**:23-50.
Schall JD (2005) Decision making. *Curr Biol* **15**:R9-R11.
Schall JD, Hanes DP, Thompson KG, King DJ (1995) Saccade target selection in frontal eye field of macaque. I. Visual and premovement activation. *J Neurosci* **15**:6905-6918.
Schlag J, Schlag-Rey M (1984) Visuomotor functions of central thalamus in monkey. II. Unit activity related to visual events, targeting, and fixation. *J Neurophysiol* **51**:1175-1195.
Schlag-Rey M, Schlag J (1984) Visuomotor functions of central thalamus in monkey. I. Unit activity related to spontaneous eye movements. *J Neurophysiol* **51**:1149-1174.
Schultz W (2001) Reward signaling by dopamine neurons. *Neuroscientist* 7:293-302.
Schultz W (2002) Getting formal with dopamine and reward. *Neuron* **36**:241-263.
Sommer MA (2003) The role of the thalamus in motor control. *Curr Opin Neurobiol* **13**:663-670.
Sommer MA, Wurtz RH (2004) What the brain stem tells the frontal cortex. I. Oculomotor signals sent from superior colliculus to frontal eye field via mediodorsal thalamus. *J Neurophysiol* **91**:1381-1402.
Sommer MA, Wurtz RH (2004) What the brain stem tells the frontal cortex. II. Role of the SC-MD-FEF pathway in corollary discharge. *J Neurophysiol* **91**:1403-1423.
Sparks DL (1986) Translation of sensory signals into commands for control of saccadic eye movements: role of primate superior colliculus. *Physiol Rev* **66**:118-171.
Sparks DL, Mays LE (1990) Signal transformations required for the generation of saccadic eye movements. *Annu Rev Neurosci* **13**:309-336.
Stanford TR (2004) Signal coding in the primate superior colliculus revealed through the use of artificial signals. In: *The superior colliculus: New approaches for studying sensorimotor signaling* (Hall WC, Moschovakis A, eds), pp 35-53: CRC Press.
Sugrue LP, Corrado GS, Newsome WT (2004) Matching behavior and the representation of value in the parietal cortex. *Science* **304**:1782-1787.

Sugrue LP, Corrado GS, Newsome WT (2005) Choosing the greater of two goods: neural currencies for valuation and decision making. *Nat Rev Neurosci* **6**:363-375.

Thompson KG, Bichot NP, Schall JD (1997) Dissociation of visual discrimination from saccade programming in macaque frontal eye field. *J Neurophysiol* **77**:1046-1050.

Thompson KG, Hanes DP, Bichot NP, Schall JD (1996) Perceptual and motor processing stages identified in the activity of macaque frontal eye field neurons during visual search. *J Neurophysiol* **76**:4040-4055.

Wyder MT, Massoglia DP, Stanford TR (2003) Quantitative assessment of the timing and tuning of visual-related, saccade-related, and delay period activity in primate central thalamus. *J Neurophysiol* **90**:2029-2052.

Wyder MT, Massoglia DP, Stanford TR (2004) Contextual modulation of central thalamic delay-period activity: representation of visual and saccadic goals. *J Neurophysiol* **91**:2628-2648.

In: Recent Breakthroughs in Basal Ganglia Research
Editor: Erwan Bezard, pp. 213

ISBN: 1-59454-880-3

PART VI: PARKINSON'S DISEASE: SYMPTOMS

Erwan Bezard
CNRS UMR 5543, University Victor Segalen, Bordeaux, France

Parkinson's disease (PD) is traditionally associated to akinesia, bradykinesia, rigidity and resting tremor. The cerebral lesions leading to PD are increasingly acknowledged to be not limited to the sole substantia nigra pars compacta and both the physiology and pathophysiology of basal ganglia have seen dramatic increase in knowledge in the past few years. Still, we do not understand the basis for the above-mentioned symptoms and besides the precursor of dopamine, L-dopa, *nihil novo sub sole* as a promising new treatment. This section is devoted to the recent explorations of the cardinal symptoms of the disease such as akinesia/bradykinesia (Burbaud), including the resting tremor that remained largely unexplained before the insightful work of Sigvardt and colleagues. Still, whatever the symptom you are considering, the information processing through basal ganglia should follow comparable rules. Franz proposes to apply information processing models developed in the cognition field that link competing inhibitory processes and focused selection of action to operations of the basal ganglia. Such proposal should help us understanding the pathophysiology of the cardinal symptoms of PD.

Smith reviewed the appealing concept of the modulation of sensitivity to neurotoxins by the environment. Here environment should be understood in its largest definition. Smith highlights the fact that different levels of enrichment of the environment, dietary restriction or casting of affected or non-affected limbs might have beneficial or deleterious effects according to the timing with the application of the neurotoxins. Such studies have direct implications on the clinical management of parkinsonian patients and on the possibility to slow down the neurodegenerative process.

Reviewing the myriad of receptor-based therapies aimed at controlling those symptoms would require a book on its own. Still, it was important to highlight the new non-receptor-based therapeutic promises. Brotchie and Nash show how tackling the NMDA receptor signalosome could lead to a symptomatic treatment, paving the way for targeting other receptor signalosomes. The understanding of the physiology and pathological reorganization of the macromolecular assemblies at the synapse level should help us defining selective treatments. This section thus call for an integrated approach of PD that would build a comprehensive understanding of BG from single corticostriatal synapses to information processing within the same theroretical framework.

In: Recent Breakthroughs in Basal Ganglia Research
Editor: Erwan Bezard, pp. 215-225
ISBN: 1-59454-880-3

Chapter 17

THE NMDA RECEPTOR SIGNALOSOME AND RELATED MACROMOLECULAR ASSEMBLIES– NOVEL MOLECULAR TARGETS FOR SYMPTOMATIC TREATMENTS IN PARKINSON'S DISEASE ?

Jonathan M. Brotchie and Joanne E. Nash
Toronto Western Research Institute, 399 Bathurst St, Toronto, ON, Canada

ABSTRACT

Macromolecular complexes of receptors and other signalling proteins represent a fundamental feature of the mechanisms regulating neurotransmission in the CNS. For instance, NMDA receptor subunits associate to form a functional receptor which can then link directly or indirectly, via scaffolding proteins, e.g. PSD-95, with other signalling molecules including kinases e.g. Fyn and phosphatases e.g. STEP or cytoskeletal elements such as actin or kinesin–like motor proteins. These interactions regulate construction, delivery, signalling and re-cycling of NMDA receptors. Disruption of these processes may underlie abnormal striatal NMDA transmission in parkinsonism and levodopa-induced dyskinesia and might be targeted by novel therapeutics.

INTRODUCTION

The purpose of this chapter is to discuss advances in the understanding of the molecular mechanisms underlying the enhanced NMDA receptor mediated transmission responsible the generation of parkinsonism and L-DOPA-induced dyskinesia (LID). We speculate that these symptoms might result from abnormalities in macromolecular assemblies complexing NMDA receptors with a range of other proteins. We, thus, highlight a concept that could, in due course, lead to novel, more effective treatments.

1 ROLE OF MACROMOLECULAR ASSEMBLIES IN REGULATING NMDA RECEPTOR FUNCTION

The last decade has witnessed a paradigm shift in the understanding of the neurobiology of how neurotransmission is regulated. It is now apparent that a fundamental feature of the regulation of these processes is the complexing of receptor subunits with both each other and with other proteins (Al-Hallaq et al., 2001; Sheng, 2001a, 2001b). This view of receptors as components of macromolecular complexes, rather than individual players, physically-isolated from all but their most direct downstream signalling cascades, has revolutionised our understanding of the intricacies of synaptic transmission. The impact on the field is akin to that in which the concept of regulated phosphorylation added another dimension to neuroscience research in the 1970s. Thus, we now appreciate that receptors are organised into functional complexes with other signalling and structural molecules at the synapse. The same receptor can have distinct functions, subcellular localisation and regulation depending upon the complement of proteins with which it is complexed. Furthermore, in addition to controlling the function of a receptor when it is at the synapse, macromolecular complexes guide and regulate the mechanisms by which receptor proteins, once synthesised, are processed, trafficked to, stabilised in and then removed from appropriate subcellular locations.

2 Diversity of Protein-Protein Interactions

Interactions between receptors and other proteins depend on many families of "Velcro"-like protein-protein-binding domains which provide the molecular basis for coordinating their association. The most widely recognised of these domains is the PDZ domain and the motif to which it binds, tSXV (Niethammer et al., 1996, though many other have been described e.g. (Niethammer et al., 1998; Sun et al., 1998; Lim et al., 1999). Several other domains and motifs, other than PDZ/tSXV, permit protein-protein interactions with relevance to NMDA receptor signalling, these include the Src homology domains (SH2 and SH3) and the guanylate kinase homology domain (GK). Many proteins have multiple protein-protein binding domains, this allows them to act as scaffolds and bind to more than one other protein. Perhaps the best-studied is the interaction between PSD-95 and NMDA receptors. PSD-95, a member of the membrane associated guanylate kinase (MAGUK) superfamily, which also includes SAP-97 and PSD-93, can act as a scaffold around which complexes are built. Not only does PSD-95 contain 3 PDZ domains, thus enabling association with up to 3, potentially different, tSXV-containing proteins, but also SH3 and GK domains which allow association with up to 2 further proteins. The complexity of such assemblies can be appreciated if one considers that PSD-95 may, itself, bind to other scaffolding proteins, this can recruit of a diverse range of molecules to a complex. As an example, the recently-described scaffold protein TANC, can bind to both PSD-95 and CaMKII directly but, because the NR2B subunit of NMDA receptors can bind to PSD-95, it can provide an indirect link between NMDA receptors and CamKII (Suzuki et al., 2005). Protein binding domains on NMDA receptors thus allow their assembly not only with scaffolding proteins but also, by direct or indirect interactions, with motor proteins, vesicular targeting proteins, cytoskeletal proteins, effector molecules, both non-NMDA glutamate receptors and receptors for other transmitters with which they may interact functionally.

3 The NMDA Signalosome

NMDA receptors are comprised of complexes of subunits of 2 families, NR1 and NR2. There are 8 splice variants of NR1 ($NR1_{a\text{-}h}$) and 4 genes encoding NR2 subunits (NR2A-D). Subunits from each family combine, probably as tetramers, of 2 NR1 and 2 NR2 subunits, to form functional receptors. With respect to subunit composition, each receptor can be either binary (i.e. containing NR1 and only one member of the NR2 family, e.g. containing NR1/2A or NR1/2B) or ternary (i.e. containing NR1 and two members of the NR2 family, e.g., NR1/2A/2B; Dunah and Standaert, 2003). Receptors with different subunit composition have different pharmacology and physiology and are expressed heterogeneously in different brain region in a manner that is regulated during development, physiologically in the adult and in disease processes (Monyer et al., 1994; van Zundert et al., 2004). In addition, to this association of NMDA subunits, interactions between NMDA subunits and scaffolding proteins, allow the formation of sub-synaptic rafts of multiple signalling and regulatory molecules. These can be assembled to form an integrated processing unit, that might include both ionotropic and G-protein coupled receptors, several downstream effectors proteins as well as proteins responsible for regulating signalling cascades or subcellular localisation, as a whole, such a complex can be termed a signalosome. An appreciation of the functioning of a receptor cannot be made in isolation from its interactions with other components of the signalosome (Iwamoto et al., 2004).

As an example of the importance of the composition of the signalosome to NMDA receptor function, it is clear that interactions between NMDA receptors and scaffolding proteins subserve indirect interactions with kinases and phosphatases. These proteins can directly regulate NMDA channel function but can also modulate the phosphorylation of other signalosome components and thus, indirectly, modulate NMDA receptor function. PSD-95 may permit the association of NR2A subunits with Src-like kinases, e.g. Fyn, and thus the phosphorylation of tyrosine residues on NR2A (Kalia and Salter, 2003; Salter and Kalia, 2004). Within the same complex, the striatally-enriched phosphatase (STEP) can bind and reverse the actions of Fyn (Nguyen et al., 2002). Similarly, the association of CaMKII and NR2A subunits is regulated by PSD-95 (Gardoni et al., 2001).

Some components of the NMDA signalosome interact with cytoskeletal elements and stabilise NMDA receptors to their synaptic localisation, e.g. direct interactions between NR1 and α-actinin (Dunah et al., 2000a). Interactions between scaffold proteins and NMDA receptors can also regulate the turnover, e.g. association with the MAGUKs PSD-95 or SAP-102 inhibits the cleavage of NR2A by calpain (Dong et al., 2004).

The signalosome might also be the molecular basis of receptor-receptor interactions, which have long been investigated solely from a pharmacological perspective (Salter, 2003). The presence of receptors within the same complex may be critical for positioning them in such a way as to mediate interactions either with each other or with downstream signalling molecules. Until recently, the concept of direct links between ionotropic and metabotropic receptors was little entertained. However, in recent years examples of such have illustrated mechanisms by which signals can be integrated within the nervous system. Thus, in the striatum, there are direct interactions between D1 receptors and NMDA receptors (Salter, 2003). In less direct ways, other receptors, for instance NMDA and adenosine A2a receptors, might become associated via interactions with intermediary proteins (Nash and Brotchie, 2000). Such assembly, to allow communication between receptor systems, is likely to be a basic, and increasingly recognised, building block of the processes mediating integration of signals within the nervous system.

4 Construction, Delivery and Regulation of the NMDA Signalosome

With respect to NMDA receptor function and the opportunities for regulation by macromolecular complexes, the following scheme can be considered.

4.1 NMDA Receptor Subunits are Synthesised in the Endoplasmic Reticulum (ER) and Transported to the Golgi Apparatus where They Associate with Other Proteins

NR1 and NR2 subunits combine in the Golgi to form functional receptors, be they binary or ternary. The mechanism regulating this association is, as yet, unclear. Extrusion of NMDA subunits from the ER-Golgi system requires association with appropriate partners. For example, a complex of NR1, PSD-95 and an NR2 subunit, is required to enable newly-synthesised NR1 subunits to leave the ER and be trafficked to the synapse (Fukaya et al., 2003). This mechanism ensures that only NMDA receptors with NR1 and NR2 subunits, i.e. only functional receptors are sent to the synapse. While the minimal components of this complex are PSD-95, NR1 and NR2 subunits, the full nature of these complexes is not appreciated. However, it could also contain motor proteins responsible for trafficking vesicles along the dendrite, vesicular proteins responsible for exocytosis and delivery to the plasma membrane and other signalling molecules such as kinases, receptors, and/or ion channels that would be involved in the functioning of the receptor once delivered.

4.2 From the Golgi, NMDA Receptor-Containing Complexes Are Trafficked along Dendrites to the Synaptic Bouton

NR1/NR2 complexes are trafficked in vesicles, as cargo along the microtubular network, from the Golgi to the vicinity of the synapse. This is dependent on their engagement with kinesin-like motor proteins. Again, macromolecular complexes regulate/are required. Thus, with respect to NR1/NR2B receptors, the motor protein KIF-17 requires interaction, via PDZ domains, with a complex containing three other proteins to promote trafficking of NR2B-containing NMDA receptors along dendrite spines (Setou et al., 2000; Guillaud et al., 2003).

4.3 At the Synaptic Bouton, NMDA Receptors Interact in a Dynamic Way with Proteins that Regulate their Internalisation and Re-cycling

The stabilisation of NMDA receptors at synaptic sites is tightly regulated (Lavezzari et al., 2004). For instance, the internalisation of synaptic NR2B receptors requires the binding to NR2B of the adaptor protein AP-2 (Lavezzari et al., 2003). AP-2 is itself a protein complex which links proteins for internalisation to clathrin, a signalling molecule driving endocytosis from the plasma membrane (Lavezzari et al., 2003). AP-2 binds to a motif, termed YEKL, adjacent to the PDZ binding domain, tSXV on NR2B. PSD-95, SAP-97 and PSD-93, by binding to this tSXV domain, can inhibit AP-2 binding to NR2B so stabilising NR2B localisation to the synapse.

5 The Neural Mechanisms of Parkinsonism and LID

Major factors in the genesis of both parkinsonism and LID are alterations in the influence of the glutamatergic inputs to the striatum (Chase et al., 1998; Calon et al., 2000). These result in imbalances in the pattern of activity of inhibitory GABAergic striatal outputs. Striatal efferents influence the outputs of the basal ganglia, from the internal segment of the globus pallidus (GPi) and substantia nigra pars reticulata (SNr) by either a "direct" monosynaptic connection or an "indirect" connection through a network involving the external pallidal segment (GPe) and the subthalamic nucleus. The interplay of these processes is responsible for selection of appropriate and suppression of inappropriate movements. Imbalances in the activities of the direct and indirect pathways lead to abnormal basal ganglia output, reflected in gross changes in firing rate, patterning and synchronisation of neuronal firing, and the suppression of appropriate movements in parkinsonism or the generation of the inappropriate movements of LID. Specifically, overactivity of the indirect pathway, driven by enhanced glutamatergic excitation of striatal output neurons projecting to GPe, is a key contributor to the mechanisms of parkinsonism (Penney and Young, 1983; Mitchell et al., 1986; Mitchell et al., 1995). Although, controversy remains (Obeso et al., 2000), and a complete understanding of LID will undoubtedly require additional and alternative concepts to be incorporated, enhanced glutamatergic excitation of striatal output neurons projecting to GPi, leading to overactivity of the direct pathway, is critical to the mechanisms of LID (Bezard et al., 2001; Brotchie et al., 2005). Thus, a common feature of both parkinsonism and LID is enhanced glutamatergic drive to striatal output neurons.

6 Role of Enhanced Striatal NMDA Receptor-Mediated Transmission in Parkinsonism and LID

Within the striatum, glutamatergic inputs can excite striatal output neurons by interactions with both ionotropic (NMDA and AMPA) and metabotropic glutamate (mGluR) receptors. There are undoubtedly important functional interactions between these receptors and abnormalities in signalling by all of them probably contribute to parkinsonism and LID. However, enhanced NMDA transmission is critical to these processes. NMDA receptor antagonists have been consistently shown, unlike either AMPA or an individual class of mGluR antagonists, to be capable of exerting both (1) anti-parkinsonian actions, without eliciting dyskinesia, and (2) anti-dyskinetic actions when administered as an adjunct to L-DOPA. Indeed, the only non-dopaminergic treatment widely employed clinically to control motor function in PD/LID, amantadine, is thought to exert its actions via blockade of NMDA transmission.

6.1 NMDA Receptors and Parkinsonism

Enhanced striatal NMDA receptor-mediated transmission is thought to be responsible for overactivity of the striatal indirect output pathway in parkinsonism. Thus, stimulation of striatal NMDA receptors elicits parkinsonism (Greenamyre and O'Brien, 1991); conversely NR2B-selective NMDA antagonists can have anti-parkinsonian actions when administered systemically (Nash et al., 2000; Steece-Collier et al., 2000; Loschmann et al., 2004) or when administered directly into the striatum (Nash and Brotchie, 2002). The anti-parkinsonian effects of NR2B-selective NMDA antagonists are remarkable in that they can demonstrate, as monotherapy, efficacy equal to dopamine replacement in the MPTP-lesioned primate. Moreover, these anti-parkinsonian actions are not accompanied by dyskinesia, even in animals previously primed so that they elicit dyskinesia when treated with L-DOPA (Loschmann et al., 2004). Overactive NR2B transmission is, thus, an attractive therapeutic target in parkinsonism. However, not all NR2B antagonists have anti-parkinsonian actions (Nash et al., 2004). On the basis of behavioural studies in the MPTP-lesioned primate, we have recently proposed that this reflects the fact the NR2B-containing NMDA receptors involved in the generation of parkinsonism are ternary, NR1/NR2A/NR2B, receptors rather than binary, NR1/NR2B (Nash et al., 2004).

The mechanism by which transmission at NR2B-containing receptors becomes overactive is not clear, and cannot be explained by simple receptor pharmacology. In striatal homogenates (containing all cellular components) there is no change in the level of NR2B subunits following dopamine depletion in the 6-OHDA-lesioned rat (Dunah et al., 2000b). Furthermore, in the MPTP-lesioned macaque, the total levels of NMDA receptors, as well as those with NR2B-like pharmacology, as assessed by subunit-selective receptor binding, are unchanged compared to normal animals (Calon et al., 2002). Similarly, in PD patients without LID, NR2B binding is normal (Calon et al., 2003). Thus, a simple up-regulation of receptor number cannot account for enhanced NR2B NMDA transmission in PD. However, what is lacking from these studies are the levels of NMDA receptors at the synapse. However, it does not appear that a change in the subcellular distribution of NMDA receptors could account for enhanced NR2B transmission since, in the 6-OHDA-lesioned rodent, there are actually reduced levels of NR2B and NR1 subunits in synaptosomal membranes (Dunah et al., 2000b). We have recently extended these findings to the MPTP-lesioned macaque where, using approaches which assess receptor subunit levels in different subcellular compartments of the neuron, we found that the synaptosomal levels of NR2B subunit protein are significantly reduced, by 40 %, in the striatum in the parkinsonian state (Hallett, 2005). NR1 subunit levels are also decreased suggesting this reduction reflects a reduction in the number of functional NMDA channels containing NR2B subunits at striatal synapses in parkinsonism.

In light of the above discussion, the question thus emerges, how can transmission at striatal NMDA receptors become enhanced in parkinsonism in a situation of reduced synaptic receptor number? We now propose that alterations in macromolecular assemblies involving NMDA receptors might be responsible for,

1. abnormal delivery and/or removal of NMDA receptors to/from the synapse, leading to decreased synaptic NR2B in PD and
2. enhanced excitatory influence of individual synaptic NMDA receptors in PD.

To date, the components of NMDA signalosomes in the striatum have not been described in any detail. In fact, it is not even clear whether striatal NMDA receptors are organised into binary or ternary complexes in parkinsonism and LID. Enhanced signalling by NMDA receptors could be driven by alterations in the binary/ternary nature of the receptors, in the receptors with they are functionally linked or changes in modulatory components of the signalosome. Abnormalities such as these could be driven by alterations in scaffolding proteins at the core of the signalosome.

Similarly, changes in the complement of scaffolding proteins, such as PSD-95, present in complexes in other sub-cellular compartments could lead to disruption of extrusion of NMDA receptors from the ER, trafficking from the Golgi or internalisation/ re-cycling from the synapse. These might underlie reduced synaptic levels of NR2B-containing NMDA receptors in parkinsonism. We have found that, in synaptosomal membrane preparations from normal primate striatum, PSD-95 co-immunoprecipitates with NR1 and NR2B subunits, suggesting that it is, as predicted, a component of at least some NMDA signalosomes. Furthermore, in the 6-OHDA-lesioned rat model of Parkinson's disease, there is a dramatic reduction in the striatal levels of PSD-95 (Picconi et al., 2004; Nash et al., 2005). Thus, in PD, other scaffolding proteins may take the place of PSD-95 in, at least, some NMDA receptor signalosomes. This could lead to NMDA receptors being associated with inappropriate molecules and underlie enhanced signalling by those NR2B-containing NMDA receptors remaining at the synapse. There are many possibilities that might underlie these changes. However, as an example of how such changes might underlie enhanced NMDA transmission in parkinsonism we suggest that a change in scaffolding protein may increase recruitment of adenosine A2a receptors to the NR2B signalosome. We have previously shown that there is a functional link between NMDA and A2a adenosine receptors in the striatum (Nash and Brotchie, 2000). This appears to involve an indirect activation of A2a receptors by NMDA receptor activation. Signalling in this pathway is enhanced in the dopamine-depleted state. Although a direct association between A2a and NMDA receptors has not been demonstrated to date, enhanced recruitment of A2a to the NMDA signalosome would be a clear mechanism by which the interaction between the two and the effects of NMDA transmission could be enhanced.

Furthermore, it is interesting to note that a reduction in PSD-95 levels (Picconi et al., 2004; Nash et al., 2005) may be responsible for the reduced numbers of synaptic NMDA receptors in parkinsonism as, PSD-95 is required for the successful extrusion of NMDA receptors from the ER and delivery to the synapse (Fukaya et al., 2003). Thus, a single change, reduction in PSD-95 expression, might contribute to both the decreased number of NR2B-containing receptors and their increased ability to influence neuronal excitability in the parkinsonian striatum.

6.2 NMDA Receptors and LID

There is also a clear role for enhanced striatal NMDA transmission in LID. Thus NMDA antagonists reduce LID when administered systemically with L-DOPA in the MPTP-lesioned primate (Papa and Chase, 1996; Blanchet et al., 1998; Blanchet et al., 1999; Chassain et al., 2003; Hill et al., 2004). Furthermore, NMDA antagonists also reduce motor fluctuations resulting from repeated L-DOPA treatment in the 6-OHDA-lesioned rat, a model sharing many mechanistic similarities with LID (Papa et al., 1995). Furthermore, the only clinically-available adjunctive therapy for LID, amantadine, is an NMDA antagonist (Verhagen-Metman et al., 1998; Metman et al., 1999; Luginger et al., 2000; Del-Dotto et al., 2001). That LID involves striatal NMDA receptors is suggested by intrastriatal injection of NMDA antagonists reducing motor fluctuations in the 6-OHDA-lesioned rat (Papa et al., 1995).

We propose that transmission at NR2A-containing NMDA receptors on neurons of the direct pathway is responsible for enhanced NMDA signalling in LID. This concept is supported by recent studies in the MPTP-lesioned macaque with LID, where, using receptor binding assays selective for NMDA receptors with different subunit compositions, Calon et al. demonstrated a dramatic increase in NR2A-like binding in the striatum of MPTP-lesioned macaques with LID (211 % of control; Calon et al., 2002). This elevation was observed in the face of non-significant (NR1) or much more limited increases in NR2B binding (+33 %). Furthermore, in the MPTP-lesioned macaque, the levels of NR2A subunits in striatal synaptosomal membranes are increased in MPTP-lesioned monkeys with LID compared to normal or MPTP-lesioned animals without LID, by 52 % and 48 % respectively (Hallett, 2005). Synaptosomal levels of NR1 are normal in LID, i.e. increased from the parkinsonian state, while NR2B remains reduced to the levels seen in untreated parkinsonism. Total cellular and endoplasmic reticulum (ER) levels of NR2A were also increased suggesting that, in LID, there is increased synthesis, and trafficking to the synapse, of NR2A-containing receptors. However, this concept, is somewhat discrepant to findings in the 6-OHDA rat where synaptosomal, and total, levels of NR2A have been reported as being normal in 6-OHDA-lesioned rats following long-term L-DOPA therapy (Dunah et al., 2000b). This may represent a species-related difference with respect to the post-translational processing of receptor proteins, and emphasises the value of employing primate models in understanding human disease, at least with respect to the fine details of neural communication at the molecular level. However, this cannot be the sole determinant of the discrepancy, a recent study in tissue from PD patients with LID, also by Calon et al. suggested that, in human LID, NR2B, not 2A, binding is, in fact, increased (Calon et al., 2003). The involvement of NR2B-containing receptors in the expression of LID, is also supported by some (Blanchet et al., 1999), but not all (Nash et al., 2004), behavioural pharmacological studies in MPTP primates. In fact, the discrepancies between studies are likely to reflect the reagents used and the fact that the NMDA receptors responsible for generating LID may contain both NR2A and NR2B subunits. Our ability to address this issue has hitherto been compromised by the limited availability of pharmacological tools with good selectivity for NR2A over NR2B, and ternary over binary, NMDA receptors to allow their investigation, especially *in vivo*. However, on the basis of behavioural studies in the MPTP primate, we have proposed that the NMDA receptors involved in the expression of LID may, like those in parkinsonism, be ternary, NR1/NR2A/NR2B receptors rather than binary, NR1/NR2A or NR1/NR2B (Nash et al., 2004).

These data collectively raise the question, *how does striatal NMDA transmission become enhanced and how does it contribute to the generation of LID?* As was the case with enhanced NMDA transmission in parkinsonism, we now hypothesise that the answer to this question may lie in abnormal interactions and complexing between NMDA subunits and other proteins that might,

1. lead to enhanced delivery, reduced internalisation or reduced breakdown, of NR2A-containing receptors to/ from/ at the synapse, leading to increased synaptic NR2A receptors in LID, and, perhaps,
2. the composition of signalosomes containing NR2A is altered to enhance their impact on cellular excitability in the dyskinetic striatum, leading to enhanced excitatory influence of individual synaptic NMDA receptors LID

At one level, enhanced NMDA transmission in LID may result from a simple up-regulation of expression of NMDA receptors and the delivery of more receptors to each synapse leading to a greater sensitivity to glutamate released from pre-synaptic vesicles. In fact, given our findings that, NR2A levels are increased across all subcellular compartments (Hallett, 2005) and those of Calon et al., (2003), that binding of NR2A-selective radioligands is increased in the striata of MPTP-lesioned macaques with LID this seems quite feasible. However, in itself this represents neither a complete nor, necessarily, the only answer. For instance, the rate of delivery of receptors could also be increased, e.g. by increased association of vesicular receptors with motor proteins, or association with different motor proteins, so contributing to enhanced NMDA receptor function. NR2A levels could also rise because turnover was decreased, this latter suggestion being consistent with our recent finding that PSD-95 levels are elevated in synaptic membranes in a rat model of LID (Nash et al., 2005) and the ability of PSD-95 to reduce calpain-mediated breakdown of NR2A (Dong et al., 2004). Additionally, enhanced NR2A signalling may result from a change in signalosome function, which could enhance the action of each individual NMDA receptor, as illustrated in the discussion above relating to enhanced NR2B function in parkinsonism.

7 Therapeutic Potential of Targeting NMDA Receptor Signalosomes

In the above discussion, we have highlighted the key role of enhanced striatal NMDA-receptor mediated transmission in both parkinsonism and LID. We have hypothesised that a major contributor to these abnormalities may relate to abnormalities of macromolecular assemblies containing NMDA receptors. This could lead to those receptors having altered excitability, increased synaptic delivery, decreased internalisation or decreased degradation. This concept remains theoretical at present. However, two recent reports do suggest that there are indeed alterations in components of the NMDA signalosome, PSD-95 and SAP-97, in animal models of PD and LID, that are consistent with this proposition (Picconi et al., 2004; Nash et al., 2005). It is thus be timely to consider whether the NMDA receptor signalosome represents an attractive therapeutic target and whether approaches to modulate its function are amenable to the development of drugs.

Firstly, it is clear that striatal NMDA receptor-mediated transmission contributes significantly to the generation of parkinsonian symptoms and LID and is thus an attractive therapeutic target. Both parkinsonism and LID can be seen as processes involving disruption of NMDA receptor-dependent synaptic plasticity (Centonze et al., 1999; Picconi et al., 2003; Picconi et al., 2004). Thus, NMDA receptor stimulation can be seen as the initiator of a cascade of other abnormalities and, thus, a potential target for correcting all. Furthermore, the molecular pharmacology of NMDA receptors offers significant opportunities for developing treatments that might show selectivity for striatal processes in PD and LID, over other critical cognitive functions involving glutamate receptor-mediated processes. NMDA receptor function is tightly regulated, the processes responsible for synthesis, assembly, trafficking and downstream signalling are subject to a diversity of molecular 'cheques and balances' involving, as described above, a multitude of interactions between NMDA receptors and other proteins. These protein-protein interactions all represent potential means to modulate abnormal NMDA receptor function in a disease process.

We have proposed that the composition of NMDA receptors involved in the generation of both parkinsonism and LID is ternary, NR1/2A/2B. Traditional pharmacological approaches could be applied to develop a therapy based around a drug that blocked these receptors while leaving transmission at binary receptors intact. While antagonists exist that are selective for binary receptors, to date, no compounds with the desired, ternary over binary, profile have been described.

Are protein-protein interactions, realistic drug targets? Conceptually the idea is attractive, a drug that bound to a specific domain important for an abnormal interaction, e.g. the PDZ domain of PSD-95 or the tSXV domain of NR2B, could have a dramatic effect of NMDA signalling. Indeed small peptides targeting such domains have been synthesised and shown to have effects with therapeutic potential in areas of neurology other than movement disorders. However, to date, no therapeutic, with a mechanism of action of disrupting protein-protein interactions in neurons has reached the marketplace. In fact, none has ever demonstrated efficacy beyond preclinical studies. In fact, no small molecule inhibitors of interactions, such as those occurring between components of the NDMA signalosome have been described. Progress thus far has relied upon the delivery of small disrupting peptides using complex, experimental technologies such as viral gene delivery or TAT-protein tagging. Given the inherent challenges in delivering peptides to CNS neurons in a manner that is efficacious, non-invasive, safe and widely-applicable as a therapy, the drugability of this approach has recently been carefully considered (Dev, 2004). Although, the development of small molecules to interfere with protein-protein interactions is associated with significant hurdles, these all appear surmountable. In fact, in the long term it is likely that small molecules represent the most likely means of developing a marketable therapeutic to disrupt such interactions.

CONCLUSION

An increased understanding of the molecular biology of NMDA transmission, both generally and specifically in Parkinson's disease, has developed, opportunities for novel symptomatic therapeutics have been highlighted. These might selectively target receptors comprising different combinations of NMDA subunits or might disrupt interactions between NMDA subunits and other adapter proteins, be those interactions direct or indirect. Much basic research and development is required to meet take advantage of these opportunities, but it is clear, that they represent not only scientifically-exciting but also realistic, in terms of therapeutic development, advances in this area.

ACKOWLEDGEMENTS

We offer our gratitude for the generous funding of our research by the Krembil Neuroscience Fund, the Michael J Fox Foundation, the Safra Foundation and the Parkinson's Society Canada.

REFERENCES

Al-Hallaq RA, Yasuda RP, Wolfe BB (2001) Enrichment of N-methyl-D-aspartate NR1 splice variants and synaptic proteins in rat postsynaptic densities. *J Neurochem* 77:110-119.

Bezard E, Brotchie JM, Gross CE (2001) Pathophysiology of levodopa-induced dyskinesia: potential for new therapies. *Nat Rev Neurosci* **2**:577-588.

Blanchet PJ, Konitsiotis S, Chase TN (1998) Amantadine reduces levodopa-induced dyskinesias in parkinsonian monkeys. Movement Disorders : *Official Journal of the Movement Disorder Society* **13**:798-802.

Blanchet PJ, Konitsiotis S, Whittemore ER, Zhou ZL, Woodward RM, Chase TN (1999) Differing effects of N-methyl-D-aspartate receptor subtype selective antagonists on dyskinesias in levodopa-treated 1-methyl-4-phenyl-tetrahydropyridine monkeys. The *Journal of Pharmacology and Experimental Therapeutics* **290**:1034-1040.

Brotchie JM, Lee J, Venderova K (2005) Levodopa-induced dyskinesia in Parkinson's disease. *J Neural Transm* **112**:359-391.

Calon F, Rajput AH, Hornykiewicz O, Bedard PJ, Di Paolo T (2003) Levodopa-induced motor complications are associated with alterations of glutamate receptors in Parkinson's disease. *Neurobiol Dis* **14**:404-416.

Calon F, Grondin R, Morissette M, Goulet M, Blanchet PJ, Di Paolo T, Bedard PJ (2000) Molecular basis of levodopa-induced dyskinesias. *Annals of Neurology* **47**:S70-78.

Calon F, Morissette M, Ghribi O, Goulet M, Grondin R, Blanchet PJ, Bedard PJ, Di PT (2002) Alteration of glutamate receptors in the striatum of dyskinetic 1-methyl-4-phenyl-1,2,3,6-tetrahydropyridine-treated monkeys following dopamine agonist treatment. *Progress in Neuro-Psychopharmacology & Biological Psychiatry* **26**:127-138.

Centonze D, Gubellini P, Picconi B, Calabresi P, Giacomini P, Bernardi G (1999) Unilateral dopamine denervation blocks corticostriatal LTP. *Journal of Neurophysiology* **82**:3575-3579.

Chase TN, Oh JD, Blanchet PJ (1998) Neostriatal mechanisms in Parkinson's disease. *Neurology* **51**:S30-35.

Chassain C, Eschalier A, Durif F (2003) Antidyskinetic effect of magnesium sulfate in MPTP-lesioned monkeys. *Experimental Neurology* **182**:490-496.

Del-Dotto P, Pavese N, Gambaccini G, Bernardini S, Metman LV, Chase TN, Bonuccelli U (2001) Intravenous amantadine improves levadopa-induced dyskinesias: an acute double-blind placebo-controlled study. Movement Disorders : *Official Journal of the Movement Disorder Society* **16**:515-520.

Dev KK (2004) Making protein interactions druggable: targeting PDZ domains. *Nat Rev Drug Discov* **3**:1047-1056.

Dong YN, Waxman EA, Lynch DR (2004) Interactions of postsynaptic density-95 and the NMDA receptor 2 subunit control calpain-mediated cleavage of the NMDA receptor. *J Neurosci* **24**:11035-11045.

Dunah AW, Standaert DG (2003) Subcellular segregation of distinct heteromeric NMDA glutamate receptors in the striatum. *Journal of Neurochemistry* **85**:935-943.

Dunah AW, Wyszynski M, Martin DM, Sheng M, Standaert DG (2000a) alpha-actinin-2 in rat striatum: localization and interaction with NMDA glutamate receptor subunits. *Brain Research Molecular Brain Research* **79**:77-87.

Dunah AW, Wang Y, Yasuda RP, Kameyama K, Huganir RL, Wolfe BB, Standaert DG (2000b) Alterations in subunit expression, composition, and phosphorylation of striatal N-methyl-D-aspartate glutamate receptors in a rat 6-hydroxydopamine model of Parkinson's disease. *Molecular Pharmacology* **57**:342-352.

Fukaya M, Kato A, Lovett C, Tonegawa S, Watanabe M (2003) Retention of NMDA receptor NR2 subunits in the lumen of endoplasmic reticulum in targeted NR1 knockout mice. *Proceedings of the National Academy of Sciences of the United States of America* **100**:4855-4860.

Gardoni F, Schrama LH, Kamal A, Gispen WH, Cattabeni F, Di_Luca M (2001) Hippocampal synaptic plasticity involves competition between Ca2+/calmodulin-dependent protein kinase II and postsynaptic density 95 for binding to the NR2A subunit of the NMDA receptor. The Journal of Neuroscience : the *Official Journal of the Society For Neuroscience* **21**:1501-1509.

Greenamyre JT, O'Brien CF (1991) N-methyl-D-aspartate antagonists in the treatment of Parkinson's disease. *Arch Neurol* **48**:977-981.

Guillaud L, Setou M, Hirokawa N (2003) KIF17 dynamics and regulation of NR2B trafficking in hippocampal neurons. *J Neurosci* **23**:131-140.

Hallett PJ, Dunah, A. W., Ravenscroft, P., Zhou, S., Bezard, E., Crossman, A. R., Brotchie, J. M., and Standaert, D. G (2005) Alterations of striatal NMDA receptor subunits accompanying parkinsonism and levodopa-induced dyskinesia in the MPTP-lesioned non-human primate model of Parkinson's disease. *Neuropharmacology*: doi: 10.1016/j.neuropharm.2004.1011.1008.

Hill MP, Ravenscroft P, Bezard E, Crossman AR, Brotchie JM, Michel A, Grimee R, Klitgaard H (2004) Levetiracetam potentiates the antidyskinetic action of amantadine in the 1-methyl-4-phenyl-1,2,3,6-tetrahydropyridine (MPTP)-lesioned primate model of Parkinson's disease. *J Pharmacol Exp Ther* **310**:386-394.

Iwamoto T, Yamada Y, Hori K, Watanabe Y, Sobue K, Inui M (2004) Differential modulation of NR1-NR2A and NR1-NR2B subtypes of NMDA receptor by PDZ domain-containing proteins. *J Neurochem* **89**:100-108.

Kalia LV, Salter MW (2003) Interactions between Src family protein tyrosine kinases and PSD-95. *Neuropharmacology* **45**:720-728.

Lavezzari G, McCallum J, Lee R, Roche KW (2003) Differential binding of the AP-2 adaptor complex and PSD-95 to the C-terminus of the NMDA receptor subunit NR2B regulates surface expression. *Neuropharmacology* **45**:729-737.

Lavezzari G, McCallum J, Dewey CM, Roche KW (2004) Subunit-specific regulation of NMDA receptor endocytosis. *J Neurosci* **24**:6383-6391.

Lim S, Naisbitt S, Yoon J, Hwang JI, Suh PG, Sheng M, Kim E (1999) Characterization of the Shank family of synaptic proteins. Multiple genes, alternative splicing, and differential expression in brain and development. *J Biol Chem* **274**:29510-29518.

Loschmann PA, De Groote C, Smith L, Wullner U, Fischer G, Kemp JA, Jenner P, Klockgether T (2004) Antiparkinsonian activity of Ro 25-6981, a NR2B subunit specific NMDA receptor antagonist, in animal models of Parkinson's disease. *Exp Neurol* **187**:86-93.

Luginger E, Wenning GK, Bosch S, Poewe W (2000) Beneficial effects of amantadine on L-dopa-induced dyskinesias in Parkinson's disease. Movement Disorders : *Official Journal of the Movement Disorder Society* **15**:873-878.

Metman LV, Del_Dotto P, LePoole K, Konitsiotis S, Fang J, Chase TN (1999) Amantadine for levodopa-induced dyskinesias: a 1-year follow-up study. *Archives of Neurology* **56**:1383-1386.

Mitchell IJ, Cross AJ, Sambrook MA, Crossman AR (1986) Neural mechanisms mediating 1-methyl-4-phenyl-1,2,3, 6-tetrahydropyridine-induced parkinsonism in the monkey: relative contributions of the striatopallidal and striatonigral pathways as suggested by 2-deoxyglucose uptake. *Neuroscience Letters* **63**:61-65.

Mitchell IJ, Hughes N, Carroll CB, Brotchie JM (1995) Reversal of parkinsonian symptoms by intrastriatal and systemic manipulations of excitatory amino acid and dopamine transmission in the bilateral 6-OHDA lesioned marmoset. *Behav Pharmacol* **6**:492-507.

Monyer H, Burnashev N, Laurie DJ, Sakmann B, Seeburg PH (1994) Developmental and regional expression in the rat brain and functional properties of four NMDA receptors. *Neuron* **12**:529-540.

Nash JE, Brotchie JM (2000) A common signaling pathway for striatal NMDA and adenosine A2a receptors: implications for the treatment of Parkinson's disease. The Journal of Neuroscience : the *Official Journal of the Society For Neuroscience* **20**:7782-7789.

Nash JE, Brotchie JM (2002) Characterisation of striatal NMDA receptors involved in the generation of parkinsonian symptoms: intrastriatal microinjection studies in the 6-OHDA-lesioned rat. Movement Disorders : *Official Journal of the Movement Disorder Society* **17**:455-466.

Nash JE, Johnston TH, Collingridge GL, Garner CC, Brotchie JM (2005) Subcellular redistribution of the synapse-associated proteins PSD-95 and SAP97 in animal models of Parkinson's disease and L-DOPA-induced dyskinesia. Faseb J:Epub ahead of print.

Nash JE, Ravenscroft P, McGuire S, Crossman AR, Menniti FS, Brotchie JM (2004) The NR2B-selective NMDA receptor antagonist CP-101,606 exacerbates L-DOPA-induced dyskinesia and provides mild potentiation of anti-parkinsonian effects of L-DOPA in the MPTP-lesioned marmoset model of Parkinson's disease. *Exp Neurol* **188**:471-479.

Nash JE, Fox SH, Henry B, Hill MP, Peggs D, McGuire S, Maneuf Y, Hille C, Brotchie JM, Crossman AR (2000) Antiparkinsonian actions of ifenprodil in the MPTP-lesioned marmoset model of Parkinson's disease. *Experimental Neurology* **165**:136-142.

Nguyen TH, Liu J, Lombroso PJ (2002) Striatal enriched phosphatase 61 dephosphorylates Fyn at phosphotyrosine 420. *J Biol Chem* **277**:24274-24279.

Niethammer M, Kim E, Sheng M (1996) Interaction between the C terminus of NMDA receptor subunits and multiple members of the PSD-95 family of membrane-associated guanylate kinases. *The Journal of Neuroscience : the Official Journal of the Society For Neuroscience* **16**:2157-2163.

Niethammer M, Valtschanoff JG, Kapoor TM, Allison DW, Weinberg TM, Craig AM, Sheng M (1998) CRIPT, a novel postsynaptic protein that binds to the third PDZ domain of PSD-95/SAP90. *Neuron* **20**:693-707.

Obeso JA, Rodriguez_Oroz MC, Rodriguez M, DeLong MR, Olanow CW (2000) Pathophysiology of levodopa-induced dyskinesias in Parkinson's disease: problems with the current model. *Annals of Neurology* **47**:S22-32; discussion S32-24.

Papa SM, Chase TN (1996) Levodopa-induced dyskinesias improved by a glutamate antagonist in Parkinsonian monkeys. *Annals of Neurology* **39**:574-578.

Papa SM, Boldry RC, Engber TM, Kask AM, Chase TN (1995) Reversal of levodopa-induced motor fluctuations in experimental parkinsonism by NMDA receptor blockade. *Brain Research* **701**:13-18.

Penney JB, Young AB (1983) Speculations on the functional anatomy of basal ganglia disorders. *Annual Review of Neuroscience* **6**:73-94.

Picconi B, Centonze D, Hakansson K, Bernardi G, Greengard P, Fisone G, Cenci MA, Calabresi P (2003) Loss of bidirectional striatal synaptic plasticity in L-DOPA-induced dyskinesia. *Nature Neuroscience* **6**:501-506.

Picconi B, Gardoni F, Centonze D, Mauceri D, Cenci MA, Bernardi G, Calabresi P, Di Luca M (2004) Abnormal Ca2+-calmodulin-dependent protein kinase II function mediates synaptic and motor deficits in experimental parkinsonism. *J Neurosci* **24**:5283-5291.

Salter MW (2003) D1 and NMDA receptors hook up: expanding on an emerging theme. *Trends in Neurosciences* **26**:235-237.

Salter MW, Kalia LV (2004) Src kinases: a hub for NMDA receptor regulation. *Nat Rev Neurosci* **5**:317-328.

Setou M, Nakagawa T, Seog DH, Hirokawa N (2000) Kinesin superfamily motor protein KIF17 and mLin-10 in NMDA receptor-containing vesicle transport. *Science* **288**:1796-1802.

Sheng M (2001a) The postsynaptic NMDA-receptor--PSD-95 signaling complex in excitatory synapses of the brain. *Journal of Cell Science* **114**:1251.

Sheng M (2001b) Molecular organization of the postsynaptic specialization. *Proc Natl Acad Sci U S A* **98**:7058-7061.

Steece-Collier K, Chambers LK, Jaw-Tsai SS, Menniti FS, Greenamyre JT (2000) Antiparkinsonian actions of CP-101,606, an antagonist of NR2B subunit-containing N-methyl-d-aspartate receptors. *Experimental Neurology* **163**:239-243.

Sun J, Tadokoro S, Imanaka T, Murakami SD, Nakamura M, Kashiwada K, Ko J, Nishida W, Sobue K (1998) Isolation of PSD-Zip45, a novel Homer/vesl family protein containing leucine zipper motifs, from rat brain. *FEBS Lett* **437**:304-308.

Suzuki T, Li W, Zhang JP, Tian QB, Sakagami H, Usada N, Kondo H, Fujii T, Endo S (2005) A novel scaffold protein, TANC, possibly a rat homolog of Drosophila rolling pebbles (rols), forms a multiprotein complex with various postsynaptic density proteins. *Eur J Neurosci* **21**:339-350.

van Zundert B, Yoshii A, Constantine-Paton M (2004) Receptor compartmentalization and trafficking at glutamate synapses: a developmental proposal. *Trends Neurosci* **27**:428-437.

Verhagen-Metman L, Del-Dotto P, van-den-Munckhof P, Fang J, Mouradian MM, Chase TN (1998) Amantadine as treatment for dyskinesias and motor fluctuations in Parkinson's disease. *Neurology* **50**:1323-1326.

In: Recent Breakthroughs in Basal Ganglia Research
Editor: Erwan Bezard, pp. 227-239
ISBN: 1-59454-880-3

Chapter 18

Converging Evidence of the Basal Ganglia's Role in Focused Action Selection and Inhibition of Competing Information

Elizabeth A. Franz*
University of Otago, Otago – New Zealand

Abstract

At any given time, we are faced with a number of potential actions and we somehow select one to be carried out over other possible competitors. This problem of action selection is of fundamental importance to studies of cognition. The present chapter highlights a recent cognitive approach to research on basal ganglia function that supports earlier cognitive models of inhibition and a more recent model that amalgamates anatomical and neurophysiological findings. Evidence from these different domains strongly supports models that link competing inhibitory processes and focused selection of action to operations of the basal ganglia.

Introduction

At any given time, we are faced with a number of potential actions and somehow select one to be carried out over other possible competitors. This problem of action selection is of fundamental importance in studies of cognition. The present chapter highlights a recent cognitive perspective on basal ganglia function. Experimental findings from patients with Parkinson's disease (PD) support the view that a basic operation of the basal ganglia is related to activation and deactivation (inhibition). According to the present author's view, this basic operation occurs on different types of information processing and is captured in earlier cognitive models on action selection and more recent anatomical-physiological models of basal ganglia function. The present aim is to present a persuasive argument that cognitive and neurophysiological models have run a parallel course in their development, and that the two have converged on a coherent framework for understanding the processes of action selection and the role of the basal ganglia.

In an earlier cognitive account of action selection, Neumann, (1987) raised the puzzling question: How does a person selectively reach for an apple from a tree that contains many apples? As Neumann suggested, a cognitive approach might consider why one apple is selected over the others that surround it. Perhaps the target apple becomes the focus of attention because it is bigger and brighter than the others. A cognitive approach might also consider how the brain organizes the goal-directed action of reaching the target apple by selecting one hand rather than the other to perform the reach. Or, it may consider the principles of organization and initiation of an appropriate sequence of movements necessary to bring the selected hand in contact with the apple. Few cognitive approaches generally deal with the possibility that the brain might

* Correspondence to Liz Franz: Action, Brain, and Cognition Laboratory, Department of Psychology, University of Otago, Box 56, Dunedin, New Zealand. E-mail address: Lfranz@psy.otago.ac.nz. Phone: 64-3-479-5269, Fax: 64-3-479-8335

not be focused solely on the goal to reach the apple, as there might be myriad other goals or subgoals (or candidate actions) that are at least partially activated simultaneously. In contrast, a cognitive approach is much less likely to be concerned with the body's maintenance of an appropriate posture to support the reaching arm so that the action can be effectively carried out. Neurophysiological approaches consider this aspect in more detail, however.

As the above example suggests, the problem of action selection entails more than initiating a reaching movement with a particular arm and hand; rather, action selection might occur on numerous levels, from properties related to the motor movements themselves, to those related to the maintenance of goals. I begin the development of the present framework with a description of some basic properties of reaction time (RT), the standard measure used in studies of cognition. RT, or the latency to respond to a stimulus to which a response is mapped, is typically viewed as a measure of cognitive processing. As shown in early seminal studies, reaction time (RT) tends to increase with the number of stimulus-response choices and is therefore often regarded as a measure of task complexity or difficulty (Hick, 1952; first documented in German by Merkel, 1885). A common interpretation is that RT reflects time taken to program a movement, which includes events preceding movement initiation (Klapp, 1975).

In what follows, I will first briefly review some early investigations that applied RT paradigms to examine performance of patients with Parkinson's disease in an attempt to elucidate whether RT effects reflect basal ganglia operations involved in response planning and preparation. I will then present an overview of recent work conducted by my colleagues and me that sheds a different light on the operations of the basal ganglia, suggesting instead that these nuclei are critically involved in activation and inhibition more generally. Following this, I will build on selected studies in the cognitive literature that view behavioral activation and inhibition as central to processes of action selection. Finally, I will argue that activation and inhibition deficits are the outcome of basal ganglia impairments and that these basic operations are captured in recent anatomical-physiological models of focused selection and inhibition of competing responses.

1 Early Investigations of RT in People with Parkinson's Disease

The literature on RT in PD patients reveals mixed findings, perhaps due to the differences in methodologies and severity of symptoms of the patients tested (King, 1959; Angel et al., 1970; Heilman et al., 1976). Some studies reported that PD patients were slower than controls on simple RT tasks (where all response information is known prior to the stimulus to move), whereas the two groups were not reliably different on choice RT tasks (Evarts et al., 1981; Bloxham et al., 1984; Sheridan et al., 1987). A corollary finding common across these studies- that simple and choice RT are not reliably different within the patient group- has led to the interpretation that PD patients are impaired in utilizing advanced information that is provided in the simple RT case (i.e., Sheridan et al., 1987; but see Rafal et al., 1984). However, it is equally plausible that the benefit of advanced information appears to be lost in PD patients because they cannot effectively hold planned information in store prior to movement initiation (Bloxham et al., 1984; see also Klapp, 1976). For example, basal ganglia dysfunction might cause the program for movement to degrade quickly due to noise in the system and to interference that arises due to an inability to inhibit the programming of other potential responses (see Sheridan et al., 1987). As suggested above, due to the broad range of methodologies employed in these studies, it has been difficult to weigh one possibility against the others.

2 A Recent Demonstration that Sheds a Different Light on the Basic Cognitive Deficits Related to Parkinson's Disease

Taking a very different approach, Franz and Miller, (2002) recently examined whether a basic function of the basal ganglia relates to activation and inhibition. Rather than using RT as a measure of task difficulty or complexity, RT was measured from analogue force profiles to examine the effects of high and low response readiness using a go no-go task. Specifically, each block of trials was equally divided into trials with color cues that corresponded to either high or low probability events. Cues in either of two colors flooded the computer screen just prior to presentation of a stimulus that signaled the subject to respond (go trial) or to withhold responding (no-go trial). To reinforce the manipulation on response readiness, participants were instructed prior to performance of each block that one color cue indicated high probability that a response is likely on that trial and the other color indicated low probability that a response is likely on that trial. Following the color cue, an arrow pointing in the direction of the response key was presented on 80 % of trials for high response readiness blocks, and an '=' was presented on the remaining 20 %. All trials with an arrow required a response (go trials), whereas trials with the '=' required withholding of the response (no-go). The opposite probability-to-response mapping was used for low probability trials (low response readiness conditions), with 80 % 'no-go' stimuli and only 20 % 'go' stimuli. Force output was measured on each trial, and RT was computed using a force threshold criterion. It was previously unknown how patients with PD would respond on high versus low readiness conditions, and the primary issue of interest was in whether they would reveal deficits in response activation, following logic provided by studies on healthy control subjects.

Previous studies on healthy control participants demonstrated that faster RTs occur with high probability compared to low probability events (i.e., Mattes et al., 1997, 2002). Mattes et al., (2002) extended similar findings to the force domain, demonstrating that high probability trials resulted in a smaller peak force than low probability trials. This latter effect, in particular, motivated the study by Franz and Miller, (2002) who noted that impairments associated with PD appear to range from too little activation (i.e., akinesia, bradykinesia) to too much activation (i.e., festination, chorea). Franz and Miller also noted that the common observation, that Parkinson's patients have difficulty stopping an ongoing movement, might be a marker of a deficit associated with inhibition (deactivation).

To capture properties of response activation in healthy controls, Mattes et al., (2002) elaborated on a framework proposed earlier by Näätänen, (1971). Specifically, Näätänen, (1971) found that mean RT lengthened with systematic increases in the interval that separates stimuli that are used to cue responses in a simple RT task. According to Näätänen, response readiness is high with expectancy of a short interval compared to expectancy of a long interval. He further asserted that with a high response readiness, only a small additional amount of response activation is required to reach a 'motor action limit', whereas with a low response readiness, a larger amount of activation is required to reach that limit. This relationship between expectancy and response activation was later elaborated by Mattes et al., (2002) to account for the effects on force that were demonstrated using direct manipulations on response readiness (described above). Franz and Miller, (2002) therefore raised the question of whether PD patients would produce different patterns of force output than controls under manipulations of response readiness, and if so, whether these patterns on force could be interpreted as deficits in response activation (and deactivation). A group of mild to moderate PD patients and age-matched controls were among the participants tested in that study.

Of primary interest, Franz and Miller, (2002) learned that the effects of response readiness on RT were similar in the patient group and their age-matched controls, although the patients showed abnormal force profiles compared with the matched controls. For the controls, peak force and force integral were smaller for high probability trials compared to low probability trials, whereas the opposite pattern was apparent in the patients. The average values of these variables collapsed across blocks performed by the left and right hands can be seen in Table 1 (see also Franz and Miller, 2002; Table 2). Note that variables for the responding hand only are shown in Table 1, although force output was concurrently measured for both hands (a point I will return to later). Further analyses revealed that, on average, the rate of force output was faster on low compared to high response readiness trials in control participants, whereas the rate of response output was approximately the same for the two conditions in the patients. Force output also occurred for a longer duration in the PD group compared to the controls, and a longer average duration for larger impulses occurred in the patients despite the lack of differences in rate. Franz and Miller, (2002) interpreted this pattern of results as reflecting an inability of the patients to modulate response activation according to the internal level of response readiness. In other words, the patients produced a constant increment of activation regardless of the level of internal readiness. It was concluded that a primary deficit of the patients is related to response activation.

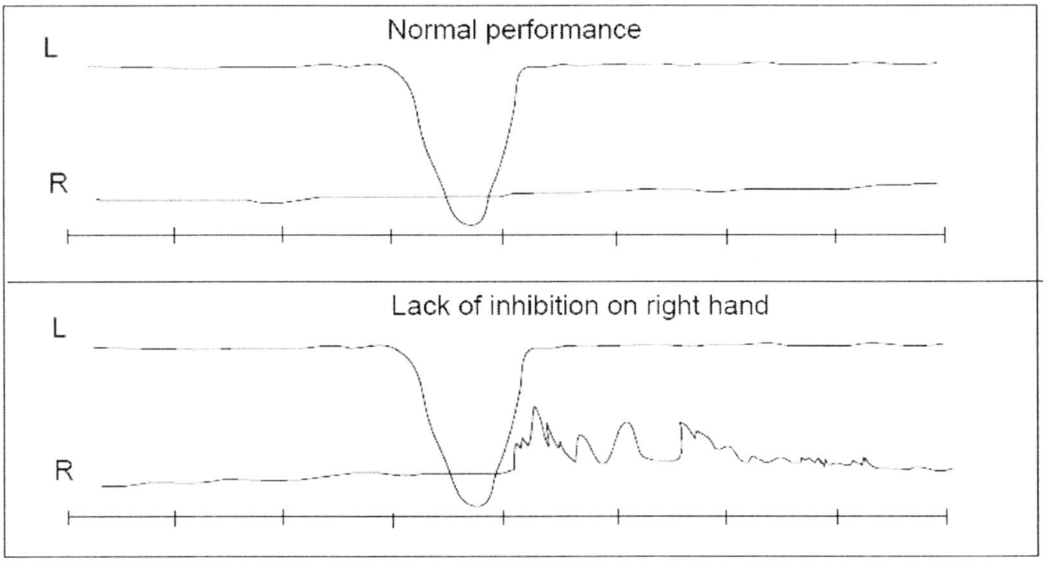

Figure 1. One representative trial of a normal 'go' response with high response readiness (top panel), and one representative trial of a response from the same condition showing evidence of a lack of inhibition (lower panel). Tick marks occur each second. L means left hand, and R means right hand. Adapted from data of (Franz and Miller, 2002).

Table 1. Mean measures of force for PD patients and age-matched (AM) controls for the responding hand on go trials with high or low readiness. Values are collapsed across the two hands.

	Peak Force (cN) Readiness condition		Impulse size (cN) Readiness condition	
	High	Low	High	Low
Group				
PD	**567**	556	27,784	26,802
AM	512	526	17,594	18,035

Adapted from Franz and Miller, (2002, Table 2).

Other related research is consistent with an inability in people with PD to appropriately modulate response activation. For example, electromyographical activity (EMG) in healthy subjects shows the characteristic triphasic pattern of agonist and antagonist activity with a normal limb movement to a target, but it has been shown that PD patients often do not generate sufficient EMG activity in the initial agonist burst to obtain the force necessary for large and fast movements (Hallett and Khoshbin, 1980). Marsden, (1984) suggested that while the selection of muscles and the sequential timing of their activations are not impaired in PD patients, perhaps the number and frequency of activated motor neurons is not adequate. It has also been suggested that the diminution of letter size known as micrographia, and a similar type of effect in continuous drawing tasks might be a manifestation of an inability to sustain an appropriate level of response activation through time (see Figure 2).

In addition to response activation, Franz and Miller, (2002) proposed that response deactivation (or inhibition) would also be impaired in the PD patients. At least two aspects of the findings of the study supported this hypothesis. First, although interest was primarily in the responding hand on each trial, force output was also measured continuously throughout each trial for the hand that was not supposed to respond (given all trials involved only a unimanual go versus no-go response, there was always one uninvolved hand). For the uninvolved hand, at least some level of force output was found on some temporal epochs following stimulus presentation, and this significant level of force occurred somewhat sporadically for control participants. In contrast, for the PD patients a significant level of force output occurred for the uninvolved hand quite often on temporal epochs that surrounded the time at which RT would normally occur (see Figure 1). In other words, response inhibition of the hand that was not supposed to respond was somewhat impaired in the PD group. Second, illustrating a possible impairment in response inhibition, the PD group also produced a larger mean force level than the control group on no-go trials, or those trials in which no response was required. Collectively, these findings are suggestive of problems in inhibiting or deactivating responses (Franz and Miller, 2002). Of additional note, given that the effects on force output were incidental (or implicit) rather than being explicit requirements of the task (subjects were instructed to press the key on 'go' trials, without any mention of force), it would seem that any explanation of response activation/deactivation deficits would have to apply both for explicit as well as implicit force demands (Franz and Miller, 2002).

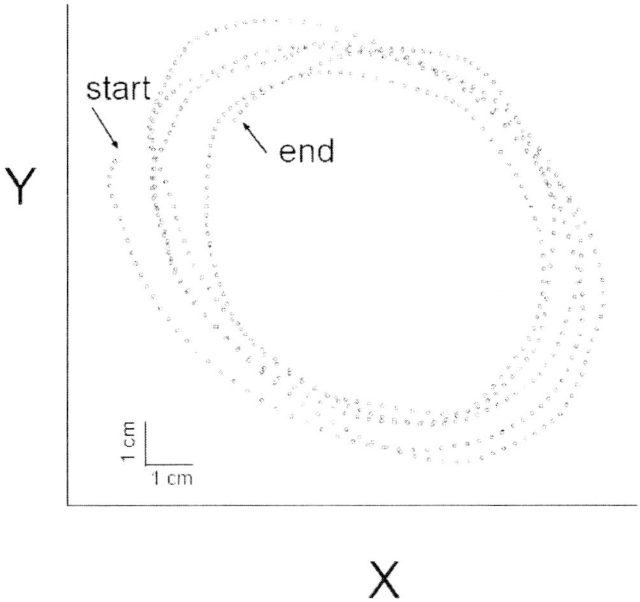

Figure 2. One representative trial of a PD patient drawing continuous circular trajectories through 10 seconds of time. Position data are plotted for the x and y dimensions for the trial. Note that the circles are getting slightly smaller through time, and that there is some evidence of variability along the trajectory. Franz data (unpublished).

With the aim of clarifying how impairments in response activation using a go no-go task might reveal a more general deficit of basal ganglia function, I now take a small leap of faith, which is to suggest that force output is an expression of a rudimentary form of cognition. By this view, the activation and inhibition deficits demonstrated in the force output profiles of Franz and Miller, (2002) reveal basic operations of abstract information processing, and these operations rely on basal ganglia motor loops. These operations, in my view, need not be different from those that operate on other forms of cognitive information, whether activating and inhibiting certain perceptual cues, cognitive sets, memories for action, or goals. It follows that, along with the operations of activation and inhibition on force output, which implicate basal ganglia motor loops, similar deficits should be revealed for cognitive processes mediated by other basal ganglia loops. To support this argument, I now refer to some elegant accounts in earlier literature on cognition that form the basis of this type of framework for action selection. I then attempt to link features of these cognitive accounts to models based on anatomy and physiology that suggest that the cognitive processes of action selection rely on the integrity of basal ganglia operations.

3 Earlier Cognitive Accounts of the Inhibition of Competing Responses

In an earlier cognitive account of action selection in the normal brain, Allport, (1980) proposed that one form of task difficulty might occur due to the demand to keep specific goals active and others inhibited. Accordingly, inhibition of specific goals or subgoals to enable activation of other more pertinent goals might entail a competitive process (Neumann, 1987). According to Allport, (1980) and echoed by Neumann, (1987), one set of inputs must be decoupled from potential command of actions, a problem of action selection that humans are extremely efficient at solving. This type of view casts a slightly different light on earlier issues examined in RT literature. Notably, it leaves open the possibility that RT increases reflect both the processes associated with response selection in the face of alternatives, and the operation of inhibitory processes on potentially active competing responses. Interestingly, a parallel development of models of action selection in the face of competing alternatives can be traced in the neurophysiological literature. What follows is an outline of that development.

4 A Parallel Development in Neurophysiology

Perhaps one of the earliest direct examples of behavioral inhibition comes from von Holst and his colleagues' work on the functional organization of drives (Von Holst and Von St. Paul, 1963; see also Mink, 1996). These researchers presented elegant examples of the competing drives in the behavior of the domestic fowl, which, due to their central internal relationships, are overcome by those behaviors that dominate in particular situations. Although Von Holst and Von St. Paul, (1963) did not specify precisely what neural mechanisms are involved in the form of behavioral inhibition they demonstrated, neurophysiological evidence in the last decade or so has converged to make a strong case that in humans, these operations are a primary function of the basal ganglia.

In this section, I will focus primarily on the work of Mink and Thach, (1991a, 1991b, 1991c), and a model of focused selection and competing inhibition that emerged from those and other related studies (Mink, 1996). For purpose of illustration, I will once again refer to the example described by Neumann, (1987), which is also cited in the opening paragraph of the present chapter—that of reaching for an apple from a tree. To perform this task, a person interrupts his locomotion by stopping and assuming a steady posture that supports the body while one arm reaches for the apple (Mink, 1996). Some of the cognitive considerations were noted above. Here, I elaborate on some of the issues that anatomico-physiological models might also consider. First, how is posture controlled so that a volitional action, such as a reach to an object, can be performed? Wilson, (1928), in his early investigations of patients with basal ganglia damage, noted that while syndromes associated with damage to the pyramidal system were not apparent, numerous problems associated with the phylogenetically-older extrapyramidal motor system could be readily observed. These included muscular rigidity, tremor, and weakness which could be associated with bradykinesia, symptoms that are now commonly associated with Parkinson's disease (see this volume). According to Wilson, the extrapyramidal system operates somewhat automatically to control posture, whereas, the pyramidal system is involved in voluntary, phasic movements. Mink, (1996) further proposed a framework based on anatomical-neurophysiological evidence that nicely captures complex tasks that involve focused selection of action and inhibition of competing responses. Central to the assumptions of this model is the time course of activation of the different structures involved during motor preparation, particularly with reference to the basal ganglia.

Since the pioneering work of DeLong, (1971), it has been known that basal ganglia activity occurs with movement. However, there has been considerable debate as to the precise role performed by the basal ganglia, and whether its nuclei actually code parameters of movement (Crutcher and DeLong, 1984). A significant amount of movement-related activity in the putamen (part of the input nuclei to basal ganglia) has been found to occur after rather than before the onset of muscle activity (Ibid), and activity of output nuclei (globus pallidus) also has been found to occur later than motor cortex activation (Anderson and Horak, 1985; Mink and Thach, 1991a, 1991b). Some have suggested that late activity occurs in more posterior portions of the striatum, whereas anterior portions might mediate earlier set-related activity (see Mink, 1996).

As Mink, (1996) describes (also using Neumann's example of reaching for an apple), multiple mechanisms involved in maintaining an upright posture (presumably extrapyramidal), both prior to the reach and during the reach, are also involved in supporting the posture of the reaching arm to perform a volitional action. These mechanisms must remain active for the rest of the body but must be selectively deactivated for the reaching arm during the reach so that the reaching action can be produced. Viewed in this way, competing motor pattern generators (MPGs: Mink's term) must be turned off so that they don't interfere with the voluntary (reaching) movement. This process of selectively inhibiting the competing

MPGs so that the MPGs for the required volitional action can be activated, is central to Mink and Thach's model of focused selection and inhibition of competing motor mechanisms (summarized in Figure 3).

As Mink, (1996) elaborates, the pattern of activity from basal ganglia circuits would result in inhibition of a large portion of thalamic and brainstem nuclei receiving projections from GPi, and disinhibition of a smaller portion of GPi nuclei, resulting in the opposite effect on their thalamic and cortical targets. According to Mink, (1996), GPi activity increases occur earlier than GPi decreases, which is consistent with excitatory STN-GPi connections being faster than the inhibitory Striatum-GPi connections (see Figure 3). Consistent with this model, studies using electrical stimulation to individual neurons in prefrontal and motor cortical regions have demonstrated that initial excitation is followed by inhibition (Nambu et al., 1990). In sum, Mink and colleagues conjectured that set-related and movement-related activity originate in the cerebral cortex and not the basal ganglia, thus calling into question the earlier-held views that the function of the basal ganglia is in movement initiation. Instead, they suggested that the basal ganglia are involved in focused selection of actions, and inhibition of competing programs.

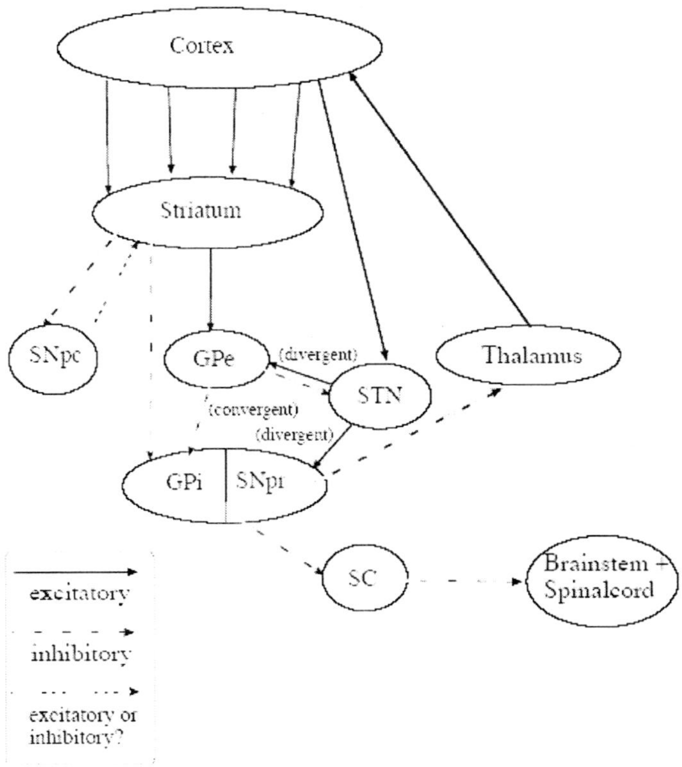

Figure 3. The striatum is one primary input that receives excitatory input from diffuse areas of the cerebral cortex. Inhibitory projections from the striatum to Gpi and on to thalamus (direct loop) and ihhibitory projections from the striatum to GPe, which in turn inhibits Gpi (referred to as part of the indirect loop) result in focused input to Gpi (and also SNpr). These projections onto Gpi are believed to be convergent, thus producing focused selection. A second input from motor and premotor areas (A4, 6) and frontal eye fields (A8) is excitatory on STN. Through divergent projections, STN is excitatory on GPe and GPi. STN also receives inhibitory input from GPe. This loop is also referred to as indirect. Importantly, the focused input from striatum to Gpi is purportedly slow, whereas the STN input to GPi and SNpr is fast. This results in fast divergent excitation and a slower focused inhibition on specific output neurons of GPi and SNpr. Note that the mechanisms intrinsic to the striatum are proposed to integrate the input from cortex and focus the output to other nuclei of the basal ganglia. A third loop involves inhibitory input from the striatum to SNpc, which in turn projects back to the striatum in either an inhibitory or excitatory manner depending on the receptors. This loop likely modulates cortical inputs to the striatum (Mink, 1996). Terminology used throughout this article: GPi (internal segment of the globus pallidus), GPe (external segment of the globus pallidus), SNpr (substantia nigra pars reticulata), STN (subthalamic nucleus), SNpc (substantia nigra pars compacta).

With respect to findings of a simple go no-go task involving reaching movements in monkeys, Romo and Schultz, (1992) found that of the neurons that responded to the instruction (approximately one fifth of the total), about half responded with sustained activity only in the "go" condition, whereas the other half responded transiently in relation to the instruction. The so-called 'late activity' changes in the basal ganglia are also consistent with the problems of deactivation reported by Franz and Miller, (2002) using their go no-go task in PD patients.

5 Integrating Cognitive and Neurophysiology Research

Is it possible that the impairments in the force profiles of PD patients on a go no-go task (Franz and Miller, 2002), the cognitive models of Neumann, (1987) and Allport, (1980), and the anatomic-physiological model of Mink and colleagues (Mink, 1996) converge on a coherent framework of action selection in which the basal ganglia play a central role? I will now attempt to fill in some of the missing pieces that hopefully will make this proposed framework more persuasive.

First, the findings of Franz and Miller, (2002) are based on the use of one of the simplest types of motor tasks, that consisting of a go versus inhibit response task (go no-go), under conditions of high or low response readiness. As suggested above, the force output on such a task might be regarded as the expression of a rudimentary form of cognition. One might assume, for example, that when preparing to make a response under high readiness, the motor system is in a state of high preparation, and therefore the initial level of internal activation is higher than in the situation in which the system is not prepared to respond (low readiness condition). Echoing a model elaborated by Mattes et al., (2002) based on an earlier proposal by Näätänen, (1971): under high readiness, the system requires only a small increase in activation to produce the response, whereas, under low readiness, much more activation is necessary. In addition to this effect on the activated responses, those responses that were to be inhibited also displayed a larger than appropriate level of force output in the PD patients (see above description). If these force output processes reflect the operation of a rudimentary form of cognition, and the basal ganglia are responsible for activation (focused selection) and inhibition (inhibition of competing actions) more generally, then similar deficits should be demonstrated in patients with basal ganglia damage on other types of cognitive tasks.

6 Other Cognitive Phenomena that Can Be Captured by this Integrated Framework: Set Switching

Another example of a fundamental task of cognition is the ability to switch from one task to another (Jersild, 1927; Allport et al., 1994; Rogers and Monsell, 1995; Hayes et al., 1998). This can be viewed in simple form as a switch between perceptual dimensions, such as color and shape. Take, for example, a task similar to that of Hayes et al., (1998), in which subjects are presented with two response keys, each labeled with a shape (circle or square) and a color (pink or yellow). On a given trial, subjects are instructed to respond on the basis of a cue word that occurs at the same time as the stimulus, with the cue word indicating to respond to "shape" or "color". One can then consider pairs of trials as Hayes et al., (1998) did, in which two "color" cues occur in succession for consecutive trials, or two "shape" cues occur in succession for consecutive trials. These trials would be considered 'no switch' trials, given the perceptual dimension is the same for both. Compare this to the pair of trials in which a "shape" cue occurs on trial 1, and a "color" cue occurs on trial 2 (or vice versa). In this case, a switch is necessary between one perceptual dimension and the other, in order to respond correctly. Hayes et al., (1998) demonstrated an increase in RT for switch compared to no switch trials of this type in healthy control subjects, and a significantly larger switch cost in patients with Parkinson's disease. Our laboratory recently replicated these findings, with further examinations of patients tested both during their normal medication cycle and with their levodopa medication withheld for approximately 12 hours (Shook et al., Under revision). The results for switching costs are shown in Figure 4. Clearly, there are increases in cognitive switch times for the patients compared to the controls. Moreover, these switch costs are exacerbated when the patients' medication is withheld, indicating that switching is dopamine-dependent.

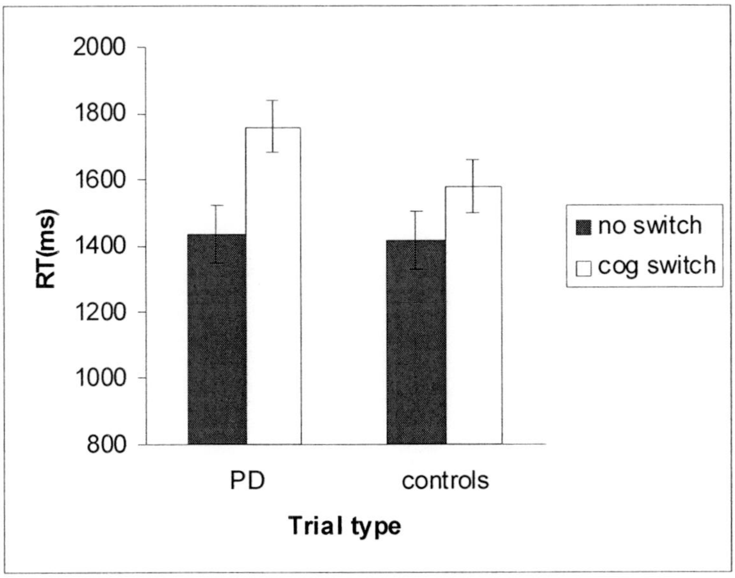

Figure 4. Cognitive switch costs (cog) compared to no switch trials, where the switch is between perceptual dimensions on consecutive trials. These data are averages from all control subjects, and all PD patients in the 'off' medication state. Standard error bars for each group are shown. Adapted from Shook et. al., 2004.

How can these findings be reconciled with the earlier-reported findings on force output? As a start point, one might consider that the perceptual cues of color and shape trigger different perceptual sets. When responding on the basis of the color cue, one's perceptual set of color is activated, and one's perceptual set of shape is deactivated (or inhibited). The reverse argument holds for processes associated with the shape cue. Thus, rather than operating on the level of force output as in the simple go no-go task of Franz and Miller, (2002), the basal ganglia operations of focused selection (activation) and inhibition of competing surround would be operating on perceptual sets. Given perceptual sets might involve vast cell assemblies in the cortex, it seems reasonable that switching from one to another would result in an increase in RT. In other words, if activation and inhibition are general deficits with PD, one might expect them to be manifest in deficits of force output on the simplest forms of motor tasks that are cued by basic perceptual shapes. However, when the task demands instead require discrimination between one perceptual dimension and another (color versus shape), these operations might reflect different cognitive loops of the basal ganglia: activating the perceptual dimension that is appropriate to responding while inhibiting the dimension that is not appropriate. In a similar manner, selecting the appropriate perceptual dimension on which to respond illustrates the process of focused selection, and inhibiting the inappropriate dimension illustrates the process of inhibiting competing responses (according to Mink's 1996 terminology).

7 Anatomical Models of Parallel Circuits within the Basal Ganglia Suggest that Similar Cognitive Operations Occur in the Different Segregated Circuits

What is required for this type of model to apply generally across numerous cognitive domains, is that different loops within the basal ganglia-cortical circuits operate in a similar manner. A remarkable feature of the basal-ganglia-cortical circuitry that has been elegantly described by neurologists, neuroscientists, and physiologists, is that there are numerous circuits that are functionally segregated and organized in parallel (Alexander et al., 1986). Figure 5 is adapted from Alexander et al., (1986), showing the different structures involved in five parallel circuits of the basal ganglia.

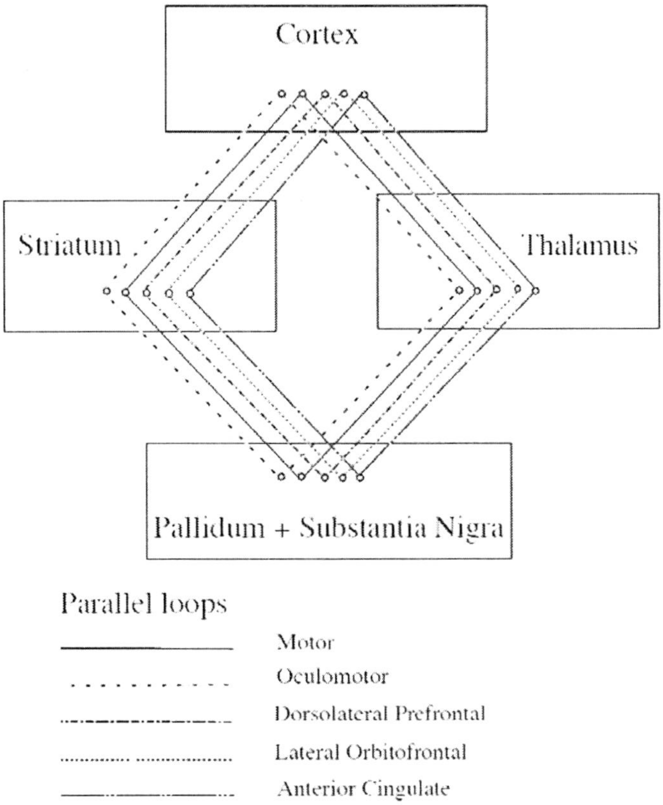

Figure 5. Schematic of parallel segregated loops involving cortical-basal ganglia-thalamic circuits. Specific nuclei at each structure are not shown. See text for details. Adapted from (Alexander et al., 1986).

Indeed, there are likely to be many more such circuits. Note that the motor circuit involves the supplementary motor area's input to the putamen (of the striatum), through the basal ganglia circuitry to the GPi and SNpr output nuclei, and back to the cortex via the VLo and VLm of the thalamus (for simplicity, the present figures do not depict specific nuclei but only general structures). As can be seen in Figure 5, the other parallel circuits are organized in a similar manner, each engaging specific regions of the structures involved. If we can imagine, therefore, that the circuits producing focused selection and inhibition of competing programs (as shown in Figure 4) also characterize each of the other functionally segregated and parallel cortical-bg-thalamic-cortical circuits, then it seems reasonable that the activation-inhibition functions that might characterize the motor loop extend to similar functions mediated by the other loops. Indeed, a number of researchers have suggested that the existence of a parallel organization of these circuits is strongly suggestive of similar basic functions occurring in all of them (Alexander et al., 1986; Hayes et al., 1998; Shook et al., Under revision).

8 Relating this Model to Symptoms Commonly Seen in Parkinson's Disease

As suggested earlier, the symptoms of PD often appear to reflect either too little or too much activation, as though the balance between activation and inhibition is awry (Franz and Miller, 2002). I now point out some of the more obvious PD symptoms that can easily be captured by the type of model described herein. Dyskinesia, or the rapid flailing movements of the limbs, might appear to be characterized by an inappropriate sequence of unwanted movements, as though some components of movement are inappropriately activated while others are inappropriately inhibited. Impairments in either the focused selection component or the inhibition component of the basal ganglia circuitry could account for these observed symptoms. Increased visual distraction is also often observed in people with PD, as though competing information is not appropriately inhibited. Indeed, as can be seen in Figures 3 and 5, one loop of the frontal-basal-ganglia circuitry comprises the frontal eye fields, caudate nucleus of the basal ganglia, and superior colliculus, which, together describe a system for eye movements that might be similar to that described above for limb movements. It is therefore plausible that symptoms such as chorea reveal similar properties as the type of distraction associated with eye movement saccades.

9 Can we Extrapolate beyond the Mink (1996) Model on the Basis of Other Cognitive Research ?

I will now briefly attempt to extrapolate beyond the model proposed by (Mink, 1996), to consider one possible function of the dopaminergic projections from the substantia nigra pars compacta on the striatum (see Figure 3). As suggested by the model of Mink, (1996), dopaminergic projections from the substantia nigra pars compacta to the striatum and back might mediate increases or decreases in striatal activity, which is also influenced by cortical projections. The model described above seems to capture activation and inhibition processes within basal ganglia circuitry, but it is somewhat undefined precisely how those circuits 'know' to focus attention on events that are significant, while inhibiting activation to those that are not.

Recent years have witnessed a rapid proliferation of findings and perspectives on the precise role of dopamine in the context of the complex frontal-basal-ganglia-thalamic circuitry described. While it seems generally accepted that dopamine plays a role in motivation and reward, the precise nature of what dopamine is doing remains debatable. Whereas early views suggested a role of dopamine as a pleasure mediator (Wise, 1996), more recent findings have led to the idea that dopamine signals that something significant is occurring, even if the stimuli these neurons responded to were not in themselves rewarding (Schultz et al., 1997). A related account holds that dopamine is involved in an attention-getting system, in which novel or unpredictable events attract the organism's attention (Horvitz et al., 1997). Although the precise role of dopamine is not yet known, an intriguing possibility is that the dopamine system is involved in both signaling significant events for action, and in maintaining current action sets until they are completely carried out. In a sense, dopamine would 'juice the system' for actions that should receive (possibly sustained) focused attention. The finding that cognitive switching costs are exacerbated when regular application of dopamine precursors are withheld temporarily from the patients, seems to be consistent with this possibility (Hayes et al., 1998; Shook et al., Under revision). Thus, although speculative, it seems possible that the dopaminergic inputs to the set-related neurons in the striatum assist in the maintenance of a particular set so that a current action can be carried out completely and without intrusion by other action selection processes. This system would enable sustained activation of networks within the striatum so that other operations of the basal ganglia loops can be carried out before another selected input goes through these loops.

Summary

The present paper was an attempt to synthesize data from recent cognitive experimentation on patients with Parkinson's disease with models of the anatomy and neurophysiology of the basal-ganglia-frontal circuitry and intrinsic circuits within the basal ganglia. One main point is that the focused selection and competing inhibition model of Mink, (1996) reflects a similar cognitive view espoused by Neumann, (1987) and others (see Allport, 1980); and these process models can account for recent findings in cognition that identify activation and inhibition as a basic function of the basal ganglia. Activation and inhibition on a force output task are proposed to reflect a rudimentary form of cognition, and a similar deficit is proposed to operate on different forms of information processing. I propose that these approaches are all converging on the same basic model of basal ganglia function, and that this basic function has already gained support from other cognitive domains. In sum, the conjecture is that the basal ganglia complex performs a function of selection on many different levels of the cognitive system (via parallel operations on segregated circuits) by focused selection (activation) and inhibition of competing information (inhibition). It is my belief that deficits of bimanual coordination, switching, focused selection, and sequencing all can be captured by basal ganglia operations involved in focused selection and inhibition of competing information.

ACKNOWLEDGEMENTS

I am grateful to Professor Jeff Miller and Professor Karen Sigvardt for their invaluable lessons and discussions on issues related to this work. Thanks also to Reuben Fischer for drawings.

REFERENCES

Alexander GE, DeLong MR, Strick PL (1986) Parallel organization of functionally segregated circuits linking basal ganglia and cortex. *Annu Rev Neurosci* **9**:357-381.

Allport DA (1980) Attention and Performance. In: Cognitive psychology-new directions. (Claxton G, ed). London: Routledge & Kegan Paul.

Allport DA, Styles EA, Hsieh S (1994) Shifting intentional set: Exploring the dynamic control of tasks. In: *Attention and performance* (C. Umilta MM, ed), pp pp. 421-452. Hillsdale, NJ: Erlbaum.

Anderson ME, Horak FB (1985) Influence of the globus pallidus on arm movements in monkeys. III. Timing of movement-related information. *J Neurophysiol* **54**:433-448.

Angel RW, Alston W, Higgins JR (1970) Control of movement in Parkinson's disease. *Brain* **93**:1-14.

Bloxham CA, Mindel TA, Frith CD (1984) Initiation and execution of predictable and unpredictable movements in Parkinson's disease. *Brain* **107** (Pt 2):371-384.

Crutcher MD, DeLong MR (1984) Single cell studies of the primate putamen. II. Relations to direction of movement and pattern of muscular activity. *Exp Brain Res* **53**:244-258.

DeLong MR (1971) Activity of pallidal neurons during movement. *J Neurophysiol* **34**:414-427.

Evarts EV, Teravainen H, Calne DB (1981) Reaction time in Parkinson's disease. *Brain* **104**:167-186.

Franz EA, Miller J (2002) Effects of response readiness on reaction time and force output in people with Parkinson's disease. *Brain* **125**:1733-1750.

Hallett M, Khoshbin S (1980) A physiological mechanism of bradykinesia. *Brain* **103**:301-314.

Hayes AE, Davidson MC, Keele SW, Rafal RD (1998) Toward a functional analysis of the basal ganglia. *J Cogn Neurosci* **10**:178-198.

Heilman KM, Bowers D, Watson RT, Greer M (1976) Reaction times in Parkinson disease. *Arch Neurol* **33**:139-140.

Hick WE (1952) On the rate of gain of information. *Quarterly Journal of Experimental Psychology* **4**:11-26.

Horvitz JC, Stewart T, Jacobs BL (1997) Burst activity of ventral tegmental dopamine neurons is elicited by sensory stimuli in the awake cat. *Brain Res* **759**:251-258.

Jersild AT (1927) Mental set and shift. Archives of Psychology 89.

King HE (1959) Defective psychomotor movement in Parkinson's disease: exploratory observations. *Perceptual and Motor Skills* **9**:326.

Klapp ST (1975) Feedback versus motor programming in the control of aimed movements. *J Exp Psychol Hum Percept Perform* **104**:161-169.

Klapp ST (1976) Short-term memory as a response preparation state. *Memory and Cognition* **4**:721-729.

Marsden CD (1984) Which motor disorder in Parkinson's disease indicates the true motor function of the basal ganglia? In: *Functions of the Basal Ganglia.* (M.O'Connor. DEa, ed), pp 225-237. London: Pitman.

Mattes S, Ulrich R, Miller J (1997) Effects of response probability on response force and simple RT. *Quarterly Journal of Experimental Psychology* **50***A*:405-420.

Mattes S, Ulrich R, Miller J (2002) Response force in RT tasks: isolating effects of stimulus probability and response probability. *Visual Cognition* **9**:477-501.

Merkel J (1885) Die Zeitlichen Verhältnisse der Wiellensthätigkeit. *Philosophische Studien* **2**:73–127.

Mink JW (1996) The basal ganglia: focused selection and inhibition of competing motor programs. *Prog Neurobiol* **50**:381-425.

Mink JW, Thach WT (1991a) Basal ganglia motor control. I. Nonexclusive relation of pallidal discharge to five movement modes. *J Neurophysiol* **65**:273-300.
Mink JW, Thach WT (1991b) Basal ganglia motor control. II. Late pallidal timing relative to movement onset and inconsistent pallidal coding of movement parameters. *J Neurophysiol* **65**:301-329.
Mink JW, Thach WT (1991c) Basal ganglia motor control. III. Pallidal ablation: normal reaction time, muscle cocontraction, and slow movement. *J Neurophysiol* **65**:330-351.
Näätänen R (1971) Non-aging fore-periods and simple reaction time. *Acta Psychologica (Amst)* **35**:316-327.
Nambu A, Yoshida S, Jinnai K (1990) Discharge patterns of pallidal neurons with input from various cortical areas during movement in the monkey. *Brain Res* **519**:183-191.
Neumann O (1987) Beyond capacity: A functional view of attention. In: *Perspectives on Perception and Action* (Sanders HHaA, ed), pp 361-394. Hillsdale, NJ: Erlbaum.
Rafal RD, Posner MI, Walker JA, Friedrich FJ (1984) Cognition and the basal ganglia. Separating mental and motor components of performance in Parkinson's disease. *Brain* **107** *(Pt 4)*:1083-1094.
Rogers RD, Monsell S (1995) The cost of a predictable switch between simple cognitive tasks. *Journal of Experimental Psychology: General* **124**:207-231.
Romo R, Schultz W (1992) Role of primate basal ganglia and frontal cortex in the internal generation of movements. III. Neuronal activity in the supplementary motor area. *Exp Brain Res* **91**:396-407.
Schultz W, Dayan P, Montague PR (1997) A neural substrate of prediction and reward. *Science* **275**:1593-1599.
Sheridan MR, Flowers KA, Hurrell J (1987) Programming and execution of movement in Parkinson's disease. *Brain* **110** *(Pt 5)*:1247-1271.
Shook S, Franz EA, Higginson CI, Sigvardt KA (Under revision) Dopamine dependency of intradimensional and extradimensional switching operations in Parkinson's patients. Neuropsychologia.
Von Holst E, Von St. Paul U (1963) On the functional organization of drives. *Animal Behavior* **11**:1-20.
Wilson SAK (1928) Modern problems in neurology. London.
Wise RA (1996) Neurobiology of addiction. *Curr Opin Neurobiol* **6**:243-251.

In: Recent Breakthroughs in Basal Ganglia Research
Editor: Erwan Bezard, pp. 241-248 ISBN: 1-59454-880-3

Chapter 19

THE AKINETIC PUZZLE: A CORTICAL VIEWPOINT (CORTICAL CORRELATES OF AKINESIA)

Pierre Burbaud
CNRS UMR 5543, University Victor Segalen, Bordeaux, France

ABSTRACT

Recent electrophysiological data collected in humans and monkeys shed new light on the involvement of the frontal cortex in the pathophysiology ofParkinson's disease (PD). In MPTP-treated monkeys, movement-related activity in the primary motor cortex (M1) appears to be more gradual, the neurons taking longer to reach their maximal activation. There is an increased synchronization among M1 cells, which could support muscle rigidity. Evidence also exists for dysfunctioning in both parts of the primate supplementary motor area. These phenomena may account for the difficulties PD patients encounter in executing self-initiated sequential movements. On the other hand, little is known about the activity of the prefrontal areas (dorsolateral prefrontal cortex, anterior cingulate cortex, orbitofrontal cortex) and the role of dopamine (DA) on cortical neurons. The future use of multiple single unit recordings coupled with more efficient imaging techniques will allow precise investigation of the nature of cortical dysfunctioning in PD.

1 PHENOMENOLOGICAL ASPECTS OF AKINESIA

The term akinesia corresponds to an impairment in the spontaneous initiation of movement. It has been improperly extended to the delayed onset of movement but also to the slowing of movement, more accurately described by the term bradykinesia. One problem is to know whether akinesia is a movement execution disorder or a defect in cognitive and/or affective processes associated with motor planning. Although patients with an advanced form of the disease might suffer from attentional problems and frequently exhibit inadapted emotional responses, it is clear that akinesia does not result from a misunderstanding of instructions or from a loss of motivation. PD patients want to move but they simply cannot (Marsden, 1982).

However, if the problem is to run the appropriate program, what is exactly dysfunctioning in the system? Is it the selection of the adapted program in space, time or both or a disorder limited to the execution of a previously selected program? If so, is it because of an ineffective activation of the cortex by sub-cortical inputs or because of an impaired processing of specific information at the cortical level?

Rigidity illustrates the problem of program selection. Instead of a selective activation of agonist muscles with an inhibition of antagonist muscles, PD patients activate both muscles in such a time range that the harmonious execution of movement is no longer possible. However, it is unlikely that rigidity alone can account for akinesia. Indeed, the severity of the rigidity in some patients might be moderate whereas they are frankly akinetic. This suggests that these two clinical symptoms account for the imbalance of the various neuronal pathways involved in tonus regulation and action planning, respectively. A poor selection in time is illustrated by the difficulty that PD patients might have in correctly chaining sequential distal movements such as those required by the opposition between the thumb and other digits. These patients have the greatest difficulty in performing a perfect sequential movement e.g. one opposition with each digit only. Usually, they tend to persevere and perform several oppositions with the same digits. Moreover, they find it difficult to initiate movement or postures at the proper time, a phenomenon which sometimes leads to falling.

PD patients are especially impaired in the self-initiation of movements. This means that the process of action initiation is a problem. Paradoxically, however, movement performance can be dramatically improved by the presence of external cues or visual landmarks such as transversal lines on the floor. Moreover, in risky situations, akinesia can suddenly abate, so the functional blockade of motor programs can be compensated by additional information or extraordinary motivation.

2 What Role Does the Cortex Play ?

During the last fifteen years, most studies have focused on the involvement of sub-cortical regions in the pathophysiology of akinesia (for review, see Berardelli et al., 2001). So far, the role of the cortex has been mainly investigated through functional imaging. However, some recent data collected with single unit recordings performed in the cortex of monkeys rendered parkinsonian by 1-methyl-4-phenyl-1,2,3,6-tetrahydropteridine (MPTP) are now available. They throw new light on the pathophysiology of akinesia.

2.1 Primary Motor Cortex

The primary motor cortex is directly involved in the control of motoneuron activity so one could expect changes at its level in PD. Several functional imaging studies have shown that metabolic activity is essentially unchanged in PD patients compared to controls during both rest periods and movement execution (Rascol et al., 1992; Playford et al., 1993; Samuel et al., 1997), whereas fMRI studies consistently reveal an increased response (Sabatini et al., 2000; Haslinger et al., 2001; Payoux et al., 2004). Moreover, a decrease in rCBF in M1 has been reported after subthalamic nucleus stimulation (Limousin et al., 1997; Ceballos-Baumann et al., 1999; Payoux et al., 2004). Such an enhanced activity within the primary motor cortex could be related to the impairment of cortico-cortical inhibition, as suggested by data obtained using in transcranial magnetic stimulation. Such a phenomenon is alleviated by levodopa (Ridding et al., 1995) dopamine agonists (Strafella et al., 2000) or subthalamic stimulation (Dauper et al., 2002).

However, which aspects of neuronal activity are reflected by indirect assessment of cortical metabolism is a matter of debate (Goldberg et al., 2002). It seems that in MPTP-treated monkeys there is no modification in M1 firing rate both at rest (Doudet et al., 1990; Goldberg et al., 2002) and during movement (Doudet et al., 1990). Movement-related activity appears to be more gradual and long-lasting during performance of simple ballistic movements than in control monkeys (Doudet et al., 1990). This is associated with an increase in behavioral reaction time and movement time but unchanged cellular reaction time. This means that the processing of responses to instructions is unaffected in M1 but that neurons take longer to reach their maximal activation before movement onset. Interestingly, bursts of activity which are sometimes observed in M1 of MPTP-treated monkeys do not necessarily trigger movement (Goldberg et al., 2002). Could the problem therefore be energy-related ?

More recently, an increased tendency to synchronization between basal ganglia and the primary motor cortex or supplementary motor area has been reported in PD patients (Brown and Marsden, 1999; Brown et al., 2001; Marsden et al., 2001; Brown, 2003). The observation that M1 neurons recorded in MPTP-treated monkeys discharged in long bursts synchronized among many cells led to the idea that such a phenomenon could result in a loss of functional segregation between cortical columns. This might underlie muscle rigidity by causing a persistent co-contraction of antagonist muscles (Goldberg et al., 2002). The loss of specificity in M1 activity is also illustrated by the increased neuronal responses to passive arm movements in MPTP versus normal monkeys (Goldberg et al., 2002). Moreover, the appearance of 10 Hz oscillations through the cortex-basal ganglia networks observed in the parkinsonian situation suggests a closer relation between basal ganglia and cortex in MPTP-treated than in normal monkeys (Goldberg et al., 2002). Thus, the parkinsonian state is associated with an increased synchronization of M1 neuronal activity, potentially leading to rigidity but also impairing the focusing of information processing. One consequence could be the difficulty for M1 neurons to reach an optimal level of activation during voluntary movement.

2.2 Supplementary Motor Areas (SMA)

As previously stated, akinesia is particularly marked in the execution of self-initiated sequential movements. Some lines of evidence suggest that the frontal mesial cortex is involved both in the initiation of complex actions and in the planning of sequential movements (Tanji, 1994; Passingham, 1996; Picard and Strick, 1996). In this respect, trancranial magnetic stimulation applied over the SMA alters the performance of sequential movements (Cunnington et al., 1996; Gerloff et al., 1997; Boylan et al., 2001). In monkeys performing sequential tasks, it has been shown that different aspects of a behavioral sequence are directly encoded by SMA neurons (Tanji, 1994; Shima and Tanji, 2000). Some neurons process the rank of movement within the sequence whereas others are more concerned by the chaining of two specific movements. Moreover, a particular class of neurons seems to encode the sequence as a whole (Shima and Tanji, 2000; Akkal et al., 2002). The more rostral part of the SMA, corresponding to the pre-SMA, is specifically devoted to the learning of sequential procedures (Nakamura et al., 1998) and the rapid adaptation of on-going movement when unexpected stimuli occur (Matsuzaka and Tanji, 1996; Shima et al., 1996), a common situation in ordinary life.

In accordance with clinical observations, functional imaging studies have reported a decreased activity within the supplementary motor area in PD patients (Jenkins et al., 1992; Playford et al., 1993; Samuel et al., 1997; Sabatini et al., 2000; Haslinger et al., 2001). This impaired motor activation can be partially normalized by various treatments such as levodopa or dopa agonists (Jenkins et al., 1992; Rascol et al., 1994), subthalamic stimulation (Limousin et al., 1997; Ceballos-Baumann et al., 1999) or pallidotomy (Ceballos-Baumann et al., 1994; Grafton et al., 1995; Samuel et al., 1997). Using the 2-DG mapping technique in a monkey model of Parkinson's disease, Bezard et al., (2001) found that metabolic activity within the basal ganglia is established before symptoms appear, whereas abnormalities in the supplementary motor area accompany the appearance of symptoms.

We found slightly different data (Escola et al., 2003). Using a very progressive model of MPTP intoxication in the monkey, we first observed a drastic reduction of activity within both parts of the SMA in full parkinsonian monkeys, in accordance with the above-mentioned authors. This was associated with a clear hypoexcitability of the same region following intra-cortical microstimulation. On the other hand, a decrease in responses to visual information that specify the expected motor response was noticed in pre-symptomatic animals without akinetic symptoms (Escola et al., 2003). This might have several consequences. First, an impaired response to instructions could promote the occurrence of errors because of an incorrect processing of instructions. Second, the problem might be that pre-SMA neurons no longer respond to visual responses of attentional value indicating when to move. This phenomenon might explain the delayed response of monkeys to imperative stimuli and consequently the increased behavioral reaction time observed in PD patients during motor tasks (Berardelli et al., 2001). Neuronal changes after a NoGo signal were also decreased (Escola et al., personal communication), a phenomenon which could be related to the difficulty of patients to stop an on-going action.

Paradoxically, we also noticed an increased response of SMAp neurons to proprioceptive information, associated with a complete disorganization of SMAp somatotopy in MPTP-treated monkeys (Escola et al., 2002). Since rigidity was very moderate in these animals, it is unlikely that this phenomenon accounts for the diffusion of proprioceptive receptive fields seen in SMAp neurons. Whatever the interpretation of these data, the net effect is an increased amount of sensory information impinging on the cortical neurons. However, at this stage the information has lost its spatial specificity and the resulting blurring message could impair cortical processing during action planning, another putative mechanism for the pathophysiology of akinesia.

2.3 Other Cortical Regions

To date no study of single unit recordings has been performed in the lateral premotor and parietal cortices of MPTP-treated monkeys. Most imaging studies report overactivity in these regions in PD patients versus normal controls during the performance of sequential finger movements (Cunnington et al., 1996; Samuel et al., 1997; Catalan et al., 1999; Cunnington et al., 1999; Sabatini et al., 2000; Haslinger et al., 2001). The lateral premotor cortex is reported to be predominantly involved in the execution of externally triggered, sensory-cued movements (Deiber et al., 1991). It is responsible for movements under visual guidance and for the retrieval of abstract plans of action that are stored in the parietal cortex. Thus, overactivity in the lateral premotor and parietal cortices is generally interpreted as a switch from the use of striato-mesial frontal to parietal-lateral premotor circuits in order to facilitate performance of complex movements (Samuel et al., 1997; Haslinger et al., 2001). Our own data suggest that the parkinsonian state is also associated with an increase in the amount of proprioceptive inputs reaching the cortex (Escola et al., 2002). This could contribute to the overactivity of the parietal cortex in these subjects.

Data is also needed concerning the activity of prefrontal regions in MPTP-treated monkeys. Abnormal metabolic changes in the dorsolateral prefrontal cortex (DLPC) have been reported in PD patients (Cools et al., 2002; Mattay et al., 2002), a finding correlated with their deficits in attentional shifting and working memory tasks (Cooper et al., 1994; Robbins et al., 1994; Bennett et al., 1995). Information from the dorsolateral cortex might reach premotor areas through the pre-SMA, the latter receiving direct inputs from this region (Bates and Goldman-Rakic, 1993; Luppino et al., 1993; Lu et al., 1994). In this respect, the hypoactivity of the pre-SMA observed in MPTP-treated monkeys could be the result of modification in neuronal activity within the DLPC.

Another region which could play a critical role in the pathophysiology of akinesia is the anterior cingulate cortex (ACC). This cortical region is part of the brain's limbic system (Bush et al., 2000). It is subdivided into affect and cognitive components (Devinsky et al., 1995). The cognitive part which is probably more directly involved in motor planning receives strong inputs from the dorsolateral prefrontal cortex, the pre-SMA, the orbito-frontal cortex and sends output to the SMAp, M1 and the spinal cord (Devinsky et al., 1995). Neuroimaging studies indicate that the ACC is involved in cognition, behavioral planning and motor control. Moreover, lesions in this region induce attentional and emotional disturbances which in some cases might lead to akinetic mutism. It seems that because of its unique anatomical position, the ACC could play the role of an interface between prefrontal, premotor and subcortical system, linking cognitive and limbic information into motor inputs. In line with this, a specific role of the ACC in the translation from intentions to actions has been proposed (Paus, 2001). Since hypometabolic activity in PD patients is consistently reported in the cingulate cortex (Jahanshahi et al., 1995) this may alter the transformation of intention into action. If so, information would not be transferred to more executive areas such as the SMA proper and M1.

Other regions of the frontal lobe such as the orbito-frontal cortex could potentially play a part in the pathophysiology of PD. This region is known to be involved in the pathophysiology of mood and compulsive disorders (Aouizerate et al., 2004). These symptoms are frequently encountered in PD patients and can be alleviated by STN stimulation (Mallet et al., 2002).

3 What Role Does Cortical Dopamine Play ?

If asked the question as to what is the biological cause of akinesia, most specialists would answer that akinesia is the result of dopamine depletion, a situation observed in PD. However, this is not totally true since it can occur with a presumably normal level of dopamine when lesions involve cortical regions such as the mesial frontal cortex in the absence of dopamine receptor blockade (Haussermann et al., 2001). It can also be observed in particular diseases (depression, schizophrenia) in which no dopaminergic loss has been reported, although a functional impairment of dopaminergic receptors has been suggested. Moreover, the precise role of dopamine is far from understood. Experimental evidence obtained in rodents and primates points to its role in training and reward prediction (for review see Schultz, 1998). However, the most spectacular effect of dopamine depletion is a motor defect.

The role of the meso-cortical tract in the regulation of cortical activity remains largely unknown (see for review, Seamans and Yang, 2004). Indeed the modes of action of dopamine in the frontal cortex are not clearly understood although most researchers agree that DA is a neuromodulator. Its subtle effect depends on a variety of factors such as the dose response profile of post-synaptic receptors, prolonged duration effects, specific stimulation of D1 and D2 receptors and history dependence of subsequent DA action. There are very few studies reporting a direct effect of DA manipulation on cortical neuron activity in primates. D1 antagonist can selectively potentiate the « memory field » of prefrontal neurons (Williams and Goldman-Rakic, 1995) whereas D2 receptor modulates the neural activity associated with memory-guided saccades (Wang et al., 2004). The influence of DA on the cellular properties of layer 3 pyramidal cells in area 46 of the macaque has been investigated in a brain slice preparation (Henze et al., 2000). DA was found to increase the excitability of prefrontal cells in response to a depolarizing pulse. However, the evidence is fragmentary and data are needed concerning the effect of DA on premotor areas such as the supplementary motor area or the anterior cingulate cortex which receive dense dopaminergic projections (Berger et al., 1991).

From a phenomenological point of view, it is tempting to think that DA might have a similar action in the cortex as it has in the striatum i.e. favoring the activity of specific cortical neurons depending on the immediate context. However, there are two major differences. First, cortical neurons receive both intra-cortical and sub-cortical information with a particular architectonic organization different from that observed in the striatum. Second, the cortex, at least in the motor and premotor areas, is much more excitable than the striatal medium spiny cells. These two arguments suggest that the action of dopamine at the cortical level is probably different from that in the striatum. From a phylogenetic viewpoint, the striatum, which is a more archaic structure, could play a part in the holistic aspects of behavioral planning such as emotional evaluation of stimuli or context detection, whereas the cortex might be involved in the instrumental, parametric and executive aspects of action.

4 Return to the Future

In summary, single unit recordings performed at the cortical level in MPTP-treated monkeys are able to shed light on the disruption of information processing at the cortical level in PD. Even though data on the particular prefrontal regions is sparse, a possible schema of cortical dysfunctioning is emerging. This is to be interpreted in the light of the phenomenological aspects of PD.

Functional imaging studies demonstrate an overactivity of the primary motor cortex in PD patients, which is partially corrected by treatment. This overactivity seems to be associated with an increased synchronization which could impair the normal segregation between cortical columns controlling the agonist and antagonist muscles. This phenomenon could play a part both in the genesis of rigidity and in the ineffective selection of movements which characterizes bradykinesia. This synchronization could be due to the loss of the focusing action of basal ganglia output through the motor thalamus. However, it is likely that direct descending pathways through the pedonculo-pontine nucleus also play a role in the pathophysiology of rigidity (Berardelli et al., 2001). Moreover, the exact role of the dopaminergic meso-cortical tract in the control of information processing at the cortical level remains unknown. Another consequence of synchronization could be that the recruitment of motor cortex neurons at the proper time when the subject has to move is ineffective. This is in accordance with the observation that neuronal activity is delayed and decreased in amplitude in MPTP-treated versus normal monkeys (Doudet et al., 1990). This insufficient activation to trigger the target motoneurons could be compensated by the number of M1 cells recruited for a given movement. However, some uncertainty remains regarding this theory.

First, the motor cortex receives few direct projections from the basal ganglia, unlike the SMA. It is thus possible that the impairment of selective processing observed in the motor cortex is the indirect result of a dysfunctioning of the SMA. If we use the well-known analogy of the M1 as the keyboard of a piano, each note corresponding to a cortical column, then the SMA can be viewed as the pianist needed to play the kinetic melody of movement (temporal planning) or chords corresponding to the coordinated activity of motor cortex cells (spatial planning). The dramatic decrease in neuronal activity observed in the SMA of MPTP-treated monkeys is in accordance with this view. In this condition, motor cortex cells are no longer under control and are overactivated, albeit in an anarchical manner.

Second, there is also an impairment of the prefrontal areas (dorsolateral prefrontal cortex, anterior cingulate cortex, orbitofrontal cortex) in PD. These regions are involved in the cognitive and affective aspects of action planning. Any alteration of their activity is likely to account for some of the intellectual and emotional symptoms of PD patients and also some aspects of akinesia. Indeed, it is likely that the anterior cingulate cortex plays a critical role in the intention to move. Lesions in this region induce an akinetic-mutism syndrome which is also associated with bradypsychia and athymormy. On the other hand, some of the behavioral (obsessional compulsive disorders) or dysthymic disorders observed in PD patients could be related to changes in the activity of the anterior cingulate and orbitofrontal cortices.

Finally, our present knowledge of cortical processing based on single unit recordings is too narrow to be able to paint a realistic picture of cortical alterations. The use in the future of multiple single unit recordings coupled with more efficient imaging techniques will allow more precise investigation of the nature of cortical dysfunctioning in Parkinson's disease and other pathological conditions.

REFERENCES

Akkal D, Bioulac B, Audin J, Burbaud P (2002) Comparison of neuronal activity in the rostral supplementary and cingulate motor areas during a task with cognitive and motor demands. *Eur J Neurosci* **15**:887-904.

Aouizerate B, Cuny E, Martin-Guehl C, Guehl D, Amieva H, Benazzouz A, Fabrigoule C, Allard M, Rougier A, Bioulac B, Tignol J, Burbaud P (2004) Deep brain stimulation of the ventral caudate nucleus in the treatment of obsessive-compulsive disorder and major depression. Case report. *J Neurosurg* **101**:682-686.

Bates JF, Goldman-Rakic PS (1993) Prefrontal connections of medial motor areas in the rhesus monkey. *J Comp Neurol* **336**:211-228.

Bennett KM, Waterman C, Scarpa M, Castiello U (1995) Covert visuospatial attentional mechanisms in Parkinson's disease. *Brain* **118** (Pt 1):153-166.

Berardelli A, Rothwell JC, Thompson PD, Hallett M (2001) Pathophysiology of bradykinesia in Parkinson's disease. *Brain* **124**:2131-2146.

Berger B, Gaspar P, Verney C (1991) Dopaminergic innervation of the cerebral cortex: unexpected differences between rodents and primates. *Trends Neurosci* **14**:21-27.

Bezard E, Crossman AR, Gross CE, Brotchie JM (2001) Structures outside the basal ganglia may compensate for dopamine loss in the presymptomatic stages of Parkinson's disease. *Faseb J* **15**:1092-1094.

Boylan LS, Pullman SL, Lisanby SH, Spicknall KE, Sackeim HA (2001) Repetitive transcranial magnetic stimulation to SMA worsens complex movements in Parkinson's disease. *Clin Neurophysiol* **112**:259-264.

Brown P (2003) Oscillatory nature of human basal ganglia activity: relationship to the pathophysiology of Parkinson's disease. *Mov Disord* **18**:357-363.

Brown P, Marsden CD (1999) Bradykinesia and impairment of EEG desynchronization in Parkinson's disease. *Mov Disord* **14**:423-429.

Brown P, Oliviero A, Mazzone P, Insola A, Tonali P, Di Lazzaro V (2001) Dopamine dependency of oscillations between subthalamic nucleus and pallidum in Parkinson's disease. *J Neurosci* **21**:1033-1038.

Bush G, Luu P, Posner MI (2000) Cognitive and emotional influences in anterior cingulate cortex. *Trends Cogn Sci* **4**:215-222.

Catalan MJ, Ishii K, Honda M, Samii A, Hallett M (1999) A PET study of sequential finger movements of varying length in patients with Parkinson's disease. *Brain* **122** (Pt 3):483-495.

Ceballos-Baumann AO, Obeso JA, Vitek JL, Delong MR, Bakay R, Linazasoro G, Brooks DJ (1994) Restoration of thalamocortical activity after posteroventral pallidotomy in Parkinson's disease. *Lancet* **344**:814.

Ceballos-Baumann AO, Boecker H, Bartenstein P, von Falkenhayn I, Riescher H, Conrad B, Moringlane JR, Alesch F (1999) A positron emission tomographic study of subthalamic nucleus stimulation in Parkinson disease: enhanced movement-related activity of motor-association cortex and decreased motor cortex resting activity. *Arch Neurol* **56**:997-1003.

Cools R, Stefanova E, Barker RA, Robbins TW, Owen AM (2002) Dopaminergic modulation of high-level cognition in Parkinson's disease: the role of the prefrontal cortex revealed by PET. *Brain* **125**:584-594.

Cooper JA, Sagar HJ, Tidswell P, Jordan N (1994) Slowed central processing in simple and go/no-go reaction time tasks in Parkinson's disease. *Brain* **117** (Pt 3):517-529.

Cunnington R, Iansek R, Bradshaw JL (1999) Movement-related potentials in Parkinson's disease: external cues and attentional strategies. *Mov Disord* **14**:63-68.

Cunnington R, Iansek R, Thickbroom GW, Laing BA, Mastaglia FL, Bradshaw JL, Phillips JG (1996) Effects of magnetic stimulation over supplementary motor area on movement in Parkinson's disease. *Brain* **119** (Pt 3):815-822.

Dauper J, Peschel T, Schrader C, Kohlmetz C, Joppich G, Nager W, Dengler R, Rollnik JD (2002) Effects of subthalamic nucleus (STN) stimulation on motor cortex excitability. *Neurology* **59**:700-706.

Deiber MP, Passingham RE, Colebatch JG, Friston KJ, Nixon PD, Frackowiak RS (1991) Cortical areas and the selection of movement: a study with positron emission tomography. *Exp Brain Res* **84**:393-402.

Devinsky O, Morrell MJ, Vogt BA (1995) Contributions of anterior cingulate cortex to behaviour. *Brain* **118** (Pt 1):279-306.

Doudet DJ, Gross C, Arluison M, Bioulac B (1990) Modifications of precentral cortex discharge and EMG activity in monkeys with MPTP-induced lesions of DA nigral neurons. *Exp Brain Res* **80**:177-188.

Escola L, Michelet T, Douillard G, Guehl D, Bioulac B, Burbaud P (2002) Disruption of the proprioceptive mapping in the medial wall of parkinsonian monkeys. *Ann Neurol* **52**:581-587.

Escola L, Michelet T, Macia F, Guehl D, Bioulac B, Burbaud P (2003) Disruption of information processing in the supplementary motor area of the MPTP-treated monkey: a clue to the pathophysiology of akinesia? *Brain* **126**:95-114.

Gerloff C, Corwell B, Chen R, Hallett M, Cohen LG (1997) Stimulation over the human supplementary motor area interferes with the organization of future elements in complex motor sequences. *Brain* **120** (Pt 9):1587-1602.

Goldberg JA, Boraud T, Maraton S, Haber SN, Vaadia E, Bergman H (2002) Enhanced synchrony among primary motor cortex neurons in the 1-methyl-4-phenyl-1,2,3,6-tetrahydropyridine primate model of Parkinson's disease. *J Neurosci* **22**:4639-4653.

Grafton ST, Waters C, Sutton J, Lew MF, Couldwell W (1995) Pallidotomy increases activity of motor association cortex in Parkinson's disease: a positron emission tomographic study. *Ann Neurol* **37**:776-783.

Haslinger B, Erhard P, Kampfe N, Boecker H, Rummeny E, Schwaiger M, Conrad B, Ceballos-Baumann AO (2001) Event-related functional magnetic resonance imaging in Parkinson's disease before and after levodopa. *Brain* **124**:558-570.

Haussermann P, Wilhelm T, Keinath S, Stolzle C, Conrad B, Ceballos-Baumann A (2001) Primary central nervous system lymphoma in the SMA presenting as rapidly progressive parkinsonism. *Mov Disord* **16**:962-965.

Henze DA, Gonzalez-Burgos GR, Urban NN, Lewis DA, Barrionuevo G (2000) Dopamine increases excitability of pyramidal neurons in primate prefrontal cortex. *J Neurophysiol* **84**:2799-2809.

Jahanshahi M, Jenkins IH, Brown RG, Marsden CD, Passingham RE, Brooks DJ (1995) Self-initiated versus externally triggered movements. I. An investigation using measurement of regional cerebral blood flow with PET and movement-related potentials in normal and Parkinson's disease subjects. *Brain* **118** (Pt 4):913-933.

Jenkins IH, Fernandez W, Playford ED, Lees AJ, Frackowiak RS, Passingham RE, Brooks DJ (1992) Impaired activation of the supplementary motor area in Parkinson's disease is reversed when akinesia is treated with apomorphine. *Ann Neurol* **32**:749-757.

Limousin P, Greene J, Pollak P, Rothwell J, Benabid AL, Frackowiak R (1997) Changes in cerebral activity pattern due to subthalamic nucleus or internal pallidum stimulation in Parkinson's disease. *Ann Neurol* **42**:283-291.

Lu MT, Preston JB, Strick PL (1994) Interconnections between the prefrontal cortex and the premotor areas in the frontal lobe. *J Comp Neurol* **341**:375-392.

Luppino G, Matelli M, Camarda R, Rizzolatti G (1993) Corticocortical connections of area F3 (SMA-proper) and area F6 (pre-SMA) in the macaque monkey. *J Comp Neurol* **338**:114-140.

Mallet L, Mesnage V, Houeto JL, Pelissolo A, Yelnik J, Behar C, Gargiulo M, Welter ML, Bonnet AM, Pillon B, Cornu P, Dormont D, Pidoux B, Allilaire JF, Agid Y (2002) Compulsions, Parkinson's disease, and stimulation. *Lancet* **360**:1302-1304.

Marsden CD (1982) The mysterious motor function of the basal ganglia: the Robert Wartenberg Lecture. *Neurology* **32**:514-539.

Marsden JF, Limousin-Dowsey P, Ashby P, Pollak P, Brown P (2001) Subthalamic nucleus, sensorimotor cortex and muscle interrelationships in Parkinson's disease. *Brain* **124**:378-388.

Matsuzaka Y, Tanji J (1996) Changing directions of forthcoming arm movements: neuronal activity in the presupplementary and supplementary motor area of monkey cerebral cortex. *J Neurophysiol* **76**:2327-2342.

Mattay VS, Tessitore A, Callicott JH, Bertolino A, Goldberg TE, Chase TN, Hyde TM, Weinberger DR (2002) Dopaminergic modulation of cortical function in patients with Parkinson's disease. *Ann Neurol* **51**:156-164.

Nakamura K, Sakai K, Hikosaka O (1998) Neuronal activity in medial frontal cortex during learning of sequential procedures. *J Neurophysiol* **80**:2671-2687.

Passingham RE (1996) Attention to action. *Philos Trans R Soc Lond B Biol Sci* **351**:1473-1479.

Paus T (2001) Primate anterior cingulate cortex: where motor control, drive and cognition interface. *Nat Rev Neurosci* **2**:417-424.

Payoux P, Remy P, Damier P, Miloudi M, Loubinoux I, Pidoux B, Gaura V, Rascol O, Samson Y, Agid Y (2004) Subthalamic nucleus stimulation reduces abnormal motor cortical overactivity in Parkinson disease. *Arch Neurol* **61**:1307-1313.

Picard N, Strick PL (1996) Motor areas of the medial wall: a review of their location and functional activation. *Cereb Cortex* **6**:342-353.

Playford ED, Jenkins IH, Passingham RE, Frackowiak RS, Brooks DJ (1993) Impaired activation of frontal areas during movement in Parkinson's disease: a PET study. Adv *Neurol* **60**:506-510.

Rascol O, Sabatini U, Chollet F, Celsis P, Montastruc JL, Marc-Vergnes JP, Rascol A (1992) Supplementary and primary sensory motor area activity in Parkinson's disease. Regional cerebral blood flow changes during finger movements and effects of apomorphine. *Arch Neurol* **49**:144-148.

Rascol O, Sabatini U, Chollet F, Fabre N, Senard JM, Montastruc JL, Celsis P, Marc-Vergnes JP, Rascol A (1994) Normal activation of the supplementary motor area in patients with Parkinson's disease undergoing long-term treatment with levodopa. *J Neurol Neurosurg Psychiatry* **57**:567-571.

Ridding MC, Inzelberg R, Rothwell JC (1995) Changes in excitability of motor cortical circuitry in patients with Parkinson's disease. *Ann Neurol* **37**:181-188.

Robbins TW, James M, Owen AM, Lange KW, Lees AJ, Leigh PN, Marsden CD, Quinn NP, Summers BA (1994) Cognitive deficits in progressive supranuclear palsy, Parkinson's disease, and multiple system atrophy in tests sensitive to frontal lobe dysfunction. *J Neurol Neurosurg Psychiatry* **57**:79-88.

Sabatini U, Boulanouar K, Fabre N, Martin F, Carel C, Colonnese C, Bozzao L, Berry I, Montastruc JL, Chollet F, Rascol O (2000) Cortical motor reorganization in akinetic patients with Parkinson's disease: a functional MRI study. *Brain* **123** (Pt 2):394-403.

Samuel M, Ceballos-Baumann AO, Blin J, Uema T, Boecker H, Passingham RE, Brooks DJ (1997) Evidence for lateral premotor and parietal overactivity in Parkinson's disease during sequential and bimanual movements. A PET study. *Brain* **120** (Pt 6):963-976.

Schultz W (1998) Predictive reward signal of dopamine neurons. *J Neurophysiol* **80**:1-27.

Seamans JK, Yang CR (2004) The principal features and mechanisms of dopamine modulation in the prefrontal cortex. *Prog Neurobiol* **74**:1-58.

Shima K, Tanji J (2000) Neuronal activity in the supplementary and presupplementary motor areas for temporal organization of multiple movements. *J Neurophysiol* **84**:2148-2160.

Shima K, Mushiake H, Saito N, Tanji J (1996) Role for cells in the presupplementary motor area in updating motor plans. *Proc Natl Acad Sci U S A* **93**:8694-8698.

Strafella AP, Valzania F, Nassetti SA, Tropeani A, Bisulli A, Santangelo M, Tassinari CA (2000) Effects of chronic levodopa and pergolide treatment on cortical excitability in patients with Parkinson's disease: a transcranial magnetic stimulation study. *Clin Neurophysiol* **111**:1198-1202.

Tanji J (1994) The supplementary motor area in the cerebral cortex. *Neurosci Res* **19**:251-268.

Wang M, Vijayraghavan S, Goldman-Rakic PS (2004) Selective D2 receptor actions on the functional circuitry of working memory. *Science* **303**:853-856.

Williams GV, Goldman-Rakic PS (1995) Modulation of memory fields by dopamine D1 receptors in prefrontal cortex. *Nature* **376**:572-575.

In: Recent Breakthroughs in Basal Ganglia Research
Editor: Erwan Bezard, pp. 249-266
ISBN: 1-59454-880-3

Chapter 20

THE DYNAMICS OF TREMOR NETWORKS IN PARKINSON'S DISEASE

José M. Hurtado, Leonid L. Rubchinsky
Center for Neuroscience, University of California, Davis, CA Department of Mathematical Sciences, Indian University School of Medicine, Indianapolis
Karen A. Sigvardt*
Center for Neuroscience, University of California, Davis, CA Department of Neurology, University of California, Davis, CA Department of Mathematical Sciences, Indian University School of Medicine, Indianapolis

ABSTRACT

Resting tremor is commonly considered the landmark feature of Parkinson's disease, yet the mechanisms underlying tremor generation remain unclear and are not encompassed within current theories regarding mechanisms giving rise to the other motor cardinal features of PD. We, as well as others, have hypothesized that the motor symptoms of PD result from specific changes in the population dynamics of the neural activity within the thalamocortical-basal ganglia pathway. In this chapter we summarize current knowledge of the spatiotemporal properties of peripheral tremor and tremor-related activity within the CNS. We then speculate on how the presence of oscillations and correlations across the networks involved in the pathophysiology of tremor might be related to the mechanisms that give rise to other parkinsonian motor symptoms, as put forward in current theories of basal ganglia function and its dysfunction in Parkinson's disease.

INTRODUCTION

In his treatise "An essay on the shaking palsy", James Parkinson (1817) recognized resting tremor as the landmark feature of the newly described disorder (Parkinson, 1817). The most important advance since then has been the discovery of its underlying neuropathology, the degeneration of SNc dopaminergic cells and the resulting loss of striatal dopamine (Hornykiewicz, 1966) This crucial finding not only gave rise to the pharmacological treatment strategy for the disease but also provided a foundation for theories of basal ganglia function in motor control. Most standard models of BG disorders (Albin et al., 1989; DeLong, 1990), however, have focused on the other symptoms of the disease –bradykinesia, akinesia and rigidity- while tremor remains largely unexplained.

The simplest hypothesis would be that tremor originates from a single pace-making nucleus, and propagates through a polysynaptic network from there towards the peripheral musculature. Indeed a similar

* Correspondence should be addressed to: Karen A. Sigvardt, Ph.D. Center for Neuroscience, University of California Davis, 1544 Newton Court, Davis, CA 95616. Phone: (530) 757-8520. Fax: (530) 757-8827. E-mail address: kasigvardt@ucdavis.edu

mechanism is known to operate in focal seizures, and in some forms of tremor such as harmaline-induced tremor, where oscillatory activity is initiated in the inferior olive (Lamarre and Mercier, 1971; De Montigny and Lamarre, 1973). This proposal has been difficult to test in the case of PD tremor. Not only has it been found that tremor-related activity encompasses almost the entire motor control network, including BG, cerebellar and spinal networks (Deuschl et al., 2000), but also that tremor can be abolished, or at least ameliorated, by lesions, or by microstimulation, in a number of different sites: STN (Rodriguez et al., 1998); GPi (Taha et al., 1997), and motor thalamus (Ohye et al., 1976; Tasker, 1990). On the other hand, some of the structures that are entrained with tremor-related oscillations do not seem to be required for the genesis of the rhythm, though they do have an effect on the dynamical aspects of tremor such as its frequency and phase (Stiles and Pozos, 1976; Rack and Ross, 1986 for peripheral afferents; Deuschl et al., 1999; for cerebellum).

We are therefore still not in a solid position to resolve to what degree PD tremor may result from pacemaking activity in a single neural structure versus the interaction between structures. However, some clues as to the mechanism of tremor can be obtained by taking into consideration several well described aspects of this phenomenon:

1. The basic architecture of the tremor generating network can be inferred from what is known from neuroanatomical studies. The neuroanatomy suggests that the structures that show tremor-related activity are connected in three basic loops (see Figure 1A) : a cortico-subthalamo-pallido thalamic loop (also called the "hyperdirect" pathway; Nambu et al., 2002), a cortico-pontocerebellar-thalamic loop and a cortico-spino-thalamic loop.
2. The optimal target for surgical treatment of tremor is in the Vim thalamus, a region that presumably receives cerebellar but not BG afferents, although ascending spinal afferents may terminate there as well (Macchi and Jones, 1997; Ilinsky and Kultas-Ilinsky, 2002). Some argue that a lesion in this region might affect pallido-thalamic fibers of passage (Pare et al., 1990). Interestingly, this lesion is effective for the relief of most types of tremor, regardless of origin.
3. PD resting tremor results from basal ganglia dopamine (DA) loss, but its severity does not correlate well with the extent of DA loss (reviewed in Deuschl et al., 2000).
4. There is a paucity of tremor-related data from striatal projection neurons. Oscillatory activity does appear in the TANs following MPTP intoxication (Raz et al., 1996) but as these are interneurons they cannot contribute to propagation of tremor to other structures. In PD as there is decreased metabolic activity in the striatum (Brooks et al., 1992) but as these are interneurons they cannot contribute to propagation of tremor to other structures.
5. Resting tremor decreases on movement onset in a majority of PD patients.
6. Patient specificity: tremor occurs in a subset (70 %) of the PD cases. The most plausible explanation is that this distribution results from a differential pattern of DA degeneration between patients. This is supported by postmortem studies suggesting that the akinetic/rigid subtype correlates better with cell loss in the lateral SNc (A9) whereas the tremulous variant results predominantly from loss in the medial region (A8; Jellinger, 1999; reviewed in Deuschl et al., 2000). In human cases of acute MPTP intoxication, where damage is likely present in both regions, tremor is prevalent (Ballard et al., 1985; Tetrud and Langston, 1992).
7. Species specificity. The MPTP application has been shown to produce parkinsonian-like tremor only in vervet monkeys (*Cercopithecus aethiops;* Bergman et al., 1998). Rhesus monkeys (*Macacca mulatta*) develop sporadic tremor of a higher frequency than that of human parkinsonism.
8. Tremor is more common in distal than in proximal musculature. An important difference between primate species is the amount of cortex devoted to distal versus proximal musculature, and the appearance of independent finger movements concomitant with an increase in the size of the cortico-motoneuronal pathway. Indeed it is the distal musculature of the fingers that is more susceptible to tremor.

Resting tremor occurs in about 70 % of PD cases, but recent theories regarding the relationship between pathophysiology and the symptoms of diseases associated with basal ganglia dysfunction have focused mostly on the hypokinetic symptoms described by James Parkinson, namely bradykinesia and akinesia and their reversal by L-Dopa therapy (Albin et al., 1989; DeLong, 1990). The reason is probably that, unlike tremor, hypokinetic symptoms can be easily understood in terms of overall increases or decreases in neuronal activity in the different structures of the BG network. Models formulated in these terms are appealing for their simplicity and provide predictions that are easily testable. However they tend to favor a steady state view of the basal ganglia, a view that has been challenged from several fronts in recent years (Obeso et al., 2000; Bar-Gad and Bergman, 2001). The study of tremor mechanisms, in contrast, forces a dynamical formulation of BG mechanisms at the outset and, therefore, provides an opportunity to complement and enrich former formulations.

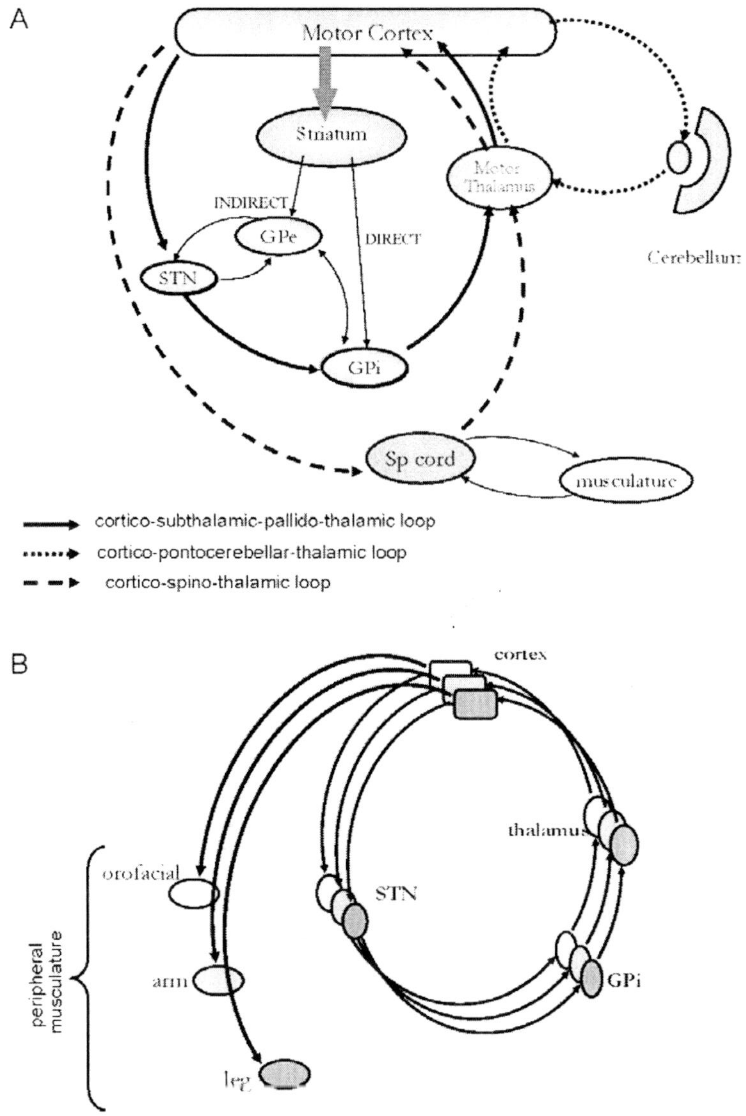

Figure 1A: The three pathways that participate in PD tremorgenesis. The cortico-subthalamo-pallido-thalamic loop (wide solid lines) is comprised of the motor cortex, STN, GPi and motor thalamus. The activity in this loop, in particular its gain, is controlled by the striatum via the direct and indirect pathways. It is proposed that striatal control of this loop is lost in PD, giving rise to tremor and other symptoms. The cortico-spinal-thalamic loop (dashed arrows) is comprised of the motor cortex, spinal cord, musculature, sensory afferents and spinal ascending systems. The cerebellar loop (dotted lines) includes cortico-pontine projections, the cerebellar nuclei, inferior olive (not shown), cerebellar cortex, motor thalamus and the cortex. The three loops are segregated at the thalamic level but can interact in the cortex. Figure 1B. Diagram illustrating the topographic organization of tremor and the principle of parallel tremor-generating oscillators. The oscillators span different structures but preserve their limb specificity, as all of the central structures shown in the diagram have segregated limb representations. This organization is supported by the patterns of coherence within and between neurons representing limb specific regions in each structure as well as the relationship between these neurons and peripheral tremor. Lateral interactions between the nuclei might be responsible for between-limb transient phase locking.

One of the current prevailing theories regarding the role of the basal ganglia in the control of movement is the "hypothesis of action selection" (Mink, 1996). This theory is based on the idea that at any point in time, several competing motor programs coexist in the CNS, as patterns of activity in the cortico-thalamic system. According to this formulation, the function of the BG is to provide a signal to select among them, and this is done by the combination of an inhibitory mechanism that suppresses those programs that interfere with the voluntary movement of highest reward value complemented by a *disinhibitory* mechanism to remove suppression of the desired movement (see also Hikosaka and Wurtz, 1985a, 1985b; Redgrave et al., 1999). This hypothesis of focused selection contains some of the basic elements of the standard model proposed by Albin and others (Albin et al., 1989; Alexander and Crutcher, 1990; Alexander et al., 1990; DeLong, 1990) to explain movement disorders of BG origin. According to that view the direct (net excitatory) and indirect (net inhibitory) circuits (see Figure 1A) are not in balance in PD; this formulation predicted that the observed disorders of movement associated with Parkinson's disease result from insufficient response activation/inhibition. However, unlike the original standard model, the hypothesis of action selection does not depend on the steady state direct/indirect pathway formulation. Indeed, alternative or complementary reformulations of this view are possible given more recent anatomical and physiological data. Anatomical data has made clear that pathways other than the "indirect" pathway can play a role in the suppression of unwanted movements. One candidate that has received wide attention in recent years is the corticosubthalamic or "hyperdirect" pathway (Nambu et al., 2002). Second, the focused selection model can easily accommodate a dynamic formulation, in which selection and suppression are not merely in balance but evolve over the course of the behavioral sequence.

While the theory of action selection has provided an explanation for the changes in firing rates within the network, in the current formulation this theory does not take into account the dynamical aspects of PD, in particular resting tremor. We, as well as others (Bergman et al., 1998; Hurtado et al., 1999; Bevan et al., 2002; Farmer, 2002), have hypothesized that the motor symptoms of PD and other diseases of BG origin, are the result of specific changes in population dynamics of the neural activity within the cortical-basal ganglia pathway. This is in contrast to the Albin-DeLong model that implicitly considers the control of on-going behavior by the BG network as a series of steady states. By population dynamics we mean that the changes are expressed not only in the firing patterns of individual cells but in the way the populations of cells interact. This idea is supported by the observation of a high incidence of oscillatory units firing at the tremor frequency in some key BG structures of PD patients with tremor (see Ohye et al., 1976; Lenz et al., 1994; Hutchison et al., 1997; Hurtado et al., 1999; Levy et al., 2000; Brown et al., 2001b; Levy et al., 2002), correlation of oscillatory neuronal activity across relatively distant (3 mm) sites within GPi (Hurtado et al., 1999), and the dynamics of the relationships (coherence and phase) between pairs of units as well as single units and muscle tremor (Hurtado et al., 1999; 2004; 2005).

For obvious reasons no microelectrode recordings from healthy human subjects are available for comparison with PD patients, but this has been compensated for by the development of the MPTP animal model of parkinsonism. It has been shown that the incidence of oscillatory neurons and the incidence of pairs with correlated activity increases in non-human primates following MPTP treatment (Filion and Tremblay, 1991; Bergman et al., 1994; Nini et al., 1995; Raz et al., 1996; Bergman et al., 1998; Raz et al., 2000). The abnormal synchronization in the pallidum following MPTP treatment can be reversed by levodopa/carbidopa treatment following a time course similar to that of the relief of PD symptoms (Heimer et al., 2002). Furthermore, in MPTP-treated monkeys with severe akinesia, neurons in M1 fire in long synchronous bursts (Goldberg et al., 2002), never observed in normal monkeys, providing evidence that akinesia may result from changes in coherence within the motor network. The dependence of coherence between different parts of the BG on dopaminergic activity has been demonstrated in humans (Brown et al., 2001b). Coherence of tremor between muscles also varies; it differs for different muscle pairs, patients and cognitive states (Hurtado et al., 2000; Raethjen et al., 2000).

In this chapter we will attempt to consolidate the idea of spatiotemporal dynamics with that of action selection. Might there be a way to link the various views of the basal ganglia – to relate the presence of oscillations and correlation in the pathophysiology of tremor with the view represented by the theory of action selection?

1 The Components of the Tremor Generating Network

Our current view of the brain regions that make up the tremor-generating network has been shaped by observations from single unit neurophysiology, functional anatomy and clinical assessment of the outcome of functional neurosurgery; invaluable additional information has been provided by anatomical tracing studies in non-human primates. It is now apparent that most structures in the limb motor network are entrained into tremor-related oscillations. Microelectrode and LFP recordings in patients and in primate models of PD tremor have shown tremor-related activity in the GPi (Taha et al., 1996; Hutchison et al., 1997; Hurtado et al., 1999), the motor thalamus (Guiot et al., 1962; Lenz et al., 1988; Lenz et al., 1994) and motor cortex (Alberts et al., 1969; Volkmann et al., 1996; Timmermann et al., 2003; Pollok et al., 2004). It has also shown that oscillatory activity in relation to tremor is present in the ipsilateral cerebellar cortex (Parker et al., 1992; Volkmann et al., 1996; Williams et al., 2002; Timmermann et al., 2003), and contralateral premotor, and somatosensory regions (Volkmann et al., 1996). A single case report of a patient with resection of the cerebellum on one side prior to PD onset suggests that PD tremor is not of cerebellar origin (Deuschl et al., 1999).

Another component of the tremor network is, of course, the corticospinal and spino-thalamo-cortical pathways. The involvement of the latter pathway has been debated since the observation by Pollock and Davis, (1930) that PD tremor persisted after dorsal rhizotomy. However, the evidence now clearly indicates that afferent activity can affect tremor (Stiles and Pozos, 1976; Rack and Ross, 1986).

Finally, another structure that has received wide attention in recent years is the pedunculopontine nucleus (PPN; Pahapill and Lozano, 2000; Mena-Segovia et al., 2004). The possibility that this nucleus forms part of the tremor network cannot be discarded, as it is closely linked to structures that participate in the tremor network such as the STN, GPi, Spinal cord and deep cerebellar nuclei. To our knowledge, no recordings have been performed there in relation to tremor; however there are reasons to believe that the PPN does not play a significant role in PD tremor generation. As a motor structure, the PPN is mostly involved in the initiation of locomotion and does not seem to play a significant role in the control of the peripheral musculature, that is the most affected in PD tremor (Stein and Aziz, 1999). In PD, PPN dysfunction is thought to be involved in such symptoms as gait disturbance and akinesia (Pahapill and Lozano, 2000).

The data on tremor-related activity combined with anatomical findings allow us to group these structures into three loops (see Figure 1A): 1) a BG loop consisting of the cortico-STN-GPi-thalamo-cortical loop which has been called the 'hyperdirect pathway' (Nambu et al., 2002) ; 2) a cerebellar loop, the cortico-cerebello-thalamo cortical loop, involving cerebellar nuclei (presumably dentate nucleus)-Vlpo (cerebellar) thalamus; and 3) the cortico-spino-thalamo-cortical loop. An important aspect of this network is that there is a topographic organization in each of the structures, that reflects on the dynamics of tremor (reviewed in Deuschl et al., 2000). Anatomical tracing studies suggest that these loops can interact in the primary motor cortex; it has been shown in macaques that pallidal and cerebellar thalamic regions provide convergent input to the hand area of primary motor cortex (Holsapple et al., 1991; Rouiller et al., 1994).

Importantly, the striatum does not seem to be part of the tremor generating network. There is a paucity of neurophysiological data showing tremor-related activity there. A few reports have shown oscillatory activity in tonically active neurons (TANs) in relation to MPTP intoxication, yet this activity is not likely to find a way out of the striatum since these are interneurons (Raz et al., 1996; Raz et al., 2001). To our knowledge, no tremor-related activity has been described in striatal output neurons so far. A distinction must, therefore, be made here between those structures that display tremor-related activity (i.e. oscillatory activity at the tremor frequency) and those that do not present that activity, but most likely participate indirectly as "controllers" of the activity of the others.

Tremor-related activity in the motor system, with the notable exception of the striatum, suggests that the latter structure, or parts of it, might exert a suppressive effect on oscillatory activity that arises spontaneously within the BG-thalamo cortical or the cerebello-thalamo-cortical network. The suppressive effect is disrupted in PD, but on voluntary movement initiation, residual striatal function might be sufficient to disable, at least temporarily, the tremor network. A lack of striatal "control" over the tremor network may be the link between the DA deficit and the appearance of tremor.

2 Dynamics of Oscillatory Activity in the Tremor-Generating Network

Casual observations of parkinsonian patients in the clinic clearly show that tremor in any one muscle can come and go over time. Furthermore, while recording from the basal ganglia during intraoperative mapping in functional neurosurgery procedure, we and others have observed that tremor-related activity in GPi would wax and wane, even though limb tremor continued, and "tremor-related" activity could exist even when then was no observable tremor. This led us to ask questions about the temporal relationships between the units in the pallidum and the limb musculature and between tremor-related units within the basal ganglia. Furthermore, the previously reported general topographic organization of the limb representations within the basal ganglia (DeLong et al., 1985) led us to question how neurons with tremor-related activity within a limb representation might be correlated in comparison to the correlations between the fields that represent different limbs.

Even though tremor-related activity occurs at several sites in the brain, there have been some reported observations suggesting that the tremor-generating network does not produce tremor-related activity that was correlated across the entire network. Early on Alberts and colleagues (Alberts et al., 1969) noted that despite the presence of tremor-related activity in the EEG, muscle tremor was not always correlated with the EEG. Several studies of multilimb EMG recordings in PD patients also indicate that tremor in different limbs is largely uncorrelated (O'Suilleabhain and Matsumoto, 1998; Hurtado et al., 2000; Raethjen et al., 2000). In addition, dual recordings in GPi tremor-related cells during stereotactic surgery have shown that although cells might be correlated to restricted portions of the musculature or to each other, uncorrelated oscillations within GPi are commonplace as well (Hurtado et al., 1999; Lemstra et al., 1999). All of these findings are consistent with the view that there is a general, though not precise, topographic organization of the individual structures that comprise the tremor generating network.

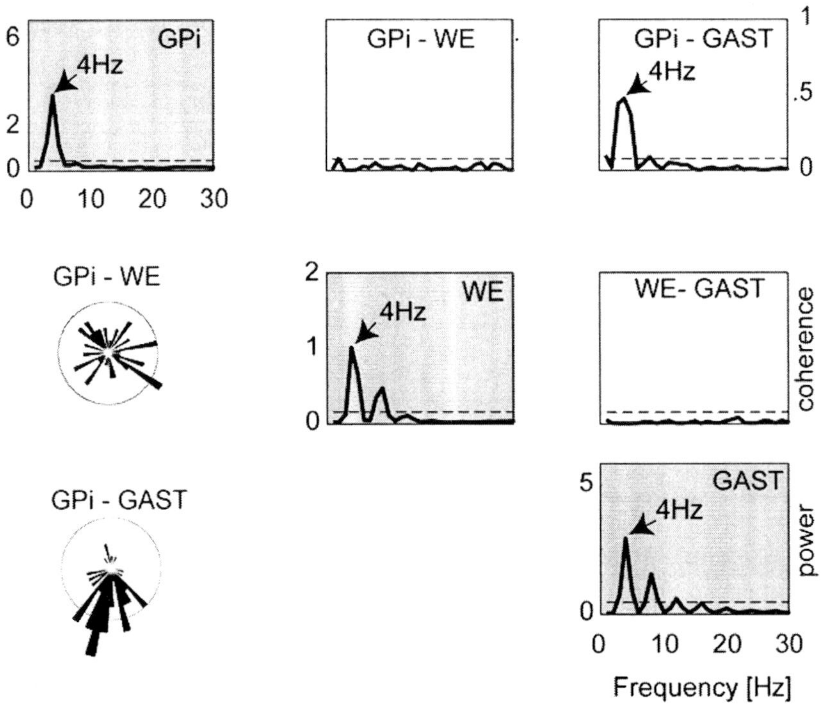

Figure 2: Tremor/GPi correlations follow a topographic order. Power spectra of GPi spike train and EMG activity of muscles from the upper and lower extremities in a 60 sec tremor episode are shown in the gray panels on the diagonal. White panels are coherence spectra for all pairs, as labeled; dashed lines, 99 % significance level. Angular histograms of the phase differences are shown for the two GPi/muscle pairs at the frequency of significant coherence. The peak at 4 Hz in the coherence spectra between the unit and the leg muscle indicate that the unit and this muscle are correlated within the tremor band; the oscillatory activity at 4Hz is not correlated with the arm muscle EMG. Color bars at right of each plot indicate the values of coherence or phase lag.

Our analysis of tremor and tremor-related neural activity suggest that although tremor activity is locally synchronized in the peripheral musculature and in central structures, the synchronization is not global as it is the case of other neural oscillations like slow wave sleep, generalized epileptic states or harmaline-induced tremor. Tremor in the peripheral musculature and tremor-related neural activity is in general not synchronized *between* different limbs or limb representations of central motor structures (see Figure 2). We can summarize this by stating that PD tremor has a *topographic organization*. The tremor-related activity propagates through a central network via parallel pathways organized in a roughly somatotopic manner (Figure 1B). This explains why sometimes we found independent oscillations; oscillating cells were in different limb representations. But because of the intermingled representations of the joints within each limb, the pathways spanning a given limb could support intra-limb interactions by divergent/convergent projections between structures or by horizontal interconnection within a structure.

3 Methods for the Evaluation of the Temporal Evolution of Oscillatory Activity and Coherence

The dynamic properties of parkinsonian tremor and the tremor-generating network are often lost in studies that examine tremor-related activity averaged over long periods of time. If tremor comes and goes over time, it is a non-stationary phenomenon and thus violates the stationarity assumption implicit in standard spectral analysis. If non-stationary signals are analyzed over long periods of time, even if they contain short segments of phase locking, the correlation could be below statistical significance (Figures 2, 3). A better assessment of the interactions would be obtained by analyzing these signals over short periods of time.

The goal of the approach we have developed (described in detail in Hurtado et al., 2004) was to find a sound statistical method for examining the temporal evolution of oscillations and phase-locking between pairs of signals. Most methods for detection of synchrony take into account both the amplitude and phase of the oscillatory signals. However, these methods can miss correlations that include just the correlations in phase. Recent modeling studies in the field of nonlinear dynamics have shown that phase synchronization of coupled nonlinear oscillators may occur even in the absence of amplitude correlations, particularly in cases where interactions are weak (reviewed in Pikovsky et al., 2001). Therefore, it has been suggested that for detection of synchrony between oscillatory neural signals *timing* correlations are better evaluated in the *phase* of the oscillators (Tass et al., 1998; Lachaux et al., 1999; Feige et al., 2000). Analyses of synchronization based on the crosscorrelogram or the coherence spectrum detect synchronization if both phase and amplitude correlations are present and can, therefore, miss the type of event where correlations exist in the timing alone, while the amplitudes vary independently. Furthermore, covariations in the amplitude envelope of two signals can lead to artifacts in the assessment of timing synchronization (Brody, 1998, 1999). We therefore used the phase of the oscillation to evaluate correlations in time (reviewed in Pikovsky et al., 2001). The first step in the analysis is to extract the oscillatory activity at the frequency of interest and calculate a phase variable from the oscillatory signals. Phase coherence is then used to assess phase locking between the two signals. In experimental data, periodic activity is often interrupted by episodes of aperiodic activity. Since detection of phase locking requires a reliable phase construction, we also had to determine whether the signal was oscillatory before proceeding with the analysis of phase correlations. We developed a criterion to discriminate between oscillatory and aperiodic segments, so we could pull out from the data segments of well behaved oscillations and perform the phase locking analysis only on the segments of activity that were oscillatory by this criterion (for details, see Hurtado et al., 2004; 2005). We also devised methods for removing phase slips (<1 cycle disruptions of an oscillation, flanked by several oscillatory periods).

For any study of oscillatory activity it is important to know the distribution of the coherence values for a pair of oscillatory signals that are independent (uncorrelated). Only values that are on the high end of the distribution (e.g. the 95th or 99th percentile) from what would be expected for independent oscillators can be considered to be phase locked. As this distribution is unknown for our data set, one approach is to generate surrogate time series that share some statistical features with the original data (Palus and Hoyer, 1998; Schreiber and Schmitz, 2000). Methods for generation of surrogate data differ in the specific features of the original data that are kept fixed across the surrogate ensemble, e.g., mean and standard deviation, or power spectrum, but the more features of the data preserved, the more strict the test. Standard surrogate methods for spike trains generate random spikes from a Poisson process with the same mean firing rate as the data, and for EMG and LFP, from a Gaussian distribution with the same mean and standard deviation. Neither method preserves the temporal features of the data. For narrow-band data it is important to preserve the temporal structure of the data because narrow-band oscillations with similar frequencies may have false positives in coherence at short observation times. Since the coherence values that we were evaluating were computed from a phase series, we generated instantaneous frequency surrogates and calculated surrogate phase series from them. To preserve some of the temporal structure in the instantaneous frequency series, we created surrogates with the same power spectrum as the instantaneous frequency series of the original data. We have compared our approach to standard methods for obtaining surrogate data by applying the methods to a pair of independent oscillators (obtained from EMGs from two different PD patients with tremor); the Gaussian method produced significant periods of phase locking (i.e., false positives) but our methods did not. Thus this method can reveal correlations over time at high temporal resolution in a statistically sound manner. Importantly, this method does not require the activity to be stationary over time and has the advantage that it resolves significant interaction on a single trial basis. An example of a data set analyzed using both standard spectral methods and the approach outlined here are presented in Figures 2 and 3.

4 The Temporal Evolution of Oscillatory Activity and Coherence in Tremor-Generating Networks

Numerous studies provide support for a shift to an oscillatory state in PD with a concomitant increase in synchrony across the basal ganglia-thalamocortical network. In Parkinson's disease tremor-related units within an area representing a single limb are coherent with peripheral tremor over long time scales (~60 sec). The general conclusions drawn from these studies have been that tremor-related activity is strongly oscillatory and correlated to the muscle tremor in a specific limb, and that there is no correlation *between* regions specific to particular limbs (Figure 2). Studies of the dynamic behavior of the tremor-generating network at high temporal resolution provide quite a different picture (Figure 3).

First, oscillations at the tremor frequency come and go in both units within GPi and in muscles; oscillatory activity in both tremor-related units and in tremulous muscles occurs intermittently over time. Second, the appearance and disappearance of oscillatory activity seems to occur independently at each site; the timing of the oscillatory signals in a single unit and a muscle are not correlated (Figures 3 and 4). The same is true for muscular tremor; the tremor in different muscles of the limb comes and goes independently (Figure 4). Third, synchrony is transient; the synchrony between an oscillatory unit in a particular field in GPi and a particular tremulous muscle within that field is transitory; the phase locking between individual units (as described by (Hurtado et al., 1999) or between individual units and specific muscles are constantly being formed and broken (Figures 3 and 4). Fourth, over-all there are two categories of interactions among the set of neuron/muscle pairs that show periods of transient synchrony. Examination of the distribution of coherence for all neuron/EMG pairs allows one to splits the pairs into natural modal components - the pairs are either coherent or non-coherent; (this classification is not simply an artifact that results from splitting the data based on some arbitrary cutoff value). For coherent pairs, phase locking occurs episodically, but the phase delay returns to similar values from one phase locking sub-episode to the next (Figure 3). Pairs that are not coherent over long periods of time also have phase locked sub-episodes, but phase locking occurs less frequently and there is no preferred phase delay (Figure 3). Fifth, tremor oscillations in GPi and EMGs are punctuated by phase slips that can be classified as synchronizing or desynchronizing depending on their effect on phase locking.

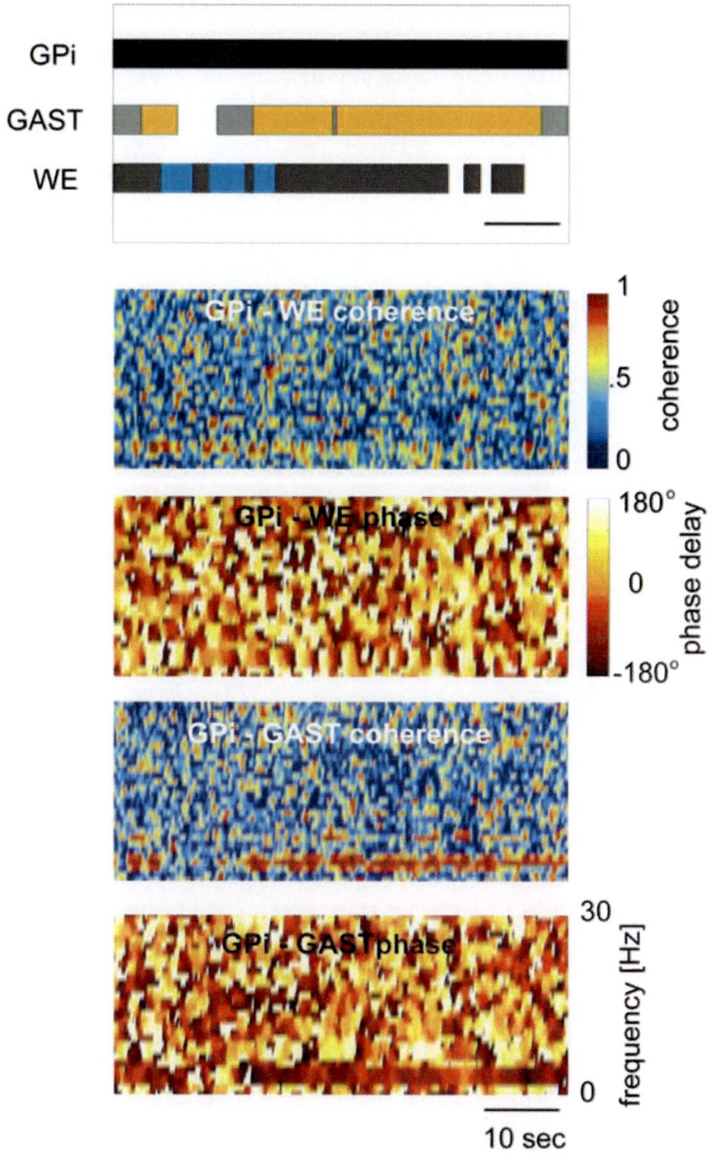

Figure 3: Temporal evolution of coherence and phase between a GPi unit which is coherent with the gastrocnemius EMG over a 60 second time window, and non-coherent with the wrist extensor. The data represented here is the same as that in Figure 2. Top panel: Representation of the statistically significant periods of tremor-related activity computed at high temporal resolution, using the methods described in Hurtado et al., 2004a. The horizontal black and gray bars represent the periods of time during which the oscillatory activity in the GPi unit and two EMG signals are each above the threshold signal-to-noise ratio. The colored sections within the bars represent the periods of time when there was significant phase-locking between the GPi unit and the two different muscles according to the criterion described in Hurtado et al 2004b. Statistically significant subepisodes of phase-locking are more prevalent in the GPi-gastrocnemius pair. Bottom four panels: High resolution time-frequency plots of computed values of coherence and phase delay between the GPi unit and EMG from muscles of the upper (wrist extensor) and lower (gastrocnemius) extremities. Each bin was computed from 10 overlapping segments of 1 sec duration, displaced in 100ms intervals. The frequency is represented on the vertical axis of each plot, as indicated by the scale to the right of the lower panel. The color scale for coherence values and phase delays are to the right of the upper two panels. The yellow/red bands in the coherence plots for both pairs are at approximate 4Hz and indicate that the GPi unit was transiently phase-locked to each of the muscles over time at this frequency. The band at 4 Hz in the GP-gastrocnemius phase plot indicates that the phase delay remained relatively constant over time and explains the presence of statistically significant coherence using standard coherence methods shown in Figure 2. The lack of a band in the GPi-wrist extensor phase plot indicates that the phase delay between the unit and the EMG is variable over time, and explains the lack of overall coherence in the plot of this pair in Figure 2.

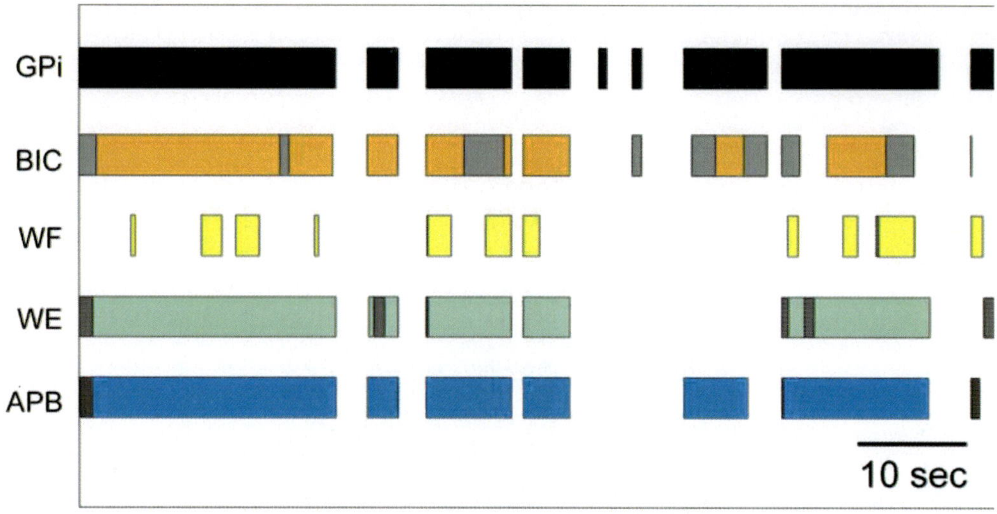

Figure 4: Temporal evolution of oscillatory activity at the tremor frequency in a GPi single unit and four muscles in a single limb and the coherence between the units and each of the four muscles. The horizontal bars represent the periods of time during which the oscillatory activity in the GPi unit and four EMG signals are each above the threshold signal-to-noise ratio. The colored sections within the bars represent the periods of time when there is significant phase-locking between the GPi unit and four different muscles. Note that the periods of occurrence of oscillatory activity in the unit and EMGs do not completely overlap and are often independent. The coherence between the unit and the four muscles is also transient and somewhat independent. Standard spectral methods show that this unit is significantly coherent with all four muscles over the time period shown. BIC, biceps; WF, wrist flexor; WE, wrist extensor; APB, extensor pollicus brevis.

5 Speculations on the Nature of Tremor-Generating Networks with Intermittent Oscillations and Transient Phase Locking

All of the interaction between neurons within the motor region of GPi and peripheral muscles appear to be transient; interactions giving rise to transient phase locking in non-coherent pairs are of a different nature than in coherent ones. That the pallido-muscular phase delay returns to a similar value across phase locking sub-episodes suggests that their locking arise from more robust interactions than in the case of those with variable delays. One possibility is that the fixed phase delay results from conduction delays within a limb-specific BG-thalamocortical network. In contrast, the variable phase delays seen in non-coherent pairs may result from weaker collateral interactions between parallel circuits. The idea of segregated circuits is not contradicted by the short periods of phase locking in non-coherent pairs, but rather suggests that interaction between parallel circuits is possible, but different from the interaction between coherent pairs within the circuit representing a topographic field. Within the field the units with tremor-related activity are always associated with the specified region of the peripheral musculature, but the neurons within the region nor the muscles of the related limb are not necessarily concurrently in the tremor state. Within this functional subcircuit, specific units and specific muscles are recruited into the tremor state relatively independently. Thus, tremor can be viewed as a property of the population of cells at each stage of the parallel pathway, not a property of the individual cells, each of which can either be in the tremor state or not at any one time.

6 Intermittent Synchrony is an Emergent Property of Coupled Nonlinear Oscillators Networks

The observed transient behavior of neuronal-muscular synchrony in the tremor-generating network is reminiscent of the behavior of a network of coupled nonlinear oscillators, which can be synchronous or close to the synchronous state when sufficiently coupled (Pecora, 1998). Intermittent synchrony is an emergent property of networks of coupled nonlinear oscillators with constant values of parameters of oscillators and couplings and is a robust dynamical phenomenon in many different kinds of systems of coupled oscillators (Heagy et al., 1995; Hasler and Maistrenko, 1997; Baker et al., 1998; Boccaletti and Valladares, 2000; Chen et al., 2001). One example of the loss of synchrony leading to the intermittency is that resulting from frequency mismatch between oscillators (Ermentrout and Rinzel, 1984). Time-dependent modulation of network properties may also contribute to the observed variability in synchrony; variations in coupling could be introduced by fluctuations in the intrinsic properties of neurons or in fluctuations in synaptic strength, a possibility is supported by recent observations of significant short-term synaptic plasticity in the basal ganglia (Fitzpatrick et al., 2001; Hanson and Jaeger, 2002). External input into neural motor-control system (including proprioceptive feedback) may also contribute to the observed variability of synchrony.

7 Experimental Evidence of Coupled Oscillators in Basal Ganglia Motor Control System

Several types of experimental data suggest that the basal ganglia-thalamocortical network for motor control contains networks of coupled nonlinear oscillators. Individual neurons within the subthalamic nucleus (see, e.g., Beurrier et al., 1999; Bevan et al., 2000) and the pallidum (Kita and Kitai, 1991; Nambu and Llinas, 1994; Cooper and Stanford, 2000) have been shown to possess nonlinear oscillatory properties, due to the nature and variety of their ionic currents. A network formed from subthalamic and GPe neurons in *in vitro* organotypic cultures has been shown to exhibit complex oscillatory behavior (Plenz and Kital, 1999). The presence of coupled oscillators is also supported by observations of tremor-related oscillatory activity in the pallidum (e.g. Miller and DeLong, 1987; Filion and Tremblay, 1991; Hutchison et al., 1997; Hurtado et al., 1999), subthalamic nucleus (see, e.g., Bergman et al., 1994), motor thalamus (Guiot et al., 1962; Lenz et al., 1988; 1994), and motor cortex (Alberts et al., 1969; Volkmann et al., 1996). These structures are known to be anatomically connected, and thus form the networks discussed here.

The idea that tremor-generating networks in Parkinson's disease are networks of coupled nonlinear oscillators is also supported by analysis of univariate time-series obtained from an accelerometer during tremor in parkinsonian patients (Timmer et al., 2000; Gao and Tung, 2002). These studies used the analysis tools of nonlinear dynamics and stochastic systems to show that an ensemble of coupled nonlinear oscillators with noise can reproduce the data recorded from patients. These studies deal with a single time-series per patient that characterizes kinematics of tremulous limb motion and, therefore, do not address the dynamics of synchrony between brain and EMG. However, the conclusions that tremor is produced by a network of coupled nonlinear oscillators with noise supports the same network organization proposed here.

8 Speculation on the Significance of the Existence of Tremor in Motor Control Networks

The suggested organization and dynamics operation of neuronal-muscular networks producing tremor may have important functional significance for motor control. While rhythmic activity at the tremor frequency is clearly pathological, the dynamics of the tremor-generating network, transformed by the loss of dopamine into a network in which more of the neurons are oscillating, may represent a feature that is not pathological – the transient nature of the correlations to the peripheral musculature. Movement control networks must be designed to generate a rich repertoire of muscle synergies to implement a large number of complex behaviors. The neural system for motor control can rely on the properties of networks of coupled oscillators to achieve this goal. An intermittent network, such as that described above, allows a large number of different patterns of transient synchronies (e.g., muscle synergies) to be formed. As a result, the perturbation needed to induce any particular pattern of synchrony can be small, because any necessary type of dynamics does not need to be created *de novo*. The advantage of this organization is that it allows for efficient formation and transitions between many possible motor synergies.

In the case of motor control networks, the unstable network dynamics, of which parkinsonian tremor is a special case, allow for effective formation of different patterns of muscle synergies, which can be oscillatory (see Murthy and Fetz, 1996; Donoghue et al., 1998). These synergies are needed to perform different movements and to make rapid transitions between them. This predicts that in healthy subjects there should be some mechanism to suppress intermittent oscillatory activity at rest. This mechanism would also minimize metabolic expense as it suppresses oscillations at rest, when they are not needed.

Following degeneration of dopaminergic innervation of the basal ganglia in Parkinson's disease this mechanism would be impaired and the motor control network will become intermittently synchronous, which clinically manifests as resting tremor. Verification of these ideas requires further experimental and modeling studies.

9 Dynamics and Motor Behavior

A number of studies in the field of motor control support a relationship between oscillatory activity and shifts in correlation in relation to motor task performance (e.g. Conway et al., 1995; Murthy and Fetz, 1996; Donoghue et al., 1998; Baker et al., 1999; Baker et al., 2001). In the frontal cortex of monkeys performing a behavioral task, neural activity can rapidly shift from being uncorrelated to correlated and back to uncorrelated in systematic relation to the task being performed (Vaadia et al., 1995). These shifts in the temporal interactions of the population of cortical neurons can occur without changes in the firing rates of the individual neurons; this seems to be a property of cortical networks in general (e.g., Riehle et al., 1997 in motor cortex; deCharms and Merzenich, 1996 in auditory cortex).

Studies by several groups have shown beta band oscillations in the sensorimotor cortex that are coherent with the contralateral EMG (Farmer et al., 1993a; 1993b; Conway et al., 1995; Baker et al., 1997; Salenius et al., 1997; Halliday et al., 1998; Kilner et al., 1999; Llinas et al., 1999). In non-human primates, coherent oscillations observed in the beta range between sites in the motor cortex, between motor cortex and EMG and between pairs of EMG are modulated during the performance of a precision grip task (Baker et al., 1997). Recent work from this group (Jackson et al., 2003), also using a precision grip task, have shown that neuron pairs from the hand regions of the motor cortex that have similar muscle fields show greater synchronization than pairs with nonoverlapping fields, and that cells with opposing effects (postspike facilitation vs. postspike suppression) in the same muscles exhibited negative synchronization.

For PD patients, implantation of deep brain stimulating (DBS) electrodes has allowed the study of correlations between cortex, subcortical structures, and the peripheral musculature. In these studies, local field potentials (LFP) from pallidum or subthalamic nucleus can be recorded from an implanted DBS electrode and compared to cortical EEG and muscle activity. Coherence in the 7-45 Hz band has been noted between STN and muscle EMG, STN and Cz electrodes (reflecting mainly supplementary motor area activity), and STN and C3 electrodes (sensorimotor cortex activity); this coherence was reported to be similar whether the patient was performing a tonic contraction or a self-paced phasic wrist movement (Marsden et al., 2001). Most of the power in this frequency band probably reflects the synchronous drive to motor units in the beta range (15-30 Hz) discovered by Wollaston, (1810) and studied more recently by Farmer et al., (1993a; 1993b), as well as the Piper rhythm (Piper, 1912; 30-60 Hz). Synchronous oscillations between STN and GPi occurred at both the tremor frequency and approximately 20 Hz, but the coherence was lower during tonic contractions than during rest (Brown et al., 2001a). Levodopa therapy resulted in the disappearance of these peaks in the coherence spectrum and the appearance of a peak at 70 Hz. Coherence in the beta range has been reported between pairs of neurons in STN (Levy et al., 2002) in patients with tremor and these oscillation were suppressed by voluntary movement and levodopa treatment. In this study, however, the oscillations at 20 Hz were not seen in patients without tremor and oscillations at 70 Hz were not reported. Results from MEG-EMG studies of motor cortex showed that voluntary tonic contraction in PD increased coherence at 5-12 Hz and reduced coherence between 15-30 Hz (Salenius et al., 2002). Levodopa reduced the 5-12 Hz coherence and enhanced 15-30 Hz and 35-60 Hz MEG-EMG coherence.

Together, these observations indicate that the mechanisms that underlie the subtle changes in synchronization required for normal motor control become disrupted in the parkinsonian brain, giving rise to the PD motor signs. The rapid association and dissociation of neuronal elements described by Vaadia and colleagues suggests the presence of a selection mechanism by which a new assembly takes over a previous one in the course of behavioral performance (Vaadia et al., 1995). In this scheme it is essential for the correlated firing that gives rise to an assembly to be terminated in a timely manner; the failure of such mechanism would possibly give rise to hypersynchrony. Indeed, one of the proposed roles of the BG in motor control is to coordinate the selection and suppression of motor programs (Mink, 1996), and it is not unlikely that selection takes the form of a bias signal for the favoring formation of a particular cortical cell assembly (i.e the "wanted movement") over others ("unwanted movement"), whereas suppression might occur in the form of disruption of the ongoing assembly. In recent years it has been proposed that the "hyperdirect" (i.e. cortico-subthalamo-pallidal) pathway might perform such a suppressive function (Nambu et al., 2002), whereas the cortico-striatal pathways participate in the selection process. Importantly, the overactivity of the hyperdirect pathway would result in oscillatory behavior involving a larger neural population, and occurring at a different frequency, than in the case of normal assembly formation. Moreover, phase locking would occur between neural populations in the different structures of the pathways, but this phase locking would be restricted to regions that are anatomically connected and that subserve a similar topographic territory of the limb motor network.

An open question in this picture is what is the pattern of synchronization, if any, within basal ganglia structures in relation to normal action selection? In total the studies cited certainly support the notion of a change in synchronization with dopamine levels, but this short term synchronization has not yet been examined in the BG during normal ongoing behavior. Single unit studies in the normal BG suggest that correlated firing is rare, while oscillatory and synchronized activity appears under DA deficit. However, significant changes in oscillatory activity in LFP have been described in relation to normal movement initiation in those same structures. Is it possible that short term neuronal synchronization is present in relation to normal movement? Since LFP most likely represents synaptic integration the possibility remains that these oscillations are not a faithful indicator of the output of a given structure and therefore do not function as a selection signal in downstream (i.e. thalamocortical) structures but represent a local integrative mechanism. We would suggest that examination at high temporal resolution across a behavioral epoch might be useful to examine this issue.

Acknowledgements

We thank our colleagues Conrad Pappas, M.D., Ph.D. and Vicki Wheelock, M.D. with whom we have had many helpful discussions about the clinical symptoms, neuroanatomy and pathophysiology of Parkinson's disease. This work was supported by NIH grant NS39121.

References

Alberts WW, Wright EW, Jr., Feinstein B (1969) Cortical potentials and Parkinsonian tremor. *Nature* **221**:670-672.

Albin RL, Young AB, Penney JB (1989) The functional anatomy of basal ganglia disorders. *Trends Neurosci* **12**:366-375.

Alexander GE, Crutcher MD (1990) Functional architecture of basal ganglia circuits: neural substrates of parallel processing. *Trends Neurosci* **13**:266-271.

Alexander GE, Crutcher MD, DeLong MR (1990) Basal ganglia-thalamocortical circuits: parallel substrates for motor, oculomotor, "prefrontal" and "limbic" functions. *Prog Brain Res* **85**:119-146.

Baker GL, Blackburn JA, Smith HJT (1998) Intermittent synchronization in a pair of coupled chaotic pendula. *Phys Rev Lett* **81**:554-557.

Baker SN, Olivier E, Lemon RN (1997) Coherent oscillations in monkey motor cortex and hand muscle EMG show task-dependent modulation. *J Physiol* **501** (Pt 1):225-241.

Baker SN, Kilner JM, Pinches EM, Lemon RN (1999) The role of synchrony and oscillations in the motor output. *Exp Brain Res* **128**:109-117.

Baker SN, Spinks R, Jackson A, Lemon RN (2001) Synchronization in monkey motor cortex during a precision grip task. I. Task-dependent modulation in single-unit synchrony. *J Neurophysiol* **85**:869-885.

Ballard PA, Tetrud JW, Langston JW (1985) Permanent human parkinsonism due to 1-methyl-4-phenyl-1,2,3,6-tetrahydropyridine (MPTP): seven cases. *Neurology* **35**:949-956.

Bar-Gad I, Bergman H (2001) Stepping out of the box: information processing in the neural networks of the basal ganglia. *Curr Opin Neurobiol* **11**:689-695.

Bergman H, Wichmann T, Karmon B, DeLong MR (1994) The primate subthalamic nucleus. II. Neuronal activity in the MPTP model of parkinsonism. *J Neurophysiol* **72**:507-520.

Bergman H, Raz A, Feingold A, Nini A, Nelken I, Hansel D, Ben-Pazi H, Reches A (1998) Physiology of MPTP tremor. *Mov Disord 13 Suppl* **3**:29-34.

Beurrier C, Congar P, Bioulac B, Hammond C (1999) Subthalamic nucleus neurons switch from single-spike activity to burst-firing mode. *J Neurosci* **19**:599-609.

Bevan MD, Wilson CJ, Bolam JP, Magill PJ (2000) Equilibrium potential of GABA(A) current and implications for rebound burst firing in rat subthalamic neurons in vitro. *J Neurophysiol* **83**:3169-3172.

Bevan MD, Magill PJ, Terman D, Bolam JP, Wilson CJ (2002) Move to the rhythm: oscillations in the subthalamic nucleus-external globus pallidus network. *Trends Neurosci* **25**:525-531.

Boccaletti S, Valladares DL (2000) Characterization of intermittent lag synchronization. *Phys Rev E Stat Phys Plasmas Fluids Relat Interdiscip Topics* **62**:7497-7500.

Brody CD (1998) Slow covariations in neuronal resting potentials can lead to artefactually fast cross-correlations in their spike trains. *J Neurophysiol* **80**:3345-3351.

Brody CD (1999) Correlations without synchrony. *Neural Comput* **11**:1537-1551.

Brooks DJ, Ibanez V, Sawle GV, Playford ED, Quinn N, Mathias CJ, Lees AJ, Marsden CD, Bannister R, Frackowiak RS (1992) Striatal D2 receptor status in patients with Parkinson's disease, striatonigral degeneration, and progressive supranuclear palsy, measured with 11C-raclopride and positron emission tomography. *Ann Neurol* **31**:184-192.

Brown P, Oliviero A, Mazzone P, Insola A, Tonali P, Di Lazzaro V (2001a) Dopamine dependency of oscillations between subthalamic nucleus and pallidum in Parkinson's disease. *J Neurosci* **21**:1033-1038.

Brown P, Marsden J, Defebvre L, Cassim F, Mazzone P, Oliviero A, Altibrandi MG, Di Lazzaro V, Limousin-Dowsey P, Fraix V, Odin P, Pollak P (2001b) Intermuscular coherence in Parkinson's disease: relationship to bradykinesia. *Neuroreport* **12**:2577-2581.

Chen JY, Wong KW, Zheng HY, Shuai JW (2001) Intermittent phase synchronization of coupled spatiotemporal chaotic systems. *Phys Rev E* **64**:016212.

Conway BA, Halliday DM, Farmer SF, Shahani U, Maas P, Weir AI, Rosenberg JR (1995) Synchronization between motor cortex and spinal motoneuronal pool during the performance of a maintained motor task in man. *J Physiol* **489** (Pt 3):917-924.

Cooper AJ, Stanford IM (2000) Electrophysiological and morphological characteristics of three subtypes of rat globus pallidus neurone in vitro. *J Physiol* **527** Pt 2:291-304.

De Montigny C, Lamarre Y (1973) Rhythmic activity induced by harmaline in the olivo-cerebello-bulbar system of the cat. *Brain Res* **53**:81-95.

deCharms RC, Merzenich MM (1996) Primary cortical representation of sounds by the coordination of action-potential timing. *Nature* **381**:610-613.

DeLong M, Crutcher M, Georgopoulos A (1985) Primate globus pallidus and subthalamic nucleus: functional organization. *Journal of Neurophysiology* **53**:530-543.

DeLong MR (1990) Primate models of movement disorders of basal ganglia origin. *Trends Neurosci* **13**:281-285.

Deuschl G, Wilms H, Krack P, Wurker M, Heiss WD (1999) Function of the cerebellum in Parkinsonian rest tremor and Holmes' tremor. *Ann Neurol* **46**:126-128.

Deuschl G, Raethjen J, Baron R, Lindemann M, Wilms H, Krack P (2000) The pathophysiology of parkinsonian tremor: a review. *J Neurol 247 Suppl* **5**:V33-48.

Donoghue JP, Sanes JN, Hatsopoulos NG, Gaal G (1998) Neural discharge and local field potential oscillations in primate motor cortex during voluntary movements. *J Neurophysiol* **79**:159-173.

Ermentrout GB, Rinzel J (1984) Beyond a pacemaker's entrainment limit: phase walk-through. *Am J Physiol* **246**:R102-106.

Farmer S (2002) Neural rhythms in Parkinson's disease. *Brain* **125**:1175-1176.

Farmer SF, Swash M, Ingram DA, Stephens JA (1993b) Changes in motor unit synchronization following central nervous lesions in man. *J Physiol* **463**:83-105.

Farmer SF, Bremner FD, Halliday DM, Rosenberg JR, Stephens JA (1993a) The frequency content of common synaptic inputs to motoneurones studied during voluntary isometric contraction in man. *J Physiol* **470**:127-155.

Feige B, Aertsen A, Kristeva-Feige R (2000) Dynamic synchronization between multiple cortical motor areas and muscle activity in phasic voluntary movements. *J Neurophysiol* **84**:2622-2629.

Filion M, Tremblay L (1991) Abnormal spontaneous activity of globus pallidus neurons in monkeys with MPTP-induced parkinsonism. *Brain Res* **547**:142-151.

Fitzpatrick JS, Akopian G, Walsh JP (2001) Short-term plasticity at inhibitory synapses in rat striatum and its effects on striatal output. *J Neurophysiol* **85**:2088-2099.

Gao JB, Tung WW (2002) Pathological tremors as diffusional processes. *Biol Cybern* **86**:263-270.

Goldberg JA, Boraud T, Maraton S, Haber SN, Vaadia E, Bergman H (2002) Enhanced synchrony among primary motor cortex neurons in the 1-methyl-4-phenyl-1,2,3,6-tetrahydropyridine primate model of Parkinson's disease. *J Neurosci* **22**:4639-4653.

Guiot G, Hardy J, Albe-Fressard D (1962) Delimitation precise des structures sus-corticales et identification de noyaux thalamiques chez l'homme par l'electrophisiologie stereotaxique. *Neurochirurgia* **5**:1-18.

Halliday DM, Conway BA, Farmer SF, Rosenberg JR (1998) Using electroencephalography to study functional coupling between cortical activity and electromyograms during voluntary contractions in humans. *Neurosci Lett* **241**:5-8.

Hanson JE, Jaeger D (2002) Short-term plasticity shapes the response to simulated normal and parkinsonian input patterns in the globus pallidus. *J Neurosci* **22**:5164-5172.

Hasler M, Maistrenko YL (1997) An introduction to the synchronization of chaotic systems: coupled skew tent maps. *IEEE Transactions on circuits and systems-I: fundamental theory and applications* **44**:856-866.

Heagy JF, Pecora LM, Carroll TL (1995) Short wavelength bifurcations and size instabilities in coupled oscillator systems. *Physical Review Letters* **74**:4185-4188.

Heimer G, Bar-Gad I, Goldberg JA, Bergman H (2002) Dopamine replacement therapy reverses abnormal synchronization of pallidal neurons in the 1-methyl-4-phenyl-1,2,3,6-tetrahydropyridine primate model of parkinsonism. *J Neurosci* **22**:7850-7855.

Hikosaka O, Wurtz RH (1985a) Modification of saccadic eye movements by GABA-related substances. I. Effect of muscimol and bicuculline in monkey superior colliculus. *J Neurophysiol* **53**:266-291.

Hikosaka O, Wurtz RH (1985b) Modification of saccadic eye movements by GABA-related substances. II. Effects of muscimol in monkey substantia nigra pars reticulata. *J Neurophysiol* **53**:292-308.

Holsapple JW, Preston JB, Strick PL (1991) The origin of thalamic inputs to the "hand" representation in the primary motor cortex. *J Neurosci* **11**:2644-2654.

Hornykiewicz O (1966) Dopamine (3-hydroxytyramine) and brain function. *Pharmacol Rev* **18**:925-964.

Hurtado JM, Rubchinsky LL, Sigvardt KA (2004) Statistical method for detection of phase-locking episodes in neural oscillations. *J Neurophysiol* **91**:1883-1898.

Hurtado JM, Gray CM, Tamas LB, Sigvardt KA (1999) Dynamics of tremor-related oscillations in the human globus pallidus: a single case study. *Proc Natl Acad Sci U S A* **96**:1674-1679.

Hurtado JM, Lachaux JP, Beckley DJ, Gray CM, Sigvardt KA (2000) Inter- and intralimb oscillator coupling in parkinsonian tremor. *Mov Disord* **15**:683-691.

Hurtado JM, Rubchinsky LL, Sigvardt KA, Wheelock VL, Pappas CT (2005) Temporal evolution of oscillations and synchrony in GPi/muscle pairs in Parkinson's disease. *J Neurophysiol* **93**:1569-1584.

Hutchison WD, Lozano AM, Tasker RR, Lang AE, Dostrovsky JO (1997) Identification and characterization of neurons with tremor-frequency activity in human globus pallidus. *Experimental Brain Research* **113**:557-563.

Ilinsky IA, Kultas-Ilinsky K (2002) Motor thalamic circuits in primates with emphasis on the area targeted in treatment of movement disorders. *Mov Disord 17 Suppl* **3**:S9-14.

Jackson A, Gee VJ, Baker SN, Lemon RN (2003) Synchrony between neurons with similar muscle fields in monkey motor cortex. *Neuron* **38**:115-125.

Jellinger KA (1999) Post mortem studies in Parkinson's disease--is it possible to detect brain areas for specific symptoms? *J Neural Transm Suppl* **56**:1-29

Kilner JM, Baker SN, Salenius S, Jousmaki V, Hari R, Lemon RN (1999) Task-dependent modulation of 15-30 Hz coherence between rectified EMGs from human hand and forearm muscles. *J Physiol* **516** (Pt 2):559-570.

Kita H, Kitai ST (1991) Intracellular study of rat globus pallidus neurons: membrane properties and responses to neostriatal, subthalamic and nigral stimulation. *Brain Res* **564**:296-305.

Lachaux JP, Rodriguez E, Martinerie J, Varela FJ (1999) Measuring phase synchrony in brain signals. *Hum Brain Mapp* **8**:194-208.

Lamarre Y, Mercier L (1971) Neurophysiological studies of harmaline induced tremor in the cat. *Can J Physiol Pharmacol* **49**:1049-1058.

Lemstra AW, Verhagen Metman L, Lee JI, Dougherty PM, Lenz FA (1999) Tremor-frequency (3-6 Hz) activity in the sensorimotor arm representation of the internal segment of the globus pallidus in patients with Parkinson's disease. *Neurosci Lett* **267**:129-132.

Lenz F, Kwan H, Martin R, Tarker R, Dostrovsky J, Lenz Y (1994) Single unit analysis of the human ventral thalamic nuclear group.Tremor-related activity in functionally identified cells. *Brain* **117**:531-543.

Lenz F, Tasker R, Kwan H, Schnider S, Kwong R, Murayama Y, Dostrovsky J, Murphy J (1988) Single unit analysis of the human ventral thalamic nuclear group: correlation of thalamic "tremor cells" with the 3-6 Hz component of parkinsonian tremor. *Journal of Neuroscience* **8**:754-764.

Levy R, Hutchison WD, Lozano AM, Dostrovsky JO (2000) High-frequency synchronization of neuronal activity in the subthalamic nucleus of parkinsonian patients with limb tremor. *J Neurosci* **20**:7766-7775.

Levy R, Ashby P, Hutchison WD, Lang AE, Lozano AM, Dostrovsky JO (2002) Dependence of subthalamic nucleus oscillations on movement and dopamine in Parkinson's disease. *Brain* **125**:1196-1209.

Llinas RR, Ribary U, Jeanmonod D, Kronberg E, Mitra PP (1999) Thalamocortical dysrhythmia: A neurological and neuropsychiatric syndrome characterized by magnetoencephalography. *Proc Natl Acad Sci U S A* **96**:15222-15227.

Macchi G, Jones EG (1997) Toward an agreement on terminology of nuclear and subnuclear divisions of the motor thalamus. *J Neurosurg* **86**:670-685.

Marsden JF, Limousin-Dowsey P, Ashby P, Pollak P, Brown P (2001) Subthalamic nucleus, sensorimotor cortex and muscle interrelationships in Parkinson's disease. *Brain* **124**:378-388.

Mena-Segovia J, Bolam JP, Magill PJ (2004) Pedunculopontine nucleus and basal ganglia: distant relatives or part of the same family? *Trends in Neurosciences* **27**:585-588.

Miller WC, DeLong MR (1987) Altered tonic activity of neurons in the globus pallidus and subthalamic nucleus in the primate MPTP model of Parkinsonism. In: *The basal ganglia II* (Jayaraman MCaA, ed), pp 415-427. New York: Plenum Press.

Mink JW (1996) The basal ganglia: focused selection and inhibition of competing motor programs. *Prog Neurobiol* **50**:381-425.

Murthy VN, Fetz EE (1996) Synchronization of neurons during local field potential oscillations in sensorimotor cortex of awake monkeys. *J Neurophysiol* **76**:3968-3982.

Nambu A, Llinas R (1994) Electrophysiology of globus pallidus neurons in vitro. *J Neurophysiol* **72**:1127-1139.

Nambu A, Tokuno H, Takada M (2002) Functional significance of the cortico-subthalamo-pallidal 'hyperdirect' pathway. *Neurosci Res* **43**:111-117.

Nini A, Feingold A, Slovin H, Bergman H (1995) Neurons in the globus pallidus do not show correlated activity in the normal monkey, but phase-locked oscillations appear in the MPTP model of parkinsonism. *J Neurophysiol* **74**:1800-1805.

Obeso JA, Rodriguez-Oroz MC, Rodriguez M, Lanciego JL, Artieda J, Gonzalo N, Olanow CW (2000) Pathophysiology of the basal ganglia in Parkinson's disease. *Trends Neurosci* **23**:S8-19.

Ohye C, Maeda T, Narabayashi H (1976) Physiologically defined VIM nucleus. Its special reference to control of tremor. *Appl Neurophysiol* **39**:285-295.

O'Suilleabhain PE, Matsumoto JY (1998) Time-frequency analysis of tremors. *Brain* **121** *(Pt 11)*:2127-2134.

Pahapill PA, Lozano AM (2000) The pedunculopontine nucleus and Parkinson's disease. *Brain* **123** (Pt 9):1767-1783.

Palus M, Hoyer D (1998) Detecting nonlinearity and phase synchronization with surrogate data. *IEEE Eng Med Biol Mag* **17**:40-45.

Pare D, Curro'Dossi R, Steriade M (1990) Neuronal basis of the parkinsonian resting tremor: a hypothesis and its implications for treatment. *Neuroscience* **35**:217-226.

Parker F, Tzourio N, Blond S, Petit H, Mazoyer B (1992) Evidence for a common network of brain structures involved in parkinsonian tremor and voluntary repetitive movement. *Brain Res* **584**:11-17.

Parkinson J (1817) An Essay on the Shaking Palsy. In: (Rowland W, ed). London.

Pecora LM (1998) Synchronization conditions and desynchronizing patterns in coupled limit-cycle and chaotic systems. *Phys Rev E* **58**:347-360.

Pikovsky A, Rosenblum M, Kurths J (2001) Synchronization: a universal concept in nonlinear sciences. Cambridge, UK: Cambridge University Press.

Piper H (1912) Electrophysiologie menschlicher Muskeln.

Plenz D, Kital ST (1999) A basal ganglia pacemaker formed by the subthalamic nucleus and external globus pallidus. *Nature* **400**:677-682.

Pollock L, Davis L (1930) Muscle tone in parkinsonian states. *Arch Neurol Psychiat* **23**:303-319.

Pollok B, Gross J, Dirks M, Timmermann L, Schnitzler A (2004) The cerebral oscillatory network of voluntary tremor. *J Physiol* **554**:871-878.

Rack P, Ross H (1986) The role of reflexes in the resting tremor of Parkinson's disease. *Brain* **109**:115-141.

Raethjen J, Lindemann M, Schmaljohann H, Wenzelburger R, Pfister G, Deuschl G (2000) Multiple oscillators are causing parkinsonian and essential tremor. *Mov Disord* **15**:84-94.

Raz A, Vaadia E, Bergman H (2000) Firing patterns and correlations of spontaneous discharge of pallidal neurons in the normal and the tremulous 1-methyl-4-phenyl-1,2,3,6-tetrahydropyridine vervet model of parkinsonism. *J Neurosci* **20**:8559-8571.

Raz A, Feingold A, Zelanskaya V, Vaadia E, Bergman H (1996) Neuronal synchronization of tonically active neurons in the striatum of normal and parkinsonian primates. *Journal of Neurophysiology* **76**:2083-2088.

Raz A, Frechter-Mazar V, Feingold A, Abeles M, Vaadia E, Bergman H (2001) Activity of pallidal and striatal tonically active neurons is correlated in mptp-treated monkeys but not in normal monkeys. *J Neurosci* **21**:RC128.

Redgrave P, Prescott TJ, Gurney K (1999) The basal ganglia: a vertebrate solution to the selection problem? *Neuroscience* **89**:1009-1023.

Riehle A, Grun S, Diesmann M, Aertsen A (1997) Spike synchronization and rate modulation differentially involved in motor cortical function. *Science* **278**:1950-1953.

Rodriguez MC, Guridi OJ, Alvarez L, Mewes K, Macias R, Vitek J, DeLong MR, Obeso JA (1998) The subthalamic nucleus and tremor in Parkinson's disease. *Mov Disord 13 Suppl* **3**:111-118.

Rouiller EM, Liang F, Babalian A, Moret V, Wiesendanger M (1994) Cerebellothalamocortical and pallidothalamocortical projections to the primary and supplementary motor cortical areas: a multiple tracing study in macaque monkeys. *J Comp Neurol* **345**:185-213.

Salenius S, Portin K, Kajola M, Salmelin R, Hari R (1997) Cortical control of human motoneuron firing during isometric contraction. *J Neurophysiol* **77**:3401-3405.

Salenius S, Avikainen S, Kaakkola S, Hari R, Brown P (2002) Defective cortical drive to muscle in Parkinson's disease and its improvement with levodopa. *Brain* **125**:491-500.

Schreiber T, Schmitz A (2000) Surrogate time series. *Physica D* **142**:346-382.

Stein JF, Aziz TZ (1999) Does imbalance between basal ganglia and cerebellar outputs cause movement disorders? *Curr Opin Neurol* **12**:667-669.

Stiles RN, Pozos RS (1976) A mechanical-reflex oscillator hypothesis for parkinsonian hand tremor. *J Appl Physiol* **40**:990-998.

Taha JM, Favre J, Baumann TK, Burchiel KJ (1996) Characteristics and somatotopic organization of kinesthetic cells in the globus pallidus of patients with Parkinson's disease. *J Neurosurg* **85**:1005-1012.

Taha JM, Favre J, Baumann TK, Burchiel KJ (1997) Tremor control after pallidotomy in patients with Parkinson's disease: correlation with microrecording findings. *Neurosurg Focus* **2**:e2.

Tasker RR (1990) Thalamotomy. *Neurosurg Clin N Am* **1**:841-864.

Tass P, Rosenblum MG, Weule J, Kurths J, Pikovsky A, Volkmann J, Schnitzler A, Freund HJ (1998) Detection of N-M phase locking from noisy data: application to magnetoencephalography. *Physical Rev Lett* **81**:3291-3294.

Tetrud J, Langston J (1992) Tremor in MPTP-induced parkinsonism. Neurology 42:407-410.

Timmer J, Haussler S, Lauk M, Lucking CH (2000) Pathological tremors: Deterministic chaos or nonlinear stochastic oscillators? *Chaos* **10**:278-288.

Timmermann L, Gross J, Dirks M, Volkmann J, Freund HJ, Schnitzler A (2003) The cerebral oscillatory network of parkinsonian resting tremor. *Brain* **126**:199-212.

Vaadia E, Haalman I, Abeles M, Bergman H, Prut Y, Slovin H, Aertsen A (1995) Dynamics of neuronal interactions in monkey cortex in relation to behavioural events. *Nature* **373**:515-518.

Volkmann J, Joliot M, Mogilner A, Ioannides A, Lado F, Fazzini E, Ribary U, Llinás R (1996) Central motor loop oscillations in parkinsonian resting tremor revealed by magnetoencephalography. *Neurology* **46**:1359-1370.

Williams D, Tijssen M, Van Bruggen G, Bosch A, Insola A, Di Lazzaro V, Mazzone P, Oliviero A, Quartarone A, Speelman H, Brown P (2002) Dopamine-dependent changes in the functional connectivity between basal ganglia and cerebral cortex in humans. *Brain* **125**:1558-1569.

Wollaston WH (1810) On the duration of muscle action. *Philos Trans Soc Lond* **MDCCCX**:1-5.

In: Recent Breakthroughs in Basal Ganglia Research
Editor: Erwan Bezard, pp. 267-281
ISBN: 1-59454-880-3

Chapter 21

INFLUENCE OF EXPERIENCE ON THE VULNERABILITY OF DA NEURONS TO OXIDATIVE STRESS: IMPLICATIONS FOR THERAPEUTIC INVENTION FOR PARKINSON'S DISEASE

Amanda D. Smith [*]
Department of Neurology, University of Pittsburgh, Pittsburgh, PA 15213 (USA)

ABSTRACT

Although the role of the loss of substantia nigral dopamine (DA) neurons in the motor symptoms of PD have been known for quite some time, treatments that do more than allay these motor symptoms have remained illusive. Additionally, L-dopa, the most efficacious drug used to treat the symptoms of PD, loses its efficacy over time as disease progression continues. Thus, it is important that new therapies be developed that not only have continuing symptomatic benefit over a long period of time, but that may also slow or halt disease progression. Thus, the following chapter examines the use of enriched environment, caloric restriction and exercise not only as symptomatic therapies but also examines their ability to protect or decrease the vulnerability of the dopaminergic nigrostriatal pathway against various toxins. The convergent mechanisms by which each of these therapies may decrease the vulnerability of DA neurons in the substantia nigra are also discussed.

INTRODUCTION

Although it is now clear that Parkinson's disease (PD) is associated with complex neuropathology, it is the loss of dopamine (DA) neurons projecting from the substantia nigra (SN) to the neostriatum that appear to be primarily responsible for the motor dysfunctions characteristic of this disease (Olanow and Tatton, 1999) Little is known about the underlying cause of this cell loss. However, the greatest known risk factor for the disease is age, with most individuals not exhibiting with frank symptoms until the age of 60. This age-dependency may provide clues into the underlying cause of the DA cell loss.

The free radical/oxidative stress theory of aging: The normal process of molecular and cellular aging is believed to be caused by oxidative stress—an hypothesis referred to as the *free radical/oxidative stress theory* of aging and first put forth by Harman in 1956 (for review see Harman, 2003; Bokov et al., 2004). This theory asserts that over time the continuing onslaught of oxidative stresses within a cell as a result of normal metabolism leads to cell death. Because of its high oxygen consumption and the presence of polyunsaturated lipids, the brain in general is more susceptible than many other tissues to damage via oxidative stress. Decreases in trophic factor support (Gage et al., 1983; Date et al., 1994; Collier et al., 1999; 2005), anti-oxidant defenses (Aksenova et al., 1998; Sasaki et al., 2001), and damaged protein

[*] Contact information: Amanda D. Smith, PhD: Department of Neurology, S-510 Biomedical Science Tower, Pittsburgh, PA 15213. Phone: (412) 383-7604. Fax: (412) 624-7327. E-mail address: smithad@pitt.edu

removal/repair processes (for review see Mattson et al., 2002b) that occur with age may render the aged brain even more vulnerable.

DA as a source of oxidative stress: Within the brain itself, DA neurons in the SN appear to be particularly sensitive to the deleterious effects of oxidative stresses. It has been reported that 5-10% of DA neurons in the SN are lost per decade after the age of 20 (Fearnley and Lees, 1991; McGeer et al., 1988; Volkow et al., 1996) and major culprit is believed to be oxidative stress (Schapira, 2001). Increases in protein oxidative products have been observed in the SN (Floor and Wetzel, 1998) and striatum (Driver et al., 2000; Dubey et al., 1996; Zhang et al., 1994; Arivazhagan et al., 2002) compared to other regions of the brain and in striatum of PD patients (Jenner 2003). That the SN appears to bear an even higher oxidative burden is probably at least in part due to its neurotransmitter—DA. Reactive oxygen species (ROS) are generated as byproducts of DA metabolism and DA also can undergo auto-oxidation (Stokes et al., 1999). Considering the highly oxidant environment in the nigrostriatal pathway created by DA metabolism, the fact that it normally takes approximately 6-7 decades to produce a 35% loss of DA neurons is evidence of the efficacy of protective mechanisms.

Although as noted, oxidative damage is a normal consequence of aging, there is considerable evidence that PD is associated with an unusually high level of oxidative damage to proteins, lipids, and DNA (Alam et al., 1997; Jenner 2003). It is reasonable to assume, therefore, that much of the neuropathology associated with PD is a consequence of pro-oxidant mechanisms overwhelming anti-oxidant, neuronal trophic support, and damaged protein removal/repair mechanisms. Indeed, a reduction in antioxidant mechanisms has been reported in the brains of patients with PD (Perry et al., 1982; Kish et al., 1985). If processes that lead to cell death in normal aging occur in PD, and the free radical/oxidative stress theory of aging is correct, the degeneration of DA neurons that occur in the SN with PD may be considered a form of accelerated aging.

Neuroprotective treatments for PD: Although it has been known for some time that the dopaminergic neurons of SN are lost in PD, treatments that halt the progression of the disease rather than only alleviating the symptoms, remain illusive. Recent animal studies using gene therapy have shown promise in preventing degeneration and for providing symptomatic relief (for review see Eberhardt and Schulz, 2004). However, translation of this approach to the clinic has been problematic and extensive clinical trials will be needed before the benefits of gene therapy can be realized in PD. If oxidative stress is indeed causative in the genesis and/or progression of PD, then it would appear that the best defense against development or continued progression of PD would be to reduce or counteract the oxidative stress, thus shifting the cellular milieu to a more anti-oxidant/protective state. Increasing anti-oxidant defense mechanisms through over-expression of superoxide dismutase or glutathione peroxidase (Callio et al., 2005; Thiruchelvam et al., 2005), exogenous administration of SOD mimetics, and small anti-oxidant proteins (Peng et al., 2005) have shown efficacy in protecting against biochemical and behavioral deficits in animal models of PD. Moreover, enriched environment, and caloric restriction, which have all been shown to combat the changes in the brain that normally occur with aging, have also been shown to increase anti-oxidant enzymes (Fernandez et al., 2004), neurotrophic factor levels (Bezard et al., 2003; Faherty et al., 2995), and DNA repair processes in the brain (Cabelof et al., 2003). The resultant effect is a decrease in oxidatively modified proteins, lipids, and DNA and, at least in the case of caloric restriction, this is accompanied by a consistent increase in lifespan (Weindruch et al., 1986; Koubova and Guarente 2003). Thus, the likelihood that enriched environment and caloric restriction can be used to not only treat and halt the normal degeneration of the DA nigrostriatal projection, but also to alter the susceptibility of the affected neuronal populations to toxic insult, should be explored more fully.

MOTOR THERAPY

Exercise and PD: Like housing in an enriched environment and caloric restriction, exercise has been shown to reduce many of the effects of aging, including the decline in antioxidant defenses (Devi et al., 2004; Navarro et al., 2004). Indeed, there is some reason to believe that the effects of an enriched environment and of caloric restriction are mediated at least in part through exercise (see below). There appears to be a consensus in the literature that exercise is beneficial in the treatment of the symptoms of PD. Indeed, physical exercise has been shown to improve the motor, cognitive and affective deficits in patients with mild to moderate PD (Palmer et al., 1986; Hurwitz, 1989; Sunvisson et al., 1997; Schenkman et al., 1998; Reuter et al 1999; Baatile et al., 2000; Lökk, 2000; and Miyai et al., 2000; Cadet et al., 2003).

Movement initiation has been shown to be significantly improved by aerobic activity (Bergen et al., 2002) and balance has been shown to be improved with resistance and balance training (Hirsch et al., 2003). Whether exercise actually prevents the development of PD or halts the progression in humans is not yet known. A few studies do report a negative correlation between PD and the level of physical activity throughout life (Sasco et al., 1992; Tsai et al., 2002; Chen et al., 2005). However, it is not yet possible to know if this is simply a reflection of the tendency toward inactivity of individuals who are in the preclinical stages of PD. Thus, although these data are encouraging, they must be viewed with caution. Because the clinical literature does not provide concrete evidence that exercise prevents the development or halts the progression of PD, we must consult the animal literature to assess the neurobiological benefits of this therapy.

Forced exercise in an animal model of PD: Tillerson and colleagues (2001, 2002) have used an animal model of PD to test the hypothesis that intensive motor therapy targeting limbs affected by degeneration of the nigrostriatal DA neurons will have a beneficial effect on motor deficits and the integrity of the DA nigrostriatal DA system. Adult male rats received injections of 6-hydroxydopamine (6-OHDA) in the medial forebrain bundle (MFB) at the level of the lateral hypothalamus, thereby gaining access to the nigrostriatal DA projection. Approximately 2 hours after surgery, the ipsilateral forelimb was immobilized by placing it in a plaster of Paris casts to force animals to rely on their impaired forelimb. 6-OHDA-treated rats that did not receive any physical therapy showed an extreme motor asymmetry in their use of limbs. Both in rearing against the wall and in landing, they greatly favored the use of those limbs that were ipsilateral to the side of the lesion.

In marked contrast, those lesioned rats that received a cast immediately following surgery through day 7 post-surgery showed no such asymmetry and, indeed, were not different from intact animals in the limb-use asymmetry task (Figure 1; Tillerson et al., 2001). Likewise, those animals that were casted on days 3-9 did not show a significant group difference when compared to sham animals. However, if the casts were not placed on the animals until 7 days after the 6-OHDA infusion, a degree of asymmetry comparable to that seen in 6-OHDA infused animals not subjected to physical therapy was observed (Tillerson et al., 2001). Similar results were obtained for forelimb kinesia, forelimb placing, and apomorphine-induced rotation. Measurement of striatal DA content and vesicular monoamine transporter-2 (VMAT-2) protein indicated that animals that underwent motor therapy either immediately after or beginning 3 days after infusion of 6-OHDA into the MFB showed a marked reduction in the loss of DA (Figure 2, Tillerson et al., 2001) and VMAT-2 as measured 30 days later.

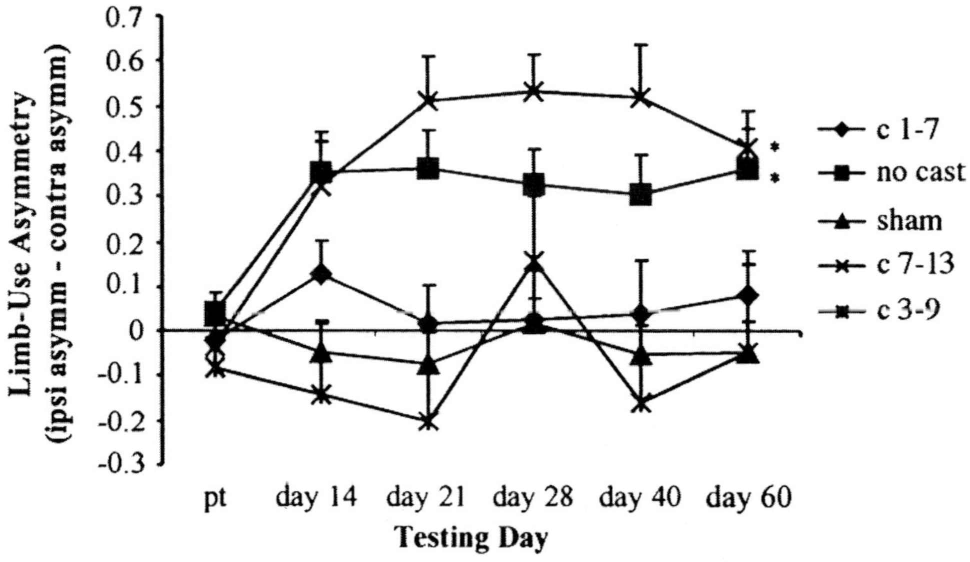

Figure 1. Effect of use-dependent therapy on subsequent limb-use in 6-OHDA-treated animals. 6-OHDA was administered unilaterally into the MFB on day 1. Behavioral measures were made prior to injury (pt, pretest). A plaster of Paris cast then was place on the forelimb ipsilateral to the site of 6-OHDA infusion on days 1–7, 3–9, or 7–13. Bonferroni post hoc analysis indicated significant differences (*) between sham controls (*sham*) and lesion + no cast (*no cast*) and lesion + late cast (*c 7–13*) groups. (From Tillerson et al., 2001).

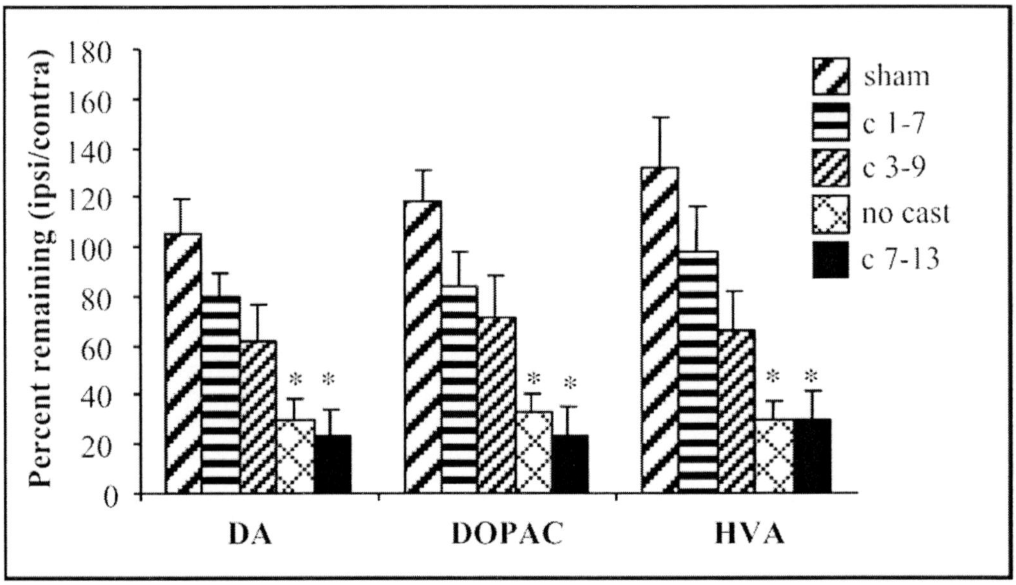

Figure 2. Effect of use-dependent therapy on DA, DOPAC and HVA content in the striatum in 6-OHDA-treated animals. 6-OHDA was administered unilaterally into the MFB on day 1. A plaster of Paris cast then was place on the forelimb ipsilateral to the site of 6-OHDA infusion on days 1–7, 3–9, or 7–13. Biochemical measures were made ≥ 2 months post-6-OHDA infusion. Bonferroni post hoc analysis indicated significant differences (*) between sham controls (*sham*) and lesion + no cast (*no cast*) and lesion + late cast (*c 7–13*) groups. (From Tillerson et al., 2001).

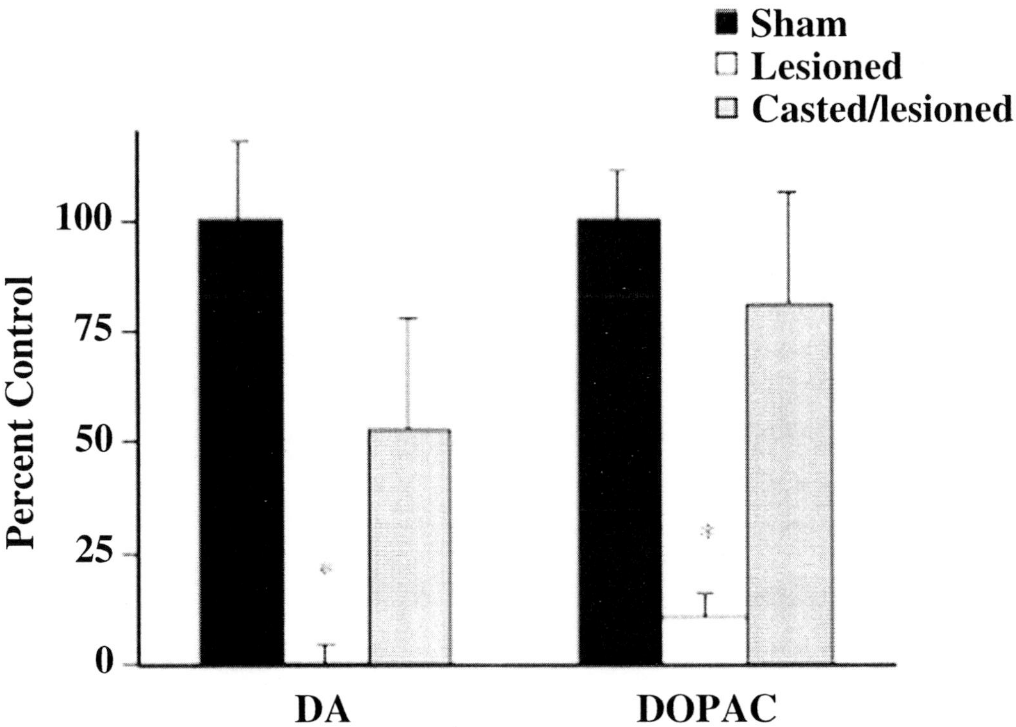

Figure 3. Effect of use-dependent therapy prior to 6-OHDA infusion on striatal DA and DOPAC content. All results are expressed as mean percent of control ± SEM. Animals in the lesioned group showed a significant loss of DA and DOPAC when compared with sham-lesioned (*$P < 0.01$) and casted/lesioned (*$P < 0.05$) animals that was attenuated in casted/lesioned animals such that no significant difference was detected between sham-lesioned and casted/lesioned animals. (From Cohen et al., 2003).

Subsequently, we have used a similar strategy to assess the benefits of motor therapy for 7 days prior to infusion of 6-OHDA into the MFB. It was observed that motor therapy *prior* to infusion of 6-OHDA also reduced the behavioral and biochemical deficits, which suggests that some of the 6-OHDA-exposed neurons are more resistant to damage incurred from the toxin (Figure 3; Cohen et al., (2003)).

Regeneration, sprouting, or reduced vulnerability? There are many possible explanations for the ability of forced motor activity to reduce the behavioral effects of 6-OHDA and the loss of DA in the striatum, including regeneration of damaged axons, sprouting of intact neurons, or even neurogenesis. Upon further investigation, we observed that animals subjected to 7 days of motor therapy prior to infusion of 6-OHDA do not display the characteristic loss of tyrosine hydroxylase (TH) immunoreactive neurons and terminals in the SN and striatum respectively.

To further examine this phenomenon we used Fluorojade B, a fluorochrome that has been shown to label degenerating neurons and their processes (Schmued et al., 1997), including neurons exposed to 6-OHDA (Zuch et al., 2000). Motor therapy via a unilateral cast completely blocked the increase in Fluorojade B staining in the striatum seen in 6-OHDA treated animals that did not undergo motor therapy (Cohen et al., 2004). This suggests that motor therapy, at least prior to toxic insult, decreases the vulnerability of DA neurons in the SN that normally succumb to oxidative stress induced by 6-OHDA.

Effects of limb "disuse" in an animal model: Interestingly, if animals previously protected against 6-OHDA toxicity, have their "protected" limb casted for a 1-week period 2-3 weeks after infusion, inducing a period of disuse, evidence of toxicity emerges (Figure 4; Tillerson et al., 2002). Indeed, these animals display behavioral and biochemical deficits comparable to animals that had a severe lesion ($\geq$ 90% loss of DA) in the absence of protective motor therapy. If in the absence of continued or additional toxin exposure, the lesion can emerge, what does this say about the status of the neurons after motor therapy immediately following 6-OHDA infusion? The data from Tillerson et al., (2002) would suggest that although the neurons have been spared from degeneration, they are not "normal", as compared to non-toxin exposed neurons, but exist in a heightened state of vulnerability. Whether the neurons remain in this heightened state indefinitely is not known. It is also not known if continued motor therapy would be required to maintain the neurons in the face of continued oxidative stress such as that which exists in the nigrostriatal pathway due to the auto-oxidation or enzymatic breakdown of DA or which occurs with normal aging considering that age is a risk factor for PD. This "disuse" experiment has not yet been done with animals forced to exercise *before* the 6-OHDA. Thus, it is possible that different mechanisms underlie the protective effects of post-6-OHDA and pre-6-OHDA motor therapy. That is to say that, nigral neurons associated with the overused limb prior to 6-OHDA infusion may actually have been "protected" from 6-OHDA toxicity altogether, whereas nigral neurons exposed to 6-OHDA prior to motor therapy may have already initiated a toxic response and remain vulnerable even if appearing to have been rescued.

Additional forms of exercise: The beneficial effects of other forms of exercise, including both voluntary and involuntary running, have also been investigated in animal models of PD. Voluntary running has been shown to provide limited behavioral protection (Mabandla et al., 2004) against 6-OHDA toxicity. However, the integrity of the dopaminergic nigrostriatal projection was not assessed in that study. Voluntary running has also been reported to protect TH immunoreactive neurons in the SN against MPTP in the absence of an effect on TH-immunoreactive terminals (Faherty et al., 2005). Although behavior was not assessed in that study, previous studies have reported continued behavioral deficits when DA neurons are protected in the absence of protection of DA terminals in the striatum (Rosenblad et al., 1999). The latest time point assessed by Faherty et al was 14 days, thus it is possible that what these investigators were observing was a hastening of the recovery of TH-immunoreactive terminals known to occur in young adult mice treated with MPTP (Saitoh et al., 1987). On the other hand, it is possible that they observed a delaying of progressive degeneration of DA neurons that would eventually have occurred. Involuntary running via treadmill ameliorates behavioral deficits (Tillerson et al., 2003; Fisher et al., 2004) and provides some degree of biochemical protection as assessed by measuring striatal DA content (Tillerson et al., 2003).

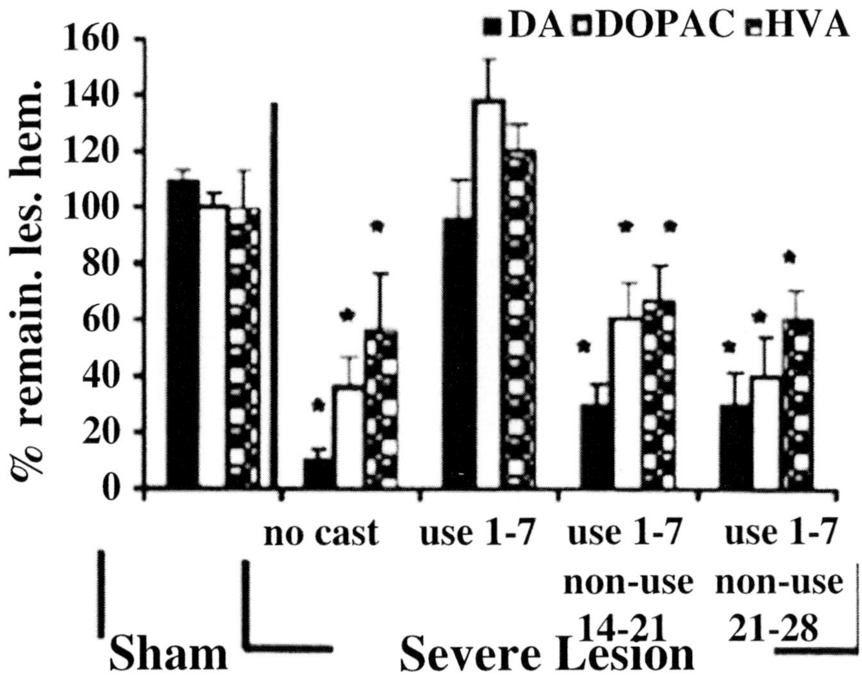

Figure 4. Effect of non-use on DA, DOPAC and HVA content in the striatum after prior use-dependent protection against severe 6-OHDA toxicity (*$P < 0.01$ versus sham). From Tillerson et al., 2002.

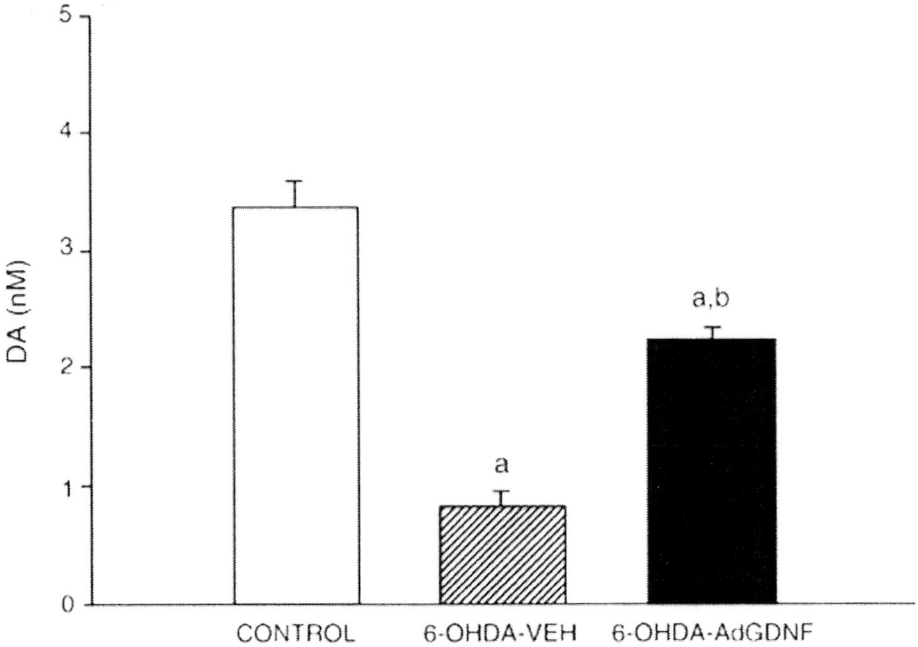

Figure 5. Basal extracellular concentration of DA in the striatum 3 weeks post-6-OHDA infusion. The extracellular concentration of DA in control animals was 3.4 ± 0.1 nM ($n = 4$). Infusion of 6-OHDA significantly decreased the concentration of DA to 0.8 ± 0.01 nM (a, $P < 0.001$ vs. control). Treatment with AdGDNF in the SN 1 week after 6-OHDA infusion increased DA levels in the striatum to 65% of control (2.2 ± 0.07 nM, $n = 9$), which was significantly higher than animals treated with vehicle [b, $P < 0.001$ vs. 6-OHDA-PBS/AdLacZ or vehicle treated animals (6-OHDA-Veh)]. However, DA levels in AdGDNF treated animals were still significantly lower than control animals (a, $P < 0.001$ vs. control). From Smith et al., 2005.

How does exercise protect? The mechanism by which motor therapy prevents/reduces the impact of oxidative stress on DA neurons may involve increased production of neurotrophic factors. Motor therapy in various forms has been shown to increase GDNF within the SN and the striatum (Cohen et al., 2003; Faherty et al., 2005). Exogenous administration of the GDNF gene (Bilang-Bleuel et al., 1997; Choi-Lundberg et al., 1998; Kordower et al., 2000; Kozlowski et al., 2000; Smith et al., 2005) or protein (Hoffer et al., 1994; Kearns and Gash, 1995) has been shown to protect the DA nigrostriatal projection in various animal models of PD (Figure 5; Smith et al., 2005).

Running alone causes a decrease in the expression of the DA transporter (DAT; Fisher et al., 2004; Faherty et al., 2005). MPTP and 6-OHDA are known to exert their toxic effects on the nigrostriatal DA system by being taken up by the DAT and DA transport inhibitors are protective against MPTP and 6-OHDA toxicity (Lees et al., 1984; 1985; Mayer et al., 1985; 1986). Thus, in some studies it is possible that DA neurons in exercised animals may actually be protected from the toxin itself. However, this seems unlikely in most of the cases reviewed above, since the exercise was initiated after toxin administration. More likely, the exercise-induced increase in trophic factors is able to trigger intracellular signaling cascades such as those involving MAP kinases that either reduce oxidative stress or enhance antioxidant defenses.

Enriched Environment

There have been many reports on the ability of enriched environments to promote synaptogenesis (Sirevaag and Greenough, 1988; Greenough and Chang, 1989; Comery et al., 1996; Kolb et al., 2003) and neurogenesis (Kempermann et al., 1998; Kempermann and Gage, 1999; Briones et al., 2005). Recently, two laboratories have described the benefits of an enrichment environment against MPTP toxicity in mice (Bezard et al., 2003; Faherty et al., 2005). Providing a social environment for ≥ 2 months with daily access to a running wheel, a small house and a sampling of 6-7 toys or tunnels decreased the degree of damage incurred from MPTP treatment. Standard housed animals had a significant loss of DA neurons that was decreased to $\leq 40\%$ in animals housed in an enriched environment (Figure 6; Bezard et al., 2003; Faherty et al., 2005). As with exercise alone, this decrease in damage was associated with a downregulation of the DAT and an increase in mRNA for BDNF (Bezard et al., 2003) and GDNF (Faherty et al., 2005). As discussed earlier, this combination of changes could play a role in the decreased vulnerability of DA neurons to MPTP.

It is not yet possible to determine which aspect of an enriched environment provides the neuroprotective stimulus. This is because animals in an enriched environment receive many types of stimulation, including social, mental, and physical. To shed some light on this issue, in addition to housing animals in an enriched environment with access to a running wheel, Faherty and colleagues (2005) housed animals in standard housing cages equipped with running wheels. As discussed in the previous section, they observed that running provided almost complete protection of DA neurons against MPTP, suggesting that running alone is sufficient to protect against toxicity in this case.

Caloric Restriction

Caloric restriction has long been known to increase lifespan in various animal species including rodents (Weindruch et al., 1986; Koubova and Guarente, 2003) and to reverse some of the damaging effects of aging on the brain (for review see Ingram et al., 2001; Mattson et al., 2001). However, only recently have investigators begun to study its potential benefits in treating age-related neurodegenerative diseases such as PD. Duan and Mattson (1999) have shown that reducing caloric intake by 30-40% in young mice for 3 months prior to MPTP administration improved motor performance as assessed by time spent on the rotarod (Figure 7; Duan and Mattson, 1999). Assessment of TH-immunoreactivity in the SN and the striatum indicated a near complete protection of DA cells and terminals respectively (Figure 8; Duan and Mattson, 1999). Caloric restriction also attenuated the behavioral and biochemical deficits in a hemiparkinsonian primate MPTP model of PD (Maswood et al., 2004). Within 24 hours after MPTP administration into the right carotid artery, monkeys maintained on normal and caloric restricted diets displayed the cardinal symptoms of PD, particularly on the side of the body contralateral to MPTP administration. Approximately 6-8 weeks after MPTP treatment movement distance and speed were significantly better in animals subjected to a reduced calorie diet. Additionally, caloric restricted animals displayed better manual dexterity in retrieving treats with the affected hand. The extent of the damage to the striatum was assessed using both positron emission tomography (PET) and post-mortem biochemical analysis of the caudate-putamen. Whereas the 6-[^{18}F]fluoro-*m*-tyrosine PET scan did not display any significant difference between groups, biochemical analysis of DA and DOPAC content in the caudate-putamen was significantly spared in monkeys maintained on a reduced calorie diet. Given that age is the greatest risk factor for PD, it is worth noting that caloric restriction was found not to protect against MPTP toxicity in 1-year-old mice (Morgan et al., 2003) suggesting that the cellular and molecular environment in the aged diseased brain may not be conducive to protective therapy.

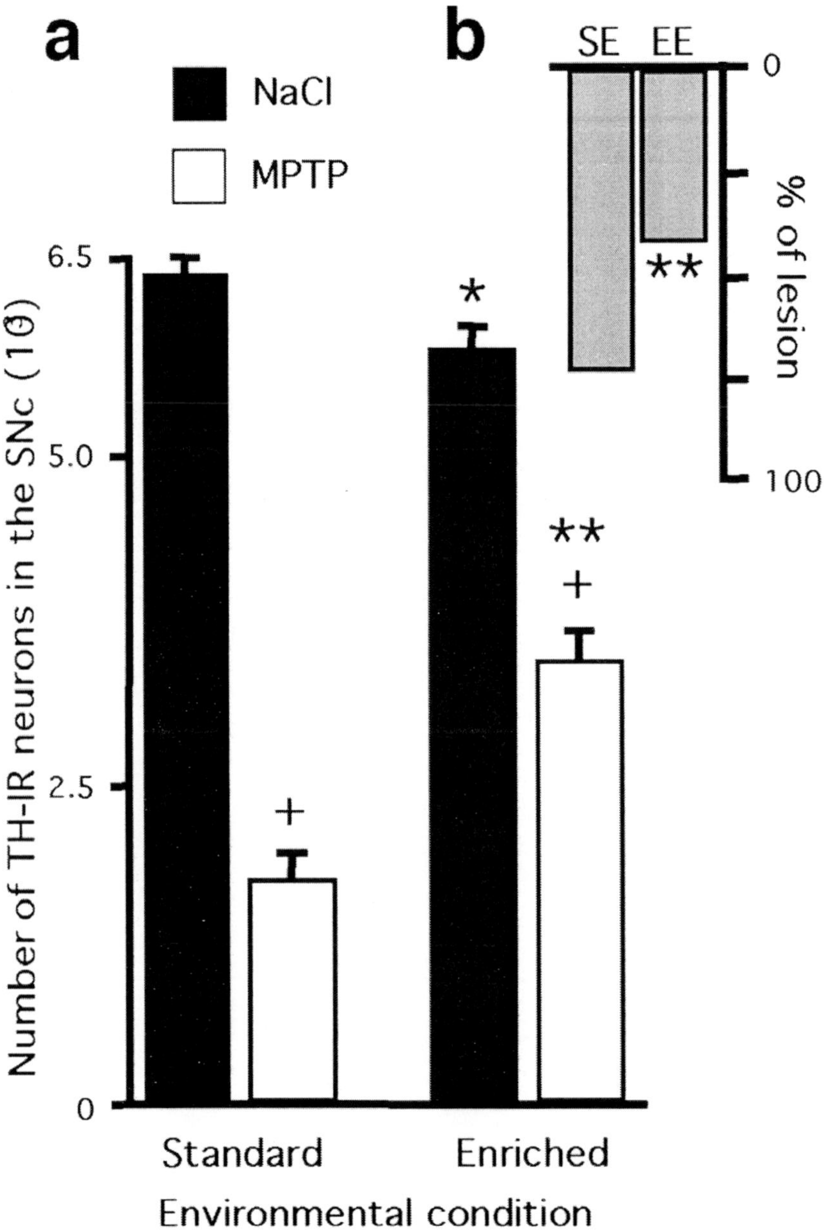

Figure 6. Effect of enriched environment on MPTP toxicity in mice (*$P < 0.05$ compared to standard housed NaCl-treated animals; +$p < 0.05$ MPTP-treated animals compared to their respective NaCl-treated control animals; **$P < 0.05$ MPTP-treated animals housed in an enriched environment compared to MPTP-treated animals housed under standard conditions). From Bezard et al., 2003.

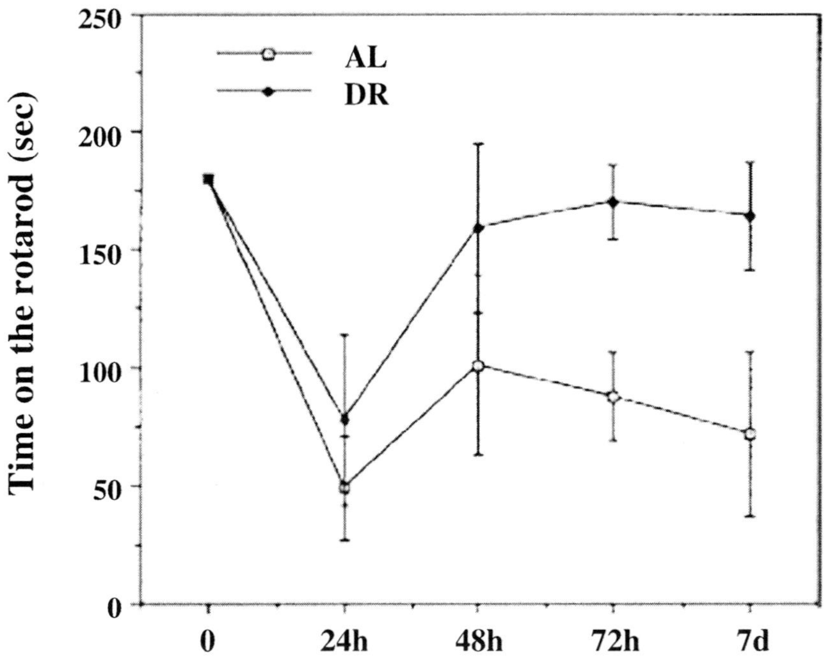

Figure 7. Time spent on the rotarod in MPTP-treated mice fed an ad libitum (AL) or caloric restricted diet (DR). From Duan and Mattson, 1999.

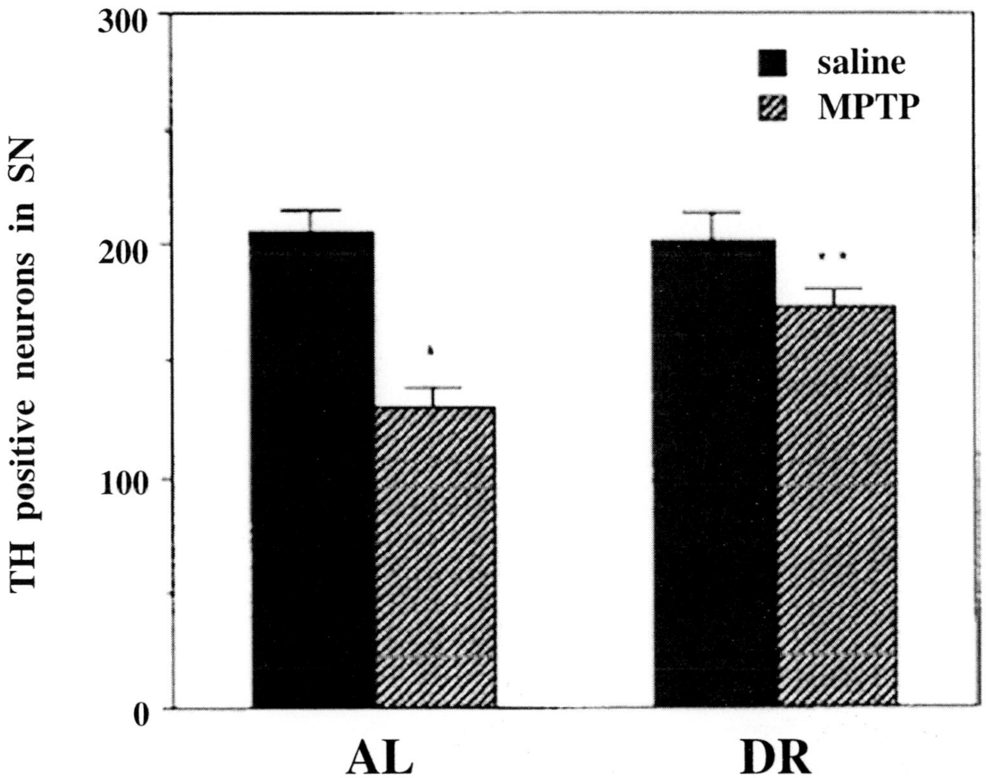

Figure 8. The number of TH immunoreactive neurons in the SN of MPTP-or saline-treated mice fed ad libitum (AL) or a caloric restricted diet (DR) 7 days after MPTP administration (*$P < 0.01$ compared to AL-saline; **$P < 0.01$ compared to AL-MPTP). From Duan and Mattson, 1999.

As with enriched environment, a consequence of the reduced calorie diet is increased motor activity, and like motor activity, caloric restriction increases the levels of protective factors within the nigrostriatal pathway. Indeed, caloric restriction increases levels of GDNF, BDNF (Duan et al., 2001a; 2001b; Maswood et al., 2004) and the stress protein, heat shock protein 70 (HSP 70) in the striatum (Duan and Mattson, 1999). Like GDNF, BDNF is protective in animal models of PD (Levivier et al., 1995; Shults et al., 1995) and delivery of the HSP 70 gene has recently been shown to protect against MPTP toxicity in mice (Dong et al., 2005). Thus, it is not completely clear, at least with regards to PD, if caloric restriction has any beneficial effect independent of that provided by increased motor activity, or if the similarities between the two therapies—indeed enriched environment as well, suggest shared mechanisms.

Role of Stress in Neuroprotection

The one common feature of the three therapies – motor therapy, enriched environment, and caloric restriction – that may be relevant to their protective effects against normal aging and PD is that they present the brain with a physiological stress (for example, Patel and Finch, 2002; Adlard and Cotman, 2004). Faced with this threat, defense mechanisms are engaged to combat the perceived danger. Primary defense mechanisms such as increased neurotrophic factor expression, heat shock proteins, and increased anti-oxidant enzyme activity, as well as secondary defense mechanisms, biomolecular repair and removal processes have been reported to be enhanced in response to these therapies (for review see Mattson et al., 2002a; 2002b; Rao, 2003; Cabelof et al., 2003). The resulting effect is a favorable shift in the pro-oxidant—anti-oxidant balance in the brain towards a more anti-oxidant status with a reduction in oxidatively modified proteins, lipids, and DNA (Sanz et al., 2005). This process has been referred to as *hormesis* or *preconditioning*; that is using the endogenous response to a mild stressor to counteract the effects of a deleterious stressor.

Conclusion

The extended preclinical phase of PD lends itself well to non-invasive therapies such as exercise, caloric restriction, and enrichment, each of which have been shown to be neuroprotective in animal models of the disease. For these interventions to have their greatest benefit, screening tests must be available with the ability to recognize those individuals who are at risk and/or who are in the preclinical phase of the disease. In the meantime, each of us is probably best advised to exercise more, remain stimulated, and eat less.

Acknowledgements

Supported in part by grants from the USPHS (NS19608 and NS045698) and the Michael J. Fox Foundation. My thanks to Michael J. Zigmond, for discussions and comments on this review.

References

Adlard PA, Cotman CW (2004) Voluntary exercise protects against stress-induced decreases in brain-derived neurotrophic factor protein expression. *Neuroscience.* **124**:985-992.

Aksenova MV, Aksenov MY, Carney JM, Butterfield DA (1998) Protein oxidation and enzyme activity decline in old brown Norway rats are reduced by dietary restriction. *Mech Ageing Dev.* **100**:157-168.

Alam ZI, Daniel SE, Lees AJ, Marsden DC, Jenner P, Halliwell B (1997). A generalised increase in protein carbonyls in the brain in Parkinson's but not incidental Lewy body disease. *J Neurochem.* **69**:1326-1329.

Arivazhagan P, Thilakavathy T, Ramanathan K, Kumaran S, Panneerselvam C (2002). Effect of DL-alpha-lipoic acid on the status of lipid peroxidation and protein oxidation in various brain regions of aged rats. *J Nutr Biochem.* **13**:619-624.

Baatile J, Langbein WE, Weaver F, Maloney C, Jost MB (2000) Effect of exercise on perceived quality of life of individuals with Parkinson's disease. *J Rehabil Res Dev.* **37**:529-534.

Bergen JL, Toole T, Elliott RG 3rd, Wallace B, Robinson K, Maitland CG (2002) Aerobic exercise intervention improves aerobic capacity and movement initiation in Parkinson's disease patients. *NeuroRehabilitation.* **17**:161-168.

Bezard E, Dovero S, Belin D, Duconger S, Jackson-Lewis V, Przedborski S, Piazza PV, Gross CE, Jaber M (2003). Enriched environment confers resistance to 1-methyl-4-phenyl-1,2,3,6-tetrahydropyridine and cocaine: involvement of dopamine transporter and trophic factors. *J Neurosci.* **23**:10999-11007.

Bilang-Bleuel A, Revah F, Colin P, Locquet I, Robert JJ, Mallet J, Horellou P (1997) Intrastriatal injection of an adenoviral vector expressing glial-cell-line-derived neurotrophic factor prevents dopaminergic neuron degeneration and behavioral impairment in a rat model of Parkinson disease. *Proc Natl Acad Sci* **94**:8818-8823.

Bokov A, Chaudhuri A, Richardson A (2004) The role of oxidative damage and stress in aging. *Mech Ageing Dev.* **125**:811-826.

Briones TL, Suh E, Hattar H, Wadowska M (2005) Dentate gyrus neurogenesis after cerebral ischemia and behavioral training. *Biol Res Nurs.* **6**:167-179.

Cabelof DC, Yanamadala S, Raffoul JJ, Guo Z, Soofi A, Heydari AR (2003) Caloric restriction promotes genomic stability by induction of base excision repair and reversal of its age-related decline. *DNA Repair* **2**:295-307.

Cadet P, Zhu W, Mantione K, Rymer M, Dardik I, Reisman S, Hagberg S, Stefano GB. (2003)

Cyclic exercise induces anti-inflammatory signal molecule increases in the plasma of Parkinson's patients. *Int J Mol Med.* **12**:485-492.

Callio J, Oury TD, Chu CT (2005). Manganese superoxide dismutase protects against 6-hydroxydopamine injury in mouse brains. *J Biol Chem.* **280**:18536-18542.

Chen H, Zhang SM, Schwarzschild MA, Hernan MA, Ascherio A. (2005) Physical activity and the risk of Parkinson disease. *Neurology* **64**:664-669.

Choi-Lundberg DL, Lin Q, Schallert T, Crippens D, Davidson BL, Chang YN, Chiang YL, Qian J, Bardwaj L, Bohn MC (1998) Behavioral and cellular protection of rat dopaminergic neurons by an adenoviral vector encoding glial cell line-derived neurotrophic factor. *Exp Neurol.* **154**:261-275.

Cohen AD, Tillerson JL, Smith AD, Schallert T, Zigmond MJ (2003) Neuroprotective effects of prior limb use in 6-hydroxydopamine-treated rats: possible role of GDNF. *J Neurochem.* **85**:299-305.

Cohen AD, El Ayadi A, Smith AD, Zigmond, MJ (2004) Forced limb use prior to 6-OHDA reduced vulnerability of dopamine neurons to oxidative stress. SFN abstract.

Collier TJ, Sortwell CE, Daley BF (1999) Diminished viability, growth, and behavioral efficacy of fetal dopamine neuron grafts in aging rats with long-term dopamine depletion: an argument for neurotrophic supplementation. *J Neurosci.* **19**:5563-5573.

Collier TJ, Dung Ling Z, Carvey PM, Fletcher-Turner A, Yurek DM, Sladek JR Jr, Kordower JH (2005) Striatal trophic factor activity in aging monkeys with unilateral MPTP-induced parkinsonism. *Exp Neurol 191 Suppl* **1**:S60-S67.

Comery TA, Stamoudis CX, Irwin SA, Greenough WT (1996) Increased density of multiple-head dendritic spines on medium-sized spiny neurons of the striatum in rats reared in a complex environment. *Neurobiol Learn Mem.* **66**:93-96.

Date I, Yoshimoto Y, Imaoka T, Miyoshi Y, Furuta T, Asari S, Ohmoto T (1994) Effect of host age upon the degree of nigrostriatal dopaminergic system recovery following cografts of adrenal medulla and pretransected peripheral nerve. *Brain Res.* **637**:50-56.

Devi SA, Kiran TR (2004) Regional responses in antioxidant system to exercise training and dietary vitamin E in aging rat brain. *Neurobiol Aging.* **25**:501-508.

Dong Z, Wolfer DP, Lipp HP, Bueler H (2005) Hsp70 gene transfer by adeno-associated virus inhibits MPTP-induced nigrostriatal degeneration in the mouse model of Parkinson disease. *Mol Ther.* **11**(1):80-88.

Driver AS, Kodavanti PR, Mundy WR (2000) Age-related changes in reactive oxygen species production in rat brain homogenates. *Neurotoxicol Teratol.* **22**:175-181.

Duan W, Mattson MP (1999) Dietary restriction and 2-deoxyglucose administration improve behavioral outcome and reduce degeneration of dopaminergic neurons in models of Parkinson's disease. *J Neurosci Res.* **57**:195-206.

Duan W, Guo Z, Mattson MP (2001a) Brain-derived neurotrophic factor mediates an excitoprotective effect of dietary restriction in mice. *J Neurochem.* **76**:619-626.

Duan W, Lee J, Guo Z, Mattson MP (2001b) Dietary restriction stimulates BDNF production in the brain and thereby protects neurons against excitotoxic injury. *J Mol Neurosci.* **16**:1-12.

Dubey A, Forster MJ, Lal H, Sohal RS (1996) Effect of age and caloric intake on protein oxidation in different brain regions and on behavioral functions of the mouse. *Arch Biochem Biophys.* **333**:189-197.

Eberhardt O, Schulz JB (2004) Gene therapy in Parkinson's disease. *Cell Tissue Res.* **318**:243-260.

Faherty CJ, Raviie Shepherd K, Herasimtschuk A, Smeyne RJ (2005). Environmental enrichment in adulthood eliminates neuronal death in experimental Parkinsonism. *Brain Res Mol Brain Res.* **134**:170-179.

Fearnley JM, Lees AJ (1991) Ageing and Parkinson's disease: substantia nigra regional selectivity. *Brain.* **114**:2283-2301.

Fernandez CI, Collazo J, Bauza Y, Castellanos MR, Lopez O (2004) Environmental enrichment-behavior-oxidative stress interactions in the aged rat: issues for a therapeutic approach in human aging. *Ann N Y Acad Sci* **1019**:53-57.

Fisher BE, Petzinger GM, Nixon K, Hogg E, Bremmer S, Meshul CK, Jakowec MW (2004). Exercise-induced behavioral recovery and neuroplasticity in the 1-methyl-4-phenyl-1,2,3,6-tetrahydropyridine-lesioned mouse basal ganglia. *J Neurosci Res.* **77**:378-390.

Floor E, Wetzel MG (1998). Increased protein oxidation in human substantia nigra pars compacta in comparison with basal ganglia and prefrontal cortex measured with an improved dinitrophenylhydrazine assay. *J Neurochem.* **70**:268-275.

Gage FH, Bjorklund A, Stenevi U, Dunnett SB (1983) Intracerebral grafting of neuronal cell suspensions. VIII. Survival and growth of implants of nigral and septal cell suspensions in intact brains of aged rats. *Acta Physiol Scand Suppl.* **522**:67-75.

Gobbo OL, O'Mara SM. (2004) Impact of enriched-environment housing on brain-derived neurotrophic factor and on cognitive performance after a transient global ischemia. *Behav Brain Res.***152**:231-41.

Greenough WT, Chang FL (1988) Dendritic pattern formation involves both oriented regression and oriented growth in the barrels of mouse somatosensory cortex. *Brain Res* **471**:148-152.

Harman D (2003) The free radical theory of aging. *Antioxid Redox Signal.* **5**:557-561.

Hirsch MA, Toole T, Maitland CG, Rider RA (2003) The effects of balance training and high-intensity resistance training on persons with idiopathic Parkinson's disease. *Arch Phys Med Rehabil.* **84**:1109-1117.

Hoffer BJ, Hoffman A, Bowenkamp K, Huettl P, Hudson J, Martin D, Lin LF, Gerhardt GA (1994) Glial cell line-derived neurotrophic factor reverses toxin-induced injury to midbrain dopaminergic neurons in vivo. *Neurosci Lett.* **182**:107-111.

Hurwitz A (1989) The benefit of a home exercise regimen for ambulatory Parkinson's disease patients. *J Neurosci Nurs.* **21**:180-184.

Ingram DK, Chefer S, Matochik J, Moscrip TD, Weed J, Roth GS, London ED,Lane MA (2001) Aging and caloric restriction in nonhuman primates: behavioral and in vivo brain imaging studies. *Ann N Y Acad Sci* **928**:316-326.

Jenner P (2003) Oxidative stress in Parkinson's disease. *Ann Neurol.* **53**:S26-S36.

Kearns CM, Gash DM (1995) GDNF protects nigral dopamine neurons against 6-hydroxydopamine in vivo. *Brain Res.* **672**:104-111.

Kempermann G, Brandon EP, Gage FH (1998) Environmental stimulation of 129/SvJ mice causes increased cell proliferation and neurogenesis in the adult dentate gyrus. *Curr Biol.* **8**:939-942.

Kempermann G, Gage FH (1999) Experience-dependent regulation of adult hippocampal neurogenesis: effects of long-term stimulation and stimulus withdrawal. *Hippocampus.* **9**:321-332.

Kish SJ, Morito C, Hornykiewicz O (1985) Glutathione peroxidase activity in Parkinson's disease brain. *Neurosci Lett.* **58**:343-346.

Kolb B, Gorny G, Soderpalm AH, Robinson TE (2003) Environmental complexity has different effects on the structure of neurons in the prefrontal cortex versus the parietal cortex or nucleus accumbens. *Synapse.* **48**:149-153.

Kordower JH, Emborg ME, Bloch J, Ma SY, Chu Y, Leventhal L, McBride J, Chen EY, Palfi S, Roitberg BZ, Brown WD, Holden JE, Pyzalski R, Taylor MD, Carvey P, Ling Z, Trono D, Hantraye P, Deglon N, Aebischer P. Neurodegeneration prevented by lentiviral vector delivery of GDNF in primate models of Parkinson's disease. *Science.* **290**: 767-773.

Koubova J, Guarente L (2003) How does calorie restriction work? *Genes Dev.* 17313-321.

Kozlowski DA, Connor B, Tillerson JL, Schallert T, Bohn MC (2000) Delivery of a GDNF gene into the substantia nigra after a progressive 6-OHDA lesion maintains functional nigrostriatal connections. *Exp Neurol* **166**:1-15.

Lees GJ, Kydd RR, Horsburgh RJ, Wright JJ (1984) Effect of nomifensine on the toxicity of 6-hydroxydopamine for mesotelencephalic dopamine neurons. *Brain Res.* **308**:155-158.

Lees GJ, Kydd RR, Wright JJ (1985) Relationship between sensorimotor neglect and the specificity, degree and locus of mesotelencephalic dopaminergic cell loss following 6-hydroxydopamine. *Psychopharmacology* **85**:115-122.

Levivier M, Przedborski S, Bencsics C, Kang UJ (1995) Intrastriatal implantation of fibroblasts genetically engineered to produce brain-derived neurotrophic factor prevents degeneration of dopaminergic neurons in a rat model of Parkinson's disease. *J Neurosci.* **15**:7810-7820.

Lökk J (2000) The effects of mountain exercise in Parkinsonian persons - a preliminary study. *Arch Gerontol Geriatr.* **31**:19-25.

Mabandla M, Kellaway L, St Clair Gibson A, Russell VA. (2004) Voluntary running provides neuroprotection in rats after 6-hydroxydopamine injection into the medial forebrain bundle. *Metab Brain Dis.* **19**:43-50.

Maswood N, Young J, Tilmont E, Zhang Z, Gash DM, Gerhardt GA, Grondin R, Roth GS, Mattison J, Lane MA, Carson RE, Cohen RM, Mouton PR, Quigley C, Mattson MP, Ingram DK. Caloric restriction increases neurotrophic factor levels and attenuates neurochemical and behavioral deficits in a primate model of Parkinson's disease. *Proc Natl Acad Sci* **101**:18171-18176.

Mattson MP, Duan W, Lee J, Guo Z (2001) Suppression of brain aging and neurodegenerative disorders by dietary restriction and environmental enrichment: molecular mechanisms. *Mech Ageing Dev.* **122**:757-778.

Mattson MP, Chan SL, Duan W (2002a) Modification of brain aging and neurodegenerative disorders by genes, diet, and behavior. *Physiol Rev.* **82**:637-672.

Mattson MP, Duan W, Chan SL, Cheng A, Haughey N, Gary DS, Guo Z, Lee J, Furukawa K (2002b) Neuroprotective and neurorestorative signal transduction mechanisms in brain aging: modification by genes, diet and behavior. *Neurobiol Aging* **23**:695-705.

Miyai I, Fujimoto Y, Ueda Y, Yamamoto H, Nozaki S, Saito T, Kang J (2000) Treadmill training with body weight support: its effect on Parkinson's disease. *Arch Phys Med Rehabil.* **81**:849-852.

Mayer RA, Jarvis MF, Wagner GC (1985) Cocaine blocks the dopamine depletion induced by MPTP. *Res Commun Chem Pathol Pharmacol.* **49**:145-148.

Mayer RA, Kindt MV, Heikkila RE (1986) Prevention of the nigrostriatal toxicity of 1-methyl-4-phenyl-1,2,3,6-tetrahydropyridine by inhibitors of 3,4-dihydroxyphenylethylamine transport. *J Neurochem.* **47**:1073-1079.

McGeer PL, Itagaki S, Akiyama H, McGeer EG (1988) Rate of cell death in parkinsonism indicates active neuropathological process. *Ann Neurol.* **24**:574-576.

Morgan WW, Richardson AG, Nelson JF (2003) Dietary restriction does not protect the nigrostriatal dopaminergic pathway of older animals from low-dose MPTP-induced neurotoxicity. *J Gerontol A Biol Sci Med Sci.* **58**:B394-B399.

Navarro A, Gomez C, Lopez-Cepero JM, Boveris A (2004) Beneficial effects of moderate exercise on mice aging: survival, behavior, oxidative stress, and mitochondrial electron transfer. *Am J Physiol Regul Integr Comp Physiol.* **286**:R505-R511.

Olanow CW, Tatton WG (1999) Etiology and pathogenesis of Parkinson's disease. *Annu Rev Neurosci.* **22**:123-144.

Palmer SS, Mortimer JA, Webster DD, Bistevins R, Dickinson GL (1986) Exercise therapy for Parkinson's disease. *Arch Phys Med Rehabil.* **67**:741-745.

Patel NV, Finch CE (2002) The glucocorticoid paradox of caloric restriction in slowing brain aging. *Neurobiol Aging.* **23**:707-717.

Peng J, Stevenson FF, Doctrow SR, Andersen JK (2005). Superoxide dismutase/catalase mimetics are neuroprotective against selective paraquat-mediated dopaminergic neuron death in the substantial nigra: Implications for Parkinson's disease. *J Biol Chem.* PMID: 15946937

Perry TL, Godin DV, Hansen S (1982) Parkinson's disease: a disorder due to nigral glutathione deficiency? *Neurosci Lett.* **33**:305-310.

Rao KS (2003) Dietary calorie restriction, DNA-repair and brain aging. *Mol Cell Biochem.* **253**:313-318.

Reuter I, Engelhardt M, Stecker K, Baas H (1999) Therapeutic value of exercise training in Parkinson's disease. *Med Sci Sports Exerc.* **31**:1544-1549.

Rosenblad C, Kirik D, Devaux B, Moffat B, Phillips HS, Bjorklund A (1999) Protection and regeneration of nigral dopaminergic neurons by neurturin or GDNF in a partial lesion model of Parkinson's disease after administration into the striatum or the lateral ventricle. *Eur J Neurosci.* **11**:1554-1566.

Saitoh T, Niijima K, Mizuno Y (1987) Long-term effect of 1-methyl-4-phenyl-1,2,3,6-tetrahydropyridine (MPTP) on striatal dopamine content in young and mature mice. *J Neurol Sci.* **77**:229-235.

Sanz A, Caro P, Ibanez J, Gomez J, Gredilla R, Barja G (2005) Dietary restriction at old age lowers mitochondrial oxygen radical production and leak at complex I and oxidative DNA damage in rat brain. *J Bioenerg Biomembr.* **37**:83-90.

Sasaki T, Senda M, Kim S, Kojima S, Kubodera A (2001) Age-related changes of glutathione content, glucose transport and metabolism, and mitochondrial electron transfer function in mouse brain. *Nucl Med Biol. 2001 Jan;***28**(1):25-31.

Sasco AJ, Paffenbarger RS Jr, Gendre I, Wing AL (1992) The role of physical exercise in the occurrence of Parkinson's disease. *Arch Neurol.* **49**:360-365.

Schapira AH (2001) Causes of neuronal death in Parkinson's disease. *Adv Neurol.* **86**: 155-162.

Schenkman M, Cutson TM, Kuchibhatla M, Chandler J, Pieper CF, Ray L, Laub KC (1998) Exercise to improve spinal flexibility and function for people with Parkinson's disease: a randomized, controlled trial. *J Am Geriatr Soc.* **46**:1207-1216.

Schmued LC, Albertson C, Slikker W Jr (1997) Fluoro-Jade: a novel fluorochrome for the sensitive and reliable Histochemical localization of neuronal degeneration. *Brain Res.* **751**:37-46.

Shults CW, Kimber T, Altar CA (1995) BDNF attenuates the effects of intrastriatal injection of 6-hydroxydopamine. *Neuroreport.* **6**:1109-1112.

Sirevaag AM, Black JE, Shafron D, Greenough WT (1988) Direct evidence that complex experience increases capillary branching and surface area in visual cortex of young rats. *Brain Res.* **471**:299-304.

Smith AD, Kozlowski DA, Bohn MC, Zigmond MJ (2005) Effect of AdGDNF on dopaminergic neurotransmission in the striatum of 6-OHDA-treated rats. *Exp Neurol.* **193**:420-426.

Stokes AH, Hastings TG, Vrana KE (1999) Cytotoxic and genotoxic potential of dopamine. *J Neurosci Res.* **55**:659-565.

Sunvisson H, Lokk J, Ericson K, Winblad B, Ekman SL (1997) Changes in motor performance in persons with Parkinson's disease after exercise in a mountain area. *J Neurosci Nurs.* **29**:255-260.

Thiruchelvam M, Prokopenko O, Cory-Slechta DA, Richfield EK, Buckley B, Mirochnitchenko O (2005). Overexpression of superoxide dismutase or glutathione peroxidase protects against the paraquat + maneb-induced Parkinson disease phenotype. *J Biol Chem.* **280**:22530-22539.

Tillerson JL, Cohen AD, Philhower J, Miller GW, Zigmond MJ, Schallert T (2001). Forced limb-use effects on the behavioral and neurochemical effects of 6-hydroxydopamine. *J Neurosci.;***21**:4427-4435.

Tillerson JL, Cohen AD, Caudle WM, Zigmond MJ, Schallert T, Miller GW (2002). Forced nonuse in unilateral parkinsonian rats exacerbates injury. *J Neurosci* **22**:6790-6799.

Tillerson JL, Caudle WM, Reveron ME, Miller GW. (2003) Exercise induces behavioral recovery and attenuates neurochemical deficits in rodent models of Parkinson's disease. *Neuroscience.* **119**:899-911.

Tsai CH, Lo SK, See LC, Chen HZ, Chen RS, Weng YH, Chang FC, Lu CS (2002) Environmental risk factors of young onset Parkinson's disease: a case-control study. *Clin Neurol Neurosurg.* **104**:328-333.

Volkow ND, Wang GJ, Fowler JS, Logan J, Gatley SJ, MacGregor RR, Schlyer DJ, Hitzemann R, Wolf AP (1996) Measuring age-related changes in dopamine D2 receptors with 11C-raclopride and 18F-N-methylspiroperidol. *Psychiatry Res.* **67**:11-16.

Weindruch R, Walford RL, Fligiel S, Guthrie D (1986) The retardation of aging in mice by dietary restriction: longevity, cancer, immunity and lifetime energy intake. *J Nutr.* **116**:641-654.

Zhang JR, Andrus PK, Hall ED (1994) Age-related phospholipid hydroperoxide levels in gerbil brain measured by HPLC-chemiluminescence and their relation to hydroxyl radical stress. *Brain Res.* **639**:275-282.

Zuch CL, Nordstroem VK, Briedrick LA, Hoernig GR, Granholm AC, Bickford PC (2000) Time course of degenerative alterations in nigral dopaminergic neurons following a 6-hydroxydopamine lesion. *J Comp Neurol.* **427**:440-454.

In: Recent Breakthroughs in Basal Ganglia Research
Editor: Erwan Bezard, pp. 283-284 ISBN: 1-59454-880-3

PART VII: NEURODEGENERATION IN BASAL GANGLIA

Miquel Vila
Department of Neurology, Movement Disorders Division, Columbia University, New York, USA

Neuron cell death in the basal ganglia is epitomized by two neurodegenerative conditions: Parkinson's disease (PD) and Huntington's disease (HD). PD is traditionally associated with the loss of pigmented dopamine-containing neurons of the substantia nigra pars compacta (SNpc) while in HD there is selective degeneration of striatal neurons, with marked atrophy of the putamen and caudate nucleus. Both disorders are characterized by the presence of intracellular proteinaceous inclusions of uncertain pathogenic significance.

This section will be devoted to the molecular mechanisms that lead to cell death in PD, with some references to HD and other neurodegenerative conditions of the basal ganglia. The cause of dopaminergic cell death in sporadic PD remains unknown. However, in recent years the identification of several genetic defects linked to familial forms of PD are providing researchers with new important clues to help unravel the molecular mechanisms underlying dopaminergic cell death (Vila and Przedborski, 2004). The different chapters in this section clearly illustrate how the new genetic discoveries are transforming the field of PD research.

Wilson and Giasson enlighten us about α-synuclein, the first identified gene directly linked to PD. This protein has not only become one of the main players in the field of PD research but is also at the origin of a new term, "synucleinopathies", that identifies neurodegenerative conditions that share common pathological lesions composed of aggregates of α-synuclein. In this context, Leonidas Stefanis discusses mounting evidence indicating that impaired protein degradation and accumulation of misfolded proteins may play a crucial role in PD-related neuron cell death. The possibility that activation of programmed cell death (PCD) molecular pathways may be activated by different cellular stressors, including protein aggregation, is addressed by Andreas Hartmann. Recent advances in the mechanisms of PCD and in the genetics of PD have strengthened the potential role for mitochondrial dysfunction in dopaminergic cell death. Kim Tieu discusses new evidence that re-situates mitochondria at the forefront of PD research. Finally, Eric Benner discusses about the role of inflammatory cells in dopaminergic cell death, as well as about the provocative possibility that vaccination strategies may be beneficial for PD patients.

The different pathogenic mechanisms discussed here are not mutually exclusive. On the contrary, all of them seem to be in play in PD-related cell death, either simultaneously or sequentially, therefore unraveling many molecular targets for potential therapeutic intervention. The elucidation of the molecular mechanisms of cell death should thus allow the development of new therapeutic strategies to slow down or stop the neurodegenerative process in PD and other neurodegenerative disorders of the basal ganglia. However, a few important considerations have to be kept in mind: (i) any therapeutic intervention aimed at targeting basic, etiopathogenetic mechanisms of neurodegeneration should be applied early in the course of the disease to be successful (preferably, even before symptoms occur, which would ask for the development of reliable disease biomarkers); (ii) due to the multiplicity of molecular pathways involved in PD-related neuron cell death, optimal neuroprotection might require administration of a pharmacological cocktail directed against several pathogenic events; (iii) neurodegeneration in PD extends well beyond SNpc dopaminergic neurons, including noradrenergic (locus coeruleus), serotonergic (raphe), and cholinergic (nucleus basalis of Meynert, dorsal motor nucleus of vagus) systems; therefore, potential novel therapies should take into consideration the widespread nature of PD-related neurodegeneration.

In: Recent Breakthroughs in Basal Ganglia Research
Editor: Erwan Bezard, pp. 285-301
ISBN: 1-59454-880-3

Chapter 22

The Janus Faces of Neuroinflammation

Eric J. Benner*[*], *R. Lee Mosley and Howard E. Gendelman
Center for Neurovirology and Neurodegenerative Disorders, Departments of Pharmacology, Pathology and Microbiology and Internal Medicine,
University of Nebraska Medical Center, Omaha, NE

Abstract

Neuroinflammation perpetrated, in significant measure, through activation of brain mononuclear phagocytes (MP; perivascular and parenchymal macrophages and microglia) is linked to the pathogenesis of diverse sets of neurodegenerative diseases including Parkinson's, Alzheimer's and Huntington's diseases, HIV-1 associated dementia (HAD) and spongiform encephalopathies. Central nervous system inflammatory processes, however, consist of complex array of cellular and molecular players of which no single cell type acts alone. This chapter serves to review how inflammatory responses are produced, regulated and how they affect a variety of neurodegenerative processes. The initial responders, or the cellular elements of innate immunity, set up a cascade that may later involve the activation and recruitment of the adaptive immune system. Both are discussed in the context of neural injury and repair.

Introduction

Over a century ago, Elie Metchnikoff discovered an inflammatory cell type in starfish larvae capable of engulfing foreign objects and cellular debris leaving him to postulate that these cells were beneficial to the host. This led to his now famous theory of phagocytosis, the first line of defense whereby macrophages protect an organism against pathogenic microbes. For this pioneering work, he was awarded, ultimately the 1908 Nobel Prize in Medicine and Physiology. Since that time a great deal of progress was made in understanding the role macrophages play in immunity. Phagocytic function is only one of the roles this cell utilizes to maintain tissue homeostasis. The macrophage also orchestrates other cellular processes, including but not limited to, intracellular killing of pathogenic microbes, antigen presentation, secretion of biologically active factors as well as playing a role as an instigator in pathological processes. What underlies such cellular functions is inflammation. Inflammation is the frontline defense of multi-cellular organisms against infection and its absence is incompatible with life. The intent of this evolutionarily conserved response is beneficial to the host. Indeed, inflammation enables the host to fend off various disease-causing microbes including bacteria, viruses and parasites. The instant potential virulent microorganisms enter the body; inflammation enables a defense, serving to eliminate the invader along with damaged tissue. All together, inflammation is a vital sensor against invasion by microbial pathogens as well as a vital component of wound healing following acute tissue infection or injury.

[*] Corresponding author: Eric J. Benner, Ph.D: Center for Neurovirology and Neurodegenerative Disorders, 985215 Nebraska Medical Center, University of Nebraska Medical Center, Omaha, NE 68198-5215. Phone 402-559-4035 / fax: 402-559-8922. E-mail address: ebenner@unmc.edu

However, this is not always true. An inflammatory response is also inexplicitly linked to a number of degenerative states including, but not limited to, cancer, arthritis, cardiovascular disease and numerous autoimmune diseases. With regards to the nervous system recent data suggests that neuroinflammation perpetrated through activation of brain mononuclear phagocytes (MP; perivascular and parenchymal macrophages and microglia) and other glial elements including astrocytes and to a lesser degree endothelial cells. These may also act in concert as a central pathway in a diverse set of neurodegenerative diseases including Parkinson's, Alzheimer's and Huntington's diseases (PD, AD, HD), HIV-1-associated dementia (HAD) and more recently for spongiform encephalopathies or prion-mediated neurodegeneration. Central nervous system (CNS) inflammatory infiltrates consist of complex array of cellular and molecular players of which no single cell type acts alone. The initial responders or the cell elements of innate immunity are glia that set up a cascade, which may later involve the activation and recruitment of the adaptive immune system.

Microglia are the primary MPs in the central nervous system that respond to injury (Kreutzberg, 1996). The cell's principal function is in brain defense. As professional phagocytes, they scavenge microbes and serve as effectors of innate immune responses and coordinate adaptive immune responses within the CNS. Activated microglia participate in inflammatory processes linked to neurodegeneration by producing neurotoxic factors including quinolinic acid, superoxide anions, matrix metalloproteinases (MMP), nitric oxide, arachidonic acid and its metabolites, chemokines, pro-inflammatory cytokines and excitotoxins including glutamate. On the other hand, neuroprotective functions of microglia have been suggested through their abilities to produce neurotrophins and to eliminate excitotoxins present in the extracellular spaces (Schwartz et al., 2003). In particular, evidence indicates that under circumstances microglia may promote neuronal survival after brain injury (Batchelor et al., 1999; 2002).

Microglia are bone marrow-derived macrophage-lineage cells which enter the brain early during embryogenesis and develop in parallel with the maturation of the nervous system itself. They are the resident phagocytes of the CNS and can react promptly in response to brain insults of various natures, ranging from pathogens to aggregated proteins and to more subtle alterations in their micro-environment such as alterations in ion homeostasis that can affect pathological processes (Kreutzberg, 1996). For this process microglial adopt macrophage-like functions and remove infected or damaged cells thus serving as a sensor in the brain. In the normal brain, microglial cells are in a resting state, their cell bodies barely visible and only few fine ramified processes are detectable. However, in pathological settings, resting microglial cells quickly proliferate, become hypertrophic, and increase or express *de novo* a large number of marker molecules such as CD11b and major histocompatibility complex (MHC) antigens (Banati et al., 1993; Kreutzberg, 1996; McGeer and McGeer, 2002) transforming to macrophage-like cells. Activated microglia are now readily visible, increase their numbers at the affected site and exhibit a "spider-like" or macrophage-like appearance. Ramified microglial change their appearance with retracted processes and enlarged cell bodies. Within the damaged area, the maximal density of activated microglia is located at the epicenter of the lesion close to injured cells (e.g., degenerating neurons). While such changes are clearly implicated in neurodegenerative processes of the CNS, the innate immune system has also been tasked with alternative functions. In addition to guarding the nervous system from invading pathogens, this system is involved in many physiological functions such as tissue remodeling during development or after damage (Wahl et al., 1997; Takeda et al., 2003), transportation of blood lipids (Wang and Munford, 1999; Takeda et al., 2003), and scavenging apoptotic cells (Callahan et al., 2003). Following activation, microglia gradually return to the morphology of the ramified type during regeneration of affected tissue. Neuroprotective responses are elicited through elimination of the ongoing infectious agents by innate immune activities and subsequently through adaptive immune functions orchestrated in the CNS by microglia and other antigen presenting cells (see below).

With regards to the adaptive immune system (humoral, cytotoxic and regulatory components), the CNS has long been regarded as an immune privileged site due to the complexities of the blood-brain barrier (BBB) and its lack of an anatomical lymphatic system. Both serve to prevent or restrict cell and macromolecule entry into the CNS. However, soluble CNS antigens do drain to the deep cervical lymph nodes by way of the olfactory system and CNS derived dendritic cells have been characterized in perivascular regions (Ling et al., 2003; Karman et al., 2004a, 2004b). Under physiological settings, dendritic cells appear sample brain antigens and may play a vital role in maintaining immunological tolerance to CNS antigens in peripheral lymphoid tissues (Karman et al., 2004a, 2004b). For example, foreign antigen injected into the rodent striatum translates into peripheral immunological tolerance. Moreover, tolerance can be transferred to naïve animals by T cells isolated from the cervical nodes suggesting that tolerance to brain antigens presented to the immune system via this route is an active process possibly involving regulatory T cell populations (Wenkel et al., 2000).

While naïve T cells are precluded from CNS entry, neuroinflammation aggressively recruits activated components of the adaptive immune system to sites of active neurodegeneration by increasing expression of cellular adhesion molecules and inducing chemokine gradients (Babcock et al., 2003). Moreover, glial cells secrete toxic factors that disrupt blood brain barrier function. Nonetheless, much evidence indicates a far more complex relationship between the CNS and immunological systems than previously thought. For example, immune molecules such as Thy-1, interleukins, and chemokines are expressed at high levels in neurons and surrounding glia and may be involved in direct communication between the CNS and immune cells (Wyss-Coray and Mucke, 2002). Signaling between neurons and glia during neuronal injury incite inflammatory responses and leukocyte migration (Babcock et al., 2003). Interestingly, molecules mediating specific antigen recognition by T lymphocytes, including major histocompatibility complex (MHC) class I and CD3ζ molecules, also have a role in axonal guidance, activity-dependent remodeling, and plasticity in the developing and mature mammalian CNS (Huh et al., 2000). Within the neuronal synapse, MHC class I molecules may participate in the refinement or elimination of synaptic connections. The cognate receptor for MHC class I peptide complexes is the αβ T cell receptor (TCR) expressed on T cells. It was recently determined that neurons, particularly in the neonatal developing CNS express mRNA transcripts for the unrearranged β subunit of the TCR (Syken and Shatz, 2003; Nishiyori et al., 2004). Functional cooperation between these two molecules in neuronal populations has yet to be determined. Interestingly, during inflammatory states, MHC class I molecules are up-regulated on neuronal surfaces yet there is no direct evidence of cytotoxic T lymphocyte (CTL) mediated neuronal damage in common neurodegerative disorders (Foster et al., 2002; Cabarrocas et al., 2003). However, MHC class I molecules alone are unstable and must associate with proteasome-derived peptides and β2-microglobulin to form stable complexes at the cell surface. Altered peptide profiles presented at the neuronal synapse during neuronal degeneration may therefore also affect neuronal plasticity and remodeling during disease states.

Further challenging this view of the "immune privileged" status of the CNS are multiple animal model systems where deficiencies in adaptive immunity translate into exacerbated neuronal loss following traumatic injuries (Stalder et al., 1998; Serpe et al., 1999; Kipnis et al., 2001; Schori et al., 2002). Such injuries are corrected in animals that receive immune reconstitution prior to experimental injury. Rodents and humans that have sustained CNS injuries also have an expanded T cell repertoire against myelin-associated antigens yet do not appear to be at an increased risk for the development of CNS autoimmunity. If there is any functional consequence of such T cell responses against CNS antigens following injury, it is not known. A more comprehensive understanding of this complex relationship and the ability to decipher the noxious roles from the beneficial aspects of neuroinflammation may lead to insightful therapeutic strategies. In this regard, we will review recent concepts for activation of the innate immune system in the CNS, discuss the role of this response in neurodegeneration with a specific emphasis on PD. Finally the beneficial aspects of this response will be discussed with recent attempts at the modification of innate immunity in neurodegenerative disease by vaccine strategies.

1 Endogenous Triggers of Innate Immunity

Neuronal cell death can occur by apoptosis or necrosis, and both appear to be operative to some extent in neurodegenerative diseases (Vila and Przedborski, 2003). Regardless of the primary insult, the way in which neurons die appears to be an important determinant in the ensuing microglial response. This is of paramount importance as phagocytes clear cells undergoing programmed cell death (PCD) during development and must do so in a manner that does not harm neighboring cells (Sastry and Rao, 2000). The rapid clearance of apoptotic cells prevents the release of toxic or immunostimulatory molecules from dead cell corpses and is thought to play an important role in maintaining homeostatic functions in surrounding tissues (Figure 1). Apoptosis leads to morphological and biochemical alterations at the cell surface making them suitable targets for resident phagocytes. Such changes include the reversal of membrane polarity, yielding a net anionic charge, and the surface expression of modified lipid (phosphatidylserine) and carbohydrate motifs. Interestingly, a component of the lipopolysaccharide (LPS) receptor (CD14) has been shown to interact with the intracellular adhesion molecule ICAM3 exposed on the cell surface of apoptotic cells and affects the phagocytic process (Elward and Gasque, 2003). Apoptosis almost always proceeds in the absence of an inflammatory response and recent data demonstrate that phagocytosis of apoptotic cells actually inhibits innate immune activation (Chan et al., 2001; Magnus et al., 2001). In keeping with this, cells undergoing apoptosis also release anti-inflammatory cytokines such as transforming growth factor beta (TGF-β) and interleukin-10 (IL-10). Moreover, animals with defects in phagocytic clearance of apoptotic cells develop autoantibodies against nuclear and lipid antigens(Asano et al., 2004).

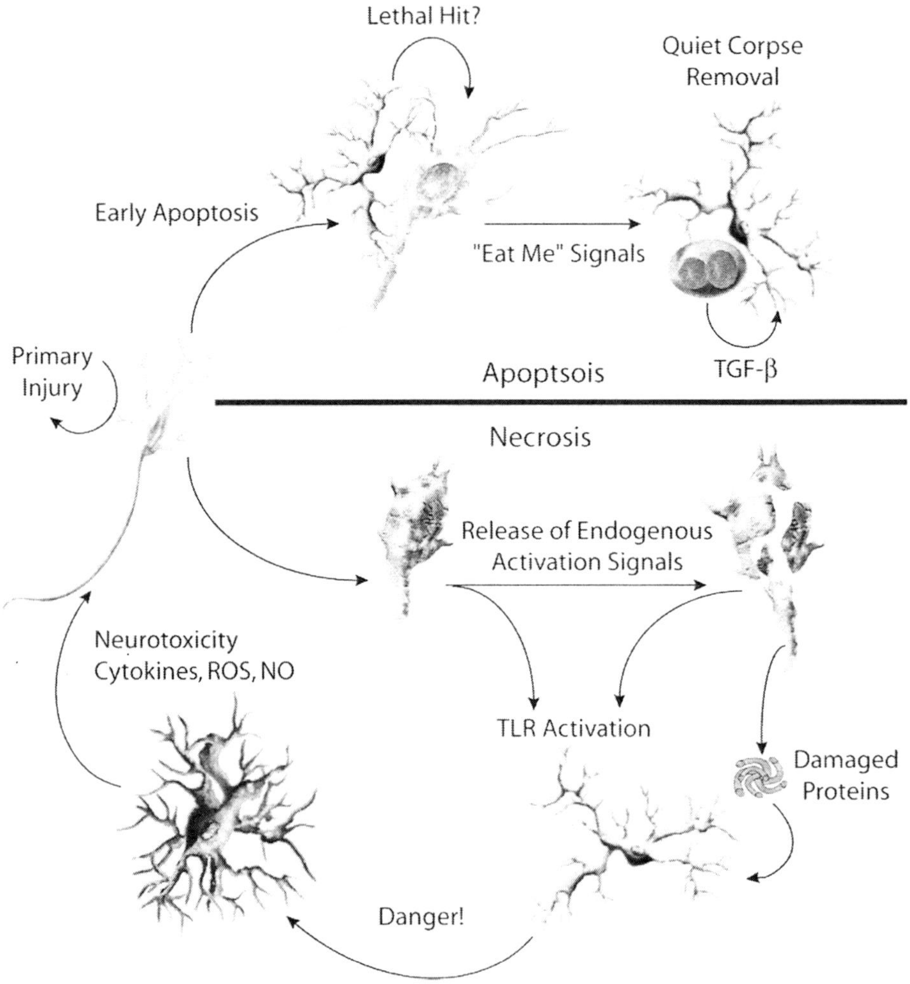

Figure 1. Innate immune responses to "appropriate" and "inappropriate" forms of cell death. In the top panel, early apoptotic neurons undergo membrane alterations and become suitable targets for phagocytes. Phagocytes may also actively participate in target cell death by delivering a lethal hit to dying cells. Apoptotic cells deliver anti-inflammatory factors to phagocytic cells to prevent immune activation. In the bottom panel, necrotic cells release intracellular organelles and damaged proteins, which alerts the innate immune system to danger and activates resident microglia. Chronic immune activation leads to further neurotoxicity through the production of inflammatory mediators and oxidative stress.

Interestingly, PCD and phagocytosis may also be linked on a more functional level. Phagocytes in the nematode *C. elegans* are vital in clearing cells that undergo physiological PCD during development. However, mutation of genes in the phagocytic pathway enhances cell survival when target cells are exposed to weak apoptotic signals (Reddien et al., 2001). It appears that phagocytes act to ensure the death cells that have displayed early signs of apoptosis prior to their engulfment and do so through target cell caspase activation (Hoeppner et al., 2001). This may have implications in neurodegenerative diseases and suggests that phagocytic cells may deliver a lethal hit to stressed cells and by doing so, may play an active role in deciding which cells survive following insult (Figure 1).

While the innate immune system removes apoptotic cells, "inappropriate" forms of cell death (i.e. necrosis) alerts this system to danger and elicits a robust inflammatory response (Gallucci et al., 1999). The molecular recognition of necrotic cells by the innate immune system is poorly understood however, recent data suggest that antigens from necrotic or stressed cells may trigger a class of pattern-recognition receptors (PRRs) designed to recognize evolutionarily distant microorganisms (Seong and Matzinger, 2004). The classical PRRs within the mammalian system are the 11 known Toll-like receptors (TLRs), which are homologues of *Drosophila* Toll. Ligands for TLRs include microbial-specific products such as LPS, lipoarabinomanan and peptidoglycan, or double stranded RNA among others. However, PRRs are likely more promiscuous than previously thought (Bell et al., 2003). In addition to the recognition of microbial products, endogenous host products associated with cellular stress also activate these receptors. While minimal evidence exists for an infectious process in classical neurodegenerative diseases, recent advances in our understanding of PRRs suggests an important role in sterile neurodegenerative states where inflammation is linked to neuronal loss.

In keeping with this, researchers have demonstrated that LPS infusion into the rat CNS produced an inflammatory reaction followed by a gradual loss of dopaminergic neurons. Moreover, LPS appears to synergistically enhance dopaminergic neurotoxicity in rotenone and MPTP models of PD (Gao et al., 2003b; 2003a). LPS signals through TLR4 and CD14 and results in a robust activation of microglial cells. There is very little evidence for an infectious process in PD however, unlike most receptors that bind very precise ligands, TLRs, including TLR4, can interact with a diverse set of ligands, some of which are endogenous host-derived molecules released from necrotic cells. Such immunostimulatory ligands include heat-shock proteins (HSP; Asea et al., 2002), high mobility group 1 protein (HMG1; Park et al., 2004), DNA (Collins et al., 2004), hyaluronan (Termeer et al., 2002), uric acid (Shi et al., 2003), and fibronectin (Okamura et al., 2001). Many of these target ligands are upregulated in stressed cells or liberated from the extracellular matrix during tissue damage. As such, they make excellent candidates for molecular triggers of immune activation following tissue injury and may therefore contribute to neurodegenerative process (Figure 1).

Interestingly, extracellular mitochondria, released during cell death, may also act as a danger signal to the innate immune system. Mitochondria likely evolved from a prokaryotic ancestor and have retained many phylogenic characteristics including double membrane structures, primitive transcription and translation machinery, and stretches of unmethylated CpG DNA motifs associated with microbial genomes. A recent report has demonstrated that mitochondrial DNA (mtDNA) and DNA, which has been modified by oxidation (8-oxodG), strongly activate cells of the MP lineage. Injection of either of these molecules into the articular joints of mice leads to a macrophage-dependent arthritic disease (Collins et al., 2004). This also has implications for neurodegenerative disease where mitochondrial pathology is associated with neuronal loss. This is particularly true in PD where defects in the electron transport chain (ETC) contribute to cellular ROS-mediated damage. mtDNA is particularly vulnerable to oxidation due to its proximity and the lack of histones which can normally buffer DNA damage. Blockade of scavenger or TLR pathways operative in chronic inflammatory disease may provide a powerful tool for therapeutic intervention in neurodegenerative diseases.

Aggregated proteins are also indicators of cellular stress and are potent activators of innate immunity. Most neurodegenerative diseases are associated with the accumulation of abnormal protein aggregates (proteopathies). Aggregated proteins can exert toxicity by disrupting intracellular transport, overwhelming the protein degradation machinery, or by sequestering normal proteins away from their physiological roles. In addition to direct toxicity, mounting evidence suggests that abnormal protein conformations can activate innate immune cells through scavenger receptors thus providing a mechanism for secondary tissue injury (Bradt et al., 1998; Yan et al., 1999; Fischer et al., 2002). Additionally, treatment of neurons with proteasome inhibitors and subsequent accumulation of ubiquinated proteins elicits prostaglandin synthesis via cyclooxygenase-2 (COX-2) suggesting that intracellular accumulation of abnormal proteins can trigger the production of proinflammatory mediators (Rockwell et al., 2000). A major pathological hallmark of PD is the presence of eosinophilic, cytoplasmic inclusions of fibrillar, misfolded proteins called Lewy bodies (LB) suggesting derangements in protein catabolism may be involved in disease pathogenesis (Dauer and Przedborski, 2003). While LBs are intracellular, the fate of associated protein aggregates following cell death is unknown and can be released following necrotic cell death or incomplete apoptosis. Moreover, neurons go to great lengths during times of cellular stress to prevent such aggregation by upregulating ATP-dependent chaparones such as HSPs that colocalize with misfolded proteins in Lbs (Muchowski, 2002). In addition to aggregated proteins, HSPs can be released during cell death and doing so may directly contribute to microglial activation (Kakimura et al., 2002).

2 DELETERIOUS ROLE OF INFLAMMATION IN NEURODEGENERATION

PD is the most common movement disorder whose cardinal features include, rigidity, bradykinesia, resting tremor, or postural instability (Fahn and Przedborski, 2000). Pathologically the disease is characterized by the profound loss of dopamine producing neurons that reside in the substantia nigra pars compacta (SNpc) and project to the caudate-putamen resulting in subsequent loss of striatal dopamine. Neuronal damage is not limited to the nigrostriatal system and is also evident in the locus coeruleus, raphe nuclei and the nucleus basalis of Meynert. With the exception of rare familial forms, the majority of PD cases are sporadic and may be due, in part, to mitochondrial defects at complex I (Dauer and Przedborski, 2003; Dawson and Dawson, 2003). Indeed, complex I inhibitors (MPTP, rotenone) recapitulate many of the pathological features of PD in humans and animal models (Dauer and Przedborski, 2003). Moreover, the presence of ubiquitinated misfolded proteins in LBs suggests derangements in protein quality control systems or degradation pathways may be involved in disease pathogenesis. Indeed recent work with proteasome inhibitors in rats recapitulates many of the PD associated hallmarks including neuronal cell death and intracellular proteinacious inclusions. A common pathological feature of the human disease is the chronic activation of microglial cells around degenerating neurons. Several clinical studies indicate a neuroinflammatory role in disease and disease progression. For example, epidemiological data suggest the use of nonsteroidal anti-inflammatory agents decrease the risk for the development of PD. There is also ample biochemical and histological evidence of oxidative stress in human and animal studies of PD including increased levels of carbonyl and nitrotyrosine protein modifications, lipid peroxidation, DNA damage, and reduction of glutathione and ferritin.

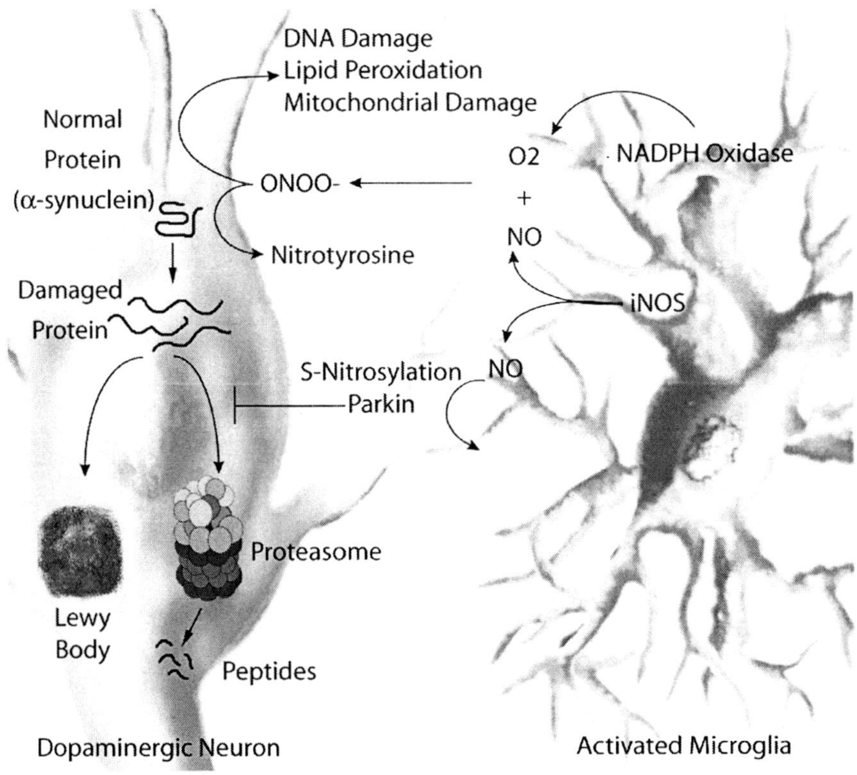

Figure 2. Neuroinflammatory factors contribute to the specific pathology seen in PD. Microglial derived NO and superoxide species react in extracellular spaces to form peroxynitrite. Peroxynitrite readily crosses cell membranes where it contributes to lipid peroxidation, DNA damage and nitrotyrosine formation in α-synuclein and other cellular proteins. Damaged proteins are targeted to cellular proteosomes for degradation via the ubiquitination pathway. Excess NO produced by activated microglia can lead to S-nitrosylation of cellular proteins, including parkin. Such modifications may diminish E3 ubiquitin ligase activity necessary for efficient protein turnover by proteosomes. Excessive protein damage caused by oxidants and disruptions in the ubiquitin pathways may overload or inhibit protein degradation quality control measures leading to the accumulation of damaged proteins in cells.

Once microglia are activated, they are capable of producing noxious factors including pro-inflammatory cytokines, chemokines, quionlinic acid, arachidonic acid and its metabolies and excitatory amino acids among others. Importantly, large amounts of reactive oxygen species (ROS) known as a respiratory burst (Locksley and Klebanoff, 1983; Sankarapandi et al., 1998) may have disastrous effects on delicate neuronal networks in the CNS. Indeed, oxidative stress is implicated as a major cause of neuronal injury in a wide range of neurological diseases including PD. It may by triggered by altered protein configuration including aggregation and results from the accumulation of reactive oxygen and nitrogen by products formed during cell processes such as oxidative phosphorylation. They are typically produced by microglia and serve to destroy invading microorganisms. Reactive oxygen species include superoxide, hyrodgen peroxide and hydroxyl free radicals as well as nitrogen intermediates (nitric oxide and peroxynitrite) and may cause damage to neurons if produced in excess such as during prolonged neuroinflammatory responses. Much of the microglial-derived ROS such as superoxide cannot efficiently transverse cellular membranes (Ischiropoulos and Beckman, 2003), making it unlikely that these extracellular ROS gain access (Wu et al., 2003) to dopaminergic neurons and trigger intraneuronal toxic events. However, superoxide can rapidly react with NO in the extracellular space to form a more stable oxidant, peroxynitrite (Wu et al., 2003). The stability of peroxynitrite facilitates its crossing of cell membranes where it can damage intracellular components in neighboring neurons. Its toxicity has been linked to the disruption of mitochondrial electron transport chain, lipid peroxidation, DNA damage, and the nitration of tyrosine residues in cellular proteins. Therefore, microglial-derived superoxide, by contributing to peroxynitrite formation, may be significant contributor to the pathogenesis of PD (Figure 2). NADPH-oxidase is a large multi-subunit complex and is the main enzyme known to produce ROS in activated macrophages and microglia. Moreover, genetic deletion of gp91, an essential subunit of NAPDH oxidase, mitigates neuronal loss in numerous models of neurodegeneration including the MPTP model of PD (Wu et al., 2003).

NO is a biological messenger molecule that has numerous physiological roles in the CNS. In addition, NO plays an important role in innate immunity and is associated with tumoricidal and bactericidal activities of macrophages (MacMicking et al., 1997). Three distinct forms of nitric oxide synthase (NOS) have been identified to date and are designated neuronal NOS (nNOS), inducible NOS (iNOS), and endothelial NOS (eNOS). In addition to the physiological roles of NO, under pathological settings with excessive production it can act as a potent neurotoxin in a number of neurodegenerative models (Samdani et al., 1997; Dawson and Dawson, 1998; Liberatore et al., 1999; Brown and Bal-Price, 2003). For example, nNOS and iNOS are both upregulated in sporadic PD and genetic ablation or pharmacological inhibition are neuroprotective in the MPTP model (Przedborski et al., 1996; Liberatore et al., 1999). Although NO may generate much of its toxicity through the formation of peroxynitrite, it also reacts with sulfur containing cysteine residues in protein (S-nitrosylation), which may modify protein structural conformations or enzymatic activities.

The generalized toxicity of these inflammatory responses provides very little insight into the selective neurodegeneration patterns observed in various disease states. However, it is tempting to speculate that the shared phenotype of multiple genetic mutations identified in familial forms of PD suggests the derangement of a common pathway may be involved. Consistent with aberrant protein accumulation in PD, malfunction of the ubiquitin-proteasome system appears to be a common link in these familial forms of PD. Indeed, many of the genes identified as of today are linked to protein misfolding and/or degradation pathways (Giasson and Lee, 2003; Vila and Przedborski, 2004). While these genetic mutations offer insight into common pathways involved in familial forms of PD, the information they offer for sporadic forms of the disease in individuals who lack these genetic lesions are not completely understood. Interestingly, recent data suggests that some of these PD associated genes are active targets of reactive nitrogen and oxygen species both of which are generated during chronic inflammation.

In keeping with this, three missense mutations (A53T, A30P, and E46K) have been identified in the gene encoding α-synuclein leading to an autosomal dominant inheritance of PD. Moreover, genomic triplication of the α–synuclein gene is associated with familial PD (Singleton et al., 2003). Transgenic overexpression of wild-type or mutant forms of α–synuclein in mice produces intraneuronal aggregates (Masliah et al., 2000), while in *drosophila*, this resulted in both aggregate formation as well as dopaminergic neuronal cell death (Auluck et al., 2002). In sporadic PD, recent studies support a role for oxidative and/or nitrative stress in α-synuclein modification and aggregation (Giasson et al., 2000). Nitrating agents such as peroxynitrite can readily nitrate α-synuclein at tyrosine residues, and generate highly stable *o,o'*-dityrosine oligomers. These biochemical lesions enhance fibril formation *in vitro*, similar to the biophysical properties of α-synuclein isolated PD brains (Norris et al., 2003). Aberrant protein conformations of modified α-synuclein can potentially overload cellular proteasomes and by doing so, may increase cellular stress associated with the accumulation of misfolded proteins in affected neurons (Figure 2; Lindersson et al., 2004).

Parkin is also a gene associated with familial PD whose protein product may be a target of nitrosative stress-associated protein modifications. Parkin is an ubiquitin E3 ligase responsible for the edition of ubiquitin to protein substrates marked for degradation by cellular proteasomes including α–synuclein and its interacting protein, synphilin-1 (Petrucelli et al., 2002). Over expression on parkin in α–synuclein transgenic flies rescues neurons from degeneration (Yang et al., 2003). Mutations in parkin result in the loss of ubiquitin E3 ligase activity and are found in juvenile PD in an autosomal recessive fashion (Kitada et al., 1998). The posttranslational modification (S-nitrosylation) of parkin also abolishes its E3 ligase activity and kills its ability to rescue cells from α–synuclein/synphilin coexpression in the presence of proteasome inhibition (Chung et al., 2004; Yao et al., 2004). Nitrosylation modifications on parkin were found in sporadic cases of PD in affected brain regions as well as both MPTP and rotenone animal models. Animal studies reveal that nitrosylation of parkin is dependent on both neuronal nitric oxide synthase (nNOS) as well as microglial derived inducible nitric oxide synthase (iNOS; Chung et al., 2004). Thus it is conceivable that inflammation contributes to oxidative modifications in parkin that in turn make affected neurons more vulnerable to cytotoxic stress caused by altered protein catabolism (Figure 2).

On a more general level, matrix metalloproteinases (MMPs) are a class of extracellular soluble or membrane bound cysteine proteases involved in remodeling of the extracellular matrix (ECM). MMPs have also been implicated in a range of neurodegenerative disease including, HAD, AD, PD and stoke. Gu et al. reported that S-nitrosylation of N-terminal cysteine residues within proMMP-9 leads to the subsequent activation of MMP-9 protease activity (Gu et al., 2002). These findings identify a potential extracellular proteolysis mechanism involved in neuronal cell death in which S-nitrosylation activates MMPs. An increase in MMP-9 expression has been determined in the MPTP model and pharmacological inhibition of MMP-9 was neuroprotective (Lorenzl et al., 2004).

3 Dual Role of Innate Immunity in the CNS

The complex role of cellular and molecular inflammatory mediators, particularly within the CNS, makes it difficult to define their specific beneficial or deleterious roles in pathophysiological states (Gendelman, 2002). Cytokines and chemokines are of particular importance in the initiation, maintenance, and resolution of inflammatory reactions within the CNS (Allan and Rothwell, 2001). Moreover, chemokines in particular, have important roles outside of inflammation including chemotactic axonal guidance during developmental stages of the CNS (Mennicken et al., 1999; Tran and Miller, 2003a). With regards to the immune system, cytokines and chemokines act as potent survival cues and their tissue levels directly modulate the activity of inflammatory cell types. Therefore, the function of these molecules within the immune system is reminiscent of the function of neurotrophins within the nervous system (Tran and Miller, 2003b). In fact, many neurotrophins are structurally related to molecules within cytokine families and these two classes of molecules appear to share functional overlap. IL-10 is a potent neuroprotective factor on purified cerebellar neurons (Bachis et al., 2001) and NGF has emerged as a potent anti-inflammatory factor with beneficial effects in the mouse model of MS (Flugel et al., 2001). Indeed, neurons themselves can produce chemokines and express many of their receptors on their surface suggesting direct dialog between these two systems. Neuronal-immune communication is not restricted to soluble mediators. The CD200 receptor (CD200R) was recently identified and is expressed on myeloid cells including microglia and its ligand, CD200, is abundant on neuronal cell surfaces (Hoek et al., 2000). Cognate CD200-CD200R interactions have potent inhibitory effects on cells of the myeloid lineage (Hoek et al., 2000; Barclay et al., 2002). Cells of the adaptive immune system can also communicate with neurons. Activated T cells express numerous neurotrophins (Ehrhard et al., 1993; Kerschensteiner et al., 1999; Moalem et al., 2000) in addition to the neurotrophic factor receptors, trkB and trkC (Besser and Wank, 1999), thus providing sufficient mechanistic means to establish direct T cell-neuron communications.

Results from transgenic mouse studies are equally confusing (Owens et al., 2001). Targeted CNS expression of numerous proinflammatory cytokines, including TNF-α, interferon-α, and IL-6 results in glial activation and subsequent neurodegeneration (Campbell et al., 1993; 1999). However, dopaminergic neurons in IL-6 deficient mice are significantly more vulnerable to MPTP-induced neurodegeneration (Bolin et al., 2002). Additionally, neuronal injury by astroglial overexpression of TNF-α was exacerbated in mice that lack functional T and B cell populations (Stalder et al., 1998) and mice lacking the receptor for TNF-α are more susceptible to neuronal loss following ischemic or excitotoxic injury (Bruce et al., 1996). These selected examples underscore the difficulty in predicting the role these factors may play in neuronal injury and warrant caution while pursuing therapies that modulate such pathways. Neuroinflammation is an "equal opportunity" field and confounding results are not limited to proinflammatory mediators. Cytokines, normally linked to beneficial functions in the CNS, can deliver equally confusing results. Knocking out TGF-β, a cytokine associated with wound repair, leads to prominent glial activation and neuronal apoptosis (Brionne et al., 2003). However, levels of mRNA for TGF-β positively correlate with the extent of amyloid pathology in postmortem AD brains. Overexpression of TGF-β in hAPP transgenic mice resulted in the deposition of Aβ in blood vessels while simultaneously decreasing Aβ load in parenchymal tissue (Wyss-Coray et al., 1997; 2001).

Functional roles for cellular immune responses can also provide conflicting studies. Virally or otherwise activated macrophages can have devastating consequences on neuronal function and survival. Depletion of peripheral monocyte/macrophage lineages in a spinal cord injury model can improve functional recovery (Popovich et al., 1999). Conversely, direct spinal injection of macrophages activated with damaged peripheral nerves *ex vivo* leads to enhanced functional recovery (Rapalino et al., 1998). Resident microglia, while contributing to noxious factors in CNS disease detailed above, also produce a plethora of neurotrophic factors upon activation. Local production of these factors can establish trophic gradients which stimulate and guide axonal sprouting following striatal injury (Batchelor et al., 1999; 2002).

The mere presence of T cells in CNS pathological settings has long been regarded as detrimental. One such case where T cells play a direct role in disease pathogenesis is multiple sclerosis (MS). MS is an inflammatory disease characterized by perivascular cuffs of mononuclear cells that include both lymphocyte and macrophage populations. Accumulation of these cells leads to significant damage to the myelin sheath and can involve the underlying axon (Kornek et al., 2000). Importantly, adoptive transfer of myelin-reactive T cells into naïve recipient mice confers a disease that shares many of the pathological and clinical signs similar to human MS. However, recent provocative data also implicate a role for adaptive immunity in otherwise non-autoimmune disorders of the CNS. Surprisingly, SCID and Rag-2 -/- mice, both deficient in adaptive immunity, show enhanced motor neuron loss within the facial nucleus following transection of the facial nerve (Serpe et al., 2000) consistent with the findings in the TNF-α transgenic neurodegeneration model. Regulatory T cells such as $CD4^{+}CD25^{+}$ play a vital role in inhibiting autoreactive T cell activation and expansion and may be of paramount importance in maintaining peripheral tolerance against CNS antigens (Zhang et al., 2004). Recent studies have shown that depletion of naturally occurring regulatory CD4+CD25+ T cells, which comprise 10 % of the total $CD4^{+}$ population, predisposes animals to development of organ-specific autoimmune diseases (Shevach, 2000). In addition, Kipnis et al. show that depletion of $CD4^{+}CD25^{+}$ regulatory T cells enhances the spontaneous T cell-dependent protective response and hence improves post-injury neuronal survival in the CNS. Conversely, an increase in the number of regulatory T cells had the opposite effect (Kipnis et al., 2002). It appears that animals capable of generating an autoimmune response to CNS antigens subsequent to injury appear to derive benefit from doing so. This suggests that the presence of naturally occurring CD4+CD25+ regulatory T cells may suppress endogenous immune activation against antigens released following CNS injury and in doing so, suppresses the benefits associated with this response.

4 THERAPEUTIC IMMUNIZATION IN NEURODEGENERATIVE DISEASE

The unexpected role of T cells in neuronal outcomes has lead to a number of studies which augment peripheral T cell immunity in an attempt to enhance immune-mediated neuroprotection. Surprisingly, multiple investigators have reported that proinflammatory T_H1 T cell responses generated through immunization with myelin-associated antigens can attenuate secondary neuronal loss following experimentally induced CNS injury (Schwartz, 2001; Weiner and Selkoe, 2002). It appears that these immunization strategies can increase local production of proinflammatory cytokines as well as multiple neurotrophic factors within the injured tissue (Moalem et al., 1999; Hammarberg et al., 2000). These provocative observations generate a new set of biological questions regarding neuro-immune interactions and it has been suggested that the observed neuroprotection can be directly linked to the proinflammatory nature of the infiltrating T cell population (Schwartz et al., 1999). However, detailed mechanistic data is lacking and the true mechanisms will likely prove as complicated as many other inflammatory mediators described above. Contrary to this, oral or nasal immunization that preferentially generates an anti-inflammatory T cell response (T_H2/T_H3) also has a demonstrated neuroprotective element and may prove more clinically feasible (Frenkel et al., 2003; Monsonego et al., 2003).

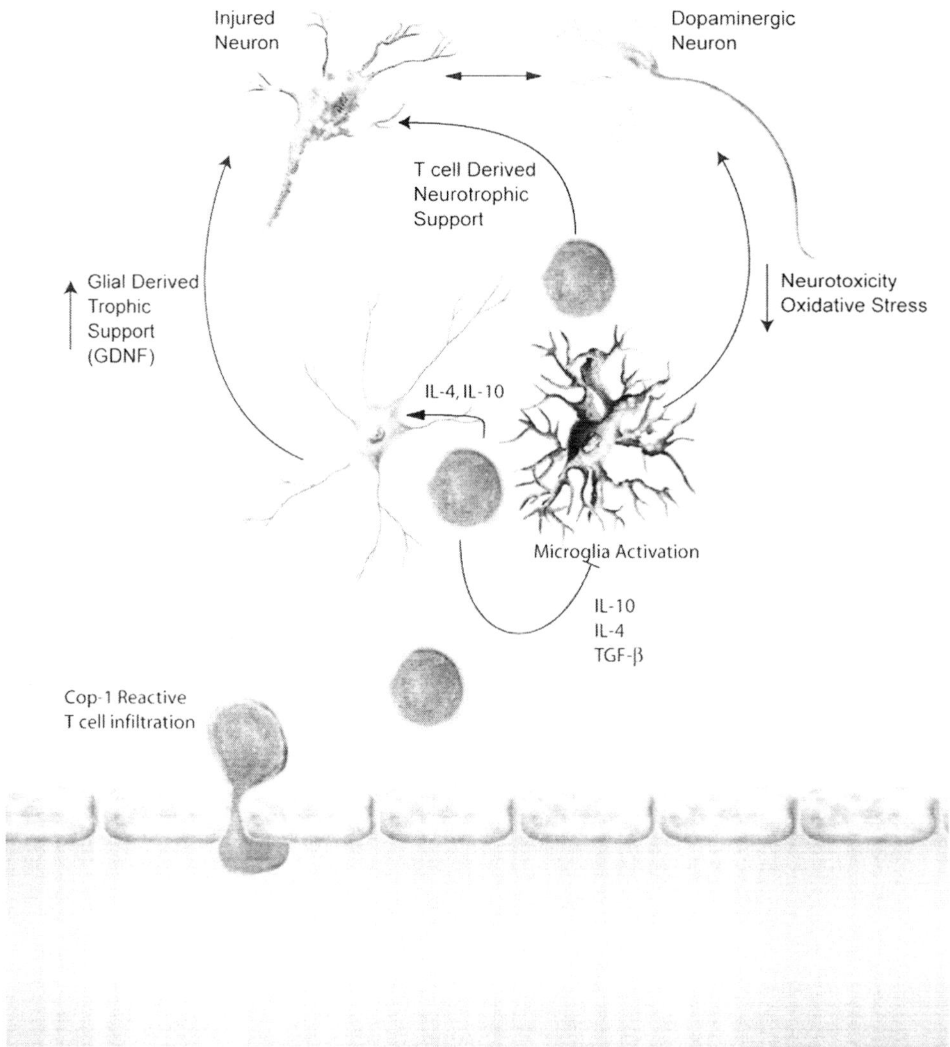

Figure 3. Proposed mechanism of Cop-1 immune mediated neuroprotection in the MPTP model of PD. TH2/TH3 Cop-1 T cells infiltrate the inflamed nigrostriatal pathway where they encounter antigens (myelin basic protein) presented by resident microglial cells in MHC. Activated T cells secrete anti-inflammatory cytokines such as IL-4, IL-10, and TGF-β that suppress toxic microglial activities. Neurotrophin expression can occur directly in T cells or T cell derived IL-4 and IL-10 may induce neurotrophin production in neighboring glia. These activities lead to passive neuroprotection by suppression of microglial responses and active neuroprotection through the local delivery of neurotrophins.

Regardless of the mechanisms involved in immunization with myelin components, their potential for devastating side effects precludes their use in a clinical setting. However, a more practical approach are the use of modified self-antigens and copolymers (Kipnis et al., 2000; Hauben et al., 2001) which can generate nonencephalitic T cell responses capable of recognizing CNS antigens. Such modified antigens, particularly copolymers such as copolymer-1 (Cop-1, Copaxone, glatiramer acetate) have a proven safety record in patients (Johnson et al., 2000; 2001). Moreover, Cop-1 immunization preferentially incites T cells with a T_H2 phenotype, which secrete anti-inflammatory cytokines such as interleukin (IL)-4, IL-10, IL-13 and TGF-β (Aharoni et al., 1997; 2000). Our own data demonstrated the neuroprotective outcomes of Cop-1 immunization in the MPTP mouse model of PD (Benner et al., 2004). We have found that the adoptive transfer of Cop-1 immune cells to MPTP recipient mice lead to a rapid T cell accumulation within the substantia nigra pars compacta, suppression of microglial activation, and increased local expression of astrocyte-associated glial cell line-derived neurotrophic factor. While the exact molecular mechanism underlying Cop-1 T cell effects remain hypothetical, it is likely that Cop-1 beneficial actions are related, at least in part, to a combined T cell local production of anti-inflammatory cytokines and neurotrophic factors within the damaged area (Figure 3). In this regard, Cop-1 vaccination is reminiscent of gene therapy (Kaspar et al., 2003). Indeed, both approaches attempt to deliver beneficial factors (e.g., trophic factors) specifically to damaged microenvironments. Still, Cop-1 immunization avoids the hurtle of gene delivery and, by also producing anti-inflammatory cytokines, may tackle the degenerative process in a more comprehensive fashion that includes the possible taming of the resident microglial responses.

Conclusion

Neurodegenerative disorders are strongly associated with innate immune activation comprised of an astonishing array of cellular and molecular players. There is little evidence for an infectious process in most neurodegenerative diseases however; multiple endogenous immune activators associated with cellular stress signals likely converge on resident glia leading to an inflammatory phenotype. The intended purpose of this response is beneficial to the host however, when regulatory controls are breached and inflammation is allowed to persist, tissue damage ensues. The identification of individual activation pathways operative in disease and the subsequent response they elicit may prove beneficial in the development of novel therapeutics, allowing us to benefit from the protective components of inflammation while inhibiting toxic responses. In addition to innate immunity, the adaptive immune system may also play a role (beneficial or toxic) in neurodegeneration, even in the absence of overt autoimmune pathology. Moreover, the adaptive immune system may be manipulated through peripheral immunization in a way that modifies innate inflammatory mechanisms in the CNS. Careful pursuit of vaccine strategies with an emphasis on safety may lead to novel treatment strategies for neurodegenerative disorders.

References

Aharoni R, Teitelbaum D, Sela M, Arnon R (1997) Copolymer 1 induces T cells of the T helper type 2 that crossreact with myelin basic protein and suppress experimental autoimmune encephalomyelitis. *Proc Natl Acad Sci U S A* **94**:10821-10826.

Aharoni R, Teitelbaum D, Leitner O, Meshorer A, Sela M, Arnon R (2000) Specific Th2 cells accumulate in the central nervous system of mice protected against experimental autoimmune encephalomyelitis by copolymer 1. *Proc Natl Acad Sci U S A* **97**: 11472-11477.

Allan SM, Rothwell NJ (2001) Cytokines and acute neurodegeneration. *Nat Rev Neurosci* **2**:734-744.

Asano K, Miwa M, Miwa K, Hanayama R, Nagase H, Nagata S, Tanaka M (2004) Masking of phosphatidylserine inhibits apoptotic cell engulfment and induces autoantibody production in mice. *J Exp Med* **200**:459-467.

Asea A, Rehli M, Kabingu E, Boch JA, Bare O, Auron PE, Stevenson MA, Calderwood SK (2002) Novel signal transduction pathway utilized by extracellular HSP70: role of toll-like receptor (TLR) 2 and TLR4. *J Biol Chem* **277**:15028-15034.

Auluck PK, Chan HY, Trojanowski JQ, Lee VM, Bonini NM (2002) Chaperone suppression of alpha-synuclein toxicity in a Drosophila model for Parkinson's disease. *Science* **295**:865-868.

Babcock AA, Kuziel WA, Rivest S, Owens T (2003) Chemokine expression by glial cells directs leukocytes to sites of axonal injury in the CNS. *J Neurosci* **23**:7922-7930.

Bachis A, Colangelo AM, Vicini S, Doe PP, De Bernardi MA, Brooker G, Mocchetti I (2001) Interleukin-10 prevents glutamate-mediated cerebellar granule cell death by blocking caspase-3-like activity. *J Neurosci* **21**:3104-3112.

Banati RB, Gehrmann J, Schubert P, Kreutzberg GW (1993) Cytotoxicity of microglia. *Glia* **7**:111-118.

Barclay AN, Wright GJ, Brooke G, Brown MH (2002) CD200 and membrane protein interactions in the control of myeloid cells. *Trends Immunol* **23**:285-290.

Batchelor PE, Liberatore GT, Wong JY, Porritt MJ, Frerichs F, Donnan GA, Howells DW (1999) Activated macrophages and microglia induce dopaminergic sprouting in the injured striatum and express brain-derived neurotrophic factor and glial cell line-derived neurotrophic factor. *J Neurosci* **19**:1708-1716.

Batchelor PE, Porritt MJ, Martinello P, Parish CL, Liberatore GT, Donnan GA, Howells DW (2002) Macrophages and Microglia Produce Local Trophic Gradients That Stimulate Axonal Sprouting Toward but Not beyond the Wound Edge. *Mol Cell Neurosci* **21**: 436-453.

Bell JK, Mullen GE, Leifer CA, Mazzoni A, Davies DR, Segal DM (2003) Leucine-rich repeats and pathogen recognition in Toll-like receptors. *Trends Immunol* **24**:528-533.

Benner EJ, Mosley RL, Destache CJ, Lewis TB, Jackson-Lewis V, Gorantla S, Nemachek C, Green SR, Przedborski S, Gendelman HE (2004) Therapeutic immunization protects dopaminergic neurons in a mouse model of Parkinson's disease. *Proc Natl Acad Sci U S A* **101**:9435-9440.

Besser M, Wank R (1999) Cutting edge: clonally restricted production of the neurotrophins brain-derived neurotrophic factor and neurotrophin-3 mRNA by human immune cells and Th1/Th2-polarized expression of their receptors. *J Immunol* **162**:6303-6306.

Bolin LM, Strycharska-Orczyk I, Murray R, Langston JW, Di Monte D (2002) Increased vulnerability of dopaminergic neurons in MPTP-lesioned interleukin-6 deficient mice. *J Neurochem* **83**:167-175.

Bradt BM, Kolb WP, Cooper NR (1998) Complement-dependent proinflammatory properties of the Alzheimer's disease beta-peptide. *J Exp Med* **188**:431-438.

Brionne TC, Tesseur I, Masliah E, Wyss-Coray T (2003) Loss of TGF-beta 1 leads to increased neuronal cell death and microgliosis in mouse brain. *Neuron* **40**:1133-1145.

Brown GC, Bal-Price A (2003) Inflammatory neurodegeneration mediated by nitric oxide, glutamate, and mitochondria. *Mol Neurobiol* **27**:325-355.

Bruce AJ, Boling W, Kindy MS, Peschon J, Kraemer PJ, Carpenter MK, Holtsberg FW, Mattson MP (1996) Altered neuronal and microglial responses to excitotoxic and ischemic brain injury in mice lacking TNF receptors. *Nat Med* **2**:788-794.

Cabarrocas J, Bauer J, Piaggio E, Liblau R, Lassmann H (2003) Effective and selective immune surveillance of the brain by MHC class I-restricted cytotoxic T lymphocytes. *Eur J Immunol* **33**:1174-1182.

Callahan MK, Halleck MS, Krahling S, Henderson AJ, Williamson P, Schlegel RA (2003) Phosphatidylserine expression and phagocytosis of apoptotic thymocytes during differentiation of monocytic cells. *J Leukoc Biol* **74**:846-856.

Campbell IL, Abraham CR, Masliah E, Kemper P, Inglis JD, Oldstone MB, Mucke L (1993) Neurologic disease induced in transgenic mice by cerebral overexpression of interleukin 6. *Proc Natl Acad Sci U S A* **90**:10061-10065.

Campbell IL, Krucker T, Steffensen S, Akwa Y, Powell HC, Lane T, Carr DJ, Gold LH, Henriksen SJ, Siggins GR (1999) Structural and functional neuropathology in transgenic mice with CNS expression of IFN-alpha. *Brain Res* **835**:46-61.

Chan A, Magnus T, Gold R (2001) Phagocytosis of apoptotic inflammatory cells by microglia and modulation by different cytokines: mechanism for removal of apoptotic cells in the inflamed nervous system. *Glia* **33**:87-95.

Chung KK, Thomas B, Li X, Pletnikova O, Troncoso JC, Marsh L, Dawson VL, Dawson TM (2004) S-nitrosylation of parkin regulates ubiquitination and compromises parkin's protective function. *Science* **304**:1328-1331.

Collins LV, Hajizadeh S, Holme E, Jonsson IM, Tarkowski A (2004) Endogenously oxidized mitochondrial DNA induces in vivo and in vitro inflammatory responses. *J Leukoc Biol* **75**:995-1000.

Dauer W, Przedborski S (2003) Parkinson's disease: mechanisms and models. *Neuron* **39**:889-909.

Dawson TM, Dawson VL (2003) Molecular pathways of neurodegeneration in Parkinson's disease. *Science* **302**:819-822.

Dawson VL, Dawson TM (1998) Nitric oxide in neurodegeneration. *Prog Brain Res* **118**:215-229.

Ehrhard PB, Erb P, Graumann U, Otten U (1993) Expression of nerve growth factor and nerve growth factor receptor tyrosine kinase Trk in activated CD4-positive T-cell clones. *Proc Natl Acad Sci U S A* **90**:10984-10988.

Elward K, Gasque P (2003) "Eat me" and "don't eat me" signals govern the innate immune response and tissue repair in the CNS: emphasis on the critical role of the complement system. *Mol Immunol* **40**:85-94.

Fahn S, Przedborski S (2000) Merritt's Neurology. New York: Lippincott Williams & Wilkins.

Fischer B, von Knethen A, Brune B (2002) Dualism of oxidized lipoproteins in provoking and attenuating the oxidative burst in macrophages: role of peroxisome proliferator-activated receptor-gamma. *J Immunol* **168**:2828-2834.

Flugel A, Matsumuro K, Neumann H, Klinkert WE, Birnbacher R, Lassmann H, Otten U, Wekerle H (2001) Anti-inflammatory activity of nerve growth factor in experimental autoimmune encephalomyelitis: inhibition of monocyte transendothelial migration. *Eur J Immunol* **31**:11-22.

Foster JA, Quan N, Stern EL, Kristensson K, Herkenham M (2002) Induced neuronal expression of class I major histocompatibility complex mRNA in acute and chronic inflammation models. *J Neuroimmunol* **131**:83-91.

Frenkel D, Huang Z, Maron R, Koldzic DN, Hancock WW, Moskowitz MA, Weiner HL (2003) Nasal vaccination with myelin oligodendrocyte glycoprotein reduces stroke size by inducing IL-10-producing CD4+ T cells. *J Immunol* **171**:6549-6555.

Gallucci S, Lolkema M, Matzinger P (1999) Natural adjuvants: endogenous activators of dendritic cells. *Nat Med* **5**:1249-1255.

Gao HM, Liu B, Zhang W, Hong JS (2003a) Synergistic dopaminergic neurotoxicity of MPTP and inflammogen lipopolysaccharide: relevance to the etiology of Parkinson's disease. *Faseb J* **17**:1957-1959.

Gao HM, Hong JS, Zhang W, Liu B (2003b) Synergistic dopaminergic neurotoxicity of the pesticide rotenone and inflammogen lipopolysaccharide: relevance to the etiology of Parkinson's disease. *J Neurosci* **23**:1228-1236.

Gendelman HE (2002) Neural immunity: Friend or foe? *J Neurovirol* **8**:474-479.

Giasson BI, Lee VM (2003) Are ubiquitination pathways central to Parkinson's disease? *Cell* **114**:1-8.

Giasson BI, Duda JE, Murray IV, Chen Q, Souza JM, Hurtig HI, Ischiropoulos H, Trojanowski JQ, Lee VM (2000) Oxidative damage linked to neurodegeneration by selective alpha-synuclein nitration in synucleinopathy lesions. *Science* **290**:985-989.

Gu Z, Kaul M, Yan B, Kridel SJ, Cui J, Strongin A, Smith JW, Liddington RC, Lipton SA (2002) S-nitrosylation of matrix metalloproteinases: signaling pathway to neuronal cell death. *Science* **297**:1186-1190.

Hammarberg H, Lidman O, Lundberg C, Eltayeb SY, Gielen AW, Muhallab S, Svenningsson A, Linda H, van Der Meide PH, Cullheim S, Olsson T, Piehl F (2000) Neuroprotection by encephalomyelitis: rescue of mechanically injured neurons and neurotrophin production by CNS-infiltrating T and natural killer cells. *J Neurosci* **20**:5283-5291.

Hauben E, Agranov E, Gothilf A, Nevo U, Cohen A, Smirnov I, Steinman L, Schwartz M (2001) Posttraumatic therapeutic vaccination with modified myelin self-antigen prevents complete paralysis while avoiding autoimmune disease. *J Clin Invest* **108**:591-599.

Hoek RM, Ruuls SR, Murphy CA, Wright GJ, Goddard R, Zurawski SM, Blom B, Homola ME, Streit WJ, Brown MH, Barclay AN, Sedgwick JD (2000) Down-regulation of the macrophage lineage through interaction with OX2 (CD200). *Science* **290**:1768-1771.

Hoeppner DJ, Hengartner MO, Schnabel R (2001) Engulfment genes cooperate with ced-3 to promote cell death in Caenorhabditis elegans. *Nature* **412**:202-206.

Huh GS, Boulanger LM, Du H, Riquelme PA, Brotz TM, Shatz CJ (2000) Functional requirement for class I MHC in CNS development and plasticity. *Science* **290**:2155-2159.

Ischiropoulos H, Beckman JS (2003) Oxidative stress and nitration in neurodegeneration: cause, effect, or association? *J Clin Invest* **111**:163-169.

Johnson KP, Brooks BR, Ford CC, Goodman A, Guarnaccia J, Lisak RP, Myers LW, Panitch HS, Pruitt A, Rose JW, Kachuck N, Wolinsky JS (2000) Sustained clinical benefits of glatiramer acetate in relapsing multiple sclerosis patients observed for 6 years. Copolymer 1 Multiple Sclerosis Study Group. *Mult Scler* **6**:255-266.

Johnson KP, Brooks BR, Cohen JA, Ford CC, Goldstein J, Lisak RP, Myers LW, Panitch HS, Rose JW, Schiffer RB, Vollmer T, Weiner LP, Wolinsky JS (2001) Extended use of glatiramer acetate (Copaxone) is well tolerated and maintains its clinical effect on multiple sclerosis relapse rate and degree of disability. 1998 [classical article]. *Neurology* **57**:S46-53.

Kakimura J, Kitamura Y, Takata K, Umeki M, Suzuki S, Shibagaki K, Taniguchi T, Nomura Y, Gebicke-Haerter PJ, Smith MA, Perry G, Shimohama S (2002) Microglial activation and amyloid-beta clearance induced by exogenous heat-shock proteins. *Faseb J* **16**:601-603.

Karman J, Ling C, Sandor M, Fabry Z (2004a) Dendritic cells in the initiation of immune responses against central nervous system-derived antigens. *Immunol Lett* **92**:107-115.

Karman J, Ling C, Sandor M, Fabry Z (2004b) Initiation of immune responses in brain is promoted by local dendritic cells. *J Immunol* **173**:2353-2361.

Kaspar BK, Llado J, Sherkat N, Rothstein JD, Gage FH (2003) Retrograde viral delivery of IGF-1 prolongs survival in a mouse ALS model. *Science* **301**:839-842.

Kerschensteiner M, Gallmeier E, Behrens L, Leal VV, Misgeld T, Klinkert WE, Kolbeck R, Hoppe E, Oropeza-Wekerle RL, Bartke I, Stadelmann C, Lassmann H, Wekerle H, Hohlfeld R (1999) Activated human T cells, B cells, and monocytes produce brain-derived neurotrophic factor in vitro and in inflammatory brain lesions: a neuroprotective role of inflammation? *J Exp Med* **189**:865-870.

Kipnis J, Yoles E, Schori H, Hauben E, Shaked I, Schwartz M (2001) Neuronal survival after CNS insult is determined by a genetically encoded autoimmune response. *J Neurosci* **21**:4564-4571.

Kipnis J, Mizrahi T, Hauben E, Shaked I, Shevach E, Schwartz M (2002) Neuroprotective autoimmunity: naturally occurring CD4+CD25+ regulatory T cells suppress the ability to withstand injury to the central nervous system. *Proc Natl Acad Sci U S A* **99**: 15620-15625.

Kipnis J, Yoles E, Porat Z, Cohen A, Mor F, Sela M, Cohen IR, Schwartz M (2000) T cell immunity to copolymer 1 confers neuroprotection on the damaged optic nerve: possible therapy for optic neuropathies. *Proc Natl Acad Sci U S A* **97**:7446-7451.

Kitada T, Asakawa S, Hattori N, Matsumine H, Yamamura Y, Minoshima S, Yokochi M, Mizuno Y, Shimizu N (1998) Mutations in the parkin gene cause autosomal recessive juvenile parkinsonism. *Nature* **392**:605-608.

Kornek B, Storch MK, Weissert R, Wallstroem E, Stefferl A, Olsson T, Linington C, Schmidbauer M, Lassmann H (2000) Multiple sclerosis and chronic autoimmune encephalomyelitis: a comparative quantitative study of axonal injury in active, inactive, and remyelinated lesions. *Am J Pathol* **157**:267-276.

Kreutzberg GW (1996) Microglia: a sensor for pathological events in the CNS. *Trends Neurosci* **19**:312-318.

Liberatore GT, Jackson-Lewis V, Vukosavic S, Mandir AS, Vila M, McAuliffe WG, Dawson VL, Dawson TM, Przedborski S (1999) Inducible nitric oxide synthase stimulates dopaminergic neurodegeneration in the MPTP model of Parkinson disease. *Nat Med* **5**:1403-1409.

Lindersson E, Beedholm R, Hojrup P, Moos T, Gai W, Hendil KB, Jensen PH (2004) Proteasomal inhibition by alpha-synuclein filaments and oligomers. *J Biol Chem* **279**:12924-12934.

Ling C, Sandor M, Fabry Z (2003) In situ processing and distribution of intracerebrally injected OVA in the CNS. *J Neuroimmunol* **141**:90-98.

Locksley RM, Klebanoff SJ (1983) Oxygen-dependent microbicidal systems of phagocytes and host defense against intracellular protozoa. *J Cell Biochem* **22**:173-185.

Lorenzl S, Calingasan N, Yang L, Albers DS, Shugama S, Gregorio J, Krell HW, Chirichigno J, Joh T, Beal MF (2004) Matrix metalloproteinase-9 is elevated in 1-methyl-4-phenyl-1,2,3,6-tetrahydropyridine-induced parkinsonism in mice. *Neuromolecular Med* **5**:119-132.

MacMicking J, Xie QW, Nathan C (1997) Nitric oxide and macrophage function. *Annu Rev Immunol* **15**:323-350.

Magnus T, Chan A, Grauer O, Toyka KV, Gold R (2001) Microglial phagocytosis of apoptotic inflammatory T cells leads to down-regulation of microglial immune activation. *J Immunol* **167**:5004-5010.

Masliah E, Rockenstein E, Veinbergs I, Mallory M, Hashimoto M, Takeda A, Sagara Y, Sisk A, Mucke L (2000) Dopaminergic loss and inclusion body formation in alpha-synuclein mice: implications for neurodegenerative disorders. *Science* **287**:1265-1269.

McGeer PL, McGeer EG (2002) Inflammatory processes in amyotrophic lateral sclerosis. *Muscle Nerve* **26**:459-470.

Mennicken F, Maki R, de Souza EB, Quirion R (1999) Chemokines and chemokine receptors in the CNS: a possible role in neuroinflammation and patterning. *Trends Pharmacol Sci* **20**:73-78.
Moalem G, Leibowitz-Amit R, Yoles E, Mor F, Cohen IR, Schwartz M (1999) Autoimmune T cells protect neurons from secondary degeneration after central nervous system axotomy. *Nat Med* **5**:49-55.
Moalem G, Gdalyahu A, Shani Y, Otten U, Lazarovici P, Cohen IR, Schwartz M (2000) Production of neurotrophins by activated T cells: implications for neuroprotective autoimmunity. *J Autoimmun* **15**:331-345.
Monsonego A, Beserman ZP, Kipnis J, Yoles E, Weiner HL, Schwartz M (2003) Beneficial effect of orally administered myelin basic protein in EAE-susceptible Lewis rats in a model of acute CNS degeneration. *J Autoimmun* **21**:131-138.
Muchowski PJ (2002) Protein misfolding, amyloid formation, and neurodegeneration: a critical role for molecular chaperones? *Neuron* **35**:9-12.
Nishiyori A, Hanno Y, Saito M, Yoshihara Y (2004) Aberrant transcription of unrearranged T-cell receptor beta gene in mouse brain. *J Comp Neurol* **469**:214-226.
Norris EH, Giasson BI, Ischiropoulos H, Lee VM (2003) Effects of oxidative and nitrative challenges on alpha -synuclein fibrillogenesis involve distinct mechanisms of protein modifications. *J Biol Chem*.
Okamura Y, Watari M, Jerud ES, Young DW, Ishizaka ST, Rose J, Chow JC, Strauss JF, 3rd (2001) The extra domain A of fibronectin activates Toll-like receptor 4. *J Biol Chem* **276**:10229-10233.
Owens T, Wekerle H, Antel J (2001) Genetic models for CNS inflammation. *Nat Med* **7**:161-166.
Park JS, Svetkauskaite D, He Q, Kim JY, Strassheim D, Ishizaka A, Abraham E (2004) Involvement of toll-like receptors 2 and 4 in cellular activation by high mobility group box 1 protein. *J Biol Chem* **279**:7370-7377.
Petrucelli L, O'Farrell C, Lockhart PJ, Baptista M, Kehoe K, Vink L, Choi P, Wolozin B, Farrer M, Hardy J, Cookson MR (2002) Parkin protects against the toxicity associated with mutant alpha-synuclein: proteasome dysfunction selectively affects catecholaminergic neurons. *Neuron* **36**:1007-1019.
Popovich PG, Guan Z, Wei P, Huitinga I, van Rooijen N, Stokes BT (1999) Depletion of hematogenous macrophages promotes partial hindlimb recovery and neuroanatomical repair after experimental spinal cord injury. *Exp Neurol* **158**:351-365.
Przedborski S, Jackson-Lewis V, Yokoyama R, Shibata T, Dawson VL, Dawson TM (1996) Role of neuronal nitric oxide in 1-methyl-4-phenyl-1,2,3,6-tetrahydropyridine (MPTP)-induced dopaminergic neurotoxicity. *Proc Natl Acad Sci U S A* **93**:4565-4571.
Rapalino O, Lazarov-Spiegler O, Agranov E, Velan GJ, Yoles E, Fraidakis M, Solomon A, Gepstein R, Katz A, Belkin M, Hadani M, Schwartz M (1998) Implantation of stimulated homologous macrophages results in partial recovery of paraplegic rats. *Nat Med* **4**:814-821.
Reddien PW, Cameron S, Horvitz HR (2001) Phagocytosis promotes programmed cell death in C. elegans. *Nature* **412**:198-202.
Rockwell P, Yuan H, Magnusson R, Figueiredo-Pereira ME (2000) Proteasome inhibition in neuronal cells induces a proinflammatory response manifested by upregulation of cyclooxygenase-2, its accumulation as ubiquitin conjugates, and production of the prostaglandin PGE(2). *Arch Biochem Biophys* **374**:325-333.
Samdani AF, Dawson TM, Dawson VL (1997) Nitric oxide synthase in models of focal ischemia. *Stroke* **28**:1283-1288.
Sankarapandi S, Zweier JL, Mukherjee G, Quinn MT, Huso DL (1998) Measurement and characterization of superoxide generation in microglial cells: evidence for an NADPH oxidase-dependent pathway. *Arch Biochem Biophys* **353**:312-321.
Sastry PS, Rao KS (2000) Apoptosis and the nervous system. *J Neurochem* **74**:1-20.
Schori H, Lantner F, Shachar I, Schwartz M (2002) Severe immunodeficiency has opposite effects on neuronal survival in glutamate-susceptible and -resistant mice: adverse effect of B cells. *J Immunol* **169**:2861-2865.
Schwartz M (2001) Harnessing the immune system for neuroprotection: therapeutic vaccines for acute and chronic neurodegenerative disorders. *Cell Mol Neurobiol* **21**:617-627.
Schwartz M, Moalem G, Leibowitz-Amit R, Cohen IR (1999) Innate and adaptive immune responses can be beneficial for CNS repair. *Trends Neurosci* **22**:295-299.
Schwartz M, Shaked I, Fisher J, Mizrahi T, Schori H (2003) Protective autoimmunity against the enemy within: fighting glutamate toxicity. *Trends Neurosci* **26**:297-302.

Seong SY, Matzinger P (2004) Hydrophobicity: an ancient damage-associated molecular pattern that initiates innate immune responses. *Nat Rev Immunol* **4**:469-478.

Serpe CJ, Sanders VM, Jones KJ (2000) Kinetics of facial motoneuron loss following facial nerve transection in severe combined immunodeficient mice. *J Neurosci Res* **62**:273-278.

Serpe CJ, Kohm AP, Huppenbauer CB, Sanders VM, Jones KJ (1999) Exacerbation of facial motoneuron loss after facial nerve transection in severe combined immunodeficient (scid) mice. *J Neurosci* **19**:RC7.

Shevach EM (2000) Regulatory T cells in autoimmmunity*. *Annu Rev Immunol* **18**:423-449.

Shi Y, Evans JE, Rock KL (2003) Molecular identification of a danger signal that alerts the immune system to dying cells. *Nature* **425**:516-521.

Singleton AB, Farrer M, Johnson J, Singleton A, Hague S, Kachergus J, Hulihan M, Peuralinna T, Dutra A, Nussbaum R, Lincoln S, Crawley A, Hanson M, Maraganore D, Adler C, Cookson MR, Muenter M, Baptista M, Miller D, Blancato J, Hardy J, Gwinn-Hardy K (2003) alpha-Synuclein locus triplication causes Parkinson's disease. *Science* **302**:841.

Stalder AK, Carson MJ, Pagenstecher A, Asensio VC, Kincaid C, Benedict M, Powell HC, Masliah E, Campbell IL (1998) Late-onset chronic inflammatory encephalopathy in immune-competent and severe combined immune-deficient (SCID) mice with astrocyte-targeted expression of tumor necrosis factor. *Am J Pathol* **153**:767-783.

Syken J, Shatz CJ (2003) Expression of T cell receptor beta locus in central nervous system neurons. *Proc Natl Acad Sci U S A* **100**:13048-13053.

Takeda K, Kaisho T, Akira S (2003) Toll-like receptors. *Annu Rev Immunol* **21**:335-376.

Termeer C, Benedix F, Sleeman J, Fieber C, Voith U, Ahrens T, Miyake K, Freudenberg M, Galanos C, Simon JC (2002) Oligosaccharides of Hyaluronan activate dendritic cells via toll-like receptor 4. *J Exp Med* **195**:99-111.

Tran PB, Miller RJ (2003a) Chemokine receptors in the brain: a developing story. *J Comp Neurol* **457**:1-6.

Tran PB, Miller RJ (2003b) Chemokine receptors: signposts to brain development and disease. *Nat Rev Neurosci* **4**:444-455.

Vila M, Przedborski S (2003) Targeting programmed cell death in neurodegenerative diseases. *Nat Rev Neurosci* **4**:365-375.

Vila M, Przedborski S (2004) Genetic clues to the pathogenesis of Parkinson's disease. *Nat Med 10 Suppl*:**S58**-62.

Wahl LM, Shankavaram U, Zhang Y (1997) Role of macrophages in vascular tissue remodelling. *Transpl Immunol* **5**:173-176.

Wang PY, Munford RS (1999) CD14-dependent internalization and metabolism of extracellular phosphatidylinositol by monocytes. *J Biol Chem* **274**:23235-23241.

Weiner HL, Selkoe DJ (2002) Inflammation and therapeutic vaccination in CNS diseases. *Nature* **420**:879-884.

Wenkel H, Streilein JW, Young MJ (2000) Systemic immune deviation in the brain that does not depend on the integrity of the blood-brain barrier. *J Immunol* **164**:5125-5131.

Wu DC, Teismann P, Tieu K, Vila M, Jackson-Lewis V, Ischiropoulos H, Przedborski S (2003) NADPH oxidase mediates oxidative stress in the 1-methyl-4-phenyl-1,2,3,6-tetrahydropyridine model of Parkinson's disease. *Proc Natl Acad Sci U S A* **100**: 6145-6150.

Wyss-Coray T, Mucke L (2002) Inflammation in neurodegenerative disease--a double-edged sword. *Neuron* **35**:419-432.

Wyss-Coray T, Masliah E, Mallory M, McConlogue L, Johnson-Wood K, Lin C, Mucke L (1997) Amyloidogenic role of cytokine TGF-beta1 in transgenic mice and in Alzheimer's disease. *Nature* **389**:603-606.

Wyss-Coray T, Lin C, Yan F, Yu GQ, Rohde M, McConlogue L, Masliah E, Mucke L (2001) TGF-beta1 promotes microglial amyloid-beta clearance and reduces plaque burden in transgenic mice. *Nat Med* **7**:612-618.

Yan SD, Roher A, Schmidt AM, Stern DM (1999) Cellular cofactors for amyloid beta-peptide-induced cell stress. Moving from cell culture to in vivo. *Am J Pathol* **155**:1403-1411.

Yang Y, Nishimura I, Imai Y, Takahashi R, Lu B (2003) Parkin suppresses dopaminergic neuron-selective neurotoxicity induced by Pael-R in Drosophila. *Neuron* **37**:911-924.

Yao D, Gu Z, Nakamura T, Shi ZQ, Ma Y, Gaston B, Palmer LA, Rockenstein EM, Zhang Z, Masliah E, Uehara T, Lipton SA (2004) Nitrosative stress linked to sporadic Parkinson's disease: S-nitrosylation of parkin regulates its E3 ubiquitin ligase activity. *Proc Natl Acad Sci U S A* **101**:10810-10814.

Zhang X, Koldzic DN, Izikson L, Reddy J, Nazareno RF, Sakaguchi S, Kuchroo VK, Weiner HL (2004) IL-10 is involved in the suppression of experimental autoimmune encephalomyelitis by CD25(+)CD4(+) regulatory T cells. *Int Immunol* **16**:249-256.

In: Recent Breakthroughs in Basal Ganglia Research
Editor: Erwan Bezard, pp. 303-312
ISBN: 1-59454-880-3

Chapter 23

PROGRAMMED CELL DEATH IN PARKINSON´S DISEASE

*Andreas Hartmann**
INSERM U 289, Hôpital de la Salpêtrière, Paris, France

ABSTRACT

Programmed cell death (PCD) encomprises all forms of cell death that involve active intracellular processes, whereas apoptosis solely relates to certain morphological criteria that may, but do not necessarly occur in PCD. Applying this broader definition to cell death mechanisms relevant to Parkinson´s disease (PD), recent research has clearly indicated that dopaminergic neurons die, at least partly, by PCD in this chronic neurodegenerative disorder. Also, PCD has been linked both to recent advances in the etiology (e.g. the familial PD forms) and the pathogenesis (chronic inflammation) of parkinsonian syndromes. Finally, PCD as such may be a fast process of cell demise, but chronic, sublethal activation of PCD signalling pathways can contribute to cellular dysfunction. Thus, PCD has been proposed as a therapeutic target in PD. The following review will therefore also adress the question whether this proposal may be translated into clinical trials in a foreseeable future.

INTRODUCTION

The question whether dopaminergic (DA) neurons in Parkinson´s disease (PD) die by apoptosis, necrosis or some alternative form of cell death has received considerable attention due to the fact that apoptosis is amenable to pharmacological inhibition, and may thus represent a therapeutic target in PD. The initial human postmortem studies related to this question showed conficting results (Hartmann and Hirsch, 2001), either because of methodological issues, especially the unreliability of 3′ DNA end labeling as a marker of apoptotic nuclear degradation or because the apoptotic changes observed might in fact reflect a perimortem effect rather than the primary disease process (Kingsbury et al., 1998). Also, innapropriate imposition of narrow criteria for the recognition of apoptosis (Tatton et al., 2003) have fuelled a debate which at times seemed almost hysterical. Surprisingly, the opponents of apoptotic cell death in PD never provided clear evidence for alternative modes of cell demise, especially necrosis.

Programmed cell death (PCD) has long been considered identical to apoptosis, a form of developmental cell death characterized morphologically by Kerr et al., (1972), and involving nuclear fragmentation and/or condensation, chromatin clumping, cell membrane preservation until very late stages, and heterophagic elimination to guarantee a safe and non-inflammatory removal of corpses. The discovery that developmental and non-developmental cell death share similar execution systems and controlling proteins has then blurred the boundary between the two processes. Moreover, it was subsequently recognized that other forms of cell death (autophagy, parapotosis) occur in non-developmental settings related to adult, neurodegenerative disorders (Anglade et al., 1997; Sperandio et al., 2000) and, importantly,

* Address for correspondence: Andreas Hartmann, MD INSERM U 289 & Centre d´Investigation Clinique, Hôpital de la Salpêtrière 47, Boulevard de l´Hôpital, 75013 Paris, France Tel. +33 (0)1 42 16 22 02 / Fax +33 (0)1 44 24 36 58. E-mail address: hartmann@ccr.jussieu.fr

are governed by signalling pathways, as "classical" apoptosis (Castro-Obregon et al., 2002; Lang-Rollin et al., 2003; Castro-Obregon et al., 2004). Consequently, in accordance with the definition given by Vila and Przedborski, (2003), I will group under the term PCD all forms of cell death that involve active intracellular processes, and use the term "apoptosis" only in reference to the morphology of cell death.

In the following review, I will focus on the role of PCD in PD, largely bypassing other basal ganglia disorders such as Huntington's disease (see, however, the paper by Turmaine et al., 2000, on a non-apoptotic form of cell death in Huntington's disease). Also, descriptive work regarding the morphologic and biochemical alterations relevant to PCD in PD will not be discussed here, as they have been extensively reviewed elsewhere (Hartmann and Hirsch, 2001; Tatton et al., 2003).

1 PCD Pathways – A Brief Reminder

Several intra- and pericellular molecular pathways of PCD have been described, in particular the so-called extrinsic and intrinsic pathways (Strasser et al., 2000; Vila and Przedborski, 2003; Figure 1). The intrinsic pathway originates from mitochondrial release of cytochrome *c* and associated activation of caspase-9 (which is a part of the so-called "apoptosome"); whereas the extrinsic pathway originates from the activation of cell surface death receptors such as Fas and TNFR1, and results in the activation of caspase-8. A role for the intrinsic pathway in PD pathogenesis has been firmly established over the last years due to gene targeting studies in PD (see below, and also rewiewed by Vila and Przedborski, 2003), whereas the involvement of the extrinsic pathway is less clear, yet plausible in the framework of pericellular inflammatory processes occuring in PD (Hunot and Hirsch, 2003). Whether the distinction between intrinsic and extrinsic pathways in PD is of therapeutical or merely pathophysiological interest, will be discussed at the end of this article. It should also be mentionned that mixed forms of apoptosis and necrosis have been observed, termed "necrosis-like PCD" or "aborted apoptosis", which occurs when a standard apoptosis programme is initiated, then blocked at the level of caspase activation and finally terminated by alternative, caspase-independent routes (Leist and Jaattela, 2001). Yet, this pathway is not passive, but seems to rely on signal transduction by the receptor-interacting protein (RIP) and free radical formation (Holler et al., 2000). Also, so-called "secondary necrosis", which can be observed after initiation of the cell death program, may occur through cleavage of plasma membrane calcium pumps by caspases, and has been suggested to link apoptosis and necrosis (Schwab et al., 2002). To complicate matters further, proteases other than caspases have been shown to participate in PCD, for instance calpains, cathepsins, and serine proteases (with regard to PD, see for instance Crocker et al., 2003).

2 Genetic Targeting of PCD Pathways in Experimental PD Models

Four studies deserve mention under this heading. Adenovirally-mediated X-chromosome-linked inhibitor of apoptosis (XIAP) gene transfer into the striatum of 1-methyl-4-phenyl-1,2,3,6-tetrahydropyridine (MPTP)-treated mice did not result in preservation of nigrostriatal terminals. However, if associated with transfer of a neurotrophic factor, glial-cell-derived neurotrophic factor (GDNF), synergistic effects on cell survival (tyrosine hydroxylase cell counts in the substantia nigra pars compacta) and function (dopamine and dopamine metabolite concentrations in the striatum) were observed (Eberhardt et al., 2000). Using a viral Apaf-1 dominant negative vector injectetd intrastriatally, Mochizuki et al., (2001) were able to protect nigral DA neurons in an MPTP mouse model of PD. Striatal dopamine (DA) and DA metabolite levels were not analysed in this study, but a behavioural improvement was reported following treatment with the Apaf-1 dominant negative vector. Vila et al., (2001) intoxicated Bax-deficient mice with MPTP and observed an almost complete protection of nigral DA neurons, and a less marked, but still significant preservation of DA and DA metabolites levels in the striatum of Bax-deficient mice compared to their wild-type littermates. Finally, Viswanath et al., (2001) have reported that mice overexpressing the baculoviral general caspase inhibitor protein p35 were significantly protected both at the nigral and striatal level against MPTP toxicity.

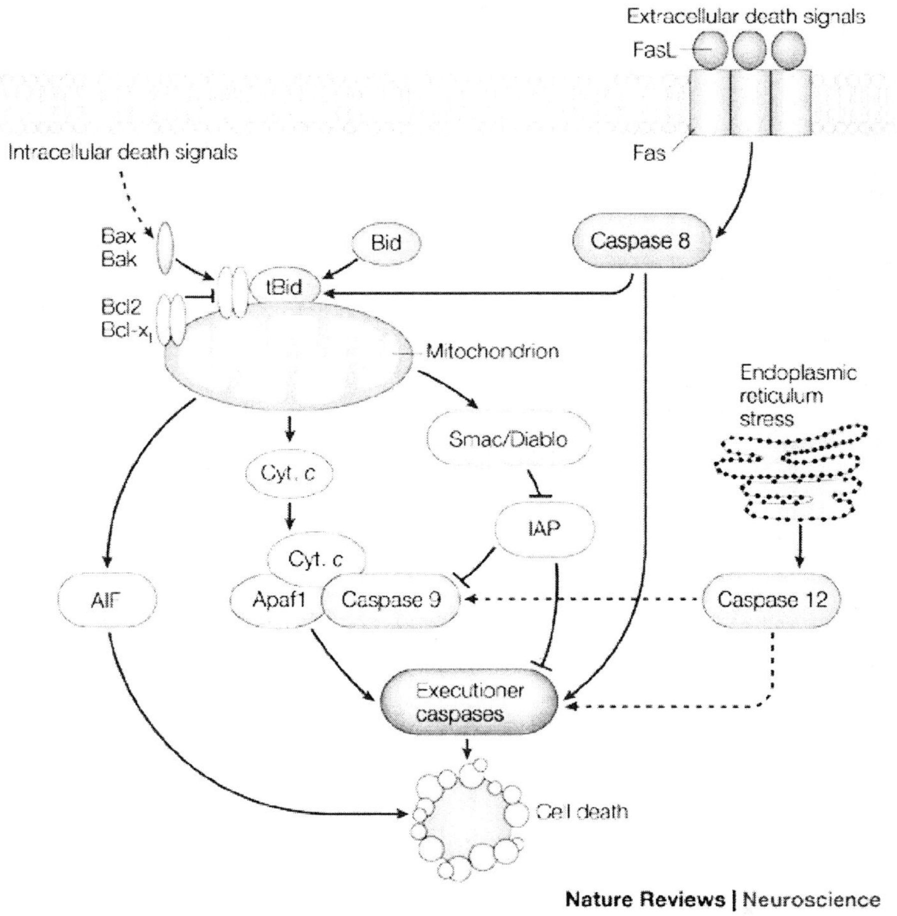

Figure 1: Molecular pathways of programmed cell death. Extracellular signals through cellular death receptors, such as Fas, and intracellular signals, including damage to subcellular constituents or endoplasmic reticulum stress, both trigger genetically programmed pathways of programmed cell death (PCD). The two main PCD pathways result in the activation of downstream executioner caspases and cell death. AIF, apoptosis-inducing factor; Cyt., cytochrome; FasL, Fas ligand; IAP, inhibitor of apoptosis; tBid, truncated Bid.

All these studies employed different MPTP intoxication protocols, different survival times, different genetic backgrounds, and showed different levels of protection against MPTP toxicity. Also, it should not be forgotten that constituent genetic alteration of biochemical pathways may induce compensatory changes which would not occur during a time-limited pharmacological intervention, as would be the case in a clinical setting. Yet, these studies are important as they provide proof-of-principle that inhibition of the PCD machinery at various levels confers at least partial resistance against DA cell death. Thus, they underline *functionally* the implication of PCD in a relevant animal model of PD.

3 PD Genetics and PCD

None of the proteins so far identified as mutated in familial forms of PD (Vila and Przedborski, 2004), e.g. alpha-synuclein (Park1), parkin (Park2), DJ-1 (Park7), UCH-L1 (Park5) or PINK1 (Park6) are molecules directly implicated in the regulation of PCD. Thus, all forms of PD characterized to date do not suggest that PCD is a primary phenomenon in DA neuronal degeneration. However, depending on the role of the pathogenic mechanism by which the proteins mentioned (gain- or loss-of-function) induce DA cell loss, most studies suggest that the mode of cell death is indeed programmed.

Alpha-synuclein mutant mice and drosophila mutants (A30P and A53T mutations, and overexpression of the human wild-type protein) show very heterogenous neuronal loss – sometimes but not always affecting the nigrostriatal system – depending on the promoters used (reviewed by Dawson et al., 2002; Shimohama et al., 2003). The mode of cell death in these models, altough usually not analysed in detail, appears to be compatible with PCD. *In vitro*, inducible expression of mutant alpha-synuclein sensitizes neuronal cell lines cells to undergo PCD (Saha et al., 2000; Tanaka et al., 2001; Zhou et al., 2002).

Neither parkin knockout mice (Goldberg et al., 2003; Itier et al., 2003; Von Coelln et al., 2004) nor drosophila mutants (Greene et al., 2003) display nigral DA cell loss, although other neuronal systems can be damaged in these animals. *In vitro*, parkin has been shown to protect against DA-, 6-hydroxydopamine (6-OHDA)- and kainate-induced PCD (Staropoli et al., 2003; Jiang et al., 2004). Since parkin is a ubiquitin E3 ligase, it has attracted interest to the potential role of the ubiquitin-proteasome system in PD pathogenesis, hypothesizing that impaired degradation of damaged or damaging proteins may contribute to DA neurodegeneration. Indeed, a recently developped rat model of PD is based on systemic proteasome inhibition, and induces PCD of nigral DA neurons (McNaught et al., 2004). PINK1 may be of special interest regarding PCD regulation in PD, since it is contains a mitochondrial localization sequence, and PINK1 overexpression protects against PCD induced by the proteasome inhibitor MG-132 *in vitro* (Shen and Cookson, 2004; Valente et al., 2004). DJ-1 appears to be also targeted to mitochondria, but is mainly localized in glial cells and does not prevent PCD in the form of staurosporine-induced cell death (Yokota et al., 2003; Canet-Aviles et al., 2004).

4 Inflammation and PCD – A Possible Link

Human caspase-1, -4 and -5, and mouse caspase-11 represent a group of caspases whose primary function is the regulation of inflammatory processes, and link these to PCD pathways. They are thus termed "inflammatory caspases" (Mariathasan et al., 2004; Martinon and Tschopp, 2004). Given the importance of inflammation as a pathogenetic mechanism in PD (Hunot and Hirsch, 2003), the role of two inflammatory caspases, caspase-1 and caspase-11, have been studied in MPTP mouse models. Using caspase-1 deleted mice, (Klevenyi et al., 1999) have demonstrated extensive protection against MPTP neurotoxicity, both striatally and nigrally. However, injecting a dominant negative viral vector for caspase-1 into the striatum of MPTP-treated mice, Mochizuki et al., (2001) were unable to protect nigral DA neurons from degenerating.

Caspase-11 is induced by stimulation with lipopolysaccharide (LPS) and mediates the activation of caspase-1 by physical interaction (Wang et al., 1998). Also, it is able to promote the activation of caspase-3 (Kang et al., 2000). This dual role of caspase-11, linking inflammation and PCD, has recently been confirmed to be relevant in nigral DA cell death. Intoxicating mice with MPTP, it was shown that caspase-11 is upregulated in the SN following treatment and that caspase-11-deficient mice are protected against an acute (but not subchronic) MPTP challenge (Furuya et al., 2004).

5 Can Chronic, but Sublethal Activation of the PCD Machinery Cause Cellular Dysfunction in PD ?

Maybe it is helpful take a fresh look at PCD and wonder whether PCD is only an acute or even subacute process. *In vitro* studies suggest that once the PCD machinery is engaged, cell death occurs very quickly, i.e. in as little as 30 minutes. However, the delicate balance between pro- and anti-apoptotic intracellular factors suggests otherwise. Neuronally expressed apoptosis inhibitor proteins buffer the caspase system (a tightly controlled system, especially in the central nervous system), and this buffering capacity might allow the localized activation of caspases (e.g. with neurites or synapses) or sequestration of active caspases without build-up of the death cascade affecting the entire neuron (Leist and Jaattela, 2001). Rather than a binary, yes-or-no decision, activation of the PCD machinery may thus be a treshold process, at some point tipping the balance over to actual cell death.

This points to a further question: Are the classical biochemical alterations of PCD (e.g. caspase activation and mitochondrial permeability pore opening) only a consequence of some upstream events (cytokines, free radical formation, excitotoxicity) or may themselves contribute to a vicious circle compromising cellular viability? In this regard, it is important to mention that cleavage of substrates essential to neurodegenerative pathologies such as amyloid-beta percursor protein (APP), presenilins (PS1,

PS2), tau, huntingtin, atrophin-1, ataxin-3 and the androgen receptor by caspases may alter their biological function (Nicotera et al., 1999). As an example, APP cleavage by caspase-3 does not result in immediate cell death but is part of the proteolytic processing of this protein; the ensuing cleavage products gain an altered biological function, i.e. exercing toxic effects on affected neurons and amplifying subsequent APP cleavage, thus creating a toxic feed-forward loop (Gervais et al., 1999). However, it is unknown how long, under *in vivo* conditions, this process requires to result in actual cell demise.

With regard to PD, a protein mutated in autosomal-recessive juvenile forms of the disease, parkin, has been shown to be a caspase-1 and -8 substrate (Kahns et al., 2002; 2003), although the biological significance of parkin cleavage remains unclear to date. In addition, parkin, which is a ubiquitin E3 ligase, has attracted interest to the potential role of the ubiquitin-proteasome system in PD pathogenesis, hypothesizing that impaired degradation of damaged or damaging proteins may contribute to DA neurodegeneration. In this respect, a recent study has suggested that three subsunits of the 19S regulatory complex of the proteasome are caspase-3 cleavage products. The net effect of this process appears to be inhibition of of the proteasomal degradation of ubiquitin-dependent, but also -independent substates, including pro-apoptotic molecules such as Smac, providing a feed-forward loop for the execution of PCD (Sun et al., 2004). With regard to the mitochondrial respiratoty chain as key locus in PD pathogenesis (Schapira et al., 1998), it is noteworthy that NDUFS1, the 75 kDa subunit of respiratory complex I, is a caspase cleavage substrate and alters both the cell's general energy and redox state, and thus its propensity to undergo PCD (Ricci et al., 2004). Finally, an experimental link between sublethal activation of apoptotic pathways and LB formation has been suggested by Hashimoto et al., (1999) who showed that release of cytochrome *c* from mitochondria into the cytosol may also function as a potent stimulator for alpha-synuclein aggregation.

In conclusion, chronic caspase activation may both reflect and result in cellular dysfunction that may well take months or years before culminating into actual cell demise.

6 Is PCD a Therapeutic Target in PD ?

In theory, PCD is an attractive candidate for drug therapy in PD. However, since PCD in PD appears to be a consequence of cellular dysfunction rather than a primary disease phenomenon, questions have been raised as to whether it truly represents a therapeutic target in PD (Hartmann and Hirsch, 2001). The major concerns raised were related to two questions: (1) Is PCD inhibition sufficient to protect *and* restore function of nigral PD neurons in PD? (2) If so, is the probable duration of anti-apoptotic treatment as a neuroprotective therapy in PD achievable without inducing major side effects (tumor formation, autoimmune disease)?

Our own *in vitro* results using primary mesecephalic cell cultures have indicated that caspase inhibition does not protect against MPP^+ toxicity, and may even accelerate cell death due to a switch from apoptosis to necrosis (Hartmann et al., 2001). This latter mechanism has not been reproduced in another study, although globally, caspase inhibition did not rescue DA neurons functionally from MPP^+ toxicity, exept when very high concentrations (300 μM) of the broad spectrum caspase inhibitor z-VAD-fmk were used (Bilsland et al., 2002). In a previous study, Eberhardt et al., (2000) were equally unable to restore function in cultured fetal mesencephalic neurons treated with MPP^+ and z-VAD-fmk. In a 6-OHDA model model using a SY5Y cell line, morphological rescue of DA neurons has been observed following treatment with broad spectrum caspase inhibitors or viral mediated expression of XIAP, but functional preservation of cells could not be achieved by these anti-PCD treatments (von Coelln et al., 2001). Recently, it has been reported that administration of a novel peptidyl broad-spectrum caspase inhibitor, Q-VD-OPH, offered significant protection against MPTP-induced neurodegeneration in mice (Yang et al., 2004). I do not wish to overemphasize the potential risks of caspase inhibition in PD or other chronic neurodegenerative disorders, but it should not be dismissed that synthetic caspase inhibitors may affect other proteins unintentionally, whose activity may be necessary for survival (Abraham and Shaham, 2004). Also, caspase inhibition *in vivo* may result in a partial execution of apoptosis resulting in the persistence of damaged cells within the tissue, or even trigger an alternative, necrosis-like form of cell death with cell membrane leakage and consequent inflammation, resulting in further disease progression (Nicotera, 2002).

Given the current state of research, the majority of arguments suggest that inhibition of the PCD machinery may offer partial (that is, at least morphological) protection against time-limited insults of the nigrostriatal DA system in toxin-based models of PD. However, functional benefit very probably implies that pathways upstream of caspase actication and cytosolic cytochrome *c* release are equally targeted. Prominent examples include the upstream regulation of PCD by the c-jun N-terminal kinase (JNK) pathway, which has been implicated in in both viral overexpression and knockout MPTP mouse models of PD to participate in degeneration of the nigrostriatal system (Xia et al., 2001; Hunot et al., 2004).

Furthermore, is dissection of precise molecular PCD pathways relevant to PD therapy? For instance, caspase-8 is traditionally considered an apical adaptor caspase belonging to the extrinsic pathway (Figure1). However, several studies have shown that caspase-8 can also be cleaved by downstream caspase considered to be executioner caspases, such as caspases-3 and –9, a mechanism also relevant in the MPTP mouse model of PD (Viswanath et al., 2001) and potentially acting a feed-forward mechanism linking the extrinsic and the intrinsic pathways in PCD.

I believe that in clinical practice, things will boil down to broad rather than specific targets of PCD pathways; and that these broad spectrum inhibitors of PCD will themselves be combined or include other pathways, upstream of caspase activation and cytosolic cytochrome *c* release.

A drug examplifying this thesis is minocycline, a tetracycline derivate, which may be active due to multiple actions, being both a caspase inhibitor and inhibitor of cytochrome *c* release, as well as an anti-inflammatory substance by virtue of blocking microglial activation and inhibiting inducible nitric oxide synthase (Thomas and Le, 2004). Other substances acting on different cell death pathways including anti-apoptotic actions in PD models include non-immunosuppressive neuroimmunophilin ligands (Tanaka and Ogawa, 2004) and rasagiline (Weinreb et al., 2004).

However, recent studies testing minocyline in experimental PD models have yielded conflicting results, judging minocycline to be either highly effective in protecting the nigrostriatal system against DA toxins (Du et al., 2001; Wu et al., 2002) or, on the contrary, potentially deleterious (Yang et al., 2003; Diguet et al., 2004b). A general problem concerning the value of current PD models to predict therapeutic sucess has been highlighted by the contradictory findings surrounding minocycline and discussed by Diguet et al., (2004a).

Obviously, a trivial question regarding the potential value of anti-PCD compounds is: if targeting events upstream activation of PCD pathways is successful, is PCD inhibition still necessary? A possible answer is that sick neurons in PD may just be more vulnerable to undergo PCD when exposed to otherwise tolerable levels of cellular (oxidative, metabolic) stress, and that PCD should therefore be regarded as an extension of the pathogenesis of cell death (Tatton et al., 2003). This pleads in favour of a simultaneous interception of pro-oxidant, pro-inflammatory and pro-apoptotic signals. In addition, as mentioned above, activation of the PCD machinery may not only be an acute process but contribute to chronic cellular dysfunction.

Finally, long term inhibition of PCD pathways may trigger the risk of autoimmune or proliferative disease unless cell-targeting strategies aimed at specific neuronal populations are used. Nevertheless, ongoing clinical Phase II and III trials using the JNK inhibitor CEP-1347 and the GAPDH ligand TCH346 in PD have, so far, not shown any evidence of systemic toxicity (Waldmeier and Tatton, 2004). However, it must be noted that both substances are not solely PCD inhitors but do display metabolic and trophic properties protective to DA (and other) neurons.

These questions, especially the systemic, long term and delayed use of PCD inhibitors, should be addressed in primate models of PD that reflect, as acurately as possible, the clinical situation. Until then, PCD inhibition remains a hopeful, but uncertain therapeutic target in PD.

REFERENCES

Abraham MC, Shaham S (2004) Death without caspases, caspases without death. *Trends Cell Biol* **14**:184-193.

Anglade P, Vyas S, Javoy-Agid F, Herrero MT, Michel PP, Marquez J, Mouatt-Prigent A, Ruberg M, Hirsch EC, Agid Y (1997) Apoptosis and autophagy in nigral neurons of patients with Parkinson's disease. *Histol Histopathol* **12**:25-31.

Bilsland J, Roy S, Xanthoudakis S, Nicholson DW, Han Y, Grimm E, Hefti F, Harper SJ (2002) Caspase inhibitors attenuate 1-methyl-4-phenylpyridinium toxicity in primary cultures of mesencephalic dopaminergic neurons. *J Neurosci* **22**:2637-2649.

Canet-Aviles RM, Wilson MA, Miller DW, Ahmad R, McLendon C, Bandyopadhyay S, Baptista MJ, Ringe D, Petsko GA, Cookson MR (2004) The Parkinson's disease protein DJ-1 is neuroprotective due to cysteine-sulfinic acid-driven mitochondrial localization. *Proc Natl Acad Sci U S A* **101**:9103-9108.

Castro-Obregon S, Del Rio G, Chen SF, Swanson RA, Frankowski H, Rao RV, Stoka V, Vesce S, Nicholls DG, Bredesen DE (2002) A ligand-receptor pair that triggers a non-apoptotic form of programmed cell death. *Cell Death Differ* **9**:807-817.

Castro-Obregon S, Rao RV, del Rio G, Chen SF, Poksay KS, Rabizadeh S, Vesce S, Zhang XK, Swanson RA, Bredesen DE (2004) Alternative, nonapoptotic programmed cell death: mediation by arrestin 2, ERK2, and Nur77. *J Biol Chem* **279**:17543-17553.

Crocker SJ, Smith PD, Jackson-Lewis V, Lamba WR, Hayley SP, Grimm E, Callaghan SM, Slack RS, Melloni E, Przedborski S, Robertson GS, Anisman H, Merali Z, Park DS (2003) Inhibition of calpains prevents neuronal and behavioral deficits in an MPTP mouse model of Parkinson's disease. *J Neurosci* **23**:4081-4091.

Dawson T, Mandir A, Lee M (2002) Animal models of PD: pieces of the same puzzle? *Neuron* **35**:219-222.

Diguet E, Gross CE, Tison F, Bezard E (2004a) Rise and fall of minocycline in neuroprotection: need to promote publication of negative results. *Exp Neurol* **189**:1-4.

Diguet E, Fernagut PO, Wei X, Du Y, Rouland R, Gross C, Bezard E, Tison F (2004b) Deleterious effects of minocycline in animal models of Parkinson's disease and Huntington's disease. *Eur J Neurosci* **19**:3266-3276.

Du Y, Ma Z, Lin S, Dodel RC, Gao F, Bales KR, Triarhou LC, Chernet E, Perry KW, Nelson DL, Luecke S, Phebus LA, Bymaster FP, Paul SM (2001) Minocycline prevents nigrostriatal dopaminergic neurodegeneration in the MPTP model of Parkinson's disease. *Proc Natl Acad Sci U S A* **98**:14669-14674.

Eberhardt O, Coelln RV, Kugler S, Lindenau J, Rathke-Hartlieb S, Gerhardt E, Haid S, Isenmann S, Gravel C, Srinivasan A, Bahr M, Weller M, Dichgans J, Schulz JB (2000) Protection by synergistic effects of adenovirus-mediated X-chromosome-linked inhibitor of apoptosis and glial cell line-derived neurotrophic factor gene transfer in the 1-methyl-4-phenyl-1,2,3,6-tetrahydropyridine model of Parkinson's disease. *J Neurosci* **20**: 9126-9134.

Furuya T, Hayakawa H, Yamada M, Yoshimi K, Hisahara S, Miura M, Mizuno Y, Mochizuki H (2004) Caspase-11 mediates inflammatory dopaminergic cell death in the 1-methyl-4-phenyl-1,2,3,6-tetrahydropyridine mouse model of Parkinson's disease. *J Neurosci* **24**:1865-1872.

Gervais FG, Xu D, Robertson GS, Vaillancourt JP, Zhu Y, Huang J, LeBlanc A, Smith D, Rigby M, Shearman MS, Clarke EE, Zheng H, Van Der Ploeg LH, Ruffolo SC, Thornberry NA, Xanthoudakis S, Zamboni RJ, Roy S, Nicholson DW (1999) Involvement of caspases in proteolytic cleavage of Alzheimer's amyloid-beta precursor protein and amyloidogenic A beta peptide formation. *Cell* **97**:395-406.

Goldberg MS, Fleming SM, Palacino JJ, Cepeda C, Lam HA, Bhatnagar A, Meloni EG, Wu N, Ackerson LC, Klapstein GJ, Gajendiran M, Roth BL, Chesselet MF, Maidment NT, Levine MS, Shen J (2003) Parkin-deficient mice exhibit nigrostriatal deficits but not loss of dopaminergic neurons. *J Biol Chem* **278**:43628-43635.

Greene JC, Whitworth AJ, Kuo I, Andrews LA, Feany MB, Pallanck LJ (2003) Mitochondrial pathology and apoptotic muscle degeneration in Drosophila parkin mutants. *Proc Natl Acad Sci U S A* **100**:4078-4083.

Hartmann A, Hirsch EC (2001) Parkinson's disease. The apoptosis hypothesis revisited. *Adv Neurol* **86**:143-153.

Hartmann A, Troadec JD, Hunot S, Kikly K, Faucheux BA, Mouatt-Prigent A, Ruberg M, Agid Y, Hirsch EC (2001) Caspase-8 is an effector in apoptotic death of dopaminergic neurons in Parkinson's disease, but pathway inhibition results in neuronal necrosis. *J Neurosci* **21**:2247-2255.

Hashimoto M, Takeda A, Hsu LJ, Takenouchi T, Masliah E (1999) Role of cytochrome c as a stimulator of alpha-synuclein aggregation in Lewy body disease. *J Biol Chem* **274**: 28849-28852.

Holler N, Zaru R, Micheau O, Thome M, Attinger A, Valitutti S, Bodmer JL, Schneider P, Seed B, Tschopp J (2000) Fas triggers an alternative, caspase-8-independent cell death pathway using the kinase RIP as effector molecule. *Nat Immunol* **1**:489-495.

Hunot S, Hirsch EC (2003) Neuroinflammatory processes in Parkinson's disease. *Ann Neurol 53 Suppl* **3**:S49-58; discussion S58-60.

Hunot S, Vila M, Teismann P, Davis RJ, Hirsch EC, Przedborski S, Rakic P, Flavell RA (2004) JNK-mediated induction of cyclooxygenase 2 is required for neurodegeneration in a mouse model of Parkinson's disease. *Proc Natl Acad Sci U S A* **101**:665-670.

Itier JM, Ibanez P, Mena MA, Abbas N, Cohen-Salmon C, Bohme GA, Laville M, Pratt J, Corti O, Pradier L, Ret G, Joubert C, Periquet M, Araujo F, Negroni J, Casarejos MJ, Canals S, Solano R, Serrano A, Gallego E, Sanchez M, Denefle P, Benavides J, Tremp G, Rooney TA, Brice A, Garcia de Yebenes J (2003) Parkin gene inactivation alters behaviour and dopamine neurotransmission in the mouse. *Hum Mol Genet* **12**:2277-2291.

Jiang H, Ren Y, Zhao J, Feng J (2004) Parkin protects human dopaminergic neuroblastoma cells against dopamine-induced apoptosis. *Hum Mol Genet* **13**:1745-1754.

Kahns S, Lykkebo S, Jakobsen LD, Nielsen MS, Jensen PH (2002) Caspase-mediated parkin cleavage in apoptotic cell death. *J Biol Chem* **277**:15303-15308.

Kahns S, Kalai M, Jakobsen LD, Clark BF, Vandenabeele P, Jensen PH (2003) Caspase-1 and caspase-8 cleave and inactivate cellular parkin. *J Biol Chem* **278**:23376-23380.

Kang SJ, Wang S, Hara H, Peterson EP, Namura S, Amin-Hanjani S, Huang Z, Srinivasan A, Tomaselli KJ, Thornberry NA, Moskowitz MA, Yuan J (2000) Dual role of caspase-11 in mediating activation of caspase-1 and caspase-3 under pathological conditions. *J Cell Biol* **149**:613-622.

Kerr JF, Wyllie AH, Currie AR (1972) Apoptosis: a basic biological phenomenon with wide-ranging implications in tissue kinetics. *Br J Cancer* **26**:239-257.

Kingsbury AE, Mardsen CD, Foster OJ (1998) DNA fragmentation in human substantia nigra: apoptosis or perimortem effect? *Mov Disord* **13**:877-884.

Klevenyi P, Andreassen O, Ferrante RJ, Schleicher JR, Jr., Friedlander RM, Beal MF (1999) Transgenic mice expressing a dominant negative mutant interleukin-1beta converting enzyme show resistance to MPTP neurotoxicity. *Neuroreport* **10**:635-638.

Lang-Rollin IC, Rideout HJ, Noticewala M, Stefanis L (2003) Mechanisms of caspase-independent neuronal death: energy depletion and free radical generation. *J Neurosci* **23**:11015-11025.

Leist M, Jaattela M (2001) Four deaths and a funeral: from caspases to alternative mechanisms. *Nat Rev Mol Cell Biol* **2**:589-598.

Mariathasan S, Newton K, Monack DM, Vucic D, French DM, Lee WP, Roose-Girma M, Erickson S, Dixit VM (2004) Differential activation of the inflammasome by caspase-1 adaptors ASC and Ipaf. *Nature* **430**:213-218.

Martinon F, Tschopp J (2004) Inflammatory caspases: linking an intracellular innate immune system to autoinflammatory diseases. *Cell* **117**:561-574.

McNaught KS, Perl DP, Brownell AL, Olanow CW (2004) Systemic exposure to proteasome inhibitors causes a progressive model of Parkinson's disease. *Ann Neurol* **56**:149-162.

Mochizuki H, Hayakawa H, Migita M, Shibata M, Tanaka R, Suzuki A, Shimo-Nakanishi Y, Urabe T, Yamada M, Tamayose K, Shimada T, Miura M, Mizuno Y (2001) An AAV-derived Apaf-1 dominant negative inhibitor prevents MPTP toxicity as antiapoptotic gene therapy for Parkinson's disease. *Proc Natl Acad Sci U S A* **98**:10918-10923.

Nicotera P (2002) Apoptosis and age-related disorders: role of caspase-dependent and caspase-independent pathways. *Toxicol Lett* **127**:189-195.

Nicotera P, Leist M, Manzo L (1999) Neuronal cell death: a demise with different shapes. *Trends Pharmacol Sci* **20**:46-51.

Ricci JE, Munoz-Pinedo C, Fitzgerald P, Bailly-Maitre B, Perkins GA, Yadava N, Scheffler IE, Ellisman MH, Green DR (2004) Disruption of mitochondrial function during apoptosis is mediated by caspase cleavage of the p75 subunit of complex I of the electron transport chain. *Cell* **117**:773-786.

Saha AR, Ninkina NN, Hanger DP, Anderton BH, Davies AM, Buchman VL (2000) Induction of neuronal death by alpha-synuclein. *Eur J Neurosci* **12**:3073-3077.

Schapira AH, Gu M, Taanman JW, Tabrizi SJ, Seaton T, Cleeter M, Cooper JM (1998) Mitochondria in the etiology and pathogenesis of Parkinson's disease. *Ann Neurol* **44**:S89-98.

Schwab BL, Guerini D, Didszun C, Bano D, Ferrando-May E, Fava E, Tam J, Xu D, Xanthoudakis S, Nicholson DW, Carafoli E, Nicotera P (2002) Cleavage of plasma membrane calcium pumps by caspases: a link between apoptosis and necrosis. *Cell Death Differ* **9**:818-831.

Shen J, Cookson MR (2004) Mitochondria and dopamine: new insights into recessive parkinsonism. *Neuron* **43**:301-304.

Shimohama S, Sawada H, Kitamura Y, Taniguchi T (2003) Disease model: Parkinson's disease. *Trends Mol Med* **9**:360-365.

Sperandio S, de Belle I, Bredesen DE (2000) An alternative, nonapoptotic form of programmed cell death. *Proc Natl Acad Sci U S A* **97**:14376-14381.

Staropoli JF, McDermott C, Martinat C, Schulman B, Demireva E, Abeliovich A (2003) Parkin is a component of an SCF-like ubiquitin ligase complex and protects postmitotic neurons from kainate excitotoxicity. *Neuron* **37**:735-749.

Strasser A, O'Connor L, Dixit VM (2000) Apoptosis signaling. *Annu Rev Biochem* **69**: 217-245.

Sun XM, Butterworth M, MacFarlane M, Dubiel W, Ciechanover A, Cohen GM (2004) Caspase activation inhibits proteasome function during apoptosis. *Mol Cell* **14**:81-93.

Tanaka K, Ogawa N (2004) Possibility of non-immunosuppressive immunophilin ligands as potential therapeutic agents for Parkinson's disease. *Curr Pharm Des* **10**:669-677.

Tanaka Y, Engelender S, Igarashi S, Rao RK, Wanner T, Tanzi RE, Sawa A, V LD, Dawson TM, Ross CA (2001) Inducible expression of mutant alpha-synuclein decreases proteasome activity and increases sensitivity to mitochondria-dependent apoptosis. *Hum Mol Genet* **10**:919-926.

Tatton WG, Chalmers-Redman R, Brown D, Tatton N (2003) Apoptosis in Parkinson's disease: signals for neuronal degradation. *Ann Neurol 53 Suppl* **3**:S61-70; discussion S70-62.

Thomas M, Le WD (2004) Minocycline: neuroprotective mechanisms in Parkinson's disease. *Curr Pharm Des* **10**:679-686.

Turmaine M, Raza A, Mahal A, Mangiarini L, Bates GP, Davies SW (2000) Nonapoptotic neurodegeneration in a transgenic mouse model of Huntington's disease. *Proc Natl Acad Sci U S A* **97**:8093-8097.

Valente EM, Abou-Sleiman PM, Caputo V, Muqit MM, Harvey K, Gispert S, Ali Z, Del Turco D, Bentivoglio AR, Healy DG, Albanese A, Nussbaum R, Gonzalez-Maldonado R, Deller T, Salvi S, Cortelli P, Gilks WP, Latchman DS, Harvey RJ, Dallapiccola B, Auburger G, Wood NW (2004) Hereditary early-onset Parkinson's disease caused by mutations in PINK1. *Science* **304**:1158-1160.

Vila M, Przedborski S (2003) Targeting programmed cell death in neurodegenerative diseases. *Nat Rev Neurosci* **4**:365-375.

Vila M, Przedborski S (2004) Genetic clues to the pathogenesis of Parkinson's disease. *Nat Med* **10** *Suppl*:S58-62.

Vila M, Jackson-Lewis V, Vukosavic S, Djaldetti R, Liberatore G, Offen D, Korsmeyer SJ, Przedborski S (2001) Bax ablation prevents dopaminergic neurodegeneration in the 1-methyl- 4-phenyl-1,2,3,6-tetrahydropyridine mouse model of Parkinson's disease. *Proc Natl Acad Sci U S A* **98**:2837-2842.

Viswanath V, Wu Y, Boonplueang R, Chen S, Stevenson FF, Yantiri F, Yang L, Beal MF, Andersen JK (2001) Caspase-9 activation results in downstream caspase-8 activation and bid cleavage in 1-methyl-4-phenyl-1,2,3,6-tetrahydropyridine-induced Parkinson's disease. *J Neurosci* **21**:9519-9528.

von Coelln R, Kugler S, Bahr M, Weller M, Dichgans J, Schulz JB (2001) Rescue from death but not from functional impairment: caspase inhibition protects dopaminergic cells against 6-hydroxydopamine-induced apoptosis but not against the loss of their terminals. *J Neurochem* **77**:263-273.

Von Coelln R, Thomas B, Savitt JM, Lim KL, Sasaki M, Hess EJ, Dawson VL, Dawson TM (2004) Loss of locus coeruleus neurons and reduced startle in parkin null mice. *Proc Natl Acad Sci U S A* **101**:10744-10749.

Waldmeier PC, Tatton WG (2004) Interrupting apoptosis in neurodegenerative disease: potential for effective therapy? *Drug Discov Today* **9**:210-218.

Wang S, Miura M, Jung YK, Zhu H, Li E, Yuan J (1998) Murine caspase-11, an ICE-interacting protease, is essential for the activation of ICE. *Cell* **92**:501-509.

Weinreb O, Bar-Am O, Amit T, Chillag-Talmor O, Youdim MB (2004) Neuroprotection via pro-survival protein kinase C isoforms associated with Bcl-2 family members. *Faseb J* **18**:1471-1473.

Wu DC, Jackson-Lewis V, Vila M, Tieu K, Teismann P, Vadseth C, Choi DK, Ischiropoulos H, Przedborski S (2002) Blockade of microglial activation is neuroprotective in the 1-methyl-4-phenyl-1,2,3,6-tetrahydropyridine mouse model of Parkinson disease. *J Neurosci* **22**:1763-1771.

Xia XG, Harding T, Weller M, Bieneman A, Uney JB, Schulz JB (2001) Gene transfer of the JNK interacting protein-1 protects dopaminergic neurons in the MPTP model of Parkinson's disease. *Proc Natl Acad Sci U S A* **98**:10433-10438.

Yang L, Sugama S, Mischak RP, Kiaei M, Bizat N, Brouillet E, Joh TH, Beal MF (2004) A novel systemically active caspase inhibitor attenuates the toxicities of MPTP, malonate, and 3NP in vivo. *Neurobiol Dis* **17**:250-259.

Yang L, Sugama S, Chirichigno JW, Gregorio J, Lorenzl S, Shin DH, Browne SE, Shimizu Y, Joh TH, Beal MF, Albers DS (2003) Minocycline enhances MPTP toxicity to dopaminergic neurons. *J Neurosci Res* **74**:278-285.

Yokota T, Sugawara K, Ito K, Takahashi R, Ariga H, Mizusawa H (2003) Down regulation of DJ-1 enhances cell death by oxidative stress, ER stress, and proteasome inhibition. *Biochem Biophys Res Commun* **312**:1342-1348.

Zhou W, Schaack J, Zawada WM, Freed CR (2002) Overexpression of human alpha-synuclein causes dopamine neuron death in primary human mesencephalic culture. *Brain Res* **926**:42-50.

In: Recent Breakthroughs in Basal Ganglia Research
Editor: Erwan Bezard, pp. 313-322
ISBN: 1-59454-880-3

Chapter 24

The Proteasome in Neurodegenerative Disorders of the Basal Ganglia

Leonidas Stefanis*
Neurobiology Laboratory, Foundation of Biomedical Research of the Academy of Athens
Soranou Efesiou 4, Athens 11527, Greece

Abstract

Accumulating data suggest that impairment of protein degradation, and in particular of the ubiquitin-proteasome system, may be a contributing, or even initiating factor in neurodegeneration. The ubiquitin-proteasome system degrades proteins that are tagged, through a series of enzymatic reactions, with multiple molecules of ubiquitin. These poly-ubiquitinayted substrates are subsequently recognized and degraded by the multicatalytic proteasome complex. Here we review recent evidence that defects in this system may play a role in Parkinson's, as well as in other Lewy body diseases, and in expanded polyglutamine repeat disorders, such as Huntington's disease. Special emphasis is placed on the association of known genetic defects in these disorders with the ubiquitin-proteasome system and on the relationship between protein misfolding-aggregation and malfunction of this system.

Introduction

Over the past few years, converging lines of evidence have led to the hypothesis that defects in the degradation of proteins through the proteasome may contribute to neurodegeneration in a variety of neurological diseases. For the purpose of this review, we will focus on those neurodegenerative diseases linked to the basal ganglia. Proteasomal dysfunction may also play a role in diseases such as Amyotrophic Lateral Sclerosis, prion diseases and Alzheimer's Disease, but because the brunt of the pathology in these disorders lies outside the basal ganglia, this evidence will not be reviewed here.

What is the proteasome? It is a multicatalytic protein complex responsible for the degradation of many types of intracellular proteins. Amongst such proteins, notable are rapidly turning over, misfolded, damaged, or oxidatively modified proteins. In a typical cell, a large bulk, often the majority, of protein degradation may occur through the proteasome. A basic mechanism of recognition of protein substrates by the proteasome is the tagging of such substrates with multiple adducts of a small molecule, ubiquitin. The process of ubiquitination involves first the activation, in an ATP-dependent fashion, of ubiquitin by a universal E1 ubiquitin-activating-enzyme, then the transfer, by an E2 ubiquitin-conjugating enzyme (UCE), of activated ubiquitin from E1 to a member of the E3 ligase family. E3 ligases also bind the substrates, to which they attach activated ubiquitin. In some cases, E4 enzymes are involved in the stacking of ubiquitin molecules on each other to form a poly-ubiquitin chain. E2s and, especially, E3s and E4s, unlike E1, are specific for a substrate or a group of substrates.

* Tel: 30-210-6597214 / Fax: 30-210-6597545, E-mail address: lstefanis@bioacademy.gr

The polyubiquitinated proteins that are formed are subsequently recognized and degraded by the proteasome. It should be noted that natively unfolded proteins may not need to be tagged by ubiquitin to be degraded by the proteasome, and thus ubiquitin-independent proteasomal degradation may also occur. The 26S proteasome contains a core catalytic component, the 20S proteasome, flanked on both sides by a 19S cap. Polyubiquitinated proteins signaled for degradation are recognized by and bind to components of the 19S cap. They then enter the cylindrical pore formed by the 20S proteasome, which consists of two identical outer rings of 7 α subunits each and two identical inner rings of 7 β subunits each, and are degraded by 3 sets of enzymes, with chymotrypsin-like, trypsin-like, and peptidylglutamyl-like activity. Proteasomal proteolysis is ATP-dependent. The polyubiquitin chains are hydrolyzed to monomeric ubiquitin through the action of deubiquitinating enzymes, which include, amongst others, ubiquitin C-terminal hydrolases (UCHs). The released monomeric ubiquitin is reactivated by E1 in order to act on further substrates (Ciechanover and Brundin, 2003).

The function of the proteasome is classically assessed by measuring in cell lysates or tissue extracts the 3 enzymatic activities present within the 20S core. Alternatively, the levels of particular substrate proteins that are known to be targeted to and degraded by the proteasome can be assessed. Such artificial substrates have been created and tagged to fluorescent proteins, so that their potential accumulation can be monitored in living cells.

The issue of proteasomal dysfunction in neurodegenerative diseases is closely related to that of aberrant protein folding. In many such diseases, abnormal proteins assume an oligomeric conformation, leading to the formation of proto-fibrils and then eventually bona-fide fibrillar structures and inclusions that characterize such disease processes. Some of these abnormal disease-related proteins appear to be targeted for degradation by the proteasome. It has now been realized that proteasomal function may be adversely affected by such proteins. Proteasomal dysfunction can in turn cause further accumulation of misfolded protein conformations, leading to further inclusion formation and thus establishing a vicious cycle (Figure 1).

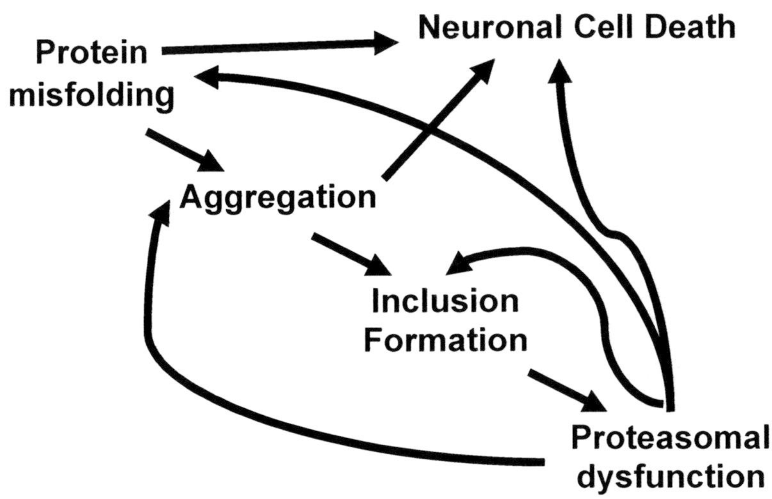

Figure 1

Thinking about how proteasomal function may be linked to neurodegeneration in disorders of the basal ganglia, one has to have a few questions in mind:

- How are the particular disease-related pathological/misfolded proteins degraded?
- What are the effects of such proteins on the ubiquitin proteasomal system (UPS)?
- Is there evidence of proteasomal/UPS dysfunction in these disorders?
- Bearing these questions in mind, we will review here the evidence linking two major classes of basal ganglia diseases, Parkinson's disease and related Lewy body diseases, and Huntington's disease and related expanded polyglutamine repeat disorders to abnormalities of the UPS of protein degradation.

1 PARKINSON'S DISEASE AND OTHER LEWY BODY-RELATED CONDITIONS

Parkinson's disease (PD) is the condition for which the largest body of evidence exists regarding involvement of the UPS in disease pathogenesis. Most important is the fact that two proteins linked genetically to the disease, Parkin and UCH-L1, are involved in the UPS. A considerable proportion of early onset autosomal recessive PD families harbor mutations in the gene encoding for Parkin. Parkin is an E3 ligase, involved in tagging specific substrates with ubiquitin, and mutations in the gene lead to diminution or loss of this activity (Hattori and Mizuno, 2004). This has led to the hypothesis that the build-up of specific Parkin substrates is responsible for disease pathogenesis in this type of autosomal recessive PD. A number of potential substrates have been identified, amongst them CDCRel-1, a glycosylated form of α-synuclein, the α-synuclein-binding protein synphilin-1, and the transmembrane protein Pael receptor but it is unclear which one, if any, of these, may play such a pathogenic role (Hattori and Mizuno, 2004). Importantly, Parkin overexpression appears to ameliorate proteasomal function (Hyun et al., 2002; Tsai et al., 2003) and to protect against proteasomal inhibition-induced death of cultured ventral midbrain dopaminergic neurons and other neuronal cells (Petrucelli et al., 2002). Furthermore, downregulation of parkin expression by antisense oligonucleotides in neuroblastoma cells led to inhibition of proteasomal-dependent protein clearance (Petrucelli et al., 2002). It has been proposed that Parkin may act, through its binding to the molecular chaperone HSP70, as a general facilitator of the degradation of proteins that misfold in the endoplasmic reticulum or the cytosol (Tsai et al., 2003). Consistent with this idea, the N-terminal ubiquitin-like domain of Parkin may bind directly to the 26S proteasome and enhance its degradative ability (Sakata et al., 2003; Tsai et al., 2003). Thus, the absence of Parkin, apart from its effects on the build-up of specific protein substrates, may lead to more generalized defects of protein degradation through the proteasome. Whether such a general defect exists in the affected regions of patients suffering from Parkin-related neurodegeneration or in the brains of mice null for Parkin has not been tested specifically. It should be noted, however, that in most, but not all, cases of PD due to Parkin mutations there is an absence of Lewy bodies (LBs) (Hattori and Mizuno, 2004). This does not exclude the possibility of the general build-up of proteasomal substrates in this condition, but suggests that, if this is the case, such a build-up does not lead to inclusion formation because of lack of ubiquitination of key substrates. Consistent with this idea, we have shown that LB-like inclusions in cultured cortical neurons treated with proteasomal inhibitors are abolished upon overexpression of a dominant negative form of the E2 enzyme Cdc34, suggesting that ubiquitination of certain substrates is required for inclusion formation in conditions of proteasomal inhibition (Rideout and Stefanis, 2002).

A mutation in UCH-L1, I93M, has been described in one family with autosomal dominant PD (Leroy et al., 1998). In addition, the presence of polymorphism S18Y has been shown in a recent meta-analysis from multiple data sets from different populations to be negatively correlated with the chance of developing PD, further arguing for the importance of this gene and the encoded protein in PD pathogenesis (Maraganore et al., 2004). UCH-L1 is an abundant neuronal-specific protein, which belongs to the family of UCHs (Chung and Baek, 1999). The cellular substrate(s) of UCH-L1 is unknown, but it may act on linear poly-ubiquitin proteins, as they are produced during translation, or on polyubiquitinated proteins following their degradation by the proteasome to a small peptide bound to the polyubiquitin chain (Pickart and Rose, 1985; Hegde et al., 1997; Larsen et al., 1998; Chung and Baek, 1999). Additional functions for UCH-L1 that have been proposed include the maintenance of sufficient levels of free ubiquitin in the cell through direct binding and inhibition of the degradation of free ubiquitin in the lysosomes (Osaka et al., 2003), or an unexpected function as an aberrant E3 ligase, that binds to protein substrates such as α-synuclein and ubiquitinates them at an aberrant site, leading to their lack of recognition by the proteasome and consequent build-up in the cell (Liu et al., 2002). It is evident that the function of this protein is still a matter of contention. I93L confers a 50 % decrease in the in vitro enzymatic activity of this protein as a deubiquitinating enzyme, leading to the hypothesis that in this particular case PD may be due to haploinsufficiency (Leroy et al., 1998; Nishikawa et al., 2003). This mutation also leads to an increase of the purported aberrant E3 ligase activity, suggesting instead a gain of function (Liu et al., 2002). The protective S18Y polymorphism leads to a slight increase of the in vitro deubiquitinating activity and to a decrease of the E3 ligase activity (Liu et al., 2002; Nishikawa et al., 2003). Lack of UCH-L1 in mice leads to sensory ataxia and axonal degeneration in the gracile nucleus. Such *gad* mice show a marked decrease in the levels of free ubiquitin (Osaka et al., 2003). Taking all this information into account, it is clear that, despite the lack of consensus on the normal function of UCH-L1 and on the changes conferred on this function by genetic alterations, this protein is somehow involved in the pathway of ubiquitination and subsequent proteasomal degradation. This provides further evidence that this pathway is linked to PD pathogenesis.

The first gene to be identified as linked to PD encodes for the presynaptic protein α-synuclein (Polymeropoulos et al., 1997; Vekrellis et al., 2004). Although α-synuclein is not as tightly linked to the proteasomal system of protein degradation as are Parkin and UCH-L1, there is some evidence that it may be influenced and may in turn impact this system. α-synuclein, a natively unfolded protein with a propensity to aggregate, constitutes a major component of LBs. It is hypothesized that aberrant folding forms of α-synuclein may be responsible for sporadic PD and related LB diseases, collectively termed "synucleinopathies" (Duda et al., 2000). Original reports (Bennett et al., 1999; Imai et al., 2000) suggested that α-synuclein is ubiquitinated and degraded by the proteasome, however we and others failed to find significant upregulation of endogenous or overexpressed α-synuclein with application of proteasomal inhibitors to cell lines or primary neurons (Ancolio et al., 2000; Paxinou et al., 2001; Rideout et al., 2001; Rideout and Stefanis, 2002). Instead, we have recently presented evidence that α-synuclein is degraded in lysosomes, via the process of chaperone-mediated autophagy (Cuervo et al., 2004). It is still possible that some α-synuclein degradation may occur in the proteasome, and this may depend on the particular conformation and localization of α-synuclein, as well as the cellular context.

A number of laboratories, including our own, have shown that overexpression of mutant α-synuclein in neuronal cell lines leads to impairment of proteasomal activity (Stefanis et al., 2001; Tanaka et al., 2001; Petrucelli et al., 2002). In another report, overexpression of WT α-synuclein was also associated with proteasomal impairment (Snyder et al., 2003). It remains to be seen whether such an impairment occurs in animal models of synucleinopathies and in the brains of patients with familial synucleinopathies. The mechanism through which proteasomal dysfunction may occur is unclear, although the general principle of "clogging" of the proteasome by aggregates may apply. We, however, failed to detect obvious aggregation of mutant α-synuclein in our cellular system (Stefanis et al., 2001), and aggregated α-synuclein appears to be degraded by lysosomes, not proteasomes (Lee et al., 2004). α-synuclein may instead have direct effects on the proteasome through binding to specific components of the 19S or the 20S proteasome (Ghee et al., 2000; Snyder et al., 2003; Lindersson et al., 2004). Alternatively, effects on the proteasome may be indirect, perhaps through the effect of aberrant α-synuclein on lysosomes.

The proteins encoded by the two other genes linked to PD, PINK-1 (Valente et al., 2004) and DJ-1 (Bonifati et al., 2003), do not have an obvious link to the proteasome system of protein degradation, although DJ-1 may serve as a SUMOlating factor, a function akin to ubiquitylation.

The fact that the ubiquitin present within LBs represents polyubiquitinated proteins (Iwatsubo et al., 1996), suggests that, if there is dysfunction of the UPS in PD and related LB diseases, this should be at the level of the proteasome. A number of studies have addressed the issue of potential dysfunction of the proteasome in sporadic PD. In a seminal paper, McNaught and Jenner, (2001) measured enzymatic activities of the proteasome in brains of patients with PD and controls, and found a specific decrease of these activities in Substantia Nigra (SN) of PD patients. This result has been confirmed by Tofaris et al., (2003), who additionally found decreased proteasomal activity in SN of DLBD patients. Other CNS regions showed normal activity in DLBD, except for frontal cortex, where activity was somewhat lower, without however reaching statistical significance (Tofaris et al., 2003). Examination of extranigral regions of PD patients by another group also showed normal proteasomal activity (Furukawa et al., 2002). It appears therefore that proteasomal activity is decreased in the SN of patients with LB-related diseases, but that most other areas of the CNS have normal activity.

It should be noted here that such assessments of proteasomal function based on the measurement of enzymatic activities derived from tissue homogenates may give both false-positive and false-negative results. The tissues studied from PD patients, with the marked proliferation of glia and microglia and the near-depletion of dopaminergic neurons, represent a markedly altered population of cells compared to control tissues. It is possible therefore that differences encountered may be due the altered cell composition, and not to a decrease per se of proteasomal activity in nigral neurons. Alternatively, if only nigral melanized neurons have decreased proteasomal activity, it may be difficult to pick that up in an assay that measures the activity of a host of cellular elements, where dopaminergic neurons are a minority. What these studies are showing therefore is a regional impairment of enzymatic activity of the proteasome at the level of the ventral midbrain in PD patients.

What could be the cause of such a decrease of proteasomal activity in the SN? McNaught et al., (2002a) have reported that there is a dramatic decrease of labeling of α subunits of the 20S proteasome only in melanized neurons of the SN of PD patients, whereas labeling for β subunits was normal. This raises the possibility of primary structural or biochemical defects of the proteasome in PD patients. Alternatively, other factors etiologically linked to PD, such as protein aggregation, aging, oxidative, nitrative or nitrosylative stress, or mitochondrial dysfunction, may be responsible.

The co-existence of proteasomal dysfunction, at least in the SN, with obvious protein aggregation in the form of LBs raises the chicken and egg dilemma. As has already been mentioned here, protein aggregation can potentially lead to proteasomal dysfunction. Can proteasomal dysfunction lead to a PD-like phenotype? We and others have shown that application of proteasomal inhibitors to cultured neuronal cells leads to formation of LB-like inclusions and neuronal death (Rideout et al., 2001; Rideout and Stefanis, 2002). Although somewhat controversial (see Kikuchi et al., 2003), it appears that proteasomal inhibition is selectively toxic to dopaminergic neurons of ventral midbrain cultures (McNaught et al., 2002c; Petrucelli et al., 2002), and therefore a regional proteasomal impairment in SN could lead to selective degeneration of dopaminergic neurons. Initial in vivo experiments with direct injection of proteasomal inhibitors in the brain, and in particular in the SN, showed that such treatment induced neuronal death that could be selective for neurons compared to glia (McNaught et al., 2002b). McNaught et al., (2004) have reported in a more recent study remarkable effects of systemic application of pharmacological proteasomal inhibitors. Such treatment appears to lead to selective neuronal degeneration and aggregation in the brainstem structures, including the SN, described by Braak et al., (2003) as predominantly affected in PD. We have recently shown that application of such inhibitors to cultured cortical neurons leads to the oligomerization of α-synuclein within LB-like inclusions, presumably because of a molecular crowding effect. In this model, α-synuclein does not influence the formation of the inclusions, but determines whether they assume a fibrillar structure (Rideout et al., 2004). On the basis of these findings, it appears that proteasomal dysfunction can lead to protein aggregation and neuronal degeneration that may be selective for regions affected in PD. Whether this is what actually occurs in PD remains to be determined.

2 Huntington's Disease and Other Expanded Polyglutamine Repeat Diseases

It is widely recognized that mutant Huntingtin and other expanded polyglutamine repeat proteins are degraded, at least in part, by the proteasome (see for example Cummings et al., 1999; Jana et al., 2001; Ravikumar et al., 2002), although alternate routes of degradation by macro-autophagy may also play a role (Ravikumar et al., 2002). It was by expressing truncated forms of such proteins in cells carrying a reporter activity for proteasomal function that Bence et al., (2001) made the seminal observation of decreased UPS function that has been confirmed by other investigators (for example, see Jana et al., 2001; Tsai et al., 2003). These models therefore provide an example of the vicious cycle of protein aggregation and UPS dysfunction depicted in Figure 1. In these cases, the initiating factor is clearly the aberrantly misfolded protein, but it may be that the ensuing proteasomal dysfunction plays a significant secondary role in disease pathogenesis. It is not known exactly how such aggregated proteins inhibit proteasomal function. A prevailing hypothesis is that the aggregated protein blocks access of other substrates either to recognition motifs or the hydrolytic core of the proteasome. This hypothesis is based on the assumption that aggregated proteins are more difficult to degrade than their WT counterparts, and that they may act as dominant negatives, binding to components of the proteasome but not being readily released. There is experimental evidence that supports these assumptions (see for example Cummings et al., 1999; Jana et al., 2001; Navon and Goldberg, 2001). Alternatively, the large amount of aggregated protein may overwhelm the system by consuming ubiquitin or other elements of the pathway, such as molecular chaperones, that smooth along the process (Sakahira et al., 2002). Proteasomal and other UPS components are present within inclusions induced by such aggregated polyglutamine repeat proteins (Cummings et al., 1998; Chai et al., 1999; Jana et al., 2001; Waelter et al., 2001), and this may lead to their relative depletion from their normal site of action, and further amplification of the loop (Jana et al., 2001). A new twist to the story has been provided by the observation that isolated mammalian proteasomes cannot degrade the regions with the polyglutamine repeats, and release them without processing them further. The accumulation of these aggregated species may have further detrimental effects, feeding into the vicious cycle (Venkatraman et al., 2004).

In addition to the above, that depend on the mode of degradation of particular polyglutamine repeat proteins, there is evidence that some of these proteins may actually normally serve functions within the UPS pathway. For example, ataxin-3 has been shown to interact directly with poly-ubiquitin, poly-ubiquitinated proteins (Burnett et al., 2003; Doss-Pepe et al., 2003; Chai et al., 2004), the proteasome (Doss-Pepe et al., 2003), as well as with certain ubiquitin-like proteins that may serve as transporters of poly-ubiquitinated proteins (Wang et al., 2000; Doss-Pepe et al., 2003). It is presumed that its normal function involves assisting in the presentation of polyubiquitinated substrates to the proteasome. In addition, ataxin-7 interacts with the S4 subunit of the 19S proteasome, and thus may serve similar functions (Matilla et al., 2001).

It should be noted that not in all cases does overexpression of polyglutamine repeats lead to inhibition of proteasomal activity. Neuroblastoma cells stably expressing expanded polyglutamine tracts did not show an impairment of enzymatic proteasomal activity at baseline, and only showed a relative impairment of induction of proteasomal activity in response to heat shock (Ding et al., 2002). Interestingly, in these cells there is an upregulation of the β subunit of the 20S proteasome LMP2. Presumably this upregulation occurrs as a response to the proteolytic stress and enables the cells to maintain normal levels of proteasomal activity at baseline.

It has been known for some time that modulation of the UPS can lead to effects on the toxicity of huntingtin or other polyglutamine repeat proteins in cellular and animal models. For example, lack of E6-AP, an E3 ligase, decreased inclusion formation but enhanced toxicity in transgenic mouse model of Spinocerebellar Ataxia Type I (SCA 1), in which mutant Ataxin-1 was overexpressed (Cummings et al., 1999). Similarly, expression of a dominant negative form of the E2 Cdc34 abrogated inclusion formation, but enhanced toxicity in a cellular model of truncated mutant huntingtin overexpression (Saudou et al., 1998). From such and other data it has been surmised that the process of ubiquitination is necessary for inclusion formation, and that inclusions may actually serve a protective function. An alternative interpretation would be that inclusions are not related positively or negatively to toxicity, but that there is something inherently protective in the process of ubiquitination itself, and that disrupting that leads to enhanced toxicity in these conditions.

UPS dysfunction, as measured by enzymatic activities of the proteasome in tissue extracts, could not be demonstrated in two separate studies of various brain regions of mice expressing mutant Huntingtin, in one case a truncated form in a conditional model (Diaz-Hernandez et al., 2003), and in the other case as a knock-in (Zhou et al., 2003). In the conditional model, like in the stable cell line mentioned above, an induction of LMP2, as well as of another inducible β subunit, LMP7, was observed, signifying a shift of normal proteasomes to the structure of the "immunoproteasome". Such immunoproteasomes have different enzymatic characteristics and may be used for antigen presentation. It is unclear what exact role this upregulation, which occurs in neuronal cells, may play in this model, but it is likely to represent an adaptive response to the proteolytic stress induced by protein aggregation. This brings us to the issue of the plasticity of the UPS and the proteasome in particular. It is clear that in the case of chronic partial insults that affect this system, there can be a number of adaptive responses. A classical one is the upregulation of certain HSPs which assist in refolding folding intermediates and thus facilitate the work of the proteasome. The proteasome itself, as mentioned, can upregulate certain subunits at the transcriptional level, and this may alter its proteolytic capability. Alternatively, there may be a general modification of proteasome biogenesis following a particular stimulus, leading, for example, to a net increase in the number of proteasomes formed in a cell.

Does UPS dysfunction actually occur in Huntington's disease (HD)? A very recent study indicates that this is the case, and provides evidence that such dysfunction is much more widespread than expected. Brain regions not affected by the disease process, and even fibroblasts from the patients had a pronounced impairment of proteasomal activity compared to controls. Interestingly, even patients from the very early stage of the disease, where there is little, if any, neuronal cell loss, showed proteasomal dysfunction. This, together with the fact that proteasomal dysfunction extended well beyond the areas with pathology and cell loss, indicates that altered cellular composition cannot account for the observed changes (Seo et al., 2004). This therefore is in marked contrast with PD, where, as mentioned, proteasomal function in extranigral tissues is largely preserved. The results in human HD appear contradictory to those in the animal models. There are two possibilities: either the adaptive, plasticity responses uncovered in the animal models do not operate in the human disease state, or they are operative initially, but are eventually overwhelmed by the proteolytic stress. Given the fact that no obvious inclusions are present in many of the tissues assessed, the inference is that prefibrillar forms of mutant Huntingtin may be responsible for such proteasomal dysfunction. These data further suggest that a decrease of proteasomal activity may not be directly toxic to some cells, and that issues of selective vulnerability or additional aggravating factors may play a role in determining whether a cell can withstand impairment of proteasomal function.

There is also other indirect evidence for the existence of UPS dysfunction in HD brains. UBB^{+1} is an aberrant form of ubiquitin caused by molecular misreading at the mRNA level. Under normal circumstances its protein levels are non-detectable, and they only accumulate in situations of proteolytic stress (Ciechanover and Brundin, 2003). Elevated protein levels of UBB^{+1} have recently been detected in brains of patients with HD or SCA3, suggesting that in these conditions the UPS is impaired. Accumulation of UBB^{+1} can actually lead to further proteasomal inhibition, and may be a contributing factor to neurotoxicity in these disorders (de Pril et al., 2004).

REFERENCES

Ancolio K, Alves da Costa C, Ueda K, Checler F (2000) Alpha-synuclein and the Parkinson's disease-related mutant Ala53Thr-alpha-synuclein do not undergo proteasomal degradation in HEK293 and neuronal cells. *Neurosci Lett* **285**:79-82.

Bence NF, Sampat RM, Kopito RR (2001) Impairment of the ubiquitin-proteasome system by protein aggregation. *Science* **292**:1552-1555.

Bennett MC, Bishop JF, Leng Y, Chock PB, Chase TN, Mouradian MM (1999) Degradation of alpha-synuclein by proteasome. *J Biol Chem* **274**:33855-33858.

Bonifati V, Rizzu P, van Baren MJ, Schaap O, Breedveld GJ, Krieger E, Dekker MC, Squitieri F, Ibanez P, Joosse M, van Dongen JW, Vanacore N, van Swieten JC, Brice A, Meco G, van Duijn CM, Oostra BA, Heutink P (2003) Mutations in the DJ-1 gene associated with autosomal recessive early-onset parkinsonism. *Science* **299**:256-259.

Braak H, Del Tredici K, Rub U, de Vos RA, Jansen Steur EN, Braak E (2003) Staging of brain pathology related to sporadic Parkinson's disease. *Neurobiol Aging* **24**:197-211.

Burnett B, Li F, Pittman RN (2003) The polyglutamine neurodegenerative protein ataxin-3 binds polyubiquitylated proteins and has ubiquitin protease activity. *Hum Mol Genet* **12**:3195-3205.

Chai Y, Berke SS, Cohen RE, Paulson HL (2004) Poly-ubiquitin binding by the polyglutamine disease protein ataxin-3 links its normal function to protein surveillance pathways. *J Biol Chem* **279**:3605-3611.

Chai Y, Koppenhafer SL, Shoesmith SJ, Perez MK, Paulson HL (1999) Evidence for proteasome involvement in polyglutamine disease: localization to nuclear inclusions in SCA3/MJD and suppression of polyglutamine aggregation in vitro. *Hum Mol Genet* **8**:673-682.

Chung CH, Baek SH (1999) Deubiquitinating enzymes: their diversity and emerging roles. *Biochem Biophys Res Commun* **266**:633-640.

Ciechanover A, Brundin P (2003) The ubiquitin proteasome system in neurodegenerative diseases: sometimes the chicken, sometimes the egg. *Neuron* **40**:427-446.

Cuervo AM, Stefanis L, Fredenburg R, Lansbury PT, Sulzer D (2004) Impaired degradation of mutant alpha-synuclein by chaperone-mediated autophagy. *Science* **305**:1292-1295.

Cummings CJ, Mancini MA, Antalffy B, DeFranco DB, Orr HT, Zoghbi HY (1998) Chaperone suppression of aggregation and altered subcellular proteasome localization imply protein misfolding in SCA1. *Nat Genet* **19**:148-154.

Cummings CJ, Reinstein E, Sun Y, Antalffy B, Jiang Y, Ciechanover A, Orr HT, Beaudet AL, Zoghbi HY (1999) Mutation of the E6-AP ubiquitin ligase reduces nuclear inclusion frequency while accelerating polyglutamine-induced pathology in SCA1 mice. *Neuron* **24**:879-892.

de Pril R, Fischer DF, Maat-Schieman ML, Hobo B, de Vos RA, Brunt ER, Hol EM, Roos RA, van Leeuwen FW (2004) Accumulation of aberrant ubiquitin induces aggregate formation and cell death in polyglutamine diseases. *Hum Mol Genet* **13**:1803-1813.

Diaz-Hernandez M, Hernandez F, Martin-Aparicio E, Gomez-Ramos P, Moran MA, Castano JG, Ferrer I, Avila J, Lucas JJ (2003) Neuronal induction of the immunoproteasome in Huntington's disease. *J Neurosci* **23**:11653-11661.

Ding Q, Lewis JJ, Strum KM, Dimayuga E, Bruce-Keller AJ, Dunn JC, Keller JN (2002) Polyglutamine expansion, protein aggregation, proteasome activity, and neural survival. *J Biol Chem* **277**:13935-13942.

Doss-Pepe EW, Stenroos ES, Johnson WG, Madura K (2003) Ataxin-3 interactions with rad23 and valosin-containing protein and its associations with ubiquitin chains and the proteasome are consistent with a role in ubiquitin-mediated proteolysis. *Mol Cell Biol* **23**:6469-6483.

Duda JE, Lee VM, Trojanowski JQ (2000) Neuropathology of synuclein aggregates. *J Neurosci Res* **61**:121-127.

Furukawa Y, Vigouroux S, Wong H, Guttman M, Rajput AH, Ang L, Briand M, Kish SJ, Briand Y (2002) Brain proteasomal function in sporadic Parkinson's disease and related disorders. *Ann Neurol* **51**:779-782.

Ghee M, Fournier A, Mallet J (2000) Rat alpha-synuclein interacts with Tat binding protein 1, a component of the 26S proteasomal complex. *J Neurochem* **75**:2221-2224.

Hattori N, Mizuno Y (2004) Pathogenetic mechanisms of parkin in Parkinson's disease. *Lancet* **364**:722-724.

Hegde AN, Inokuchi K, Pei W, Casadio A, Ghirardi M, Chain DG, Martin KC, Kandel ER, Schwartz JH (1997) Ubiquitin C-terminal hydrolase is an immediate-early gene essential for long-term facilitation in Aplysia. *Cell* **89**:115-126.

Hyun DH, Lee M, Hattori N, Kubo S, Mizuno Y, Halliwell B, Jenner P (2002) Effect of wild-type or mutant Parkin on oxidative damage, nitric oxide, antioxidant defenses, and the proteasome. *J Biol Chem* **277**:28572-28577.

Imai Y, Soda M, Takahashi R (2000) Parkin suppresses unfolded protein stress-induced cell death through its E3 ubiquitin-protein ligase activity. *J Biol Chem* **275**:35661-35664.

Iwatsubo T, Yamaguchi H, Fujimuro M, Yokosawa H, Ihara Y, Trojanowski JQ, Lee VM (1996) Purification and characterization of Lewy bodies from the brains of patients with diffuse Lewy body disease. *Am J Pathol* **148**:1517-1529.

Jana NR, Zemskov EA, Wang G, Nukina N (2001) Altered proteasomal function due to the expression of polyglutamine-expanded truncated N-terminal huntingtin induces apoptosis by caspase activation through mitochondrial cytochrome c release. *Hum Mol Genet* **10**:1049-1059.

Kikuchi S, Shinpo K, Tsuji S, Takeuchi M, Yamagishi S, Makita Z, Niino M, Yabe I, Tashiro K (2003) Effect of proteasome inhibitor on cultured mesencephalic dopaminergic neurons. *Brain Res* **964**:228-236.

Larsen CN, Krantz BA, Wilkinson KD (1998) Substrate specificity of deubiquitinating enzymes: ubiquitin C-terminal hydrolases. *Biochemistry* **37**:3358-3368.

Lee HJ, Khoshaghideh F, Patel S, Lee SJ (2004) Clearance of alpha-synuclein oligomeric intermediates via the lysosomal degradation pathway. *J Neurosci* **24**:1888-1896.

Leroy E, Boyer R, Auburger G, Leube B, Ulm G, Mezey E, Harta G, Brownstein MJ, Jonnalagada S, Chernova T, Dehejia A, Lavedan C, Gasser T, Steinbach PJ, Wilkinson KD, Polymeropoulos MH (1998) The ubiquitin pathway in Parkinson's disease. *Nature* **395**:451-452.

Lindersson E, Beedholm R, Hojrup P, Moos T, Gai W, Hendil KB, Jensen PH (2004) Proteasomal inhibition by alpha-synuclein filaments and oligomers. *J Biol Chem* **279**:12924-12934.

Liu Y, Fallon L, Lashuel HA, Liu Z, Lansbury PT, Jr. (2002) The UCH-L1 gene encodes two opposing enzymatic activities that affect alpha-synuclein degradation and Parkinson's disease susceptibility. *Cell* **111**:209-218.

Maraganore DM, Lesnick TG, Elbaz A, Chartier-Harlin MC, Gasser T, Kruger R, Hattori N, Mellick GD, Quattrone A, Satoh J, Toda T, Wang J, Ioannidis JP, de Andrade M, Rocca WA (2004) UCHL1 is a Parkinson's disease susceptibility gene. *Ann Neurol* **55**:512-521.

Matilla A, Gorbea C, Einum DD, Townsend J, Michalik A, van Broeckhoven C, Jensen CC, Murphy KJ, Ptacek LJ, Fu YH (2001) Association of ataxin-7 with the proteasome subunit S4 of the 19S regulatory complex. *Hum Mol Genet* **10**:2821-2831.

McNaught KS, Jenner P (2001) Proteasomal function is impaired in substantia nigra in Parkinson's disease. *Neurosci Lett* **297**:191-194.

McNaught KS, Perl DP, Brownell AL, Olanow CW (2004) Systemic exposure to proteasome inhibitors causes a progressive model of Parkinson's disease. *Ann Neurol* **56**:149-162.

McNaught KS, Belizaire R, Jenner P, Olanow CW, Isacson O (2002a) Selective loss of 20S proteasome alpha-subunits in the substantia nigra pars compacta in Parkinson's disease. *Neurosci Lett* **326**:155-158.

McNaught KS, Bjorklund LM, Belizaire R, Isacson O, Jenner P, Olanow CW (2002b) Proteasome inhibition causes nigral degeneration with inclusion bodies in rats. *Neuroreport* **13**:1437-1441.

McNaught KS, Mytilineou C, Jnobaptiste R, Yabut J, Shashidharan P, Jennert P, Olanow CW (2002c) Impairment of the ubiquitin-proteasome system causes dopaminergic cell death and inclusion body formation in ventral mesencephalic cultures. *J Neurochem* **81**:301-306.

Navon A, Goldberg AL (2001) Proteins are unfolded on the surface of the ATPase ring before transport into the proteasome. *Mol Cell* **8**:1339-1349.

Nishikawa K, Li H, Kawamura R, Osaka H, Wang YL, Hara Y, Hirokawa T, Manago Y, Amano T, Noda M, Aoki S, Wada K (2003) Alterations of structure and hydrolase activity of parkinsonism-associated human ubiquitin carboxyl-terminal hydrolase L1 variants. *Biochem Biophys Res Commun* **304**:176-183.

Osaka H, Wang YL, Takada K, Takizawa S, Setsuie R, Li H, Sato Y, Nishikawa K, Sun YJ, Sakurai M, Harada T, Hara Y, Kimura I, Chiba S, Namikawa K, Kiyama H, Noda M, Aoki S, Wada K (2003) Ubiquitin carboxy-terminal hydrolase L1 binds to and stabilizes monoubiquitin in neuron. *Hum Mol Genet* **12**:1945-1958.

Paxinou E, Chen Q, Weisse M, Giasson BI, Norris EH, Rueter SM, Trojanowski JQ, Lee VM, Ischiropoulos H (2001) Induction of alpha-synuclein aggregation by intracellular nitrative insult. *J Neurosci* **21**:8053-8061.

Petrucelli L, O'Farrell C, Lockhart PJ, Baptista M, Kehoe K, Vink L, Choi P, Wolozin B, Farrer M, Hardy J, Cookson MR (2002) Parkin protects against the toxicity associated with mutant alpha-synuclein: proteasome dysfunction selectively affects catecholaminergic neurons. *Neuron* **36**:1007-1019.

Pickart CM, Rose IA (1985) Ubiquitin carboxyl-terminal hydrolase acts on ubiquitin carboxyl-terminal amides. *J Biol Chem* **260**:7903-7910.

Polymeropoulos MH, Lavedan C, Leroy E, Ide SE, Dehejia A, Dutra A, Pike B, Root H, Rubenstein J, Boyer R, Stenroos ES, Chandrasekharappa S, Athanassiadou A, Papapetropoulos T, Johnson WG, Lazzarini AM, Duvoisin RC, Di Iorio G, Golbe LI, Nussbaum RL (1997) Mutation in the alpha-synuclein gene identified in families with Parkinson's disease. *Science* **276**:2045-2047.

Ravikumar B, Duden R, Rubinsztein DC (2002) Aggregate-prone proteins with polyglutamine and polyalanine expansions are degraded by autophagy. *Hum Mol Genet* **11**:1107-1117.

Rideout HJ, Stefanis L (2002) Proteasomal inhibition-induced inclusion formation and death in cortical neurons require transcription and ubiquitination. *Mol Cell Neurosci* **21**:223-238.

Rideout HJ, Larsen KE, Sulzer D, Stefanis L (2001) Proteasomal inhibition leads to formation of ubiquitin/alpha-synuclein-immunoreactive inclusions in PC12 cells. *J Neurochem* **78**:899-908.

Rideout HJ, Dietrich P, Wang Q, Dauer WT, Stefanis L (2004) alpha-synuclein is required for the fibrillar nature of ubiquitinated inclusions induced by proteasomal inhibition in primary neurons. *J Biol Chem* **279**:46915-46920.

Sakahira H, Breuer P, Hayer-Hartl MK, Hartl FU (2002) Molecular chaperones as modulators of polyglutamine protein aggregation and toxicity. *Proc Natl Acad Sci U S A* **99** *Suppl* **4**:16412-16418.

Sakata E, Yamaguchi Y, Kurimoto E, Kikuchi J, Yokoyama S, Yamada S, Kawahara H, Yokosawa H, Hattori N, Mizuno Y, Tanaka K, Kato K (2003) Parkin binds the Rpn10 subunit of 26S proteasomes through its ubiquitin-like domain. *EMBO Rep* **4**:301-306.

Saudou F, Finkbeiner S, Devys D, Greenberg ME (1998) Huntingtin acts in the nucleus to induce apoptosis but death does not correlate with the formation of intranuclear inclusions. *Cell* **95**:55-66.

Seo H, Sonntag KC, Isacson O (2004) Generalized brain and skin proteasome inhibition in Huntington's disease. *Ann Neurol* **56**:319-328.

Snyder H, Mensah K, Theisler C, Lee J, Matouschek A, Wolozin B (2003) Aggregated and monomeric alpha-synuclein bind to the S6' proteasomal protein and inhibit proteasomal function. *J Biol Chem* **278**:11753-11759.

Stefanis L, Larsen KE, Rideout HJ, Sulzer D, Greene LA (2001) Expression of A53T mutant but not wild-type alpha-synuclein in PC12 cells induces alterations of the ubiquitin-dependent degradation system, loss of dopamine release, and autophagic cell death. *J Neurosci* **21**:9549-9560.

Tanaka Y, Engelender S, Igarashi S, Rao RK, Wanner T, Tanzi RE, Sawa A, V LD, Dawson TM, Ross CA (2001) Inducible expression of mutant alpha-synuclein decreases proteasome activity and increases sensitivity to mitochondria-dependent apoptosis. *Hum Mol Genet* **10**:919-926.

Tofaris GK, Razzaq A, Ghetti B, Lilley KS, Spillantini MG (2003) Ubiquitination of alpha-synuclein in Lewy bodies is a pathological event not associated with impairment of proteasome function. *J Biol Chem* **278**:44405-44411.

Tsai YC, Fishman PS, Thakor NV, Oyler GA (2003) Parkin facilitates the elimination of expanded polyglutamine proteins and leads to preservation of proteasome function. *J Biol Chem* **278**:22044-22055.

Valente EM, Abou-Sleiman PM, Caputo V, Muqit MM, Harvey K, Gispert S, Ali Z, Del Turco D, Bentivoglio AR, Healy DG, Albanese A, Nussbaum R, Gonzalez-Maldonado R, Deller T, Salvi S, Cortelli P, Gilks WP, Latchman DS, Harvey RJ, Dallapiccola B, Auburger G, Wood NW (2004) Hereditary early-onset Parkinson's disease caused by mutations in PINK1. *Science* **304**:1158-1160.

Vekrellis K, Rideout HJ, Stefanis L (2004) Neurobiology of alpha-synuclein. *Mol Neurobiol* **30**:1-21.

Venkatraman P, Wetzel R, Tanaka M, Nukina N, Goldberg AL (2004) Eukaryotic proteasomes cannot digest polyglutamine sequences and release them during degradation of polyglutamine-containing proteins. *Mol Cell* **14**:95-104.

Waelter S, Boeddrich A, Lurz R, Scherzinger E, Lueder G, Lehrach H, Wanker EE (2001) Accumulation of mutant huntingtin fragments in aggresome-like inclusion bodies as a result of insufficient protein degradation. *Mol Biol Cell* **12**:1393-1407.

Wang G, Sawai N, Kotliarova S, Kanazawa I, Nukina N (2000) Ataxin-3, the MJD1 gene product, interacts with the two human homologs of yeast DNA repair protein RAD23, HHR23A and HHR23B. *Hum Mol Genet* **9**:1795-1803.

Zhou H, Cao F, Wang Z, Yu ZX, Nguyen HP, Evans J, Li SH, Li XJ (2003) Huntingtin forms toxic NH2-terminal fragment complexes that are promoted by the age-dependent decrease in proteasome activity. *J Cell Biol* **163**:109-118.

In: Recent Breakthroughs in Basal Ganglia Research
Editor: Erwan Bezard, pp. 323-336

ISBN: 1-59454-880-3

Chapter 25

Mitochondrial Dysfunction and Neurodegenerative Disorders in the Basal Ganglia

Kim Tieu *

Department of Environmental Medicine and Center for Aging and Developmental Biology, School of Medicine and Dentistry, University of Rochester, Rochester, New York, 14642, USA.

Abstract

The involvement of mitochondrial dysfunction in Parkinson's disease (PD) and Huntington's disease (HD) has been a topic of interest and debate for decades. Although bioenergetic defects can be clearly demonstrated in models of PD and HD using mitochondrial toxins, the extent to which mitochondrial impairment plays a role in the pathogenesis of these disorders in humans is controversial. However, recent discoveries of genetic mutations in PD such as α-synuclein, parkin, DJ-1 and PINK-1, and the demonstration that these mutant proteins, as well as mutant huntingtin in HD, do induce mitochondrial abnormalities, have revitalized the interest and excitement in the role of mitochondria in these disorders. In this chapter we will examine the evidence for the involvement of mitochondrial impairment in PD and HD as well as the complex and dynamic features of mitochondrial biology, which may explain why this common pathogenic feature leads to such diversified forms of neurodegenerative diseases.

Introduction

Neurodegenerative disorders represent a large group of diseases with heterogeneous clinical and pathological expressions characterized by the death of specific subsets of neurons within specific anatomical areas of the nervous system. Currently, hundreds of different neurodegenerative disorders are recognized. In the basal ganglia, Parkinson's disease (PD) and Huntington's disease (HD) are the two most prevalent and publicized neurodegenerative disorders. Although mutated genes have been discovered in subsets of patients with these diseases, the mechanisms of neurodegeneration leading to specific cell loss in these disorders still remain elusive.

In recent years, the pathogenic role of mitochondrial dysfunction in these disorders has emerged as a topic of interest and debate. Mitochondria are organelles that serve as a "powerhouse" to supply most of the energy required by cells to maintain normal physiological functions and homeostasis. These multifunctional organelles are involved in pyruvate oxidation, in generation of intermediates from the tricarboxylic acid (TCA) cycle, in metabolism of amino acids, fatty acids, steroids, and in oxidative phosphorylation (OXPHOS)-an aerobic metabolic process that generates adenosine triphosphate (ATP). In

* To whom correspondence and reprint requests should be addressed, University of Rochester, 575 Elmwood Avenue, Department of Environmental Medicine, BOX EHSC, E-mail: kim_tieu@urmc.rochester.edu

contrast, abnormal mitochondria act as a major source of reactive oxygen species (ROS) and apoptogenic proteins. Equally detrimental to cells, mitochondria also serve as a pivotal and central station where different apoptotic cascades interweave and converge. Abnormality in mitochondria, therefore, can potentially lead to oxidative stress, loss of ATP production and apoptosis. As an organ that has an extremely high demand for energy, the brain is particularly vulnerable to the effects of mitochondrial dysfunction.

In this chapter we will first review some unique mitochondrial features which may account for the heterogeneous forms of cell death in neurodegenerative disorders. We then will discuss the mechanisms by which neurons may die through mitochondrial dysfunction. Lastly, we will examine the evidence for the involvement of mitochondrial dysfunction in two basal ganglia neurodegenerative disorders: PD and HD. Please refer to a recent review (Tieu and Przedborski, 2004) for a more comprehensive review of the roles of mitochondrial dysfunction in other neurodegenerative disorders, including Leber Hereditary Optic Neuropathy, Leigh Syndrome, 12SrRNA Mutation-Linked Parkinsonism, Friedreich's Ataxia, Hereditary Spastic Paraplegia, Deafness–Dystonia Syndrome, Wilson's Disease, Alzheimer's Disease, and Amyotrophic Lateral Sclerosis.

1 Mitochondrial Biology

The mitochondrion is an organelle whose structural composition and function are governed by laws and principles distinct from those that apply to the rest of the cell. The following mitochondrial features may provide an explanation for some of the most intriguing features of neurodegenerative disorders, including heterogeneous phenotypic expression among patients with mitochondrial dysfunction and the consistent adult-onset and the progressive nature of these neurological disorders.

1.1 Mitochondrial Genetics

Mitochondria are under the genetic control of two genomes: nuclear DNA (nDNA) and mitochondrial DNA (mtDNA). Besides the nucleus, the mitochondrion is the only other organelle in the cell that possesses its own DNA and genetic machinery. The human mitochondrial genome is a small 16.6 kb circular, double stranded DNA. It encodes only 13 out of the approximately 80 proteins in the five complexes of the human respiratory chain (Poyton and McEwen, 1996; Wallace, 1999). Most of the approximately 850 mitochondrial proteins, which account for about 10 % of the total cellular proteins (Di Donato, 2000), are encoded by nDNA. These proteins are synthesized in the cytoplasm prior to being imported into the mitochondria. Any mutation resulting in defects of mitochondrial proteins would potentially lead to cell death. Since all mtDNAs of a zygote originate from the ovum, diseases linked to mtDNA mutations are maternally inherited. That is, mothers who carry mtDNA mutations pass them to the children, and then only the daughters may pass them to their children. A complete list of known mtDNA mutations is available through MITOMAP database (www.mitomap.org). Recently, the prevalence of individuals having or at risk of developing mtDNA disease is estimated to be 10-15 per 100,000 persons (Chinnery and Turnbull, 2001).

There are thousands of mtDNAs in each cell, with about five mtDNAs per mitochondrion, and in patients, a pathogenic mitochondrial mutation may be present in some, but not necessarily in all cells. This mixture of mutated and wild-type mtDNA (*heteroplasmy*) can vary greatly among patients, organs and even cells. The replication of mtDNA occurs constantly in all cells, even in postmitotic, non-dividing cells such as neurons (*relaxed replication*), and upon cell division, the mitochondria are partitioned randomly among the daughter cells (*mitotic segregation*). Over time, these phenomena may alter the proportion of mutated mtDNA in the cell and influence the expression of the disease. Specific nuclear factors can also influence mitochondrial segregation (Battersby et al., 2003), supporting a role for nuclear-mitochondrial interactions in determining the degree of heteroplasmy. Before an oxidative phosphorylation deficit occurs and symptoms arise, a minimal proportion of mutated mtDNA is required (*threshold effect*). The higher the proportion of mutated mtDNA, the more severe the phenotype of the diseases. Because mtDNA encodes only for polypeptide subunits of the respiratory chain, all pathogenic mtDNA mutations affect oxidative phosphorylation. The nervous system and one of its cellular constituents, neurons, are highly dependent on oxidative phosphorylation. Thus, this specific organ and this specific cell type are especially vulnerable to the effects of mtDNA mutations.

1.2 Mitochondrial Dynamics: Fission, Fusion and Translocation

Mitochondria are dynamic, double membrane organelles that undergo frequent changes in shape, size, number and location. Morphologically, mitochondria appear in various shapes and sizes, including the traditional cigar-shaped structures that appear in most standard textbooks, as well as short oval tubules or reticular networks. These diverse mitochondrial morphologies vary not only between cell types, but also within individual cells. Mitochondrial morphology is maintained and controlled mainly by the key processes of *fission* and *fusion*. Mitochondrial fission leads to multiple smaller mitochondria which are more motile within the cell as compared to the larger ones (Skulachev, 2001; Rintoul et al., 2003). In contrast to fission, the process of fusion results in larger mitochondria, which could offer a larger ATP supply than the smaller ones do (Skulachev, 2001). Under normal physiological conditions, these two processes ensure proper mitochondrial morphology to maximize their ATP storage and to facilitate their translocation to various cellular sites where ATP is needed. In addition, these processes of fission and fusion may promote complementation of damaged and undamaged mitochondrial DNA and thereby minimize the local accumulation of defective mitochondria. On the other hand, this protective mechanism may promote the spreading of mutations to healthy mitochondria. In short, a balance of fusion and fission is crucial not only to mitochondrial morphology but also to function and hence, cell survival. Together, these dynamic features of mitochondria, coupled with their unique genetics, highlight the complexity and heterogeneity of mitochondrial biology and function.

2 Mechanisms of Cell Death Induced by Mitochondrial Dysfunction

One of the most intriguing and unresolved questions that pertain to mitochondrial abnormality and neurodegeneration is the mechanism by which selective cell death occurs. As emphasized above, most of the definite *mitochondrial* neurodegenerative diseases are linked to a defect in the respiratory chain. In keeping with this, it can be predicted that a complete loss of mitochondrial respiration would be lethal as suggested by the embryonic death of mutant mice deficient in mitochondrial transcription factor A (Larsson et al., 1998). In contrast, the chronic nature of PD and HD discussed in this chapter suggests that the observed mitochondrial defects are probably more subtle, causing a progressive increase in mitochondrial and cellular dysfunction which ultimately will culminate in cell death.

2.1 ROS Overproduction, ATP Deficit and Excitotoxicity in Abnormal Mitochondria

The extremely high demand for energy by the brain is highlighted by the fact that while the brain represents only about 1/50 of the body weight, it consumes about 1/5 of the total oxygen inhaled. Of this amount of oxygen, about 90 % is utilized by mitochondria to produce ATP. Although through the course of evolution, mitochondria have provided eukaryotic cells with a very efficient method of aerobic metabolism, an estimated 1-2 % of this consumed oxygen is converted to ROS rather than to water (Richter, 1992). The inability of the respiratory chain to completely reduce oxygen to water, coupled with its high rate of oxygen consumption, contributes to high ROS production by the mitochondrial respiratory chain. Under a state of mitochondrial dysfunction, however, this rate may increase even more dramatically. There are two main sites in the electron transport chain where ROS are generated: complex I and complex III (see Figure 1). A dysfunction in complex I activity as a result of either a mutation or a chemical blockade by the mitochondrial poisons 1-methyl-4-phenyl-1,2,3,6-tetrahydropyridine (MPTP) or rotenone can dramatically increase ROS production. This elevation is a result of "leakage" of electrons from complex I and the subsequent reduction of oxygen to superoxide. Increased production of ROS is also seen when complex III is blocked in a manner that leads to the accumulation of ubisemiquinone as seen with antimycin, a complex III inhibitor. Intrinsically, mitochondria are equipped with defense mechanisms against ROS including antioxidant enzymes such as manganese superoxide dismutase, glutathione peroxidase and catalase. Therefore, any imbalance between the production of ROS and the antioxidant capacity of these enzymes leads to local oxidative stress, possibly causing serious functional and structural damage to the mitochondria. However, how much of this mitochondrial oxidative stress can also damage cytoplasmic molecules and other organelles is still an unresolved question. The view that mitochondrial ROS *leak* out and inflict extramitochondrial oxidative damage is commonly found in the literature. However, the majority of ROS produced inside the mitochondria are not membrane permeant and not very reactive.

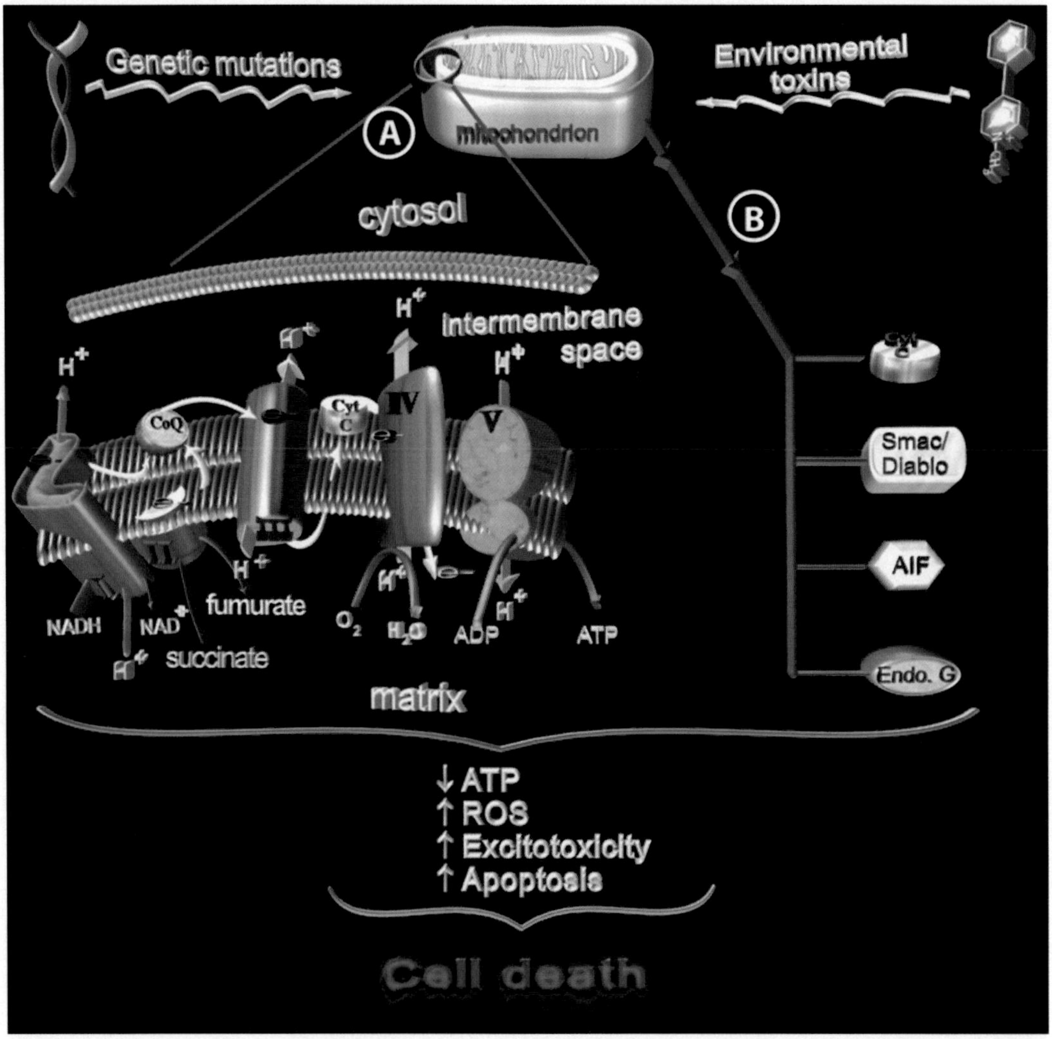

Figure 1. (A) Schematic representation of the electron transport chain (ETC). Electrons (e^-) donated from NADH to Complex I or from succinate to Complex II are passed to Coenzyme Q (CoQ), ubisemiquinone, ubiquinol, Complex III, cytochrome c, Complex IV and finally to O_2 to produce H_2O. The energy released from this ETC is used to pump protons out of the inner mitochondrial membrane into the intermembrane space, thus creating a membrane potential gradient. This potential energy then drives the flow of protons back into the mitochondrial matrix via Complex V (ATP synthase) and thereby produces ATP. Complex I and complex III are the two major sites where ROS are produced. Dysfunction in the ETC, in general, as a result of genetic mutations or blockade by specific inhibitors of the complexes, would lead to increase in ROS production, reduction in ATP production or increase in excitotoxicity. These abnormalities can potentially result in cell death. (B) Mitochondria also play a major role in apoptotic cascades. Shown here are the various apoptogenic molecules released from mitochondria, leading to caspase 3 activation and DNA fragmentation and ultimately, cell death. Environmental toxins, genetic mutations, or both, may lead to mitochondrial abnormalities, which, in turn, contribute to cell death in neurodegenerative disorders such as PD and HD.

Defective mitochondria also result in reduced production of ATP, which can in turn disrupt the proper functioning of many critical cellular processes. However, it is unclear how low ATP levels must be prior to the occurrence of cellular damage. Studies performed in cultured cells expressing mitochondrial mutations that impair ATP synthesis show that cellular stores of ATP drop to pathologically meaningful levels only when those mutant cells are placed under conditions of high ATP demand (James et al., 1999). This suggests that cells such as neurons, which have variable ATP demands related to changes in ion fluxes across the membrane in response to external stimuli, would experience an energy crisis only when ATP demands rise above basal levels. Thus, any prolonged energy crisis due to a sustained increase in ATP demand not only may impair basic ATP-dependent neuronal functions such as neurotransmission, but also may increase the probability for that neuron to die.

ATP deficiency may also lead to Na^+/K^+-ATPase failure and subsequent neuronal membrane depolarization, which, in turn, activates the voltage-dependent NMDA receptors (Greene and Greenamyre, 1996). According to this scenario, once NMDA receptors are activated, a massive influx of calcium occurs, leading to excitotoxicity (Albin and Greenamyre, 1992; Dugan and Choi, 1994). Consistent with this, NMDA antagonists such as MK-801 can attenuate neurotoxicity induced by neurotoxins that block the electron transport chain (Greene and Greenamyre, 1996; Schulz et al., 1996; Massieu et al., 2001).

2.2 Mitochondria and Apoptosis

In addition to ROS, mitochondria also act as a major supplier of apoptogenic molecules (see Figure 1). Apoptosis is an active form of cell death regulated by multiple molecular pathways (Vila and Przedborski, 2003). In addition to representing the "intrinsic" apoptotic pathway, mitochondria also serve as a central hub to regulate and interact with other pathways of apoptosis (Vila and Przedborski, 2003). The exponential interest in the role of mitochondria in apoptosis is primarily sparked by the initial discoveries that apoptosis could be induced by cytochrome *c* released from mitochondria (Kluck et al., 1997; Yang et al., 1997). After being released into the cytosol, cytochrome *c* interacts with Apaf1 and pro-caspase 9 to form a complex known as apoptosome, which subsequently activates caspase 3 (Li et al., 1997). Caspases are a group of proteolytic enzymes which cleave their substrates at specific aspartic acid residues. Activation of caspases potentially leads to apoptotic death. A second mitochondrial apoptogenic protein released into the cytosol is Smac/Diablo. In the cytosol, this protein inhibits the inhibitors of apoptosis proteins (IAPs), again leading to activation of caspases. Two other mitochondrial proteins, apoptosis inducing factor (AIF) and endonuclease G, are released and translocated from mitochondria to nuclei where they induce DNA fragmentation. In addition to these apoptogenic molecules, mitochondria also contain anti-apoptotic proteins such as Bcl-2 and Bcl-X_L. A balance between these pro-and anti-apoptotic molecules must be tightly controlled to ensure proper execution of apoptosis.

3 Mitochondrial Dysfunction and Neurodegeneration in PD and HD

The link between mitochondrial dysfunction and neurodegeneration can be divided into three categories: (1) neurodegenerative diseases resulting from mtDNA mutations, (2) neurodegenerative diseases resulting from nDNA mutations affecting mitochondrial targeted proteins and (3) neurodegenerative diseases with an unknown cause or linked to mutations in non-mitochondrial proteins. The role of mitochondrial dysfunction in PD and HD belongs to this third category, which by far is most controversial since an unequivocal mitochondrial origin has largely not been documented for these disorders. Nevertheless, cumulating evidence does support the role of mitochondrial dysfunction in the pathogenesis of PD and HD. Recent developments in these fields have sparked interest in the relationship between mitochondrial dysfunction and the pathogenesis of these disorders and have been met with an unquestionably significant degree of enthusiasm within the scientific and clinical communities.

3.1 Parkinson's Disease

Parkinson's disease (PD) is the second most common neurodegenerative disorder, after Alzheimer's disease. PD is characterized by the loss of dopaminergic neurons in the substantia nigra pars compacta (SNpc), leading to a reduction of dopamine in the putamen (Dauer and Przedborski, 2003). When this putamenal DA is depleted about 80 %, the symptoms of PD emerge (Dauer and Przedborski, 2003). Abnormal movements such as resting tremor, rigidity, and postural instability are common in PD patients (Dauer and Przedborski, 2003).

The etiology of PD is currently unknown. Environmental toxins have been hypothesized to play a dominant role based on the observations that parkinsonism can be caused by postencephalitic infection and by accidental MPTP injection (Dauer and Przedborski, 2003); however, from more recent discoveries of various familial forms of early-onset PD, a genetic-based theory is gaining momentum. Perhaps both mechanisms are involved and mitochondrial dysfunction could represent a common denominator that unifies these two theories. The theory of mitochondrial dysfunction as a pathogenic mechanism in PD originates from the initial discovery that MPTP induced a parkinsonian syndrome in humans and various animal species (Dauer and Przedborski, 2003). Subsequently, 1-methyl-4-phenylpyridinium ion (MPP^+), the active neurotoxic metabolite of MPTP, was identified as an inhibitor of complex I of the mitochondrial electron transport chain (Nicklas et al., 1985). This latter finding prompted many investigators to search for mitochondrial defects in PD patients. Shortly thereafter, numerous publications reported a reduction in complex I activity in platelets (Parker et al., 1989) and brain tissues, including the SNpc (Schapira et al., 1990; Haas et al., 1995) of PD patients. In cytoplasmic hybrid cell lines (cybrid cells), which contain mitochondria derived from platelets of PD patients, complex I activity was also reduced (Swerdlow et al., 1996), and this finding led the authors to conclude that complex I deficiency seen in PD is mitochondrially inherited. Consistent with this, the same authors subsequently reported that complex I activity was reduced in maternal descendants of families with PD inherited exclusively through maternal lines (Swerdlow et al., 1998). Furthermore, mutations and polymorphisms in subunits of complex I have been proposed as susceptibility in subsets of PD patients (Kosel et al., 1998; van der Walt et al., 2003). It has to be emphasized, however, that complex I deficiency only occurs in a subset of PD patients. This may partly explain the negative finding of a complex I defect in a more recent study of cybrids in PD (Aomi et al., 2001). Nevertheless, together, all of these studies suggest that either by genetic mutations, environmental toxins, or both, complex I deficiency may play a pathogenic role in some PD patients. Unquestionably, the demonstration that exposure of rats to rotenone, a lipophilic complex I inhibitor, reproduces some of the key features of PD (Betarbet et al., 2000), has also revived interest in the relationship between complex I defects and sporadic PD. Furthermore, mitochondrial substrates, such as the ketone body, d-β-hydroxybutyrate (Tieu et al., 2003b), or creatine (Matthews et al., 1999), by improving oxidative phosphorylation, attenuate dopaminergic neurodegeneration caused by the parkinsonian toxin MPTP.

The discoveries of various genes involved in early onset PD have also provided insights into the mitochondrial mechanism by which PD-causing mutations may induce neurodegeneration. Recently, mutations in PTEN-induced putative kinase 1 (PINK1)/BRPK gene have been reported in autosomal recessive (Hatano et al., 2004; Rohe et al., 2004; Valente et al., 2004b) and sporadic early onset PD (Valente et al., 2004a). The function of this ubiquitously expressed protein is currently unknown but it is upregulated by PTEN (Phosphatase and Tensin homolog deleted on chromosome Ten) (Unoki and Nakamura, 2001). Although PTEN is a tumor suppressor gene (Li et al., 1997; Risinger et al., 1997; Steck et al., 1997; Obata et al., 1998), PINK1 does not have a suppressive effect on cell growth (Unoki and Nakamura, 2001). Based on its high degree of homology with the serine-threonine kinases of the Ca^{2+}/calmodulin family, PINK1 is suggested to have putative serine/threonine protein kinase catalytic activity (Unoki and Nakamura, 2001; Nakajima et al., 2003). PINK1 has a mitochondrial targeting motif and consistent with this observation, it localizes to the mitochondria in transfected cell lines (Valente et al., 2004b). The mutations thus far identified are expected to impair PINK1 kinase activity or substrate recognition and to cause PD by a loss of function. Impaired phosphorylation of a PINK1 substrate, especially in mitochondria, is a likely scenario to explain the pathogenic mechanism of these mutations. In neuroblastoma cells transiently transfected with either wild-type or mutant PINK1, no alteration in base-line cell viability was detectable (Valente et al., 2004b). In contrast, when these cells were challenged with the proteasome inhibitor MG132, those overexpressing wild-type PINK1 survived the toxic insult, whereas those overexpressing mutant PINK1 died to the same extent as the non-transfected neuroblastoma control cells (Valente et al., 2004b). These results suggest that wild-type mitochondrial PINK1 confers greater resistance to the mitochondria, and thus to the cell, against cytotoxic insults. Future studies leading to the elucidation of the substrates of PINK1 and its altered physiological activity induced by the mutations will unquestionably yield significant insights into the mechanism of cell death, mediated by mitochondria, in PD.

Eleven different DJ-1 mutations, encompassing missense, truncation, splice site mutations and large deletions, have been linked to an autosomal recessive form of PD (Abou-Sleiman et al., 2003; Bonifati et al., 2003; Hague et al., 2003). Under normal physiological conditions, wild-type DJ-1 is in the cytosol; however, in the oxidative stress environment, it translocates to the mitochondria (Bonifati et al., 2003; Canet-Aviles et al., 2004). Oxidation of one of the cysteine residues (C106) in DJ-1 is critical for this mitochondrial translocation (Canet-Aviles et al., 2004). The localization of DJ-1 in mitochondria is critical for cell survival following toxic insults, as preventing DJ-1 from accumulating in the mitochondria through mutation of C106 renders cells more susceptible to the MPP^+ toxicity (Canet-Aviles et al., 2004). Down regulation of endogenous DJ-1 by siRNA in a cell culture model also enhances cell death induced by oxidative stress (Yokota et al., 2003). Some PD-causing mutations appear to also impair DJ-1 mitochondrial translocation (Bonifati et al., 2003). Although the exact molecular mechanism of DJ-1 remains to be determined, these studies suggest that a lack of DJ-1 translocation to mitochondria may be pivotal in the neurodegenerative process of PD.

In addition to PINK1 and DJ-1, other genes identified in familial early onset of PD have also been demonstrated to affect mitochondrial function. For example, parkin, an E3 ubiquitin protein ligase found in a recessively inherited form of PD (Kitada et al., 1998), was recently demonstrated to localize in the outer mitochondrial membrane where it delays ceramide-mediated mitochondrial swelling, subsequent cytochrome *c* release and caspase-3 activation (Darios et al., 2003). Consistent with this report, in drosophila lacking parkin, the appearance of abnormal mitochondrial pathology precedes muscle degeneration, locomotor defects and the subsequent premature death (Greene et al., 2003). Furthermore, the levels of some proteins involved in mitochondrial function and oxidative stress are reduced in the ventral midbrain of parkin-deficient mice (Palacino et al., 2004). Consistent with this, polarographical measurements indicate reduced oxygen consumption mediated by complexes I, II and IV, suggesting an overall reduction in respiratory capacity. These mice also exhibit reduced antioxidant capacity and increased oxidative damage (Palacino et al., 2004).

Missense mutations in α-synuclein are linked to a dominantly inherited form of PD (Polymeropoulos et al., 1997; Kruger et al., 1998; Zarranz et al., 2004). Although the exact normal physiological function of α-synuclein is uncertain, this prevalent presynaptic protein may be involved in the modulation of synaptic vesicle function (Dauer and Przedborski, 2003). Recently, injection of MPTP into transgenic mice overexpressing α-synuclein was shown to result in abnormal mitochondrial morphology (Song et al., 2004). Although this study does not assess whether this abnormal feature would result in enhanced dopaminergic cell death, previous studies show that overexpression of α-synuclein *in vitro* in cells with impaired mitochondrial function does result in increased levels of free radicals (Hsu et al., 2000). In addition, α-synuclein-knockout mice are more resistant to MPTP toxicity (Dauer et al., 2002). At present, the mechanism by which α-synuclein induces mitochondrial abnormality is unknown.

Although current knowledge of these parkinsonian genes and their proteins is far from being complete, the studies discussed above have two implications: 1) Genetic makeup predisposes one to susceptibility to environmental mitochondrial toxins; 2) Genetic evidence for mitochondrial dysfunction in PD now exists.

3.2 Huntington's Disease

Huntington's disease (HD) is a progressive neurodegenerative disorder characterized clinically by chorea, psychiatric disturbances and a decline in cognitive function. The symptoms usually begin in the 4th or 5th decade of life and worsen rapidly thereafter. Typically, the length of time from the onset of symptoms to death is about 18 years. HD is an autosomal dominant disease caused by an expansion of a trinucleotide CAG repeat that encodes the non-essential amino acid glutamine in exon 1 of the *IT15* gene on chromosome 4 (The Huntington's Disease Collaborative Research Group, 2001). The normal range of CAG repeats – and thus of the glutamine tail in the encoded protein called huntingtin – is between six and 35. The age at onset is inversely correlated to the length of the repeat, but apparently not the severity or the rate of progression (Brinkman et al., 1997). Environmental factors, however, also seem to play a major role in the pathogenesis of HD as demonstrated in animal models (van Dellen et al., 2000; Hockly et al., 2002) and in HD patients (Wexler et al., 2004). Pathologically, there is a selective neurodegeneration of medium spiny neurons in the striatum. Nuclear inclusions of protein aggregates are also a common feature in the striatum and cortex.

Since the discovery of the HD mutation (The Huntington's Disease Collaborative Research Group, 1993), the normal function of huntingtin remains unknown and the pathogenic mechanism leading to neurodegeneration in HD is elusive. Nonetheless, a defect in bioenergetics has been proposed as a major pathogenic factor. Positron emission tomography studies with [^{18}F]-fluorodeoxyglucose have demonstrated a marked reduction in glucose utilization in the caudate and putamen of HD patients (Kuhl et al., 1982). This hypometabolism precedes the detection of striatal atrophy, indicating that cell death occurs after metabolic impairment (Kuhl et al., 1982; 1984). These results also suggest a dysfunction in nerve terminals since they are the sites of the greatest energy consumption by ATP-dependent pumps to restore ionic gradients following synaptic transmission (Erecinska and Silver, 1989). The early damage to the terminals in HD is consistent with the findings of decreases in various types of neurotransmitter receptors observed in these patients (Penney and Young, 1982; Dure et al., 1991) and in a transgenic mouse model of HD (Cha et al., 1998). Because mitochondria are highly concentrated in nerve terminals (Knull, 1978), it is tempting to speculate that the abnormalities described above result from a mitochondrial dysfunction, and indeed, several lines of evidence are consistent with this idea. Biochemically, in muscle biopsies of HD patients, there is a reduced ratio of phosphocreatine to inorganic phosphate (Koroshetz et al., 1997), suggesting a buildup of precursor products as a result of decreased ATP production. Nuclear magnetic resonance spectroscopy shows elevated lactate levels in the cortex and basal ganglia of symptomatic HD patients (Jenkins et al., 1993), indicating an impairment of oxidative phosphorylation. In well-coupled mitochondria, a higher rate of oxygen consumption correlates with a higher rate of ATP production (Chance and Williams, 1955; Hagihara, 1961). Using polarography to monitor the rate of oxygen consumption, a significant reduction in respiration was detected in mitochondrial preparations of caudate and cortex postmortem tissue from HD patients (Brennan et al., 1985). This alteration is in agreement with the previously presumed reduction in enzymatic activity of complexes II-III in these brain regions of HD patients (Mann et al., 1990; Browne et al., 1997). Complex II is an enzyme that serves as a common component of both the tricarboxylic acid cycle and the electron transport chain. It catalyzes the oxidation of succinate to fumarate and transfers electrons to coenzyme Q. In animal models, inhibition of this enzyme by malonate or 3-nitropropionic acid (3-NP) reproduces a pattern of striatal lesions similar to that seen in HD patients (Brouillet et al., 1999).

Despite a wealth of data showing a mitochondrial abnormality in HD, the link between mutated huntingtin and mitochondria is still largely speculative. Huntingtin is expressed ubiquitously in all brain cells and is localized mostly in the cytoplasm and to a smaller extent, in the nucleus (Sapp et al., 1997). However, immuno-electron microscopy recently shows that the protein is also localized in the neuronal mitochondrial membrane (Panov et al., 2002) where it seems to accumulate at the level of the outer membrane (Choo et al., 2004). Functionally, brain mitochondria from HD mice have a lower membrane potential and depolarize at lower calcium concentrations than the wild-type control (Panov et al., 2002). Given these results, it has been proposed that mutant huntingtin either induces the opening of the mitochondrial permeability transition (MPT) pore or decreases the threshold for calcium to induce MPT pore opening (Choo et al., 2004). Of note, the release of cytochrome *c* as a result of huntingtin-induced MPT pore opening is inhibited by cyclosporin A (Choo et al., 2004). In addition, mutant huntingtin also increases the vulnerability of a striatal cell line, created from Hdh Q111 mouse embryos, to the mitochondrial toxin 3-NP (Ruan et al., 2004). This latter study implies that individuals with mutated huntingtin would be more susceptible to environmental mitochondrial toxins. Lastly, mutated huntingtin can also affect mitochondrial function indirectly through transcription factors. Huntingtin binds to and modifies the activity of transcription factors such as TATA-binding protein, cAMP-responsive element-binding protein (CREB)-binding protein (CBP), specificity protein 1 (Sp1), and p53 (Beal and Ferrante, 2004; Li and Li, 2004). These factors regulate the transcription of some mitochondrial proteins involved in energy metabolism and apoptosis. For example, CREB activation has been shown to be involved in mitochondrial dysfunction (Arnould et al., 2002). Huntingtin can also alter the activity of transcription factors through histone acetylation, which is regulated by both histone acetyltransferase and histone deacetylases. Indeed, huntingtin binds to and reduces the activity of histone acetyltransferase. The reduction in acetylation of histones leads to alterations in chromatin structure which, in turn, alters gene expression (Gregory et al., 2001). It is thus conceivable that mutated huntingtin can perturb mitochondrial function by interfering with the activities of transcription factors; however, a HD transgenic mouse study fails to confirm that these transcription factors are affected (Yu et al., 2002). More studies are needed to resolve this issue.

Further reinforcing the hypothesis of metabolic defect as a pathogenic mechanism in HD are the studies which show that administration of mitochondrial substrates are neuroprotective in animal models of HD. For example, creatine, a mitochondrial energy improving compound, has been shown in various HD animal models to prevent body weight loss and brain atrophy, improve motor performance, and extend life expectancy (Matthews et al., 1998; Ferrante et al., 2000; Andreassen et al., 2001; Dedeoglu et al., 2003). Coenzyme Q10, an electron carrier and a cofactor of the electron transport chain, also has similar protective effects in these animals (Beal et al., 1994; Schilling et al., 2001; Ferrante et al., 2002). Another mitochondrial substrate, D-β-hydroxybutyrate, has also been demonstrated to confer neuroprotection in an animal model of HD (Tieu et al., 2003a). The efficacies of these compounds in animals, however, have not been replicated in human HD clinical trials (The Huntington's Disease Collaborative Research Group, 2001). Formation of protein aggregates has also been attributed to mitochondrial dysfunction. In keeping with this, administration of congo red intraperitoneally (Sanchez et al., 2003) and of trehalose orally (Tanaka et al., 2004) inhibited protein aggregation and improved motor performance and survival in HD transgenic mice, supporting a very contentious argument that proteinacious intranuclear inclusions in HD are cytotoxic.

Conclusion

Emerging studies appear to indicate that the diverse causes of neurodegenerative diseases, including PD and HD, may share one common pathogenic mechanism: mitochondrial impairment. Although defects in bioenergetics in PD and HD had been suggested for many years and inhibitors of the mitochondrial respiratory chain have been used to model these disorders, until recently, it was difficult to reconcile the role of mitochondrial dysfunction in the familial forms of PD and HD. However, recent studies show that mutated genes such as PINK1, parkin, DJ-1 and α-synuclein in PD as well as huntingtin in HD, to varying degrees, do cause mitochondrial abnormalities.

In the past decade, we have learned that through depletion of ATP, the generation of ROS, and the release of apoptogenic proteins, mitochondrial abnormalities may hold a key role in neurodegenerative processes. Initiation of these events, either individually or more likely in combination, would potentially lead to neuronal death. We also understand that mitochondrial abnormalities can arise from a variety of causes: mutations in mtDNA, mutations in nDNA encoding for mitochondrial proteins, mutations in non-mitochondrial proteins and from those without definitive genetic basis. Together, these pathogenic mitochondrial alterations lead to a wide range of neurodegenerative disorders, including PD and HD.

What we still have not fully comprehended, however, is how can this one common pathogenic feature - the mitochondrial impairment - lead to such diversified forms of neurodegenerative diseases? For example, if mitochondrial dysfunction is a common pathogenic feature in PD and HD, then it is not clear how this abnormality would lead to specific cell loss in specific brain regions, not only within, but also between these disorders. Many aspects of the biology of the mitochondria that were reviewed in this chapter such as heteroplasmy, mitotic segregation, the threshold effect and other dynamic features of mitochondria may contribute to our difficulty in unequivocally linking specific cell death in these neurodegenerative diseases to a mitochondrial defect. Alternatively, perhaps mitochondrial dysfunction does not play an etiological role, but rather, is a major contributor to the pathogenesis of these diseases. Nevertheless, the recent discoveries that mutated proteins in PD and HD can induce mitochondrial defects open new and exciting avenues of research into both basic mitochondrial biology and these prevalent neurodegenerative diseases of the basal ganglia.

Acknowledgments

The authors wish to thank Dr. Paula Ashe for her assistance in preparing this chapter.

References

Abou-Sleiman PM, Healy DG, Quinn N, Lees AJ, Wood NW (2003) The role of pathogenic DJ-1 mutations in Parkinson's disease. *Ann Neurol* **54**:283-286.

Albin RL, Greenamyre JT (1992) Alternative excitotoxic hypotheses. *Neurology* **42**:733-738.

Andreassen OA, Ferrante RJ, Huang HM, Dedeoglu A, Park L, Ferrante KL, Kwon J, Borchelt DR, Ross CA, Gibson GE, Beal MF (2001) Dichloroacetate exerts therapeutic effects in transgenic mouse models of Huntington's disease. *Ann Neurol* **50**:112-117.

Aomi Y, Chen CS, Nakada K, Ito S, Isobe K, Murakami H, Kuno SY, Tawata M, Matsuoka R, Mizusawa H, Hayashi JI (2001) Cytoplasmic transfer of platelet mtDNA from elderly patients with Parkinson's disease to mtDNA-less HeLa cells restores complete mitochondrial respiratory function. *Biochem Biophys Res Commun* **280**:265-273.

Arnould T, Vankoningsloo S, Renard P, Houbion A, Ninane N, Demazy C, Remacle J, Raes M (2002) CREB activation induced by mitochondrial dysfunction is a new signaling pathway that impairs cell proliferation. *Embo J* **21**:53-63.

Battersby BJ, Loredo-Osti JC, Shoubridge EA (2003) Nuclear genetic control of mitochondrial DNA segregation. *Nat Genet* **33**:183-186.

Beal MF, Ferrante RJ (2004) Experimental therapeutics in transgenic mouse models of Huntington's disease. *Nat Rev Neurosci* **5**:373-384.

Beal MF, Henshaw DR, Jenkins BG, Rosen BR, Schulz JB (1994) Coenzyme Q10 and nicotinamide block striatal lesions produced by the mitochondrial toxin malonate. *Ann Neurol* **36**:882-888.

Betarbet R, Sherer TB, MacKenzie G, Garcia-Osuna M, Panov AV, Greenamyre JT (2000) Chronic systemic pesticide exposure reproduces features of Parkinson's disease. *Nat Neurosci* **3**:1301-1306.

Bonifati V, Rizzu P, van Baren MJ, Schaap O, Breedveld GJ, Krieger E, Dekker MC, Squitieri F, Ibanez P, Joosse M, van Dongen JW, Vanacore N, van Swieten JC, Brice A, Meco G, van Duijn CM, Oostra BA, Heutink P (2003) Mutations in the DJ-1 gene associated with autosomal recessive early-onset parkinsonism. *Science* **299**:256-259.

Brennan WA, Jr., Bird ED, Aprille JR (1985) Regional mitochondrial respiratory activity in Huntington's disease brain. *J Neurochem* **44**:1948-1950.

Brinkman RR, Mezei MM, Theilmann J, Almqvist E, Hayden MR (1997) The likelihood of being affected with Huntington disease by a particular age, for a specific CAG size. *Am J Hum Genet* **60**:1202-1210.

Brouillet E, Conde F, Beal MF, Hantraye P (1999) Replicating Huntington's disease phenotype in experimental animals. *Prog Neurobiol* **59**:427-468.

Browne SE, Bowling AC, MacGarvey U, Baik MJ, Berger SC, Muqit MM, Bird ED, Beal MF (1997) Oxidative damage and metabolic dysfunction in Huntington's disease: selective vulnerability of the basal ganglia. *Ann Neurol* **41**:646-653.

Canet-Aviles RM, Wilson MA, Miller DW, Ahmad R, McLendon C, Bandyopadhyay S, Baptista MJ, Ringe D, Petsko GA, Cookson MR (2004) The Parkinson's disease protein DJ-1 is neuroprotective due to cysteine-sulfinic acid-driven mitochondrial localization. *Proc Natl Acad Sci U S A* **101**:9103-9108.

Cha JH, Kosinski CM, Kerner JA, Alsdorf SA, Mangiarini L, Davies SW, Penney JB, Bates GP, Young AB (1998) Altered brain neurotransmitter receptors in transgenic mice expressing a portion of an abnormal human huntington disease gene. *Proc Natl Acad Sci U S A* **95**:6480-6485.

Chance B, Williams GR (1955) A simple and rapid assay of oxidative phosphorylation. *Nature* **175**:1120-1121.

Chinnery PF, Turnbull DM (2001) Epidemiology and treatment of mitochondrial disorders. *Am J Med Genet* **106**:94-101.

Choo YS, Johnson GV, MacDonald M, Detloff PJ, Lesort M (2004) Mutant huntingtin directly increases susceptibility of mitochondria to the calcium-induced permeability transition and cytochrome c release. *Hum Mol Genet* **13**:1407-1420.

Darios F, Corti O, Lucking CB, Hampe C, Muriel MP, Abbas N, Gu WJ, Hirsch EC, Rooney T, Ruberg M, Brice A (2003) Parkin prevents mitochondrial swelling and cytochrome c release in mitochondria-dependent cell death. *Hum Mol Genet* **12**:517-526.

Dauer W, Przedborski S (2003) Parkinson's disease: mechanisms and models. *Neuron* **39**:889-909.

Dauer W, Kholodilov N, Vila M, Trillat AC, Goodchild R, Larsen KE, Staal R, Tieu K, Schmitz Y, Yuan CA, Rocha M, Jackson-Lewis V, Hersch S, Sulzer D, Przedborski S, Burke R, Hen R (2002)

Resistance of alpha -synuclein null mice to the parkinsonian neurotoxin MPTP. *Proc Natl Acad Sci U S A* **99**:14524-14529.

Dedeoglu A, Kubilus JK, Yang L, Ferrante KL, Hersch SM, Beal MF, Ferrante RJ (2003) Creatine therapy provides neuroprotection after onset of clinical symptoms in Huntington's disease transgenic mice. *J Neurochem* **85**:1359-1367.

Di Donato S (2000) Disorders related to mitochondrial membranes: pathology of the respiratory chain and neurodegeneration. *J Inherit Metab Dis* **23**:247-263.

Dugan LL, Choi DW (1994) Excitotoxicity, free radicals, and cell membrane changes. *Ann Neurol 35 Suppl*:S17-21.

Dure LSt, Young AB, Penney JB (1991) Excitatory amino acid binding sites in the caudate nucleus and frontal cortex of Huntington's disease. *Ann Neurol* **30**:785-793.

Erecinska M, Silver IA (1989) ATP and brain function. *J Cereb Blood Flow Metab* **9**:2-19.

Ferrante RJ, Andreassen OA, Dedeoglu A, Ferrante KL, Jenkins BG, Hersch SM, Beal MF (2002) Therapeutic effects of coenzyme Q10 and remacemide in transgenic mouse models of Huntington's disease. *J Neurosci* **22**:1592-1599.

Ferrante RJ, Andreassen OA, Jenkins BG, Dedeoglu A, Kuemmerle S, Kubilus JK, Kaddurah-Daouk R, Hersch SM, Beal MF (2000) Neuroprotective effects of creatine in a transgenic mouse model of Huntington's disease. *J Neurosci* **20**:4389-4397.

Greene JC, Whitworth AJ, Kuo I, Andrews LA, Feany MB, Pallanck LJ (2003) Mitochondrial pathology and apoptotic muscle degeneration in Drosophila parkin mutants. *Proc Natl Acad Sci U S A* **100**:4078-4083.

Greene JG, Greenamyre JT (1996) Manipulation of membrane potential modulates malonate-induced striatal excitotoxicity in vivo. *J Neurochem* **66**:637-643.

Gregory PD, Wagner K, Horz W (2001) Histone acetylation and chromatin remodeling. *Exp Cell Res* **265**:195-202.

Haas RH, Nasirian F, Nakano K, Ward D, Pay M, Hill R, Shults CW (1995) Low platelet mitochondrial complex I and complex II/III activity in early untreated Parkinson's disease. *Ann Neurol* **37**:714-722.

Hagihara B (1961) Techniques for the application of polarography to mitochondrial respiration. *Biochim Biophys Acta* **46**:134-142.

Hague S, Rogaeva E, Hernandez D, Gulick C, Singleton A, Hanson M, Johnson J, Weiser R, Gallardo M, Ravina B, Gwinn-Hardy K, Crawley A, St George-Hyslop PH, Lang AE, Heutink P, Bonifati V, Hardy J (2003) Early-onset Parkinson's disease caused by a compound heterozygous DJ-1 mutation. *Ann Neurol* **54**:271-274.

Hatano Y, Li Y, Sato K, Asakawa S, Yamamura Y, Tomiyama H, Yoshino H, Asahina M, Kobayashi S, Hassin-Baer S, Lu CS, Ng AR, Rosales RL, Shimizu N, Toda T, Mizuno Y, Hattori N (2004) Novel PINK1 mutations in early-onset parkinsonism. *Ann Neurol* **56**:424-427.

Hockly E, Cordery PM, Woodman B, Mahal A, van Dellen A, Blakemore C, Lewis CM, Hannan AJ, Bates GP (2002) Environmental enrichment slows disease progression in R6/2 Huntington's disease mice. *Ann Neurol* **51**:235-242.

Hsu LJ, Sagara Y, Arroyo A, Rockenstein E, Sisk A, Mallory M, Wong J, Takenouchi T, Hashimoto M, Masliah E (2000) alpha-synuclein promotes mitochondrial deficit and oxidative stress. *Am J Pathol* **157**:401-410.

James AM, Sheard PW, Wei YH, Murphy MP (1999) Decreased ATP synthesis is phenotypically expressed during increased energy demand in fibroblasts containing mitochondrial tRNA mutations. *Eur J Biochem* **259**:462-469.

Jenkins BG, Koroshetz WJ, Beal MF, Rosen BR (1993) Evidence for impairment of energy metabolism in vivo in Huntington's disease using localized 1H NMR spectroscopy. *Neurology* **43**:2689-2695.

Kitada T, Asakawa S, Hattori N, Matsumine H, Yamamura Y, Minoshima S, Yokochi M, Mizuno Y, Shimizu N (1998) Mutations in the parkin gene cause autosomal recessive juvenile parkinsonism. *Nature* **392**:605-608.

Kluck RM, Bossy-Wetzel E, Green DR, Newmeyer DD (1997) The release of cytochrome c from mitochondria: a primary site for Bcl-2 regulation of apoptosis. *Science* **275**:1132-1136.

Knull HR (1978) Association of glycolytic enzymes with particulate fractions from nerve endings. *Biochim Biophys Acta* **522**:1-9.

Koroshetz WJ, Jenkins BG, Rosen BR, Beal MF (1997) Energy metabolism defects in Huntington's disease and effects of coenzyme Q10. *Ann Neurol* **41**:160-165.

Kosel S, Grasbon-Frodl EM, Mautsch U, Egensperger R, von Eitzen U, Frishman D, Hofmann S, Gerbitz KD, Mehraein P, Graeber MB (1998) Novel mutations of mitochondrial complex I in pathologically proven Parkinson disease. *Neurogenetics* **1**:197-204.

Kruger R, Kuhn W, Muller T, Woitalla D, Graeber M, Kosel S, Przuntek H, Epplen JT, Schols L, Riess O (1998) Ala30Pro mutation in the gene encoding alpha-synuclein in Parkinson's disease. *Nat Genet* **18**:106-108.

Kuhl DE, Metter EJ, Riege WH, Markham CH (1984) Patterns of cerebral glucose utilization in Parkinson's disease and Huntington's disease. *Ann Neurol 15 Suppl*:S119-125.

Kuhl DE, Phelps ME, Markham CH, Metter EJ, Riege WH, Winter J (1982) Cerebral metabolism and atrophy in Huntington's disease determined by 18FDG and computed tomographic scan. *Ann Neurol* **12**:425-434.

Larsson NG, Wang J, Wilhelmsson H, Oldfors A, Rustin P, Lewandoski M, Barsh GS, Clayton DA (1998) Mitochondrial transcription factor A is necessary for mtDNA maintenance and embryogenesis in mice. *Nat Genet* **18**:231-236.

Li P, Nijhawan D, Budihardjo I, Srinivasula SM, Ahmad M, Alnemri ES, Wang X (1997) Cytochrome c and dATP-dependent formation of Apaf-1/caspase-9 complex initiates an apoptotic protease cascade. *Cell* **91**:479-489.

Li SH, Li XJ (2004) Huntingtin-protein interactions and the pathogenesis of Huntington's disease. *Trends Genet* **20**:146-154.

Mann VM, Cooper JM, Javoy-Agid F, Agid Y, Jenner P, Schapira AH (1990) Mitochondrial function and parental sex effect in Huntington's disease. *Lancet* **336**:749.

Massieu L, Del Rio P, Montiel T (2001) Neurotoxicity of glutamate uptake inhibition in vivo: correlation with succinate dehydrogenase activity and prevention by energy substrates. *Neuroscience* **106**:669-677.

Matthews RT, Yang L, Jenkins BG, Ferrante RJ, Rosen BR, Kaddurah-Daouk R, Beal MF (1998) Neuroprotective effects of creatine and cyclocreatine in animal models of Huntington's disease. *J Neurosci* **18**:156-163.

Matthews RT, Ferrante RJ, Klivenyi P, Yang L, Klein AM, Mueller G, Kaddurah-Daouk R, Beal MF (1999) Creatine and cyclocreatine attenuate MPTP neurotoxicity. *Exp Neurol* **157**:142-149.

Nakajima A, Kataoka K, Hong M, Sakaguchi M, Huh NH (2003) BRPK, a novel protein kinase showing increased expression in mouse cancer cell lines with higher metastatic potential. *Cancer Lett* **201**:195-201.

Nicklas WJ, Vyas I, Heikkila RE (1985) Inhibition of NADH-linked oxidation in brain mitochondria by 1-methyl-4-phenyl-pyridine, a metabolite of the neurotoxin, 1-methyl-4-phenyl-1,2,5,6-tetrahydropyridine. *Life Sci* **36**:2503-2508.

Obata K, Morland SJ, Watson RH, Hitchcock A, Chenevix-Trench G, Thomas EJ, Campbell IG (1998) Frequent PTEN/MMAC mutations in endometrioid but not serous or mucinous epithelial ovarian tumors. *Cancer Res* **58**:2095-2097.

Palacino JJ, Sagi D, Goldberg MS, Krauss S, Motz C, Wacker M, Klose J, Shen J (2004) Mitochondrial dysfunction and oxidative damage in parkin-deficient mice. *J Biol Chem* **279**:18614-18622.

Panov AV, Gutekunst CA, Leavitt BR, Hayden MR, Burke JR, Strittmatter WJ, Greenamyre JT (2002) Early mitochondrial calcium defects in Huntington's disease are a direct effect of polyglutamines. *Nat Neurosci* **5**:731-736.

Parker WD, Jr., Boyson SJ, Parks JK (1989) Abnormalities of the electron transport chain in idiopathic Parkinson's disease. *Ann Neurol* **26**:719-723.

Penney JB, Jr., Young AB (1982) Quantitative autoradiography of neurotransmitter receptors in Huntington disease. *Neurology* **32**:1391-1395.

Polymeropoulos MH, Lavedan C, Leroy E, Ide SE, Dehejia A, Dutra A, Pike B, Root H, Rubenstein J, Boyer R, Stenroos ES, Chandrasekharappa S, Athanassiadou A, Papapetropoulos T, Johnson WG, Lazzarini AM, Duvoisin RC, Di Iorio G, Golbe LI, Nussbaum RL (1997) Mutation in the alpha-synuclein gene identified in families with Parkinson's disease. *Science* **276**:2045-2047.

Poyton RO, McEwen JE (1996) Crosstalk between nuclear and mitochondrial genomes. *Annu Rev Biochem* **65**:563-607.

Richter C (1992) Reactive oxygen and DNA damage in mitochondria. *Mutat Res* **275**:249-255.

Rintoul GL, Filiano AJ, Brocard JB, Kress GJ, Reynolds IJ (2003) Glutamate decreases mitochondrial size and movement in primary forebrain neurons. *J Neurosci* **23**:7881-7888.

Risinger JI, Hayes AK, Berchuck A, Barrett JC (1997) PTEN/MMAC1 mutations in endometrial cancers. *Cancer Res* **57**:4736-4738.

Rohe CF, Montagna P, Breedveld G, Cortelli P, Oostra BA, Bonifati V (2004) Homozygous PINK1 C-terminus mutation causing early-onset parkinsonism. *Ann Neurol* **56**:427-431.

Ruan Q, Lesort M, MacDonald ME, Johnson GV (2004) Striatal cells from mutant huntingtin knock-in mice are selectively vulnerable to mitochondrial complex II inhibitor-induced cell death through a non-apoptotic pathway. *Hum Mol Genet* **13**:669-681.

Sanchez I, Mahlke C, Yuan J (2003) Pivotal role of oligomerization in expanded polyglutamine neurodegenerative disorders. *Nature* **421**:373-379.

Sapp E, Schwarz C, Chase K, Bhide PG, Young AB, Penney J, Vonsattel JP, Aronin N, DiFiglia M (1997) Huntingtin localization in brains of normal and Huntington's disease patients. *Ann Neurol* **42**:604-612.

Schapira AH, Cooper JM, Dexter D, Clark JB, Jenner P, Marsden CD (1990) Mitochondrial complex I deficiency in Parkinson's disease. *J Neurochem* **54**:823-827.

Schilling G, Coonfield ML, Ross CA, Borchelt DR (2001) Coenzyme Q10 and remacemide hydrochloride ameliorate motor deficits in a Huntington's disease transgenic mouse model. *Neurosci Lett* **315**:149-153.

Schulz JB, Matthews RT, Henshaw DR, Beal MF (1996) Neuroprotective strategies for treatment of lesions produced by mitochondrial toxins: implications for neurodegenerative diseases. *Neuroscience* **71**:1043-1048.

Skulachev VP (2001) Mitochondrial filaments and clusters as intracellular power-transmitting cables. *Trends Biochem Sci* **26**:23-29.

Song DD, Shults CW, Sisk A, Rockenstein E, Masliah E (2004) Enhanced substantia nigra mitochondrial pathology in human alpha-synuclein transgenic mice after treatment with MPTP. *Exp Neurol* **186**:158-172.

Steck PA, Pershouse MA, Jasser SA, Yung WK, Lin H, Ligon AH, Langford LA, Baumgard ML, Hattier T, Davis T, Frye C, Hu R, Swedlund B, Teng DH, Tavtigian SV (1997) Identification of a candidate tumour suppressor gene, MMAC1, at chromosome 10q23.3 that is mutated in multiple advanced cancers. *Nat Genet* **15**:356-362.

Swerdlow RH, Parks JK, Miller SW, Tuttle JB, Trimmer PA, Sheehan JP, Bennett JP, Jr., Davis RE, Parker WD, Jr. (1996) Origin and functional consequences of the complex I defect in Parkinson's disease. *Ann Neurol* **40**:663-671.

Swerdlow RH, Parks JK, Davis JN, 2nd, Cassarino DS, Trimmer PA, Currie LJ, Dougherty J, Bridges WS, Bennett JP, Jr., Wooten GF, Parker WD (1998) Matrilineal inheritance of complex I dysfunction in a multigenerational Parkinson's disease family. *Ann Neurol* **44**:873-881.

Tanaka M, Machida Y, Niu S, Ikeda T, Jana NR, Doi H, Kurosawa M, Nekooki M, Nukina N (2004) Trehalose alleviates polyglutamine-mediated pathology in a mouse model of Huntington disease. *Nat Med* **10**:148-154.

The Huntington's Disease Collaborative Research Group (1993) A novel gene containing a trinucleotide repeat that is expanded and unstable on Huntington's disease chromosomes. The Huntington's Disease Collaborative Research Group. *Cell* **72**:971-983.

The Huntington's Disease Collaborative Research Group (2001) A randomized, placebo-controlled trial of coenzyme Q10 and remacemide in Huntington's disease. *Neurology* **57**:397-404.

Tieu K, Przedborski S (2004) Mitochondrial dysfunction and neurodegenerative disorders. In: *Mitochondrial Medicine* (DiMauro S HM, Schon E, ed), p in press. Lancaster: The Parthenon Publishing Group.

Tieu K, Wu DC, Przedborski S (2003a) D-hydroxybutyrate prevents striatal lesion in the 3-nitropropionic acid animal model of Hungtington's disease. In: *Society for Neuroscience* (Online AVIP, ed). Washington, DC.

Tieu K, Perier C, Caspersen C, Teismann P, Wu DC, Yan SD, Naini A, Vila M, Jackson-Lewis V, Ramasamy R, Przedborski S (2003b) D-beta-hydroxybutyrate rescues mitochondrial respiration and mitigates features of Parkinson disease. *J Clin Invest* **112**:892-901.

Unoki M, Nakamura Y (2001) Growth-suppressive effects of BPOZ and EGR2, two genes involved in the PTEN signaling pathway. *Oncogene* **20**:4457-4465.

Valente EM, Salvi S, Ialongo T, Marongiu R, Elia AE, Caputo V, Romito L, Albanese A, Dallapiccola B, Bentivoglio AR (2004a) PINK1 mutations are associated with sporadic early-onset parkinsonism. *Ann Neurol* **56**:336-341.

Valente EM, Abou-Sleiman PM, Caputo V, Muqit MM, Harvey K, Gispert S, Ali Z, Del Turco D, Bentivoglio AR, Healy DG, Albanese A, Nussbaum R, Gonzalez-Maldonado R, Deller T, Salvi S, Cortelli P, Gilks WP, Latchman DS, Harvey RJ, Dallapiccola B, Auburger G, Wood NW (2004b) Hereditary early-onset Parkinson's disease caused by mutations in PINK1. *Science* **304**:1158-1160.

van Dellen A, Blakemore C, Deacon R, York D, Hannan AJ (2000) Delaying the onset of Huntington's in mice. *Nature* **404**:721-722.

van der Walt JM, Nicodemus KK, Martin ER, Scott WK, Nance MA, Watts RL, Hubble JP, Haines JL, Koller WC, Lyons K, Pahwa R, Stern MB, Colcher A, Hiner BC, Jankovic J, Ondo WG, Allen FH, Jr., Goetz CG, Small GW, Mastaglia F, Stajich JM, McLaurin AC, Middleton LT, Scott BL, Schmechel DE, Pericak-Vance MA, Vance JM (2003) Mitochondrial polymorphisms significantly reduce the risk of Parkinson disease. *Am J Hum Genet* **72**:804-811.

Vila M, Przedborski S (2003) Targeting programmed cell death in neurodegenerative diseases. *Nat Rev Neurosci* **4**:365-375.

Wallace DC (1999) Mitochondrial diseases in man and mouse. *Science* **283**:1482-1488.

Wexler NS, Lorimer J, Porter J, Gomez F, Moskowitz C, Shackell E, Marder K, Penchaszadeh G, Roberts SA, Gayan J, Brocklebank D, Cherny SS, Cardon LR, Gray J, Dlouhy SR, Wiktorski S, Hodes ME, Conneally PM, Penney JB, Gusella J, Cha JH, Irizarry M, Rosas D, Hersch S, Hollingsworth Z, MacDonald M, Young AB, Andresen JM, Housman DE, De Young MM, Bonilla E, Stillings T, Negrette A, Snodgrass SR, Martinez-Jaurrieta MD, Ramos-Arroyo MA, Bickham J, Ramos JS, Marshall F, Shoulson I, Rey GJ, Feigin A, Arnheim N, Acevedo-Cruz A, Acosta L, Alvir J, Fischbeck K, Thompson LM, Young A, Dure L, O'Brien CJ, Paulsen J, Brickman A, Krch D, Peery S, Hogarth P, Higgins DS, Jr., Landwehrmeyer B (2004) Venezuelan kindreds reveal that genetic and environmental factors modulate Huntington's disease age of onset. *Proc Natl Acad Sci U S A* **101**:3498-3503.

Yang J, Liu X, Bhalla K, Kim CN, Ibrado AM, Cai J, Peng TI, Jones DP, Wang X (1997) Prevention of apoptosis by Bcl-2: release of cytochrome c from mitochondria blocked. *Science* **275**:1129-1132.

Yokota T, Sugawara K, Ito K, Takahashi R, Ariga H, Mizusawa H (2003) Down regulation of DJ-1 enhances cell death by oxidative stress, ER stress, and proteasome inhibition. *Biochem Biophys Res Commun* **312**:1342-1348.

Yu ZX, Li SH, Nguyen HP, Li XJ (2002) Huntingtin inclusions do not deplete polyglutamine-containing transcription factors in HD mice. *Hum Mol Genet* **11**:905-914.

Zarranz JJ, Alegre J, Gomez-Esteban JC, Lezcano E, Ros R, Ampuero I, Vidal L, Hoenicka J, Rodriguez O, Atares B, Llorens V, Gomez Tortosa E, del Ser T, Munoz DG, de Yebenes JG (2004) The new mutation, E46K, of alpha-synuclein causes Parkinson and Lewy body dementia. *Ann Neurol* **55**:164-173.

In: Recent Breakthroughs in Basal Ganglia Research
Editor: Erwan Bezard, pp. 337-349
ISBN: 1-59454-880-3

Chapter 26

Potential Role of α-Synuclein in Neurodegenerative Diseases of the Basal Ganglia: Influence of Genetic and Environmental Factors

***Christina A. Wilson*[1] *and Benoit I. Giasson*[2]**
[1]Department of [1]Pathology and Laboratory Medicine and [2]Pharmacology, University of Pennsylvania School of Medicine, Philadelphia, Pennsylvania 19104

Abstract

α-Synuclein is a small abundant presynaptic neuronal protein expressed throughout the brain. By an aberrant process, α-synuclein can polymerize in ~10 nm fibrils that accumulate to from various forms of pathological inclusions associated with neurodegenerative diseases. These inclusions can be abundant in the basal ganglia and they may contribute to neuronal dysfunctions. Point mutations in α-synuclein or increased expression of α-synuclein, due to short chromosomal duplication/trisomy that include the α-synuclein gene, can promote the formation of α-synuclein inclusions. Furthermore, defects or mutations in several other genes have also been implicated in affecting the aggregation in α-synuclein. Finally, environmental factors in the form of toxins may also precipitate the aggregation of α-synuclein leading to disease. In the chapter, the recent progress in this area of research with focus on the basal ganglia will be reviewed.

Introduction

α-Synuclein is a small abundant presynaptic neuronal protein expressed throughout the brain. By an aberrant process, α-synuclein can polymerize in ~10 nm fibrils that accumulate to from various forms of pathological inclusions associated with neurodegenerative diseases. These inclusions can be abundant in the basal ganglia and they may contribute to neuronal dysfunctions. Point mutations in α-synuclein or increased expression of α-synuclein, due to short chromosomal duplication/trisomy that include the *α-synuclein* gene, can promote the formation of α-synuclein inclusions. Furthermore, defects or mutations in several other genes have also been implicated in affecting the aggregation in α-synuclein. Finally, environmental factors in the form of toxins may also precipitate the aggregation of α-synuclein leading to disease. In the chapter, the recent progress in this area of research with focus on the basal ganglia will be reviewed.

1 α-SYNUCLEIN AND PARKINSON'S DISEASE

Parkinson's disease (PD) is the most common neurodegenerative movement disease and it affects ~ 1 % of the population by 65 years of age. Clinically, PD is characterized by resting tremor, rigidity, bradykinesia, postural instability and responds to levodopa therapy (Simuni and Hurtig, 2000). A definite diagnosis of PD is rendered by a post-mortem analyses showing an extensive loss of dopaminergic neurons and gliosis in the substantia nigra pars compacta (SNpc), and the presence of round, eosinophilic intracytoplasmic inclusions known as Lewy bodies (LBs) in some of these neurons (McGeer et al., 1988; Fearnley and Lees, 1991; Cornford et al., 1995; Forno, 1996).

The neuronal circuitry involved in coordinated movement is complex (Obeso et al., 2000), but it is widely accepted that the degeneration of the dopaminergic neurons in the SNpc is the cause of many of the symptoms of Parkinson's disease (Pakkenberg et al., 1991; Damier et al., 1999). The demise of these neurons, which project to the striatum, results in a profound reduction in striatal dopamine content disturbing the normal activity of the "motor circuit" (Obeso et al., 2000).

Table 1. Identified genes causal of familial parkinsonism. In all cases the mode of genetic transmission is autosomal, but can be recessive or dominant.

Locus (location)	Gene	Inheritance
PARK1 (4q21-22)	a-synuclein	dominant
PARK2 (6q25-27)	parkin	recessive
PARK4 (4q21-22)	a-synuclein	dominant
PARK5 (4p14)	UCH-L1	dominant
PARK6 (1p35-36)	PINK1	recessive
PARK7 (1p36)	DJ-1	recessive
PARK8 (12p11-q13)	Dardarin/LRRK2	dominant

α-synuclein was the first gene identified directly linked to Parkinson's disease (PD) (Polymeropoulos et al., 1997). In the seminal report by Polymeropoulos and colleagues, the Ala53Thr mutation resulting from a G to A transition at position 209 of the *α-synuclein* gene (also known as *PARK1* gene; see Table 1) was identified in a large Italian family (the Contursi kindred) and three small possibly related Greek families with autosomal dominant PD (Polymeropoulos et al., 1997). Shortly following the report of the Ala53Thr mutation in α-synuclein, a series of studies convincingly demonstrated that α-synuclein is the building block of the ~10 nm filamentous polymers that form Lewy bodies (LBs; Fig. 1A; see Duda et al., 2000a; Goedert, 2001 for reviews). Moreover, the Ala53Thr mutation in α-synuclein has now been identified in at least 8 additional families (Athanassiadou et al., 1999; Markopoulou and Langston, 1999; Papadimitriou et al., 1999; Spira et al., 2001). Two other mutations in α-synuclein, Ala30Pro and Glu46Lys, are also found in rare kindreds with PD and the related disorder dementia with LBs (Kruger et al., 1998; Zarranz et al., 2004). The trisomy of a short region in chromosome 4 that includes the *α-synuclein* gene, resulting in increased expression of α-synuclein, has also been identified in some families with the related disease dementia with LBs (DLB; Singleton et al., 2003; Farrer et al., 2004). The inability to identify missense mutations in *α-synuclein* in this kindred resulted in the naming of this locus as PARK4 (see Table 1). Duplication of short regions of chromosome 4 that include the *α-synuclein* gene can result in classical PD (Chartier-Harlin et al., 2004).

Although LBs are often emphasized in the description of PD for historical and histological reasons, the formation of α-synuclein inclusions in neuronal processes, known as Lewy neurites (LNs) are much more abundant than LBs (Duda et al., 2002a; Neumann et al., 2002), and they likely play a much greater role in perturbing neuronal function. These aberrant inclusions in neuronal processes were first described as dystrophic ubiquitin-positive neurites (LNs; Dickson et al., 1991; Braak et al., 1994). Moreover, filamentous inclusions of α-synuclein is the major component of several types of pathological inclusions characteristic of many disorders resulting in the coining of the term "synucleinopathies" (Spillantini et al., 1997; Irizarry et al., 1998; Spillantini et al., 1998b; 1998a; Tu et al., 1998; Takeda et al., 2000).

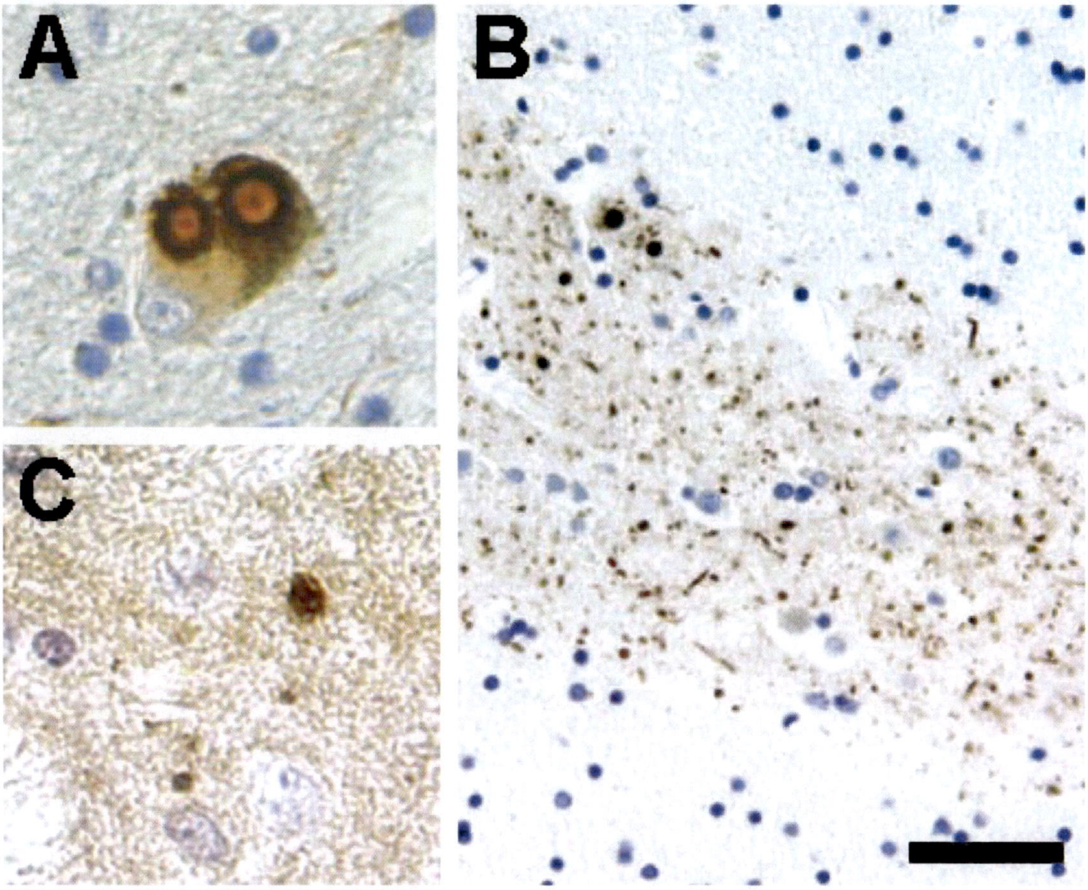

Figure 1. α-Synuclein pathological inclusions in the SN and basal ganglia. (A) Classical LBs in a dopaminergic neuron in the SNpc of a PD patient immunolabeled with an anti-α-synuclein antibody. (B) Immunostaining with an antibody to α-synuclein in the putamen (bordered below and above by internal capsule) of a patient with DLB demonstrating abundant inclusions in the form of Lewy neurites and grain-like aggregates. (C) Similar grain-like α-synuclein inclusions in the striatum of human A53T α-synuclein transgenic mice detected with antibody specific for α-synuclein. Scale bar = 200 μm in A, 250 μm in B and 40 μm in C.

2 α-Synuclein Pathology in the Basal Ganglia

In PD research, there has been an emphasis on the reduction of dopamine levels in the striatum resulting from the loss of dopaminergic neurons in SNpc. However, recently it has also been suggested that the impairment of the normal function of dopaminergic neurons due to α-synuclein inclusions in their axons and axonal terminal may also contribute to dysfunction (Duda et al., 2002a). α-Synuclein inclusions in the form of Lewy neurites, neuroaxonal spheroids or "grain-like" aggregates in dopaminergic neuronal processes in the basal ganglia is a common finding in PD (see Figure 1B), as well as in related diseases such as DLB (Shoji et al., 2000; Duda et al., 2002a). Furthermore, similar inclusions are observed in patients with the A53T α-synuclein mutation (Duda et al., 2002b) and in α-synuclein transgenic mouse models (Fig. 1C; Giasson et al., 2002). It is likely that these aggregates contribute to impairing nigrostriatal projections by acting as "plugs" in axons and axonal terminal, diminishing transport and neuronal connections, eventually leading to neurodegeneration.

The presence of α-synuclein inclusions is also an important finding in patients with neurodegeneration with brain iron accumulation type 1 (NBIA-1; previously known as Hallervorden-Spatz disease). This rare progressive neurodegenerative disorder occurs in both a familial and sporadic form, and it is clinically characterized by parkinsonism features, pyramidal signs, seizures, and mental deterioration that can culminate in dementia (Dooling et al., 1974). The histopathological findings defining NBIA-1 are abundant neuroaxonal spheroids (i.e. 20-100 μm wide axonal swelling), which appear to consist mainly of α-synuclein and intense iron deposits in the globus pallidus and SNpc (Arawaka et al., 1998; Tu et al., 1998; Newell et al., 1999; Wakabayashi et al., 1999; Galvin et al., 2000; Saito et al., 2000). The recessive familial form of this disease has been attributed to the defects in the *pantothenate kinase 2* (PK2) gene (Zhou et al., 2001). PK2 is one of 4 human genes that express a pantothenate kinase, but PK2 is unique in that it is targeted to the mitochondria (Hortnagel et al., 2003; Johnson et al., 2004). Although pantothenate kinase activity is required for the biosynthesis of coenzyme A from pantothenate (vitamin B5), it is unclear how disruption of PK2 activity leads to disease. Although widespread LBs have been reported in the majority of the brains of NBIA-1 patients, some brains do not appear to display these lesions (see Saito et al., 2000 and references therein) and the association between α-synuclein pathological inclusions and loss of PK2 activity has not been assessed.

Multiple system atrophy (MSA) is an adult onset neurodegenerative disease characterized by varying degrees of parkinsonism, cerebellar ataxia, and autonomic dysfunction (Wenning et al., 1994; Wenning et al., 1997; Gilman et al., 1999). MSA is characterized histologically by the accumulation of filamentous α-synuclein inclusions in the form of neuronal cytoplasmic inclusions (NCIs) and glial cytoplasmic inclusions (GCIs) in oligodendrocytes (Papp et al., 1989; Kato et al., 1991; Spillantini et al., 1998b; Tu et al., 1998; Wakabayashi et al., 1998). MSA brains show varying degrees of atrophy and demyelination of the cerebellum, pons, and medulla, as well as the loss of pigmented neurons of the substantia nigra. GCIs can be found throughout the white matter, but the greatest abundance of these inclusions occurs in the basal ganglia, the substantia nigra, the pontine nucleus, medulla and cerebellum (Arima et al., 1992; Lantos, 1997; Duda et al., 2000b).

3 α-Synuclein Protein and Mechanisms of Pathogenesis

α-Synuclein is a relatively small protein of 140 amino acids which includes 6 imperfect repeats (KTKEGV) distributed throughout most of the amino-terminal half of the protein and an acidic carboxyl-terminal region (see Figure 2; Jakes et al., 1994). α-Synuclein is a heat-stable protein that is "natively unfolded", such that it has no apparent defined structure (Weinreb et al., 1996; Davidson et al., 1998), and it is predominantly expressed in the neurons of the central nervous system (CNS), where it is localized at presynaptic terminals (Jakes et al., 1994; George et al., 1995; Iwai et al., 1995; Withers et al., 1997), but its function remains unknown (Chandra et al., 2004).

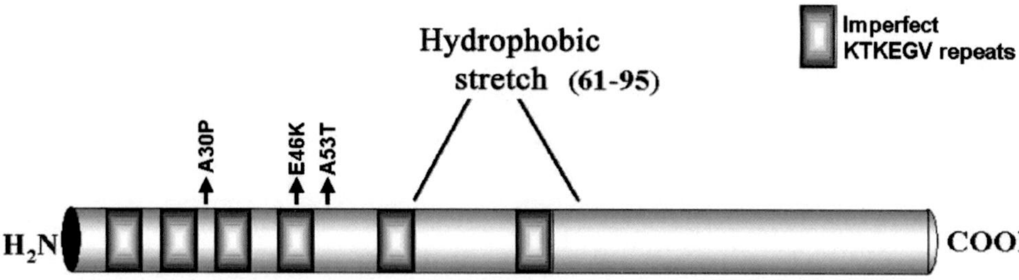

Figure 2. Schematic of the α-synuclein molecule. α-synuclein has six degenerate KTKEGV repeats throughout the N-terminal region and a hydrophobic stretch that spans the middle region. The locations of the known missense mutations causal of disease are indicated.

Although the formation of α-synuclein inclusions is clearly associated with the etiology of many neurodegenerative diseases, the nature of the toxic polymerized species of α-synuclein is still debated. Some studies in cultured cells indicate that simple overexpression of α-synuclein may have toxicity (Saha et al., 2000; Zhou et al., 2000; Xu et al., 2002). However, most evidence support the notion that the aberrant polymerization of α-synuclein into ~10 nm-wide amyloidogenic filaments, which eventually form large intracytoplasmic inclusions, that can cause the dysfunction and demise of neurons or oligodendrocytes (Goedert, 2001; Dawson et al., 2002; Giasson et al., 2002). For example, the most common disease-causing point mutation in α-synuclein (Ala53Thr) significantly increases the propensity of the mutant of α-synuclein to fibrillize *in vitro* and results in the widespread and abundant distribution of α-synuclein pathology in individuals (Conway et al., 1998; Giasson et al., 1999; Narhi et al., 1999; Duda et al., 2002b). Similarly, expression of Ala53Thr human α-synuclein in mice results in the formation of abundant α-synuclein inclusions that is accompanied by severe neuronal dysfunction and axonal degeneration leading to death (Dawson et al., 2002; Giasson et al., 2002; Lee et al., 2002). The density of cortical α-synuclein pathological inclusions in patients also correlates with dementia severity (Lennox et al., 1989; Samuel et al., 1996; Haroutunian et al., 2000; Hurtig et al., 2000; Mattila et al., 2000). Therefore, it is likely that α-synuclein inclusions can impair cellular function by obstructing normal cellular trafficking, disrupting cell morphology, and by trapping other cellular components, all of which eventually can lead to cell death. However, Lansbury and colleagues have supported an alternative, but not mutually exclusive, concept that protofibrils may be toxic (Goldberg and Lansbury, 2000; Caughey and Lansbury, 2003).

4 Other Genetic Defects Associated with Parkinsonism that May Influence α-Synuclein Inclusion Formation

Recently, several genes have been identified in kindred with parkinsonism and it is possible that some of these proteins may be involved in the aggregation of α-synuclein. Mutations in the *DJ-1* gene (also knows as *PARK7*) were found to be responsible for disease in some families with autosomal recessive parkinsonism (Bonifati et al., 2003). The pathological features of patients with DJ-1 mutations are not clear since post-mortem analyses are not available. The first PARK7 kindred identified had a homozygous deletion of a large region within the *DJ-1* gene (encompassing the promoter and first five exons) that resulted in the complete loss of DJ-1 expression (Bonifati et al., 2003). In addition, several homozygous point mutations in *DJ-1* (Leu166Pro, Glu64Asp, and Met26Ile) have been identified in families with an autosomal PD-like disorder (Abou-Sleiman et al., 2003; Bonifati et al., 2003; Gorner et al., 2004), but no post-mortem analysis has been reported.

DJ-1 is a small 189-amino acid protein that exists in vivo as a dimer (Honbou et al., 2003; Huai et al., 2003; Lee et al., 2003; Tao and Tong, 2003; Wilson et al., 2003) and the Leu166Pro mutation has been shown to prevent the dimerization of DJ-1, resulting in its rapid degradation (Miller et al., 2003; Gorner et al., 2004; Olzmann et al., 2004). The function of DJ-1 remains mostly unknown, but it exerts some ability to protect against oxidative stress (Yokota et al., 2003). Furthermore, exposure of cultured cells to oxidative challenges results in the oxidation of DJ-1, as seen by a shift in its isoelectric property, suggesting that its may act as a sensor of oxidative stress (Mitsumoto et al., 2001). In cultured cells DJ-1 has been shown to have the ability to reduce the aggregation of α-synuclein resulting form oxidative damage (Shendelman et al., 2004).

Mutations in the *PARK6* gene on chromosome 1p35-p36 that encodes a putative mitochondrial protein kinase PINK1 (PTEN-induced kinase 1) were recently shown to be responsible for some kindreds with an autosomal recessive form of parkinsonism (Hatano et al., 2004; Valente et al., 2004). Homozygous nonsense (W437X, R246X) and missense mutations (G309A, H271Q, E417G, L347P), as well as compound nonsense mutations (Q239X/R492X) have been linked to the disease , but autopsies have not been performed to determine the extent of neurodegeneration or the presence of proteinaceous inclusions. PINK1 (also known as BRPK) is a putative serine/threonine protein kinase, but its substrates are unknown (Nakajima et al., 2003). PINK1 has a mitochondrial localization sequence and has been shown to be targeted to the mitochondria (Valente et al., 2004), but it might also function in the cytoplasm. The effects of PINK1 on α-synuclein aggregation have not been studies, but it may protect neurons against stress-induced mitochondrial dysfunction and apoptosis (Valente et al., 2004).

As much as 50 % of patients with early onset (<40 years of age) parkinsonism harbor mutations in the *parkin* gene (also knows as *PARK2*) and a range of nonsense, missense and insertion/deletion mutations have been shown to be causal of the majority of cases of autosomal recessive juvenile parkinsonism (AR-JP; Kitada et al., 1998; Giasson and Lee, 2001). Patients with *parkin* mutations exhibit the typical clinical features of parkinsonism and are markedly responsive to levodopa therapy. Neuropathological examination demonstrates that neuronal loss and gliosis are restricted to the SN and locus coeruleus (Mori et al., 1998;

Hayashi et al., 2000). In most cases, Lewy pathology was not observed in patients with parkin mutation (Mori et al., 1998; Hayashi et al., 2000), but some brains present with LBs (Farrer et al., 2001; Sasaki et al., 2004).

The *parkin* gene encodes a 465-amino acid protein with known E3 ligase activity such that it can target specific proteins for ubiquitination and destruction by the proteasome (Shimura et al., 2000; Zhang et al., 2000; Chung et al., 2001; Staropoli et al., 2003). Many parkin substrates have been identified (see Giasson and Lee, 2001 for review) and several findings suggest that parkin may also play an important role in maintaining mitochondrial integrity (Darios et al., 2003; Goldberg et al., 2003; Greene et al., 2003; Palacino et al., 2004; Pesah et al., 2004). The role of parkin in idiopathic PD and α-synuclein aggregation are still unclear, but a partial loss of activity may contribute to pathogenesis (Chung et al., 2004).

A nucleotide substitution resulting in the Ile93Met amino acid change in the enzyme ubiquitin carboxy-terminal hydrolase L1 (UCH-L1; also knows as *PARK5*) was identified in two siblings with parkinsonism (Leroy et al., 1998). It is unclear if these individuals have typical PD since no autopsy has been performed. Furthermore, there are several factors that put in question whether the Ile93Met substitution is a mutation causal of disease. For example, the mode of transmission of this substitution seems to be autosomal dominant since only one UCH-L1 allele is altered. However the parents of the affected sibling were unaffected, which indicates that this substitution could be a rare polymorphism that segregates with disease, or that mutation has incomplete penetrance (Liu et al., 2002). UCH-L1 is a thiol protease that has been characterized for its ability to hydrolyze C-terminal bonds between ubiquitin and small adducts, and the Ile93Met substitution results in only a partial loss (~50 %) of catalytic activity (Leroy et al., 1998). Therefore, it is also unclear how the simple reduction of the activity of one of the two copies of the UCH-L1 gene products could have a dominant effect. UCH-L1 has been shown to also have an ubiquityl ligase-like activity that act to increase poly-ubiquitination of mono- or di-ubiquitinated α-synuclein and the Ile93Met substitution also results in reduced (~50 %) ligase activity (Liu et al., 2002).

PARK8 (also termed dardarin or leucine-rich repeat kinase-2) was recently cloned in several kindreds with autosomal dominant parkinsonism characterized by various forms of protein aggregates, including α-synuclein inclusions (Liu et al., 2002; Wszolek et al., 2004; Zimprich et al., 2004). PARK8 is a large 2527 amino acid protein with a leucine-rich repeat, a tyrosine kinase catalytic domain and a WD40 domain. Most disease-causing mutations in PARK8 are missense mutations and it is unclear what function they impair, but they appear to lead to α-synuclein aggregation (Zimprich et al., 2004).

Tau gene haplotypes are associated with PD (Pastor et al., 2000; Maraganore et al., 2001) and specific mutations in the *tau* gene causes the disease frontotemporal dementia with parkinsonism linked to chromosome 17 (FTDP-17; Forman et al., 2000; Lee et al., 2001). Degeneration of the nigrostriatal pathway is common in patients with FTDP-17 (Reed et al., 2001). Tau is a microtubule binding protein that is also the major component of the fibrils that constitute many types of pathological inclusions such as neurofibrillary tangles in Alzheimer's disease (Forman et al., 2000; Lee et al., 2001). α-Synuclein can interact with tau to initiate tau filament formation, and the association between these two proteins can enhance the fibrillogenesis of each other (Giasson et al., 2003). These finding are consistent with the frequent co-existence of α-synuclein and tau inclusions in the same brains and often in the same cells (Giasson et al., 2003; Lee et al., 2004). For example, concurrent α-synuclein and tau pathological inclusions in the globus pallidus of patients with NBIA-1 is common (Saito et al., 2000).

5 Effects of Environmental Toxins on α-Synuclein Aggregation and Degeneration of the Basal Ganglia

Environmental factors (Rajput et al., 1986; 1987; Tanner et al., 1987; Koller et al., 1990; Hubble et al., 1993) and exposure to toxins, like 1-methyl-4-phenyl-1,2,3,6-tetrahydropyridine (MPTP; Langston et al., 1999; Beal, 2001), have been suggested in the etiology of PD and related neurodegenerative diseases. However, there is still little direct evidence for the involvement of these factors in idiopathic PD, although exposure to MPTP can cause parkinsonism due to the degeneration of dopaminergic striatal nigra pathway, but without the involvement of proteinaceous inclusions (Langston et al., 1999; Beal, 2001). The most provocative finding supporting the involvement of environmental factors in PD is an experimental model where rats are exposed to rotenone by intravenous injection or subcutaneous administration (Betarbet et al.,

2000; Sherer et al., 2003a). Rotenone is a pesticide that inhibits mitochondrial complex I. The administration of rotenone to rats results in many features that closely parallel those seen in PD patients, including bradykinesia, posture instability and unsteady gait. Moreover, it was reported that rotenone causes the selective degeneration of the nigrostriatal dopaminergic system, with injury being more profound in the striatum than the SN. However, this specificity of neuronal degeneration by rotenone has been challenged by others that noted degeneration in other brain regions including striatal neurons (Hoglinger et al., 2003). Some SN neurons that survive rotenone exposure develop intracytoplasmic α-synuclein-rich inclusions that resembled LBs by light and electron microscopy (Betarbet et al., 2000). Furthermore, some inclusions have a distinct core surrounded by a fibrillar halo similar to authentic human LBs, and other inclusions resembled "pale bodies", presumptive LB precursor lesions. Moreover, it is likely that increased generation of free radicals resulting from mitochondrial inhibition, and not energy (i.e. ATP) depletion, is the predominant mechanism of cellular damage leading α-synuclein aggregation (Sherer et al., 2003b).

6 Closing Remarks

In the last few years, rapid progress has been realized in understanding the importance of α-synuclein aggregation in neurodegenerative diseases. Recent studies have emphasize the previously underappreciated extent of smaller α-synuclein inclusions in many brain regions including in the basal ganglia, which likely contribute to the clinical presentations in PD and related diseases. Moreover, the recent identification of new defective genes associated with parkinsonism suggests that these may act as modifiers of α-synuclein aggregation, but few studies in the area have yet been completed. Environmental factors and neurotoxins may also act alone or in concert with these gene products to influence α-synuclein induced pathogenesis. Future analyses will be needed to assess the importance of these recently identified gene defects on the formation of α-synuclein inclusion as well as the biological effects of α-synuclein inclusions in the basal ganglia.

References

Abou-Sleiman PM, Healy DG, Quinn N, Lees AJ, Wood NW (2003) The role of pathogenic DJ-1 mutations in Parkinson's disease. *Ann Neurol* **54**:283-286.

Arawaka S, Saito Y, Murayama S, Mori H (1998) Lewy body in neurodegeneration with brain iron accumulation type 1 is immunoreactive for alpha-synuclein. *Neurology* **51**:887-889.

Arima K, Murayama S, Mukoyama M, Inose T (1992) Immunocytochemical and ultrastructural studies of neuronal and oligodendroglial cytoplasmic inclusions in multiple system atrophy. 1. Neuronal cytoplasmic inclusions. *Acta Neuropathol (Berl)* **83**:453-460.

Athanassiadou A, Voutsinas G, Psiouri L, Leroy E, Polymeropoulos MH, Ilias A, Maniatis GM, Papapetropoulos T (1999) Genetic analysis of families with Parkinson disease that carry the Ala53Thr mutation in the gene encoding alpha-synuclein. *Am J Hum Genet* **65**:555-558.

Beal MF (2001) Experimental models of Parkinson's disease. *Nat Rev Neurosci* **2**:325-334.

Betarbet R, Sherer TB, MacKenzie G, Garcia-Osuna M, Panov AV, Greenamyre JT (2000) Chronic systemic pesticide exposure reproduces features of Parkinson's disease. *Nat Neurosci* **3**:1301-1306.

Bonifati V, Rizzu P, van Baren MJ, Schaap O, Breedveld GJ, Krieger E, Dekker MC, Squitieri F, Ibanez P, Joosse M, van Dongen JW, Vanacore N, van Swieten JC, Brice A, Meco G, van Duijn CM, Oostra BA, Heutink P (2003) Mutations in the DJ-1 gene associated with autosomal recessive early-onset parkinsonism. *Science* **299**:256-259.

Braak H, Braak E, Yilmazer D, de Vos RA, Jansen EN, Bohl J, Jellinger K (1994) Amygdala pathology in Parkinson's disease. *Acta Neuropathol (Berl)* **88**:493-500.

Caughey B, Lansbury PT (2003) Protofibrils, pores, fibrils, and neurodegeneration: separating the responsible protein aggregates from the innocent bystanders. *Annu Rev Neurosci* **26**:267-298.

Chandra S, Fornai F, Kwon HB, Yazdani U, Atasoy D, Liu X, Hammer RE, Battaglia G, German DC, Castillo PE, Sudhof TC (2004) Double-knockout mice for alpha- and beta-synucleins: effect on synaptic functions. *Proc Natl Acad Sci U S A* **101**:14966-14971.

Chartier-Harlin MC, Kachergus J, Roumier C, Mouroux V, Douay X, Lincoln S, Levecque C, Larvor L, Andrieux J, Hulihan M, Waucquier N, Defebvre L, Amouyel P, Farrer M, Destee A (2004) Alpha-synuclein locus duplication as a cause of familial Parkinson's disease. *Lancet* **364**:1167-1169.

Chung KK, Thomas B, Li X, Pletnikova O, Troncoso JC, Marsh L, Dawson VL, Dawson TM (2004) S-nitrosylation of parkin regulates ubiquitination and compromises parkin's protective function. *Science* **304**:1328-1331.

Chung KK, Zhang Y, Lim KL, Tanaka Y, Huang H, Gao J, Ross CA, Dawson VL, Dawson TM (2001) Parkin ubiquitinates the alpha-synuclein-interacting protein, synphilin-1: implications for Lewy-body formation in Parkinson disease. *Nat Med* **7**:1144-1150.

Conway KA, Harper JD, Lansbury PT (1998) Accelerated in vitro fibril formation by a mutant alpha-synuclein linked to early-onset Parkinson disease. *Nat Med* **4**:1318-1320.

Cornford ME, Chang L, Miller BL (1995) The neuropathology of parkinsonism: an overview. *Brain Cogn* **28**:321-341.

Damier P, Hirsch EC, Agid Y, Graybiel AM (1999) The substantia nigra of the human brain. II. Patterns of loss of dopamine-containing neurons in Parkinson's disease. *Brain* **122** (Pt 8):1437-1448.

Darios F, Corti O, Lucking CB, Hampe C, Muriel MP, Abbas N, Gu WJ, Hirsch EC, Rooney T, Ruberg M, Brice A (2003) Parkin prevents mitochondrial swelling and cytochrome c release in mitochondria-dependent cell death. *Hum Mol Genet* **12**:517-526.

Davidson WS, Jonas A, Clayton DF, George JM (1998) Stabilization of alpha-synuclein secondary structure upon binding to synthetic membranes. *J Biol Chem* **273**:9443-9449.

Dawson T, Mandir A, Lee M (2002) Animal models of PD: pieces of the same puzzle? *Neuron* **35**:219-222.

Dickson DW, Ruan D, Crystal H, Mark MH, Davies P, Kress Y, Yen SH (1991) Hippocampal degeneration differentiates diffuse Lewy body disease (DLBD) from Alzheimer's disease: light and electron microscopic immunocytochemistry of CA2-3 neurites specific to DLBD. *Neurology* **41**:1402-1409.

Dooling EC, Schoene WC, Richardson EP, Jr. (1974) Hallervorden-Spatz syndrome. *Arch Neurol* **30**:70-83.

Duda JE, Lee VM, Trojanowski JQ (2000a) Neuropathology of synuclein aggregates. *J Neurosci Res* **61**:121-127.

Duda JE, Giasson BI, Mabon ME, Lee VM, Trojanowski JQ (2002a) Novel antibodies to oxidized synuclein show abundant striatal pathology in Lewy body diseases. *Ann Neurol* **52**:205-210.

Duda JE, Giasson BI, Mabon ME, Miller DC, Golbe LI, Lee VM, Trojanowski JQ (2002b) Concurrence of alpha-synuclein and tau brain pathology in the Contursi kindred. *Acta Neuropathol (Berl)* **104**:7-11.

Duda JE, Giasson BI, Gur TL, Montine TJ, Robertson D, Biaggioni I, Hurtig HI, Stern MB, Gollomp SM, Grossman M, Lee VM, Trojanowski JQ (2000b) Immunohistochemical and biochemical studies demonstrate a distinct profile of alpha-synuclein permutations in multiple system atrophy. *J Neuropathol Exp Neurol* **59**:830-841.

Farrer M, Kachergus J, Forno L, Lincoln S, Wang DS, Hulihan M, Maraganore D, Gwinn-Hardy K, Wszolek Z, Dickson D, Langston JW (2004) Comparison of kindreds with parkinsonism and alpha-synuclein genomic multiplications. *Ann Neurol* **55**:174-179.

Farrer M, Chan P, Chen R, Tan L, Lincoln S, Hernandez D, Forno L, Gwinn-Hardy K, Petrucelli L, Hussey J, Singleton A, Tanner C, Hardy J, Langston JW (2001) Lewy bodies and parkinsonism in families with parkin mutations. *Ann Neurol* **50**:293-300.

Fearnley JM, Lees AJ (1991) Ageing and Parkinson's disease: substantia nigra regional selectivity. *Brain* **114** (Pt 5):2283-2301.

Forman MS, Lee VM, Trojanowski JQ (2000) New insights into genetic and molecular mechanisms of brain degeneration in tauopathies. *J Chem Neuroanat* **20**:225-244.

Forno LS (1996) Neuropathology of Parkinson's disease. *J Neuropathol Exp Neurol* **55**:259-272.

Galvin JE, Giasson B, Hurtig HI, Lee VM, Trojanowski JQ (2000) Neurodegeneration with brain iron accumulation, type 1 is characterized by alpha-, beta-, and gamma-synuclein neuropathology. *Am J Pathol* **157**:361-368.

George JM, Jin H, Woods WS, Clayton DF (1995) Characterization of a novel protein regulated during the critical period for song learning in the zebra finch. *Neuron* **15**:361-372.

Giasson BI, Lee VM (2001) Parkin and the molecular pathways of Parkinson's disease. *Neuron* **31**:885-888.

Giasson BI, Uryu K, Trojanowski JQ, Lee VM (1999) Mutant and wild type human alpha-synucleins assemble into elongated filaments with distinct morphologies in vitro. *J Biol Chem* **274**:7619-7622.

Giasson BI, Duda JE, Quinn SM, Zhang B, Trojanowski JQ, Lee VM (2002) Neuronal alpha-synucleinopathy with severe movement disorder in mice expressing A53T human alpha-synuclein. *Neuron* **34**:521-533.

Giasson BI, Forman MS, Higuchi M, Golbe LI, Graves CL, Kotzbauer PT, Trojanowski JQ, Lee VM (2003) Initiation and synergistic fibrillization of tau and alpha-synuclein. *Science* **300**:636-640.

Gilman S, Low PA, Quinn N, Albanese A, Ben-Shlomo Y, Fowler CJ, Kaufmann H, Klockgether T, Lang AE, Lantos PL, Litvan I, Mathias CJ, Oliver E, Robertson D, Schatz I, Wenning GK (1999) Consensus statement on the diagnosis of multiple system atrophy. *J Neurol Sci* **163**:94-98.

Goedert M (2001) Alpha-synuclein and neurodegenerative diseases. *Nat Rev Neurosci* **2**:492-501.

Goldberg MS, Lansbury PT, Jr. (2000) Is there a cause-and-effect relationship between alpha-synuclein fibrillization and Parkinson's disease? *Nat Cell Biol* **2**:E115-119.

Goldberg MS, Fleming SM, Palacino JJ, Cepeda C, Lam HA, Bhatnagar A, Meloni EG, Wu N, Ackerson LC, Klapstein GJ, Gajendiran M, Roth BL, Chesselet MF, Maidment NT, Levine MS, Shen J (2003) Parkin-deficient mice exhibit nigrostriatal deficits but not loss of dopaminergic neurons. *J Biol Chem* **278**:43628-43635.

Gorner K, Holtorf E, Odoy S, Nuscher B, Yamamoto A, Regula JT, Beyer K, Haass C, Kahle PJ (2004) Differential effects of Parkinson's disease-associated mutations on stability and folding of DJ-1. *J Biol Chem* **279**:6943-6951.

Greene JC, Whitworth AJ, Kuo I, Andrews LA, Feany MB, Pallanck LJ (2003) Mitochondrial pathology and apoptotic muscle degeneration in Drosophila parkin mutants. *Proc Natl Acad Sci U S A* **100**:4078-4083.

Haroutunian V, Serby M, Purohit DP, Perl DP, Marin D, Lantz M, Mohs RC, Davis KL (2000) Contribution of Lewy body inclusions to dementia in patients with and without Alzheimer disease neuropathological conditions. *Arch Neurol* **57**:1145-1150.

Hatano Y, Li Y, Sato K, Asakawa S, Yamamura Y, Tomiyama H, Yoshino H, Asahina M, Kobayashi S, Hassin-Baer S, Lu CS, Ng AR, Rosales RL, Shimizu N, Toda T, Mizuno Y, Hattori N (2004) Novel PINK1 mutations in early-onset parkinsonism. *Ann Neurol* **56**:424-427.

Hayashi S, Wakabayashi K, Ishikawa A, Nagai H, Saito M, Maruyama M, Takahashi T, Ozawa T, Tsuji S, Takahashi H (2000) An autopsy case of autosomal-recessive juvenile parkinsonism with a homozygous exon 4 deletion in the parkin gene. *Mov Disord* **15**: 884-888.

Hoglinger GU, Feger J, Prigent A, Michel PP, Parain K, Champy P, Ruberg M, Oertel WH, Hirsch EC (2003) Chronic systemic complex I inhibition induces a hypokinetic multisystem degeneration in rats. *J Neurochem* **84**:491-502.

Honbou K, Suzuki NN, Horiuchi M, Taira T, Niki T, Ariga H, Inagaki F (2003) Crystallization and preliminary crystallographic analysis of DJ-1, a protein associated with male fertility and parkinsonism. *Acta Crystallogr D Biol Crystallogr* **59**:1502-1503.

Hortnagel K, Prokisch H, Meitinger T (2003) An isoform of hPANK2, deficient in pantothenate kinase-associated neurodegeneration, localizes to mitochondria. *Hum Mol Genet* **12**:321-327.

Huai Q, Sun Y, Wang H, Chin LS, Li L, Robinson H, Ke H (2003) Crystal structure of DJ-1/RS and implication on familial Parkinson's disease. *FEBS Lett* **549**:171-175.

Hubble JP, Cao T, Hassanein RE, Neuberger JS, Koller WC (1993) Risk factors for Parkinson's disease. *Neurology* **43**:1693-1697.

Hurtig HI, Trojanowski JQ, Galvin J, Ewbank D, Schmidt ML, Lee VM, Clark CM, Glosser G, Stern MB, Gollomp SM, Arnold SE (2000) Alpha-synuclein cortical Lewy bodies correlate with dementia in Parkinson's disease. *Neurology* **54**:1916-1921.

Irizarry MC, Growdon W, Gomez-Isla T, Newell K, George JM, Clayton DF, Hyman BT (1998) Nigral and cortical Lewy bodies and dystrophic nigral neurites in Parkinson's disease and cortical Lewy body disease contain alpha-synuclein immunoreactivity. *J Neuropathol Exp Neurol* **57**:334-337.

Iwai A, Masliah E, Yoshimoto M, Ge N, Flanagan L, de Silva HA, Kittel A, Saitoh T (1995) The precursor protein of non-A beta component of Alzheimer's disease amyloid is a presynaptic protein of the central nervous system. *Neuron* **14**:467-475.

Jakes R, Spillantini MG, Goedert M (1994) Identification of two distinct synucleins from human brain. *FEBS Lett* **345**:27-32.

Johnson MA, Kuo YM, Westaway SK, Parker SM, Ching KH, Gitschier J, Hayflick SJ (2004) Mitochondrial localization of human PANK2 and hypotheses of secondary iron accumulation in pantothenate kinase-associated neurodegeneration. *Ann N Y Acad Sci* **1012**:282-298.

Kato S, Nakamura H, Hirano A, Ito H, Llena JF, Yen SH (1991) Argyrophilic ubiquitinated cytoplasmic inclusions of Leu-7-positive glial cells in olivopontocerebellar atrophy (multiple system atrophy). *Acta Neuropathol (Berl)* **82**:488-493.

Kitada T, Asakawa S, Hattori N, Matsumine H, Yamamura Y, Minoshima S, Yokochi M, Mizuno Y, Shimizu N (1998) Mutations in the parkin gene cause autosomal recessive juvenile parkinsonism. *Nature* **392**:605-608.

Koller W, Vetere-Overfield B, Gray C, Alexander C, Chin T, Dolezal J, Hassanein R, Tanner C (1990) Environmental risk factors in Parkinson's disease. *Neurology* **40**:1218-1221.

Kruger R, Kuhn W, Muller T, Woitalla D, Graeber M, Kosel S, Przuntek H, Epplen JT, Schols L, Riess O (1998) Ala30Pro mutation in the gene encoding alpha-synuclein in Parkinson's disease. *Nat Genet* **18**:106-108.

Langston JW, Forno LS, Tetrud J, Reeves AG, Kaplan JA, Karluk D (1999) Evidence of active nerve cell degeneration in the substantia nigra of humans years after 1-methyl-4-phenyl-1,2,3,6-tetrahydropyridine exposure. *Ann Neurol* **46**:598-605.

Lantos P (1997) Multiple system atrophy. *Brain Pathol* **7**:1293-1297.

Lee MK, Stirling W, Xu Y, Xu X, Qui D, Mandir AS, Dawson TM, Copeland NG, Jenkins NA, Price DL (2002) Human alpha-synuclein-harboring familial Parkinson's disease-linked Ala-53 --> Thr mutation causes neurodegenerative disease with alpha-synuclein aggregation in transgenic mice. *Proc Natl Acad Sci U S A* **99**:8968-8973.

Lee SJ, Kim SJ, Kim IK, Ko J, Jeong CS, Kim GH, Park C, Kang SO, Suh PG, Lee HS, Cha SS (2003) Crystal structures of human DJ-1 and Escherichia coli Hsp31, which share an evolutionarily conserved domain. *J Biol Chem* **278**:44552-44559.

Lee VM, Goedert M, Trojanowski JQ (2001) Neurodegenerative tauopathies. *Annu Rev Neurosci* **24**:1121-1159.

Lee VM, Giasson BI, Trojanowski JQ (2004) More than just two peas in a pod: common amyloidogenic properties of tau and alpha-synuclein in neurodegenerative diseases. *Trends Neurosci* **27**:129-134.

Lennox G, Lowe J, Landon M, Byrne EJ, Mayer RJ, Godwin-Austen RB (1989) Diffuse Lewy body disease: correlative neuropathology using anti-ubiquitin immunocytochemistry. *J Neurol Neurosurg Psychiatry* **52**:1236-1247.

Leroy E, Boyer R, Auburger G, Leube B, Ulm G, Mezey E, Harta G, Brownstein MJ, Jonnalagada S, Chernova T, Dehejia A, Lavedan C, Gasser T, Steinbach PJ, Wilkinson KD, Polymeropoulos MH (1998) The ubiquitin pathway in Parkinson's disease. *Nature* **395**:451-452.

Liu Y, Fallon L, Lashuel HA, Liu Z, Lansbury PT, Jr. (2002) The UCH-L1 gene encodes two opposing enzymatic activities that affect alpha-synuclein degradation and Parkinson's disease susceptibility. *Cell* **111**:209-218.

Maraganore DM, Hernandez DG, Singleton AB, Farrer MJ, McDonnell SK, Hutton ML, Hardy JA, Rocca WA (2001) Case-Control study of the extended tau gene haplotype in Parkinson's disease. *Ann Neurol* **50**:658-661.

Markopoulou K, Langston JW (1999) Candidate genes and Parkinson's disease: where to next? *Neurology* **53**:1382-1383.

Mattila PM, Rinne JO, Helenius H, Dickson DW, Roytta M (2000) Alpha-synuclein-immunoreactive cortical Lewy bodies are associated with cognitive impairment in Parkinson's disease. *Acta Neuropathol (Berl)* **100**:285-290.

McGeer PL, Itagaki S, Akiyama H, McGeer EG (1988) Rate of cell death in parkinsonism indicates active neuropathological process. *Ann Neurol* **24**:574-576.

Miller DW, Ahmad R, Hague S, Baptista MJ, Canet-Aviles R, McLendon C, Carter DM, Zhu PP, Stadler J, Chandran J, Klinefelter GR, Blackstone C, Cookson MR (2003) L166P mutant DJ-1, causative for recessive Parkinson's disease, is degraded through the ubiquitin-proteasome system. *J Biol Chem* **278**:36588-36595.

Mitsumoto A, Nakagawa Y, Takeuchi A, Okawa K, Iwamatsu A, Takanezawa Y (2001) Oxidized forms of peroxiredoxins and DJ-1 on two-dimensional gels increased in response to sublethal levels of paraquat. *Free Radic Res* **35**:301-310.

Mori H, Kondo T, Yokochi M, Matsumine H, Nakagawa-Hattori Y, Miyake T, Suda K, Mizuno Y (1998) Pathologic and biochemical studies of juvenile parkinsonism linked to chromosome 6q. *Neurology* **51**:890-892.

Nakajima A, Kataoka K, Hong M, Sakaguchi M, Huh NH (2003) BRPK, a novel protein kinase showing increased expression in mouse cancer cell lines with higher metastatic potential. *Cancer Lett* **201**:195-201.

Narhi L, Wood SJ, Steavenson S, Jiang Y, Wu GM, Anafi D, Kaufman SA, Martin F, Sitney K, Denis P, Louis JC, Wypych J, Biere AL, Citron M (1999) Both familial Parkinson's disease mutations accelerate alpha-synuclein aggregation. *J Biol Chem* **274**:9843-9846.

Neumann M, Kahle PJ, Giasson BI, Ozmen L, Borroni E, Spooren W, Muller V, Odoy S, Fujiwara H, Hasegawa M, Iwatsubo T, Trojanowski JQ, Kretzschmar HA, Haass C (2002) Misfolded proteinase K-resistant hyperphosphorylated alpha-synuclein in aged transgenic mice with locomotor deterioration and in human alpha-synucleinopathies. *J Clin Invest* **110**:1429-1439.

Newell KL, Boyer P, Gomez-Tortosa E, Hobbs W, Hedley-Whyte ET, Vonsattel JP, Hyman BT (1999) Alpha-synuclein immunoreactivity is present in axonal swellings in neuroaxonal dystrophy and acute traumatic brain injury. *J Neuropathol Exp Neurol* **58**:1263-1268.

Obeso JA, Rodriguez-Oroz MC, Rodriguez M, Lanciego JL, Artieda J, Gonzalo N, Olanow CW (2000) Pathophysiology of the basal ganglia in Parkinson's disease. *Trends Neurosci* **23**:S8-19.

Olzmann JA, Brown K, Wilkinson KD, Rees HD, Huai Q, Ke H, Levey AI, Li L, Chin LS (2004) Familial Parkinson's disease-associated L166P mutation disrupts DJ-1 protein folding and function. *J Biol Chem* **279**:8506-8515.

Pakkenberg B, Moller A, Gundersen HJ, Mouritzen Dam A, Pakkenberg H (1991) The absolute number of nerve cells in substantia nigra in normal subjects and in patients with Parkinson's disease estimated with an unbiased stereological method. *J Neurol Neurosurg Psychiatry* **54**:30-33.

Palacino JJ, Sagi D, Goldberg MS, Krauss S, Motz C, Wacker M, Klose J, Shen J (2004) Mitochondrial dysfunction and oxidative damage in parkin-deficient mice. *J Biol Chem* **279**:18614-18622.

Papadimitriou A, Veletza V, Hadjigeorgiou GM, Patrikiou A, Hirano M, Anastasopoulos I (1999) Mutated alpha-synuclein gene in two Greek kindreds with familial PD: incomplete penetrance? *Neurology* **52**:651-654.

Papp MI, Kahn JE, Lantos PL (1989) Glial cytoplasmic inclusions in the CNS of patients with multiple system atrophy (striatonigral degeneration, olivopontocerebellar atrophy and Shy-Drager syndrome). *J Neurol Sci* **94**:79-100.

Pastor P, Ezquerra M, Munoz E, Marti MJ, Blesa R, Tolosa E, Oliva R (2000) Significant association between the tau gene A0/A0 genotype and Parkinson's disease. *Ann Neurol* **47**:242-245.

Pesah Y, Pham T, Burgess H, Middlebrooks B, Verstreken P, Zhou Y, Harding M, Bellen H, Mardon G (2004) Drosophila parkin mutants have decreased mass and cell size and increased sensitivity to oxygen radical stress. *Development* **131**:2183-2194.

Polymeropoulos MH, Lavedan C, Leroy E, Ide SE, Dehejia A, Dutra A, Pike B, Root H, Rubenstein J, Boyer R, Stenroos ES, Chandrasekharappa S, Athanassiadou A, Papapetropoulos T, Johnson WG, Lazzarini AM, Duvoisin RC, Di Iorio G, Golbe LI, Nussbaum RL (1997) Mutation in the alpha-synuclein gene identified in families with Parkinson's disease. *Science* **276**:2045-2047.

Rajput AH, Uitti RJ, Stern W, Laverty W (1986) Early onset Parkinson's disease in Saskatchewan--environmental considerations for etiology. *Can J Neurol Sci* **13**:312-316.

Rajput AH, Uitti RJ, Stern W, Laverty W, O'Donnell K, O'Donnell D, Yuen WK, Dua A (1987) Geography, drinking water chemistry, pesticides and herbicides and the etiology of Parkinson's disease. *Can J Neurol Sci* **14**:414-418.

Reed LA, Wszolek ZK, Hutton M (2001) Phenotypic correlations in FTDP-17. *Neurobiol Aging* **22**:89-107.

Saha AR, Ninkina NN, Hanger DP, Anderton BH, Davies AM, Buchman VL (2000) Induction of neuronal death by alpha-synuclein. *Eur J Neurosci* **12**:3073-3077.

Saito Y, Kawai M, Inoue K, Sasaki R, Arai H, Nanba E, Kuzuhara S, Ihara Y, Kanazawa I, Murayama S (2000) Widespread expression of alpha-synuclein and tau immunoreactivity in Hallervorden-Spatz syndrome with protracted clinical course. *J Neurol Sci* **177**:48-59.

Samuel W, Galasko D, Masliah E, Hansen LA (1996) Neocortical lewy body counts correlate with dementia in the Lewy body variant of Alzheimer's disease. *J Neuropathol Exp Neurol* **55**:44-52.

Sasaki S, Shirata A, Yamane K, Iwata M (2004) Parkin-positive autosomal recessive juvenile Parkinsonism with alpha-synuclein-positive inclusions. *Neurology* **63**:678-682.

Shendelman S, Jonason A, Martinat C, Leete T, Abeliovich A (2004) DJ-1 is a redox-dependent molecular chaperone that inhibits alpha-synuclein aggregate formation. *PLoS Biol* **2**:e362.

Sherer TB, Kim JH, Betarbet R, Greenamyre JT (2003a) Subcutaneous rotenone exposure causes highly selective dopaminergic degeneration and alpha-synuclein aggregation. *Exp Neurol* **179**:9-16.

Sherer TB, Betarbet R, Testa CM, Seo BB, Richardson JR, Kim JH, Miller GW, Yagi T, Matsuno-Yagi A, Greenamyre JT (2003b) Mechanism of toxicity in rotenone models of Parkinson's disease. *J Neurosci* **23**:10756-10764.

Shimura H, Hattori N, Kubo S, Mizuno Y, Asakawa S, Minoshima S, Shimizu N, Iwai K, Chiba T, Tanaka K, Suzuki T (2000) Familial Parkinson disease gene product, parkin, is a ubiquitin-protein ligase. *Nat Genet* **25**:302-305.

Shoji M, Harigaya Y, Sasaki A, Ueda K, Ishiguro K, Matsubara E, Watanabe M, Ikeda M, Kanai M, Tomidokoro Y, Shizuka M, Amari M, Kosaka K, Nakazato Y, Okamoto K, Hirai S (2000) Accumulation of NACP/alpha-synuclein in lewy body disease and multiple system atrophy. *J Neurol Neurosurg Psychiatry* **68**:605-608.

Simuni T, Hurtig H (2000) Parkinson's disease: the clinical picture. In: Neurodegenerative dementias (Clark C, Trojanoswki J, eds). New York: McGraw-Hill.

Singleton AB, Farrer M, Johnson J, Singleton A, Hague S, Kachergus J, Hulihan M, Peuralinna T, Dutra A, Nussbaum R, Lincoln S, Crawley A, Hanson M, Maraganore D, Adler C, Cookson MR, Muenter M, Baptista M, Miller D, Blancato J, Hardy J, Gwinn-Hardy K (2003) alpha-Synuclein locus triplication causes Parkinson's disease. *Science* **302**:841.

Spillantini MG, Crowther RA, Jakes R, Hasegawa M, Goedert M (1998a) alpha-Synuclein in filamentous inclusions of Lewy bodies from Parkinson's disease and dementia with lewy bodies. *Proc Natl Acad Sci U S A* **95**:6469-6473.

Spillantini MG, Schmidt ML, Lee VM, Trojanowski JQ, Jakes R, Goedert M (1997) Alpha-synuclein in Lewy bodies. *Nature* **388**:839-840.

Spillantini MG, Crowther RA, Jakes R, Cairns NJ, Lantos PL, Goedert M (1998b) Filamentous alpha-synuclein inclusions link multiple system atrophy with Parkinson's disease and dementia with Lewy bodies. *Neurosci Lett* **251**:205-208.

Spira PJ, Sharpe DM, Halliday G, Cavanagh J, Nicholson GA (2001) Clinical and pathological features of a Parkinsonian syndrome in a family with an Ala53Thr alpha-synuclein mutation. *Ann Neurol* **49**:313-319.

Staropoli JF, McDermott C, Martinat C, Schulman B, Demireva E, Abeliovich A (2003) Parkin is a component of an SCF-like ubiquitin ligase complex and protects postmitotic neurons from kainate excitotoxicity. *Neuron* **37**:735-749.

Takeda A, Hashimoto M, Mallory M, Sundsumo M, Hansen L, Masliah E (2000) C-terminal alpha-synuclein immunoreactivity in structures other than Lewy bodies in neurodegenerative disorders. *Acta Neuropathol (Berl)* **99**:296-304.

Tanner CM, Chen B, Wang WZ, Peng ML, Liu ZL, Liang XL, Kao LC, Gilley DW, Schoenberg BS (1987) Environmental factors in the etiology of Parkinson's disease. *Can J Neurol Sci* **14**:419-423.

Tao X, Tong L (2003) Crystal structure of human DJ-1, a protein associated with early onset Parkinson's disease. *J Biol Chem* **278**:31372-31379.

Tu PH, Galvin JE, Baba M, Giasson B, Tomita T, Leight S, Nakajo S, Iwatsubo T, Trojanowski JQ, Lee VM (1998) Glial cytoplasmic inclusions in white matter oligodendrocytes of multiple system atrophy brains contain insoluble alpha-synuclein. *Ann Neurol* **44**:415-422.

Valente EM, Abou-Sleiman PM, Caputo V, Muqit MM, Harvey K, Gispert S, Ali Z, Del Turco D, Bentivoglio AR, Healy DG, Albanese A, Nussbaum R, Gonzalez-Maldonado R, Deller T, Salvi S, Cortelli P, Gilks WP, Latchman DS, Harvey RJ, Dallapiccola B, Auburger G, Wood NW (2004) Hereditary early-onset Parkinson's disease caused by mutations in PINK1. *Science* **304**:1158-1160.

Wakabayashi K, Yoshimoto M, Tsuji S, Takahashi H (1998) Alpha-synuclein immunoreactivity in glial cytoplasmic inclusions in multiple system atrophy. *Neurosci Lett* **249**:180-182.

Wakabayashi K, Yoshimoto M, Fukushima T, Koide R, Horikawa Y, Morita T, Takahashi H (1999) Widespread occurrence of alpha-synuclein/NACP-immunoreactive neuronal inclusions in juvenile and adult-onset Hallervorden-Spatz disease with Lewy bodies. *Neuropathol Appl Neurobiol* **25**:363-368.

Weinreb PH, Zhen W, Poon AW, Conway KA, Lansbury PT, Jr. (1996) NACP, a protein implicated in Alzheimer's disease and learning, is natively unfolded. *Biochemistry* **35**:13709-13715.

Wenning GK, Ben Shlomo Y, Magalhaes M, Daniel SE, Quinn NP (1994) Clinical features and natural history of multiple system atrophy. An analysis of 100 cases. *Brain* **117** (Pt 4):835-845.

Wenning GK, Tison F, Ben Shlomo Y, Daniel SE, Quinn NP (1997) Multiple system atrophy: a review of 203 pathologically proven cases. *Mov Disord* **12**:133-147.

Wilson MA, Collins JL, Hod Y, Ringe D, Petsko GA (2003) The 1.1-A resolution crystal structure of DJ-1, the protein mutated in autosomal recessive early onset Parkinson's disease. *Proc Natl Acad Sci U S A* **100**:9256-9261.

Withers GS, George JM, Banker GA, Clayton DF (1997) Delayed localization of synelfin (synuclein, NACP) to presynaptic terminals in cultured rat hippocampal neurons. *Brain Res Dev Brain Res* **99**:87-94.

Wszolek ZK, Pfeiffer RF, Tsuboi Y, Uitti RJ, McComb RD, Stoessl AJ, Strongosky AJ, Zimprich A, Muller-Myhsok B, Farrer MJ, Gasser T, Calne DB, Dickson DW (2004) Autosomal dominant parkinsonism associated with variable synuclein and tau pathology. *Neurology* **62**:1619-1622.

Xu J, Kao SY, Lee FJ, Song W, Jin LW, Yankner BA (2002) Dopamine-dependent neurotoxicity of alpha-synuclein: a mechanism for selective neurodegeneration in Parkinson disease. *Nat Med* **8**:600-606.

Yokota T, Sugawara K, Ito K, Takahashi R, Ariga H, Mizusawa H (2003) Down regulation of DJ-1 enhances cell death by oxidative stress, ER stress, and proteasome inhibition. *Biochem Biophys Res Commun* **312**:1342-1348.

Zarranz JJ, Alegre J, Gomez-Esteban JC, Lezcano E, Ros R, Ampuero I, Vidal L, Hoenicka J, Rodriguez O, Atares B, Llorens V, Gomez Tortosa E, del Ser T, Munoz DG, de Yebenes JG (2004) The new mutation, E46K, of alpha-synuclein causes Parkinson and Lewy body dementia. *Ann Neurol* **55**:164-173.

Zhang Y, Gao J, Chung KK, Huang H, Dawson VL, Dawson TM (2000) Parkin functions as an E2-dependent ubiquitin- protein ligase and promotes the degradation of the synaptic vesicle-associated protein, CDCrel-1. *Proc Natl Acad Sci U S A* **97**:13354-13359.

Zhou B, Westaway SK, Levinson B, Johnson MA, Gitschier J, Hayflick SJ (2001) A novel pantothenate kinase gene (PANK2) is defective in Hallervorden-Spatz syndrome. *Nat Genet* **28**:345-349.

Zhou W, Hurlbert MS, Schaack J, Prasad KN, Freed CR (2000) Overexpression of human alpha-synuclein causes dopamine neuron death in rat primary culture and immortalized mesencephalon-derived cells. *Brain Res* **866**:33-43.

Zimprich A, Biskup S, Leitner P, Lichtner P, Farrer M, Lincoln S, Kachergus J, Hulihan M, Uitti RJ, Calne DB, Stoessl AJ, Pfeiffer RF, Patenge N, Carbajal IC, Vieregge P, Asmus F, Muller-Myhsok B, Dickson DW, Meitinger T, Strom TM, Wszolek ZK, Gasser T (2004) Mutations in LRRK2 cause autosomal-dominant parkinsonism with pleomorphic pathology. *Neuron* **44**:601-607.

In: Recent Breakthroughs in Basal Ganglia Research
Editor: Erwan Bezard, pp. 351-360
ISBN: 1-59454-880-3

Part VIII: Mechanisms of L-DOPA-Induced Dyskinesia in Parkinson´s Disease: What do Animal Models Tell Us?

M. Angela Cenci
Basal Ganglia Pathophysiology Unit, Dept. of Experimental Medical Sciences, Lund University, BMC F11, S.221 84 Lund (Sweden)

Abstract

L-DOPA pharmacotherapy causes dyskinesia (abnormal involuntary movements) in the vast majority of patients with Parkinson´s disease (PD). Current notions attribute the development of dyskinesia to two main factors, viz. the loss of nigrostriatal dopamine (DA) projections and the maladaptive changes produced by L-DOPA at sites post-synaptic to the nigrostriatal neuron. Animal models allow us to dissect the impact of these two factors under strictly controlled conditions. Studies in animal models have taught us that DA denervation dramatically reduces the threshold for the dyskinetic action of L-DOPA, and that the chronic administration of L-DOPA causes an increasing incidence and severity of dyskinesia over the time. The development of dyskinesia during chronic L-DOPA treatment is closely linked with changes in gene and protein expression in striatal neurons. These changes are intimately related to an overactive signalling through DA D1 receptors. Recent experimental work shows that also presynaptic factors, involving the regulation of L-DOPA/DA release and metabolism in the brain, may show plasticity during L-DOPA treatment. Present and future studies in animal models of L-DOPA-induced dyskinesia will ultimately allow us to compile a unifying pathophysiological model taking into account both presynaptic and postsynaptic changes.

Abbreviations:

DA	Dopamine
L-DOPA	L-3,4-dihydroxyphenylalanine
LTP	Long-term potentiation
MPTP	1-methyl-4-phenyl-1,2,3,6-tetrahydropyridine
NMDA	N-methyl-D-aspartate
6-OHDA	6-hydroxydopamine
PD	Parkinson´s disease

Clinical Introduction

The finding of severe dopamine (DA) deficiency in the brains of Parkinson´s disease (PD) patients paved the way for the use of L-DOPA as a replacement therapy (Birkmayer and Hornykiewicz, 1961; Cotzias et al., 1967). Thanks to its symptomatic efficacy, L-DOPA revolutionized the treatment of PD and considerably improved the life quality of the patients. However, it soon became apparent that L-DOPA pharmacotherapy was not devoid of complications. If the amino acid was administered at high doses, and/or for a sufficiently long time, it did not just restore normal mobility, it also induced excessive abnormal movement. Already in the beginning of the seventies, the induction of abnormal involuntary movements (dyskinesia) was recognized as the most common, dose-limiting side effect of L-DOPA pharmacotherapy in PD (Cotzias et al., 1969; McDowell et al., 1970; Schwarz and Fahn, 1970; Calne et al., 1971). L-DOPA-induced dyskinesia usually affected all parts of the body in a choreiform pattern (which consists in rapid, nonrythmic movements flowing from one body part to another). These movements were most common and most severe at the time when a dose of L-DOPA was producing the maximal relief of parkinsonian motor symptoms (hence the term "peak-of-dose dyskinesia"). They could occur very early after the initiation of L-DOPA pharmacotherapy as a sign of L-DOPA overdosage, but their incidence and severity increased with continuing treatment (Cotzias et al., 1969; Calne et al., 1971; Duvoisin, 1974). Although early studies reported that dyskinesias would subside with a lowering of the L-DOPA dosage (Cotzias et al., 1969), it is now well established parkinsonian dyskinesias may have a more complex relationship to the timing and dosage of L-DOPA. Indeed, the current classification of L-DOPA-induced dyskinesias in PD distinguishes between three main patterns: (i) "On" dyskinesias, which appear when central levels of DA are high, and are ameliorated by downward dose adjustments; (ii) diphasic dyskinesias, which are seen when the central levels of DA are rising or declining, i.e. at the beginning and at the end the L-DOPA action cycle; (iii) "off" periods dyskinesia, which appear when central levels of DA are low, and which mainly affect the legs and feet (Luquin et al., 1992). These different patterns are usually associated with somewhat different physical manifestations (Luquin et al., 1992). By now, a number of clinical and epidemiological studies have identified the main risk factors for L-DOPA-induced dyskinesia, which are, young age at PD onset, severity and duration of PD, and dose and duration of L-DOPA treatment (see chapter by Manson and Schrag in this book).

1 Animal Models of L-DOPA-Induced Dyskinesia

The available animal models of L-DOPA-induced dyskinesia invariably reproduce the "on" or "peak-of-dose" pattern, which is due to high levels of DA receptor stimulation. To the best of our knowledge, animal models of diphasic and "off" dyskinesias have never been reported. In both nonhuman primate and rodent species, models of L-DOPA-induced dyskinesia are generally obtained by first lesioning the nigrostriatal DA projection, and then exposing the animals to single or bidaily administration of L-DOPA. The nigrostriatal lesion is produced using the neurotoxins, 1-methyl-4-phenyl-1,2,3,6-tetrahydropyridine (MPTP) and 6-hydroxydopamine (6-OHDA) in nonhuman primates and rodents, respectively (Bezard et al., 2001; Cenci et al., 2002a; Lundblad et al., 2004). Our knowledge about the possible mechanisms of dyskinesia is largely derived from these animal models. Both rodent (Winkler et al., 2002; Lundblad et al., 2004; Stefanova et al., 2004) and nonhuman primate studies (Jenner, 2000; Langston et al., 2000) concur to indicate that the incidence of L-DOPA-induced dyskinesia is conditioned by the extent of nigrostriatal DA damage and by the daily dose of L-DOPA. However, the nigrostriatal lesion does not seem to be an absolute requirement for the development of dyskinetic movements. Choreiform dyskinesias physically identical to those seen in DA denervated-animals have been reported in squirrel monkeys treated with therapeutic doses of L-DOPA (Togasaki et al., 2001), or in macaque monkeys treated with high doses of the drug (Zeng et al., 2000; Pearce et al., 2001). Thus, the available experimental evidence indicates that the nigrostriatal DA lesion increases the sensitivity to L-DOPA, but that dyskinesia would occur even in the presence of an intact DA innervation if the drug were to be administered at sufficiently high doses. The available evidence also indicates that the threshold dose of L-DOPA that is required to induce dyskinesia may vary greatly between phylogenetically related species (cf. Pearce et al., 2001 and Togasaki et al., 2001), and even among genetically similar individuals that belong to the same species and sustain the same type of nigrostriatal lesion (Cenci et al., 1998; Andersson et al., 1999; Lundblad et al., 2004).

Animal models allow us to clearly distinguish two sides to the dyskinetic action of L-DOPA. One side is the acute, "triggering" effect of single L-DOPA doses on the expression of dyskinetic movements. The other side is a delayed effect, which accounts both for the increasing incidence and severity of dyskinesia during chronic L-DOPA treatment and for the persistence of this movement disorder once it has developed. Indeed, an animal who has been rendered dyskinetic by a course of L-DOPA treatment will continue to exhibit abnormal movements when re-exposed to L-DOPA, despite prolonged periods of treatment discontinuation. This delayed effect is often referred to as a "priming" for dyskinesia. Distiguishing between the immediate dyskinetic effect and the priming action L-DOPA is very important. Indeed, treatments that suppress the physical expression of dyskinesia may not interfere with the neural plastic changes that sustain dyskinesia development. Conversely, some treatments that prevent the priming for dyskinesia would not necessarily attenuate its expression once this movement disorder is already established.

Our knowledge of the factors that determine either the expression or the development of dyskinesia is only partial. Research in the past years has identified several bits in a complex puzzle, but it is fair to say that an integrated picture is still lacking. In the following, we shall briefly review the most important findings obtained in animal models of L-DOPA-induced dyskinesia, and we shall also highlight important gaps in knowledge that will have to be addressed by further experimental work.

2 Presynaptic Factors

When discussing about the possible causes of dyskinesia it is common to distinguish between factors that act presynaptically or postsynaptically with respect to the dopaminoceptive neurons in the brain. Research in the past years had placed a lot of emphasis on the postsynaptic changes produced by L-DOPA, but more recent work is bringing presynaptic factors back into the limelight. In particular, it is becoming increasingly clear that large fluctuations in central levels of L-DOPA and DA have a direct hand in both the expression of dyskinesia and the priming process that is set in motion by L-DOPA. In a recent microdialysis study (Carta et al., 2005) we have compared the peripheral and central levels of L-DOPA in 6-OHDA lesioned rats that had previously sustained a course of chronic L-DOPA treatment. During the treatment, some rats developed abnormal involuntary movements, whereas others did not. On the last day of L-DOPA treatment, we measured the levels of this amino acid both in the venous blood and in the striatal extracellular fluid following an intraperitoneal injection of L-DOPA. Blood levels of this aminoacid did not differ between dyskinetic and nondyskinetic cases. However, L-DOPA levels in the striatal extracellular fluid were several-fold larger in the dyskinetic group, and their temporal rise and decline matched perfectly the time course of the L-DOPA-induced abrnormal involuntary movements. Moreover, even the rats previously classified as "non-dyskinetic" exhibited AIMs when a sufficiently large concentration of L-DOPA was infused in the striatum through the microdialysis probe. This study demonstrates the causal link between a surge in striatal L-DOPA levels and the expression of dyskinesia. In keeping with these data, several lines of evidence indicate that fluctuations in central L-DOPA levels are implicated in neural plastic changes at the basis of the dyskinesia-priming process. One line of evidence comes from studies comparing the biochemical effects of bidaily L-DOPA injections with those of continuous i.p. infusion of L-DOPA in 6-OHDA lesioned rats. In one study, continuous L-DOPA infusion for 21 days could reverse changes in brain peptide levels produced by the 6-OHDA lesion without causing additional biochemical abnormalities (Engber et al., 1992). By contrast, rats receiving intermittent injections of L-DOPA exhibited an upregulation of dynorphin in the striatonigral pathway (Engber et al., 1992). We now know that this type of biochemical change is linked to the development of dyskinesia (see below). A more direct demonstration of a causal role for central L-DOPA fluctuations in the dyskinesia priming process has recently been obtained using a viral vector-based approach for continuous L-DOPA delivery (Carlsson et al., 2005). The dyskinetic response to peripheral L-DOPA administration, and the associated molecular changes, were completely reversed in rats receiving intrastriatal injections of viral vectors that achieved a local production of L-DOPA at high and stable levels (Carlsson et al., 2005). Pharmacological studies in MPTP-intoxicated nonhuman primates further support the crucial role of intermittent DA receptor stimulation in the genesis of dyskinesia. These studies have shown that long-acting DA receptor agonists used in the treatment of PD do not induce dyskinesia if they are administered to L-DOPA-naive animals (Grondin et al., 1996; Pearce et al., 1998). It was initially debated whether the large dyskinesiogenic potential of L-DOPA versus the DA agonists is due to the fact that the former stimulates all subtypes of DA receptors whereas the DA agonists currently used in the treatment of PD are generally more active on the D2/D3 receptor class. However, experiments performed by Bedard and collaborators have shown that even D2/3 agonists can induce dyskinesia in monkeys if their half-life is very short (Bedard et al., 1992; Blanchet et al., 1995a; 1995b).

3 THE DOPAMINE RECEPTORS

Denervation-induced supersensitivity of striatal DA receptors has been suggested to play a causal role in L-DOPA-induced dyskinesia (Klawans et al., 1977; Marconi et al., 1994), and the involvement of D1- or D2-type DA receptors has been a matter of ample debate (Bedard et al., 1992; Blanchet et al., 1995a; 1995b). Post-mortem studies on striatal D1 and D2 binding sites in PD have yielded conflicting results (Lee et al., 1978; Seeman et al., 1987; Cortes et al., 1989; Rinne et al., 1991). In a positron-emission tomography study (Turjanski et al., 1997), in vivo binding densities at striatal D1 and D2 receptors were found not to differ between PD patients suffering from L-DOPA-induced dyskinesia compared to non-dyskinetic cases. Studies in patients cannot however address functional changes in the DA receptor status that may have a great impact on the postsynaptic response to L-DOPA. In particular, DA receptor supersensitivity may involve an increased efficiency of signal-transduction in the absence of receptor upregulation. It is known, for example, that DA-denervating lesions cause supersensitivity of D1 receptor-mediated gene induction in striatal neurons (Berke et al., 1998), although the striatal expression of D1 receptors is either unaltered or downregulated (Gerfen et al., 1990). Recently, Aubert et al., (2005) carried out a very systematic and elegant study of striatal D1 and D2 receptor expression in a monkey model of L-DOPA-induced dyskinesia. Binding densities at D1 and D2 receptors did not differ significantly between dyskinetic and nondyskinetic animals. However, the sensitivity of D1 receptors, measured by D1 agonist-induced [35S]GTPγS binding, was greatly augmented in the dyskinetic group (Aubert et al., 2005). These results are in agreement with data obtained in the rat model of L-DOPA-induced dyskinesia, which indicate that changes in gene expression associated with dyskinesia are particularly prominent in the D1 receptor-rich population of striatal efferent neurons (the so-called "direct pathway" neurons; Andersson et al., 1999; Konradi et al., 2004). Another type of regulation of D1 receptor function occurs at the level of intracellular trafficking. It has been shown that treatment with L-DOPA induces an internalization of D1 receptors in striatal neurons (Muriel et al., 1999). By contrast, the subcellular localization of D1 receptors in the striatum is not affected by long-acting DA agonists, which do not induce dyskinesia (Muriel et al., 2002). The functional consequences of D1 receptor internalization are presently unclear, and the relevance of this phenomenon to the motor complications of L-DOPA pharmacotherapy remains a matter of debate. Dopamine D3 receptors have also been suggested to play an important role in dyskinesia. In 6-OHDA lesioned rats, high doses of L-DOPA induce an ectopic expression of D3 receptors in the dorsal striatum, and this effect parallels the development of rotational sensitization (Bordet et al., 1997). Bezard et al., (2003) found a large upregulation of D3 receptor binding in the putamen and the internal pallidal segment in MPTP-intoxicated macaques affected by L-DOPA-induced dyskinesia. In this animal model, dyskinesia was attenuated by a D3 receptor-selective partial agonist that did not cause a worsening of PD-like motor symptoms (Bezard et al., 2003). These data implicated D3 receptor overexpression as an important factor in the genesis of L-DOPA-induced dyskinesia. However, other nonhuman primates models of L-DOPA-induced dyskinesia do not show striatal upregulation of the D3 receptor above normal levels (Hurley et al., 1996a; Quik et al., 2000).

4 GLUTAMATE-DEPENDENT PLASTICITY

Studies in 6-OHDA-lesioned rats have shown that extracellular glutamate levels are increased in the striatum following DA denervation and dyskinesiogenic L-DOPA treatment (Robelet et al., 2004). The mechanisms underlying this increase may be both local and circuit-related. The striatum receives a massive, topographically organized glutamatergic input from the cerebral cortex (Alexander and Crutcher, 1990), and different lines of evidence point to cortical overactivity as a major pathophysiological factor in L-DOPA-induced dyskinesia (Rascol et al., 1998; Bezard et al., 2001; Johansson et al., 2001). A functional imaging study in PD patients showed that the activation of motor and premotor cortices during the execution of simple movements is greatly enhanced in patients suffering from L-DOPA-induced dyskinesia compared to non-dyskinetic cases (Rascol et al., 1998). In addition to its immediate impact on the coding of movement-related information, an enhanced phasic discharge of cortical neurons may cause long-lasting changes in the efficiency of corticostriatal transmission (Calabresi et al., 1996). Prompted by this hypothesis, Picconi et al., (2003) examined for alterations in the inducibility and/or expression of corticostriatal LTP in dyskinetic rats. LTP was induced by high frequency stimulation of the corticostriatal pathway in a striatal slice preparation (Picconi et al., 2003). The 6-OHDA lesion caused a loss of LTP, and chronic treatment with L-DOPA restored the inducibility of LTP in both dyskinetic and nondyskinetic rats. However, dyskinetic animals showed a specific alteration of synaptic plasticity, whereby LTP could not be reversed by low-frequency stimulation of the corticostriatal pathway (Picconi et al., 2003). This study indicates that L-DOPA-induced dyskinesia is associated with an irreversible, activity-dependent potentiation of corticostriatal synapses. It is tempting to speculate that such an abnormality disrupts the ability of striatal neurons to sort out relevant signals, and discard irrelevant ones, during the coding of movement-related information.

The importance of glutamate transmission in the pathophysiology of L-DOPA-induced dyskinesia is supported by the antidyskinetic efficacy of ionotropic glutamate receptor antagonists both in animal models of PD and in PD patients (Chase et al., 2000; Brown et al., 2002). Indeed, amantadine, which exerts non-competitive antagonism at the NMDA receptor channel (Kornhuber et al., 1994) is currently used for the treatment of L-DOPA-induced dyskinesia in many clinics (Blanchet et al., 2003). Additional compounds with antagonistic properties at NMDA (Papa and Chase, 1996) or AMPA receptors (Konitsiotis et al., 2000) have been found to exert antidyskinetic actions in animal models of PD and/or in small clinical trials in PD patients (Verhagen Metman et al., 1998a; 1998b). The antidyskinetic action of glutamate receptor antagonists has traditionally been interpreted as being due to a normalization of the striatal response to glutamate (Chase, 1998; Chase and Oh, 2000a). Indeed, striatal neurons show an increased phosphorylation of NMDA receptor subunits in chronically L-DOPA treated animals, possibly leading to a potentiation of NMDA receptor-gated currents (Chase, 1998; Chase and Oh, 2000a). The striatum is obviously the main dopaminoceptive component of the basal ganglia, and any functional abnormality at this level is bound to have an impact on downstream targets. However, very little work has been done in order to try and ascertain whether chronic L-DOPA treatment causes maladaptive neuroplasticity also in the basal ganglia output nuclei and/or in the subthalamic nucleus. These structures show prominent alterations in neural firing in L-DOPA-induced dyskinesia (see chapter by Chakraborty et al., this book) and are themselves targets of dense glutamatergic projections (Monakow et al., 1978; Parent and Hazrati, 1995).

5 Changes in Striatal Gene and Protein Expression

The recent introduction of a model of L-DOPA-induced abnormal involuntary movements (AIMs) in rodents has greatly facilitated studying the mechanisms of L-DOPA-induced dyskinesia at the molecular level (for review see Cenci et al., 2002a; Cenci and Lundblad, 2005). One powerful feature of the rodent AIMs model is the possibility to define different response profiles to L-DOPA among the treated subjects in a way similar to that used in clinical studies. Thus, the model allows for a comparison between subjects afflicted by abnormal movements and subjects that show motor improvement in response to L-DOPA in the absence of dyskinetic adverse effects. Thus far, we have disclosed a strong association between L-DOPA-induced dyskinesia and the upregulation of prodynorphin mRNA (also called preproenkephalin-B mRNA) and ΔFosB-like transcription factor proteins in the motor part of the striatum (Cenci et al., 1998; Andersson et al., 1999; Andersson et al., 2001; Westin et al., 2001; Winkler et al., 2002). These changes in gene and protein expression are very long-lived, possibly accounting for the persistence of a dyskinesia-prone state even after discontinuation of L-DOPA treatment (Andersson et al., 2003). Interestingly, similar changes have been detected both in nonhuman primate models of L-DOPA-induced dyskinesia (Doucet et al., 1996; Tel et al., 2002; Henry et al., 2003) and in the basal ganglia of dyskinetic PD patients as examined post-mortem (Cenci et al., 2002b; Henry et al., 2003). In the rat model, we were able to establish that the L-DOPA-induced upregulation of ΔFosB is causal to the development of dyskinesia. Indeed, the gradual increase in AIM severity during a course of chronic L-DOPA treatment was prevented by antisense-mediated knockdown of ΔFosB in the striatum, which also blocked the L-DOPA-induced upregulation of prodynorphin mRNA (Andersson et al., 1999). Protein-DNA binding activities in the dyskinetic striatum were studied using electrophoretic mobility-shift and supershift assays (Andersson et al., 2001). This analysis showed that ΔFosB and JunD were the main proteins binding to both cyclic AMP-response elements (CRE) and activator protein 1 (AP-1) enhancers, which are present in the promoter sequence of many genes. These data suggested that a potentially large number of genes was likely to show altered levels of expression in dyskinetic rats. In order to obtain a complete picture of gene expression profile associated with dyskinesia we carried out a gene chip array analysis of striatal mRNA from dyskinetic rats and non-dyskinetic cases (Konradi et al., 2004). The dyskinetic group exhibited significant alterations in the expression of over 500 genes, among which, twice as many genes were upregulated than downregulated. Some functional categories of genes showed particular prominent upregulations, in particular, genes coding for ion-transporting ATPases; genes coding for synaptic components (vesicles and scaffold proteins), and genes coding for GABA receptor subunits and GABA transporters. By contrast, consistent downregulations were seen among genes involved in energy production (e.g. creatine pathaway), and among genes coding for ribosomal subunits. Taken together, the results of this study indicated that striatal GABAergic neurons had an increased transcriptional and synaptic activity in dyskinetic animals, and that genes involved in ion homeostasis, Ca2+ -dependent signaling, and structural plasticity were upregulated. Unexpectedly, the gene expression patterns also implied that the dyskinetic striatum had a decreased capacity for energy production and reduced cellular defenses against oxidative stress. We are now verifying this unheralded findings using alternative approaches. We are also attempting to define the protein expression profile associated with dyskinesia using proteomics screening methods, such as two dimensional gel electrophoresis followed by mass-spectrometric analyses.

6 Unresolved Issues and Future Developments

This short review was not intended to offer a coherent pathophysiological explanation of L-DOPA-induced dyskinesia, which wouldn´t be so easy to propose. At present we can only hold for certain that L-DOPA-induced dyskinesia is due to fluctuations in central DA levels, with ensuing pulsatile stimulation of brain DA receptors. The cascade of molecular and cellular adaptations that translate this event into a dyskinetic motor output is largely unknown. Much more work will be required in order to clarify (i) why the sensitivity to DA receptor stimulation is augmented in dyskinetic animals; (ii) how changes in gene and protein expression come about in the "dyskinetic" striatum; (iii) which molecular changes play a key role in causing altered patterns of firing and oscillatory activities in the motor loops of the basal ganglia. (cf Chakraborty et al., this book). Finally, clinical (De la Fuente-Fernandez et al., 2004a) as well as experimental observations call for further studies on the central pharmacokinetics and turnover of L-DOPA in dyskinesia. These studies will hopefully provide a handle to the different individual susceptibility to L-DOPA-induced dyskinesia that is encountered not only among patients, but also under the strictly controlled conditions of our experimental models.

References

Alexander GE, Crutcher MD (1990) Functional architecture of basal ganglia circuits: neural substrates of parallel processing. *Trends Neurosci* **13**:266-271.

Andersson M, Hilbertson A, Cenci MA (1999) Striatal fosB expression is causally linked with l-DOPA-induced abnormal involuntary movements and the associated upregulation of striatal prodynorphin mRNA in a rat model of Parkinson's disease. *Neurobiol Dis* **6**:461-474.

Andersson M, Konradi C, Cenci MA (2001) cAMP response element-binding protein is required for dopamine-dependent gene expression in the intact but not the dopamine-denervated striatum. *J Neurosci* **21**:9930-9943.

Andersson M, Westin JE, Cenci MA (2003) Time course of striatal DeltaFosB-like immunoreactivity and prodynorphin mRNA levels after discontinuation of chronic dopaminomimetic treatment. *Eur J Neurosci* **17**:661-666.

Aubert I, Guigoni C, Hakansson K, Li Q, Dovero S, Barthe N, Bioulac BH, Gross CE, Fisone G, Bloch B, Bezard E (2005) Increased D1 dopamine receptor signaling in levodopa-induced dyskinesia. *Ann Neurol* **57**:17-26.

Bedard PJ, Mancilla BG, Blanchette P, Gagnon C, Di Paolo T (1992) Levodopa-induced dyskinesia: facts and fancy. What does the MPTP monkey model tell us? *Can J Neurol Sci* **19**:134-137.

Berke JD, Paletzki RF, Aronson GJ, Hyman SE, Gerfen CR (1998) A complex program of striatal gene expression induced by dopaminergic stimulation. *J Neurosci* **18**:5301-5310.

Bezard E, Brotchie JM, Gross CE (2001) Pathophysiology of levodopa-induced dyskinesia: potential for new therapies. *Nat Rev Neurosci* **2**:577-588.

Bezard E, Ferry S, Mach U, Stark H, Leriche L, Boraud T, Gross C, Sokoloff P (2003) Attenuation of levodopa-induced dyskinesia by normalizing dopamine D3 receptor function. *Nat Med* **9**:762-767.

Birkmayer W, Hornykiewicz O (1961) Der L-Dioxyphenylalanin (=DOPA)-Effekt bei der Parkinson-akinase. *Wien Klin Wschr* **73**:787-788.

Blanchet PJ, Gomez-Mancilla B, Bedard PJ (1995a) DOPA-induced "peak dose" dyskinesia: clues implicating D2 receptor-mediated mechanisms using dopaminergic agonists in MPTP monkeys. *J Neural Transm Suppl* **45**:103-112.

Blanchet PJ, Metman LV, Chase TN (2003) Renaissance of amantadine in the treatment of Parkinson's disease. *Adv Neurol* **91**:251-257.

Blanchet PJ, Gomez-Mancilla B, Di Paolo T, Bedard PJ (1995b) Is striatal dopaminergic receptor imbalance responsible for levodopa-induced dyskinesia? *Fundam Clin Pharmacol* **9**:434-442.

Bordet R, Ridray S, Carboni S, Diaz J, Sokoloff P, Schwartz JC (1997) Induction of dopamine D3 receptor expression as a mechanism of behavioral sensitization to levodopa. *Proc Natl Acad Sci U S A* **94**:3363-3367.

Brown T, de Groote C, Brotchie JM (2002) Recent advances in the treatment of L-DOPA-induced dyskinesia. *Drugs* **5**:454-468.

Calabresi P, Pisani A, Mercuri NB, Bernardi G (1996) The corticostriatal projection: from synaptic plasticity to dysfunctions of the basal ganglia. *Trends Neurosci* **19**:19-24.

Calne DB, Reid JL, Vakil SD, Pallis C (1971) Problems with L-dopa therapy. *Clin Med* **78**:21-23.

Carlsson T, Winkler C, Burger C, Muzyczka N, Mandel RJ, Cenci A, Bjorklund A, Kirik D (2005) Reversal of dyskinesias in an animal model of Parkinson's disease by continuous L-DOPA delivery using rAAV vectors. *Brain* **128**:559-569.

Carta M, Lundblad M, Stancampiano R, Fadda F, Cenci MA (2005) High striatal levels of L-DOPA are associated with the appearance of abnormal involuntary movements in the rat 6-hydroxydopamine model. manuscript.

Cenci MA, Lundblad M (2005) Utility of 6-hydroxydopamine lesioned rats in the preclinical screening of novel treatments for Parkinson disease. In: *Animal Models of Movement Disorders* (LeDoux M, ed), pp 193-208. San Diego, CA: Elsevier Academic Press.

Cenci MA, Lee CS, Bjorklund A (1998) L-DOPA-induced dyskinesia in the rat is associated with striatal overexpression of prodynorphin- and glutamic acid decarboxylase mRNA. *Eur J Neurosci* **10**:2694-2706.

Cenci MA, Whishaw IQ, Schallert T (2002a) Animal models of neurological deficits: how relevant is the rat? *Nat Rev Neurosci* **3**:574-579.

Cenci MA, Andersson M, Daniel SE, Kingsbury AE, Kilford L, Lees AJ (2002b) Striatal expression of FosB-related transcription factors in human Parkinson's disease: Relation to L-DOPA-induced dyskinesia. *Mov Disord 17/Suppl.* **5**:P284.

Chase TN (1998) Levodopa therapy: consequences of the nonphysiologic replacement of dopamine. *Neurology* **50**:S17-25.

Chase TN, Oh JD (2000a) Striatal dopamine- and glutamate-mediated dysregulation in experimental parkinsonism. *Trends Neurosci* **23**:S86-91.

Chase TN, Oh JD, Konitsiotis S (2000) Antiparkinsonian and antidyskinetic activity of drugs targeting central glutamatergic mechanisms. *J Neurol* **247***(Suppl 2)*:1136-1142.

Cortes R, Camps M, Gueye B, Probst A, Palacios JM (1989) Dopamine receptors in human brain: autoradiographic distribution of D1 and D2 sites in Parkinson syndrome of different etiology. *Brain Res* **483**:30-38.

Cotzias G, Van Woert M, Schiffer L (1967) Aromatic amino acids and modification of parkinsonism. *N Engl J Med* **276**:374-379.

Cotzias G, Papavasiliou P, Gellene R (1969) Modification of parkinsonism-chronic treatment with L-dopa. *N Engl J Med* **280**:337-345.

De la Fuente-Fernandez R, Schulzer M, Mak E, Calne DB, Stoessl AJ (2004a) Presynaptic mechanisms of motor fluctuations in Parkinson's disease: a probabilistic model. *Brain* **127**:888-899.

Doucet JP, Nakabeppu Y, Bedard PJ, Hope BT, Nestler EJ, Jasmin BJ, Chen JS, Iadarola MJ, St-Jean M, Wigle N, Blanchet P, Grondin R, Robertson GS (1996) Chronic alterations in dopaminergic neurotransmission produce a persistent elevation of deltaFosB-like protein(s) in both the rodent and primate striatum. *Eur J Neurosci* **8**:365-381.

Duvoisin RC (1974) Hyperkinetic reactions with L-DOPA. In: Current concepts in the treatment of parkinsonism (Yahr MD, ed), pp 203-210. New York: Raven Press.

Engber TM, Boldry RC, Kuo S, Chase TN (1992) Dopaminergic modulation of striatal neuropeptides: differential effects of D1 and D2 receptor stimulation on somatostatin, neuropeptide Y, neurotensin, dynorphin and enkephalin. *Brain Res* **581**:261-268.

Gerfen CR, Engber TM, Mahan LC, Susel Z, Chase TN, Monsma FJ, Jr., Sibley DR (1990) D1 and D2 dopamine receptor-regulated gene expression of striatonigral and striatopallidal neurons. *Science* **250**:1429-1432.

Grondin R, Goulet M, Di Paolo T, Bedard PJ (1996) Cabergoline, a long-acting dopamine D2-like receptor agonist, produces a sustained antiparkinsonian effect with transient dyskinesias in parkinsonian drug-naive primates. *Brain Res* **735**:298-306.

Henry B, Duty S, Fox SH, Crossman AR, Brotchie JM (2003) Increased striatal pre-proenkephalin B expression is associated with dyskinesia in Parkinson's disease. *Exp Neurol* **183**:458-468.

Hurley MJ, Jolkkonen J, Stubbs CM, Jenner P, Marsden CD (1996a) Dopamine D3 receptors in the basal ganglia of the common marmoset and following MPTP and L-DOPA treatment. *Brain Res* **709**:259-264.

Jenner P (2000) Factors influencing the onset and persistence of dyskinesia in MPTP-treated primates. *Ann Neurol* **47**:S90-99; discussion S99-104.

Johansson PA, Andersson M, Andersson KE, Cenci MA (2001) Alterations in cortical and basal ganglia levels of opioid receptor binding in a rat model of l-DOPA-induced dyskinesia. *Neurobiol Dis* **8**:220-239.

Klawans HL, Goetz C, Nausieda PA, Weiner WJ (1977) Levodopa-induced dopamine receptor hypersensitivity. *Trans Am Neurol Assoc* **102**:80-83.

Konitsiotis S, Blanchet PJ, Verhagen Metman L, Lamers E, Chase TN (2000) AMPA receptor blockade improves levodopa-induced dyskinesia in MPTP monkeys. *Neurology* **54**:1589-1595.

Konradi C, Westin JE, Carta M, Eaton ME, Kuter K, Dekundy A, Lundblad M, Cenci MA (2004) Transcriptome analysis in a rat model of L-DOPA-induced dyskinesia. *Neurobiol Dis* **17**:219-236.

Kornhuber J, Weller M, Schoppmeyer K, Riederer P (1994) Amantadine and memantine are NMDA receptor antagonists with neuroprotective properties. *J Neural Transm Suppl* **43**:91-104.

Langston JW, Quik M, Petzinger G, Jakowec M, Di Monte DA (2000) Investigating levodopa-induced dyskinesias in the parkinsonian primate. *Ann Neurol* **47**:S79-89.

Lee T, Seeman P, Rajput A, Farley IJ, Hornykiewicz O (1978) Receptor basis for dopaminergic supersensitivity in Parkinson´s disease. *Nature* **273**:59-61.

Lundblad M, Picconi B, Lindgren H, Cenci MA (2004) A model of L-DOPA-induced dyskinesia in 6-hydroxydopamine lesioned mice: relation to motor and cellular parameters of nigrostriatal function. *Neurobiol Dis* **16**:110-123.

Luquin MR, Scipioni O, Vaamonde J, Gershanik O, Obeso JA (1992) Levodopa-induced dyskinesias in Parkinson's disease: clinical and pharmacological classification. *Mov Disord* **7**:117-124.

Marconi R, Lefebvre-Caparros D, Bonnet AM, Vidailhet M, Dubois B, Agid Y (1994) Levodopa-induced dyskinesias in Parkinson's disease phenomenology and pathophysiology. *Mov Disord* **9**:2-12.

McDowell F, Lee J, Swift T, al. e (1970) Treatment of Parkinson´s syndrome with dihydroxyphenylalanine (levodopa). *Ann Intern Med* **72**:29-35.

Monakow KH, Akert K, Kunzle H (1978) Projections of the precentral motor cortex and other cortical areas of the frontal lobe to the subthalamic nucleus in the monkey. *Exp Brain Res* **33**:395-403.

Muriel MP, Orieux G, Hirsch EC (2002) Levodopa but not ropinirole induces an internalization of D1 dopamine receptors in parkinsonian rats. *Mov Disord* **17**:1174-1179.

Muriel MP, Bernard V, Levey AI, Laribi O, Abrous DN, Agid Y, Bloch B, Hirsch EC (1999) Levodopa induces a cytoplasmic localization of D1 dopamine receptors in striatal neurons in Parkinson's disease. *Ann Neurol* **46**:103-111.

Papa SM, Chase TN (1996) Levodopa-induced dyskinesias improved by a glutamate antagonist in Parkinsonian monkeys. *Ann Neurol* **39**:574-578.

Parent A, Hazrati LN (1995) Functional anatomy of the basal ganglia. II. The place of subthalamic nucleus and external pallidum in basal ganglia circuitry. *Brain Res Brain Res Rev* **20**:128-154.

Pearce RK, Banerji T, Jenner P, Marsden CD (1998) De novo administration of ropinirole and bromocriptine induces less dyskinesia than L-dopa in the MPTP-treated marmoset. *Mov Disord* **13**:234-241.

Pearce RK, Heikkila M, Linden IB, Jenner P (2001) L-dopa induces dyskinesia in normal monkeys: behavioural and pharmacokinetic observations. *Psychopharmacology (Berl)* **156**:402-409.

Picconi B, Centonze D, Hakansson K, Bernardi G, Greengard P, Fisone G, Cenci MA, Calabresi P (2003) Loss of bidirectional striatal synaptic plasticity in L-DOPA-induced dyskinesia. *Nat Neurosci* **6**:501-506.

Quik M, Police S, He L, Di Monte DA, Langston JW (2000) Expression of D(3) receptor messenger RNA and binding sites in monkey striatum and substantia nigra after nigrostriatal degeneration: effect of levodopa treatment. *Neuroscience* **98**:263-273.

Rascol O, Sabatini U, Brefel C, Fabre N, Rai S, Senard JM, Celsis P, Viallard G, Montastruc JL, Chollet F (1998) Cortical motor overactivation in parkinsonian patients with L-dopa-induced peak-dose dyskinesia. *Brain* **121** *(Pt 3)*:527-533.

Rinne JO, Laihinen A, Lonnberg P, Marjamaki P, Rinne UK (1991) A post-mortem study on striatal dopamine receptors in Parkinson's disease. *Brain Res* **556**:117-122.

Robelet S, Melon C, Guillet B, Salin P, Kerkerian-Le Goff L (2004) Chronic L-DOPA treatment increases extracellular glutamate levels and GLT1 expression in the basal ganglia in a rat model of Parkinson's disease. *Eur J Neurosci* **20**:1255-1266.

Schwarz GA, Fahn S (1970) Newer medical treatments in parkinsonism. *Med Clin North Am* **54**:773-785.

Seeman P, Bzowej NH, Guan HC, Bergeron C, Reynolds GP, Bird ED, Riederer P, Jellinger K, Tourtellotte WW (1987) Human brain D1 and D2 dopamine receptors in schizophrenia, Alzheimer's, Parkinson's, and Huntington's diseases. *Neuropsychopharmacology* **1**:5-15.

Stefanova N, Lundblad M, Tison F, Poewe W, Cenci MA, Wenning GK (2004) Effects of pulsatile L-DOPA treatment in the double lesion rat model of striatonigral degeneration (multiple system atrophy). *Neurobiol Dis* **15**:630-639.

Tel BC, Zeng BY, Cannizzaro C, Pearce RK, Rose S, Jenner P (2002) Alterations in striatal neuropeptide mRNA produced by repeated administration of L-DOPA, ropinirole or bromocriptine correlate with dyskinesia induction in MPTP-treated common marmosets. *Neuroscience* **115**:1047-1058.

Togasaki DM, Tan L, Protell P, Di Monte DA, Quik M, Langston JW (2001) Levodopa induces dyskinesias in normal squirrel monkeys. *Ann Neurol* **50**:254-257.

Turjanski N, Lees AJ, Brooks DJ (1997) In vivo studies on striatal dopamine D1 and D2 site binding in L-dopa-treated Parkinson's disease patients with and without dyskinesias. *Neurology* **49**:717-723.

Verhagen Metman L, Del Dotto P, Natte R, van den Munckhof P, Chase TN (1998b) Dextromethorphan improves levodopa-induced dyskinesias in Parkinson's disease. *Neurology* **51**:203-206.

Verhagen Metman L, Blanchet PJ, van den Munckhof P, Del Dotto P, Natte R, Chase TN (1998a) A trial of dextromethorphan in parkinsonian patients with motor response complications. *Mov Disord* **13**:414-417.

Westin JE, Andersson M, Lundblad M, Cenci MA (2001) Persistent changes in striatal gene expression induced by long-term L-DOPA treatment in a rat model of Parkinson's disease. *Eur J Neurosci* **14**:1171-1176.

Winkler C, Kirik D, Bjorklund A, Cenci MA (2002) L-DOPA-induced dyskinesia in the intrastriatal 6-hydroxydopamine model of parkinson's disease: relation to motor and cellular parameters of nigrostriatal function. *Neurobiol Dis* **10**:165-186.

Zeng BY, Pearce RK, MacKenzie GM, Jenner P (2000) Alterations in preproenkephalin and adenosine-2a receptor mRNA, but not preprotachykinin mRNA correlate with occurrence of dyskinesia in normal monkeys chronically treated with L-DOPA. *Eur J Neurosci* **12**:1096-1104.

In: Recent Breakthroughs in Basal Ganglia Research
Editor: Erwan Bezard, pp. 361-367
ISBN: 1-59454-880-3

Chapter 28

THE PATHOPHYSIOLOGY OF DYSKINESIA: ELECTROPHYSIOLOGICAL RECORDINGS AND SYSTEM-LEVEL CHANGES IN CORTICO-BASAL GANGLIA CIRCUIT

Subojit Chakraborty, Celine Guigoni, Christian E. Gross, Erwan Bezard and Thomas Boraud
CNRS UMR 5543, Université Victor Segalen, 146 rue Léo Saignat, 33076 Bordeaux, France

ABSTRACT

The mechanisms underlying the evolution of L-dopa-induced dyskinesia (LID) in Parkinson's disease (PD) are only beginning to be uncovered and are far from being fully understood. Although of paramount importance, electrophysiological recordings and system-level metabolic changes in the cortico-basal ganglia circuit have been parsimoniously used. Still, the few studies that have been conducted yield important clues about the pathophysiology of dyskinesia, and beyond, that, about the information processing by the basal ganglia circuit.

ABBREVIATIONS

2-DG	2-deoxyglucose
DA	Dopamine
Gpe	external part of the Globus Pallidus
Gpi	internal part of the Globus Pallidus
L-dopa	L-3,4-dihydroxyphenylalanine
LFP	local field potential
LID	L-dopa-induced dyskinesia
MPTP	1-methyl-4-phenyl-1,2,3,6-tetrahydropyridine
6-OHDA	6-hydroxydopamine
PD	Parkinson's disease
SNc	substantia nigra pars compacta
SNr	substantia nigra pars reticulata
STN	subthalamic nucleus

The mechanisms underlying the evolution of L-dopa-induced dyskinesia (LID) in Parkinson's disease (PD) are only beginning to be uncovered and are far from being fully understood. Recent evidence reveals that dyskinesias are an outcome of disruption of the basal ganglia network following a sequence of events that include pulsatile stimulation of striatal dopamine receptors, downstream alterations of intracellular protein and gene signals, and abnormal changes in the non-dopamine transmitter systems (Bezard et al.,

2001). These events collectively lead to altered neuronal discharge patterns between the basal ganglia output neurons and the cerebral cortex.

Current understanding of the pathophysiological mechanisms of LID stems primarily from experimental studies conducted in 6-OHDA lesioned rats and in MPTP lesioned non-human primates (Engber et al., 1989; Brotchie and Fox, 1999; Henry et al., 1999). Indeed, the latter model has been very useful for the study of LID because primates show the various types of LID akin to those seen in parkisonian patients. In addition, recent electrophysiological and metabolic studies of the basal ganglia have shown similar abnormalities associated with LID in both humans and primates.

In one such study, Mitchell et al., (1992) reported that the regional brain uptake of 2-DG in the ventral anterior (VA) and the ventrolateral (VL) thalamic nuclei in MPTP-lesioned primates was significantly reduced in LID compared to the following 3 groups: normal, untreated Parkinsonian and non-dyskinetic L-dopa-treated animals. This suggests a decrease in metabolic activity in the region of the thalamus that mainly receives input from GPi. In a different, yet compatible study based on *in situ* hybridization for cytochrome oxidase subunit I (COI) – a molecular marker for functional neuronal activity, Vila et al., (1997) found attenuation in COI mRNA expression in the GPi in L-dopa-treated subjects. However, it should be noted that the temporal characteristics of the methods of these two studies run from a few minutes to several hours and that neither are capable of accounting for transient and/or subtle changes in neuronal discharge patterns that might not necessarily alter the firing rate over the time epoch studied. Nonetheless, these results suggest that LID is linked with under-activity of basal ganglia output nuclei (GPi and SNr). Furthermore, a recent study of metabolic tracing with 2-DG from our group has shown that there are significant changes in the accumulation of 2-DG observed specifically in the limbic and association domains of basal ganglia and thalamic nuclei in LID. Thus, LID has an additional affective, motivational and cognitive component (Guigoni et al., 2004).

Further insight into the functioning of basal ganglia comes from studies using single unit recording techniques in rats, primates and humans (Filion, 1979; Filion and Tremblay, 1991; Filion et al., 1991; Hutchinson et al., 1997; Stefani et al., 1997; Papa et al., 1999; Ruskin et al., 1999; Allers et al., 2000; Lozano et al., 2000; Boraud et al., 2001; Heimer et al., 2002; Ruskin et al., 2003). To summarize, under normal conditions, phasically active neurons dominate in the striatum while tonically active neurons predominate in the rest of the basal ganglia structures (Kaneoke and Vitek, 1996; Boraud et al., 2002). The tonically active neurons exhibit three different patterns of discharge: random, regular and bursting. Normal GPe neurons are tonic in nature, firing at the rate of 50-70 spikes/s with almost an equal representation of the three discharge patterns and only a few cells showing an oscillatory autocorrelogram in the range of 5-20 Hz (DeLong, 1971, 1972; Filion and Tremblay, 1991; Boraud et al., 1998; Raz et al., 2000). Normal GPi neurons, the major output structure of the basal ganglia besides the SNr, are also tonic with a firing rate of 60-80 spike/s and an equal distribution of random, regular and bursting patterns (DeLong, 1971, 1972; Filion and Tremblay, 1991; Nini et al., 1995; Boraud et al., 1998; Boraud et al., 2001). These neurons, however, exhibit negligible oscillation or correlated pairing (Raz et al., 1996; 2000; 2001).

Extensive single-unit studies have also been undertaken in the MPTP induced primate model of Parkinsonism and to see the effect of DA and DA agonists especially in pallidal neurons (both GPe and GPi). GPe neurons in MPTP treated monkeys yield conflicting results. Some studies report a decrease in the firing frequency in the vicinity of 30-50 spikes/s (Miller and DeLong, 1987; Filion and Tremblay, 1991; Raz et al., 2000) while others show no significant change at all (Bezard et al., 1999; Boraud et al., 2001). Firing patterns show a marked increase in the number of bursting neurons with a significant rise of oscillatory cells and of correlated pairs (Raz et al., 2000). In addition, therapy with DA agonists increases the firing rate of GPe neurons accompanied by alterations in the firing pattern (Filion and Tremblay, 1991; Boraud et al., 1998; Lozano et al., 2000; Boraud et al., 2001). These GPe neurons, however, have no distinct correlation to clinical improvement of symptoms or to the onset of dyskinesia (Boraud et al., 2001).

In the Parkinsonian state, GPi neurons significantly increase their discharge rates (Miller and DeLong, 1987; Boraud et al., 1996; 1998; Lozano et al., 2000; Boraud et al., 2001 ; but also see Wichmann and DeLong, 1999; Raz et al., 2000). Similarly, recordings from human Parkinsonian subjects during surgery show increased firing rates in the GPi (Hutchinson et al., 1994; 1997; Merello et al., 1999; Lozano et al., 2000). The characteristic neuronal discharge of these cells in Parkisonism, of both primates and humans, is predominantly the bursting type with some studies showing phase-locked oscillations in about half of the neural population recorded (Filion and Tremblay, 1991; Bergman et al., 1994; Hutchinson et al., 1994; Nini et al., 1995; Boraud et al., 1996; Hutchinson et al., 1997; Boraud et al., 1998; Hurtado et al., 1999; Lemstra et al., 1999; Raz et al., 2000; Boraud et al., 2001). Following therapy with L-dopa or DA agonists, extracellular recordings reveal the firing rate of GPi neurons to be drastically attenuated and in some cases even abolished (Filion et al., 1991; Hutchinson et al., 1997; Stefani et al., 1997; Boraud et al., 1998; Papa et al., 1999; Boraud et al., 2001) with a simultaneous diminution in the number of bursting neurons (Boraud et al., 2001). Thus, the picture that emerges from the GPi recording studies is twofold: 1) the reduction in the firing rate is probably coupled to improvement in Parkisonian symptoms, and 2) dyskinesia is a likely consequence of an excessive decrease in firing rate and an alteration of firing pattern with the latter having a crucial role in its generation. Indeed, the beneficial effects observed following pallidotomy or pallidal deep brain stimulation is hypothesized to act by reversing such abnormal discharge patterns in GPi neurons. Moreover, both D1 and D2 dopamine receptor agonists tend to shift most neurons to a random firing pattern suggesting that irregular firing could be a cause of dyskinesia. On the other hand, the mixed D1/D2 agonist apomorphine does not explicitly transform the GPi neuronal firing pattern to a random one in spite of inducing an equivalent level of dyskinesia. GPi neurons having an irregular discharge pattern before administration of apomorhine might even show a regular or bursting pattern with the advent of dyskinesia following apomorphine treatment (Boraud et al., 2001). The changes observed in the discharge characteristics highlight the notion that in the genesis of dyskinesia a change to a different firing pattern could be vital rather than the irregular pattern itself.

Parkinsonian primate recordings have demonstrated synchronized oscillations within structures like GPi, GPe and STN (Bergman et al., 1994; Raz et al., 1996; 2000). Furthermore, Raz et al., (2001) has for the first time shown synchronization between GPi and tonically active striatal neurons in MPTP treated monkeys. Similarly, work done on Parkinsonian patients by Levy et al., (2002) has also shown oscillatory synchronization both at low (3-6 Hz) and at higher (15-30 Hz) frequencies in pallidal neurons. Such results have led Bergman and co-workers to hypothesize that the primary function of basal ganglia nuclei could merely be to reduce dimensionality by de-correlation of activity provided by the feeding cortical inputs, and the role of SNc is to actively reinforce relevant events being fed into this complex network. As a result, basal ganglia would not actively participate in initiation and control of movements but would rather be a conduit for providing necessary information for such motor functions. Breakdown in synchronization of neuronal activity within structures in the basal ganglia, like the STN or GPi, would consequently lead to failure of the network to process relevant information reaching the appropriate structures and hence culminate in Parkinsonism. In the absence of relevant feedback of information, cortical areas involved in motor planning and execution would function abnormally.

Additionally, basal ganglia circuitry has been shown to exhibit robust beta oscillatory activity in the Parkinsonian state (Brown et al., 2001; Raz et al., 2001). Under healthy conditions such oscillations are generally absent with very slow oscillations (period of many seconds) seen instead (Ruskin et al., 1999; Wichmann et al., 2002). This raises the possibility that some of the observed Parkinsonian symptoms as well as LID might have a common basis as reflected by abnormal synchronization/desynchronization within the network. In fact, a recent study by Heimer et al., (2002) using cross-correlograms with extended time scales has demonstrated increased neuronal synchronization among pallidal neurons in MPTP induced Parkinsonism. This synchronization was maintained during 'off' periods but reversed during 'on' periods of chronic L-dopa therapy. Unlike other previous studies (Raz et al., 2000; Levy et al., 2002), the correlated activity was for most part non-oscillatory in nature. In GPe though they reported a distinct increase in correlation in the 'off' state as well as a significant decrease in the firing rate. Thus, such modulations (hypersynchronization) could be involved in the generation of LID.

The molecular and cellular mechanisms underlying abnormal synchronization of neuronal discharge are still obscure. Recent studies by Grace and colleagues (Moore and Grace, 2002) have implicated gap junctions as potential candidates to induce abnormal synchronization of striatal efferents. Electrotonic coupling is known to be directly involved in mediating neuronal network activity and rhythmic behavioral patterning (Jefferys, 1995; Rothwell, 1998; Perez Velazquez and Carlen, 2000). It is now well-established that striatal cells express connexins, proteins which are the building blocks of gap junctions, and that striatal gap junction communication is modulated by DA and glutamate transmission (Bennett, 1997; O'Donnell and Grace, 1997; Bennett et al., 1999; Belluardo et al., 2000). Gap junctions formed by connexin proteins on the surface of two adjacent cell membranes allow cross-flow of electricity and other cell to cell communication (Bruzzone et al., 1996). Recent evidence has shown that the direct flow of electrical current between coupled neurons can even transmit bursting activity between them (Skinner et al., 1999). Hence, it is reasonable to postulate that gap junction coupling might have a significant role in synchronization and stabilization of the synchronized state in neuronal network activity. Studies using gap junction inactivators have just begun to address this issue. Several gap junction channel blockers or inactivators are known but most of them are non-selective in nature (Spray et al., 2002). The most selective of them all, so far, is carbenoxolone, a derivitive of glycyrrhetinic acid, known to act directly on gap junctions and able to reduce conductance by up to 80 % (Davidson and Baumgarten, 1988). Work by Moore and Grace, (2002) has shown carbenoxolone to reduce apomorphine-induced stereotypies in rats. Although the neural mechanisms underlying DA receptor- mediated stereotypies and that of LID are not similar, they are comparable as both are thought to be hyperkinetic movement disorders, and hence, likely have overlapping pathophysiology. In fact, recent observations by Brotchie and colleagues (Lee et al., 2004) suggests that increased gap junction communication has a role in the priming but not in the expression of LID.

That neuronal synchronization in the basal ganglia network has a significant hand in the evolution of LID can be deduced indirectly by the action of the drug Levetiracetam. As a corollary to uniquely reducing epileptiform (hyper)synchronisation in rat hippocampal slices, Levetiracetam has been postulated to be a potential candidate in alleviating LID (Margineanu and Klitgaard, 2000; Klitgaard, 2003; Margineanu and Klitgaard, 2003). Recent evidence indicates that the drug Levetiracetam does effectively improve the choreic form of LID in MPTP lesioned marmosets and also potentiates the antidyskinetic action of Amantadine in such Parkinsonian marmosets (Bezard et al., 2004; Hill et al., 2004a; 2004b). Further experiments done by our group have found Levetiracetam to decrease local field potential (LFP) power in the SNr of non-primed (with L-dopa) 6-OHDA lesioned rats in comparison to sham lesioned primed/non-primed and 6-OHDA lesioned primed rats (Meissner et al., 2004). Paradoxically however, the effect of Levetiracetam on synchronization was observed to be increased in both the SNr of primed 6-OHDA lesioned dyskinetic rats and the GP of primed MPTP lesioned non-dyskinetic monkeys. This particular finding needs to be further investigated in light of recent findings identifying the synaptic vesicle protein SV2A as the Levetiracetam binding site and SV2A being involved in exocytosis (Lynch et al., 2004). Thus, Levetiracetam through modulation of SV2A and thereby exocytosis, should tune up or down the gain of some basal ganglia connections and thus regulate the synchronization process. Which synapses are modulated and what are the exact mechanisms of modulation remain to be determined, but cortico-striatal synapses are a good candidate because i) SV2A is present in the cortico-striatal synapse and ii) it has been proposed that oscillation in the cortex-basal ganglia network would result from a decrease in the strength of cortico-striatal connections (Leblois et al., submitted).

The electrophysiological recording studies so far have been conducted following systemic administration of dopaminergic drugs. As a result, dyskinesia can only be interpreted as a consequence of hyperstimulation of dopamine receptors in the striatum. Further studies, using single and multi-single unit recordings as well as LFP techniques, need to be undertaken to delineate the effects of systemic versus local (eg. striatal) administration of dopamimetic agents on neurons of the various structures of the basal ganglia network before we can with confidence tackle the problem of LID. The multitude of target mechanisms likely to be involved in the genesis of LID described to date just shows the complexity of the problem and we have only begun to address them earnestly.

REFERENCES

Allers KA, Kreiss DS, Walters JR (2000) Multisecond oscillations in the subthalamic nucleus: Effects of apomorphine and dopamine cell lesion. *Synapse* **38**:38-50.

Belluardo N, Mud inverted question marko G, Trovato-Salinaro A, Le Gurun S, Charollais A, Serre-Beinier V, Amato G, Haefliger J, Meda P, Condorelli D (2000) Expression of connexin36 in the adult and developing rat brain. *Brain Res* **865**:121-138.

Bennett M (1997) Gap junctions as electrical synapses. *J Neurocytol* **26**:349-366.

Bennett S, Arnold J, Chen J, Stenger J, Paul D, Roberts D (1999) Long-term changes in connexin32 gap junction protein and mRNA expression following cocaine self-administration in rats. *Eur J Neurosci* **11**:3329-3338.

Bergman H, Wichmann T, Karmon B, DeLong MR (1994) The primate subthalamic nucleus. II. Neuronal activity in the MPTP model of parkinsonism. *J Neurophysiol* **72**:507-520.

Bezard E, Brotchie JM, Gross CE (2001) Pathophysiology of levodopa-induced dyskinesia : Potential for new therapies. *Nature Rev Neurosci* **2**:577-588.

Bezard E, Boraud T, Bioulac B, Gross C (1999) Involvement of the subthalamic nucleus in glutamatergic compensatory mechanisms. *Eur J Neurosci* **11**:2167-2170.

Bezard E, Hill M, Crossman A, Brotchie J, Michel A, Grimee R, Klitgaard H (2004) Levetiracetam improves choreic levodopa-induced dyskinesia in the MPTP-treated macaque. *Eur J Pharmacol* **485**:159-164.

Boraud T, Bezard E, Bioulac B, Gross C (1996) High frequency stimulation of the internal globus pallidus reduces the higher firing frequency of its neurons in the MPTP-treated monkey model of Parkinson's disease. *Neurosci Lett* **215**:17-20.

Boraud T, Bezard E, Bioulac B, Gross C (2001) Dopamine agonist-induced dyskinesias are correlated to both firing pattern and frequency alteration of pallidal neurons in the MPTP-treated monkey. *Brain* **124**:546-557.

Boraud T, Bezard E, Bioulac B, Gross CE (2002) From single extracellular unit recording in experimental and human Parkinsonism to the development of a functional concept of the role played by the basal ganglia in motor control. *Prog Neurobiol* **66**:265-283.

Boraud T, Bezard E, Guehl D, Bioulac B, Gross C (1998) Effects of L-DOPA on neuronal activity of the Globus Pallidus externalis (GPe) and Globus Pallidus internalis (GPi) in the MPTP-treated monkey. *Brain Res* **787**:157-160.

Brotchie JM, Fox SH (1999) Quantitative assessment of dyskinesias in subhuman primates. *Mov Disord* **14**:40-47.

Brown P, Oliviero A, Mazzone P, Insola A, Tonali P, Di Lazzaro V (2001) Dopamine dependency of oscillations between subthalamic nucleus and pallidum in Parkinson's disease. *J Neurosci* **21**:1033-1038.

Bruzzone R, White T, Paul D (1996) Connections with connexins: the molecular basis of direct intercellular signaling. *Eur J Biochem* **238**:1-27.

Davidson J, Baumgarten I (1988) Glycyrrhetinic acid derivatives: a novel class of inhibitors of gap-junctional intercellular communication. Structure-activity relationships. *J Pharmacol Exp Ther* **246**:1104-1107.

DeLong MR (1971) Activity of pallidal neurons during movement. *J Neurophysiol* **34**: 414-427.

DeLong MR (1972) Activity of basal ganglia neurons during movement. *Brain Res* **40**: 127-135.

Engber TM, Susel Z, Juncos JL, Chase TN (1989) Continuous and intermittent levodopa differentially affect rotation induced by D-1 and D-2 dopamine agonists. *Eur J Pharmacol* **168**:291-298.

Filion M (1979) Effects of interruption of the nigrostriatal pathway and of dopaminergic agents on the spontaneous activity of globus pallidus neurons in the awake monkey. *Brain Res* **547**:142-151.

Filion M, Tremblay L (1991) Abnormal spontaneous activity of globus pallidus neurons in monkeys with MPTP-induced parkinsonism. *Brain Res* **547**:142-151.

Filion M, Tremblay L, Bedard PJ (1991) Effects of dopamine agonists on the spontaneous activity of globus pallidus neurons in monkeys with MPTP-induced parkinsonism. *Brain Res* **547**:152-161.

Guigoni C, Qin L, Aubert I, Dovero S, Bioulac BH, Bloch B, Crossman AR, Gross CE, Bezard E (2004) Involvement of sensorimotor, limbic, and associative basal ganglia domains in levodopa-induced dyskinesia. J Neurosci in revision.

Heimer G, Bar-Gad I, Goldberg JA, Bergman H (2002) Dopamine replacement therapy reverses abnormal synchronization of pallidal neurons in the 1-methyl-4-phenyl-1,2,3,6-tetrahydropyridine primate model of parkinsonism. *J Neurosci* **22**:7850-7855.

Henry B, Crossman AR, Brotchie JM (1999) Effect of repeated L-DOPA, bromocriptine, or lisuride administration on preproenkephalin-A and preproenkephalin-B mRNA levels in the striatum of the 6-hydroxydopamine-lesioned rat. *Exp Neurol* **155**:204-220.

Hill M, Brotchie J, Crossman A, Bezard E, Michel A, Grimee R, Klitgaard H (2004a) Levetiracetam Interferes With the L-dopa priming process in MPTP-lesioned drug-naïve marmosets. *Clin Neuropharmacol* **27**:171-177.

Hill M, Ravenscroft P, Bezard E, Crossman A, Brotchie J, Michel A, Grimee R, Klitgaard H (2004b) Levetiracetam potentiates the antidyskinetic action of amantadine in the 1-methyl-4-phenyl-1,2,3,6-tetrahydropyridine (MPTP)-lesioned primate model of Parkinson's disease. *J Pharmacol Exp Ther* **310**:386-394.

Hurtado JM, Gray CM, Tamas LB, Sigvardt KA (1999) Dynamics of tremor-related oscillations in the human globus pallidus: A single case study. *Proc Natl Acad Sci U S A* **96**:1674-1679.

Hutchinson WD, Levy R, Dostrovsky JO, Lozano AM, Lang AE (1997) Effects of apomorphine on Globus Pallidus neurons in parkinsonian patients. *Ann Neurol* **42**:767-775.

Hutchinson WD, Lozano AM, Davis KD, Saint-Cyr JA, Lang AE, Dostrovsky JO (1994) Differential neuronal activity in segments of Globus Pallidus in Parkinson's disease patients. *Neuroreport* **5**:1533-1537.

Jefferys J (1995) Nonsynaptic modulation of neuronal activity in the brain: electric currents and extracellular ions. *Physiol Rev* **75**:689-723.

Kaneoke Y, Vitek JL (1996) Burst and oscillations as disparate neuronal properties. *J Neurosci Meth* **68**:211-223.

Klitgaard H (2003) Levetiracetam: the preclinical profile of a new class of antiepileptic drugs? *Epilepsia* **42**:13-18.

Leblois A, Boraud T, Bergman H, Hansel D (submitted) A model of the Basal Ganglia physiology and pathophysiology based on a competition between direct and hyper-direct pathways.

Lee JI, Carlen P, lang AET, Brotchie JM (2004) blockade of gap junction communication as an approach to prevent priming for L-DOPA-induced dyskinesia in parkinson's disease: effects on striatal preporenkephaline-B expression. society for Neuroscience abstract:308.320.

Lemstra AW, Metman LV, Lee JI, Dougherty PM, Lenz FA (1999) Tremor-frequency (3-6 Hz) activity in the sensorimotor arm representation of the internal segment of the globus pallidus in patients with Parkinson's disease. *Neurosci Lett* **267**:129-132.

Levy R, Hutchison WD, Lozano AM, Dostrovsky JO (2002) Synchronized neuronal discharge in the basal ganglia of parkinsonian patients is limited to oscillatory activity. *J Neurosci* **22**:2855-2861.

Lozano AM, Lang AE, Levy R, Hutchison W, Dostrovsky J (2000) Neuronal recordings in Parkinson's disease Patients with dyskinesias induced by apomorphine. *Ann Neurol* **47**:S141-S146.

Lynch B, Lambeng N, Nocka K, Kensel-Hammes P, Bajjalieh S, Matagne A, Fuks B (2004) The synaptic vesicle protein SV2A is the binding site for the antiepileptic drug levetiracetam. *Proc Natl Acad Sci U S A* **101**:9861-9866.

Margineanu D, Klitgaard H (2000) Inhibition of neuronal hypersynchrony in vitro differentiates levetiracetam from classical antiepileptic drugs. *Pharmacol Res* **42**:281-285.

Margineanu D, Klitgaard H (2003) Levetiracetam has no significant gamma-aminobutyric acid-related effect on paired-pulse interaction in the dentate gyrus of rats. *Eur J Pharmacol* **466**:255-261.

Meissner W, Reese R, Klitgaard H, Bioulac B, Gross C, Boraud T, Bezard E (2004) effect of levetiracetam on synchronisation and L-DOPA-induced decrease in firing frequency of basal ganglia output neurons in the primed 6-OHDA rat and the primed MPTP-treated monkey. society for Neuroscience abstract:678.671.

Merello M, Balej J, Delfino M, Cammarota A, Betti O, Leiguarda R (1999) Apomorphine induces changes in GPi spontaneous outflow in patients with Parkinson's disease. *Mov Disord* **14**:45-49.

Miller WC, DeLong MR (1987) Altered tonic activity of neurons in the globus pallidus and subthalamic nucleus in the primate MPTP model of parkinsonism. In: *Basal Ganglia II* (Carpenter MB, Jayaraman A, eds), pp 415-427. New York: Plenum Press.

Mitchell IJ, Boyce S, Sambrook MA, Crossman AR (1992) A 2-deoxyglucose study of the effects of dopamine agonists on the parkinsonian primate brain. Implications for the neural mechanisms that mediate dopamine agonist-induced dyskinesia. *Brain* **115**:809-824.

Moore H, Grace AA (2002) A Role for Electrotonic Coupling in the Striatum in the Expression of Dopamine Receptor-mediated Stereotypies. *Neuropsychopharmacology* **27**:980-992.

Nini A, Feingold A, Slovin H, Bergman H (1995) Neurons in the globus pallidus do not show correlated activity in the normal monkey, but phase-locked oscillations appear in the MPTP model of parkinsonism. *J Neurophysiol* **74**:1800-1805.

O'Donnell P, Grace A (1997) Cortical afferents modulate striatal gap junction permeability via nitric oxide. *Neuroscience* **76**:1-5.

Papa SM, Desimone R, Fiorani M, Oldfield EH (1999) Internal globus pallidus discharge is nearly suppressed during levodopa-induced dyskinesias. *Ann Neurol* **46**:732-738.

Perez Velazquez J, Carlen P (2000) Gap junctions, synchrony and seizures. *trends neurosci* **23**:68-74.

Raz A, Vaadia E, Bergman H (2000) Firing patterns and correlations of spontaneous discharge of pallidal neurons in the normal and the tremulous 1-methyl-4-phenyl-1,2,3,6-tetrahydropyridine vervet model of parkinsonism. *J Neurosci* **20**:8559-8571.

Raz A, Feingold A, Zelanskaya V, Vaadia E, Bergman H (1996) Neuronal synchronization of tonically active neurons in the striatum of normal and parkinsonian primates. *J Neurophysiol* **76**:2083-2088.

Raz A, Frechter-Mazar V, Feingold A, Abeles M, Vaadia E, Bergman H (2001) Activity of pallidal and striatal tonically active neurons is correlated in MPTP-treated monkeys but not in normal monkeys. *J Neurosci* **21**:RC128(121-125).

Rothwell J (1998) Physiology and anatomy of possible oscillators in the central nervous system. *Mov Disord* **13**:24-28.

Ruskin DN, Bergstrom DA, Tierney PL, Walters JR (2003) Correlated multisecond oscillations in firing rate in the basal ganglia: modulation by dopamine and the subthalamic nucleus. *Neuroscience* **117**:427-438.

Ruskin DN, Bergstrom DA, Kaneoke Y, Patel BN, Twery MJ, Walters JR (1999) Multisecond oscillations in firing rate in the basal ganglia: Robust modulation by dopamine receptor activation and anesthesia. *J Neurophysiol* **81**:2046-2055.

Skinner F, Zhang L, Velazquez J, Carlen P (1999) Bursting in inhibitory interneuronal networks: A role for gap-junctional coupling. *J Neurophysiol* **81**:1274-1283.

Spray D, Rozental R, Srinivas M (2002) Prospects for rational development of pharmacological gap junction channel blockers. *Curr Drug Targets* **3**:455-464.

Stefani A, Stanzione P, Bassi A, Mazzone P, Vangelista T, Bernardi G (1997) Effects of increasing doses of apomorphine during stereotaxic neurosurgery in Parkinson's disease : clinical score and internal globus pallidus activity. *J Neural Transm* **104**:895-904.

Vila M, Levy R, Herrero M-T, Ruberg M, Faucheux B, Obeso JA, Agid Y, Hirsch EC (1997) Consequences of nigrostriatal denervation on the functioning of the basal ganglia in human and nonhuman primates: an *in situ* hybridization study of cytochrome oxidase subunit I mRNA. *J Neurosci* **17**:765-773.

Wichmann T, DeLong MR (1999) Neurobiology - Oscillations in the basal ganglia. *Nature* **400**:621-622.

Wichmann T, Kliem MA, Soares J (2002) Slow oscillatory discharge in the primate basal ganglia. *J Neurophysiol* **87**:1145-1148.

In: Recent Breakthroughs in Basal Ganglia Research
Editor: Erwan Bezard, pp. 369-380
ISBN: 1-59454-880-3

Chapter 29

Levodopa-Induced-Dyskinesias - The Clinical Problem: Clinical Features, Incidence, Risk Factors, Management and Impact on Quality of Life

Alice Manson*
Wessex Neurological Centre, Southampton General Hospital, Southampton
Anette Schrag
Royal Free and University College London Medical School, London, UK

Abstract

Levodopa-induced dyskinesias (LID) belong to the most common dose-limiting adverse effects of levodopa therapy in Parkinson´s disease (PD). The majority of patients will have developed dyskinesias after 10 years of treatment, and about 40% after 5 years. The main clinical factors associated with LID are, duration of treatment with levodopa, and severity and duration of disease. In addition, patients with younger age of onset and genetically determined subtypes of PD have a higher rate of LID. More continuous dopaminergic stimulation with the longer acting dopamine agonists has been shown to reduce and delay the onset of dyskinesias . Numerous treatment strategies to overcome existing LID are being employed, including stereotactic surgery in the deep basal ganglia nuclei, and some pharmacological interventions (amantadine). Studies on the subjective impact of dyskinesias so far have failed to show consistently that dyskinesias have a significant adverse effect on quality of life on the overall population of patients with PD. However, while LID may not be disabling in many patients with early disease, further studies need to assess the impact of this complication, and of its treatment, on the overall quality of life of patients with advanced PD.

Summary

Levodopa-induced dyskinesias (LID), the abnormal involuntary movements associated with dopaminergic therapy in Parkinson's disease (PD), present a significant problem in the advanced stages of the disease and belong to the most common dose-limiting adverse effects of levodopa therapy. The most common type: "peak-dose" LID occur at the peak effect of dopaminergic therapy, whilst 'diphasic dyskinesias' have a "beginning- and end-of-dose" pattern, and the typically dystonic "off-period dyskinesia" occur during off-periods, most frequently in the early mornings. A wide range of latencies to onset of LID has been reported in the literature, but the majority of patients (especially those with younger onset disease) will have developed dyskinesias after 10 years of treatment, and about 40 % after 5 years.

* Correspondence to Dr Alice Manson: Wessex Neurological Centre, Southampton General Hospital, Tremona Road, Southampton, UK. Tel: (+44) 7929625766. Fax: (+44) 2380848788. E-mail address: anilandalice@hotmail.com

The main clinical factors associated with development of LID appear to be duration of treatment with levodopa and severity and duration of disease. In addition, those with younger age of onset and genetically determined subtypes have been reported to have a higher rate of LID. This corresponds to conclusions from animal research, which has suggested that the degree of striatal dopaminergic depletion is one of the main factors for the development of LID. More recently the important role of pulsatile stimulation of dopaminergic receptors has become recognized and it has been postulated that more continuous dopaminergic stimulation is preferable to pulsatile treatment. More continuous dopaminergic stimulation with the longer acting dopamine agonists have been shown to reduced and delay the onset of dyskinesias. Numerous other treatment strategies to overcome existing LID are being employed, including stereotactic surgery both with deep brain stimulation and lesional procedures as well as a number of pharmacological interventions, including amantadine. As more treatment options become available, increasingly sophisticated measures for LID are being developed to attempt accurate assessment of efficacy, although all of these have limitations. Studies on the subjective impact of dyskinesias so far have failed to show consistently that dyskinesias have a significant adverse effect on quality of life on the overall population of patients with PD. However, while LID may not be disabling in many patients with early disease, further studies need to assess the impact of this complication, and of its treatment, on the overall quality of life of patients with advanced disease.

1 PHENOMENOLOGY

Soon after the introduction of levodopa for the treatment of PD, Cotzias, (1971) it was noted that this treatment could be associated with undesirable, involuntary movements or 'dyskinesia' (derived from 'dys' difficult and 'kinesis' movement). These seemed to occur at the peak of the anti-parkinsonian effect of treatment, thus leading to the current term '*peak-dose*' dyskinesias, the most commonly recognized pattern of dyskinesia (Nutt, 1990; Fahn, 2000; Schrag and Quinn, 2000), and could be modified by lowering the dose of medication, suggesting a dose-relationship. Muenter et al., (1977) termed this pattern of dyskinesia the 'improvement - dystonia - improvement' or 'IDI' response, and established that it occurred around the time of peak plasma levels of medication (Muenter and Tyce, 1971). However, pharmacokinetic studies of various dopaminergic agents demonstrate that although their presence and duration is somewhat dose dependent, the severity of 'peak-dose' dyskinesias does not correspond well to plasma levels, suggesting a more complex mechanism to be involved in their generation (Nutt, 1990; Manson et al., 2001a). Indeed, LID often occurs as a patient-specific 'all or nothing' response to adequate levodopa-dosing (also termed 'square-wave dyskinesia'), and whilst their duration may be dose-responsive, it is also other factors such as emotional stress that affect severity (Mones et al., 1970; Nutt, 1990; Durif et al., 1999a; Nutt, 2000). Peak dose LID is typified by usually generalized, choreiform type movements: ie fleeting, non-rhythmic, purposeless fidgety-type movements. Often these are exaggerated by stress or activity and are typically asymmetric, being most prominent in the most affected side by the underlying disease (Mones et al., 1970; Murphy, 1979; Nutt, 1990; Marconi et al., 1994). In addition to chorea, several other manifestations of peak-dose LID are seen: namely ballism (wild, proximal, flinging movements of the limbs), choreoathetoid (writhing-type) movements and more dystonic posturing (which will often begin in the lower limb or foot and herald the onset of the LID). Pure dystonic movements, although occasionally occurring as peak-dose phenomena, are more often seen in the 'wearing on or off' phase or as an 'off'-period manifestation (see below).

As well as the 'IDI' response, Muenter et al., (1977) also described a 'DID' (dystonia-improvement-dystonia) response, whereby patients would experience drug-induced dyskinesia as the effects of levodopa were wearing on and off. They noted that these dyskinesias affected predominantly younger patients and occurred when the concentration of levodopa in plasma passed through a critical but relatively low level, whereas it remained absent as long as the concentration remained above that level. As noted above, this type of dyskinesia is predominantly dystonic in character and, due to its occurrence at the onset and/or end of action of levodopa, is now more commonly referred to as '*diphasic*' or '*biphasic*' dyskinesia.

The third pattern of dyskinesia in levodopa treated patients with PD is that of '*off*'*-period dystonia*. This is a recognized complication of drug therapy and is separate from the dystonia of untreated parkinsonism: it commonly occurs early in the morning, and is often painful, distressing and disabling. It frequently affects the lower limbs but can involve any region such as facial, oral, lingual and laryngeal muscles. Indeed, virtually any pattern of LID can be seen and any body part be affected, including the respiratory and extra-ocular muscles (Mones et al., 1970; LeWitt, 1998; Fahn, 2000).

2 Incidence

Estimation of rate of LID occurrence as a function of disease duration is difficult, partly due to the uncertainty of establishing the exact time of disease onset; in addition, dyskinesias only occur after antiparkinsonian treatment is started, with the exception of dystonia, which can occasionally occur in untreated PD (Bozi and Bhatia, 2003). Therefore latency from onset of treatment to onset of dyskinesias often provides a more useful measure. LID have been reported to occur in about 10 % of patients per year in the first seven years of levodopa treatment (Grandas et al., 1999), but the reported incidence of LID varies greatly between studies. It is widely accepted that it is rare to develop LID within the first year of treatment, but reports vary from 0 % (Barbeau, 1980) to 17 % (Parkinson study group, 1996) to 66 % of patients with dyskinesia in the first year of treatment (Brannan and Yahr, 1995). A recent prospective practice -based study found that LID occurred in 16 % of patients after 3 years and 32 % at 5 years (Garcia Ruiz et al., 2004), and in prospective randomized controlled trials (RCTs) approximately 40-50 % of patients are found to develop LID after 5 years of treatment with levodopa (Montastruc et al., 1994; Dupont et al., 1996; Rinne et al., 1998; Rascol et al., 2000a; Ahlskog and Muenter, 2001; Holloway et al., 2004), although some earlier RCT have reported lower rates (Block et al., 1997; Koller et al., 1999). Similarly, in a clinic-based retrospective analysis of 35 patients with PD treated with levodopa, Poewe et al., (1986) documented dyskinesia in 54 % of patients at 6 years of therapy, whereas in another retrospective study of movement disorders clinic based records, Blanchet et al., (1996) reported a LID prevalence of 56 % developing in a mean of under 3 years of levodopa therapy. Such differences may partly reflect differences in ascertainment if dyskinesias are not reported by patients, who are often unaware especially of mild dyskinesias (Mones et al., 1970; Goetz et al., 1997; Schrag and Quinn, 2000; Vitale et al., 2001), and if dyskinesias are not present during a consultation. However, they are also likely to reflect variations in population characteristics, particularly with regard to age and disease duration at start of treatment, disease severity, differences in levodopa doses employed and types of concomitant medication. Thus, in patients with onset of PD before age 40 years up to 90 % have developed dyskinesias after 5 years of treatment, and up to 100 % after ten years (Kostic et al., 1991; Schrag et al., 1998). In addition, patients in studies with lower rates of dyskinesias were often taking lower doses of levodopa or other antiparkinsonian medication (Schrag and Quinn, 2000). Such differences are therefore also likely to explain the lower rate of dyskinesias in community-based epidemiological studies, where levodopa doses are generally low and mean age older (Tandberg et al., 1995; Schrag and Quinn, 2000). Thus, one community based study reported only 13 % had dyskinesias after 5 years of treatment with levodopa or less and only 36 % after 6-9 years of treatment (Schrag and Quinn, 2000), but levodopa dose was on average only 423 mg after a mean disease duration of 7 years.

Although it is largely accepted that dyskinesias occur almost inevitably at the later stages of PD, the appearance of LID undoubtedly varies from patient to patient (Miyawaki et al., 1997), and there may be a small group of patients who do not develop dyskinesias. Although the reasons for this are unclear, a number of risk factors for the early development of dyskinesias have been identified (Miyawaki et al., 1997; Schrag and Quinn, 2000).

3 Risk Factors

Typically, LID only occur in patients with a good response to levodopa, ie mainly in patients with idiopathic PD on an effective dose of levodopa, although patients with multiple system atrophy may develop severe, and usually atypical, LID early in the disease. As mentioned above, there also appears to be a dose-dependence to the occurrence and severity of LID in PD and reduction in levodopa dose often improves or abolishes these. A number of other factors have been implicated in influencing the development of dyskinesias in PD. These include age at onset (Kostic et al., 1991; Grandas et al., 1999; Schrag and Quinn, 2000; Kumar et al., 2004), disease duration (Lesser et al., 1979; Parkinson study group, 1996; Ahlskog and Muenter, 2001), gender (Parkinson study group, 1996; Lyons et al., 1998), family history of PD (Schrag and Quinn, 2000; Papapetropoulos et al., 2004), duration of treatment (Miyawaki et al., 1997), and dosage of levodopa (Barbeau, 1980; Lees and Stern, 1981; Tanner, 1986; Grandas et al., 1999), and severity of disease (Onofrj et al., 1998; Schrag and Quinn, 2000; Ahlskog and Muenter, 2001; Garcia Ruiz et al., 2004). After controlling for correlations between many of these factors, *disease duration* and, more strongly, *duration of levodopa treatment* is the most consistently demonstrated factor implicated in LID generation (Lesser et al., 1979; Parkinson study group, 1996; Grandas et al., 1999; Schrag and Quinn, 2000; Ahlskog and Muenter, 2001). Although duration of treatment has been demonstrated to be an independent risk factor in generating LID (Grandas et al., 1999; Schrag and Quinn, 2000), it is possible that this, at least partly, also reflects *disease severity*, which has been shown to correlate highly with treatment duration as well as rate of dyskinesias (Baas, 2000; Schrag and Quinn, 2000). This is further illustrated by cases of very late presenting Parkinson's disease developing LID after only a few doses of levodopa treatment (Onofrj et al., 1998), and cases of MPTP induced parkinsonism where striatal denervation is severe and marked LID develop within months of levodopa treatment (Ballard et al., 1985). This is also consistent with data from animal models of PD, whereby LID correspond to the degree and rate of neuronal loss in the striatum (Schneider, 1989; 2003). While recently it has been demonstrated that in animal models LID can also be induced in the absence of denervation (Jenner, 2000; Langston et al., 2000), these data taken together strongly support the importance of striatal neuronal loss in the development of dyskinesias in PD (see discussion elsewhere in this book).

Apart from duration of treatment with levodopa, the *dose of levodopa* has been reported as a risk factor for LID in several studies (Barbeau, 1980; Lees and Stern, 1981; Tanner, 1986; Grandas et al., 1999). One prospective RCT demonstrated atypically low rates of LID of 20 % at 5 years of treatment with levodopa, and it was postulated that this was a result of low doses of levodopa being prescribed (Koller et al., 1999).

A number of RCT trials were initiated to investigate the effect of alternative antiparkinsonian medication on the occurrence of dyskinesias (Hundemer et al., 2000; Musch and group, 2000; Parkinson study group, 2000; Rascol et al., 2000a; Bracco et al., 2004). These demonstrated that dopamine agonists not only allow reduction of levodopa dose and reduce the severity of existing dyskinesias but also delay the need for levodopa treatment and reduce the rate of new onset dyskinesias. In addition, there is increasing evidence from animal models that more continuous stimulation of striatal dopamine receptors leads to a reduced rate of dyskinesias (Jenner, 2004; Smith et al., 2004). This suggests that the intermittent treatment with levodopa doses leading to fluctuating plasma levels of levodopa and pulsatile stimulation of dopamine receptors also plays a role in the generation of dyskinesias, and new approaches to more continuous dopamine receptor stimulation are being explored (Manson et al., 2002; Deane et al., 2004; Stocchi and Olanow, 2004).

As discussed above, it has also been recognized that LID tend to be more severe and occur more frequently and sooner in younger onset (before the age of 50) patients (Kostic et al., 1991; Denny and Behari, 1999; Grandas et al., 1999; Schrag and Quinn, 2000; Kumar et al., 2004). In particular, patients with some forms of genetically determined parkinsonism have a higher risk of developing LID (Quinn et al., 1987; Schrag et al., 1998; Schrag and Quinn, 2000; Papapetropoulos et al., 2004) and a higher prevalence of dyskinesias has been reported in patients with a family history of PD than those without (Schrag and Quinn, 2000; Papapetropoulos et al., 2004). Thus, patients with mutations in the parkin genes have a high rate and early onset of (often atypical) dyskinesias in addition to other distinguishing characteristics, including a young mean age at onset (Khan et al., 2004). However, it is unknown whether patients with these rare genetic forms of parkinsonism represent small sub-groups within the overall population of PD patients or whether they represent a relative "pure" example of genetic risk factors for PD with dyskinesias is unknown.

Female gender has also been suggested to be associated with a higher risk for dyskinesias (Parkinson study group, 1996; Lyons et al., 1998) and but more recent studies have not confirmed the finding of increased dyskinesia in women (Grandas et al., 1999; Schrag and Quinn, 2000; Martinelli et al., 2003). It could be hypothesized that the threshold for reporting LID in the initial studies was lower in women who may find them more socially unacceptable and thus less tolerable.

4 Assessment of LID

One of the challenges facing the clinician or researcher dealing with LID is their accurate assessment (Brown and Manson, 1999). The fluctuating nature of Parkinson's disease itself as well as the variability of LID makes accurate measurement extremely difficult. Patients may vary markedly in the duration and severity of LID from day to day and even within the same day without any predictable pattern. Often assessment is by way of subjective questionnaire based scales only, such as in the Unified Parkinson's disease rating scale or UPDRS (Fahn et al., 1987), or visual analogue based diary cards, which are of limited sensitivity and have a number of additional drawbacks (Goetz, 1999). Firstly, it relies on patient compliance and capability of not only recognizing LID but also accurately recalling and documenting them. Secondly, some scales separate severity and duration of LID but as many patients are only aware of their LID above a certain threshold of severity, it could be argued that these are in fact measuring the same thing. Lastly a purely subjective measure poses obvious limitations on any treatments trials, particularly as a significant placebo effect is known to exist in PD (Goetz, 2000). In these cases, a more objective measure is usually utilized in the form of a video dopaminergic challenge, whereby a controlled dose of levodopa (or sometimes apomorphine) is given under standardized conditions after which the patient is videoed and the LID can be rated by and independent observer at a later stage (Melamed et al., 1999). Various rating scales are used (Goetz, 1999), from the modified AIMS scale, (which has the drawback of initially being developed for assessing anti-psychotic induced movement disorders rather than LID specifically) to specifically adapted scales whereby tasks to specifically provoke LID are included (Widner and Defer, 1999). However, video challenges are highly labour-intensive, often unpleasant for the patients and will still only provide a 'snapshot' of LID, which may not be truly representative or reliably reproducible (Damier et al., 1999), and whilst standard protocol dictates that it should be performed first thing in the morning (Melamed et al., 1999), many patients actually notice their LID build up towards the end of the day (Nutt, 1990). Therefore many trials end up using a rather complicated combination of several measures, including subjective rating scales, patient kept diaries and video challenges. For these reasons there is an ongoing quest to develop a means of accurately measuring LID by purely objective methods (Brown and Manson, 1999). Complex laboratory-based techniques involving techniques such as camera-assisted motion analysis and accelerometry can provide fairly accurate assessment of LIDs in controlled settings but may not capture the actual pattern or severity of LIDs in the setting of every day life. Therefore attempts have been made to extend their use outside the laboratory and although some promise was shown with an ambulatory device using triaxial accelerometry and specially designed computer software (Manson et al., 2000a), other groups using similar techniques have not demonstrated adequate efficacy (Hoff et al., 2004). Therefore, although research continues in this area, a video challenge currently remains the gold standard for assessment of LID, despite the problems associated (Damier et al., 1999).

5 Management of LID

The simplest and most obvious strategy to reduce LID is usually simply to reduce the dose of dopaminergic therapy, but this inevitably leads to an unacceptable worsening of parkinsonian symptoms. Thus, the clinical challenge is to find an alternative method, which does not worsen parkinsonism. After it was noted that patients on dopamine agonists tended to develop less LID than those on levodopa (Lees and Stern, 1981), it was initially postulated that the different receptor subtype stimulation might play a key role in the generation of LID. However, it is now widely believed that one of the major factors in promoting motor complications such as LID in PD is the artificial pulsatile dopaminergic stimulation which occurs as a result of pharmacological treatments (Chase, 1998a), and thus there is now a move towards more continuous dopaminergic stimulation as a means of avoiding or minimizing these (Chase, 1998b; Hadj Tahar et al., 2003). To try to achieve this, many clinicians have employed sustained release preparations of levodopa but there is little evidence to show that this strategy is effective (Dupont et al., 1996; Block et al., 1997; Pahwa et al., 1997; Koller et al., 1999; Durif, 1999b). Other attempts to minimize LID by simply manipulating the dose regimes of levodopa are made but can be of limited value, indeed the previously employed strategy of fractionating doses is now thought possibly to be more detrimental as low plasma troughs of levodopa level may increase the pulsatility of dopamine receptor stimulation. Therefore, even when peak dose LID are a problem, fewer larger doses of levodopa may minimize LID more effectively

(Verhagen Metman et al., 1997; Nutt, 2000). Instead, it has now become widely accepted practice to use dopamine agonists in place of levodopa where possible (Stocchi et al., 1997; Rascol, 1999; Olanow and Obeso, 2000), as these have a longer duration of action and thus theoretically provide more continuous stimulation. This can be a useful strategy both for patients in whom dyskinesias have already developed, when dopamine agonists can be introduced as a 'dopa-sparing' technique with a smoothening of response, and in levodopa-naïve patients in order to delay the appearance of LID. Whilst many studies have demonstrated that starting with a dopamine agonist delays the emergence of dyskinesia (Lees and Stern, 1981; Brannan and Yahr, 1995; Rinne et al., 1998; Parkinson study group, 2000; Schrag and Quinn, 2000; Rascol et al., 2000a; Holloway et al., 2004). It should be noted that some studies do not support the idea that later prescription of levodopa therapy promotes dyskinesia (Blin et al., 1988; Blanchet et al., 1996), and there is little evidence to suggest that it has a bearing on the emergence of severe troublesome LID, which seem to occur at the same stage whether levodopa treatment is delayed or not (Katzenschlager and Lees, 2002; Holloway et al., 2004). However, it has also been argued that dopamine agonists have a neuroprotective effect (Marek et al., 2002; Whone et al., 2003), and although this remains highly controversial (Morrish, 2003), it is widely believed that it is preferable to start de novo patients on a dopamine agonist as first-line therapy, particularly if they are young and are therefore at increased risk of developing this complication. It should however be noted that evidence suggests that levodopa is more effective in controlling the symptoms of PD, is often is better tolerated and may be associated with a better quality of life than comparators (Rascol et al., 2000a; Katzenschlager and Lees, 2002; Holloway et al., 2004).

When LID becomes severe enough to present a major management problem there are several additional strategies available, many targeting the non-dopaminergic pathways to avoid interfering with control of other aspects of the disease (Brotchie, 1998; Rascol, 2000b).

The drug amantadine has been used to treat Parkinson's Disease for many years but its anti-dyskinetic properties, which are thought mediated via its glutamate NMDA receptor antagonism, have only been well demonstrated fairly recently (Verhagen Metman et al., 1998; Luginger et al., 2000; Snow et al., 2000). However there is some doubt as to the true benefit and duration of effect (Crosby et al., 2003; Thomas et al., 2004) and although relatively easy to use, it does produce side-effects which means it is not suitable for all patients.

Other non-dopaminergic pharmacological agents, which have been investigated, include 5HT re-uptake inhibitors such as fluoxetine (Durif et al., 1995), opioid receptor antagonists such as naloxone and naltrexone (Sandyk and Snider, 1986), and alpha-2 adrenergic antagonists, such as yohimbine and idazoxan (Rascol et al., 1997), topiramate (Silverdale et al., 2004), levetiracetam (Hill et al., 2003). Although early animal studies and pilot studies with patients proved promising in many (Trabucchi et al., 1982; Henry and Brotchie, 1996; Rascol et al., 1997; Henry et al., 1999), a clear clinical benefit has not since been confirmed (Rascol et al., 1994; Manson et al., 2000c; Manson et al., 2001b; Colosimo and Craus, 2003).

A different approach is to try to selectively target the dopaminergic system to reduce dyskinesias without worsening parkinsonism and clozapine, a newer 'atypical' neuroleptic has shown promise in this respect (Bennett et al., 1993; Bennett et al., 1994; Durif et al., 1997; Pierelli et al., 1998; Durif et al., 2004). However, its potential side-effect profile means that it cannot be used routinely and unfortunately controlled trials of similar but less toxic drugs, such as olanzapine and quetiapine have not shown equal benefit (Manson et al., 2000b; Katzenschlager et al., 2004).

The administration of continuous subcutaneous infusions of the potent dopamine agonist apomorphine has been shown to significantly reduce LID, particularly when additional daytime medication is discontinued and apomorphine is given as monotherapy during the waking day (Colzi et al., 1998; Kanovsky et al., 2002; Manson et al., 2002; Katzenschlager et al., 2005). However, this treatment is complex and expensive and requires a motivation and cooperation from both patients and carers to enable success (Lees, 1993; Manson et al., 2002).

When severe, and when patients have exhausted available medical treatments, neurosurgical procedures, either by lesioning or deep brain stimulation (DBS) provide a further and effective option for treatment of LID (Pollak, 1999; Lang, 2000). The pallidum is the most effective target for treatment of LID (Kumar et al., 1998) although thalamotomy has been shown to have some anti-dyskinetic effect (Lang, 2000) and, more recently, subthalamotomy has been reported to reduce dyskinesias in some cases (Benabid et al., 2000). In recent years, DBS, which is largely reversible and allows adjustment of settings according to clinical need, have replaced lesioning techniques in many centres. DBS of the subthalamic nucleus enables a significant reduction in levodopa dosage and thus results in a marked reduction in LID (Benabid et al., 2000; Lang, 2000). However, it should be remembered that these neurosurgical techniques carry a significant risk with them and are still relatively experimental (Quinn and Bhatia, 1998). While lesioning techniques are irreversible, DBS, although largely reversible requires a high intensity of medical aftercare and is thus costly.

6 EFFECT OF LID ON QUALITY OF LIFE (QOL)

It is clear that LID are disabling and carry with them significant co-morbidity especially when severe. Patients complain that they are socially embarrassing, and prevent or impede their daily activities including vital tasks such as eating and drinking (Katzenschlager et al., 2003). In addition, they are associated with weight loss (Manson et al., 2002), which can present a major problem in PD, and other symptoms such as breathlessness and even vomiting (Steiger et al., 1991). Therefore one would expect LID to have a significant effect on QoL and some studies have reported worse QoL scores in patients with dyskinesias, particularly biphasic dyskinesias (Rubenstein et al., 1998; Damiano and McGrath MM, 2000; Kuopio et al., 2000; Pechevis et al., 2001; Siderowf et al., 2002; Chapuis et al., 2004). However, once confounding factors such as disease severity and treatment duration were accounted for, surprising clinical studies have mostly failed to show such association between dyskinesias and poor QoL (Schrag and Quinn, 2000; Marras et al., 2004). One explanation for this result could be that often dyskinesias are mild with little impact on daily living, whereas only a smaller group of patients with advanced disease have severe, disabling dyskinesia. In addition, if only patients with a good response to levodopa develop dyskinesia they would be expected to have a better quality of life than those receiving little anti-parkinsonian benefit from their medication (Schrag and Quinn, 2000). This hypothesis is supported by results by (Marras et al., 2004) who analyzed the association between dyskinesia and QoL in a prospective RCT of the dopamine agonist pramipexole versus levodopa in the treatment of de novo PD. They found that LID were associated with a better QoL in the first few years of treatment even after accounting for the difference in UPDRS scores of parkinsonism, suggesting that the presence of dyskinesia in this early disease group did not impact on patients' QoL. However, it is likely that more severe, disabling dyskinesia in more advanced disease are associated with poorer QoL, but longer follow-up studies with appropriate multivariate analyses are required to answer this question. Indeed as the trend for more quality of life measures to be introduced into clinical trials we will gain more information in this important area, to ensure that improvement in objective rating scale scores are translated in subjective benefit for the patients.

REFERENCES

Ahlskog JE, Muenter MD (2001) Frequency of levodopa-related dyskinesias and motor fluctuations as estimated from the cumulative literature. *Mov Disord* **16**:448-458.

Baas H (2000) Dyskinesia in Parkinson's disease. Pathophysiology and clinical risk factors. *J Neurol* 247 *Suppl* **4**:IV/12-16.

Ballard PA, Tetrud JW, Langston JW (1985) Permanent human parkinsonism due to 1-methyl-4-phenyl-1,2,3,6-tetrahydropyridine (MPTP): seven cases. *Neurology* **35**: 949-956.

Barbeau A (1980) High level levodopa therapy in severly akinetic Parkinsonism patients: twelve years later. In: *Parkinson's diseas: current progress, problems and management*. (Rinne VK KM, Stamm G, ed), pp 229-239. Amsterdam: Elsevier.

Benabid AL, Benazzouz A, Limousin P, Koudsie A, Krack P, Piallat B, Pollak P (2000) Dyskinesias and the subthalamic nucleus. *Ann Neurol* **47**:S189-192.

Bennett JP, Jr., Landow ER, Schuh LA (1993) Suppression of dyskinesias in advanced Parkinson's disease. II. Increasing daily clozapine doses suppress dyskinesias and improve parkinsonism symptoms [see comments]. *Neurology* **43**:1551-1555.

Bennett JP, Jr., Landow ER, Dietrich S, Schuh LA (1994) Suppression of dyskinesias in advanced Parkinson's disease: moderate daily clozapine doses provide long-term dyskinesia reduction. *Mov Disord* **9**:409-414.

Blanchet PJ, Allard P, Gregoire L, Tardif F, Bedard PJ (1996) Risk factors for peak dose dyskinesia in 100 levodopa-treated parkinsonian patients. *Can J Neurol Sci* **23**:189-193.

Blin J, Bonnet AM, Agid Y (1988) Does levodopa aggravate Parkinson's disease? *Neurology* **38**:1410-1416.

Block G, Liss C, Reines S, Irr J, Nibbelink D (1997) Comparison of immediate-release and controlled release carbidopa/levodopa in Parkinson's disease. A multicenter 5-year study. The CR First Study Group. *Eur Neurol* **37**:23-27.

Bozi M, Bhatia KP (2003) Paroxysmal exercise-induced dystonia as a presenting feature of young-onset Parkinson's disease. *Mov Disord* **18**:1545-1547.

Bracco F, Battaglia A, Chouza C, Dupont E, Gershanik O, Marti Masso JF, Montastruc JL (2004) The long-acting dopamine receptor agonist cabergoline in early Parkinson's disease: final results of a 5-year, double-blind, levodopa-controlled study. *CNS Drugs* **18**:733-746.

Brannan T, Yahr MD (1995) Comparative study of selegiline plus L-dopa-carbidopa versus L-dopa-carbidopa alone in the treatment of Parkinson's disease. *Ann Neurol* **37**:95-98.

Brotchie JM (1998) Adjuncts to dopamine replacement: a pragmatic approach to reducing the problem of dyskinesia in Parkinson's disease. *Mov Disord* **13**:871-876.

Brown P, Manson A (1999) Dyskinesias assessment and ambulatory devices. *Mov Disord 14 Suppl* **1**:67-68.

Chapuis S, Ouchchane L, Metz O, Gerbaud L, Durif F (2004) Impact of the motor complications of Parkinson's disease on the quality of life. Mov Disord 2004 Aug 31; [Epub ahead of print].

Chase TN (1998a) Levodopa therapy: consequences of the nonphysiologic replacement of dopamine. *Neurology* **50**:S17-25.

Chase TN (1998b) The significance of continuous dopaminergic stimulation in the treatment of Parkinson's disease. *Drugs* **55**:1-9.

Colosimo C, Craus A (2003) Noradrenergic drugs for levodopa-induced dyskinesia. *Clin Neuropharmacol* **26**:299-305.

Colzi A, Turner K, Lees AJ (1998) Continuous subcutaneous waking day apomorphine in the long term treatment of levodopa induced interdose dyskinesias in Parkinson's disease. *J Neurol Neurosurg Psychiatry* **64**:573-576.

Cotzias GC (1971) Levodopa in the treatment of Parkinsonism. *Jama* **218**:1903-1908.

Crosby NJ, Deane KH, Clarke CE (2003) Amantadine for dyskinesia in Parkinson's disease. *Cochrane Database Syst Rev*:CD003467.

Damiano AM, McGrath MM WM, SnyderCF, LeWitt PA, Richter RR (2000) Evaluation of a measurement strategy for Parkinson's disease: patient health-related quality of life. *Quality of Life Research* **9**:87-100.

Damier P, Jaillon C, Clavier I, Arnulf I, Bonnet AM, Bejjani BP, Agid Y (1999) Dyskinesias assessment in Phase II studies. *Mov Disord 14 Suppl* **1**:54-59.

Deane K, Spieker S, Clarke C (2004) Catechol-O-methyltransferase inhibitors for levodopa-induced complications in Parkinson's disease. *Cochrane Database Syst Rev* **4**.

Denny AP, Behari M (1999) Motor fluctuations in Parkinson's disease. *J Neurol Sci* **165**: 18-23.

Dupont E, Andersen A, Boas J, Boisen E, Borgmann R, Helgetveit AC, Kjaer MO, Kristensen TN, Mikkelsen B, Pakkenberg H, Presthus J, Stien R, Worm-Petersen J, Buch D (1996) Sustained-release Madopar HBS compared with standard Madopar in the long-term treatment of de novo parkinsonian patients. *Acta Neurol Scand* **93**:14-20.

Durif F (1999b) Treating and preventing levodopa-induced dyskinesias: current and future strategies [In Process Citation]. *Drugs Aging* **14**:337-345.

Durif F, Vidailhet M, Debilly B, Agid Y (1999a) Worsening of levodopa-induced dyskinesias by motor and mental tasks [In Process Citation]. *Mov Disord* **14**:242-245.

Durif F, Vidailhet M, Bonnet AM, Blin J, Agid Y (1995) Levodopa-induced dyskinesias are improved by fluoxetine. *Neurology* **45**:1855-1858.

Durif F, Vidailhet M, Assal F, Roche C, Bonnet AM, Agid Y (1997) Low-dose clozapine improves dyskinesias in Parkinson's disease. *Neurology* **48**:658-662.

Durif F, Debilly B, Galitzky M, Morand D, Viallet F, Borg M, Thobois S, Broussolle E, Rascol O (2004) Clozapine improves dyskinesias in Parkinson disease: a double-blind, placebo-controlled study. *Neurology* **62**:381-388.

Fahn S (2000) The spectrum of levodopa-induced dyskinesias. *Ann Neurol* **47**:S2-9; discussion S9-11.

Fahn S, Elton R, Committee. amotUD (1987) Unified Parkinson's Disease rating Scale. In: *Recent Developments in Parkinson's Disease* (Fahn S MC, Calne DB, Goldstein M, ed), pp 153-164. Florham Park, NJ: macmillan Health Care Information.

Garcia Ruiz PJ, Meseguer E, Del Val J, Vazquez A, Sanchez Bernardos V (2004) Motor complications in Parkinson disease: a prospective follow-up study. *Clin Neuropharmacol* **27**:49-52.

Goetz CG (1999) Rating scales for dyskinesias in Parkinson's disease. *Mov Disord* **14**:48-53.

Goetz CG, Stebbins GT, Blasucci LM, Grobman MS (1997) Efficacy of a patient-training videotape on motor fluctuations for on- off diaries in Parkinson's disease. *Mov Disord* **12**:1039-1041.

Goetz CGea (2000) Objective changes in motor function during placebo treatment in PD. *Neurology* **54**:710-714.

Grandas F, Galiano ML, Tabernero C (1999) Risk factors for levodopa-induced dyskinesias in Parkinson's disease. *J Neurol* **246**:1127-1133.

Hadj Tahar A, Grondin R, Gregoire L, Calon F, Di Paolo T, Bedard PJ (2003) New insights in Parkinson's disease therapy: can levodopa-induced dyskinesia ever be manageable. *Adv Neurol* **91**:51-64.

Henry B, Brotchie JM (1996) Potential of opioid antagonists in the treatment of levodopa-induced dyskinesias in Parkinson's disease. *Drugs Aging* **9**:149-158.

Henry B, Fox SH, Peggs D, Crossman AR, Brotchie JM (1999) The alph2-adrenergic receptor antagonist Idazoxan reduces dyskineia and enhances anti-parkinsonian actions of l-dopa in the MPTP-lesioned primate model of Parkinson's disease. *Mov Disord* **14**, No.5:744-753.

Hill MP, Bezard E, McGuire SG, Crossman AR, Brotchie JM, Michel A, Grimee R, Klitgaard H (2003) Novel antiepileptic drug levetiracetam decreases dyskinesia elicited by L-dopa and ropinirole in the MPTP-lesioned marmoset. *Mov Disord* **18**:1301-1305.

Hoff JI, van der Meer V, van Hilten JJ (2004) Accuracy of objective ambulatory accelerometry in detecting motor complications in patients with Parkinson disease. *Clin Neuropharmacol* **27**:53-57.

Holloway RG, Shoulson I, Fahn S, Kieburtz K, Lang A, Marek K, McDermott M, Seibyl J, Weiner W, Musch B, Kamp C, Welsh M, Shinaman A, Pahwa R, Barclay L, Hubble J, LeWitt P, Miyasaki J, Suchowersky O, Stacy M, Russell DS, Ford B, Hammerstad J, Riley D, Standaert D, Wooten F, Factor S, Jankovic J, Atassi F, Kurlan R, Panisset M, Rajput A, Rodnitzky R, Shults C, Petsinger G, Waters C, Pfeiffer R, Biglan K, Borchert L, Montgomery A, Sutherland L, Weeks C, DeAngelis M, Sime E, Wood S, Pantella C, Harrigan M, Fussell B, Dillon S, Alexander-Brown B, Rainey P, Tennis M, Rost-Ruffner E, Brown D, Evans S, Berry D, Hall J, Shirley T, Dobson J, Fontaine D, Pfeiffer B, Brocht A, Bennett S, Daigneault S, Hodgeman K, O'Connell C, Ross T, Richard K, Watts A (2004) Pramipexole vs levodopa as initial treatment for Parkinson disease: a 4-year randomized controlled trial. *Arch Neurol* **61**:1044-1053.

Hundemer HP, Lledo A, van Laar T, Quail D, Oertel W, Schwarz J, Wolters E (2000) The safety of pergolide monotherapy in early-stage Parkinson's disease. One-year interim analysis of a 3-year double-blind, randomized study of pergolide versus levodopa. *Mov Disord* **15** (suppl 3):115.

Jenner P (2000) Factors influencing the onset and persistence of dyskinesia in MPTP-treated primates. *Ann Neurol* **47**:S90-99; discussion S99-104.

Jenner P (2004) Avoidance of dyskinesia: preclinical evidence for continuous dopaminergic stimulation. *Neurology* **62**:S47-55.

Kanovsky P, Kubova D, Bares M, Hortova H, Streitova H, Rektor I, Znojil V (2002) Levodopa-induced dyskinesias and continuous subcutaneous infusions of apomorphine: results of a two-year, prospective follow-up. *Mov Disord* **17**:188-191.

Katzenschlager R, Lees AJ (2002) Treatment of Parkinson's disease: levodopa as the first choice. *J Neurol 249 Suppl* **2**:II19-24.

Katzenschlager R, Manson AJ, Evans A, Watt H, Lees AJ (2004) Low dose quetiapine for drug induced dyskinesias in Parkinson's disease: a double blind cross over study. *J Neurol Neurosurg Psychiatry* **75**:295-297.

Katzenschlager R, Schrag A, Hobart J, Manson A, Evans A, AJ. L (2003) The impact of dyskinesias in Parkinson's disease: development of a patient-based outcome measure. *J Neurol* **250**:32.

Katzenschlager R, Hughes A, Evans A, Manson AJ, Hoffman M, Swinn L, Watt H, Bhatia K, Quinn N, Lees AJ (2005) Continuous subcutaneous apomorphine therapy improves dyskinesias in Parkinson's disease: A prospective study using single-dose challenges. *Mov Disord* **20**:151-157.

Khan NL, Katzenschlager R, Watt H, Bhatia KP, Wood NW, Quinn N, Lees AJ (2004) Olfaction differentiates parkin disease from early-onset parkinsonism and Parkinson disease. *Neurology* **62**:1224-1226.

Koller WC, Hutton JT, Tolosa E, Capilldeo R (1999) Immediate-release and controlled-release carbidopa/levodopa in PD: a 5-year randomized multicenter study. Carbidopa/Levodopa Study Group. *Neurology* **53**:1012-1019.

Kostic V, Przedborski S, Flaster E, Sternic N (1991) Early development of levodopa-induced dyskinesias and response fluctuations in young-onset Parkinson's disease. *Neurology* **41**:202-205.

Kumar N, Van Gerpen JA, Bower JH, Ahlskog JE (2004) Levodopa-dyskinesia incidence by age of Parkinson's disease onset. Mov Disord.

Kumar R, Lozano AM, Montgomery E, Lang AE (1998) Pallidotomy and deep brain stimulation of the pallidum and subthalamic nucleus in advanced Parkinson's disease. *Mov Disord* **13**:73-82.

Kuopio AM, Marttila R, Helenius H, Toivonen M, UK. R (2000) The quality of life in Parkinson's disease. *Mov Disord* **15**:216-223.

Lang AE (2000) Surgery for levodopa-induced dyskinesias. *Ann Neurol* **47**:S193-199; discussion S199-202.

Langston JW, Quik M, Petzinger G, Jakowec M, Di Monte DA (2000) Investigating levodopa-induced dyskinesias in the parkinsonian primate. *Ann Neurol* **47**:S79-89.

Lees AJ (1993) Dopamine agonists in Parkinson's disease: a look at apomorphine. *Fundam Clin Pharmacol* **7**:121-128.

Lees AJ, Stern GM (1981) Sustained bromocriptine therapy in previously untreated patients with Parkinson's disease. *J Neurol Neurosurg Psychiatry* **44**:1020-1023.

Lesser RP, Fahn S, Snider SR, Cote LJ, Isgreen WP, Barrett RE (1979) Analysis of the clinical problems in parkinsonism and the complications of long-term levodopa therapy. *Neurology* **29**:1253-1260.

LeWitt PA (1998) Conjugate eye deviations as dyskinesias induced by levodopa in Parkinson's disease. *Mov Disord* **13**:731-734.

Luginger E, Wenning GK, Bosch S, Poewe W (2000) Beneficial effects of amantadine on L-dopa-induced dyskinesias in Parkinson's disease [In Process Citation]. *Mov Disord* **15**:873-878.

Lyons KE, Hubble JP, Troster AI, Pahwa R, Koller WC (1998) Gender differences in Parkinson's disease. *Clin Neuropharmacol* **21**:118-121.

Manson AJ, Schrag A, Lees AJ (2000b) Low-dose olanzapine for levodopa induced dyskinesias. *Neurology* **55**:795-799.

Manson AJ, Iakovidou E, Lees AJ (2000c) Idazoxan is ineffective for levodopa-induced dyskinesias in Parkinson's disease. *Mov Disord* **15**:336-337.

Manson AJ, Turner K, Lees AJ (2002) Apomorphine monotherapy in the treatment of refractory motor complications of Parkinson's disease: long-term follow-up study of 64 patients. *Mov Disord* **17**:1235-1241.

Manson AJ, Katzenschlager R, Hobart J, Lees AJ (2001b) High dose naltrexone for dyskinesias induced by levodopa. *J Neurol Neurosurg Psychiatry* **70**:554-556.

Manson AJ, Brown P, O'Sullivan JD, Asselman P, Buckwell D, Lees AJ (2000a) An ambulatory dyskinesia monitor. *J Neurol Neurosurg Psychiatry* **68**:196-201.

Manson AJ, Hanagasi H, Turner K, Patsalos PN, Carey P, Ratnaraj N, Lees AJ (2001a) Intravenous apomorphine therapy in Parkinson's disease: clinical and pharmacokinetic observations. *Brain* **124**:331-340.

Marconi R, Lefebvre-Caparros D, Bonnet AM, Vidailhet M, Dubois B, Agid Y (1994) Levodopa-induced dyskinesias in Parkinson's disease phenomenology and pathophysiology. *Mov Disord* **9**:2-12.

Marek K, Seibyl J, Shoulson I, Holloway R, Kieburtz K, McDermott M, Kamp C, Shinaman A, Fahn S, Lang A, Weiner W, Welsh M (2002) Dopamine transporter brain imaging to assess the effects of pramipexole vs levodopa on Parkinson disease progression. *Jama* **287**:1653-1661.

Marras C, Lang A, Krahn M, Tomlinson G, Naglie G (2004) Quality of life in early Parkinson's disease: impact of dyskinesias and motor fluctuations. *Mov Disord* **19**:22-28.

Martinelli P, Contin M, Scaglione C, Riva R, Albani F, Baruzzi A (2003) Levodopa pharmacokinetics and dyskinesias: are there sex-related differences? *Neurol Sci* **24**:192-193.

Melamed E, Olanow CW, Nutt JG, Lang AE (1999) Dyskinesias assessment workshop: reports from the working groups. *Mov Disord* **14**:69-73.

Miyawaki E, Lyons K, Pahwa R, Troster AI, Hubble J, Smith D, Busenbark K, McGuire D, Michalek D, Koller WC (1997) Motor complications of chronic levodopa therapy in Parkinson's disease. *Clin Neuropharmacol* **20**:523-530.

Mones RJ, Elizan TS, Siegel G (1970) L-dopa induced dyskinesias in 152 patients with Parkinson's disease. *Trans Am Neurol Assoc* **95**:286-287.

Montastruc JL, Rascol O, Senard JM, Rascol A (1994) A randomised controlled study comparing bromocriptine to which levodopa was later added, with levodopa alone in previously untreated patients with Parkinson's disease: a five year follow up. *J Neurol Neurosurg Psychiatry* **57**:1034-1038.

Morrish PK (2003) REAL and CALM: what have we learned? *Mov Disord* **18**:839-840.

Muenter MD, Tyce GM (1971) L-dopa therapy of Parkinson's disease: plasma L-dopa concentration, therapeutic response, and side effects. *Mayo Clin Proc* **46**:231-239.

Muenter MD, Sharpless NS, Tyce GM, Darley FL (1977) Patterns of dystonia ("I-D-I" and "D-I-D-") in response to l-dopa therapy for Parkinson's disease. *Mayo Clin Proc* **52**: 163-174.

Murphy MJ (1979) L-dopa induced dyskinesias in asymmetrical parkinsonism. *Acta Neurol Latinoam* **25**:47-52.

Musch B, group TIs (2000) Evaluation of patients with Parkinson's disease treated with cabergoline alone in a 5-year comparative study of cabergoline versus levodopa: A subgroup analysis. *Mov Disord* **15** (suppl 3):121.

Nutt JG (1990) Levodopa-induced dyskinesia: review, observations, and speculations. *Neurology* **40**:340-345.

Nutt JG (2000) Clinical pharmacology of levodopa-induced dyskinesia. *Ann Neurol* **47**:S160-164; discussion S164-166.

Olanow CW, Obeso JA (2000) Preventing levodopa-induced dyskinesias. *Ann Neurol* **47**:S167-176; discussion S176-168.

Onofrj M, Paci C, Thomas A (1998) Sudden appearance of invalidating dyskinesia-dystonia and off fluctuations after the introduction of levodopa in two dopaminomimetic drug naive patients with stage IV Parkinson's disease [letter]. *J Neurol Neurosurg Psychiatry* **65**:605-606.

Pahwa R, Lyons K, McGuire D, Silverstein P, Zwiebel F, Robischon M, Koller WC (1997) Comparison of standard carbidopa-levodopa and sustained-release carbidopa-levodopa in Parkinson's disease: pharmacokinetic and quality-of-life measures. *Mov Disord* **12**:677-681.

Papapetropoulos S, Argyriou AA, Ellul J, Chroni E (2004) Comparison of motor fluctuations and L-dopa-induced dyskinesias in patients with familial and sporadic Parkinson's disease. *Eur J Neurol* **11**:115-119.

Parkinson study group (1996) Impact of deprenyl and tocopherol treatment on Parkinson's disease in DATATOP patients requiring levodopa. Parkinson Study Group. *Ann Neurol* **39**:37-45.

Parkinson study group (2000) Pramipexole vs levodopa as initial treatment for Parkinson disease: A randomized controlled trial. Parkinson Study Group. *Jama* **284**:1931-1938.

Pechevis M, Clarke C, VIeregge P, al. e (2001) Direct and indirect costs of Parkinson's Disease and L-dopa induced dyskinesias: A prospective European study. *Parkinsonism Relat Disord* **7**(suppl):106.

Pierelli F, Adipietro A, Soldati G, Fattapposta F, Pozzessere G, Scoppetta C (1998) Low dosage clozapine effects on L-dopa induced dyskinesias in parkinsonian patients. *Acta Neurol Scand* **97**:295-299.

Poewe WH, Lees AJ, Stern GM (1986) Low-dose L-dopa therapy in Parkinson's disease: a 6-year follow-up study. *Neurology* **36**:1528-1530.

Pollak P (1999) Neurosurgical treatment of dyskinesias: pathophysiological consideration. *Mov Disord* 14 *Suppl* **1**:33-39.

Quinn N, Bhatia K (1998) Functional neurosurgery for Parkinson's disease. Has come a long way, though much remains experimental [editorial]. *Bmj* **316**:1259-1260.

Quinn N, Critchley P, Marsden CD (1987) Young onset Parkinson's disease. *Mov Disord* **2**:73-91.

Rascol O (1999) L-dopa-induced peak-dose dyskinesias in patients with Parkinson's disease: a clinical pharmacologic approach. *Mov Disord* **14**:19-32.

Rascol O (2000b) Medical treatment of levodopa-induced dyskinesias. *Ann Neurol* **47**: S179-188.

Rascol O, Arnulf I, Agid Y, al e (1997) L-dopa-Induced dyskinesias improvement by an alpha-2 antagonist, idazoxan in patients with Parkinson's disease. *Mov Disord* **12** (suppl 1):111

Rascol O, Brooks DJ, Korczyn AD, De Deyn PP, Clarke CE, Lang AE (2000a) A five-year study of the incidence of dyskinesia in patients with early Parkinson's disease who were treated with ropinirole or levodopa. 056 Study Group. *N Engl J Med* **342**:1484-1491.

Rascol O, Fabre N, Blin O, Poulik J, Sabatini U, Senard JM, Ane M, Montastruc JL, Rascol A (1994) Naltrexone, an opiate antagonist, fails to modify motor symptoms in patients with Parkinson's disease. *Mov Disord* **9**:437-440.

Rinne UK, Bracco F, Chouza C, Dupont E, Gershanik O, Marti Masso JF, Montastruc JL, Marsden CD (1998) Early treatment of Parkinson's disease with cabergoline delays the onset of motor complications. Results of a double-blind levodopa controlled trial. *The PKDS009 Study Group. Drugs* **55**:23-30.
Rubenstein LM, Voelker M, Chrischilles E, Glenn D, Wallace R, Rodnitzky R (1998) The usefulness of the Functional Status Questionnaire and Medical Outcomes Study Short Form in Parkinson's disease research. *Qual Life Res* 7:27-90.
Sandyk R, Snider SR (1986) Naloxone treatment of L-dopa-induced dyskinesias in Parkinson's disease [letter]. *Am J Psychiatry* **143**:118.
Schneider JS (1989) Levodopa-induced dyskinesias in parkinsonian monkeys: relationship to extent of nigrostriatal damage. *Pharmacol Biochem Behav* **34**:193-196.
Schneider JS, Gonczi H, Decamp E (2003) Development of levodopa-induced dyskinesias in parkinsonian monkeys may depend upon rate of symptom onset and/or duration of symptoms. *Brain Res* **990**:38-44.
Schrag A, Quinn N (2000) Dyskinesias and motor fluctuations in Parkinson's disease. A community- based study. *Brain* **123**:2297-2305.
Schrag A, Ben-Shlomo Y, Brown R, Marsden CD, Quinn N (1998) Young-onset Parkinson's disease revisited--clinical features, natural history, and mortality. *Mov Disord* **13**:885-894.
Siderowf A, Ravina B, Glick HA (2002) Preference-based quality-of-life in patients with Parkinson's disease. *Neurology* **59**:103-108.
Silverdale MA, Nicholson SL, Crossman AR, Brotchie JM (2004) Topiramate reduces levodopa-induced dyskinesia in the MPTP-lesioned marmoset model of Parkinson's disease. Mov Disord.
Smith LA, Jackson MJ, Al-Barghouthy G, Rose S, Kuoppamaki M, Olanow W, Jenner P (2004) Multiple small doses of levodopa plus entacapone produce continuous dopaminergic stimulation and reduce dyskinesia induction in MPTP-treated drug-naive primates. *Mov Disord Oct* **15**; [Epub ahead of print].
Snow BJ, Macdonald L, McAuley D, Wallis W (2000) The effect of amantadine on levodopa-induced dyskinesias in Parkinson's disease: a double-blind, placebo-controlled study. *Clin Neuropharmacol* **23**:82-85.
Steiger MJ, Quinn NP, Marsden CD (1991) Sickness due to levodopa-induced neck dyskinesias in Parkinson's disease [letter]. *Mov Disord* **6**:91-92.
Stocchi F, Olanow C (2004) Continuous dopaminergic stimulation in early and advanced Parkinson's disease. *Neurology* **62** (suppl1):S56-63.
Stocchi F, Nordera G, Marsden CD (1997) Strategies for treating patients with advanced Parkinson's disease with disastrous fluctuations and dyskinesias. *Clin Neuropharmacol* **20**:95-115.
Tandberg E, Larsen JP, Nessler EG, Riise T, Aarli JA (1995) The epidemiology of Parkinson's disease in the county of Rogaland, Norway. *Mov Disord* **10**:541-549.
Tanner C (1986) Drug-induced movement disorders (tardive dyskinesia and dopa-induced dyskinesa). In: *Extrapyramidal disorders. Handbook of clinical neurology* (PJ Ninken GB, HL Klawans, ed), pp 185-204. Amsterdam: Elsevier Science.
Thomas A, Iacono D, Luciano AL, Armellino K, Di Iorio A, Onofrj M (2004) Duration of amantadine benefit on dyskinesia of severe Parkinson's disease. *J Neurol Neurosurg Psychiatry* **75**:141-143.
Trabucchi M, Bassi S, Frattola L (1982) Effect of naloxone on the "on-off' syndrome in patients receiving long- term levodopa therapy. *Arch Neurol* **39**:120-121.
Verhagen Metman L, van den Munckhof P, Klaassen AA, Blanchet P, Mouradian MM, Chase TN (1997) Effects of supra-threshold levodopa doses on dyskinesias in advanced Parkinson's disease [see comments]. *Neurology* **49**:711-713.
Verhagen Metman L, Del Dotto P, van den Munckhof P, Fang J, Mouradian MM, Chase TN (1998) Amantadine as treatment for dyskinesias and motor fluctuations in Parkinson's disease [see comments]. *Neurology* **50**:1323-1326.
Vitale C, Pellecchia MT, Grossi D, Fragassi N, Cuomo T, Di Maio L, Barone P (2001) Unawareness of dyskinesias in Parkinson's and Huntington's diseases. *Neurol Sci* **22**:105-106.
Whone AL, Watts RL, Stoessl AJ, Davis M, Reske S, Nahmias C, Lang AE, Rascol O, Ribeiro MJ, Remy P, Poewe WH, Hauser RA, Brooks DJ (2003) Slower progression of Parkinson's disease with ropinirole versus levodopa: The REAL-PET study. *Ann Neurol* **54**:93-101.
Widner H, Defer GL (1999) Dyskinesias assessment: from CAPIT to CAPSIT. Core Assessment Program for Intracerebral Transplantations. Core Assessment Program for Surgical Interventional Therapies. *Mov Disord* **14** Suppl 1:60-66.

In: Recent Breakthroughs in Basal Ganglia Research
Editor: Erwan Bezard, pp. 381-382
ISBN: 1-59454-880-3

PART IX: DEEP BRAIN STIMULATION: THE PANACEA TO BASAL GANGLIA DISEASES?

Abdelhamid Benazzouz
Laboratoire de Neurophysiologie, CNRS UMR 5543,
Université Victor Segalen Bordeaux 2, 33076 Bordeaux, France.

High frequency stimulation (HFS), named also deep brain stimulation (DBS), is actually considered as an effective therapeutical approach in several movement disorders like advanced stages of Parkinson's disease and dystonia. DBS is also experienced in non motor pharmacoresistant pathologies like epilepsy and obsessive convulsive disorders.

The idea that DBS may constitute an effective therapy of chronic disorders came in the 1950s as an alternative to ablative neurosurgical procedures. The first user of DBS was Pool (1954), who implanted a stimulating electrode in the caudate nucleus of a patient suffering from depression, anorexia and Parkinson's disease. The author reported that DBS improved mood and appetite of the patient during the 8th weeks of stimulation. In the 1960s, electrical stimulation of the thalamus was used to determine with precision the functional location prior to ablation of the VIM (ventro-median nucleus) of the thalamus for the treatment of parkinsonian rest tremor (Hassler et al., 1960). Since VIM stimulation reversed tremor only at high frequencies (Nashold and Slaughter, 1969 ; Ohye et al., 1977) and as it is a reversible technique, it was proposed that chronic HFS of the VIM could be used as an adjuvent therapy to contralateral thalamotomy in order to avoid the side effects of bilateral thalamotomies (Benabid et al., 1987). In the 1990s, bilateral DBS of the VIM became a therapy of choice instead of the unilateral thalamotomy (Benabid et al., 1991, 1996; Blond et al., 1992). As the VIM has been reported to be implicated in only tremor without any effect on akinesia and rigidity, which are the most invalidant symptoms, the subthalamic nucleus (STN) has been shown to be implicated in the pathophysiology of Parkinson's disease by the hyperactivity of its neurons (Miller and De Long, 1987). Moreover, experimental surgical lesions of the STN have been reported to alleviate the three cardinal motor symptoms, akinesia, rigidity and tremor in non human primate rendered parkinsonian by systemic injections of MPTP (Bergman et al., 1990; Aziz et al., 1991). These authors have shown that the improvement of motor symptoms was accompanied by the appearance of abnormal movements. Given these results, we proposed to replace lesions by HFS of the STN and we have shown that 100 Hz stimulations induced a dramatic improvement of rigidity and motor performances in MPTP-treated monkeys (Benazzouz et al., 1993). In addition we have shown that the degree of improvement was similar to that induced by L-Dopa and that the current intensity inducing positive effects did not induce side effects (Benazzouz et al., 1993, 1996). Regarding these results, it was logical to apply this therapeutical approach to human patients suffering from advanced stages of Parkinson's disease. This new therapeutical approach was transfered to invalid parkinsonian patients, in which HFS was able to reverse the motor symptoms with a dramatic improvement of quality of life (Benabid et al., 1994; Limousin et al., 1995, 1996, 1998; Krack et al., 1998, 2000, 2004). The main interest of DBS has been the demonstration that this technique has a similar effect to that obtained by lesion procedure except the advantage that with DBS side effects can be controled by varying the stimulating voltage. After our successfull experience, HFS of the STN was used in a large number of centers in the word (Bejjani et al., 2000; Houetto et al., 2002; Lozano et al. 2003).

Despite the striking clinical benefits seen with the application of chronic stimulation, the mechanism through which DBS acts has puzzled both clinicians and neuroscientists. From a functional point of view, electrical stimulation at high frequency usually mimics the effect of ablation procedures by inducing the same beneficial effects. From these observations, it was suggested that HFS of the STN may modify the activity of the target nuclei by exercising an inhibitory effect. However, from the recent findings using different techniques (electrophysiology, microdialysis, functional anatomy, etc ...) and different preparations (in vivo and in vitro) it appears that the mechanism of DBS is more complex than expected.

The following chapters of this book section provide an overview of recent data on different pathologies for which DBS is now considered as the main effective therapy. The pathologies detailed here are Parkinson's disease, dystonia, obsessive convulsive disorder and epilepsy.

In: Recent Breakthroughs in Basal Ganglia Research
Editor: Erwan Bezard, pp. 383-392 ISBN: 1-59454-880-3

Chapter 30

Psychosurgery of Obsessive-Compulsive Disorder a New Indication for Deep Brain Stimulation ?

Dominique Guehl, Abdelhamid Benazzouz, Bernard Bioulac and Pierre Burbaud

Service de Neurophysiologie Clinique, Université Victor Segalen Bordeaux 2, CNRS UMR 5543, Centre Hospitalier Pellegrin, Place Amélie-Raba Léon, 33076 Bordeaux, France.

Emmanuel Cuny and Alain Rougier

Service de Neurochirurgie, Université Victor Segalen Bordeaux 2, Centre Hospitalier Pellegrin, Place Amélie-Raba Léon, 33076 Bordeaux, France.

Jean Tignol

Service de Psychiatrie d'Adultes, Université Victor Segalen Bordeaux 2, Centre Hospitalier Charles Perrens, Centre Carreire, 121 rue de la Béchade, 33076 Bordeaux, France.

Bruno Aouizerate

Service de Psychiatrie d'Adultes, Université Victor Segalen Bordeaux 2, Centre Hospitalier Charles Perrens, Centre Carreire, 121 rue de la Béchade, 33076 Bordeaux, France. Service de Neurophysiologie Clinique, Université Victor Segalen Bordeaux 2, CNRS UMR 5543, Centre Hospitalier Pellegrin, Place Amélie-Raba Léon, 33076 Bordeaux, France.

Abstract

Obsessive-compulsive disorder (OCD) is a relatively common psychiatric condition with an estimated lifetime prevalence of 2-3 % of the general population. It is generally characterized by a chronic course leading to a profound impairment in psychosocial functioning and quality of life. Although its pathophysiology is still far from resolved, an ample body of literature suggests the role of an overactivity in the frontal-subcortical loops originating in the orbitofrontal cortex and the anterior cingulate cortex, respectively. Today, the well-established efficacy of antidepressants, acting preferentially by blocking serotonin reuptake, in addition to psychological treatments, have considerably changed the poor prognosis of the illness. However, both conventional therapeutic approaches failed to substantially alleviate obsessive-compulsive symptoms in 20-30% of cases. From these considerations, several surgical strategies have been successfully proposed for treating the resistant forms of OCD, including anterior capsulotomy, anterior cingulotomy, subcaudate tractotomy and limbic leucotomy. They have been performed in order to interrupt the orbitofrontal and anterior cingulate loops at either cortical or subcortical level, which have been found as disrupted in OCD. Despite these lesion-producing techniques have relatively low incidence of serious complications or undesirable effects, deep brain stimulation, as a reversible procedure, has recently been introduced in the field of OCD, primarily targeting the limb part of the internal capsule or the ventral caudate nucleus. It has been shown to be of promising benefit that should be considered with caution and confirmed by further research in this area.

INTRODUCTION

Obsessive-compulsive disorder (OCD) is a relatively common anxiety disorder that affects 2 to 3 % of the general population (Robins et al., 1984; Karno et al., 1988). According to this estimated lifetime prevalence rate, OCD is the fourth most frequent psychiatric condition following the phobias, substance abuse and major depression. OCD, which is characterized by recurrent intrusive thoughts and repetitive, time-consuming behaviors, is a severely disabling mental illness owing to its intensity, the continuous and unchanging or deteriorative course of its symptoms and the disturbance in psychosocial functioning that they cause (Rasmussen and Tsuang, 1984; Koran et al., 1996; Antony et al., 1998; Skoog and Skoog, 1999).

Although the pathophysiology of OCD is still far from resolved, the existence of a biological basis for OCD is now clearly established and will be discussed from a phenomenological point of view, on the one hand, and in the light of our increasing knowledge of the physiology of prefrontal cortex-striatal-thalamic-cortical functional loops, on the other. The present chapter will then examine the therapeutic effects of various psychosurgical approaches used for the management of OCD forms, which were unresponsive to the prototypical treatments classically proposed in this psychiatric condition. All these procedures have consisted in performing lesions in order to produce an interruption of fibers connecting the thalamus to the prefrontal cortex through the anterior limb of the internal capsule. Although they have not been shown to generate serious complications and/or untoward effects in most cases, deep brain stimulation (DBS) has recently been considered as a potential alternative for treating resistant OCD.

1 BRIEF OVERETVIEW OF THE PATHOPHYSIOLOGY OF OCD

1.1 Phenomenological Aspects

The heart of the obsessional process is the subject's underlying impression that "something is wrong" (Schwartz, 1998, 1999). In other words, obsessions may be thought of as the permanent perception of a mistake and/or error in certain behavioral situations. Compulsions occur as behavioral responses aiming to relieve the tensions or anxiety generated by the situation. If obtained, this relief may be felt to be a form of reward. Nevertheless, it is only transient, thereby creating a feeling of considerable anxiety. This leads to immediately reproduce the behavior in a cyclic manner on the basis of an internal motivational state through an expectation of the reward. Although it is undoubtedly simplistic, this phenomenological view is of interest because it suggests that several malfunctioning processes are altered within the OCD: 1) error recognition; and, 2) emotion and motivation and their activational aspects, e.g. activations for initiation and sustaining behavioral reactions and tendency to work for reward. Therefore, it can be expected in OCD that there is a dysfunction of the brain regions that mediate the processes of error detection and/or management of the reward and emotional systems, including orbitofrontal and anterior cingulate cortices and ventral striatum, respectively (Devinsky et al., 1995; Le Moal, 1995; Piazza and Le Moal, 1996; Koob and Le Moal, 1997; Piazza and Le Moal, 1997; Schwartz, 1998, 1999; Bush et al., 2000; Paus, 2001; Bush et al., 2002).

1.2 Anatomo-functional Data

Several studies have found a dysregulation in the frontal-subcortical circuits originating in the orbitofrontal and anterior cingulate cortices, respectively, in the pathogenesis of OCD. These loops concern discrete regions of the striatum and the pallidum and receive clearly delimited inputs from the cortex, which are subsequently integrated as far as the thalamus. Although they are part of a parallel model of organization, these circuits are permanently interrelated (Cummings, 1993).

The orbitofrontal cortex (OFC) and ventromedial areas are involved in appraisal in determining the emotional and motivational values of environmental information, and in integrating the subject's prior experience, which is crucial in decision-making (Charney and Bremner, 1999; Tremblay and Schultz, 1999, 2000a, 2000b; Krawczyk, 2002; Ramnani and Owen, 2004). The OFC also contributes to the selection, comparison and judgment of stimuli and error detection process (Rosenkilde et al., 1981; Thorpe et al., 1983; Ramnani and Owen, 2004).

The anterior cingulate cortex (ACC) is comprised of 1) a ventral or affective region which is intimately connected to the orbitofrontal cortex, ventral striatum, amygdala, hippocampus and hypothalamus. This subdivision could keep attention on the internal emotional and motivational status and participates to the regulation of autonomic responses; and, 2) a dorsal or cognitive region which is interconnected to the dorsolateral prefrontal cortex, parietal cortex, and premotor and supplementary motor areas. This cortical area serves a wide range of cognitive functions including attention, working memory, error detection, conflict monitoring and response selection with decisional and planning aspects in situations with high likelihood of making an error, and anticipation of incoming information (Niki and Watanabe, 1979; Devinsky et al., 1995; Shima and Tanji, 1998; Schwartz, 1999; Bush et al., 2000; Akkal et al., 2002; Bush et al., 2002; Shidara and Richmond, 2002).

Ventral striatum that is received important inputs from the OFC and ACC, is primarily implicated in the preparation, initiation and execution of behavioral responses oriented toward reward delivery after cognitive and emotional integration of behaviorally relevant information at the cortical level (Hollerman et al., 1998; Tremblay et al., 1998; Hassani et al., 2001).

The medial dorsal nucleus (MD) is the principal thalamic nucleus related to the prefrontal cortex. It is composed of several subdivisions, the main portion being centrally located and represented by the parvicellular part (MDpc), and its most medial portion corresponding to the magnocellular part (MDmc; Goldman-Rakic and Porrino, 1985). The MDpc receives inputs from the substantia nigra, which has a major role in cognition, and projects preferentially to the ACC (Barbas, 2000; McFarland and Haber, 2002). On the other hand, the MDmc, which receives inputs from the limbic portions of the basal ganglia (ventral striatum) and from the amygdala, projects mainly to the OFC (Barbas, 2000; McFarland and Haber, 2002).

1.3 Functional Neuroimagery

Functional imaging research has shown that several brain structures in OCD patients have abnormal functional activity. Most teams have reported an increased functional activity in the OFC, ACC, head of the caudate nucleus and thalamus (Saxena et al., 1998; Baxter, 1999; Aouizerate et al., 2004a). These functional abnormalities have been found in basal conditions and during provocation tests. Moreover, the therapeutic efficacy of antidepressants and cognitive-behavioral therapies seems to be associated with a progressive reduction in activity of the OFC, ACC and the caudate nucleus, thereby strengthening other data suggesting a functional deterioration in the orbitofrontal and anterior cingulate cortex system in the pathogenesis of OCD (Saxena et al., 1998; Baxter, 1999).

In conclusion, it seems in OCD that there is a dysfunction of the brain regions that belong to the orbitofrontal and anterior cingulate loops in view of evidences obtained from separate and complementary approaches

2 Psychosurgical Treatments

Over the last 30 years, distinct therapeutic approaches have been developed ranging from psychotropic drugs to psychotherapies and are now considered as effective for the management of OCD. In this respect, an ample body of literature has emerged and proven the efficacy of antidepressant agents for treating OCD. They preponderantly block the serotonin transporter located in the outer membrane of presynaptic axonal terminals of the brainstem raphe neurons, and thereby inhibit neurotransmitter reuptake (Flament and Bisserbe, 1997; Piggot and Seay, 1998; Goodman, 1999; McDougle, 1999; Pigott and Seay, 1999). Psychological treatments primarily based on cognitive-behavioral therapy with the current procedure termed "exposure with self-imposed response prevention" also appear to be effective in the treatment of OCD, alone or in combination with pharmacotherapy (Greist, 1994; Foa et al., 1998; 2002). However, at least 20-30 % of patients with OCD fail to respond adequately, with minimal to no change during these conventional treatments (Pigott and Seay, 1998; Pallanti et al., 2002). Therefore, psychosurgical approaches have been considered for the management of these chronic, severely distressing forms of OCD resistant to all currently available traditional treatments. The most frequently described procedures, which are anterior capsulotomy, anterior cingulotomy, subcaudate tractotomy and limbic leucotomy, have shown to be of significant benefit in the treatment of refractory OCD.

2.1 Anterior Capsulotomy

Developed in Sweden, anterior capsulotomy targets the fiber bundles in the anterior limb of the internal capsule connecting the medial dorsal nucleus of the thalamus and the frontal lobes. Two surgical techniques have been used, the radiofrequency thermolesion, or thermocapsulotomy and the newer radiosurgical, or gamma knife capsulotomy techniques (Mindus and Jenike, 1992; Jenike, 1998; Binder and Iskandar, 2000; Greenberg et al., 2000; Greenberg et al., 2003). The first report by Herner (1961) showed that at follow-up ranging from 24 to 80 months there were either good or fair results in 78 % of the 18 patients with OCD who underwent this surgical procedure. Bingley et al. (1977) found that all 35 patients with OCD responded favorably to this type of operation. Sixteen (46 %) were free of symptoms, 9 (26 %) were much improved, and 10 (28 %) were slightly improved at the mean 35-month follow-up. There was an improvement in the degree of social rehabilitation and working capacity. In the study by Kullberg (1977), general improvement, reflecting symptom reduction and social capacity, was considered as "excellent" in 1 (12.5 %), "good" in 2 (25 %), "moderate" in 2 (25 %), "slight" in 1 and "none" in 2 (25 %) of the 8 patients treated with capsulotomy at the end of the follow-up period (1 to 9 years). Fodstad et al. (1982) reported that two patients with OCD were much improved during the 24-month period following capsulotomy. In the review by Mindus et al. (1994), it has been mentioned that 5 (23 %) of the 22 patients with OCD exhibited a clinical aggravation at the 8-year follow-up after capsulotomy while 2 (9 %) were improved by 1 % to 25 %, 5 (23 %) improved by 26 % to 50 %, 3 (13 %) improved by 51 % to 75 %, and 7 (32 %) improved by 75 % to 100 %. Adverse effects of anterior capsulotomy in the initial series of 116 patients observed by Herner (1961) may be summarized as follows: three patients died 11, 28 and 33 months, respectively after the intervention; two from suicide and the third from a cardiac infarction. Operative complications were considerably few: stenosis of the aqueduct of the mid-brain in one case and epilepsy in four (3.4 %). The side effects on the emotional, the volitional and the intellectual status, as found in 55.5 % of the 18 patients with OCD, were mild and transient. Other untoward effects were urinary (and fecal) incontinence and weight gain but were not serious. In the study by Bingley et al. (1977), transient hemiparesis due to small subcortical hemorrhage was described in 2 (6 %) of the 35 OCD cases treated with capsulotomy. EEG showed abnormalities localized mainly to the frontal regions in 28 of these patients at 10-day follow-up before normalization after a more prolonged post-operative period. Post-operative mental confusion also occurred in some cases and lasted less than 1-3 weeks after the operation. Other mental changes consisted of diminution of inhibition, elevation of mood, and loss of initiative, but were not disturbing in most cases (Bingley et al., 1977; Kullberg, 1977; Chiocca and Martuza, 1990). At the 1-year follow-up, capsulotomy contributed to reverse personality changes observed pre-operatively (Mindus et al., 1988; Mindus and Nyman, 1991; Mindus et al., 1999).

2.2 Anterior Cingulotomy

Developed in the United States and Canada, anterior cingulotomy was the most currently neurosurgical procedure applied to the treatment of anxiety disorders including refractory OCD. Bilateral thermolesions are created by the radiofrequency in the anterior cingulate cortex (Broadman areas 24 and 32; Mindus and Jenike, 1992; Jenike, 1998; Binder and Iskandar, 2000; Greenberg et al., 2000; Greenberg et al., 2003). Whitty et al. (1952) first reported the satisfactory results of this surgical procedure in two patients with OCD. The therapeutic effects were much less favorable in the study by Kullberg (1977) showing that cingulotomy produced a slight reduction in OCD symptoms severity only in 1 (33 %) of the three OCD cases at the end of the follow-up period (1-8 years). In a larger sample, Ballantine et al. (1987) found that 8 (25 %) of the 32 patients with OCD were slightly improved, 10 (31.3 %) were markedly improved, 4 (12.5 %) were functionally normal on medication and/or psychotherapy maintenance, and 4 (12.5 %) were essentially well without any treatment. The remaining 6 patients (18.7 %) showed unchanged or worsened

OCD symptoms at the mean follow-up of 8.6 years. Jenike et al. (1991), using Y-BOCS for assessment of OCD symptom severity, described that 6 (43 %) of the 14 patients with OCD treated with cingulotomy showed no improvement, while the remaining 8 (57 %) exhibited moderate to marked improvement at the mean 13.1-year follow-up, defined by a 50 % or greater reduction in Y-BOCS scores. However, 2 of these 8 patients considered their improvement related to the current treatment with SRI (fluoxetine) and CBT rather than cingulotomy. In the most recent prospective study, Dougherty et al. (2002) reported favorable effects of this surgical procedure in 44 patients with intractable OCD. At a mean follow-up of 32 months, 14 (32 %) were considered as responders according to the conservative criteria for treatment response and 6 others (14 %) were partial responders after one or more cingulotomies. Serious operative complications were relatively few including hemiplegias (0.03 %) and controllable seizure disorders (1 %) in the series of 696 bilateral cingulotomies (Ballantine et al., 1987). Death by suicide was observed in 18 patients over the mean 8.6-year follow-up, but all 18 of these patients had preoperative suicidal ideation and 13 had made suicide attempts before the operation. Transient mental confusion was reported in 2 (15.4 %) of the 13 patients treated with cingulotomy during the first post-operative days (Kullberg, 1977). Memory deficits were described in two (5 %) of the 44 patients who underwent this type of surgical intervention, but had resolved within 6 to 12 months. Apathy with loss of energy was noted in one patient (2 %) following cingulotomy. Urinary disturbances were reported in three (7 %) patients, which resolved within the first days postoperatively (Dougherty et al., 2002).

2.3 Subcaudate Tractotomy

Developed in the United Kingdom, subcaudate tractotomy is characterized by bilateral lesions made by radioactive yttrium-90 seeds placed in the substantia innominata, which is located close to the ventral part of the anterior capsule beneath the head of the caudate nucleus. The goal is to interrupt connections from orbitofrontal cortex to subcortical regions such as ventral caudate, nucleus accumbens, which then send projections to the ventromedial pallidum and thalamus (Mindus and Jenike, 1992; Jenike, 1998; Binder and Iskandar, 2000; Greenberg et al., 2000; 2003). There were relatively few reports of the efficacy of this surgical procedure in the treatment of refractory OCD. In the series of Strom-Olsen and Carlisle (1971), it has been found that at the follow-up ranging from 16 months to 8 years, 7 (35 %) of the 20 patients with OCD had no symptoms, 3 (15 %) were improved but with slight residual symptoms for which no treatment was needed, 3 (15 %) were improved but with persistent symptoms requiring treatment. The remaining 7 (35 %) patients exhibited unchanged or worsened symptoms. In most cases, the reducing effects on OCD symptom severity were delayed within 4-5 months after the operation. Bridges et al. (1973) also reported favorable effects of subcaudate tractotomy in 24 patients intractable OCD. Sixteen (67 %) of these patients showed a good outcome, as defined by no symptoms or mild residual symptoms after three years of follow-up. The remaining eight (33 %) patients had poor outcome with persistence of significant symptoms responsible for functional impairment in the patient's life. These results were confirmed by Goktepe et al. (1975) showing that nine (50 %) of the 18 patients with OCD had satisfactory response after a 2.5-year period of follow-up. In the most recent report by Hodgkiss et al. (1995), good outcome was only observed in 5 (33.3 %) of the 15 patients with OCD within one year after surgery. There was no post-operative mortality in the series of Strom-Olsen and Carlisle (1971), conducted on 150 patients with psychiatric conditions operated. However, there were three deaths by suicide in the series of 208 cases of Goktepe et al. (1975). The incidence of post-operative epilepsy in this series was 2.2 % (Goktepe et al., 1975). Weight gain was observed (Strom-Olsen and Carlisle, 1971). Psychological and behavioral changes occurred (Strom-Olsen and Carlisle, 1971; Goktepe et al., 1975). The commonly reported symptoms were tiredness and lethargy lasting for up to three months or irritability, outspokenness and volubility, which most often failed to have repercussions on family and social relationships. Neuropsychological testing revealed frontal lobe dysfunction 6 months after the operation (Kartsounis et al., 1991).

2.4 Limbic Leucotomy

Introduced in the United Kingdom, limbic leucotomy is based on bilateral lesions of cingulotomy, in addition to those of the original subcaudate tractotomy (Mindus and Jenike, 1992; Jenike, 1998; Binder and

Iskandar, 2000; Greenberg et al., 2000; Greenberg et al., 2003). In the early report of Kelly et al. (Kelly and Mitchell-Heggs, 1973; Kelly et al., 1973), 13 (76 %) of the 17 patients with OCD were clinically improved within six weeks after the intervention. Of these 13 patients, 1 (8 %) was rated as "symptom free", 6 (46 %) as "much improved" and 6 (46 %) as "improved". The favorable effects even persisted after a mean 16- or 17-month period of follow-up (Kelly and Mitchell-Heggs, 1973; Kelly et al., 1973; Mitchell-Heggs et al., 1976). These data confirmed the efficacy of this surgical procedure in the treatment of OCD, as reported in the early controlled retrospective study by Tan et al. (1971) showing a more profound reduction in OCD symptom severity in 24 patients with OCD operated compared with their 13 controls at three months after the operation, and at the end of the five years of follow-up. Post-operative complications were as follows: subarachnoid hemorrhage in one patient (4 %) and death in three patients (12.5 %; of whom one committed suicide; Tan et al., 1971). There also were transient confusion, headache, apathy, laziness, incontinence of urine, and rarely feces, which currently lasted from one to several days. Weight gain was observed. Poorer memory and concentration capacities, outspokenness or irritability were inconsistently reported (Tan et al., 1971; Kelly and Mitchell-Heggs, 1973; Kelly et al., 1973; Mitchell-Heggs et al., 1976).

Therefore, functional neurosurgical procedures, leading to an interruption of reciprocal connections between the frontal lobes and subcortical structures, have been demonstrated to be effective in the management of OCD unresponsive to both pharmacological and psychological strategies typically used in this condition.

2.5 Deep Brain Stimulation

Although the lesion techniques have been shown to cause relatively few serious undesirable effects, the reversible procedure deep brain stimulation (DBS) in the context of recent technological advances is becoming an important strategy in the treatment of severe forms of Parkinson's disease and other movement disorders (Malhi and Sachdev, 2002). This approach has recently been proposed as a potential therapeutic alternative in resistant OCD. When applied bilaterally in the anterior limb of internal capsule instead of bilateral anterior capsulotomy, DBS has been found to markedly improve OCD symptoms in three of the four OCD patients who underwent this procedure. The beneficial effects occurred almost immediately after the beginning of DBS (Nuttin et al., 1999). Headache was experienced during the first 10 seconds of stimulation and then disappeared. Long-term follow-up studies have confirmed the favorable effects of chronic anterior capsular stimulation. There was a sustained 35 % and greater reduction in Y-BOCS scores in two of the three cases of long-standing treatment-resistant OCD during the 33-month period of follow-up. At 1-year DBS, there was no deleterious effect or harmful consequence on neuropsychological functioning and personality traits (Gabriels et al., 2003). It has also been shown that a gradual decline in OCD symptom severity with recovery defined by scores in Y-BOCS below 8, which was obtained within the first 3 months of DBS application and remained stable over the seven following months. This was paralleled by a dramatic improvement in global psychosocial functioning (Anderson and Ahmed, 2003). Other brain regions have been proposed as target for DBS. In this context, a reduction in OCD manifestations by approximately 60 % was reported in two Parkinsonian patients with comorbid OCD after 6-month DBS of the subthalamic nucleus. However, symptomatic improvement was observed immediately after discharge from the neurosurgery unit (Mallet et al., 2002). The nucleus accumbens that represents the ventral and limbic part of the caudate nucleus has also been chosen as target for DBS. There was a clinical amelioration in three of the four patients with OCD and related anxiety disorders, which was achieved a few days to several weeks after beginning of DBS. No side effect was reported during the 24- to 30-month period of follow-up (Sturm et al., 2003). We also found favorable effects of DBS of the ventral caudate nucleus, including the nucleus accumbens and ventromedial head of the caudate nucleus in one patient suffering from resistant OCD with comorbid major depression (Aouizerate et al., 2004b). Reduction in depressive symptom severity occurred early during the first three months of DBS until remission obtained at 6 months while there was a delayed onset of the anti-obsessional effects, which was observed at 12-month DBS. This was parallel by a progressive improvement in global functioning. No adverse effect of any kind was noted. There was no deterioration of neuropsychological testing exploring specifically executive and mnemonic functions.

Conclusion

To summarize, a disturbance in frontal-subcortical loops originating in the OFC and ACC has extensively been documented to play a major role in the pathogenesis of OCD. Numerous neuroimaging studies have consistently shown abnormalities in functional activity in the OFC, ACC, ventral striatum and medial thalamus. These brain regions have been found implicated in cognitive and emotional processes that can be thought disrupted in OCD in the light of phenomenological considerations. Data collected from psychosurgery proposed in resistant OCD forms, which failed to successfully respond to both conventional pharmacological treatments and psychotherapies, confirm the involvement of such brain regions in the pathophysiology of OCD. The most widely used approaches (anterior capsulotomy, anterior cingulotomy, subcaudate tractotomy and limbic leucotomy) intended to perform lesions within the frontal-subcortical loops have been shown to produce an improvement in OCD symptoms in 40 to 60 % of cases. They were relatively well-tolerated. However, chronic DBS has been proposed as reversible procedure for the management of refractory OCD. This surgical technique has benefited from methodological advances made in the field of Parkinson's disease and other movement disorders. Different target regions have been considered for DBS in OCD, such as the anterior limb of the internal capsule, the subthalamic nucleus or the nucleus accumbens. Although DBS seems to be the first-line choice since it is the least invasive, there are still many unknown factors. The most critical issue is to determine the optimal targets both in terms of their possible efficacy and their risks and benefits. In this perspective, further preclinical studies in laboratory animals would be helpful in order to provide more precise information about the anatomo-functional bases of OCD. The promising anti-obsessional effects and safety of DBS remain to be confirmed in larger populations of patients with intractable OCD.

References

Akkal D, Bioulac B, Audin J, Burbaud P (2002) Comparison of neuronal activity in the rostral supplementary and cingulate motor areas during a task with cognitive and motor demands. *Eur J Neurosci* **15**:887-904.

Anderson D, Ahmed A (2003) Treatment of patients with intractable obsessive-compulsive disorder with anterior capsular stimulation. Case report. *J Neurosurg* **98**:1104-1108.

Antony MM, Downie F, Swinson RP (1998) Diagnostic issues and epidemiology in obsessive-compulsive disorder. In: *Obsessive-compulsive disorder*. Theory, research, and treatment (Swinson RP, Antony MM, Rachman S, Richter MA, eds), pp 3-32. New-York: The Guilford Press.

Aouizerate B, Guehl D, Cuny E, Rougier A, Bioulac B, Tignol J, Burbaud P (2004a) Pathophysiology of obsessive-compulsive disorder: a necessary link between phenomenology, neuropsychology, imagery and physiology. *Prog Neurobiol* **72**:195-221.

Aouizerate B, Cuny E, Martin-Guehl C, Guehl D, Amieva H, Benazzouz A, Fabrigoule C, Allard M, Rougier A, Bioulac B, Tignol J, Burbaud P (2004b) Deep brain stimulation of the ventral caudate nucleus in the treatment of obsessive-compulsive disorder and major depression. Case report. *J Neurosurg* **101**:682-686.

Ballantine HT, Jr., Bouckoms AJ, Thomas EK, Giriunas IE (1987) Treatment of psychiatric illness by stereotactic cingulotomy. *Biol Psychiatry* **22**:807-819.

Barbas H (2000) Connections underlying the synthesis of cognition, memory, and emotion in primate prefrontal cortices. *Brain Res Bull* **52**:319-330.

Baxter LR (1999) Functional imaging of brain systems mediating obsessive-compulsive disorder: Clinical studies. In: *Neurobiology of mental illness* (Charney DS, Nestler EJ, Bunney BS, eds), pp 534-547. New York: Oxford University Press.

Binder DK, Iskandar BJ (2000) Modern neurosurgery for psychiatric disorders. *Neurosurgery* **47**:9-21; discussion 21-23.

Bingley T, Leksell L, Meyerson BA, Rylander G (1977) Long-term results of stereotactic anterior capsulotomy in chronic obsessive-compulsive neurosis. In: *Neurosurgical treatment in Psychiatry, Pain and Epilepsy* (Sweet WH, Obrador S, Martin-Rodriguez JG, eds), pp 287-299. Baltimore: University Park Press.

Bridges PK, Goktepe EO, Maratos J (1973) A comparative review of patients with obsessional neurosis and with depression treated by psychosurgery. *Br J Psychiatry* **123**:663-674.

Bush G, Luu P, Posner MI (2000) Cognitive and emotional influences in anterior cingulate cortex. *Trends Cogn Sci* **4**:215-222.

Bush G, Vogt BA, Holmes J, Dale AM, Greve D, Jenike MA, Rosen BR (2002) Dorsal anterior cingulate cortex: a role in reward-based decision making. *Proc Natl Acad Sci U S A* **99**:523-528.

Charney DS, Bremner JD (1999) The neurobiology of anxiety disorders: Clinical studies. In: *Neurobiology of mental illness* (Charney DS, Nestler EJ, Bunney BS, eds), pp 494-517. New York: Oxford University Press.

Chiocca EA, Martuza RL (1990) Neurosurgical therapy of obsessive-compulsive disorder. In: *Obsessive-compulsive disorders: Theory and management* (Jenike MA, Baer L, Minichiello WE, eds), pp 283-294. Chicago: Year Book Medical Publishers.

Cummings JL (1993) Frontal-subcortical circuits and human behavior. *Arch Neurol* **50**:873-880.

Devinsky O, Morrell MJ, Vogt BA (1995) Contributions of anterior cingulate cortex to behaviour. *Brain* **118** (Pt 1):279-306.

Dougherty DD, Baer L, Cosgrove GR, Cassem EH, Price BH, Nierenberg AA, Jenike MA, Rauch SL (2002) Prospective long-term follow-up of 44 patients who received cingulotomy for treatment-refractory obsessive-compulsive disorder. *Am J Psychiatry* **159**:269-275.

Flament MF, Bisserbe JC (1997) Pharmacologic treatment of obsessive-compulsive disorder: comparative studies. *J Clin Psychiatry* **58** Suppl 12:18-22.

Foa EB, Franklin ME, Kozac MJ (1998) Psychosocial treatments for obsessive-compulsive disorder. Literature review. In: *Obsessive-compulsive disorder. Theory, research, and treatment* (Swinson RP, Antony MM, Rachman S, Richter MA, eds), pp 258-276. New-York: The Guilford Press.

Foa EB, Franklin ME, Moser J (2002) Context in the clinic: how well do cognitive-behavioral therapies and medications work in combination? *Biol Psychiatry* **52**:987-997.

Fodstad H, Strandman E, Karlsson B, West KA (1982) Treatment of chronic obsessive compulsive states with stereotactic anterior capsulotomy or cingulotomy. *Acta Neurochir (Wien)* **62**:1-23.

Gabriels L, Cosyns P, Nuttin B, Demeulemeester H, Gybels J (2003) Deep brain stimulation for treatment-refractory obsessive-compulsive disorder: psychopathological and neuropsychological outcome in three cases. *Acta Psychiatr Scand* **107**:275-282.

Goktepe EO, Young LB, Bridges PK (1975) A further review of the results of sterotactic subcaudate tractotomy. *Br J Psychiatry* **126**:270-280.

Goldman-Rakic PS, Porrino LJ (1985) The primate mediodorsal (MD) nucleus and its projection to the frontal lobe. *J Comp Neurol* **242**:535-560.

Goodman WK (1999) Obsessive-compulsive disorder: diagnosis and treatment. *J Clin Psychiatry 60 Suppl* **18**:27-32.

Greenberg BD, Murphy DL, Rasmussen SA (2000) Neuroanatomically based approaches to obsessive-compulsive disorder. Neurosurgery and transcranial magnetic stimulation. *Psychiatr Clin North Am* **23**:671-686, xii.

Greenberg BD, Price LH, Rauch SL, Friehs G, Noren G, Malone D, Carpenter LL, Rezai AR, Rasmussen SA (2003) Neurosurgery for intractable obsessive-compulsive disorder and depression: critical issues. *Neurosurg Clin N Am* **14**:199-212.

Greist JH (1994) Behavior therapy for obsessive compulsive disorder. *J Clin Psychiatry* **55** Suppl:60-68.

Hassani OK, Cromwell HC, Schultz W (2001) Influence of expectation of different rewards on behavior-related neuronal activity in the striatum. *J Neurophysiol* **85**:2477-2489.

Herner T (1961) Treatment of mental disorders with frontal stereotactic thermo-lesions. A follow-up study of 116 cases. *Acta Psychiatr Scand Suppl*. **36**.

Hodgkiss AD, Malizia AL, Bartlett JR, Bridges PK (1995) Outcome after the psychosurgical operation of stereotactic subcaudate tractotomy, 1979-1991. *J Neuropsychiatry Clin Neurosci* **7**:230-234.

Hollerman JR, Tremblay L, Schultz W (1998) Influence of reward expectation on behavior-related neuronal activity in primate striatum. *J Neurophysiol* **80**:947-963.

Jenike MA (1998) Neurosurgical treatment of obsessive-compulsive disorder. Br J Psychiatry Suppl:79-90.

Jenike MA, Baer L, Ballantine T, Martuza RL, Tynes S, Giriunas I, Buttolph ML, Cassem NH (1991) Cingulotomy for refractory obsessive-compulsive disorder. A long-term follow-up of 33 patients. *Arch Gen Psychiatry* **48**:548-555.

Karno M, Golding JM, Sorenson SB, Burnam MA (1988) The epidemiology of obsessive-compulsive disorder in five US communities. *Arch Gen Psychiatry* **45**:1094-1099.

Kartsounis LD, Poynton A, Bridges PK, Bartlett JR (1991) Neuropsychological correlates of stereotactic subcaudate tractotomy. *A prospective study. Brain* **114** (Pt 6):2657-2673.

Kelly D, Mitchell-Heggs N (1973) Stereotactic limbic leucotomy--a follow-up study of thirty patients. *Postgrad Med J* **49**:865-882.

Kelly D, Richardson A, Mitchell-Heggs N, Greenup J, Chen C, Hafner RJ (1973) Stereotactic limbic leucotomy: a preliminary report on forty patients. *Br J Psychiatry* **123**:141-148.

Koob GF, Le Moal M (1997) Drug abuse: hedonic homeostatic dysregulation. *Science* **278**:52-58.

Koran LM, Thienemann ML, Davenport R (1996) Quality of life for patients with obsessive-compulsive disorder. *Am J Psychiatry* **153**:783-788.

Krawczyk DC (2002) Contributions of the prefrontal cortex to the neural basis of human decision making. *Neurosci Biobehav Rev* **26**:631-664.

Kullberg G (1977) Differences in effects of capsulotomy and cingulotomy. In: *Neurosurgical treatment in Psychiatry, Pain and Epilepsy* (Sweet WH, Obrador S, Martin-Rodriguez JG, eds), pp 301-308. Baltimore: University Park Press.

Le Moal M (1995) Mesocorticolimbic dopaminergic neurons. Functional and regulatory roles. In: *Psychopharmacology: The fourth generation of progress* (Bloom FE, Kupfer DJ, eds), pp 283-294. New York: Raven Press.

Malhi GS, Sachdev P (2002) Novel physical treatments for the management of neuropsychiatric disorders. *J Psychosom Res* **53**:709-719.

Mallet L, Mesnage V, Houeto JL, Pelissolo A, Yelnik J, Behar C, Gargiulo M, Welter ML, Bonnet AM, Pillon B, Cornu P, Dormont D, Pidoux B, Allilaire JF, Agid Y (2002) Compulsions, Parkinson's disease, and stimulation. *Lancet* **360**:1302-1304.

McDougle CJ (1999) The neurobiology and treatment of obsessive-compulsive disorder. In: *Neurobiology of mental illness* (Charney DS, Nestler EJ, Bunney BS, eds), pp 518-533. New York: Oxford University Press.

McFarland NR, Haber SN (2002) Thalamic relay nuclei of the basal ganglia form both reciprocal and nonreciprocal cortical connections, linking multiple frontal cortical areas. *J Neurosci* **22**:8117-8132.

Mindus P, Nyman H (1991) Normalization of personality characteristics in patients with incapacitating anxiety disorders after capsulotomy. Acta Psychiatr Scand 83:283-291.

Mindus P, Jenike MA (1992) Neurosurgical treatment of malignant obsessive compulsive disorder. *Psychiatr Clin North Am* **15**:921-938.

Mindus P, Rasmussen SA, Lindquist C (1994) Neurosurgical treatment for refractory obsessive-compulsive disorder: implications for understanding frontal lobe function. J Neuropsychiatry *Clin Neurosci* **6**:467-477.

Mindus P, Edman G, Andreewitch S (1999) A prospective, long-term study of personality traits in patients with intractable obsessional illness treated by capsulotomy. *Acta Psychiatr Scand* **99**:40-50.

Mindus P, Nyman H, Rosenquist A, Rydin E, Meyerson BA (1988) Aspects of personality in patients with anxiety disorders undergoing capsulotomy. *Acta Neurochir Suppl (Wien)* **44**:138-144.

Mitchell-Heggs N, Kelly D, Richardson A (1976) Stereotactic limbic leucotomy--a follow-up at 16 months. *Br J Psychiatry* **128**:226-240.

Niki H, Watanabe M (1979) Prefrontal and cingulate unit activity during timing behavior in the monkey. *Brain Res* **171**:213-224.

Nuttin B, Cosyns P, Demeulemeester H, Gybels J, Meyerson B (1999) Electrical stimulation in anterior limbs of internal capsules in patients with obsessive-compulsive disorder. *Lancet* **354**:1526.

Pallanti S, Quercioli L, Koran LM (2002) Citalopram intravenous infusion in resistant obsessive-compulsive disorder: an open trial. *J Clin Psychiatry* **63**:796-801.

Paus T (2001) Primate anterior cingulate cortex: where motor control, drive and cognition interface. *Nat Rev Neurosci* **2**:417-424.

Piazza PV, Le Moal ML (1996) Pathophysiological basis of vulnerability to drug abuse: role of an interaction between stress, glucocorticoids, and dopaminergic neurons. *Annu Rev Pharmacol Toxicol* **36**:359-378.

Piazza PV, Le Moal M (1997) Glucocorticoids as a biological substrate of reward: physiological and pathophysiological implications. *Brain Res Brain Res Rev* **25**:359-372.

Piggot TA, Seay SM (1998) Biological treatments for obsessive-compulsive disorder. Literature review. In: *Obsessive-compulsive disorder. Theory, research, and treatment* (Eds R. P. Swinson MMA, S. Rachman, M. A. Richter., ed), pp pp. 298-326.: The Guilford Press: New-York.

Pigott TA, Seay SM (1999) A review of the efficacy of selective serotonin reuptake inhibitors in obsessive-compulsive disorder. *J Clin Psychiatry* **60**:101-106.

Ramnani N, Owen AM (2004) Anterior prefrontal cortex: insights into function from anatomy and neuroimaging. *Nat Rev Neurosci* **5**:184-194.

Rasmussen SA, Tsuang MT (1984) The epidemiology of obsessive compulsive disorder. *J Clin Psychiatry* **45**:450-457.

Robins LN, Helzer JE, Weissman MM, Orvaschel H, Gruenberg E, Burke JD, Jr., Regier DA (1984) Lifetime prevalence of specific psychiatric disorders in three sites. *Arch Gen Psychiatry* **41**:949-958.

Rosenkilde CE, Bauer RH, Fuster JM (1981) Single cell activity in ventral prefrontal cortex of behaving monkeys. *Brain Res* **209**:375-394.

Saxena S, Brody AL, Schwartz JM, Baxter LR (1998) Neuroimaging and frontal-subcortical circuitry in obsessive-compulsive disorder. *Br J Psychiatry Suppl*:**26**-37.

Schwartz JM (1998) Neuroanatomical aspects of cognitive-behavioural therapy response in obsessive-compulsive disorder. An evolving perspective on brain and behaviour. *Br J Psychiatry Suppl*:**38**-44.

Schwartz JM (1999) A role of volition and attention in the generation of new brain circuitry. Toward a neurobiology of mental force. *J Consciousness studies* **6**:115-142.

Shidara M, Richmond BJ (2002) Anterior cingulate: single neuronal signals related to degree of reward expectancy. *Science* **296**:1709-1711.

Shima K, Tanji J (1998) Role for cingulate motor area cells in voluntary movement selection based on reward. *Science* **282**:1335-1338.

Skoog G, Skoog I (1999) A 40-year follow-up of patients with obsessive-compulsive disorder [see commetns]. *Arch Gen Psychiatry* **56**:121-127.

Strom-Olsen R, Carlisle S (1971) Bi-frontal stereotactic tractotomy. A follow-up study of its effects on 210 patients. *Br J Psychiatry* **118**:141-154.

Sturm V, Lenartz D, Koulousakis A, Treuer H, Herholz K, Klein JC, Klosterkotter J (2003) The nucleus accumbens: a target for deep brain stimulation in obsessive-compulsive- and anxiety-disorders. *J Chem Neuroanat* **26**:293-299.

Tan E, Marks IM, Marset P (1971) Bimedial leucotomy in obsessive-compulsive neurosis: a controlled serial enquiry. *Br J Psychiatry* **118**:155-164.

Thorpe SJ, Rolls ET, Maddison S (1983) The orbitofrontal cortex: neuronal activity in the behaving monkey. *Exp Brain Res* **49**:93-115.

Tremblay L, Schultz W (1999) Relative reward preference in primate orbitofrontal cortex. *Nature* **398**:704-708.

Tremblay L, Schultz W (2000a) Modifications of reward expectation-related neuronal activity during learning in primate orbitofrontal cortex. *J Neurophysiol* **83**:1877-1885.

Tremblay L, Schultz W (2000b) Reward-related neuronal activity during go-nogo task performance in primate orbitofrontal cortex. *J Neurophysiol* **83**:1864-1876.

Tremblay L, Hollerman JR, Schultz W (1998) Modifications of reward expectation-related neuronal activity during learning in primate striatum. *J Neurophysiol* **80**:964-977.

Whitty CW, Duffield JE, Tov PM, Cairns H (1952) Anterior cingulectomy in the treatment of mental disease. *Lancet* **1**:475-481.

In: Recent Breakthroughs in Basal Ganglia Research
Editor: Erwan Bezard, pp. 393-401
ISBN: 1-59454-880-3

Chapter 31

DEEP BRAIN STIMULATION IN ADULT AND PEDIATRIC DYSTONIA

Laura Cif, Simone Hemm, Nathalie Vayssiere and Philippe Coubes*
Research Group on Movement Disorders, Department of Neurosurgery (Professor Philippe Coubes, Montpellier University Hospital, 34295 MONTPELLIER, CEDEX 5, FRANCE

ABSTRACT

In numerous medical conditions, abnormal movements (dystonia, dyskinesia, myoclonus and tremor) are present, isolated or associated to other neurological deficits. Deep brain stimulation has been validated as a safe and efficient neurosurgical procedure in treating Parkinson's disease and essential tremor. Following the experience with ablative surgery for dystonia and high frequency stimulation for Parkinson's disease, chronic electrical stimulation of the internal globus pallidus (GPi) has been proposed to treat dystonic syndromes. Based on etiology and spread of symptoms, 82 patients were divided in subgroups separating, for each of them, children and adults: primary DYT1 dystonia, primary dystonia without DYT1 mutation, primary segmental dystonia, post anoxic cerebral palsy, pantothenate kinase associated neurodegeneration (PKAN) and mitochondrial diseases. Assessment was performed pre- and postoperatively by the Burke-Fahn's Marsden Dystonia Rating Scale, motor and disability sections. Bilateral electrode implantation was performed under general anaesthesia. High frequency (130 Hz), 450 microseconds pulse width and monopolar stimulation were used. Clinical improvement at three years was comparable for the groups of primary dystonia (82% for the motor scores) and for secondary dystonia and degenerative diseases, improvement was less important but still interesting (40%). The originality of the method consisted in the use of stereotactic MR imaging for the target selection without microelectrode recordings. This reduced the duration of the procedure and the risk of haemorrhage. Chronic high frequency stimulation of the GPi can be considered as an efficient and lasting treatment in primary dystonia and proposed in selected cases of secondary dystonia and degenerative disease.

1 BACKGROUND AND PURPOSE

In a wide range of medical conditions, a movement disorder can be the only symptom or associated with other neurological deficits. Most of them are of unknown origin. Dystonia, dyskinesia, myoclonus, tremor, atethosis are abnormal movements susceptible to be influenced by high frequency modulation of the basal ganglia. Nevertheless, the control of these symptoms is highly depending on the etiology. Movement disorders (primary dystonia, myoclonus dystonia, tardive dyskinesia), which are not associated to other neurologic deficits are more likely to be controlled by high frequency stimulation.

The first validated medical condition successfully treated by DBS was Parkinson's disease (Benabid et al., 1987). Gradually, the indications of this therapy were enlarged to include other diseases associating

* Correspondence should be addressed to Philippe Coubes: Tel/Fax: +33 (0) 4.67.33.74.64. E-mail address: p-coubes@chu-montpellier.fr or urmae@chu-montpellier.fr

movement disorders as dystonia and dyskinesia of various etiologies. Based on the experience of our group (Gros et al., 1976) with brain lesioning surgery (pallidotomy) in a young patient with idiopathic generalized dystonia (Iacono et al., 1996) and deep brain stimulation in Parkinson's disease dyskinesias (Benabid et al., 1987; 1993), we reported, in 1996 the efficacy of DBS in a young lady presenting with severe life threatening primary generalized dystonia after chronic bilateral stimulation of the GPi (Coubes et al., 1999). Since then, 105 dystonic patients have been treated for bilateral stimulation of the Internal Globus Pallidus (GPi) in the department of Neurosurgery in Montpellier.

The originality of our method consists in the use of stereotactic MR imaging for the target determination, without microelectrode recordings. This considerably reduces the procedure's duration, which is important especially in children, as well as for the risk of hemorrhage (0 cases). Furthermore, the surgery, performed under general anesthesia is more adapted in patients with permanent involuntary movements.

We will present the population, the surgical procedure, the clinical management of the patients and the results obtained by DBS. In order to discuss the criteria for selecting the patients with dystonia-dyskinesia, we will briefly remember this medical condition. Whatever, others movement disorders being influenced by the DBS are sometimes associated to dystonia in patients to be operated.

Dystonia has been defined as a neurological syndrome characterized by involuntary, sustained muscle contractions, causing twisting and repetitive movements or abnormal postures (Fahn, 1988).

Its classification is a difficult task and can be done in several ways: according to the spread of the symptoms to different body parts, to the age of onset or to etiology.

According to distribution, dystonia is classified into one of the following categories: focal, segmental, multifocal, hemidystonia and generalized dystonia.

The age of onset is another important criterion for characterizing a patient. In case of early onset, the disease is more likely to generalize and severely worsen than for adult-onset dystonias.

Dystonia can also be classified by etiology. In primary dystonia, dystonia is the only symptom and can be sporadic or inherited (DYT1 dystonia; Ozelius et al., 1997). Dystonia-plus is a group of syndromes where dystonia is usually associated to another neurological condition such as parkinsonism or myoclonus (Dopa responsive dystonia, myoclonus-dystonia syndrome). The group of heredodegenerative dystonia includes numerous diseases and in this group, dystonia is typically not pure. Amino acid disorders, lipid disorders, Lesch-Nyhan disease, pantothenate kinase-associated neurodegeneration (PKAN), mitochondrial diseases, Wilson's disease, Huntington's disease, Juvenile parkinsonism-dystonia (Dwork et al., 1993; Ishikawa and Takahashi, 1998) and many others are heredodegenerative dystonias. Secondary dystonias are generated by insults such as drugs, strokes, tumors, infections (Fahn, 1988) and this subgroup includes also dystonia-dyskinesia secondary to cerebral palsy (CP).

In 40 % of the patients with early-onset dystonia a specific cause can be found. The etiologic diagnosis is established by clinical evaluation, neuroimaging and molecular analysis in primary dystonias. When the history of the disease, the clinical examination and the brain MRI suggest a heredogenerative dystonia, further investigations are performed such as blood work-up, urine sample analysis, CSF testing, electrophysiologic studies, muscle, skin, liver biopsies, PET, eye slit-lamp examination.

The identification of the etiology is very important for the prognosis of the disease and because of the existence of very few medical conditions having a specific treatment such as Dopa-Responsive dystonia, creatine deficiency or Wilson's disease.

Several drugs are available to treat dystonic symptoms but their efficacy is often limited, transient and difficult to assess (often because of the fluctuation of the symptoms). The most important drugs to be used are Levodopa (also as a diagnosis trial in Dopa responsive dystonia), anticholinergics, benzodiazepines, baclofen (Ford et al., 1998) and Botulinum toxin injections especially for the treatment of focal dystonias (Tsui et al., 1987)

Because of the poor efficacy of the pharmaceutical treatment in front of the very severe forms of movement disorders (especially generalized dystonias with onset in childhood) we were led to propose deep brain stimulation as another therapeutic strategy in order to control the symptoms generating life-threatening complications.

2 Patients (Population)

In the following report, we will present the results of 82 patients presenting with segmentary or generalized dystonia treated by bilateral chronic electrical stimulation of the GPi. The population was divided in subgroups separating, for each of them, children from adults. Group 1 included patients with primary generalized DYT1 dystonia, group 2 primary generalized dystonia without DYT1 mutation, group 3 primary segmentary dystonia (cervico-axial dystonia), group 4 generalized dystonia-dyskinesia due to postanoxic cerebral palsy, group 5 generalized dystonia secondary to PKAN and group 6 one generalized dystonia secondary to mitochondrial disease. For the reported patients, the follow-up was at least of 6 months.

3 Clinical Evaluation

Dystonic movements and abnormal postures were evaluated using the Burke-Fahn-Marsden-Dystonia-Rating scale (BFMDRS, motor and disability part; Burke et al., 1985) before the surgical procedure, several times during the post-operative hospital stay, every month during the first year and every three months afterwards.

4 Surgical Procedure

Bilateral electrode implantation was performed in a single surgical session under general anesthesia (Vayssiere et al., 2000; Coubes et al., 2002a; Vayssiere et al., 2002). The MR-compatible Leksell stereotactic frame was applied and a 3D-SPGR (spoiled gradient recall) acquisition was performed. The postero-ventral part of the GPi was located through axial, sagittal and coronal MRI studies (Figure 1A). The target coordinates (x, y, z) and the trajectory angles (α,β) were calculated using a dedicated software.

Two four contact electrodes (DBS 3389, Medtronic, Minneapolis) were implanted under strict profile radioscopic control. Immediate postoperative control MRI (Figure 1B) was obtained with the stereotactic frame on. Electrodes were connected to a pulse generator five days later (Itrel II or III, Kinetra and Soletra Medtronic, Minneapolis, USA), which was subcutaneously introduced in the abdominal area. In each patient final placement of the implanted devices and their connections were checked by a radiographic control during the postoperative hospital stay (Figure 2).

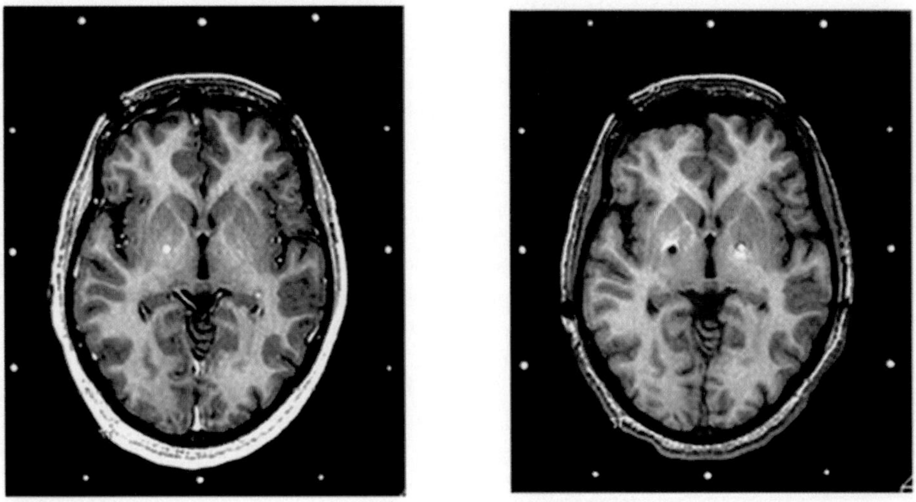

Figure 1: Stereotactic MRI. A. Pre-operative planning. B. Post-operative MRI to control the final electrode position (black points = electrode artefacts).

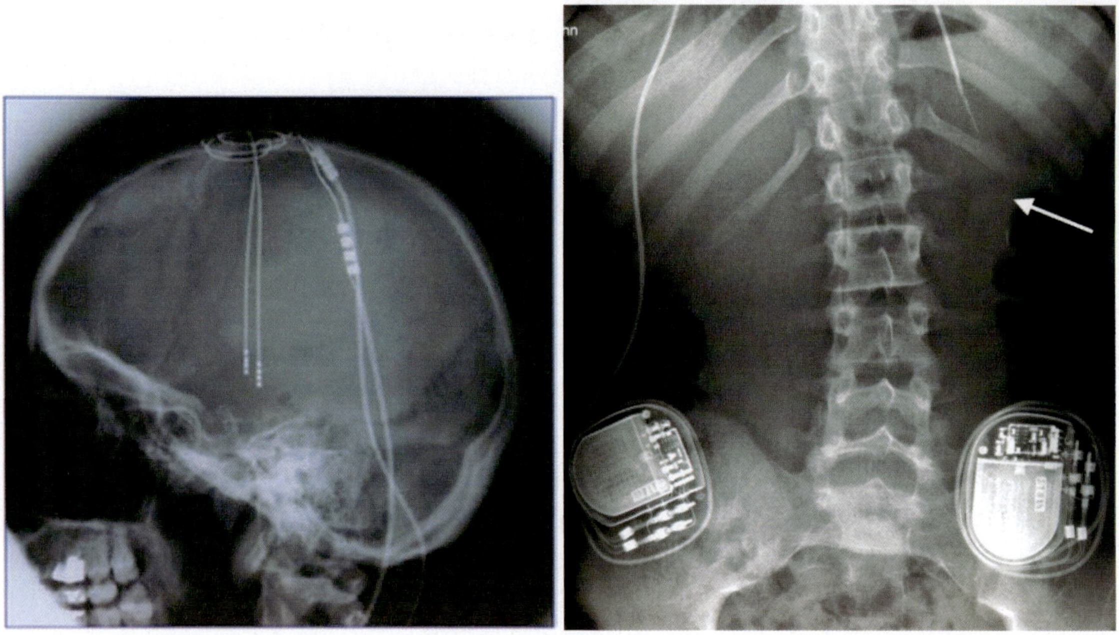

A B

Figure 2: Radiographic control of the implanted devices. A. Electrodes connected to the extensions. B. Neurostimulators with extension length reserve. Arrow indicates an extension fracture during sports.

5 Electric Parameter Settings

After implantation, stimulators were switched on. Electrical variables were set at high frequency (130 Hz), 450 μsec for the pulse width with one contact negatively activated. Intensity was progressively increased according to the needs of each patient (Hemm et al., 2004) and the clinical evolution. Usually the first levels for the voltage were between 0.5 V-0.8 V. Over the time, we modified the electric parameters in several patients, by increasing the voltage or activating a second contact when requested. The mean steady state value of the voltage was 1.6 ± 0.3 V (Hemm et al., 2002).

6 Clinical Results

Motor and disability scores' evolution (BMFDRS) for primary and secondary dystonia is presented in tables I and II. Within each group, the improvement was progressive over time. With more than 3 years of follow-up, the clinical improvement was comparable for the two groups of primary dystonia (82 % of improvement on the motor scale). The results obtained in the group of secondary dystonia and heredodegenerative diseases are less important but yet around 40 % with 3 years of follow-up. After three years, the disability score improvement was superior in the group of primary DYT1 dystonia (80 %) compared to non-DYT1 primary dystonia (56 %) and to secondary dystonia (19 %). The group of secondary dystonia and heredodegenerative diseases is a very heterogeneous group. This is why results for all etiologies should be presented separately.

Table I. Motor scores (BMFDRS) of dystonic children treated by DBS. Mean values ±SD. The values in brackets represent the mean improvement in percent. A reduction in the score indicates an improvement in function. BMFDRS denotes Burke-Marsden Fahn's Dystonia Rating Scale.

Motor score		6 months after surgery/120 (%)[1]	1 year after surgery/120 (%)[1]	2 years after surgery/120 (%)[1]	>3 years after surgery/120 (%)[1]
PGD[2] DYT1+[3]	Children	80±20 n[5]=14	77±28 n[5]=13	76±25 n[5]=12	82±18 n[5]=10
	Adults	75±17 n[5]=10	75±21 n[5]=8	70±16 n[5]=5	73±13 n[5]=2
PGD[2] DYT1-	Children	72±15 n[5]=12	74±15 n[5]=12	66±26 n[5]=11	72±21 n[5]=9
	Adults	53±31 n[5]=19	64±33 n[5]=19	73±26 n[5]=16	71±26 n[5]=7
Cervico-axial dystonia	Children	77±0 n[5]=1	93±0 n[5]=1	93±12 n[5]=1	100±10 n[5]=1
	Adults	70±34 n[5]=7	88±20 n[5]=7	85±21 n[5]=7	84±12 n[5]=4
Post-anoxic cerebral palsy	Children	31±8 n[5]=5	37±6 n[5]=2	45±0 n[5]=1	51±0 n[5]=1
	Adults	41±25 n[5]=5	38±10 n[5]=5	45±15 n[5]=3	62±0 n[5]=1
PKAN[4]	Children	51±29 n[5]=4	59±25 n[5]=4	22±5 n[5]=2	33±13 n[5]=2
	Adults	71±12 n[5]=2	73±10 n[5]=2	74±1 n[5]=2	88±0 n[5]=1
Mitochondrial disease	Children	27±17 n[5]=3	46±30 n[5]=2	25±10 n[5]=2	42±6 n[5]=2

[1]The improvement in percent is calculated based on the maximal possible gain $(Score_{preop}-Score_{postop})/(Score_{preop})$.
[2]Primary generalized dystonia
[3]DYT1 mutation
[4]Pantothenate kinase-associated neurodegeneration
[5]Number of patients

7 Discussion

We report here our experience with bilateral chronic electrical stimulation of the GPi in the treatment of primary DYT1 positive and DYT1 negative generalized dystonia (Coubes et al., 1999; 2000; 2002b; Cif et al., 2004; Coubes et al., 2004), generalized dystonia–dyskinesia secondary to post-anoxic cerebral palsy, generalized dystonia secondary to PKAN syndrome and to mitochondrial diseases. We also report the results obtained by DBS in primary segmentary dystonia. Since the first child has been operated, 126 other patients underwent surgery for DBS in our department (Dystonia + Parkinson).

While at the beginning DBS was proposed in children with primary generalized dystonia (with or without DYT1 mutation), selection criteria were revisited and enlarged for including now other types of dystonia in children and adults. Several patients with generalized dystonia associating myoclonus underwent surgery for chronic electrical stimulation of the GPi and we could see an early and complete control of myoclonus. These findings led us to propose this treatment in a child with genetically proven myoclonus-dystonia syndrome (Cif et al., 2004) MDS, DYT11, mutation in the epsilon-amino sarcoglycan gene, (Gasser, 1998; Zimprich et al., 2001) and we obtained a very satisfactory improvement of his symptoms. We confirmed in a second patient with MDS the efficiency of GPi stimulation.

Table II. Disability scores (BMFDRS) of dystonic children treated by DBS. Mean values ±SD. The values in brackets represent the mean improvement in percent. A reduction in the score indicates an improvement in function. BMFDRS denotes Burke-Marsden Fahn's Dystonia Rating Scale.

Disability score		6 months after surgery/120 (%)[1]	1 year after surgery/120 (%)[1]	2 years after surgery/120 (%)[1]	>3 years after surgery/120 (%)[1]
PGD[2] DYT1+[3]	Children	58±39 n[5]=14	62±39 n[5]=13	63±27 n[5]=12	80±14 n[5]=10
	Adults	61±25 n[5]=10	64±29 n[5]=8	74±14 n[5]=52	58±3 n[5]=2
PGD[2] DYT1-	Children	42±23 n[5]=12	52±21 n[5]=12	54±24 n[5]=11	49±29 n[5]=9
	Adults	35±34 n[5]=19	52±31 n[5]=19	67±30 n[5]=16	58±36 n[5]=7
Cervico-axial dystonia	Children	100±0 n[5]=1	92±0 n[5]=1	92±0 n[5]=1	100±0 n[5]=1
	Adults	55±36 n[5]=7	74±22 n[5]=7	86±18 n[5]=7	74±12 n[5]=4
Cerebral palsy	Children	4±3 n[5]=5	11±4 n[5]=2	19±0 n[5]=1	19±0 n[5]=1
	Adults	21±22 n[5]=5	28±21 n[5]=5	17±12 n[5]=3	27±0 n[5]=1
PKAN[4]	Children	35±32 n[5]=4	28±28 n[5]=4	2±2 n[5]=2	7±14 n[5]=2
	Adults	31±20 n[5]=2	43±10 n[5]=2	62±7 n[5]=2	65±0 n[5]=1
Mitochondrial disease	Children	8±5 n[5]=3	10±3 n[5]=2	11±3 n[5]=2	8±1 n[5]=2

[1]The improvement in percent is calculated based on the maximal possible gain $(Score_{preop}-Score_{postop})/(Score_{preop})$.

[2]Primary generalized dystonia

[3]DYT1 mutation

[4]Pantothenate kinase-associated neurodegeneration

[5]Number of patients

Being confronted with very severe clinical conditions in patients with secondary dystonia and heredodegenerative diseases, in which the efficacy of the medical treatment was poor and in which dystonic movements and postures were comparable with those met in primary dystonia, we proposed DBS in several selected patients of these groups.

The criteria for patient selection in this group were clinical, electrophysiological and based on brain imaging as well as on etiology.

We performed surgery for DBS in patients in whom dystonia and dyskinesia were prominent compared to other neurological deficits (especially motor deficit and spasticity), the motor pattern was preserved and in patients presenting with severe or life threatening symptoms due to dystonia-dyskinesia (swallowing difficulties, permanent opisthotonos, painful muscle spasms).

Electroencephalogram, electroretinogram, visual and brainstem, somatosensory evoked responses were obtained. Motor evoked potentials were performed in elder children to identify pyramidal tract impairment. In order to exclude brain abnormalities contraindicating surgery (major cortical atrophy, severe periventicular leucomalacia especially met in cerebral palsy, basal ganglia and thalamic lesions), brain MR under general anesthesia was performed in all patients.

As shown in tables I and II, best results were obtained within the group of patients with DYT1 mutation (Coubes et al., 1999; 2000; Roubertie et al., 2000; Coubes et al., 2004). The surgery of abnormal movements should intervene before the occurrence of skeletal deformities, which always diminish the outcome. Within the population of non-DYT1 dystonia and especially in secondary and heredodegenerative dystonias, results are not so predictable. The improvement of dysarthria is variable and often very few influenced by DBS.

Stimulation's switch-off systematically causes the recurrence of symptoms within some hours or days. Whatever, in several patients we could see a long lasting preservation of the clinical improvement for several weeks without stimulation.

In the secondary dystonia group, the efficacy of stimulation is far more limited. We were led to propose it for very handicapped patients for whom other therapeutic strategies failed to improve dystonia, as already mentioned. The clinical and etiological heterogeneity among this group almost prevents any global interpretation of these results. In this group, a frequently associated hypertonia of pyramidal origin influences the dystonic component.

An important negative prognostic factor under stimulation is the existence of a permanent hypertonia at rest, whatever may be its origin. Although we are not yet able to predict the long-term prognosis, we observed an interesting improvement with dyskinesia-dystonia secondary to a perinatal anoxia (dyskinetic forms of cerebral palsy accounting for less than 10 % of all forms CP; Lin, 2003), PKAN and mitochondrial diseases treated by GPi stimulation. A constant control of pain associated with muscle spasms was obtained in patients suffering from secondary dystonia.

The progression of the causal disease is of critical importance for patient's prognosis.

Using this surgical method based on MRI alone (Vayssiere et al., 2000; Coubes et al., 2002a; Vayssiere et al., 2002), the associated morbidity is low. We didn't observe hemorrhage due to the intracerebral tracts as reported before in other series (our experience reaches 236 electrodes; Starr et al., 1999). Secondary infection of the stimulation system remains the major complication of this technique and was observed in 4 patients. We summarize the complications observed in our population in table III.

Table III. Complications

Complications	Number
Hemorrage	0
Infection	4
Lead fracture	3
Extension fracture	3

The remarkable tolerance of the internal pulse generator must be also emphasized in children. We never observed any complication due to displacement with growth (loss of efficacy linked to the displacement of the electrode). As shown in figure 2 and 3, a residual length was enrolled around the battery and the electrode in order to compensate for growth in children and to provide some flexibility with movements in the system (Figure 2). Furthermore, it appears that growth does not interfere with stimulation, and the implantation of a single 90 cm extension compensates adequately for the growth of the child. We observed in two patients (1 child, 1 adult) an extension fracture due to adherences fixing the extension and limiting its mobility and flexibility.

The children's physical development (height, body weight) was followed as well as hormonal levels (Insuline-like Growth Factor-IGF1, Insuline–like Growth Factor Binding Protein 3-IGF-BP3, Estradiol, Testosterone, Folliculine Stimulating Hormone-FSH and Luteinizing Hormone-LH) in order to check puberty development.

8 Conclusion

Despite cost and complexity of the follow-up, bilateral chronic electrical stimulation can be proposed as first line treatment for early onset primary generalized dystonia when pharmacologically intractable and also early considered in segmentary dystonia in adults and in well-selected cases of secondary dystonia. It is conservative, adaptable, reversible and well tolerated by the whole population. It must be applied soon, especially in primary dystonia before neuro-ortopaedic sequels occur. The complication rate remains low.

For secondary dystonia, pallidal stimulation can partially improve dystonic syndromes with important control of pain and swallowing difficulties.

References

Benabid AL, Pollak P, Louveau A, Henry S, de Rougemont J (1987) Combined (thalamotomy and stimulation) stereotactic surgery of the VIM thalamic nucleus for bilateral Parkinson disease. *Appl Neurophysiol* **50**:344-346.

Benabid AL, Pollak P, Seigneuret E, Hoffmann D, Gay E, Perret J (1993) Chronic VIM thalamic stimulation in Parkinson's disease, essential tremor and extra-pyramidal dyskinesias. *Acta Neurochir Suppl (Wien)* **58**:39-44.

Burke RE, Fahn S, Marsden CD, Bressman SB, Moskowitz C, Friedman J (1985) Validity and reliability of a rating scale for the primary torsion dystonias. *Neurology* **35**:73-77.

Cif L, Valente EM, Hemm S, Coubes C, Vayssiere N, Serrat S, Di Giorgio A, Coubes P (2004) Deep brain stimulation in myoclonus-dystonia syndrome. *Mov Disord* **19**:724-727.

Coubes P, Roubertie A, Vayssiere N, Hemm S, Echenne B (2000) Treatment of DYT1-generalised dystonia by stimulation of the internal globus pallidus. *Lancet* **355**: 2220-2221.

Coubes P, Vayssiere N, El Fertit H, Hemm S, Cif L, Kienlen J, Bonafe A, Frerebeau P (2002a) Deep brain stimulation for dystonia. Surgical technique. *Stereotact Funct Neurosurg* **78**:183-191.

Coubes P, Echenne B, Roubertie A, Vayssiere N, Tuffery S, Humbertclaude V, Cambonie G, Claustres M, Frerebeau P (1999) [Treatment of early-onset generalized dystonia by chronic bilateral stimulation of the internal globus pallidus. Apropos of a case]. *Neurochirurgie* **45**:139-144.

Coubes P, Cif L, El Fertit H, Hemm S, Vayssiere N, Serrat S, Picot MC, Tuffery S, Claustres M, Echenne B, Frerebeau P (2004) Electrical stimulation of the globus pallidus internus in patients with primary generalized dystonia: long-term results. *J Neurosurg* **101**:189-194.

Coubes P, Cif L, Azais M, Roubertie A, Hemm S, Diakonoya N, Vayssiere N, Monnier C, Hardouin E, Ganau A, Tuffery S, Claustre M, Echenne B (2002b) [Treatment of dystonia syndrome by chronic electric stimulation of the internal globus pallidus]. *Arch Pediatr* **9** *Suppl* **2**:84s-86s.

Dwork AJ, Balmaceda C, Fazzini EA, MacCollin M, Cote L, Fahn S (1993) Dominantly inherited, early-onset parkinsonism: neuropathology of a new form. Neurology 43:69-74.

Fahn S (1988) Concept and classification of dystonia. *Adv Neurol* **50**:1-8.

Ford B, Greene PE, Louis ED, Bressman SB, Goodman RR, Brin MF, Sadiq S, Fahn S (1998) Intrathecal baclofen in the treatment of dystonia. *Adv Neurol* **78**:199-210.

Gasser T (1998) Inherited myoclonus-dystonia syndrome. Adv Neurol 78:325-334.

Gros C, Frerebeau P, Perez-Dominguez E, Bazin M, M PJ (1976) Long term results of stereotaxic surgery for infantile dystonia and dyskinesia. *Neurochirurgia (Stuttg)* **19**: 171-178.

Hemm S, N D, G M, N V, L C, P C (2002) Stimulated volume and energy consumption in improved dystonic patients treated by high frequency GPi stimulation. *Movement Disorders* **17**:S302.

Hemm S, Vayssiere N, Mennessier G, Cif L, Zanca M, Ravel P, Frerebeau P, P C (2004) Evolution of brain impedance in dystonic patients treated by GPi electrical stimulation. *Neuromodulation* 7:67-75.

Iacono RP, Kuniyoshi SM, Lonser RR, Maeda G, Inae AM, Ashwal S (1996) Simultaneous bilateral pallidoansotomy for idiopathic dystonia musculorum deformans. *Pediatr Neurol* **14**:145-148.

Ishikawa A, Takahashi H (1998) Clinical and neuropathological aspects of autosomal recessive juvenile parkinsonism. *J Neurol* 245:P4-9.

Lin JP (2003) The cerebral palsies: a physiological approach. *J Neurol Neurosurg Psychiatry* **74** Suppl 1:i23-29.

Ozelius LJ, Hewett JW, Page CE, Bressman SB, Kramer PL, Shalish C, de Leon D, Brin MF, Raymond D, Corey DP, Fahn S, Risch NJ, Buckler AJ, Gusella JF, Breakefield XO (1997) The early-onset torsion dystonia gene (DYT1) encodes an ATP-binding protein. *Nat Genet* **17**:40-48.

Roubertie A, Echenne B, Cif L, Vayssiere N, Hemm S, Coubes P (2000) Treatment of early-onset dystonia: update and a new perspective. *Childs Nerv Syst* **16**:334-340.

Starr PA, Vitek JL, DeLong M, Bakay RA (1999) Magnetic resonance imaging-based stereotactic localization of the globus pallidus and subthalamic nucleus. *Neurosurgery* **44**:303-313; discussion 313-304.

Tsui JK, Fross RD, Calne S, Calne DB (1987) Local treatment of spasmodic torticollis with botulinum toxin. *Can J Neurol Sci* **14**:533-535.

Vayssiere N, Hemm S, Zanca M, Picot MC, Bonafe A, Cif L, Frerebeau P, Coubes P (2000) Magnetic resonance imaging stereotactic target localization for deep brain stimulation in dystonic children. *J Neurosurg* **93**:784-790.

Vayssiere N, Hemm S, Cif L, Picot MC, Diakonova N, El Fertit H, Frerebeau P, Coubes P (2002) Comparison of atlas- and magnetic resonance imaging-based stereotactic targeting of the globus pallidus internus in the performance of deep brain stimulation for treatment of dystonia. *J Neurosurg* **96**:673-679.

Zimprich A, Grabowski M, Asmus F, Naumann M, Berg D, Bertram M, Scheidtmann K, Kern P, Winkelmann J, Muller-Myhsok B, Riedel L, Bauer M, Muller T, Castro M, Meitinger T, Strom TM, Gasser T (2001) Mutations in the gene encoding epsilon-sarcoglycan cause myoclonus-dystonia syndrome. *Nat Genet* **29**:66-69.

In: Recent Breakthroughs in Basal Ganglia Research
Editor: Erwan Bezard, pp. 403-411
ISBN: 1-59454-880-3

Chapter 32

DEEP BRAIN STIMULATION IN EPILEPSY

Colin Deransart*[*], *Karine Bressand*, *Philippe Kahane and Antoine Depaulis
INSERM U.704 - UJF "Dynamique des Réseaux Neuronaux"

ABSTRACT

Epilepsy refers to several neurological disorders affecting approximately 50 million people worldwide. Many patients suffering from drug-resistant epileptic seizures cannot be treated by conventionnal resective surgical approaches. Deep brain stimulation appears as a new promising therapy for the treatment of these medically-intractable seizures. Although still experimental, it received considerable attention in the past few years, and has been applied either to the epileptic focus itself or within different structures thought to be somehow related to the epileptic focus. Since the pioneering studies of Gale et al. supporting the existence of a gate control of epileptic seizures, accumulating evidence over the past 25 years supported that the basal ganglia system could remotely control several kinds of epileptic seizures. More recently, emphasis has been put on the subthalamic nucleus, according to experimental data and to neurological and neurosurgical approach developped in the treatment of movement disorders since the early 90's. The purpose of this review is to summarize and discuss these experimental data, in prospect to match them with the emerging clinical data.

INTRODUCTION

Epilepsy refers to several neurological disorders affecting approximately 50 million people worldwide. It is characterized by the spontaneous recurrence of seizures, either convulsive or non-convulsive. Seizures result from the sudden transient oversynchronization and overactivation of a neuronal population within the brain. They are currently divided into partial (focal) and generalized epilepsies (Commission on Classification and Terminology of the International League Against Epilepsy, 1989), although this classification has been frequently challenged by both experimental and clinical data (Meeren et al., 2002; Holmes et al., 2004). Partial seizures initially involve a limited part of neurones within one hemisphere. They are characterized by the focal aspect of their symptoms and electroencephalographic changes. Consciousness may be impaired and partial seizures may spread to other parts of the brain, thus progressing to a secondarily generalized seizure. Primary generalized seizures simultaneously involve both cerebral hemispheres from the beginning, displaying no evidence of local onset. Electroencephalographic changes are bilateral and synchronous. The clinical expression is either convulsive (bilateral tonic and/or clonic movements) or non-convulsive (staring, motor inhibition with behavioral arrest). Epileptic syndromes are built on several single factors: seizure type, severity and recurrence of seizures, aetiology, age of onset, anatomical data, precipitating factors, evolution. Thus, in addition to a distinction between partial and

[*] Corresponding author: INSERM U.704 - UJF "Dynamique des Réseaux Neuronaux", Université Joseph Fourier, UFR Biologie - Bât. B, Domaine Universitaire. 2280 rue de la Piscine - BP 53, 38041 GRENOBLE cedex 9. Tel: +33 (0)4 76 63 54 16 / Fax: +33 (0)4 76 63 54 15. E-mail address: colin.deransart@ujf-grenoble.fr

generalized epilepsies, the epileptic syndromes are classified as idiopathic or non idiopathic disorders according to the relative contribution of genetic and acquired pathologic factors. Determination of the epileptic syndrome determine to a great extent the type of evaluation and therapy the patient will receive (Holmes, 1997). Each kind of seizure responds differently to pharmacological treatments. However, these treatments are mainly symptomatic and it is assumed that up to 30 % of seizures are resistant to antiepileptic drugs, despite accurate diagnosis and carefully monitored treatments. Resective surgery can be proposed in patients suffering from drug-resistant seizures, but it is restricted to patients suffering from partial seizures in which an epileptic focus is clearly delineated outside "eloquent" brain regions (Rosenfeld, 2002). Other surgical treatment, named palliative, can be proposed (multiple subpial transections, callosotomy, vagus nerve stimulation), the aim of which is not to suppress the seizures but to reduce their frequency and/or severity (for a review, see Polkey, 2003). It remains that many patients suffering from drug-resistant seizures cannot be helped by theses different surgical approaches, and this explain why alternative treatment such as deep brain stimulation are under evaluation.

1 Targeted Structures

Since the seventies, different brain structures thought to modulate seizures have been targeted for deep brain stimulation (DBS) in order to improve severely affected epileptic patients. Basic mechanisms of DBS are thought to involve both silencing of neurons and generation of a new pattern of activity (see Benazzouz and Dostrovsky in this book). Instead of trying to interfere directly with the generating circuits underlying seizures, these approaches target remote structures thought to be somehow involved in the modulation of seizures. To this respect, the first trial of epilepsy treatment by brain stimulation was made by Cooper et al., (1973) who reported decreased seizure frequency after electrical stimulation of the cerebellum (Cooper et al., 1973). Because these results where not confirmed by further favourable controlled trials, chronic cerebellar stimulation in epilepsy was no longer considered, although its potential has probably not been fully exploited (Polkey, 2003).

Stimulation targets for epilepsy also included the anterior and centromedian nuclei of the thalamus. Due to their tight connections to the limbic system, the anterior nuclei have been considered in patients suffering from partial and secondarily generalized seizures. Although bilateral stimulation of the anterior nuclei was reported to significantly decrease seizure frequency in 5 patients, the observed benefits did not differ between stimulation-on and stimulation-off periods, raising the possibility that the implantation itself could account for some of the benefit (Hodaie et al., 2002). Conversely, in a recent report from 5 implanted patients, worsening of seizures was observed in 4 patients when the stimulation was discontinued (Kerrigan et al., 2004). Although up to 20 epileptic patients have been implanted so far for DBS of the anterior nuclei, the collected data preclude from any definite conclusion concerning the reliability of this approach. With their sparse connections to the cerebral cortex, the centromedian nuclei of the thalamus are part of the reticulothalamocortical system mediating cerebral cortex excitability. The group of Velasco has carried out the major work in this area. Improvement following DBS of the centromedian nuclei was highly significant for generalised tonic-clonic seizures and atypical absences but not for either tonic seizures or complex partial seizures (Velasco et al., 1995; 2000).

3 The Basal Ganglia in Epilepsy

The use of DBS of basal ganglia (BG) structures arises mainly from animal experimental data obtained from several groups during the past 25 years. Some of these data are summerized thereafter. The basal ganglia process information arising from the entire cerebral cortex (including the hippocampus), as well as from non-specific thalamic nuclei and the amygdala. In turn, via their efferent pathways, they control the excitability of brainstem premotor networks and thalamic nuclei connected to the motor, premotor, prefrontal and limbic areas of the cerebral cortex. Although there is still no consensus on the functional architecture of the basal ganglia, many aspects of their anatomical and the physiological organisation support that these structures can dynamically control seizures through their input-output relationships (Slaght et al., 2002). Although these data primarily derive from animal studies, clinical studies also largely contributed to our current understanding of the role of the basal ganglia in motor, cognitive and motivational components of behavior. In line with electrophysiological studies in primates, imaging studies in human support that mental processes associated with these different components of behavior are represented differentially within distinct cortico-basal ganglia circuits. For instance, the premotor circuit which connects the premotor cortex to the anterior part of the putamen, is thought to subserve preparation function of movements. The concomitant activation of the posterior putamen and sensorimotor cortex supports that the sensorimotor circuit has a primary executive role in the control of movement. The associative circuit through the prefrontal cortex to the caudate nucleus and large parts of the putamen rostral to the anterior commissure are thought to subserve several cognitive aspects of movements. The motivational aspects of behavior are processed by limbic circuits through the limbic and paralimbic cortex, hippocampal formation and amygdala to the ventral striatum (Lehericy and Gerardin, 2002). Although evidence for a role of such clearly delineated cortico-basal loops in human epilepsies is still lacking, several clinical aspect of seizures have been suggested to be compatible with the involvement of basal ganglia (Vercueil and Hirsch, 2002). Interestingly, Dematteis et al., (2003) recently reported transient bilateral striatal alterations secondary to an inflammatory process, which in turn could have disrupted a striatal inhibitory control over seizures originating from the left temporal lobe (Dematteis et al., 2003).

Over the past decades, the role of the basal ganglia system among the circuits involved in the expression of different types of seizures has been mainly addressed in animal models. Several animal models of seizures and epilepsy have been developed to understand the pathophysiological mechanisms underlying the different types of epileptic seizures (for review see Loscher and Schmidt, 1988; Fischer, 1989; Jobe et al., 1991; Sperk, 1994; Danober et al., 1998; Fisher, 1998). Recently, preference has been given to models of recurrent seizures whether they are genetic or induced by an initial manipulation, because they better reproduce the chronicity of this disease and the sudden occurrence and interruption of the seizures.

The possibility that the basal ganglia could be considered as a generator of epileptic seizures seems unlikely. Although spontaneous oscillatory activities have been recorded from these structures (Cassim et al., 2002) and that electrophysiological as well as metabolic changes occur concomittantly to seizing activities, the difficulty to initiate seizure through chemical or electrical stimulation of BG structures do not support their involvement in the triggering of seizure. Furthermore, lesion of the BG structures generally fails to abolish seizures, thus supporting the view that BG structures are not part of the seizure initiating circuit but rather exert some remote control. Changes in the activity of the BG circuits suggest that seizures propagate to or through these structures and may have consequences on the development of the on-going seizures (Deransart and Depaulis, 2002). Thus, although the BG are not involved per se in the generation of seizures, they appear to have a role in the modulation of phenotypic expression and perhaps in seizure control as suggested by the pioneering work of Karen Gale's group on the substantia nigra.

3.1 Experimental Evidence Supporting a Role of the BG in the Control of Seizures

Evidence supporting a role of the BG as a control circuit for epileptic seizures has been mainly collected from pharmacological or electrical modulation of these circuits and from electrophysiological analysis of neuronal activity. Data reported thereafter mainly focus on the involvement of the substantia nigra reticulata (SNR) and the subthalamic nucleus (STN).

3.2 Pharmacological Data

In 1980, Gale and Iadarola were the first to correlate an increase of GABA in SNR (induced by the systemic injection of an inhibitor of GABA-transaminase) with antiepileptic effects (Gale and Iadarola, 1980). Later, they showed that the potentiation of the GABAergic neurotransmission within the SNR, by bilateral microinjections of GABAmimetic drugs, suppressed convulsions in various models of generalized seizures in the rat (Iadarola and Gale, 1982). Local pharmacological inhibition of the SNR could also be obtained by blockade of the glutamatergic input which mainly arises from the STN. In particular, injections of different NMDA antagonists in the SNR suppressed absence-seizures in a genetic model of absence epilepsy in the Wistar rat (GAERS) whereas injections of non-NMDA antagonists were without effects (Deransart et al., 1996). Since then, several studies have confirmed that the pharmacological inhibition of

SNR suppresses both behavioural and EEG expression of various forms of generalized epileptic seizures. Recently, it was shown that deep brain stimulation applied to the SNR, also suppressed generalized convulsive seizures induced by fluorothyl inhalation (Velisek et al., 2002) absence seizures in GAERS (Feddersen et al., in preparation) and focal seizures in kainate treated mice (Deransart et al., in preparation). In GAERS and kainate-treated mice, bilateral and biphasic acute stimulation of SNR with a frequency of 130 Hz and a pulse width of 60 μs interrupt absence seizures and focal seizures without motor side effects. Repetition of such stimulation leads to a loss of the antiepileptic effects. Continuous chronic stimulation has no significant effect. These results suggest that continuous stimulation protocols need to be further evaluated and support the need for further investigation in the field of closed-loop stimulation devices to achieve optimal suppression.

Pharmacological inhibition of nigral neurons appears, however, to be without effects on tonic seizures involving brainstem structures and it has been suggested that such manipulations are more efficient on seizures involving forebrain circuits (Deransart et al., 2001). Furthermore, the neuronal population involved within the SNR appears critical and a differential involvement of anterior and posterior populations of this structure has been demonstrated in adult male rats (Veliskova et al., 2002). The circuits involved in the antiepileptic effects resulting from the inhibition of the SNR may differ according to the type of seizure under control and remain to be characterized. However, disinhibition of neurons of the superior colliculus, one of the main output of the SNR, seems to play a significant role in the mediation of these antiepileptic effects in some forms of seizures (Garant and Gale, 1987; Depaulis et al., 1990). Within this structure, specific populations of neurons appear to be involved in the control of different forms of seizures (Shehab et al., 1992; 1995; Nail-Boucherie et al., 2002). Similarly, some projections from the superior colliculus to the intralaminar thalamic nuclei are likely to interfere in this remote control circuit and thus act on the cortical oscillatory activities (Nail-Boucherie et al., 2002).

The glutamatergic input from the subthalamic nucleus (STN) appears critical in the maintenance of the SNR neuronal activity (Slaght et al., 2002). Bilateral injection of a GABA agonist into the STN, or high-frequency stimulation of this structure, decrease the activity of nigral neurons (Féger and Robledo, 1991; Benazzouz et al., 1995). We were the first to show that both ways of reducing the excitatory influence of the STN can suppress epileptic seizures (Deransart et al., 1996; Vercueil et al., 1998). In both cases, the suppression of absence seizures in GAERS could be dissociated from behavioral effects such as dystonia or stereotypies. These antiepileptic effects of STN inhibition were also confirmed in generalized convulsive seizures induced by fluorothyl inhalation (Veliskova et al., 1996), amygdala kindling (Deransart et al., 1998a), intravenous bicuculline or by its focal application into the anterior piriform cortex (Dybdal and Gale, 2000). More recently, high frequency stimulation of the subthalamic nucleus has been reported to protect against seizures induced by local kainate injection in the amygdala (Bressand et al., 1999; Loddenkemper et al., 2001). It is also in agreement with previous findings showing that intranigral injection of NMDA antagonists have antiepileptic effects (for review see Depaulis et al., 1994). The involvement of the GABAergic input from the pallidum to the subthalamic nucleus was investigated only in few models of epilepsy (Sato et al., 1976; Kelsey and Belluzzi, 1982; McLachlan and Bihari, 1990; Zhang et al., 1997). In GAERS, disinhibition of the globus pallidus or the ventral pallidum by injection of a GABAA antagonist suppresses absence seizures at doses deprived of behavioral effects. This antiepileptic effect is correlated with a decrease of glutamate levels in the SNR measured by microdialysis, suggesting that this effect is mediated, at least in part, through a reduction of activity of the subthalamic nucleus (Deransart et al., 1999).

3.3 Electrophysiological Data

These pharmacological data raised the possibility that some basal ganglia circuits may act as an endogenous control system able to interrupt epileptic seizures and/or to modulate their occurrence (Depaulis et al., 1994; Deransart et al., 1998b). According to such hypothesis, changes of neuronal activity associated with the occurrence and/or interruption of seizures should be detected in some circuits of the basal ganglia. There is good evidence from electrophysiological recordings in both human patients and animal models, that the epileptic activity is transmitted to the basal ganglia (Spiegel et al., 1956; Kaniff et al., 1983; Bonhaus et al., 1986; 1991; Rektor et al., 2001; see also Semah, 2002; Vercueil and Hirsch, 2002). However, in animal models, data were mainly collected during pharmacologically- or electrically-induced seizures and the endogenous nature of the observed changes of activity is thus difficult to determine. To investigate the possible correlations between unit activities in the basal ganglia during ictal and interictal periods, the genetic model of absence seizures was preferred for its spontaneous occurrence of frequent seizures and for the lack of motor components that could interfere with the recording. Initial results obtained by single unit recordings in GAERS maintained under neuroleptanalgesia indicated that, concomitantly with the occurrence of spike-and-wave discharges at the cortical level, SNR and STN neurons increase their firing rate whereas dopaminergic neurons of the substantia nigra pars compacta tend to decrease their activity. Conversely, at the end of the seizure, SNR and STN neurons decrease their firing rate whereas dopaminergic neurons of the substantia nigra pars compacta display a transient overactivity (Lücking et al., 2002). Because fentanyl and haloperidol used for neurolept-analgesia are known to interfere with neurotransmissions involved in the basal ganglia circuits, multi-unit recordings with chronically implanted movable tetrode were developed in the GAERS. The relative amplitudes of unit signals on the four channels, which are in part determined by the spatial locations of the spike generators relative to each recording surfaces of the tetrode, were used to allow unit isolation (McNaughton et al., 1983). Combined to off-line sorting of the tetrode data, this approach allowed us to access single unit activities in the SNR of freely-behaving epileptic animals (Rebrik et al., 1998; Deransart et al., 2003). Using this method, no particular organization of the discharge mode of the SNR neurons was observed during interictal periods, as compared to recordings performed in non-epileptic rats in previous studies (Gulley et al., 1999). However, when an absence seizure occurred on the cortical EEG, the frequency of discharge of the SNR neurons increases and action potentials become organized in bursts. These bursts always followed by a few milliseconds the spike component of each spike-and-wave element. In addition, a few seconds before the end of the seizure, a decrease of SNR neuronal activity is observed. Concomitant with the end of the seizure, some SNR neurons become silent for about 500–600 ms, after which they resume their regular firing mode. Changes could also be observed in SNR neurons synchrony. During interictal periods no clear synchrony between SNR neurons could be observed. As soon as spike-and-waves appear on the EEG and thereafter, synchrony between SNR neurons is increased. At the end of the seizure, before the seizure stops, some neurons stop firing in burst and their activity is no more phase-locked to the spike-and-waves on the EEG. Altogether, the decrease in the firing rate and synchronisation of SNR neurons before the complete end of the seizure, suggests that inhibition of SNR neurons could participate in the blockade of an ongoing seizure (Deransart et al., 2003). The nature and the temporal pattern of these changes are in agreement with what could be predicted from the pharmacological data and suggest that these events could contribute to the antiepileptic effects resulting from a physiological triggering of the control system. Whether these changes of SNR unit activity during seizures result from the propagation of cortical activity via the STN and/or via the striatum remain to be determined (Slaght et al., 2004).

4 Clinical Data

In 1978, Chkhenkeli et al. were the first to report therapeutic benefit of striatal low frequency stimulation in patients with pharmaco-resistant temporal lobe epilepsies. Conversely, high frequency stimulations of the caudate nucleus could worsen epileptic activity (Chkhenkeli, 1978; Chkhenkeli and Chkhenkeli, 1997). Emphasis on the STN has been put more recently, according to experimental data afore-mentioned and to neurological a neurosurgical skills developed in the treatment of movement disorders since the early 1990. The first STN stimulation was performed in 1998 in a 5-year-old girl with pharmacologically resistant, inoperable epilepsy caused by focal centroparietal dysplasia (Benabid et al., 2002). Thereafter, among 4 other patients treated in Grenoble, continuous high frequency stimulation of the STN induced a significant reduction in the seizures frequency from 67 % to 80 % for 3 of them (Chabardes et al., 2002). These results are in agreement with those of two other studies : two of the five patients who underwent stimulation at the Cleaveland Clinic Foundation had a reduction in seizure frequency of 60 to 80 % after 16 and 10 month of follow-up, respectively (Loddenkemper et al., 2001). The other three patients, including one with focal status epilepticus, did not show any improvement. An additional patient presenting Lennox-Gastaut syndrome and who was stimulated in Beirut showed a complete suppression of generalized tonic-clonic seizures and reduction of myoclonic and atypical absences of over 75 % after a 1 year follow-up (Alaraj et al., 2001).

Conclusion

Experimental data and preliminary clinical results suggest that neuromodulation of the BG, particularly of the STN, can be effective in patients with pharmaco-resistant epilepsy who are not suitable for conventional resective surgery. DBS of anterior and centromedian nuclei of the thalamus or of the STN have been proven to be feasible, safe and possibly effective in patients suffering from different kinds of epilepsies. The fact that responders experienced a seizure reduction, but not a complete remission of seizure means that surgical resection remains the gold standard treatment of drug-resistant epilepsies whenever this option is possible. At the present time, it is not clear which brain region, when stimulated, will be the most effective for preventing or stopping seizures. Also unknown is which types of seizures are most likely to respond to DBS as well as the optimal stimulation parameters to be used. The mechanism by which electrical stimulation might suppress seizures also remains to be understood as well as the possible difference in organisation of neuronal networks thought to be involved in DBS in parkinsonian and epileptic patients. In this respect, interaction between clinicians and researchers appears essential and controlled clinical trials will be necessary to further document and validate this new therapeutic approach.

References

Alaraj A, Commair Y, Mikati M, Wakim J, Louak E, Atweh S (2001) Subthalamic nucleus deep brain stimulation: a novel method for the treatment for non-focal intractable epilepsy. In: Neuromodulation: defining the future. Cleveland, OH.

Benabid AL, Minotti L, Koudsie A, de Saint Martin A, Hirsch E (2002) Antiepileptic effect of high-frequency stimulation of the subthalamic nucleus (corpus luysi) in a case of medically intractable epilepsy caused by focal dysplasia: a 30-month follow-up: technical case report. *Neurosurgery* **50**:1385-1391; discussion 1391-1382.

Benazzouz A, Piallat B, Pollak P, Benabid AL (1995) Responses of substantia nigra pars reticulata and globus pallidus complex to high frequency stimulation of the subthalamic nucleus in rats: electrophysiological data. *Neurosci Lett* **189**:77-80.

Bonhaus DW, Walters JR, McNamara JO (1986) Activation of substantia nigra neurons: role in the propagation of seizures in kindled rats. *J Neurosci* **6**:3024-3030.

Bonhaus DW, Russell RD, McNamara JO (1991) Activation of substantia nigra pars reticulata neurons: role in the initiation and behavioral expression of kindled seizures. *Brain Res* **545**:41-48.

Bressand K, Dematteis M, Kahane P, Benazzouz A, Benabid AL (1999) Involvement of the subthalamic nucleus in the control of temporal lobe epilepsy: study by high frequency stimulation in rats. In: *Soc Neurosci* (Abstract), p 1656.

Cassim F, Labyt E, Devos D, Defebvre L, Destee A, Derambure P (2002) Relationship between oscillations in the basal ganglia and synchronization of cortical activity. *Epileptic Disord 4 Suppl* **3**:S31-45.

Chabardes S, Kahane P, Minotti L, Koudsie A, Hirsch E, Benabid AL (2002) Deep brain stimulation in epilepsy with particular reference to the subthalamic nucleus. *Epileptic Disord 4 Suppl* **3**:S83-93.

Chkhenkeli SA (1978) The inhibitory influence of the nucleus caudatus electrostimulation on the human's amygdalar and hippocampal activity at temporal lobe epilepsy. *Bull Georgian Acad Sci 4/***6**:406-411.

Chkhenkeli SA, Chkhenkeli IS (1997) Effects of therapeutic stimulation of nucleus caudatus on epileptic electrical activity of brain in patients with intractable epilepsy. *Stereotact Funct Neurosurg* **69**:221-224.

Commission on Classification and Terminology of the International League Against Epilepsy (1989) Proposal for revised classification of epilepsies and epileptic syndromes. Commission on Classification and Terminology of the International League Against Epilepsy. *Epilepsia* **30**:389-399.

Cooper IS, Amin I, Gilman S (1973) The effect of chronic cerebellar stimulation upon epilepsy in man. *Trans Am Neurol Assoc* **98**:192-196.

Danober L, Deransart C, Depaulis A, Vergnes M, Marescaux C (1998) Pathophysiological mechanisms of genetic absence epilepsy in the rat. *Prog Neurobiol* **55**:27-57.

Dematteis M, Kahane P, Vercueil L, Depaulis A (2003) MRI evidence for the involvement of basal ganglia in epileptic seizures: an hypothesis. *Epileptic Disord* **5**:161-164.

Depaulis A, Vergnes M, Marescaux C (1994) Endogenous control of epilepsy: the nigral inhibitory system. *Prog Neurobiol* **42**:33-52.
Depaulis A, Vergnes M, Liu Z, Kempf E, Marescaux C (1990) Involvement of the nigral output pathways in the inhibitory control of the substantia nigra over generalized non-convulsive seizures in the rat. *Neuroscience* **39**:339-349.
Deransart C, Depaulis A (2002) The control of seizures by the basal ganglia? A review of experimental data. *Epileptic Disord 4 Suppl* **3**:S61-72.
Deransart C, Marescaux C, Depaulis A (1996) Involvement of nigral glutamatergic inputs in the control of seizures in a genetic model of absence epilepsy in the rat. *Neuroscience* **71**:721-728.
Deransart C, Le BT, Marescaux C, Depaulis A (1998a) Role of the subthalamo-nigral input in the control of amygdala-kindled seizures in the rat. *Brain Res* **807**:78-83.
Deransart C, Vercueil L, Marescaux C, Depaulis A (1998b) The role of basal ganglia in the control of generalized absence seizures. *Epilepsy Res* **32**:213-223.
Deransart C, Le-Pham BT, Hirsch E, Marescaux C, Depaulis A (2001) Inhibition of the substantia nigra suppresses absences and clonic seizures in audiogenic rats, but not tonic seizures: evidence for seizure specificity of the nigral control. *Neuroscience* **105**:203-211.
Deransart C, Riban V, Le BT, Hechler V, Marescaux C, Depaulis A (1999) Evidence for the involvement of the pallidum in the modulation of seizures in a genetic model of absence epilepsy in the rat. *Neurosci Lett* **265**:131-134.
Deransart C, Hellwig B, Heupel-Reuter M, Leger JF, Heck D, Lucking CH (2003) Single-unit analysis of substantia nigra pars reticulata neurons in freely behaving rats with genetic absence epilepsy. *Epilepsia* **44**:1513-1520.
Deransart et al. (in preparation).
Dybdal D, Gale K (2000) Postural and anticonvulsant effects of inhibition of the rat subthalamic nucleus. *J Neurosci* **20**:6728-6733.
Feddersen et al. (in preparation).
Féger J, Robledo P (1991) The effects of activation or inhibition of the subthalamic nucleus on the metabolic and electrophysiological activities within the pallidal complex and the substantia nigra in the rat. *Eur J Neurosci* **3**:947-952.
Fischer RS (1989) Animal models of the epilepsies. *Brain Res Brain Res Rev* **14**:245-278.
Fisher RS (1998) Epilepsy. In: Pharmacological Management of Neurological and Psychiatric Disorders (Enna SJ, Coyle JT, eds): McGraw-Hill.
Gale K, Iadarola MJ (1980) Seizure protection and increased nerve-terminal GABA: delayed effects of GABA transaminase inhibition. *Science* **208**:288-291.
Garant DS, Gale K (1987) Substantia nigra-mediated anticonvulsant actions: role of nigral output pathways. *Exp Neurol* **97**:143-159.
Gulley JM, Kuwajima M, Mayhill E, Rebec GV (1999) Behavior-related changes in the activity of substantia nigra pars reticulata neurons in freely moving rats. *Brain Res* **845**:68-76.
Hodaie M, Wennberg RA, Dostrovsky JO, Lozano AM (2002) Chronic anterior thalamus stimulation for intractable epilepsy. *Epilepsia* **43**:603-608.
Holmes GL (1997) Classification of seizures and the epilepsies. In: The comprehensive evaluation and treatment of epilepsy. (Schachter SC, Schomer DL, eds), pp 1-36. New York: Academic Press, Inc.
Holmes MD, Brown M, Tucker DM (2004) Are "generalized" seizures truly generalized? Evidence of localized mesial frontal and frontopolar discharges in absence. *Epilepsia* **45**:1568-1579.
Iadarola MJ, Gale K (1982) Substantia nigra: site of anticonvulsant activity mediated by gamma-aminobutyric acid. *Science* **218**:1237-1240.
Jobe PC, Mishra PK, Ludvig N, Dailey JW (1991) Scope and contribution of genetic models to an understanding of the epilepsies. *Crit Rev Neurobiol* **6**:183-220.
Kaniff TE, Chuman CM, Neafsey EJ (1983) Substantia nigra single unit activity during penicillin-induced focal cortical epileptiform discharge in the rat. *Brain Res Bull* **11**:11-13.
Kelsey JE, Belluzzi JD (1982) Endorphin mediation of post-ictal effects of kindled seizures in rats. *Brain Res* **253**:337-340.
Kerrigan JF, Litt B, Fisher RS, Cranstoun S, French JA, Blum DE, Dichter M, Shetter A, Baltuch G, Jaggi J, Krone S, Brodie M, Rise M, Graves N (2004) Electrical stimulation of the anterior nucleus of the thalamus for the treatment of intractable epilepsy. *Epilepsia* **45**:346-354.

Lehericy S, Gerardin E (2002) Normal functional imaging of the basal ganglia. *Epileptic Disord 4 Suppl* **3**:S23-30.

Loddenkemper T, Pan A, Neme S, Baker KB, Rezai AR, Dinner DS, Montgomery EB, Jr., Luders HO (2001) Deep brain stimulation in epilepsy. *J Clin Neurophysiol* **18**:514-532.

Loscher W, Schmidt D (1988) Which animal models should be used in the search for new antiepileptic drugs? A proposal based on experimental and clinical considerations. *Epilepsy Res* **2**:145-181.

Lücking CH, Hellwig B, Deransart C (2002) The role of the basal ganglia in the control of tremor and epilepsy. *Advances in Clinical Neurophysiology (Supplements to Clinical Neurophysiology)* **54**:23-30.

McLachlan RS, Bihari F (1990) Secondary generalization of seizures from a cortical penicillin focus following stimulation of the basal forebrain. *Exp Neurol* **109**:237-242.

McNaughton BL, O'Keefe J, Barnes CA (1983) The stereotrode: a new technique for simultaneous isolation of several single units in the central nervous system from multiple unit records. *J Neurosci Methods* **8**:391-397.

Meeren HK, Pijn JP, Van Luijtelaar EL, Coenen AM, Lopes da Silva FH (2002) Cortical focus drives widespread corticothalamic networks during spontaneous absence seizures in rats. *J Neurosci* **22**:1480-1495.

Nail-Boucherie K, Le-Pham BT, Marescaux C, Depaulis A (2002) Suppression of absence seizures by electrical and pharmacological activation of the caudal superior colliculus in a genetic model of absence epilepsy in the rat. *Exp Neurol* **177**:503-514.

Polkey CE (2003) Alternative surgical procedures to help drug-resistant epilepsy - a review. *Epileptic Disord* **5**:63-75.

Rebrik S, Tzonev S, Miller KD (1998) Analysis of Tetrode Recordings in Cat Visual System. In: *Proceedings of CNS97* (Computation and Neural Systems Meeting (Bower JM, ed). Big Sky Montana: Plenum Press.

Rektor I, Bares M, Kubova D (2001) Movement-related potentials in the basal ganglia: a SEEG readiness potential study. *Clin Neurophysiol* **112**:2146-2153.

Rosenfeld JV (2002) Epilepsy surgery, hypothalamic hamartomas and the quest for a cure. *J R Coll Surg Edinb* **47**:653-659.

Sato M, Onishi T, Nakashima T, Otsuki S (1976) [Experimental study on epilepsy using a "kindling preparation". Correlation between septal seizure development and psychomotor seizure]. *No To Shinkei* **28**:667-679.

Semah F (2002) PET imaging in epilepsy: basal ganglia and thalamic involvement. *Epileptic Disord 4 Suppl* **3**:S55-60.

Shehab S, Coffey P, Dean P, Redgrave P (1992) Regional expression of fos-like immunoreactivity following seizures induced by pentylenetetrazole and maximal electroshock. *Exp Neurol* **118**:261-274.

Shehab S, Simkins M, Dean P, Redgrave P (1995) The dorsal midbrain anticonvulsant zone--I. Effects of locally administered excitatory amino acids or bicuculline on maximal electroshock seizures. *Neuroscience* **65**:671-679.

Slaght SJ, Paz T, Mahon S, Maurice N, Charpier S, Deniau JM (2002) Functional organization of the circuits connecting the cerebral cortex and the basal ganglia: implications for the role of the basal ganglia in epilepsy. *Epileptic Disord 4 Suppl* **3**: S9-22.

Slaght SJ, Paz T, Chavez M, Deniau JM, Mahon S, Charpier S (2004) On the activity of the corticostriatal networks during spike-and-wave discharges in a genetic model of absence epilepsy. *J Neurosci* **24**:6816-6825.

Sperk G (1994) Kainic acid seizures in the rat. *Prog Neurobiol* **42**:1-32.

Spiegel EA, Wycis HT, Baird HW, 3rd, Szekely EG (1956) Functional state of basal ganglia in extrapyramidal and convulsive disorders; an electrographic study. *AMA Arch Neurol Psychiatry* **75**:167-174.

Velasco F, Velasco M, Velasco AL, Jimenez F, Marquez I, Rise M (1995) Electrical stimulation of the centromedian thalamic nucleus in control of seizures: long-term studies. *Epilepsia* **36**:63-71.

Velasco F, Velasco M, Jimenez F, Velasco AL, Brito F, Rise M, Carrillo-Ruiz JD (2000) Predictors in the treatment of difficult-to-control seizures by electrical stimulation of the centromedian thalamic nucleus. *Neurosurgery* **47**:295-304; discussion 304-295.

Velisek L, Veliskova J, Moshe SL (2002) Electrical stimulation of substantia nigra pars reticulata is anticonvulsant in adult and young male rats. *Exp Neurol* **173**:145-152.

Veliskova J, Velsek L, Moshe SL (1996) Subthalamic nucleus: a new anticonvulsant site in the brain. *Neuroreport* **7**:1786-1788.

Veliskova J, Claudio OI, Galanopoulou AS, Kyrozis A, Lado FA, Ravizza T, Velisek L, Moshe SL (2002) Developmental aspects of the basal ganglia and therapeutic perspectives. *Epileptic Disord 4 Suppl* **3**:S73-82.

Vercueil L, Hirsch E (2002) Seizures and the basal ganglia: a review of the clinical data. *Epileptic Disord 4 Suppl* **3**:S47-54.

Vercueil L, Benazzouz A, Deransart C, Bressand K, Marescaux C, Depaulis A, Benabid AL (1998) High-frequency stimulation of the subthalamic nucleus suppresses absence seizures in the rat: comparison with neurotoxic lesions. *Epilepsy Res* **31**:39-46.

Zhang X, Le Gal La Salle G, Ridoux V, Yu PH, Ju G (1997) Prevention of kainic acid-induced limbic seizures and Fos expression by the GABA-A receptor agonist muscimol. *Eur J Neurosci* **9**:29-40.

In: Recent Breakthroughs in Basal Ganglia Research
Editor: Erwan Bezard, pp. 413-418

ISBN: 1-59454-880-3

Chapter 33

Clinical Results of STN DBS in Parkinson's Disease

Frédéric Torny
Chef de clinique, Service de Neurologie. CHU de Limoges, 2 avenue Martin-Luther-King, 87042 Limoges Cedex

Paul Krack
Service de Neurologie. CHU de Grenoble, B.P. 217, 38043 Grenoble Cedex 09

Abstract

In advanced Parkinson's disease, motor complications of dopaminergic treatment become disabling in many patients. Subthalamic Nucleus Deep Brain Stimulation (STN DBS) is a surgical treatment of motor fluctuations and levodopa-induced dyskinesias. Overall outcome of STN DBS depends on patient selection, precision of targeting, surgical complications and postoperative management of drugs and stimulation parameters. This treatment is most useful in relatively young patients with severe levodopa-induced motor complications, a good motor score in the best on-medication state and the absence of levodopa-resistant signs, cognitive deterioration, and severe comorbidity. Bilateral stimulation and correctly placed electrodes allow an improvement in parkinsonian symptoms (akinesia, rigidity, and tremor) in off-medication condition. After surgery a decrease in medication is possible allowing for a marked decrease in levodopa-induced dyskinesias. This amelioration persists at five years but the symptomatic treatment does not prevent the progression of the underlying neurodegenerative disease with progressive appearance of signs that do not respond to either L-dopa or subthalamic stimulation. Surgical complications are relatively rare but can be severe. Regular follow-up is necessary to prevent treatment-related adverse effects and to adapt the therapy during evolution.

Keywords: Parkinson's disease, subthalamic stimulation

Introduction

Levodopa is the mainstay of treatment for Parkinson's disease (PD). However, long-term therapy is inevitably complicated by motor fluctuations in which patients cycle between a state of severe parkinsonism (the "off" period, when the medication is not working, also referred to as "off medication") and a state of improved mobility when the medication is working (the "on" period, also referred to as "on medication"). In addition, patients are often impaired by dyskinesias during the on period and by painful dystonia during the off period. These complications result in disability that cannot be satisfactorily controlled by medical therapy in many patients. Advances in understanding of the functional organization of the basal ganglia (Alexander and Crutcher, 1990), and development of stereotaxic surgery and deep brain stimulation (DBS) provided opportunities for new therapeutic strategies. Lesioning the subthalamic nucleus (STN) in monkeys rendered parkinsonian by 1-methyl-4-phenyl-1,2,3,6-tetrahydropyridine (MPTP) intoxication was shown to be very effective in relieving parkinsonian motor symptoms (Bergman et al., 1990; Aziz et al., 1991). STN DBS for Parkinson's disease was proposed in the 90's based on this animal model of Levodopa-responsive parkinsonism (Pollak et al., 1993).

1 Patient Selection

The results of STN DBS depend on patient selection. The main criteria rest on the type of disease the patient suffers from, his general health state, cognitive and behavioural state, results of brain MRI and other factors (Charles et al., 2002; Lang and Widner, 2002; Welter et al., 2002). Mainly off-motor symptom respond to STN stimulation, proportionally to their presurgical response to levodopa or apomorphine (Pinter et al., 1999), so patients who do not have a good response to levodopa or with postsynaptic dopaminergic lesion will show poor improvement (Krack et al., 2000; Kleiner-Fisman et al., 2003). A good motor score in the best 'on' state (on medication before surgery) with few non-responsive levodopa symptoms (good axial score: gait and postural stability) will predict optimal DBS results. In order to justify the risks of surgical procedure, the stimulation should be proposed only in disabled patients, who are unable to cope with their activities of daily living and requiring some assistance, after failure of all drugs strategies, disability being related to off-period parkinsonian symptoms or to dyskinesia. Patients with a biological age over 70 years, severe comorbidity (unstable angina pectoris, severe cerebral macroangiopathy...), uncontrolled affective or behavioural disorders should be excluded to avoid general or surgical complications in the perioperative period. Although overall neuropsychological status including the executive function remains unchanged after surgery (Ardouin et al., 1999; Jahanshahi et al., 2000; Pillon et al., 2000), older patients and patients with poor neuropsychological performance can be aggravated (Hariz et al., 2000; Saint-Cyr et al., 2000). Personal, social and professional considerations should also be discussed, in order to decide the most suitable time for surgery. STN DBS seems most useful for relatively young patients who have severe levodopa-induced motor complication and who are independent in activities of daily living on their best on-medication state, without levodopa-resistant signs or cognitive deterioration (Krack et al., 2003). Given these considerations, it is clear that surgery addresses only a minority of PD patients, estimates varying between 10 and 15 percent of the total population.

2 Effects on Parkinsonian Symptom

The beneficial effects of STN stimulation are significant providing that the electrodes are correctly placed into the target. STN DBS mainly affects the controlateral hemibody. Bilateral STN DBS improves parkinsonism considerably more than unilateral STN DSB and is the most appropriate surgical option for most patients with bilateral as well as asymmetrical symptom (Limousin et al., 1995; Kumar et al., 1999). There are few reports on unilateral STN DBS, but as DBS and medication interfere it is problematic to treat a patient with DBS and drugs on one side and with drugs only on the other side of the body. When patients are evaluated off medication, stimulation results in improved scores for akinesia, rigidity and tremor: the Unified Parkinson's Disease Rating Scale (UPDRS) motor score is improved by about 50 to 60 % and off-motor phases tend to disappear so patients need no more help in activities of daily living (Kumar et al., 1998; Limousin et al., 1998; Houeto et al., 2000; The PD study group, 2001; Pollak et al., 2002). Levodopa-responsive axial signs also improve (Bejjani et al., 2000). The small benefits that have been noted with stimulation in the on-medication state by some authors do not necessarily translate by improved function and may be partially related to evaluation artefacts (dose used for the Levodopa test, improvement in dyskinesia; Bejjani et al., 2000; The PD study group, 2001). Usually, stimulation amplitude has to be increased in all patients in the first one to three months following the electrode implantation. This could be related to the time of healing of the lesion-like effect induced by the implantation. In the long-term, the voltage necessary to obtain a motor benefit remains relatively stable (no relevant tolerance effect). At five years, the parkinsonian motor signs off medication are still significantly improved: tremor and rigidity remains stable, akinesia slightly increases, and most patients remain independent in their activities of daily living (Krack et al., 2003). Deterioration of axial symptoms (including speech, postural stability, freezing of gait), characteristic of natural history of Parkinson's disease (Markham and Diamond, 1986) and attributed to the increasing severity of nondopaminergic lesions (Bonnet et al., 1987), do not seem to be prevented by STN DBS. The STN being glutamatergic, and glutamate being neurotoxic, a potential neuroprotective effect of STN surgery has been proposed (Rodriguez et al., 1998). Indeed, in the 6-OHDA rat model of parkinsonism neuroprotective effects of STN lesions have been found (Piallat et al., 1996), but these results remain controversial (Paul et al., 2004). It remains to be seen whether neuroprotective effects can be demonstrated in the MPTP model, which is a model much closer to human pathology. Whether STN DBS can influence the natural history of the disease or not cannot be deduced from the data in humans, as the available studies were done without controls.

3 Effects on Dyskinesia

Postoperatively, off-period painful dystonia is immediately alleviated, synchronously with stimulation by a specific effect (Krack et al., 1999). The on-period dyskinesias are decreased by 60 to 80 % (Krack et al., 1997; Fraix et al., 2000; Houeto et al., 2000; The PD study group, 2001; Pollak et al., 2002). The mechanism associates the replacement of a pulsatile stimulation of dopaminergic receptors with drugs by the continuous electrical stimulation of the target (mimicking a lesion), and the decrease in dopaminergic medication (Limousin et al., 1998; Fraix et al., 2000; The PD study group, 2001; Pollak et al., 2002) responsible for a partial reversibility of the sensitization phenomenon (Bejjani et al., 2000; Wenzelburger et al., 2002). Reduction of dopaminergic medication is generally obtained in the first 3 months and kept relatively stable thereafter (Krack et al., 2003). However, some patients may need a secondary increase of drug dosage (for example to treat aggravation of apathy), which can lead to a reincrease in dyskinesia.

4 Adverse Effects

The most severe surgical complication is intracranial hematoma, an inherent risk of stereotaxic surgery, estimated to occur in 2 to 8 % of patients (Obeso et al., 1997; Alkhani and Lozano, 2001; The PD study group, 2001; Pollak et al., 2002; Krack et al., 2003), and that seems to be increased by increased numbers of microelectrodes passes (The PD study group, 2001). Intracranial hematoma is less frequent in non-microelectrode guided surgery (Alkhani and Lozano, 2001; Hariz, 2002). Whether the benefit of more precise targeting using microelectrodes prevails over the benefit of having less hematoma is a debate that is not closed. Other complications are mostly temporary. Psychiatric problems, including depression, mania , hallucinations, psychosis, were reported by several groups (Berney et al., 2002; Doshi et al., 2002; Houeto et al., 2002), and may be related to many factors: preexisting psychiatric illness, changes in medication, surgery-related stress, social readaptation, impact of STN DBS on the limbic circuit (Krack et al., 2003; Funkiewiez et al., 2004). Weight gain, already described after pallidotomy (Lozano et al., 1995; Limousin et al., 1998; Gironell et al., 2002), is frequent in the first postoperative months and must be prevented by nutritional counseling. Weight increase is most likely related to the reduction in energy expenditure parallel to the reduction in dyskinesias and off periods as also seen also with continuous apomorphine infusions (Manson et al., 2002). A regular follow-up of the patients in the peri-operative period is necessary to prevent these complications. In the long-term, some patients developed apathy, often responding to reincreasing dopaminergic medication. Apathy developing in the long term and being resistant to dopaminergic treatment may reflect the natural progression of the disease in the same way as progressive deterioration of cognitive function, speech, freezing, handwriting or balance (Krack et al., 2003). The small increase with time in stimulation parameters may play a role in deterioration of some motor signs, especially dysarthria, which can be related to current diffusion to corticobulbar fibers in the close vicinity of the STN. Adaptation of drugs and electrical parameters are sometimes necessary and justify, like other therapeutics, a regular follow-up of the patients.

Conclusion

STN DBS is an adaptative, relatively safe, and highly efficient therapy of levodopa-responsive parkinsonian symptoms. The outcome depends on patient selection, the precision of the surgical procedure, and the quality of the neurological follow-up.

References

Alexander GE, Crutcher MD (1990) Functional architecture of basal ganglia circuits: neural substrates of parallel processing. *Trends Neurosci* **13**:266-271.

Alkhani A, Lozano AM (2001) Pallidotomy for parkinson disease: a review of contemporary literature. *J Neurosurg* **94**:43-49.

Ardouin C, Pillon B, Peiffer E, Bejjani P, Limousin P, Damier P, Arnulf I, Benabid AL, Agid Y, Pollak P (1999) Bilateral subthalamic or pallidal stimulation for Parkinson's disease affects neither memory nor executive functions: a consecutive series of 62 patients. *Ann Neurol* **46**:217-223.

Aziz TZ, Peggs D, Sambrook MA, Crossman AR (1991) Lesion of the subthalamic nucleus for the alleviation of 1-methyl-4-phenyl-1,2,3,6-tetrahydropyridine (MPTP)-induced parkinsonism in the primate. *Mov Disord* **6**:288-292.

Bejjani BP, Gervais D, Arnulf I, Papadopoulos S, Demeret S, Bonnet AM, Cornu P, Damier P, Agid Y (2000) Axial parkinsonian symptoms can be improved: the role of levodopa and bilateral subthalamic stimulation. *J Neurol Neurosurg Psychiatry* **68**:595-600.

Bergman H, Wichmann T, DeLong MR (1990) Reversal of experimental parkinsonism by lesions of the subthalamic nucleus. *Science* **249**:1436-1438.

Berney A, Vingerhoets F, Perrin A, Guex P, Villemure JG, Burkhard PR, Benkelfat C, Ghika J (2002) Effect on mood of subthalamic DBS for Parkinson's disease: a consecutive series of 24 patients. *Neurology* **59**:1427-1429.

Bonnet AM, Loria Y, Saint-Hilaire MH, Lhermitte F, Agid Y (1987) Does long-term aggravation of Parkinson's disease result from nondopaminergic lesions? *Neurology* **37**:1539-1542.

Charles PD, Van Blercom N, Krack P, Lee SL, Xie J, Besson G, Benabid AL, Pollak P (2002) Predictors of effective bilateral subthalamic nucleus stimulation for PD. *Neurology* **59**:932-934.

Doshi PK, Chhaya N, Bhatt MH (2002) Depression leading to attempted suicide after bilateral subthalamic nucleus stimulation for Parkinson's disease. *Mov Disord* **17**:1084-1085.

Fraix V, Pollak P, Van Blercom N, Xie J, Krack P, Koudsie A, Benabid AL (2000) Effect of subthalamic nucleus stimulation on levodopa-induced dyskinesia in Parkinson's disease. *Neurology* **55**:1921-1923.

Funkiewiez A, Ardouin C, Caputo E, Krack P, Fraix V, Klinger H, Chabardes S, Foote K, Benabid AL, Pollak P (2004) Long term effects of bilateral subthalamic nucleus stimulation on cognitive function, mood, and behaviour in Parkinson's disease. *J Neurol Neurosurg Psychiatry* **75**:834-839.

Gironell A, Pascual-Sedano B, Otermin P, Kulisevsky J (2002) [Weight gain after functional surgery for Parkinsons disease]. *Neurologia* **17**:310-316.

Hariz MI (2002) Safety and risk of microelectrode recording in surgery for movement disorders. *Stereotact Funct Neurosurg* **78**:146-157.

Hariz MI, Johansson F, Shamsgovara P, Johansson E, Hariz GM, Fagerlund M (2000) Bilateral subthalamic nucleus stimulation in a parkinsonian patient with preoperative deficits in speech and cognition: persistent improvement in mobility but increased dependency: a case study. *Mov Disord* **15**:136-139.

Houeto JL, Damier P, Bejjani PB, Staedler C, Bonnet AM, Arnulf I, Pidoux B, Dormont D, Cornu P, Agid Y (2000) Subthalamic stimulation in Parkinson disease: a multidisciplinary approach. *Arch Neurol* **57**:461-465.

Houeto JL, Mesnage V, Mallet L, Pillon B, Gargiulo M, du Moncel ST, Bonnet AM, Pidoux B, Dormont D, Cornu P, Agid Y (2002) Behavioural disorders, Parkinson's disease and subthalamic stimulation. *J Neurol Neurosurg Psychiatry* **72**:701-707.

Jahanshahi M, Ardouin CM, Brown RG, Rothwell JC, Obeso J, Albanese A, Rodriguez-Oroz MC, Moro E, Benabid AL, Pollak P, Limousin-Dowsey P (2000) The impact of deep brain stimulation on executive function in Parkinson's disease. *Brain* **123** (Pt 6):1142-1154.

Kleiner-Fisman G, Fisman DN, Kahn FI, Sime E, Lozano AM, Lang AE (2003) Motor cortical stimulation for parkinsonism in multiple system atrophy. *Arch Neurol* **60**:1554-1558.

Krack P, Limousin P, Benabid AL, Pollak P (1997) Chronic stimulation of subthalamic nucleus improves levodopa-induced dyskinesias in Parkinson's disease. *Lancet* **350**:1676.

Krack P, Pollak P, Limousin P, Benazzouz A, Deuschl G, Benabid AL (1999) From off-period dystonia to peak-dose chorea. The clinical spectrum of varying subthalamic nucleus activity. *Brain* **122** *(Pt 6)*:1133-1146.

Krack P, Dowsey PL, Benabid AL, Acarin N, Benazzouz A, Kunig G, Leenders KL, Obeso JA, Pollak P (2000) Ineffective subthalamic nucleus stimulation in levodopa-resistant postischemic parkinsonism. *Neurology* **54**:2182-2184.

Krack P, Batir A, Van Blercom N, Chabardes S, Fraix V, Ardouin C, Koudsie A, Limousin PD, Benazzouz A, LeBas JF, Benabid AL, Pollak P (2003) Five-year follow-up of bilateral stimulation of the subthalamic nucleus in advanced Parkinson's disease. *N Engl J Med* **349**:1925-1934.

Kumar R, Lozano AM, Sime E, Halket E, Lang AE (1999) Comparative effects of unilateral and bilateral subthalamic nucleus deep brain stimulation. *Neurology* **53**:561-566.

Kumar R, Lozano AM, Kim YJ, Hutchison WD, Sime E, Halket E, Lang AE (1998) Double-blind evaluation of subthalamic nucleus deep brain stimulation in advanced Parkinson's disease. *Neurology* **51**:850-855.

Lang AE, Widner H (2002) Deep brain stimulation for Parkinson's disease: patient selection and evaluation. *Mov Disord 17 Suppl* **3**:S94-101.

Limousin P, Pollak P, Benazzouz A, Hoffmann D, Broussolle E, Perret JE, Benabid AL (1995) Bilateral subthalamic nucleus stimulation for severe Parkinson's disease. *Mov Disord* **10**:672-674.

Limousin P, Krack P, Pollak P, Benazzouz A, Ardouin C, Hoffmann D, Benabid AL (1998) Electrical stimulation of the subthalamic nucleus in advanced Parkinson's disease. *N Engl J Med* **339**:1105-1111.

Lozano AM, Lang AE, Galvez-Jimenez N, Miyasaki J, Duff J, Hutchinson WD, Dostrovsky JO (1995) Effect of GPi pallidotomy on motor function in Parkinson's disease. *Lancet* **346**:1383-1387.

Manson AJ, Turner K, Lees AJ (2002) Apomorphine monotherapy in the treatment of refractory motor complications of Parkinson's disease: long-term follow-up study of 64 patients. *Mov Disord* **17**:1235-1241.

Markham CH, Diamond SG (1986) Long-term follow-up of early dopa treatment in Parkinson's disease. *Ann Neurol* **19**:365-372.

Obeso JA, Rodriguez MC, Gorospe A, Guridi J, Alvarez L, Macias R (1997) Surgical treatment of Parkinson's disease. *Baillieres Clin Neurol* **6**:125-145.

Paul G, Meissner W, Rein S, Harnack D, Winter C, Hosmann K, Morgenstern R, Kupsch A (2004) Ablation of the subthalamic nucleus protects dopaminergic phenotype but not cell survival in a rat model of Parkinson's disease. *Exp Neurol* **185**:272-280.

Piallat B, Benazzouz A, Benabid AL (1996) Subthalamic nucleus lesion in rats prevents dopaminergic nigral neuron degeneration after striatal 6-OHDA injection: behavioural and immunohistochemical studies. *Eur J Neurosci* **8**:1408-1414.

Pillon B, Ardouin C, Damier P, Krack P, Houeto JL, Klinger H, Bonnet AM, Pollak P, Benabid AL, Agid Y (2000) Neuropsychological changes between "off" and "on" STN or GPi stimulation in Parkinson's disease. *Neurology* **55**:411-418.

Pinter MM, Alesch F, Murg M, Helscher RJ, Binder H (1999) Apomorphine test: a predictor for motor responsiveness to deep brain stimulation of the subthalamic nucleus. *J Neurol* **246**:907-913.

Pollak P, Fraix V, Krack P, Moro E, Mendes A, Chabardes S, Koudsie A, Benabid AL (2002) Treatment results: Parkinson's disease. *Mov Disord 17 Suppl* **3**:S75-83.

Pollak P, Benabid AL, Gross C, Gao DM, Laurent A, Benazzouz A, Hoffmann D, Gentil M, Perret J (1993) [Effects of the stimulation of the subthalamic nucleus in Parkinson disease]. *Rev Neurol (Paris)* **149**:175-176.

Rodriguez MC, Obeso JA, Olanow CW (1998) Subthalamic nucleus-mediated excitotoxicity in Parkinson's disease: a target for neuroprotection. *Ann Neurol* **44**:S175-188.

Saint-Cyr JA, Trepanier LL, Kumar R, Lozano AM, Lang AE (2000) Neuropsychological consequences of chronic bilateral stimulation of the subthalamic nucleus in Parkinson's disease. *Brain* **123** (Pt 10):2091-2108.

The PD study group (2001) Deep-brain stimulation of the subthalamic nucleus or the pars interna of the globus pallidus in Parkinson's disease. *N Engl J Med* **345**:956-963.

Welter ML, Houeto JL, Tezenas du Montcel S, Mesnage V, Bonnet AM, Pillon B, Arnulf I, Pidoux B, Dormont D, Cornu P, Agid Y (2002) Clinical predictive factors of subthalamic stimulation in Parkinson's disease. *Brain* **125**:575-583.

Wenzelburger R, Zhang BR, Poepping M, Schrader B, Muller D, Kopper F, Fietzek U, Mehdorn HM, Deuschl G, Krack P (2002) Dyskinesias and grip control in Parkinson's disease are normalized by chronic stimulation of the subthalamic nucleus. *Ann Neurol* **52**:240-243.

In: Recent Breakthroughs in Basal Ganglia Research
Editor: Erwan Bezard, pp. 419-432 ISBN: 1-59454-880-3

Chapter 34

BEHAVIORAL DISORDERS IN PATIENTS WITH PARKINSON'S DISEASE TREATED BY BILATERAL STIMULATION OF THE SUBTHALAMIC NUCLEUS

M. L. Welter
Centre d'investigation clinique, Fédération de Neurologie
Hopital de la Salpétrière
Paris, France

ABSTRACT

Deep brain stimulation of the subthalamic nucleus (STN) has for a number of years provided a real option for the treatment of patients with severe Parkinson's disease (PD), with a 60 to 80% reduction in motor disability and levodopa-related motor complications. The peri-operative morbidity and mortality rates associated with this neurosurgical procedure are relatively low (1 to 3%) and are influenced by the patients' clinical characteristics. Recently, psychic disturbances occurring either immediately after the operation or later in time have been reported. While depression syndromes and anxiety symptoms are improved or unchanged, major depression episodes and sometimes suicide may occur after surgery. Conversely, many cases of manic episodes occurring after surgery have been observed, usually transient, which could sometimes resolve spontaneously within a few weeks, but generally require antipsychotic medication or mood stabilizers. The most common psychotic manifestation of STN stimulation is confusional syndrome, transient and benign in most cases. Lastly, apathy and emotional lability have been frequently described with severe impact on quality of life in some cases. Finally, some authors suggest that 20 to 25% of PD patients experience postoperative emotional or behavioral disorders. This suggests that global management of these patients, consisting of a complete motor and psychological evaluation, is necessary in order to anticipate and prevent severe psychic disturbances after the surgery. That stimulation of deep brain structures like the basal ganglia has psychic effects supports the hypothesis that the basal ganglia are involved in the control of emotion and behavior.

INTRODUCTION

Deep brain stimulation of the subthalamic nucleus (STN) has for a number of years provided a real option for the treatment of patients with severe Parkinson's disease (PD; Krack et al., 1998; Limousin et al., 1998; Houeto et al., 2000). This surgical procedure results in a 60 to 80 % reduction in motor disability and in the complications induced by the dopaminergic treatment (Moro et al., 1999; Houeto et al., 2000; Molinuevo et al., 2000). The peri-operative morbidity and mortality rates associated with this neurosurgical procedure are relatively low (1 to 3 %; Krack et al., 2003; Lyons et al., 2004; Volkmann, 2004) and are influenced by the patients' clinical characteristics (Saint-Cyr et al., 2000). Hemorrhage resulting in death or sequelae has been reported but, unlike transient post-operative confusional episodes, is rare (Krack et al.,

1998; Houeto et al., 2000; Lyons et al., 2004). Stimulation of the STN induces little or no change in cognitive functions (Ardouin et al., 1999; Fields and Troster, 2000; Pillon et al., 2000; Alegret et al., 2001; Perozzo et al., 2001), although a reduction in verbal fluency may sometimes be observed, particularly in elderly patients and patients with mild cognitive decline (Saint-Cyr et al., 2000; Vingerhoets et al., 2002; Daniele et al., 2003). More recently, there have been reports of psychic disturbances occurring either immediately after the operation or later in time. Most of the reported observations are single cases, and there are very few data in the literature on the frequency of these symptoms. Some authors have suggested that 20 to 25 % of patients experience postsurgical emotional or behavioral disorders (Saint-Cyr, 2004). However, the impact of bilateral stimulation of the STN on the psychic state of patients and the prevalence of these disorders remains controversial. It could be that, in addition to inducing abnormal movements or dyskinesia, deep brain stimulation triggers 'abnormal movements' of the mind? If so, how? Are these disorders a psychological response to changes in motor disability, or are they cause by the effect of high frequency electrical stimulation on the activity of STN neurons or adjacent structures? In some patients the clinical picture may begin before the stimulator is switched on remain unaffected when it is switched off, suggesting that the behavioral disorders are not related to STN stimulation. In others, modifying or stopping stimulation can reduce or even abolish the abnormal behavior, suggesting that electrical stimulation of the STN is the cause of these symptoms. Anatomical studies in animals suggest that there are non-motor limbic and associative circuits in the subthalamic nucleus that may be involved in the control of emotions and mood (Baron et al., 2002; Baunez et al., 2002; Yelnik, 2002; Karachi et al., 2004). A change in their activity under high-frequency stimulation could then precipitate postoperative psychic symptoms. The first part of this chapter describes the psychological disturbances reported in patients undergoing bilateral stimulation of the STN. In the second part, we will consider the physiopathological hypotheses regarding the origin of these symptoms.

1 Behavioral Disorders in Patients with High-Frequency Subthalamic Stimulation

1.1 Mood Disorders and Anxiety

Depression is the most common mental disorder in Parkinson's disease, with a prevalence of around 50 %. It is characterized by a combination of a loss of self-esteem, feelings of hopelessness and worthlessness, pessimism, irritability and sadness (Brandstädter and Oertel, 2003). Suicidal thoughts are frequent, and episodes of major depression occur in over 20 % of cases of depression (Cummings and Masterman, 1999). The suicide rate in PD patients is low, however. Furthermore, PD patients frequently experience symptoms of anxiety and panic attacks, in association with depression. Generalized anxiety disorders, social phobias, phobic disorders, obsessive-compulsive disorders and unspecified anxiety have been reported (Walsh and Bennett, 2001). Depression and anxiety have a great impact on quality of life of PD patients, but their relationship to motor disability is still a matter of debate (Brown et al., 1988; Starkstein et al., 1992). Depressive syndromes in PD patients with dopaminergic treatment may improve as parkinsonian motor disability decreases (Jouvent et al., 1983; Maricle et al., 1995). Since the effects of high-frequency stimulation of the STN and levodopa treatment on parkinsonian motor disability are highly correlated (Charles et al., 2002; Welter et al., 2002), STN stimulation might also ameliorate depressive syndromes in PD patients. Some authors have indeed reported a significant improvement in depressive syndromes (Houeto et al., 2002; Lagrange et al., 2002; Daniele et al., 2003; Funkiewiez et al., 2004) and anxious disorders (Mallet et al., 2002; Daniele et al., 2003; Schneider et al., 2003; Dujardin et al., 2004) in PD patients after surgery. In the study by Funkiewiez et al., (2004), 47 % of patients had no depressive symptoms 3 years after surgery, compared to only 20 % before surgery. However, major depression (Houeto et al., 2002; Herzog et al., 2003), sometimes with suicide or attempted suicide, has also been observed in several patients immediately after the operation (Houeto et al., 2000; Berney et al., 2002; Doshi et al., 2002; Lagrange et al., 2002; Stefurak et al., 2003), and the incidence of depression in the first month after surgery can reach 25 % (Volkmann et al., 2001; Berney et al., 2002). A case of acute melancholia was reported in a patient during stimulation of the left substantia nigra pars reticulata (SNr; Bejjani et al., 1999). Depression was not induced by chronic stimulation in the STN, however, and motor disability was improved by 51 % 1 year after surgery. At a later stage, in particular when the effect of STN allows a antiparkinsonian treatment to be reduced 60 to 70 %, a depressive syndrome may also appear or the level of depression remain unchanged (Houeto et al., 2002; Krack et al., 2003; Dujardin et al., 2004). In summary, while depression syndromes and anxiety symptoms are improved or unchanged, major depression episodes and sometimes suicide may occur after surgery, but this is rare. In most cases of post-surgical depression, patients either had a history of depression or were depressed at the time of surgery (Houeto et al., 2002), suggesting that the recurrence of postoperative depression reflects the decompensation of a preexisting mood disorder. Candidates for deep brain stimulation should therefore undergo a thorough psychiatric evaluation prior to surgery.

As depression, there have been many reported cases of manic episodes occurring after surgery. The patients classically show elation, overactivity, hypersexuality and grandiose delusions without any disturbance of consciousness or hallucinations (Houeto et al., 2002; Romito et al., 2002; Daniele et al., 2003). These mood disturbances occur in the first days of stimulation (Kulisevsky et al., 2002; Romito et al., 2002; Herzog et al., 2003a) and sometimes before it starts (Romito et al., 2002). Two patients had bouts of laughter under STN stimulation during the postoperative period, accompanied by an improvement in motor disability, that disappeared when the stimulator was switched off (Krack et al., 2001). The stimulation contacts were located within the STN. One of the patients presented hypomanic behavior in the weeks that followed, suggesting that the laughter was a symptom of an incipient mood disorder. Changing the stimulation parameters (i.e. changing the contact) does not always produce an immediate regression of manic symptoms (Kulisevsky et al., 2002; Sensi et al., 2004). However, although the manic episodes may be spectacular, they are usually transient. They sometimes resolve spontaneously within a few weeks, but generally require antipsychotic medication (clozapine) or mood stabilizers (Vingerhoets et al., 2002; Herzog et al., 2003a).

1.2 Psychosis, Delusions and Aggressive Behaviors

Visual hallucinations are the most prevalent postoperative psychotic symptom in PD. Visual hallucinations with a confusional syndrome may occur in the days following surgery, but are usually transient (Krack et al., 1998; Houeto et al., 2000; Vingerhoets et al., 2002; Herzog et al., 2003a). The hallucinations are characteristically nocturnal and take the form of animals, people or objects. The rapid reduction or withdrawal of dopaminergic agonists has been incriminated in some cases (Houeto et al., 2002). The presence of a bifrontal aeroma, which may be extensive in some patients, may also contribute, but this remains to be determined. Some patients may experience hallucinations well after surgery, in the absence of any notable change in therapy. In such cases, the hallucinations may be, at least partly, related to cognitive impairment (Funkiewiez et al., 2004). These patients often presented similar episodes while receiving high doses of dopaminergic or anticholinergic agonists prior to surgery, suggesting that inclusion for STN stimulation of patients with previous history of hallucinations must be carefully weighted. Although postoperative confusional states are frequent, acute delusional states are more rare. Paranoid delusions, persecutory complexes or jealousy have been reported in some patients (Herzog et al., 2003; Funkiewiez et al., 2004). Aggressive behavior or rage have also been described during stimulation (Bejjani et al., 2002; Funkiewiez et al., 2004; Okun et al., 2004). In one case, the attack happened during surgery while the most medial part of the STN or the adjacent region were being stimulated (Bejjani et al., 2002). In another case the attacks occurred during stimulation of the left STN (Okun et al., 2004). In both cases, the disturbances disappeared within minutes when stimulation was stopped, but the second patient required antipsychotic medication (quetiapine) since bilateral STN stimulation was needed in order to improve his motor disability (Okun et al., 2004).

A confusional syndrome is therefore the most common psychotic manifestation of STN stimulation and is transient and benign in most cases. When this symptom occurs in a patient with a moderate cognitive impairment and is persistent, it could be the sign of an incipient dementia. Paranoid or aggressive states are rare and usually result from bad localization of the stimulation contact.

Table 1. Treatment of Parkinson's disease by bilateral stimulation of the subthalamic nucleus: behavioral results in published studies.

	Number	Age (years)	Disease duration (years)	Evaluation	Motor results	Psychic results
Krack (1998)	8 patients	51 ± 4	**16 ± 8**	UPDRS III	UPDRS III= -70%/ -40%	3/8 mental confusion 2/8: apathy, lack of initiative, anxiety
Bejjani (1999)	1 patient	65	30	-	-	Acute melancholia: contact 0 left: SNr
Houeto (2000)	23 patients	53 ± 2	15 ± 1	UPDRS III	UPDRS III: -67%	3 with transient confusional state 4 with depression (2 major depression and 1 attempted suicide
Alegret (2001)	15 patients	61 ± 8	16 ± 8	UPDRS Declarative memory Visual-spatial function Prefrontal functions (fluency) Obsessive–compulsive traits	UPDRS III: -57%	Moderate deterioration in verbal memory and visuospatial functions Moderate improvement in prefrontal functions and obsessive–compulsive symptoms
Krack (2001)	2 patients	47 and 54				Laughter with elated mood: Patient 1: contact 0 bilaterally, with a dramatic antiparkinsonian effect; Patient 2: right electrode contact 3, with a dramatic antiparkinsonian effect
Houeto (2002)	24 patients	59	16 ± 4.5	UPDRS III, IV Mattis dementia rating scale, 'frontal score' SAS, MINI, IOWA	UPDRS III: -69.5% UPDRS IV: -75%	Social Adjustment Scale: Work: adjustment: 2/2 Social life: adjustment: 8/24; maladjustment: 16/24 (slight to moderate: 14; marked to severe: 2) Family life: adjustment: 14/24; maladjustment: 10/24 (slight to moderate: 8; marked to severe: 2) Married life: adjustment: 11/22; maladjustment: 11/22 (slight to moderate: 9; marked to severe: 2) Children relation: adjustment: 14/21; maladjustment: 7/21 (slight to moderate: 6; marked to severe: 1) Global: adjustment: 9/24; maladjustment: 15/24 (slight to moderate: 14; marked to severe: 1)

Table 1. Continued

	Number	Age (years)	Disease duration (years)	Evaluation	Motor results	Psychic results
Houeto (2002) Continued						MINI: Suicidal risk: 3/12 with history of depression, 4 after surgery Hypomania: 1/24, 18/24: euphoria Agoraphobia: 4/24, postoperative worsening in 2 Anxiety: 17/24 preoperative; 18/24 postoperative Addiction: 2/24Psychosis: 1 postoperative (with dopaminergic agonist treatment) IOWA No change overall: 3 groups: same (n=7)/+ (n=8)/- (n=8): lack of initiative, perseveration, lability, irritability, apathy, vulnerability Emotional hyperreactivity: 15/24
Berney (2002)	24 patients	64 ± 8	13 ± 5	UPDRS MADRS, HDRS	UPDRS III: -45%	6 patients: depression; neuropsychological status: same except for verbal fluency
Doshi (2002)	31 patients	-	-	-	-	3 postoperative depressions with suicidal ideation (n=2) and history of preoperative depression (n=2). Improvement with change in electrode settings (n=2)
Bejjani (2002)	1 patient	58	13	-	Intraoperative	Agitated state with fury and shouting. Medial intraoperative trajectory: triangle of Sano
Lagrange (2002)	60 patients	56 ± 10	14 ± 8	UPDRS PDQL BDI	UPDRS III: -55% PDQL: +43%	Improvement of mood on BDI 5 transient psychiatric complications: mania (n=1), delusion (n=1), depression (n=3; 2 suicide attempts)
Martinez-Martin (2002)	17 patients	61 ± 7.7	16.5 ± 8.5	UPDRS HAD PDQ-39	UPDRS III: -62% PDQ-39.SI: -50%	3 cases of transient depression Emotional well-being: -22%, shame: -29.5%

Table 1. Continued

	Number	Age (years)	Disease duration (years)	Evaluation	Motor results	Psychic results
Vingerhoets (2002)	20 patients	63 ± 8	16 ± 5	ADL, UPDRS III	ADL: -37% (20 months) UPDRS III: -45%	3/20: agitation with postoperative confusion needed a treatment by clozapine 4/20: post-operative antidepressive treatment Neuropsychology: same, except for verbal fluency
Kulisevsky (2002)	15 patients	-	-	-	-	3 patients: manic episode 48 h after implantation, contact 0 bilaterally; Improvement of mood disorders: contacts 2 and 3 without specific treatment
Romito (2002)	30 patients	-	-	-	-	5 cases of mood disorder with hypersexuality, beginning 2 days after surgery, before the stimulation was turned 'on' or just after the start of stimulation. Stabilization after several months
Mallet (2002)	2 patients			UPDRS YBOCS	UPDRS III: -27% and -80%	Compulsion: -100% and obsession: -78%
Funkiewicz (2003)	50 patients	54.5 ± 8		UPDRS ARCI: psychoactive drugs (49 items)	UPDRS III: -75%	Levodopa and STN stimulation: subjective feelings of well-being, euphoria, increase in motivation; decrease in fatigue, anxiety and tension.
Herzog (2003b)	48 patients	60 ± 6	15 ± 5	UPDRS	UPDRS III: -51%	Perioperative confusional state (7), psychotic syndrome (2), hypomanic behavior (2), depression (5): rapid spontaneous improvement except for depression and 1 case of manic psychosis (clozapine)

Table 1. Continued

	Number	Age (years)	Disease duration (years)	Evaluation	Motor results	Psychic results
Schneider (2003)	12 patients	62 ± 6	17 ± 6.5	UPDRS III, finger-tapping, walking HAMD, BPRS, Well-Being Scale Mood induction: 40 facial expressions Emotional memory (Wechsler) Emotional discrimination (PENN facial discrimination test) Verbal fluency	UPDRS III: -70.5%	Mood induction: same results either with stimulation or with levodopa alone Emotional memory: improvement with the 'on' stimulation condition compared to the 'off' stimulation condition Emotional discrimination: no effect of stimulation Verbal fluency: no clear effect HAMD and BPRS: both worsened in the 'off' stimulation condition
Okun (2003)	5 patients	46-73	14	UPDRS III Visual Analogue Mood Scale (VAMS): 30' MDRS, HAMD, HAS	UPDRS III: STN: -25% GPi: -24%	Stimulation motor site: 4/5 STN: +; 2/5 STN: -; 2/4 Gpi: +; 1/4 Gpi: - Dorsal stim: 2/5 STN: +; 2/5 STN: -; 2/4 Gpi: +; 1/4 Gpi: - Ventral stim: 0/5 STN: +; 2/5 STN: -; 1/4 Gpi: +; 2/4 Gpi: -
Daniele (2003)	20 patients	57 ± 8	14 ± 5.5	UPDRS III, ADL, PDQL Cognition Anxiety and depression, interview	UPDS III: -47%	Decline in verbal fluency: -15% after 1 year; reduction in perseverative errors and improvement on MMS; Depressive symptoms: -16% after 1 year; anxiety symptoms: -17% after 1 year 3/20: transient manic symptoms with hypersexuality
Sensi (2004)	1 patient	64	8	UPDRS III Cognitive		Second day after surgery: spontaneous aggressive outbursts, kleptomania; alternation between moments of well-being and euphoria and moments of aggression and impulsivity: left electrode, 4 contacts. Clozapine treatment required

Table 1. Continued

	Number	Age (years)	Disease duration (years)	Evaluation	Motor results	Psychic results
Schroeder (2004)	10 patients	61 ± 11	16 ± 3	UPDRS MMS, IQ Visual perception (VOSP) Facial perception (Benton recognition test)	UPDRS III: -50%	Impaired recognition of anger while under stimulation
Dujardin (2004)	12 patients	58 ± 7	13 ± 3	UPDRS MDRS, MMS EFE decoding (1 or 0) Executive functions MADRS, AMDP-AT (anxiety)	UPDRS III: -46% MMS, MDRS: no change	Reduced EFE decoding: sadness, anger and disgust Reduced verbal fluency Reduced anxiety
Funkiewicz (2004)	77 patients	55 ± 8	15 ± 5	UPDRS MDRS, Wisconsin card sorting test, frontal score, Grober and Buschke BDI	UPDRS III: -59% Cognition: no major change	Improvement in BDI Apathy: 25%, 3 years after surgery 4 cases of depression with attempted suicides; 1 suicide 5 cases of hypomania; 2 cases with impulsive aggressive behavior with a postoperative delirium in 1 case; 3 cases of psychosis, hallucinations or delusions
Fontaine (2004)	1 patient	49	14	UPDRS Y-BOCS, MADRS MDRS, Frontal score, Trail-making test	UPDRS III: -76%	Y-BOCS: -97%, compulsion: -94%, obsessions: -100% Improvement of OCD 1 week after introduction of bilateral STN stimulation

1.3 Apathy

Many cases of postsurgical apathy have been reported (Krack et al., 1998; Dujardin et al., 2004; Funkiewiez et al., 2004). Apathy, sometimes referred to as aboulia or lack of motivation, is described as tiredness by the patient, who is neither depressed nor demented. Patients do not express sadness, they have projects but feel too exhausted to undertake new projects. They are described by their entourage as being physically capable of activity, but do nothing. They can spend hours sitting in an armchair without doing anything. In serious cases, this can result in a severe social maladjustment (Houeto et al., 2002). In our center, 25 % of a series of 40 consecutive patients who underwent surgery for bilateral deep brain stimulation, who had never been apathetic before, presented apathy 5 years after surgery (not shown). Similar results have been reported by others (Funkiewiez et al., 2004). Apathy is usually improved by starting or increasing dopaminergic treatment even if the patient's motor state does not require it (Funkiewiez et al., 2004). In some cases it persists in spite this type of treatment and may be associated with a frontal lobe-like cognitive decline, suggesting that this symptom results from the neurodegenerative process.

Apathy seems to be the most common disabling psychic side effect of bilateral stimulation of the STN. Because of its highly negative effect on quality of life of both patients and their entourage, it should be detected by interviewing the patient. Since apathy can be improved by increasing dopaminergic treatment, it is probably related to dysfunction of forebrain dopaminergic systems (Rogers et al., 1987).

1.4 Emotional Lability, Pseudobulbar Syndrome

Emotional lability, or difficulty in controlling emotions, has been frequently reported in PD patients, before surgery (Baruch et al., 1992). After surgery, it was observed in 75 % of patients, without any pyramidal or pseudobulbar signs, and with no response to dopaminergic treatment (Houeto et al., 2002). A case of postoperative pseudobulbar syndrome was recently observed after unilateral stimulation of the right STN in a patient who had undergone a left pallidotomy 8 years earlier (Okun et al., 2004). Crying without a change in mood was induced by stimulation through each of the contacts, and stopped when the stimulator was turned off. It was accompanied by a feeling of fear when the ventral-most lead was stimulated (cathode lead 0) and anxiety when the dorsal-most lead was stimulated (cathode lead 3). Stimulation through the same contacts improved motor disability. Treatment with sertraline, a serotonin reuptake inhibitor, prevented crying during STN stimulation.

Emotional lability occurs frequently but usually in isolation and without any major impact on the patient's disability. Cases of complete pseudobulbar syndrome remain anecdotal.

2 Anatomo-functional Basis of Psychic Disorders in Patients with Bilateral High-frequency Timulation of the STN

Mood disturbances or behavioral changes occurring after bilateral STN stimulation might be a psychological reaction to the improvement in the patient's motor state, a consequence of the decrease in dopaminergic treatment or directly induced by the high frequency stimulation of the STN or adjacent structures.

Stimulation of the STN provides a rapid and marked reduction in motor disability (Krack et al., 1998; Limousin et al., 1998; Houeto et al., 2000; Krack et al., 2003). The suddenness of the improvement in the patient's motor state could well lead to psychological disturbances, as reported in similar circumstances such as after a cardiac transplant (Miller, 2002; Brown et al., 2004). Almost overnight, patients who were previously impaired and not fully autonomous, suddenly regained their independence, as if they had been released from a 'prison'. It is easy to imagine the potential impact that the situation could have on their psychic state and on their relationships with their entourage. Numerous cases of severe postoperative familial, social and professional maladjustment have in fact been reported (Houeto et al., 2002). Another possibility is that the actual postsurgical motor improvement does not fulfill the patient's expectations of improvement or at least the patient's perception of his or her motor state. Indeed, it is not uncommon for some patients to complain of a lack of any improvement or insufficient improvement of their motor state even though there is a very real, objective motor improvement. These various observations suggest the need for psychological management prior to surgery, both for the patient and his or her entourage, to prepare them for the psychic and motor postsurgical changes.

The reduction in dopaminergic treatment after surgery could contribute to the appearance of depressive symptoms and apathy. 1) Mood disorders are improved by the introduction of dopaminergic treatment in de novo patients (Cummings, 1992) and when levodopa treatment produces its maximal effect in patients with motor fluctuations (Maricle et al., 1995). 2) The apathy or aboulia described in some patients after surgery was improved by introducing or increasing dopaminergic treatment (Funkiewiez et al., 2004). 3) Some patients became addicted postoperatively to levodopa treatment in spite of the improvement in their parkinsonian disability (Houeto et al., 2002). This suggests that a nigrostriatal dopaminergic deficit was compensated by high-frequency STN stimulation, but not a mesolimbic dopaminergic deficit (Houeto et al., 2002).

There are also evidences that these postoperative emotional changes are directly induced by STN high-frequency stimulation. Animal studies suggest that the STN is subdivided into three regions with different functions: a dorsolateral part involved in the regulation of movement, a ventromedial 'limbic' part involved in the regulation of emotions and an intermediate 'associative' part involved in the regulation of cognitive functions (Baron et al., 2002; Baunez et al., 2002; Yelnik, 2002; Karachi et al., 2004). High frequency stimulation of the associative and limbic part of the STN could modulate the activity of the limbic circuit (Okun et al., 2003), comprising the ventral striatum (including the accumbens nucleus), the thalamus (center median-parafascicular nucleus), the prefrontal orbitofrontal cortex, the anterior cingulate cortex, the parahippocampal gyrus, the insula and the amygdala (Alexander et al., 1986). In studies with positron emission tomography evaluating the regional blood flow, an activation of orbitofrontal and anterior cingulate cortices have been observed during STN stimulation (Sestini et al., 2002; Schroeder et al., 2003). Studies based on recognition of facial expression indeed suggest that STN stimulation affects the functioning of the amygdala (Dujardin et al., 2004; Schroeder et al., 2004). The amygdala has been implicated in the perception and production of negative affects and in associative aversive learning (Morris et al., 1996). As in patients with lesions of the amygdala (Calder, 1996), STN stimulation impaired decoding of facial expressions for negative affect. The recognition of faces expressing anger, disgust (Dujardin et al., 2004; Schroeder et al., 2004) and sadness (Dujardin et al., 2004) were reduced, whereas executive functions were unaltered (Dujardin et al., 2004). Similar results were reported with antidepressant treatment that inhibit serotonin or norepinephrine reuptake in healthy subjects (Harmer et al., 2004). This suggests that bilateral stimulation of the STN could induce changes in the emotional state of the patients, for example improvement of a depressive syndrome, by modulating the activity of serotonin or noradrenergic circuits. The fact that the improvement of depression after surgery is not correlated with the improvement in quality of life (Lagrange et al., 2002; Martinez-Martin et al., 2002) or with parkinsonian motor disability (Martinez-Martin et al., 2002) is compatible with this hypothesis. In conclusion, it is highly probable that high-frequency stimulation of the STN can induce behavioral disorders by modulating the activity of the limbic circuit (Sestini et al., 2002; Stefurak et al., 2003). The neuronal mechanisms underlying these behavioral modifications remain unknown. High-frequency stimulation of the STN induces an inhibition of neuronal activity, but also a change in the firing pattern of STN neurons (Tai et al., 2003; Welter et al., 2004). The appearance of behavioral disorders such as elation, disinhibition, hypersexuality and euphoria could then result from the inhibition of the spontaneous neuronal activity in the limbic portion of the STN in PD patients (Welter et al., 2004). The fact that lesions of the STN induce a state close to hypomania, with disinhibition, logorrhea and euphoria and without mental confusion or signs of frontal dysfunction (Trillet et al., 1995), support this hypothesis. It also suggests that during the perioperative period, the microsubthalamotomy induced by the positioning of the electrode, could result in a disinhibition, as sometimes observed in patients before the stimulation has been started (Romito et al., 2002) and which can decreased spontaneously within a few days. Furthermore, the combination of STN stimulation and dopaminergic treatment immediately after surgery could result in a combination of psychotropic effects (Funkiewiez et al., 2003; Schneider et al., 2003), and leading to hypomania. This suggests that, after the operation, dopaminergic treatment should be reintroduced with caution. Lastly, a modulation of neuronal activity of the basal ganglia output (i.e. in the internal part of the globus pallidus and in the SNr) has been observed in animal studies during high-frequency STN stimulation (Hashimoto et al., 2003; Maurice et al., 2003; Tai et al., 2003). High-frequency stimulation of the left SNr induced an acute depressive state in a parkinsonian patient (Bejjani et al., 1999), associated with an increase in cerebral blood flow in the right parietal lobe, the left orbitofrontal cortex, the left globus pallidus, the left amygdala and the anterior thalamus. Modification of the neuronal activity of the SNr might therefore be responsible for depressive syndromes observed after surgery.

Finally, electrical stimulation may affect axons passing through or close to the STN, and modulate the activity of neurons in structures with limbic and associative functions that neither project to or are innervated by the STN, such as the medial forebrain bundle, the zona incerta (Stefurak et al., 2003) and the hypothalamus. The majority of dopaminergic neurons of the ventral tegmental area project to the nucleus accumbens via the medial forebrain bundle. This circuit is activated by psychostimulant drugs, food and sexual activity (Koob, 1992). Stimulation of these fibers could induce an 'overactivation' of the nucleus accumbens resulting in the production of positive affect and addictive behaviors (Koob, 1992; Koch et al., 1996) after surgery. Stimulation of the lateral hypothalamus might result in emotional disorders after STN stimulation, as suggested by the description of a 'pressure to laugh' with mirth in patients with hypothalamic hamartoma medial and anterior to the STN (Sturm et al., 2000).

Conclusion

The risk of behavioral disorders occurring after bilateral stimulation of the STN in parkinsonian patients thus appears to be fairly low. With the exception of apathy (Krack et al., 1998; Funkiewiez et al., 2004), these disturbances occur mostly in patients with a previous history of psychiatric disorders or who were psychologically disturbed at the time of surgery (Houeto et al., 2002). Several anxious disorders are alleviated after surgery, however (Mallet et al., 2002; Daniele et al., 2003; Schneider et al., 2003; Dujardin et al., 2004). These data suggest that global management of these patients, consisting of a complete motor and psychological evaluation by a team including psychiatrists and psychologists trained to recognize both the psychiatric pathologies in parkinsonian patients and the effects of deep brain stimulation on social and affective life, is necessary in order to anticipate and prevent severe psychic disturbances after the surgery.

That stimulation of deep brain structures like the basal ganglia has psychic effects supports the hypothesis that the basal ganglia are involved in the control of emotion and behavior. This raises the possibility that this reversible and adaptable therapy, with its low morbidity, could be applied to some psychiatric diseases. Recently, obsessive–compulsive symptoms that existed prior to the surgery were spectacularly improved in several parkinsonian patients receiving STN stimulation for motor disability (Mallet et al., 2002; Fontaine et al., 2004). These results suggest that STN stimulation might be considered for patients with severe obsessive–compulsive disorders. This topic is dealt with in another chapter .

AMDP-AT: Association of Methodology and Documentation in Psychiatry-Anxiety scale; ADL: Activites of Daily Living; ARCI : Addiction Research Center Inventory; BDI: Beck Depression Inventory; BPRS : Brief Psychiatric Rating Scale; EFE: Emotional Facial Expression; HAD: Hospital Anxiety and Depression scale ; HAMD: Hamilton Depression scale; HAS : Hamilton Anxiety Scale; IOWA: Scale of personality change ; IQ: Intellectual Quotient; MADRS: Mongtomery and Asberg Depression rating Scale; MDRS: Mattis dementia rating scale; MINI-500: Mini international neuropsychiatric inventory; MMS: Mini-Mental Status; SAS: Social Adjustment Scale, PDQL : Parkinson's Disease Quality of Life; UPDRS : Unified Parkinson's Disease Rating Scale; Y-BOCS: Yale-Brown Obsessive Compulsive Scale.

References

Alegret M, Junque C, Valldeoriola F, Vendrell P, Pilleri M, Rumia J, Tolosa E (2001) Effects of bilateral subthalamic stimulation on cognitive function in Parkinson disease. *Arch Neurol* **58**:1223-1227.

Alexander GE, DeLong MR, Strick PL (1986) Parallel organization of functionally segregated circuits linking basal ganglia and cortex. *Annu Rev Neurosci* **9**:357-381.

Ardouin C, Pillon B, Peiffer E, Bejjani P, Limousin P, Damier P, Arnulf I, Benabid AL, Agid Y, Pollak P (1999) Bilateral subthalamic or pallidal stimulation for Parkinson's disease affects neither memory nor executive functions: a consecutive series of 62 patients. *Ann Neurol* **46**:217-223.

Baron MS, Wichmann T, Ma D, DeLong MR (2002) Effects of transient focal inactivation of the basal ganglia in parkinsonian primates. *J Neurosci* **22**:592-599.

Baruch P, Pourcher E, Jouvent R (1992) Parkinson's disease and mood components. Clin Neuropharmacol 15:8B.

Baunez C, Amalric M, Robbins TW (2002) Enhanced food-related motivation after bilateral lesions of the subthalamic nucleus. *J Neurosci* **22**:562-568.

Bejjani BP, Damier P, Arnulf I, Thivard L, Bonnet AM, Dormont D, Cornu P, Pidoux B, Samson Y, Agid Y (1999) Transient acute depression induced by high-frequency deep-brain stimulation. *N Engl J Med* **340**:1476-1480.

Bejjani BP, Houeto JL, Hariz M, Yelnik J, Mesnage V, Bonnet AM, Pidoux B, Dormont D, Cornu P, Agid Y (2002) Aggressive behavior induced by intraoperative stimulation in the triangle of Sano. *Neurology* **59**:1425-1427.

Berney A, Vingerhoets F, Perrin A, Guex P, Villemure JG, Burkhard PR, Benkelfat C, Ghika J (2002) Effect on mood of subthalamic DBS for Parkinson's disease: a consecutive series of 24 patients. *Neurology* **59**:1427-1429.

Brandstädter D, Oertel WH (2003) Depression in Parkinson' disease. *Adv Neurol* **91**:371-382.

Brown PA, Launius BK, Mancini MC, Cush EM (2004) Depression and anxiety in the heart transplant patient: a case study. *Crit Care Nurs Q* **27**:92-95.

Brown RG, MacCarthy B, Gotham AM, Der GJ, Marsden CD (1988) Depression and disability in Parkinson's disease: a follow-up of 132 cases. *Psychol Med* **18**:49-55.

Calder AJ (1996) Facial emotion recognition after bilateral amygdala damage: differentially severe impairment of fear. *Cognit Neuropsychol* **13**:699-745.

Charles PD, Van Blercom N, Krack P, Lee SL, Xie J, Besson G, Benabid AL, Pollak P (2002) Predictors of effective bilateral subthalamic nucleus stimulation for PD. *Neurology* **59**:932-934.

Cummings JL (1992) Depression and Parkinson's disease: a review. *Am J Psychiatry* **149**:443-454.

Cummings JL, Masterman DL (1999) Depression in patients with Parkinson's disease. Int *J Geriatr Psychiatry* **14**:711-718.

Daniele A, Albanese A, Contarino MF, Zinzi P, Barbier A, Gasparini F, Romito LM, Bentivoglio AR, Scerrati M (2003) Cognitive and behavioural effects of chronic stimulation of the subthalamic nucleus in patients with Parkinson's disease. *J Neurol Neurosurg Psychiatry* **74**:175-182.

Doshi PK, Chhaya N, Bhatt MH (2002) Depression leading to attempted suicide after bilateral subthalamic nucleus stimulation for Parkinson's disease. *Mov Disord* **17**:1084-1085.

Dujardin K, Blairy S, Defebvre L, Krystkowiak P, Hess U, Blond S, Destee A (2004) Subthalamic nucleus stimulation induces deficits in decoding emotional facial expressions in Parkinson's disease. *J Neurol Neurosurg Psychiatry* **75**:202-208.

Fields JA, Troster AI (2000) Cognitive outcomes after deep brain stimulation for Parkinson's disease: a review of initial studies and recommendations for future research. *Brain Cogn* **42**:268-293.

Fontaine D, Mattei V, Borg M, von Langsdorff D, Magnie MN, Chanalet S, Robert P, Paquis P (2004) Effect of subthalamic nucleus stimulation on obsessive-compulsive disorder in a patient with Parkinson disease. Case report. *J Neurosurg* **100**:1084-1086.

Funkiewiez A, Ardouin C, Krack P, Fraix V, Van Blercom N, Xie J, Moro E, Benabid AL, Pollak P (2003) Acute psychotropic effects of bilateral subthalamic nucleus stimulation and levodopa in Parkinson's disease. *Mov Disord* **18**:524-530.

Funkiewiez A, Ardouin C, Caputo E, Krack P, Fraix V, Klinger H, Chabardes S, Foote K, Benabid AL, Pollak P (2004) Long term effects of bilateral subthalamic nucleus stimulation on cognitive function, mood, and behaviour in Parkinson's disease. *J Neurol Neurosurg Psychiatry* **75**:834-839.

Harmer CJ, Shelley NC, Cowen PJ, Goodwin GM (2004) Increased positive versus negative affective perception and memory in healthy volunteers following selective serotonin and norepinephrine reuptake inhibition. *Am J Psychiatry* **161**:1256-1263.

Hashimoto T, Elder CM, Okun MS, Patrick SK, Vitek JL (2003) Stimulation of the subthalamic nucleus changes the firing pattern of pallidal neurons. *J Neurosci* **23**:1916-1923.

Herzog J, Reiff J, Krack P, Witt K, Schrader B, Muller D, Deuschl G (2003a) Manic episode with psychotic symptoms induced by subthalamic nucleus stimulation in a patient with Parkinson's disease. *Mov Disord* **18**:1382-1384.

Herzog J, Volkmann J, Krack P, Kopper F, Potter M, Lorenz D, Steinbach M, Klebe S, Hamel W, Schrader B, Weinert D, Muller D, Mehdorn HM, Deuschl G (2003) Two-year follow-up of subthalamic deep brain stimulation in Parkinson's disease. *Mov Disord* **18**:1332-1337.

Houeto JL, Damier P, Bejjani PB, Staedler C, Bonnet AM, Arnulf I, Pidoux B, Dormont D, Cornu P, Agid Y (2000) Subthalamic stimulation in Parkinson disease: a multidisciplinary approach. *Arch Neurol* **57**:461-465.

Houeto JL, Mesnage V, Mallet L, Pillon B, Gargiulo M, du Moncel ST, Bonnet AM, Pidoux B, Dormont D, Cornu P, Agid Y (2002) Behavioural disorders, Parkinson's disease and subthalamic stimulation. *J Neurol Neurosurg Psychiatry* **72**:701-707.

Jouvent R, Abensour P, Bonnet AM, Widlocher D, Agid Y, Lhermitte F (1983) Antiparkinsonian and antidepressant effects of high doses of bromocriptine. An independent comparison. *J Affect Disord* **5**:141-145.

Karachi C, Yelnik J, Tande D, Tremblay L, Hirsch EC, Francois C (2004) The pallidosubthalamic projection: An anatomical substrate for nonmotor functions of the subthalamic nucleus in primates. *Mov Disord* **20**:20.

Koch M, Schmid A, Schnitzler HU (1996) Pleasure-attenuation of startle is disrupted by lesions of the nucleus accumbens. *Neuroreport* **7**:1442-1446.

Koob GF (1992) Neurobiological mechanisms in cocaine and opiate dependence. *Res Publ Assoc Res Nerv Ment Dis* **70**:79-92.

Krack P, Pollak P, Limousin P, Hoffmann D, Xie J, Benazzouz A, Benabid AL (1998) Subthalamic nucleus or internal pallidal stimulation in young onset Parkinson's disease. *Brain* **121**:451-457.

Krack P, Kumar R, Ardouin C, Dowsey PL, McVicker JM, Benabid AL, Pollak P (2001) Mirthful laughter induced by subthalamic nucleus stimulation. *Mov Disord* **16**:867-875.

Krack P, Batir A, Van Blercom N, Chabardes S, Fraix V, Ardouin C, Koudsie A, Limousin PD, Benazzouz A, LeBas JF, Benabid AL, Pollak P (2003) Five-year follow-up of bilateral stimulation of the subthalamic nucleus in advanced Parkinson's disease. *N Engl J Med* **349**:1925-1934.

Kulisevsky J, Berthier ML, Gironell A, Pascual-Sedano B, Molet J, Pares P (2002) Mania following deep brain stimulation for Parkinson's disease. *Neurology* **59**:1421-1424.

Lagrange E, Krack P, Moro E, Ardouin C, Van Blercom N, Chabardes S, Benabid AL, Pollak P (2002) Bilateral subthalamic nucleus stimulation improves health-related quality of life in PD. *Neurology* **59**:1976-1978.

Limousin P, Krack P, Pollak P, Benazzouz A, Ardouin C, Hoffmann D, Benabid AL (1998) Electrical stimulation of the subthalamic nucleus in advanced Parkinson's disease. *N Engl J Med* **339**:1105-1111.

Lyons KE, Wilkinson SB, Overman J, Pahwa R (2004) Surgical and hardware complications of subthalamic stimulation: a series of 160 procedures. *Neurology* **63**:612-616.

Mallet L, Mesnage V, Houeto JL, Pelissolo A, Yelnik J, Behar C, Gargiulo M, Welter ML, Bonnet AM, Pillon B, Cornu P, Dormont D, Pidoux B, Allilaire JF, Agid Y (2002) Compulsions, Parkinson's disease, and stimulation. *Lancet* **360**:1302-1304.

Maricle RA, Nutt JG, Carter JH (1995) Mood and anxiety fluctuation in Parkinson's disease associated with levodopa infusion: preliminary findings. *Mov Disord* **10**:329-332.

Martinez-Martin P, Valldeoriola F, Tolosa E, Pilleri M, Molinuevo JL, Rumia J, Ferrer E (2002) Bilateral subthalamic nucleus stimulation and quality of life in advanced Parkinson's disease. *Mov Disord* **17**:372-377.

Maurice N, Thierry AM, Glowinski J, Deniau JM (2003) Spontaneous and evoked activity of substantia nigra pars reticulata neurons during high-frequency stimulation of the subthalamic nucleus. *J Neurosci* **23**:9929-9936.

Miller M (2002) Depression after cardiac transplant treated with interpersonal psychotherapy and paroxetine. *Am J Psychother* **56**:555-561.

Molinuevo JL, Valldeoriola F, Tolosa E, Rumia J, Valls-Sole J, Roldan H, Ferrer E (2000) Levodopa Withdrawal After Bilateral Subthalamic Nucleus Stimulation in Advanced Parkinson Disease. *Arch Neurol* **57**:983-988.

Moro E, Scerrati M, Romito LM, Roselli R, Tonali P, Albanese A (1999) Chronic subthalamic nucleus stimulation reduces medication requirements in Parkinson's disease. *Neurology* **53**:85-90.

Morris JS, Frith CD, Perrett DI, Rowland D, Young AW, Calder AJ, Dolan RJ (1996) A differential neural response in the human amygdala to fearful and happy facial expressions. *Nature* **383**:812-815.

Okun MS, Green J, Saben R, Gross R, Foote KD, Vitek JL (2003) Mood changes with deep brain stimulation of STN and GPi: results of a pilot study. *J Neurol Neurosurg Psychiatry* **74**:1584-1586.

Okun MS, Raju DV, Walter BL, Juncos JL, DeLong MR, Heilman K, McDonald WM, Vitek JL (2004) Pseudobulbar crying induced by stimulation in the region of the subthalamic nucleus. *J Neurol Neurosurg Psychiatry* **75**:921-923.

Perozzo P, Rizzone M, Bergamasco B, Castelli L, Lanotte M, Tavella A, Torre E, Lopiano L (2001) Deep brain stimulation of the subthalamic nucleus in Parkinson's disease: comparison of pre- and postoperative neuropsychological evaluation. *J Neurol Sci* **192**:9-15.
Pillon B, Ardouin C, Damier P, Krack P, Houeto JL, Klinger H, Bonnet AM, Pollak P, Benabid AL, Agid Y (2000) Neuropsychological changes between "off" and "on" STN or GPi stimulation in Parkinson's disease. *Neurology* **55**:411-418.
Rogers D, Lees AJ, Smith E, Trimble M, Stern GM (1987) Bradyphrenia in Parkinson's disease and psychomotor retardation in depressive illness. An experimental study. *Brain* **110**:761-776.
Romito LM, Scerrati M, Contarino MF, Bentivoglio AR, Tonali P, Albanese A (2002) Long-term follow up of subthalamic nucleus stimulation in Parkinson's disease. *Neurology* **58**:1546-1550.
Saint-Cyr J (2004) La stimulation du NST peut être toxique pour les fonctions cognitives ou psychologiques chez les patients parkinsoniens. *Mouvements Neurosciences cliniques* **5**:12-15.
Saint-Cyr JA, Trepanier LL, Kumar R, Lozano AM, Lang AE (2000) Neuropsychological consequences of chronic bilateral stimulation of the subthalamic nucleus in Parkinson's disease. *Brain* **123**:2091-2108.
Schneider F, Habel U, Volkmann J, Regel S, Kornischka J, Sturm V, Freund HJ (2003) Deep brain stimulation of the subthalamic nucleus enhances emotional processing in Parkinson disease. *Arch Gen Psychiatry* **60**:296-302.
Schroeder U, Kuehler A, Lange KW, Haslinger B, Tronnier VM, Krause M, Pfister R, Boecker H, Ceballos-Baumann AO (2003) Subthalamic nucleus stimulation affects a frontotemporal network: a PET study. *Ann Neurol* **54**:445-450.
Schroeder U, Kuehler A, Hennenlotter A, Haslinger B, Tronnier VM, Krause M, Pfister R, Sprengelmeyer R, Lange KW, Ceballos-Baumann AO (2004) Facial expression recognition and subthalamic nucleus stimulation. *J Neurol Neurosurg Psychiatry* **75**:648-650.
Sensi M, Eleopra R, Cavallo MA, Sette E, Milani P, Quatrale R, Capone JG, Tugnoli V, Tola MR, Granieri E, Data PG (2004) Explosive-aggressive behavior related to bilateral subthalamic stimulation. *Parkinsonism Relat Disord* **10**:247-251.
Sestini S, Scotto di Luzio A, Ammannati F, De Cristofaro MT, Passeri A, Martini S, Pupi A (2002) Changes in regional cerebral blood flow caused by deep-brain stimulation of the subthalamic nucleus in Parkinson's disease. *J Nucl Med* **43**:725-732.
Starkstein SE, Mayberg HS, Leiguarda R, Preziosi TJ, Robinson RG (1992) A prospective longitudinal study of depression, cognitive decline, and physical impairments in patients with Parkinson's disease. *J Neurol Neurosurg Psychiatry* **55**:377-382.
Stefurak T, Mikulis D, Mayberg H, Lang AE, Hevenor S, Pahapill P, Saint-Cyr J, Lozano A (2003) Deep brain stimulation for Parkinson's disease dissociates mood and motor circuits: a functional MRI case study. *Mov Disord* **18**:1508-1516.
Sturm JW, Andermann F, Berkovic SF (2000) "Pressure to laugh": an unusual epileptic symptom associated with small hypothalamic hamartomas. *Neurology* **54**:971-973.
Tai CH, Boraud T, Bezard E, Bioulac B, Gross C, Benazzouz A (2003) Electrophysiological and metabolic evidence that high-frequency stimulation of the subthalamic nucleus bridles neuronal activity in the subthalamic nucleus and the substantia nigra reticulata. *Faseb J* **17**:1820-1830.
Trillet M, Vighetto A, Croisile B, Charles N, Aimard G (1995) [Hemiballismus with logorrhea and thymo-affective disinhibition caused by hematoma of the left subthalamic nucleus]. *Rev Neurol* **151**:416-419.
Vingerhoets FJ, Villemure JG, Temperli P, Pollo C, Pralong E, Ghika J (2002) Subthalamic DBS replaces levodopa in Parkinson's disease: two-year follow-up. *Neurology* **58**:396-401.
Volkmann J (2004) Deep brain stimulation for the treatment of Parkinson's disease. *J Clin Neurophysiol* **21**:6-17.
Volkmann J, Allert N, Voges J, Weiss PH, Freund HJ, Sturm V (2001) Safety and efficacy of pallidal or subthalamic nucleus stimulation in advanced PD. *Neurology* **56**:548-551.
Walsh K, Bennett G (2001) Parkinson's disease and anxiety. *Postgrad Med J* **77**:89-93.
Welter ML, Houeto JL, Bonnet AM, Bejjani PB, Mesnage V, Dormont D, Navarro S, Cornu P, Agid Y, Pidoux B (2004) Effects of high-frequency stimulation on subthalamic neuronal activity in parkinsonian patients. *Arch Neurol* **61**:89-96.
Welter ML, Houeto JL, Tezenas du Montcel S, Mesnage V, Bonnet AM, Pillon B, Arnulf I, Pidoux B, Dormont D, Cornu P, Agid Y (2002) Clinical predictive factors of subthalamic stimulation in Parkinson's disease. *Brain* **125**:575-583.
Yelnik J (2002) Functional anatomy of the basal ganglia. *Mov Disord* **17**:S15-21.

INDEX

A

B

C

D

E

F

G

H

I

O

P

Q

R

S

T

U

V

W

Y

Z